Farid Amirouche

Fundamentals of Multibody Dynamics

Theory and Applications

Birkhäuser
Boston • Basel • Berlin

Farid M. L. Amirouche
The University of Illinois at Chicago
Department of Mechanical and
Industrial Engineering
Chicago, IL 60607-7022
U.S.A.

Cover design by Alex Gerasev.

AMS Subject Classifications: 70Exx, 70E05, 70E15, 70E17, 70E18, 70E20, 70E45, 70E50, 70E55, 70E60, 70E99, 70Fxx, 70F07, 70F10, 70F16, 70F20, 70F25, 70F35, 70G10, 70G40, 70G45, 70G55, 70G60, 70G65, 70H03, 70H05, 70H08, 70H09, 70H20, 70H25, 70H40, 70H45, 70Jxx, 70J10, 70J50, 70K43, 70M20, 93Dxx

Library of Congress Cataloging-in-Publication Data
Amirouche, Farid M. L.
Fundamentals of multibody dynamics : theory and applications / Farid Amirouche.
p. cm.
Includes bibliographical references and index.
ISBN 0-8176-4236-6 (acid-free paper)
1. Dynamics. 2. Kinematics. I. Title.

QA845.A54 2004
531'.11–dc22 2004048503

ISBN-10 0-8176-4236-6
ISBN-13 978-0-8176-4236-5
eISBN 0-8176-4406-7
Printed on acid-free paper.

Birkhäuser ®

Based on the author's previous edition, *Computational Methods in Multibody Dynamics*, Prentice-Hall, Englewood Cliffs, NJ, 1992.

Printed in the United States of America. (KeS/EB)

9 8 7 6 5 4 3 2 1 SPIN 10832352

www.birkhauser.com

To my son and daughter Larby and Anissa

Contents

Preface

Multibody dynamics has grown in the past two decades to become an important tool in the design, prototyping and simulation of complex articulated mechanical systems. Its versatility in analyzing a broad range of applications has made it an attractive feature in our teaching curriculum. Here at our university we offer two courses in multibody dynamics that are given in subsequent semesters. The first course deals mainly with rigid-body dynamics whereas the second course is devoted to constrained systems and the effects of flexibility in multibody dynamics.

This book brings together brilliant concepts of dynamics that combine the efforts of many researchers in the field of mechanics. The book's strength lies in its use of matrices in generating the kinematic coefficients associated with the formulation of the governing equations of motion. The book outlines the most effective ways of handling constraints and discusses in great detail how to use both finite-element and boundary-element methods to study the effects of deformation and vibration of structures during the course of certain maneuvers. A large number of examples worked out in their entirety follow every section. This is a learning-by-example approach that has been very instrumental in my teaching and research over the years.

This book is the result of many years of research and teaching of multibody dynamics. In essence, it is a revision of my previous book entitled *Computational Methods in Multibody Dynamics* published in 1991 by Prentice Hall. I have added a new chapter on particle dynamics, which serves as a review of previous dynamical principles that the student needs to be familiar with. An extensive review of the kinematics of a rigid body is presented in Chapter 2 together with a Matlab listing at the end of the chapter that deals with Euler angles and Euler parameters. The remainder of the chapters deal directly with treelike structures with open and closed loops and prescribed motions including flexibility effects. The book bridges the gap between dynamics and engineering applications such as robotics, mechanics, and biosystems and puts into perspective the importance of modeling in the solution of these problems.

The first seven chapters form the first part of a course in multibody dynamics where students are only expected to learn how to apply the developed procedures and principles to obtain the equations of motion. The remaining four chapters form the second part of the course where students are exposed to some concepts in vibration and finite-element methods. The first semester is aimed at junior/senior-level students and first-year graduate students, whereas the second is primarily for advanced seniors and graduate students.

The book is divided into twelve chapters. Chapter 3 is dedicated to the kinematics of rigid bodies and of interconnected multibody systems, in particular with treelike structure. The kinematic expressions are derived from the assumptions that each body can undergo all possible relative motions, including three rotations and three translations. Each chapter is divided into subsections in which several worked-out problems are presented to highlight the use of matrices in the representation of the kinematics of more than a one-body system.

Chapter 4 is devoted to the analysis of forces; this includes springs and dampers and concludes with some inertial properties. The chapter includes some key examples on the contribution of body forces and torques to the formation of generalized active forces.

Chapter 5 presents the formulation of the equations of motion making use of previously derived kinematic expressions from previous chapters and Kane's equations, Huston's tensor approach and matrix procedures developed by the author. The end of the chapter provides a special application using the techniques developed in this book. Furthermore, a flowchart used in the Matlab program is given for students who are interested in writing their own code. Student projects of this sort are useful and suggested as part of their homework. In Chapter 6 the energy equations derived by Lagrange, Hamilton and Gibbs equations are presented with examples to show the effectiveness of all the methods. The governing equations of motion of treelike multibody systems are given both in matrix form and in tensor (an index type of notation) form ready for computer implementation.

In Chapter 7 holonomic or nonholonomic constraints are studied extensively in the context of constrained multibody systems, where several methods for coordinate reduction through orthogonal complementary arrays are introduced. The pseudo-upper triangular method, the zero-eigenvalue theorem, and embedding methods are studied, together with some useful engineering examples. Prescribed motions and constraint force evaluations are also presented. The stability of the constraints as they change from 2D to 3D and the handling of zero constraints form the basis of Chapter 8.

Chapter 9 is used as a review and preparation for the analysis of flexible multibody systems. Linearization of the equations of motion and the eigenvalue problem solution to continuous and discrete systems are studied.

The next two chapters are used for the formulation of the equations of motion using both the procedures derived in Chapter 5 combined with finite-element methods. This includes treelike systems with terminal flexible links and systems with more than one flexible body, respectively.

The final chapter introduces boundary-element methods to multibody dynamics. We developed this concept back in 1996 and it was the subject of a Ph.D. thesis of one of my former students, M. Kerdjouj. It is an excellent idea that needs to be explored further by our researchers and students as it may lead to a more satisfactory way of handling surface flexibility and body deformation in multibody systems (MBS).

This book would not have been possible without the help of current and former students during the past fifteen years. I am indebted to Jia Tongyia (Tony) and Kemal

S. Ider, Mohamed Kerdjoudj, Shareef Nazer, and Tung C.W (Roger), Rick Tong, Xie Mingjun and Tajiri Gordon for their stimulating thoughts through the years of their Ph.D. thesis research. I am especially thankful for the help of my students Francisco Romero, Carlos Lopez, Pedro Gonzalez, David McNeil, Ravikumar Vardrajan, Dr. Mark Gonzalez, Joe Solomon and Nikhil Kulkarni during the course of the multibody dynamics course. I would like to acknowledge Mrs. Irena Zivkovic for her help in the initial text change from Microsoft Word to Latex. It is certainly the dedication of Surya Pratap Rai in helping with the final manuscript organization and editing that made the book possible. I would like to acknowledge the help of my current students Carlos Lopez, Jude Martin, Khurram Mahmudi and Ivan Zivkovic for their attentive help with the figures in Chapters 3–5.

Furthermore, I want to extend my sincere thanks to the reviewers who have taken the time out of their busy schedules to help with the review of some of the chapters of the book. My special thanks to Paul Mitiguy, Lazslo Palkovics, Moustafa El-Gindy, Kurt Anderson, Arun Banerjee, Jan Leuridan and Peter Brett as well as others who have contributed directly or indirectly to the content of the book.

Lastly, I would like to thank my family for their continuous love and support and the joy they bring into my life. I would like to mention that my father, whose memory lives on, is still an inspiration to my devotion and work. I am also blessed to have my mother's timeless moral support.

To all those who crave knowledge in dynamics, I hope this book will stimulate your thoughts and ideas; and for that I wish you success and prosperity.

Farid Amirouche
Chicago, Illinois 2004

Fundamentals of Multibody Dynamics

1

Particle Dynamics: The Principle of Newton's Second Law

1.1 Introduction

The dynamics of mechanical systems was first stated by Isaac Newton in his *Principia* of 1687. Newton's laws form the basis for the derivation of the equations of motion for particles. Modern dynamics is introduced through the use of vectors, free-body diagrams and reference frames. The use of particle mass to represent a body is an idealized concept that provides the simplest model in dynamics. It is important to note that Newton's second law, in its present form, has been used to derive current dynamic principles such as work and energy. Dynamics is made easy by transforming its vectorial representation to scalar forms by means of using dot products. This was the basis for variational principles. We shall review the dot product and cross product of vectors to see how some of these rules apply. The aim of this chapter is to review the basic principles in Newton's second law to formulate the equation of motion of problems involving system of particles.

1.2 Vectors

There are three types of vectors, namely, free vector, sliding vector and fixed vector. A free vector is a vector whose point of application and direction are free to move in space. In general, a free vector is a vector that can move without any constraint. A sliding vector is a vector whose point of application is not fixed but can slide along its line of action. A bound vector is a fixed vector.

Consider Figure 1.1 where $\bar{V}$ is a vector defined as:

$$\bar{V} = V_x\mathbf{i} + V_y\mathbf{j} + V_z\mathbf{k} \qquad (1.1)$$

where **i**, **j**, **k** are unit vectors along x, y and z, respectively. V_x, V_y and V_z are the components of the vector along the respective axis. For example, given a vector $\bar{V}$ along a direction $\bar{n}$, then we can write

$$\bar{V} = |V|\,\bar{n} \qquad (1.2)$$

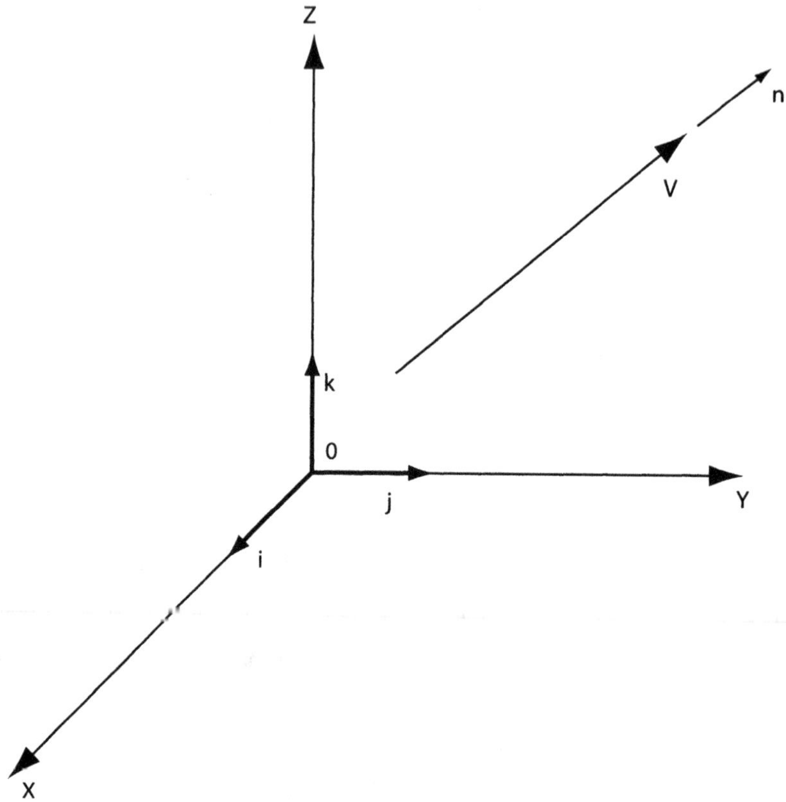

Fig. 1.1. Defining a vector $\bar{V}$ and its unit vector $\bar{n}$ with respect to a reference (x, y, z)

where $\bar{n}$ defines a unit vector (length of one). So if a vector is divided by its magnitude, it results into a unit vector; both $\bar{n}$ and $\bar{V}$ have the same direction.

The sum of two vectors is commutative.

$$\bar{A} + \bar{B} = \bar{B} + \bar{A} = \bar{C} \tag{1.3}$$

Vector addition can be used to group vectors of any size to form inner addition.

$$\begin{aligned} \bar{A} + \bar{B} + \bar{C} + \bar{D} &= \bar{A} + (\bar{B} + \bar{C}) + \bar{D} \\ &= (\bar{A} + \bar{B}) + (\bar{C} + \bar{D}) = (\bar{A} + \bar{B} + \bar{C}) + \bar{D} \end{aligned} \tag{1.4}$$

When adding vectors, one needs to keep in mind that convenience is an important factor as grouping is performed to better achieve the end result. This is useful when we have closed form polygon of vectors.

1.2.1 Scalar Product

The scalar product of two vectors is denoted by

$$\bar{A} \cdot \bar{B} = AB \cos\theta \tag{1.5}$$

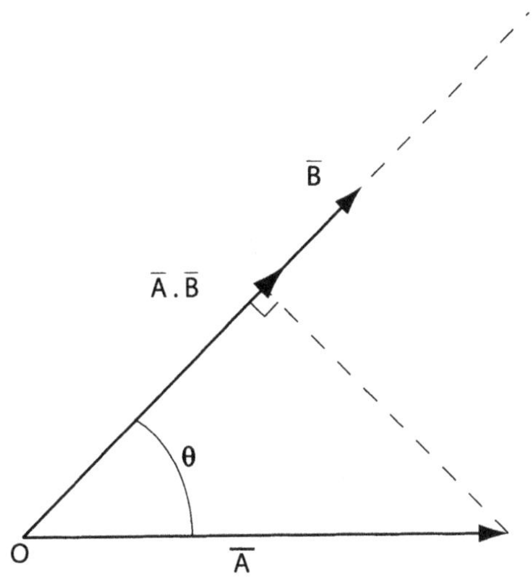

Fig. 1.2. Two vectors $\bar{A}$ and $\bar{B}$ in space and the projection of $\bar{A}$ along $\bar{B}$

The dot product of any two vectors represents the projection of one vector onto the other. The projection of $\bar{A}$ onto $\bar{B}$ or $\bar{B}$ onto $\bar{A}$ yields a scalar with a magnitude of $AB\cos\theta$; AB represents the product of the magnitude of vectors $\bar{A}$ and $\bar{B}$, respectively.

Consider two vectors fixed in an inertial frame R with unit vectors **i**, **j**, **k** . These vectors can be expressed as

$$\begin{aligned} \bar{V} &= V_x\mathbf{i} + V_y\mathbf{j} + V_z\mathbf{k} \\ \bar{W} &= W_x\mathbf{i} + W_y\mathbf{j} + W_z\mathbf{k} \end{aligned} \tag{1.6}$$

The dot product of the above two vectors is then found to be

$$\begin{aligned} \bar{V}\cdot\bar{W} &= (V_x\mathbf{i} + V_y\mathbf{j} + V_z\mathbf{k})\cdot(W_x\mathbf{i} + W_y\mathbf{j} + W_z\mathbf{k}) \\ \bar{V}\cdot\bar{W} &= V_xW_x + V_yW_y + V_zW_z \end{aligned} \tag{1.7}$$

Recall that

$$\begin{cases} \mathbf{i}\cdot\mathbf{i} = 1 & \mathbf{j}\cdot\mathbf{j} = 1 & \mathbf{k}\cdot\mathbf{k} = 1 \\ \mathbf{i}\cdot\mathbf{j} = 0 & \mathbf{i}\cdot\mathbf{k} = 0 & \mathbf{j}\cdot\mathbf{k} = 0 \end{cases} \tag{1.8}$$

In most cases, the dot product is used to project a vector quantity along a certain direction, hence eliminating its vector representation. This is a very useful concept and is used quite extensively in dynamics.

1.2.2 Vector Product

Another important feature in physics is the cross product of vectors. This, of course, is important when we define dynamic principles that are the result of a cross product

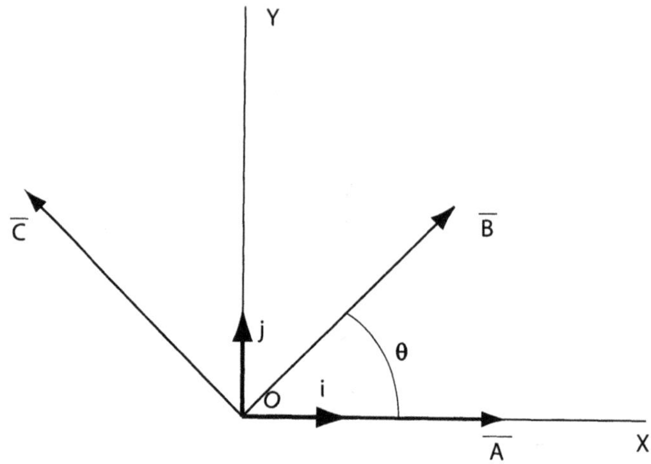

Fig. 1.3. C is the cross product of A and B

operation such as the moment of a force or an angular momentum. A vector normal to the plane of the two vectors is the result of the product of two vectors. For instance, let the product of two vectors be defined by

$$\bar{C} = \bar{A} \times \bar{B} \tag{1.9}$$

The vector $\bar{C}$ can then be expressed as

$$\bar{C} = |C| \; \mathbf{k} \tag{1.10}$$

and its magnitude is given by

$$|C| = AB \sin\theta \tag{1.11}$$

The rule of vector cross product can be understood through a simple schematic denoted by

$$\begin{matrix} \mathbf{i} \circlearrowright \mathbf{j} \\ \mathbf{k} \end{matrix} \tag{1.12}$$

where

$$\begin{aligned} \mathbf{i} \times \mathbf{j} &= \mathbf{k} \\ \mathbf{k} \times \mathbf{i} &= \mathbf{j} \end{aligned} \tag{1.13}$$

As long as the product of two vectors is along the clockwise direction, the third term defines the result. If the cross product were in the opposite direction of the arrow, then it would result in a negative sign to be introduced such as

$$\mathbf{j} \times \mathbf{i} = -\mathbf{k}$$

$$\mathbf{k} \times \mathbf{j} = -\mathbf{i} \tag{1.14}$$

Conventionally, mathematical representation of the cross product of two vectors in Cartesian coordinates is by calculating the determinant of a 3×3 matrix that is given by two vectors A and B where

$$\bar{A} = A_x\mathbf{i} + A_y\mathbf{j} + A_z\mathbf{k} \quad \text{and} \quad \bar{B} = B_x\mathbf{i} + B_y\mathbf{j} + B_z\mathbf{k}$$

The cross product is then derived as follows:

$$\bar{A} \times \bar{B} = \begin{vmatrix} i & j & k \\ A_x & A_y & A_z \\ B_x & B_y & B_z \end{vmatrix} \tag{1.15}$$

$$= (A_y B_z - B_y A_z)\,\mathbf{i} - (A_x B_z - B_x A_z)\,\mathbf{j} + (A_x B_y - B_x A_y)\,\mathbf{k}$$

When more than two vectors are involved in a cross product, we need to show the proper grouping of vectors using parentheses. For example, consider the following example

$$\bar{A} \times (\bar{B} \times \bar{C}) \times (\bar{D} \times \bar{E}) = \bar{F} \tag{1.16}$$

This is different from the following grouping of vectors

$$(\bar{A} \times \bar{B}) \times \bar{C} \times (\bar{D} \times \bar{E}) \neq \bar{F} \tag{1.17}$$

Also note that we can write the cross product of three vectors as a sum of two vectors involving the dot product, which is simpler to compute:

$$\bar{A} \times (\bar{B} \times \bar{C}) = (\bar{A} \cdot \bar{C})\bar{B} - (\bar{A} \cdot \bar{B})\bar{C} \tag{1.18}$$

The latter becomes handy when developing equations that contain a cross product of more than two vectors.

1.3 Derivative of a Vector

Suppose a particle m is traveling along a path C as shown in Figure 1.4. Let $\bar{r}$ define the position of the particle with respect to the fixed frame R. At time t, the position of the particle is $\bar{r}(t)$ and at $(t + \Delta t)$ it is $\bar{r}(t + \Delta t)$. The derivative of $\bar{r}$ with respect to time is known as the velocity and is given by:

$$\frac{d\bar{r}}{dt} = \lim_{\Delta t \to 0} \frac{\bar{r}(t + \Delta t) - \bar{r}(t)}{\Delta t} \tag{1.19}$$

$$\bar{r}(t + \Delta t) = \bar{r}(t) + \Delta \bar{r} \tag{1.20}$$

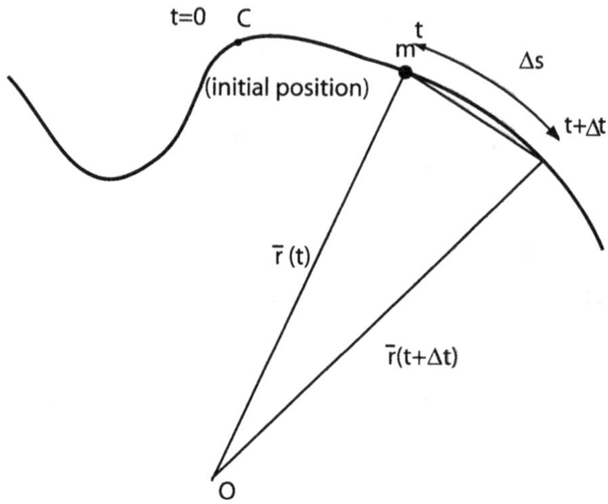

Fig. 1.4. A particle m traveling along a path C

Substituting the above equation into equation (1.19), we obtain

$$\frac{d\bar{r}}{dt} = \lim_{\Delta t \to 0} \frac{\Delta \bar{r}}{\Delta t} \tag{1.21}$$

We can write the above equation further as

$$\bar{v} = \lim_{\Delta t \to 0} \left(\frac{\Delta r}{\Delta s}\right) \left(\frac{\Delta s}{\Delta t}\right) = \left(\frac{d\bar{r}}{ds}\right) \frac{ds}{dt} \tag{1.22}$$

Note how

$$\left|\frac{\Delta \bar{r}}{\Delta s}\right| \to 1 \quad \text{as} \quad \Delta t \to 0$$

Hence $d\bar{r}/ds$ is a unit tangent vector, and we can refer to it as $\bar{e}_t$ defined as

$$\frac{d\bar{r}}{ds} = \bar{e}_t \tag{1.23}$$

Furthermore, the term $\dfrac{ds}{dt}$ or $\dot{s}$ is a scalar term defining the speed; hence the velocity can be expressed as

$$\bar{v} = \dot{s}\bar{e}_t \tag{1.24}$$

This defines the velocity of a particle which is the product of its speed along a vector $\bar{e}_t$ tangent to the path of travel C. Further differentiation of the velocity yields the acceleration. Hence,

$$\bar{a} = \frac{d\bar{v}}{dt} = \ddot{s}\bar{e}_t + \dot{s}\dot{\bar{e}}_t \tag{1.25}$$

If the path does not change, $\bar{e}_t$ remains constant and $\dot{\bar{e}}_t$ is zero. On the other hand, if the path C is such that $\bar{e}_t$ changes direction then $\dot{\bar{e}}_t \neq 0$. The question then is what is $\dot{\bar{e}}_t$? To answer this question we need to define the concept of angular velocity.

1.4 Velocity and Acceleration in Several Coordinates

Cartesian Coordinates

Let the reference frame R shown in Figure 1.5 define a set of Cartesian coordinates X, Y and Z with $\mathbf{i}, \mathbf{j}, \mathbf{k}$ as their respective unit vectors. The position $\bar{r}$ of a particle m moving on a path C and its velocity are

$$\begin{aligned} \bar{r} &= x\mathbf{i} + y\mathbf{j} + z\mathbf{k} \\ \bar{v} &= \dot{x}\mathbf{i} + \dot{y}\mathbf{j} + \dot{z}\mathbf{k} \end{aligned} \tag{1.26}$$

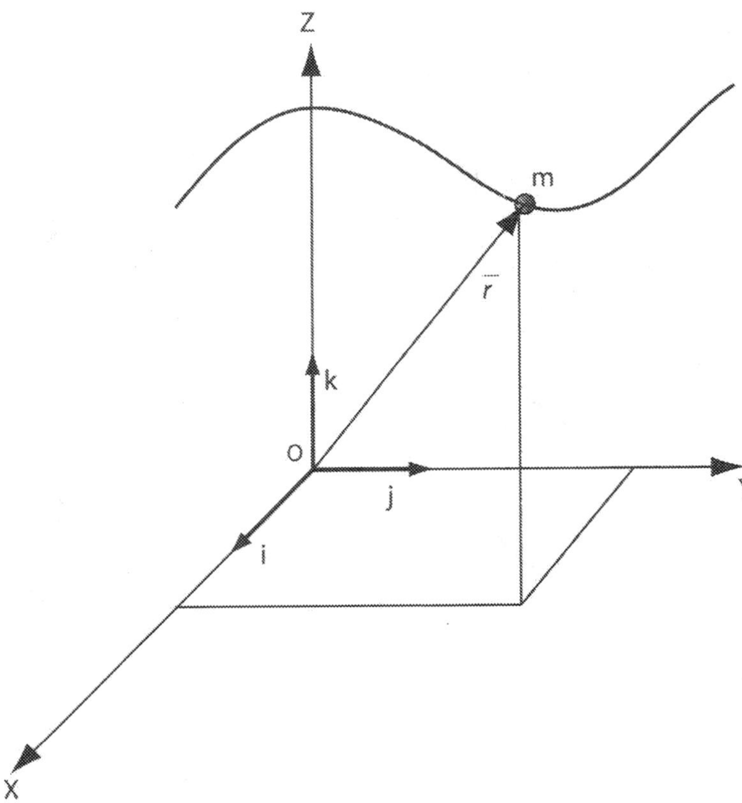

Fig. 1.5. Position of a particle m defined in a Cartesian coordinate

where $\dot{\mathbf{i}}$, $\dot{\mathbf{j}}$ and $\dot{\mathbf{k}}$ are zero and do not appear in the velocity equation. Similarly we obtain the acceleration of particle m with respect to R as:

$$\bar{a} = \frac{d\bar{v}}{dt} = \ddot{x}\mathbf{i} + \ddot{y}\,\mathbf{j} + \ddot{z}\mathbf{k} \tag{1.27}$$

Example 1.4.1

The piston is connected to the flywheel of radius R by a rod of length l. (a) Find the position of the piston as a function of time if the angular displacement is defined as $\theta = \omega t$. (b) Derive an expression for the piston center mass velocity.

Solution

From the triangle shown in the Figure 1.6 we define a geometrical relationship based on the sides of $\Delta\ Rly$ where:

$$l^2 = y^2 + R^2 - 2Ry\cos\theta \tag{1.28}$$

This can be expressed further as

$$y^2 - (2R\cos\theta)y - (l^2 - R^2) = 0 \tag{1.29}$$

Solving for y which defines the position of the piston, we get

$$y = R\cos\theta \pm \sqrt{l^2 - R^2\sin^2\theta}$$

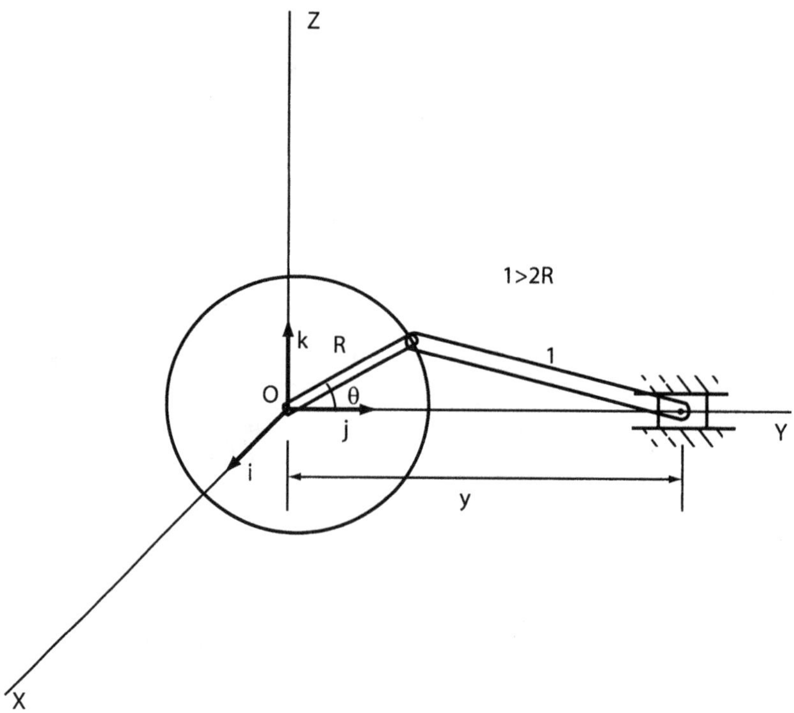

Fig. 1.6. A piston connected to a flywheel of radius R

or

$$y = R\cos\omega t + \sqrt{l^2 - R^2\sin^2\omega t}$$

So the position of the piston with respect to an inertia frame using the Cartesian coordinates X,Y,Z is

$$\bar{r} = yj$$

and its velocity is

$$\bar{v} = \dot{y}j$$

where

$$\dot{y} = -R\omega\sin\omega t + \frac{1}{2}u^{\frac{-1}{2}}\dot{u}$$

where

$$u = l^2 - R^2\sin^2\omega t$$

and

$$\dot{u} = -2R^2\omega\sin\omega t\cos\omega t$$

Cylindrical Coordinates

In this problem, we have modeled the position as a point mass. Hence l represents the distance from its mass center to the point of attachment on the flywheel (Figure 1.6). Let x, y and z be axes of an inertial frame of reference denoted by R. We now introduce three new parameters r, z and θ to define the position of a particle m in space (Figure 1.7). We define unit vectors $\bar{e}_r$, $\bar{e}_\theta$ and $\bar{e}_z$ along the respective axes as shown. The position of m with respect to R is given by:

$$\bar{r} = r\bar{e}_r + z\bar{e}_z \tag{1.30}$$

and

$$\frac{d\bar{r}}{dt} = \dot{r}\bar{e}_r + r\dot{\bar{e}}_r + \dot{z}\bar{e}_z \tag{1.31}$$

Note that $\bar{e}_r$ and $\bar{e}_\theta$ can be written as a function of the unit vectors **i**, **j** and **k**

$$\begin{aligned} \bar{e}_r &= \cos\theta\,\mathbf{i} + \sin\theta\,\mathbf{j} \\ \bar{e}_\theta &= -\sin\theta\,\mathbf{i} + \cos\theta\,\mathbf{j} \end{aligned} \tag{1.32}$$

Differentiation of the above equations leads to

$$\begin{aligned} \frac{d\bar{e}_r}{dt} &= \dot{\theta}(-\sin\theta\,\mathbf{i} + \cos\theta\,\mathbf{j}) \\ \frac{d\bar{e}_\theta}{dt} &= \dot{\theta}(-\cos\theta\,\mathbf{i} - \sin\theta\,\mathbf{j}) \end{aligned} \tag{1.33}$$

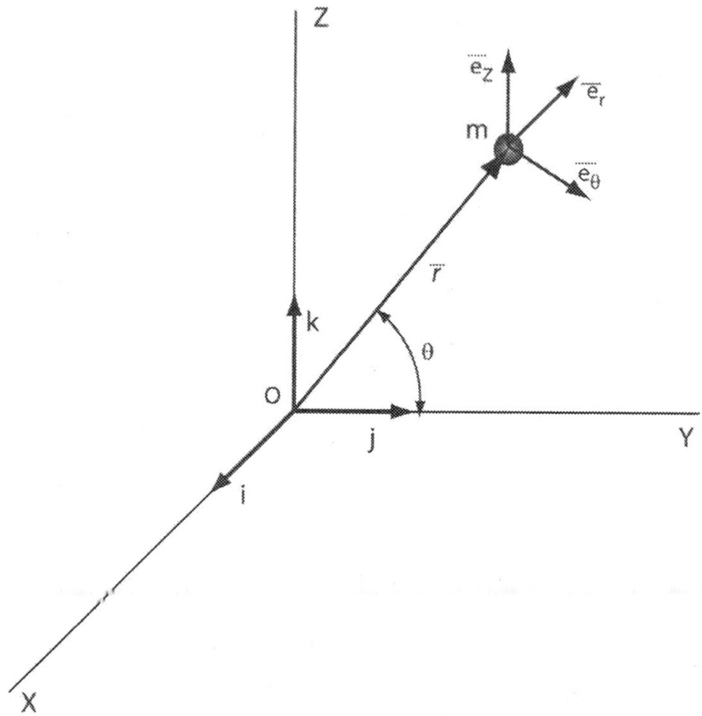

Fig. 1.7. A projection of $\bar{e}_r$ and $\bar{e}_\theta$ in the ij plane

Noting that the above two expressions in terms of $\bar{e}_r$ and $\bar{e}_\theta$ are

$$\frac{d\bar{e}_r}{dt} = \dot{\theta}\,\bar{e}_\theta$$

and

$$\frac{d\bar{e}_\theta}{dt} = -\dot{\theta}\,\bar{e}_r \tag{1.34}$$

Substituting these two terms of equation (1.34) in the velocity equation, we obtain the velocity expression in the cylindrical coordinates

$$\bar{v} = \dot{r}\bar{e}_r + r\dot{\theta}\,\bar{e}_\theta + \dot{z}\bar{e}_z \tag{1.35}$$

Differentiating further, we obtain the acceleration vector:

$$\bar{a} = \ddot{r}\bar{e}_r + \dot{r}\dot{\bar{e}}_r + \dot{r}\dot{\theta}\,e_\theta + r\ddot{\theta}\,\bar{e}_\theta + r\dot{\theta}\,\dot{\bar{e}}_\theta + \ddot{z}\,\bar{e}_z \tag{1.36}$$

Simplifying the above we arrive at the acceleration of a point in space in terms of the cylindrical coordinates.

$$\bar{a} = (\ddot{r} - r\dot{\theta}^2)\bar{e}_r + (\dot{r}\ddot{\theta} + 2\dot{r}\dot{\theta})\bar{e}_\theta + \ddot{z}\,\bar{e}_z \tag{1.37}$$

Note: We obtained the velocity and acceleration strictly in terms of r, z and θ and their time derivatives.

Example 1.4.2

A mass m slides along a frictionless track. Consider a circular track, which is part of a disk that rotates in a horizontal plane at a constant angular velocity (ω_0 = constant) as shown in the Figure 1.8. The initial conditions are such that at time $r = r_o$ and $v = v_o$ and the radial acceleration a_r is zero. Show that the position of the particle is given by

$$r(t) = r_0 \sinh \omega_0 t \quad \text{where} \quad r_0 = \frac{v_0}{\omega_0}$$

Solution

We can use the cylindrical coordinates in defining the position of the particle with respect to an inertial frame R, where

$$\bar{r} = r\bar{e}_r + z\bar{e}_z$$

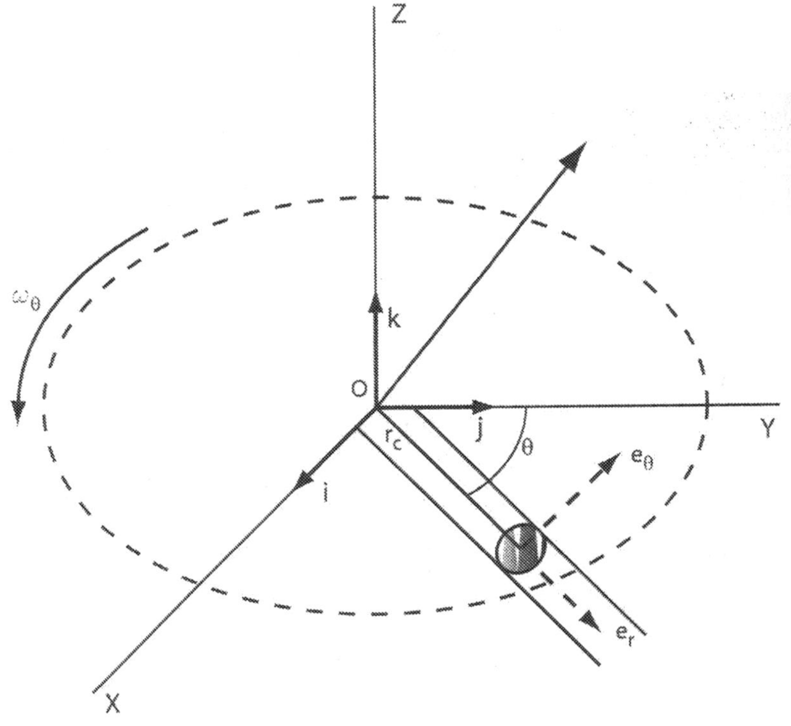

Fig. 1.8. A particle sliding on a rotating frictionless track

Differentiating the position frame twice we get the acceleration

$$\bar{a} = (\ddot{r} - r\dot{\theta}^2)\bar{e}_r + (r\ddot{\theta} + 2\dot{r}\dot{\theta})\bar{e}_\theta$$

Since the mass rotates in a planar motion there is no z component and the radial acceleration is zero (ω = constant). Hence

$$\ddot{r} - r\dot{\theta}^2 = 0$$

or

$$\ddot{r} = r\omega_0^2$$

Differentiation of the given position $\bar{r}(t)$ yields

$$\ddot{r} = \omega_0 v_0 \sinh \omega_0 t$$

Substituting in the radial component of the acceleration the above equation becomes

$$\omega_0 v_0 \sinh \omega_0 t - \omega_0^2 \left(\frac{v_0}{\omega_0}\right) \sinh \omega_0 t = 0$$

which of course satisfies the given equation, as the two terms are equal. The initial conditions are also satisfied for $t = 0$.

Cylindrical Coordinates: Alternative Derivation

Consider Figure 1.7 where the position vector of particle m is defined strictly in terms of r and z.

$$\bar{r} = r\bar{e}_r + z\bar{e}_z$$

Furthermore, let the angular velocity of the mass m with respect to the fixed inertia frame R be given by

$$\bar{\omega} = \dot{\phi}\bar{e}_z \tag{1.38}$$

Again the velocity of particle m with respect to an inertial fixed frame R in terms of the cylindrical coordinates obtained by differentiation of $\bar{r}$:

$$\frac{d\bar{r}}{dt} = \bar{v} = \dot{r}\bar{e}_r + \dot{z}\bar{e}_z + r\dot{\bar{e}}_r + z\dot{\bar{e}}_z = \dot{r}\bar{e}_r + \dot{z}\bar{e}_z + r\dot{\phi}\bar{e}_\phi \tag{1.39}$$

where the unit vector derivatives are obtained by making use of the angular velocity as follows:

$$\dot{e}_r = \bar{\omega} \times \bar{e}_r = \dot{\phi}\bar{e}_\phi \tag{1.40}$$

and

$$\dot{e}_z = \bar{\omega} \times \bar{e}_z = 0 \tag{1.41}$$

Note how equations (1.39) and (1.35) are identical. The above derivation makes use of the angular velocity definition given by equation (1.38), whereas equation (1.35) is obtained through strict differentiation.

Spherical Coordinates

In spherical coordinates the position of a point in space is defined in terms of three parameters r, θ and ϕ (see Figure 1.9). Essentially, this is a case where the most convenient way to represent a position is done through two orientation angles and a length measured from a fixed point O. The angle θ represents the angle between the ZOP plane and the XZ plane, and ϕ is the angle between the Z axis and line OP.

The position of a point particle P is given by

$$\bar{r} = r\bar{e}_r \tag{1.42}$$

The velocity of point P is obtained by differentiation of the above equation, yielding

$$\bar{v} = \dot{r}\bar{e}_r + r\dot{\bar{e}}_r \tag{1.43}$$

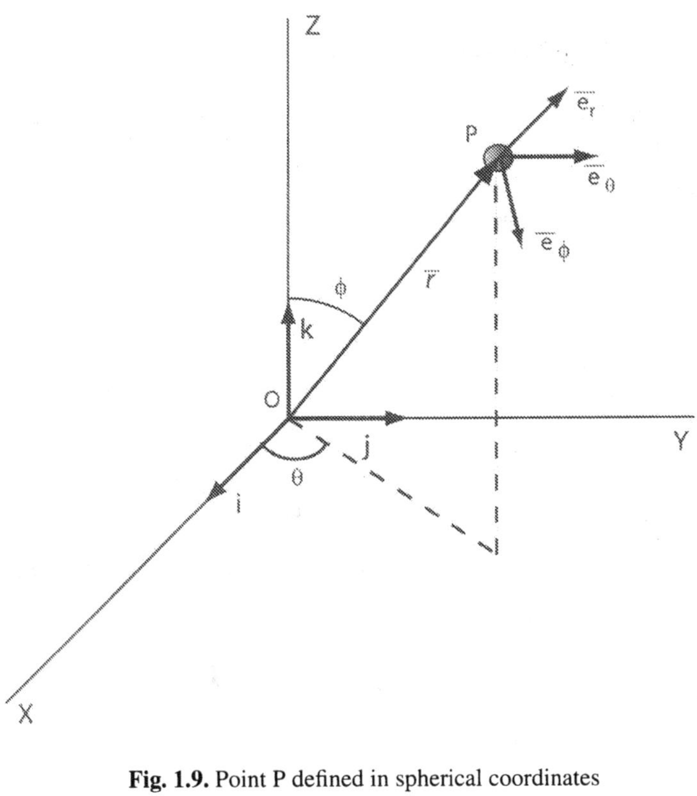

Fig. 1.9. Point P defined in spherical coordinates

Now we need to find an expression for $\dot{\bar{e}}_r$. Note that $\bar{e}_r$ can be expressed as follows

$$\bar{e}_r = \sin\phi\cos\theta\ \mathbf{i} + \sin\phi\sin\theta\ \mathbf{j} + \cos\phi\ \mathbf{k} \tag{1.44}$$

Then, through simple calculus we obtain its derivative as

$$\begin{aligned}\dot{\bar{e}}_r &= (\dot{\phi}\cos\phi\cos\theta - \dot{\theta}\sin\phi\sin\theta)\mathbf{i} \\ &\quad + (\dot{\phi}\cos\phi\sin\theta + \dot{\theta}\sin\phi\cos\theta)\,\mathbf{j} + (-\dot{\phi}\sin\phi)\,\mathbf{k}\end{aligned} \tag{1.45}$$

In addition, we can write the unit vectors $\bar{e}_\phi$ and $\bar{e}_\theta$ in terms of **i**, **j** and **k**

$$\begin{aligned}\bar{e}_\phi &= \cos\phi\cos\theta\ \mathbf{i} + \cos\phi\sin\theta\ \mathbf{j} - \sin\phi\ \mathbf{k} \\ \bar{e}_\theta &= -\sin\theta\ \mathbf{i} + \cos\theta\ \mathbf{j}\end{aligned} \tag{1.46}$$

Substituting the above equation into (1.43) we get the velocity of point P in spherical coordinates

$$\bar{v} = \dot{\bar{r}}\bar{e}_r + r\dot{\phi}\,e_\phi + r\dot{\theta}\sin\phi\,\bar{e}\theta \tag{1.47}$$

In similar fashion, the acceleration through differentiation of equation (1.47) is found to be

$$\begin{aligned}\bar{a} &= (\ddot{r} - r\dot{\phi}^2 - r\dot{\theta}^2\sin^2\phi)\bar{e}_r \\ &\quad + (2\dot{r}\dot{\phi} + r\ddot{\phi} - r\dot{\theta}^2\sin\phi\cos\phi)\bar{e}_\phi \\ &\quad + (2\dot{r}\dot{\theta}\sin\phi + 2r\dot{\phi}\dot{\theta}c\cos\phi + r\ddot{\theta}\sin\phi)\bar{e}_\theta\end{aligned} \tag{1.48}$$

Of course one can see that the final equation is expressed strictly in terms of i, ϕ and θ along $\bar{e}_r$, $\bar{e}_\phi$, and $\bar{e}_\theta$.

Alternative Derivation for Spherical Coordinates

The derivation of the velocity and acceleration of point m in spherical coordinates could be obtained making use of the angular velocity definition as follows. Let the angular velocity be defined by

$$\bar{\omega} = \dot{\phi}\bar{e}_z + \dot{\theta}\bar{e}_\phi \tag{1.49}$$

Then the velocity of particle P is found through differentiation of $\bar{r}$, where

$$\begin{aligned}\frac{d\bar{r}}{dt} &= \bar{v} = \dot{r}\bar{e}_r + r\dot{\bar{e}}_r \\ &= \dot{r}\bar{e}_r + r(\bar{\omega}\times\bar{e}_r) \\ &= \dot{r}e_r + r[(\dot{\phi}\,e_z + \dot{\theta}\,e_\phi)\times\bar{e}_r]\end{aligned} \tag{1.50}$$

Note that

$$\bar{e}_z = \cos\theta\,\bar{e}_r - \sin\theta\,\bar{e}_\phi \tag{1.51}$$

Substituting $\bar{e}_z$ in the angular velocity and velocity equation we obtain

$$\bar{\omega} = \dot{\phi}\,(\cos\theta\,\bar{e}_r - \sin\theta\,\bar{e}_\phi) + \dot{\theta}\,\bar{e}_\phi \tag{1.52}$$

and

$$\bar{v} = \dot{r}\bar{e}_r + (r\dot{\phi}\,\sin\theta)\bar{e}_\phi + r\dot{\theta}\,\bar{e}_\theta \tag{1.53}$$

Differentiating further, we arrive at

$$\begin{aligned}\bar{a} = \frac{d\bar{v}}{dt} &= (\ddot{r} - r\dot{\theta}^2 - r\dot{\phi}^2\sin^2\theta)\bar{e}_r \\ &+ (r\ddot{\theta} + 2\dot{r}\dot{\theta} - r\dot{\phi}^2\sin\theta\,\cos\theta)\bar{e}_\theta + r\ddot{\phi}\sin\theta \\ &+ 2\dot{r}\dot{\phi}\sin\theta + 2r\dot{\theta}\,\dot{\phi}\cos\theta)\bar{e}_\phi\end{aligned} \tag{1.54}$$

The results are the same as equation (1.48). In the above derivation we made use of the definition of the angular velocity, which states that

$$\frac{d\bar{r}}{dt} = \bar{\omega} \times \bar{r} \tag{1.55}$$

where $\bar{\omega}$ defines the angular velocity of $\bar{r}$ in the fixed inertial frame R.

1.5 Dynamics of a System of Particles

This section deals with the dynamics of n number of particles. It will serve as a motivation to gain some insight into rigid-body dynamics and as well as the derivation of dynamical principles such as work and energy. Particles can be viewed individually or as a group forming a system.

Each mass is subjected to forces of action and reaction, these internal forces denoted by $\bar{f}_{ij}$ are colinear and $\bar{f}_{ij} = -\bar{f}_{ji}$; so $\bar{f}_{ii} = 0$ $(i, j = 1 \ldots n)$ (from Figure 1.10).

Let F_i represent the external forces applied to m_i; then for a particle i using Newton's second law, we have

$$m_i\ddot{\bar{r}}_i = \bar{F}_i + \sum_{j=1}^{n} \bar{f}_{ij} \tag{1.56}$$

For a system of particles, we get

$$\sum_{i=1}^{n} m_i\ddot{\bar{r}}_i = \sum_{i=1}^{n} \bar{F}_i + \sum_{i=1}^{n}\sum_{j=1}^{n} \bar{f}_{ij} \tag{1.57}$$

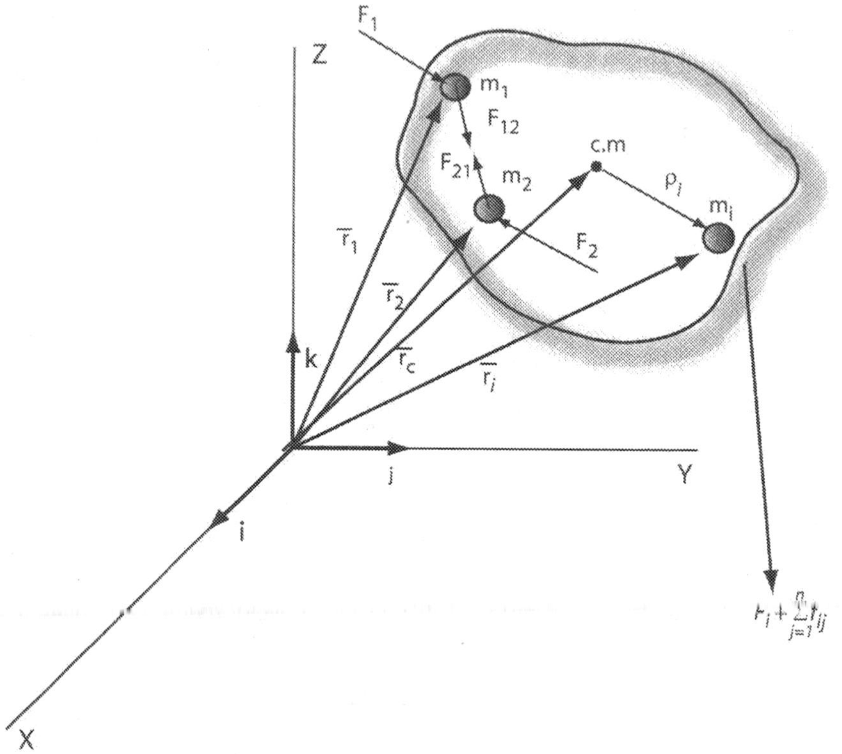

Fig. 1.10. A system of n particles subjected to external and internal forces

From the law of action and reaction we see that

$$\sum_{i=1}^{n}\sum_{j=1}^{n} \bar{f}_{ij} = 0 \tag{1.58}$$

and the total mass is

$$\sum m_i = m \tag{1.59}$$

We define the center of mass of the system as follows

$$\frac{1}{m}\sum_{i=1}^{n} m_i r_i = r_c \tag{1.60}$$

where r_c is the position from O to the center of mass (C.M.) of the system. The total external force acting on the center of mass can be represented by a resultant force

$$\bar{F} = \sum_{i=1}^{n} \bar{F}_i \tag{1.61}$$

The equations of motion for the system of particles then becomes

$$\bar{F} = m\ddot{\bar{r}}_c \tag{1.62}$$

where $\ddot{\bar{r}}_c$ defines the center of mass acceleration and F is the resultant external force applied at C.

1.6 Work and Kinetic Energy

Some of the important tools in mechanics and dynamics in general are principles that use scalar quantities and can be used to measure such things as work and energy. The latter are the outcome of Newton's second law when it is subjected to the projection (dot product) of these forces along certain vectors. For example, observe how the concept of work and energy is derived. For a single particle m_i moving in space, we can write the following:

$$\bar{F}_i = m_i \bar{a}_i \tag{1.63}$$

Then, let us take the dot product of the inertial component $m\bar{a}$ with the velocity vector $\bar{v}$. This is shown in the following equation:

$$\bar{F}_i \cdot \bar{v} = m_i \bar{a}_i \cdot \bar{v} \tag{1.64}$$

The above equation can be expressed further as

$$\bar{F}_i \cdot \frac{d\bar{r}_i}{dt} = m_i \frac{d\bar{v}_i}{dt} \cdot \bar{v}_i \tag{1.65}$$

Using the chain rule, we note that

$$\frac{d}{dt}(\bar{v} \cdot \bar{v}) = \frac{d\bar{v}}{dt} \cdot \bar{v} + \bar{v} \cdot \frac{d\bar{v}}{dt} \tag{1.66}$$

which is equivalent to

$$\frac{d}{dt}(\bar{v}_i \cdot \bar{v}_i) = 2\frac{d\bar{v}_i}{dt} \cdot \bar{v}_i \tag{1.67}$$

Substituting the above expression into equation (1.65) we get

$$\bar{F}_i \cdot d\bar{r}_i = \frac{1}{2} m_i d\left(\bar{v}_i \cdot \bar{v}_i\right) \tag{1.68}$$

The dot product yields a square term for the velocity, therefore

$$\bar{F}_i \cdot d\bar{r}_i = \frac{1}{2} md\left|v\right|^2 \tag{1.69}$$

For the system of n particles, we integrate equation (1.69) which results into

$$\int_{r1}^{r2} \bar{F} \cdot d\bar{r} = \frac{1}{2} mv^2 |_{v1}^{v2} \tag{1.70}$$

or

$$\int_{r1}^{r2} \bar{F} \cdot d\bar{r} = \frac{1}{2}mv_2^2 - \frac{1}{2}mv_1^2 \tag{1.71}$$

Observe how equation (1.71) has scalar components on both sides of the equation and defines two new concepts, namely, work on the left-hand side and kinetic energy on the other side.

The above equation states that work performed by $\bar{F}$ on a particle, which moves from position 1 to position 2, is equal to the change in kinetic energy between the two positions. The above equation of work and energy applies to a system of particles whose center of mass is subject to a resultant force.

Let's check the contribution of the work performed by the internal forces. We write the work equation W_i as follows

$$W_i = \int \left(\bar{F}_i + \sum_{j=1}^{n} \bar{f}_{ij} \right) d\bar{r} \quad \begin{cases} i = 1, \ldots, n \\ j = 1, \ldots, m \end{cases} \tag{1.72}$$

where f_{ij} denote the internal force acting on m_i, and $\bar{F}_i$ the external force.

In Figure 1.10 the position of a particle m_i be given by

$$\bar{r}_i = \bar{r}_c + \bar{\rho}_i \tag{1.73}$$

where $\bar{r}_c$ is the position vector of the center of mass with reference to R and ρ_i is the position vector of m_i with reference to the center of mass.

The sum of the work performed by all n particles is then found to be

$$W = \sum_{i=1}^{n} W_i = \sum \int \left(\bar{F}_i + \sum_{j=1}^{n} \bar{f}_{ij} \right) \cdot (d\bar{r}_c + d\bar{\rho}_i) \tag{1.74}$$

Performing the dot product and expanding the equation yields

$$W = \sum_{i=1}^{n} \int \left(\bar{F}_i + \sum_{j=1}^{m} f_{ij} \right) \cdot d\bar{r}_c + \sum_{i=1}^{n} \int \left(\bar{F}_i + \sum_{j=1}^{m} \bar{f}_{ij} \right) \cdot d\bar{\rho}_i \tag{1.75}$$

For each particle, the kinetic energy can be expressed as

$$\begin{aligned} \text{KE}_i &= \left[\frac{1}{2} m \dot{\bar{r}}_i \cdot \dot{\bar{r}}_i \right]_1^2 = \left[\frac{1}{2} m (\dot{\bar{r}}_c + \dot{\bar{\rho}}_i) \cdot (\dot{\bar{r}}_c + \dot{\bar{\rho}}_i) \right]_1^2 \\ &= \left[\frac{1}{2} m_i (\dot{\bar{r}}_c \cdot \dot{\bar{r}}_c + 2\dot{\bar{r}}_c \cdot \dot{\bar{\rho}}_i + \dot{\bar{\rho}}_i \cdot \dot{\bar{\rho}}_i) \right]_1^2 \end{aligned} \tag{1.76}$$

The kinetic energy of the system of particles is the sum of the kinetic energy associated with each particle.

$$\mathrm{KE} = \sum_{i=1}^{n} W_i = \left[\frac{1}{2} m v_c^2\right]_1^2 + \left[\bar{r}_c \cdot \sum_{i=1}^{n} m_i \bar{\dot{\rho}}_i\right]_1^2 + \left[\frac{1}{2} \sum m_i \bar{\dot{\rho}}_i^2\right]_1^2 \tag{1.77}$$

Note that

$$\sum m_i \dot{\rho}_i = 0$$

Hence, equation (1.77) reduces to

$$\mathrm{KE} = \sum_{i=1}^{n} W_i = \left[\frac{1}{2} m v^2\right]_1^2 + \left[\frac{1}{2} \sum_{i=1}^{n} m_i \dot{\rho}_i^2\right]_1^2 \tag{1.78}$$

Equalizing the work and kinetic energy, we obtain the following general expression for the work equation

$$\sum_{i=1}^{n} \int (\bar{F}_i \cdot d\bar{r}_c) + \sum_{i=1}^{n} \int (\bar{F}_i + \sum_{j=1}^{n} f_{ij}) \cdot d\rho_i = \left[\frac{1}{2} m v_c^2\right] + \left[\frac{1}{2} \sum_{i=1}^{n} m_i \dot{\rho}^2\right] \tag{1.79}$$

Note that the kinetic energy associated with $\dot{\rho}$ is due to the rigid-body rotation. Of course, we see that the total kinetic energy is a combination of energy associated with the center mass, v_c (due to translation) and the second term depicts the contribution from rotation. Note also how equations (1.71) and (1.79) are equivalent if $\rho =$ const. (no rotation).

Further Differentiation of Work

Let's consider an object in curvilinear motion relative to an inertial reference frame and specify its position by coordinates measured along its path from a reference point O in terms of the tangential unit vector $\bar{n}_t$ (see Figure 1.11). The object's velocity is

$$\bar{v} = \frac{ds}{dt} \bar{n}_t \tag{1.80}$$

Because $\bar{v} = d\bar{r}/dt$ we can multiply the velocity by dt to obtain an expression for the vector $d\bar{r}$ describing an infinitesimal displacement along the path:

$$d\bar{r} = \bar{v}\, dt = ds \bar{n}_t \tag{1.81}$$

The work done by the external forces acting on the object as a result of the displacement $d\bar{r}$ is

$$\sum \bar{F} \cdot d\bar{r} = \left(\sum \bar{F} \cdot \bar{n}_t\right) ds = \sum F_t ds \tag{1.82}$$

where $\sum F_t$ is the tangential component of the total force. Therefore, as the object moves from a position s_1 to a position s_2 the work is

$$W_{12} = \int_{s_1}^{s_2} \sum F_t ds \tag{1.83}$$

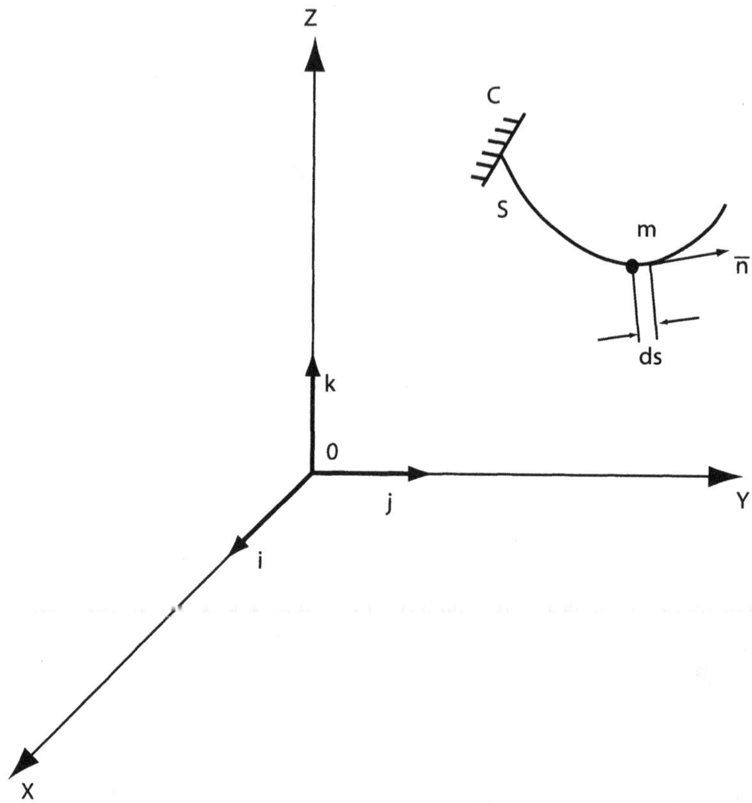

Fig. 1.11. A mass in curvilinear motion

This shows how work is a scalar quantity and that only the force component along the position vector contributes to the work.

Example 1.6.1
A device shown in Figure 1.12 is used for impact analysis. A 40 kg person sits on a rigid massless platform which is lifted to position 1 and released from rest. It falls and strikes a hard floor when it is in position 2. The spring constant $k = 1500$ N/m and the tension in each spring is 150 N when the person is in position 2. Neglect friction. What is the velocity of the person/seat just before it strikes the hard floor?

Solution
(a) Let r_0 be the unstretched length of one of the springs. In position 2, the tension in the spring is 150 N and its length is 3 m. Assuming the spring force is linear, we can write the following:

$$F = k(r - r_0)$$

or

$$150 = k(3 - r_0) = (1500)(3 - r_0)$$

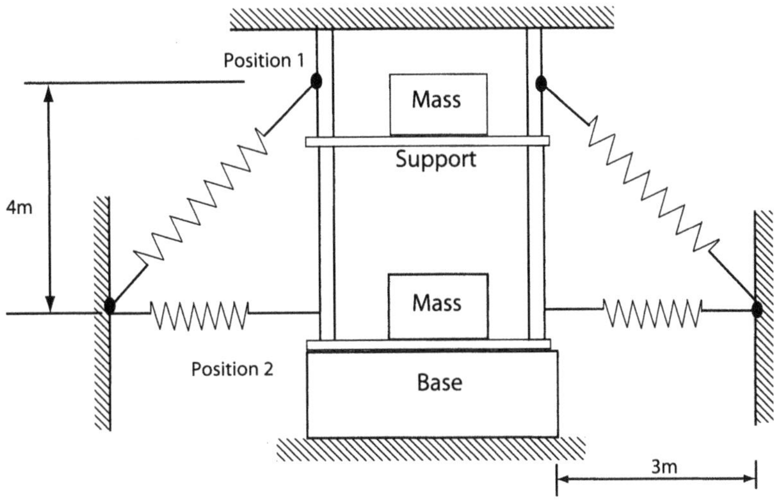

Fig. 1.12. A bouncing mechanism

We obtain $r_0 = 2$ m. The values of the stretch of each spring in positions 1 and 2 are easily obtained as $r_1 = \sqrt{4^2 + 3^2} - 2 = 3$ m and $r_2 = 1$ m. The total work performed by the forces acting on the center-piece mass are equal to the work done by the two springs from position 1 to position 2 is

$$W_{\text{springs}} = 2\left[-\frac{1}{2}k(r_2^2 - r_1^2)\right] = -(1500)[1^2 - 3^2] = 12{,}000\,\text{N-m}.$$

The work done by the body weight from position 1 to position 2 following our definition is

$$W = \int F ds = \int_0^h mgds = mgh$$

$$W_{\text{weight}} = mg(4\text{m}) = 1570\ \text{N-m}.$$

From the principle of work and energy, we can deduce the velocity when the person reaches position 2. That is

$$\sum W = W_{\text{springs}} + W_{\text{weight}} = \frac{1}{2}mv_2^2 - \frac{1}{2}mv_1^2$$

Solving for v_2 we obtain

$$v_2 = 8.32\ \text{m/s}$$

Note: The rigid mass contribution to work is neglected.

1.6.1 Conservation of Energy

The total work performed on an object by the external forces can be expressed as the change of a function of the object's position, called the potential energy. When all the forces that do work have this property, we can state the principle of work and energy as the conservation law or simply the law of conservation of energy: The sum of the kinetic and potential energies is then constant.

When we derived the principle of work and energy from Newton's second law, we were able to evaluate the integral on one side of the equation, obtaining the change in the kinetic energy. The work performed by the external force F can only be evaluated if $\bar{F}$ is given and expressed in a way consistent with $\bar{r}$. This is shown by

$$W_{12} = \int_{r_1}^{r_2} \sum \bar{F} \cdot d\bar{r} = \frac{1}{2}mv_2^2 - \frac{1}{2}mv_1^2 \tag{1.84}$$

Suppose we introduce a scalar function of position such that

$$dV = -\sum \bar{F} \cdot d\bar{r} \tag{1.85}$$

Then we evaluate the integral defining the work as follows:

$$W_{12} = \int_{r_1}^{r_2} \sum \bar{F} \cdot d\bar{r} = \int_{V_1}^{V_2} -dV = -(V_2 - V_1) \tag{1.86}$$

where V_1 and V_2 represent the values of V at the positions $\bar{r}_1$ and $\bar{r}_2$. The principle of work and energy would then have the simple form

$$\frac{1}{2}mv_1^2 + V_1 = \frac{1}{2}mv_2^2 + V_2 \tag{1.87}$$

which simply states that the sum of the kinetic energy and the function V is constant:

$$\frac{1}{2}mv^2 + V = \text{constant} \tag{1.88}$$

If the kinetic energy increases, V must decrease, and vice versa, in order for the total energy to remain constant. For this reason, we refer to V as the potential energy or energy required to produce the work needed.

Hence, if a potential energy exists for a given force $\bar{F}$, which means that a function of position V exists such that $dV = -\bar{F} \cdot d\bar{r}$, then $\bar{F}$ is said to be conservative. If all the forces that do work on a system are conservative, the total energy — the sum of the kinetic energy and the potential energies of the forces — is constant, or conserved. In that case, the system is said to be conservative, and you can use conservation of energy instead of the principle of work and energy to relate a change in its position to the change in its kinetic energy. The two approaches are equivalent.

1.6.2 Conservative Forces

You can apply conservation of energy only if the forces doing work on an object or system are conservative and their potential energies can be evaluated. In what follows, determine the potential energies of some conservative forces and use the results to demonstrate applications of conservation of energy.

Note that the work done by a conservative force as an object moves from position 1 to position 2 is independent of the object's path. This result follows from equation (1.84), which states that the work depends only on the values of the potential energy at positions 1 and 2. Equation (1.84) also implies that if the object moves along a closed path, returning to position 1, the work done by a conservative force is zero.

If we express V in terms of a Cartesian coordinate system,

$$V = V(x, y, z)$$

its differential dV is

$$dV = \frac{\partial V}{\partial x}dx + \frac{\partial V}{\partial y}dy + \frac{\partial V}{\partial z}dz \tag{1.89}$$

Expressing $\bar{F}$ and $d\bar{r}$ in terms of their Cartesian components, their dot product is

$$\begin{aligned} \bar{F} \cdot d\bar{r} &= (F_x\mathbf{i} + F_y\mathbf{j} + F_z\mathbf{k}) \cdot (dx\mathbf{i} + dy\mathbf{j} + dy\mathbf{k}) \\ &= F_x dx + F_y dy + F_z dz \end{aligned} \tag{1.90}$$

Substituting this expression into equation (1.89) we obtain

$$\frac{\partial V}{\partial x}dx + \frac{\partial V}{\partial y}dy + \frac{\partial V}{\partial z}dz = -(F_x dx + F_y dy + F_z dz) \tag{1.91}$$

which implies that

$$F_x = \frac{\partial V}{\partial x},\ F_y = \frac{\partial V}{\partial y},\ F_z = \frac{\partial V}{\partial z} \tag{1.92}$$

Given a potential energy V expressed in Cartesian coordinates, you can use the relations to determine the corresponding force. The force $\bar{F}$ is

$$\bar{F} = -\left(\frac{\partial V}{\partial x}\mathbf{i} + \frac{\partial V}{\partial y}\mathbf{j} + \frac{\partial V}{\partial z}\mathbf{k}\right) = -\nabla V, \tag{1.93}$$

where ∇V is the gradient of the potential energy V.

If a force $\bar{F}$ is conservative, its curl $\nabla \times \bar{F}$ is zero. The expression for the curl of $\bar{F}$ in Cartesian coordinates is

$$\nabla \times \bar{F} = \begin{vmatrix} \mathbf{i} & \mathbf{j} & \mathbf{k} \\ \dfrac{\partial}{\partial x} & \dfrac{\partial}{\partial y} & \dfrac{\partial}{\partial z} \\ F_x & F_y & F_z \end{vmatrix} \tag{1.94}$$

Substituting equation (1.92) into this expression confirms that $\nabla \times \bar{F} = 0$ when $\bar{F}$ is conservative. The converse is also true: A force $\bar{F}$ is conservative if its curl is zero. You can use this condition to determine whether a given force is conservative.

Example 1.6.2
Suppose that a mass m rests on a surface and you push it horizontally so that it slides along a path of length L. The magnitude of the friction force is $\mu_k mg$, and it points opposite to the direction of the mass motion. Show whether the friction force is conservative.

Solution
Let us compute the work performed by the friction force, that is

$$W_{12} = \int_0^L -\mu_k mgds = -\mu_k mgL$$

The work is proportional to the length of the path and therefore it is not independent of the object's path. This concludes that the friction forces are not conservative.

Example 1.6.3
Derive an expression for the work performed on a point mass m when it is subjected to an arbitrary force in terms of spherical coordinates. Given $\bar{F} = F_x\bar{\mathbf{i}} + F_y\bar{\mathbf{j}} + F_z\bar{\mathbf{k}}$ Evaluate its curl $\nabla \times \bar{F}$ and show whether $\bar{F}$ is conservative.

Solution
Let the position and angular velocity of a particle m in spherical coordinates be given by

$$\bar{r} = r\bar{e}_r$$
$$\bar{\omega} = \dot{\theta}\bar{e}_z + \dot{\phi}\bar{e}_\theta$$

Then

$$\frac{d\bar{r}}{dt} = \dot{r}\bar{e}_r + r\left[(\dot{\theta}\bar{e}_z + \dot{\phi}\bar{e}_\theta) \times \bar{e}_r\right]$$

We can express $\bar{e}_z$ in terms of $\bar{e}_r$ and $\bar{e}_\phi$ as:

$$\bar{e}_z = \cos\phi\bar{e}_r - \sin\phi\bar{e}_\phi$$

Expanding the derivative of $\bar{r}$ equation results into

$$\frac{d\bar{r}}{dt} = \dot{r}\bar{e}_r + r\left[(\dot{\theta}\cos\phi\bar{e}_r - \dot{\theta}\sin\phi\bar{e}_\phi + \dot{\phi}\bar{e}_\theta) \times \bar{e}_r\right]$$

The derivative of $\bar{r}$ then becomes

$$\frac{d\bar{r}}{dt} = \frac{dr}{dt}\bar{e}_r + r\frac{d\theta}{dt}\sin\phi\bar{e}_\theta - r\frac{d\phi}{dt}\bar{e}_\phi$$

which yields the differential of $\bar{r}$ as

$$d\bar{r} = dr\bar{e}_r + rd\theta \sin\phi \bar{e}_\theta - rd\phi \bar{e}_\phi$$

Recall the definition of work performed by $\bar{F}$. We obtain

$$\bar{F} \cdot d\bar{r} = (F_r\bar{e}_r + F_\theta\bar{e}_\theta + F_\phi\bar{e}_\phi) \cdot (dr\bar{e}_r + rd\theta \sin\phi e_\theta - rd\phi\bar{e}_\phi)$$

which can be simplified further to yield

$$\bar{F} \cdot d\bar{r} = F_r dr + r \sin\phi F_\theta d\theta - rF_\phi d\phi$$

The potential energy related to work can be obtained through equation (1.86). Substituting the above equation in the potential energy equation results into

$$\bar{F} \cdot d\bar{r} = -dV = F_r dr + r \sin\phi F_\theta d\theta - rF_\phi d\phi = \left(\frac{\partial V}{\partial r}dr + \frac{\partial V}{\partial \theta}d\theta + \frac{\partial V}{\partial \phi}d\phi\right)$$

The above equation defines the relationship between the force and the potential energy. The components of the forces in spherical coordinates are

$$\begin{aligned} F_r &= -\frac{\partial V}{\partial r} \\ F_\theta &= -\frac{1}{r \cdot \sin\phi}\frac{\partial V}{\partial \theta} \\ F_\phi &= \frac{1}{r}\frac{dV}{d\phi} \end{aligned}$$

Therefore, we write the force expression in the spherical coordinates as

$$\bar{F} = -\left(\frac{\partial \bar{e}_r}{\partial r} + \frac{1}{r \cdot \sin\phi}\frac{\partial \bar{e}_\theta}{\partial \theta} - \frac{1}{r}\frac{\partial \bar{e}_\phi}{\partial \phi}\right)V$$

The curl of $\bar{F}$ is defined by

$$\text{curl } \bar{F} = \nabla \times \bar{F}$$

where

$$\text{curl } \bar{F} = \begin{vmatrix} \bar{e}_r & \bar{e}_\theta & \bar{e}_\phi \\ \dfrac{-\partial}{\partial r} & \dfrac{-1}{r \sin\phi}\dfrac{\partial}{\partial \theta} & \dfrac{1}{r}\dfrac{\partial}{\partial \phi} \\ F_r & F_\theta & F_\phi \end{vmatrix}$$

the result of which is

$$\text{curl } \bar{F} = \frac{1}{r}\left[-\frac{1}{\sin\phi}\frac{\partial F_\phi}{\partial \theta} - \frac{\partial F_\theta}{\partial \phi}\right]\bar{e}_r - \left[-\frac{\partial F_\phi}{\partial r} - \frac{1}{r}\frac{\partial F_r}{\partial \phi}\right]\bar{e}_\theta$$
$$+ \left[-\frac{\partial F_\theta}{\partial r} + \frac{1}{r \sin\phi}\frac{\partial F_r}{\partial \theta}\right]\bar{e}_\phi$$

In order for $\bar{F}$ to be conservative, the curl of $\bar{F}$ must be zero. Hence each term must be set to zero to form a relationship between the force components which satisfy the conditions of $\bar{F}$ to be conservative.

Example 1.6.4
A particle m moves in a plane (x, y) and is acted on by a force, which depends on position only. This force is defined as follows:

$$\bar{F} = \frac{F_0}{a^3}(xy^2\mathbf{i} + x^2y\mathbf{j})$$

The particle moves from point A to point B following a line $y = x$ and a parabola $y = \frac{x^2}{a}$. Let the speed of m at point A be υ_0. Find the speed at point B in both cases.

Solution
Use the work–energy principle to define υ_B. First we write

$$\int_A^B \bar{F}.d\bar{r} = \frac{1}{2} \cdot m\upsilon^2|_A^B$$

$$\int \bar{F} \cdot d\bar{r} = \frac{1}{2}m\upsilon_B^2 - \frac{1}{2}m\upsilon_A^2$$

We need to define the work–energy equation by defining an expression for $d\bar{r}$. From the Cartesian coordinates, we define the position vector

$$\bar{r} = x\mathbf{i} + y\mathbf{j}$$

Hence

$$d\bar{r} = dx\mathbf{i} + dy\mathbf{j}$$

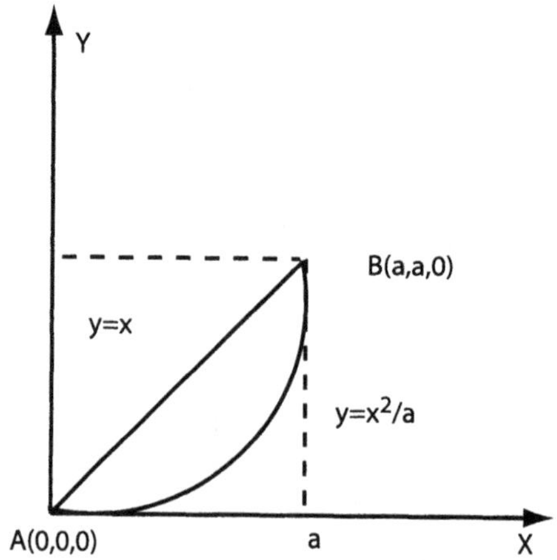

Fig. 1.13. Two distinct paths to follow between A and B

The work equation is then found to be

$$W_{AB} = \int_A^B \bar{F} \cdot dr = \frac{F_0}{a^3} \int_A^B (xy^2\mathbf{i} + x^2y\mathbf{j}) \cdot (dx\mathbf{i} + dy\mathbf{j})$$

$$= \frac{F_0}{a^3} \int_A^B (xy^2dx + x^2ydy)$$

(a) Consider that the path from A to B given by $y = x$, then $\mathrm{d}y = \mathrm{d}x$. The work done by F is found to be

$$W_{AB} = \frac{F_0}{a^3} \left[\int_0^a xx^2dx + \int_0^a x^2xdx \right]$$

$$W_{AB} = \frac{2F_0}{a^3} \int_0^a (x^3dx)$$

$$= \frac{2F_0}{a^3} \frac{x^4}{4} \cdot |_0^a$$

$$= \frac{F_0a}{2}$$

The velocity v_B is then found by equating the kinetic energy with the work done, resulting in

$$\frac{1}{2}mv_B^2 - \frac{1}{2}mv_0^2 = \frac{F_0a}{2}$$

$$v_B^2 - v_0^2 = \frac{F_0a}{m}$$

The initial velocity is known, therefore we can solve for v_B and obtain

$$v_B = \sqrt{\left(\frac{F_0a}{2} + v_0^2\right)}$$

(b) Consider the second path, which is given by $y = \frac{x^2}{a}$. We obtain $dy = \frac{2xdx}{a}$ from which we get the work along the parabola AB

$$W_{AB} = \frac{F_0}{a^3} \int_0^a \left(\frac{x^5}{a^2} + \frac{2x^5}{a^2} \right) dx$$

$$= \frac{F_0a}{2}$$

Note: The work done along the parabola is the same as the work along the line $y = x$. One should note that a force which yields the same work from A to B is said to be conservative.

Example: 1.6.5
For example 1.6.4 show that $\bar{F}$ is conservative and find the potential energy associated with $\bar{r}$.

Solution
Let us apply the curl, which states $\nabla \times \bar{F} = 0$

$$\nabla \times \bar{F} = \begin{vmatrix} \bar{i} & \bar{j} & \bar{k} \\ \dfrac{\partial}{\partial x} & \dfrac{\partial}{\partial y} & \dfrac{\partial}{\partial z} \\ \dfrac{F_0 xy^2}{a^3} & \dfrac{F_0 x^2 y}{a^3} & 0 \end{vmatrix} = \vec{i}(0) - \vec{j}(0) + \vec{k}\left(\frac{F_0}{a^3}2xy - \frac{F_0}{a^3}2xy\right)$$

Thus $\nabla \times \bar{F} = 0$ assumes that the potential energy V exists. We also know that $\bar{F} = -\nabla V$ or

$$\frac{F_0}{a^3}xy^2 = -\frac{\partial V}{\partial x},\ \frac{F_0}{a^3}x^2y = -\frac{\partial V}{\partial y},\ 0 = -\frac{\partial V}{\partial z}$$

From the above relations integrating with respect to x, this first term yields

$$\frac{F_0}{a^3}\left(\frac{x^2}{2}y^2\right) + f(y, z) = -V$$

We can check and see whether the above potential function satisfies $\frac{\partial V}{\partial y} = -\frac{F_0}{a^3}x^2y$. We see that

$$\frac{F_0}{a^3}x^2y + \frac{\partial f}{\partial y} = -\frac{\partial V}{\partial y}$$

which will hold to be true if $\frac{\partial V}{\partial y} = 0$. Furthermore, we notice that $\frac{\partial V}{\partial z} = 0$, to satisfy the last equation. Therefore, we conclude that

$$V = -\frac{F_0}{2a^3}\left(x^2y^2\right) + C$$

Observe how we can check further that V is related to work and in fact can be used to find the work W.

$$W_{AB} = -(V_B - V_A) = V_A - V_B$$

With the initial conditions for A (0, 0, 0) and B $(a,\ a,\ 0)$ we obtain

$$W_{AB} = C - \left(-\frac{F_0}{2a^3}a^4 + C\right) = \frac{F_0}{2a}$$

1.7 Principle of Impulse and Momentum

We have shown how Newton's second law was used to derive the principle of work and energy. Similarly we can derive another useful tool for the analysis of motion by integrating Newton's second law with respect to time. First, let us express Newton's second law as follows:

$$\sum \bar{F}_i = m\frac{d\bar{v}}{dt} \tag{1.95}$$

where $\sum \bar{F}_i$ represent the external force applied to the mass m. Let $\bar{F}$ represent the resultant of all forces acting on m. Then we integrate equation (1.95) with respect to time to obtain

$$\int \bar{F}dt = \int md\bar{v} = m\bar{v}_2 - m\bar{v}_1 \tag{1.96}$$

where $\bar{v}_1$ and $\bar{v}_2$ are the velocities of the center of mass of the object at the times t_1 and t_2. The term on the left is called the linear impulse, and the term on the right is the linear momentum. This result is called the principle of impulse and momentum. The impulse applied to an object during an interval of time is equal to the change in its linear momentum.

The average with respect to time of the total force acting on an object from t_1 to t_2 is

$$\sum \bar{F}_{av} = \frac{1}{t_2 - t_1}\int_{t_1}^{t_2} \bar{F}dt \tag{1.97}$$

So we can write the above equation (1.97) as

$$(t_2 - t_1)\sum \bar{F}_{av} = m\bar{v}_2 - m\bar{v}_1 \tag{1.98}$$

With this equation we can determine the average value of the total force acting on an object during a given interval of time, if we know the change in its velocity during that time.

Impacts occur in many situations of concern in engineering. In this section we consider a basic question: If the velocities of two objects before they collide is known, how do we determine their velocities afterward? In other words, what is the effect of the impact on their motion?

If colliding objects are not subjected to external forces, their total linear momentum must be the same before and after the impact. Even when they are subjected to external forces, the force of the impact is in most cases a large force which acts in a short period of time, that the effect of external forces on their motion during the impact is negligible. Suppose that objects A and B of mass m_A and m_B respectively moving with velocities $\bar{v}_A$ and $\bar{v}_B$ collide, and let $\bar{v}'_A$ and $\bar{v}'_B$ be their velocities after the impact. If the effects of external forces are negligible, their total linear momentum is conserved and we can write the following:

$$m_A\bar{v}_A + m_B\bar{v}_B = m_A\bar{v}'_A + m_B\bar{v}'_B \tag{1.99}$$

Furthermore, if A and B happen to move together after impact, then their center of mass velocity can be derived from the above equation where,

$$\bar{v}' = \frac{m_A\bar{v}_A + m_B\bar{v}_B}{m_A + m_B} \tag{1.100}$$

If A and B remain together after they collide, the collision is defined as a perfectly plastic impact. An important feature of this result is that we can determine the velocity following the impact without considering the physical nature of the impact. The vectors v'_A and v'_B define the direction the objects take after the impact.

If A and B do not stick together, linear momentum conservation alone does not provide enough equations to determine their velocities after the impact, if there is loss of energy.

As a result of the impact, part of the object's kinetic energy can be lost due to dissipation of energy to a variety of material conditions. As a consequence, the impulse they impart to each other during the restitution phase of the impact, from a given time t_C to t_2, when the objects separate from one another, are in general smaller than the impulse they impart from a given time t_1 to t_C, which is the time it takes for the two bodies to deform. The ratio of these impulses is called the coefficient of restitution:

$$e = \frac{\int_{t_C}^{t_2} \bar{F} dt}{\int_{t_1}^{t_C} \bar{F} dt} \tag{1.101}$$

From the above equation, applying the concept of linear momentum to each interval of time, we obtain the restitution equation

$$e = \frac{v'_B - v'_A}{v_A - v_B} \tag{1.102}$$

Thus, the coefficient of restitution is related in a simple way to the relative velocities of the objects before and after the impact. If e is known as is the case of a number of materials, you can use equation (1.102) together with the equation of conservation of linear momentum, equation (1.99), to determine v'_A and v'_B.

If $e = 0$, equation (1.102) indicates that $v'_B = v'_A$. The objects remain together after the impact, and the impact is perfectly plastic. If $e = 1$, it can be shown that the total kinetic energy is conserved and is the same before and after the impact:

$$\frac{1}{2} m_A v_A^2 + \frac{1}{2} m_A v_B^2 = \frac{1}{2} m_A (v'_A)^2 + \frac{1}{2} m_A (v'_B)^2 \tag{1.103}$$

An impact in which kinetic energy is conserved is called perfectly elastic. Although this is sometimes an useful approximation, energy is lost in any impact in which material objects come into contact. If you can hear a collision, kinetic energy has been converted into sound. Permanent deformations of the colliding objects after the impact also represent losses of kinetic energy.

1.8 Angular Momentum

The angular momentum was derived from the basis of examining the effects of forces and linear momentum about the vector generated during the angular rotation. Obviously the magnitude and direction involves the position from a given point to the line of action of the force, velocity and their magnitude. What follows is the derivation of the angular momentum as it leads to a unique discovery, which is the formulation and derivation of the equations of motion due to rotation. Consider a system of particles as denoted in Figure 1.14. We can write the position vector associated with the center of mass as:

$$\bar{r}_i = \bar{r}_c + \bar{\rho}_i \tag{1.104}$$

Let O be the origin of an inertial frame R and $\bar{r}_1$ the position from the fixed origin O to a particle m_1, $\bar{r}_2$ define the position to m_2 and r_n the position to m_n. Let the resultant force acting on each particle mass m_i be divided into two parts

$$\bar{F'}_i = \bar{F}_i + \sum_{j=1}^{n} \bar{f}_{ij} \tag{1.105}$$

where F_i denotes the resultant external force applied to m_i and $\bar{f}_{ij}$ represents the internal forces acting on m_i. We compute the moment resultant at O of all forces acting on the system as follows:

$$\bar{M}_0 = \sum_{i=1}^{n} \bar{r}_i \times \bar{F}_i \tag{1.106}$$

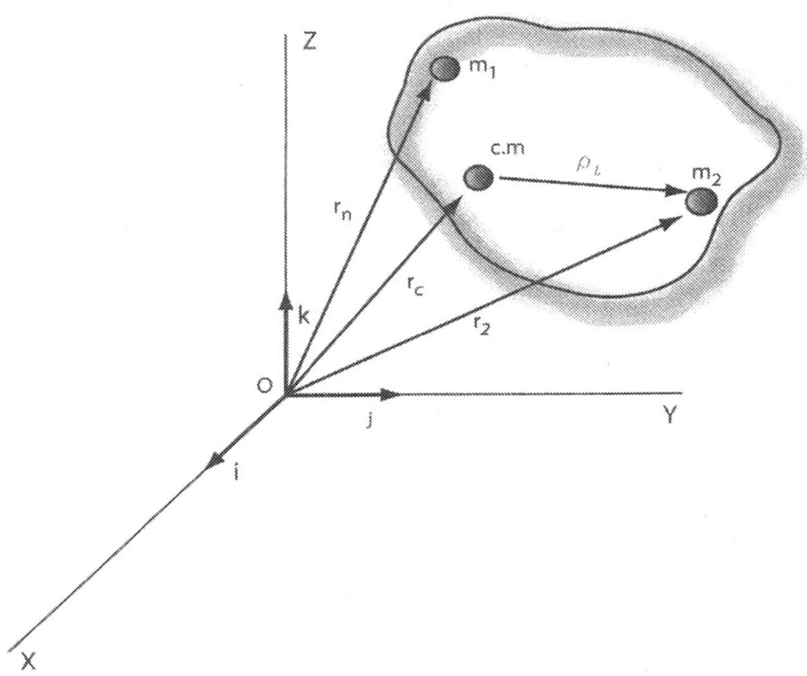

Fig. 1.14. A system of *n* particles

Again the internal forces do not contribute to the work or moment at point O. We define the moment of linear momentum as

$$\bar{H}_0 = \sum_{i=1}^{n} \bar{r} \times \bar{p}_i \tag{1.107}$$

where the linear momentum is denoted by p_i and is defined by

$$\bar{p}_i = m\bar{v}_i = m\dot{\bar{r}} \tag{1.108}$$

We wish to show that the rate of change of $\bar{H}$ is

$$\dot{\bar{H}}_0 = \bar{M}_0 \tag{1.109}$$

The velocity $\dot{\bar{r}}$ is found following equation (1.52). Equation (1.109) further illustrates how the moment about O of all external forces results in an equation, which involve ω and $\dot{\omega}$, therefore leading to an equation of motion of the system due to rotation.

From equation (1.107) we obtain the derivative of $\bar{H}$, where

$$\dot{H}_0 = \sum_{i=1}^{n} \dot{\bar{r}}_i \times \bar{p}_i + \sum_{i=1}^{n} \bar{r}_i \times \dot{\bar{p}} \tag{1.110}$$

The first term drops out because $\dot{\bar{r}}$ and $\bar{p}$ are along the same direction, what remains is

$$\dot{H}_0 = \sum_{i=1}^{n} \bar{r}_i \times \dot{\bar{p}} = \bar{M}_o \tag{1.111}$$

It is obvious that equation (1.109) is true, because

$$\dot{\bar{p}} = m\bar{a} = \bar{F} \tag{1.112}$$

Observe how $\dot{\bar{r}}$ which defines the position from 0 to m_i can be expressed as

$$\bar{r}_i = \bar{r}_G + \bar{r}_{i/G} \tag{1.113}$$

where $\bar{r}_G$ denotes the position to the center mass and $\bar{r}_{i/G}$ defines the relative position from m_i to G. After substitution of equation (1.113) into equation (1.111) we obtain

$$\bar{H}_o = \bar{H}_G + \bar{H}_{i/G} \tag{1.114}$$

The angular momentum about a fixed point is given by the angular momentum about the center mass G plus the angular momentum of point i with respect to G.

1.9 Principle of Virtual Work

Consider a particle m of a system of N particles with respect to an inertial frame R and let the virtual displacement δr be defined such that it doesn't violate any geometrical constraints. Let F represent the resultant of external forces applied to m. Since the particle is in equilibrium, then $F = 0$. Therefore

$$\bar{F} \cdot \delta\bar{r} = 0 \tag{1.115}$$

If a system of particles is in static equilibrium under the action of forces of constraints and applied forces, then in any virtual displacement of the system, with generalized coordinates $q_1, q_2, \ldots, q_N$, the position of a particle m in R does not violate work constraints, the applied forces is zero. Let a system be specified by N coordinates, then we write the position of a particle m in R as

$$\bar{r} = r(q_1, q_2, \ldots, q_n, t) \tag{1.116}$$

where q_k denotes the generalized coordinates and t time. Then we define

$$\delta\bar{r} = \frac{\partial r}{\partial q_1}\delta q_1 + \frac{\partial r}{\partial q_2}\delta q_2 + \cdots + \frac{\partial r}{\partial q_n}\delta q_n \tag{1.117}$$

In this case the variation δq_i comprising the virtual displacement $\delta\bar{r}$ may be completely arbitrary and independent and the constraints wouldn't be violated. The derivative of $\bar{r}$ in R with respect to time is

$$\dot{\bar{r}} = \frac{\partial\bar{r}}{\partial q_1}\dot{q}_1 + \frac{\partial\bar{r}}{\partial q_2}\dot{q}_2 + \cdots + \frac{\partial\bar{r}}{\partial q_n}\dot{q}_n + \frac{\partial\bar{r}}{\partial t} \tag{1.118}$$

Notice from the equation above that we can deduce the following relation

$$\frac{\partial\dot{\bar{r}}}{\partial\dot{q}_k} = \frac{\partial\bar{r}}{\partial q_k} \tag{1.119}$$

So the virtual displacement can be written as

$$\delta\bar{r} = \sum_{k=1}^{n} \frac{\partial\dot{r}}{\partial\dot{q}}\delta q_k \tag{1.120}$$

Example 1.9.1
Consider a slider crank mechanism as shown in Figure 1.15. Point P denotes the center mass of the piston, which slides on a frictionless surface along the horizontal. Let $\bar{F}$ be a horizontal force of magnitude $|\bar{F}|$ applied to m. Find the value of θ for which the slider crank mechanism shown is in equilibrium.

Solution
The velocity of point P can be written as a function of the velocity of point B with respect to O and the relative velocity of point P with respect to B. That is

$$\bar{\upsilon}_p = \bar{\upsilon}_B + \bar{\upsilon}_{P/B}$$

or

$$\bar{\upsilon}_p = \bar{\upsilon}_B + \bar{\omega}_{L_2} \times \bar{r}_{P/B}$$

The velocity of point B is simply given by

$$\bar{\upsilon}_B = \bar{\omega}_{L_1} \times \bar{r}_{B/0}$$

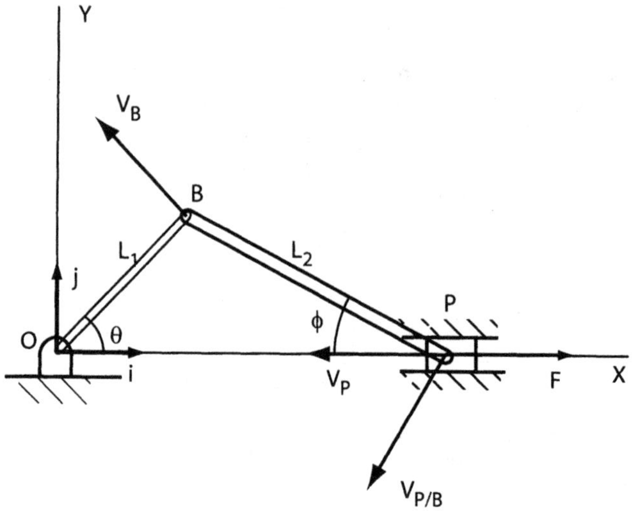

Fig. 1.15. A slider crank mechanism

where $\bar{r}_{B/0}$ is the position of B with respect to O. Then,

$$\bar{v}_B = \dot{\theta}\,\mathbf{k} \times [L_1 \cos\theta\,\mathbf{i} + L_1 \sin\theta\,\mathbf{j}]$$

which yields two components

$$\bar{v}_B = L_1\dot{\theta}\cos\theta\,\mathbf{j} - L_1\dot{\theta}\sin\theta\,\mathbf{i}$$

Similarly we find an expression for $\bar{v}_{P/B}$.

$$\begin{aligned}\bar{v}_{P/B} &= \bar{\omega}_{L_2} \times \bar{r}_{P/B} = \dot{\phi}\,\mathbf{k} \times [L_2 \cos\theta\,\mathbf{i} - L_2 \sin\theta\,\mathbf{j}] \\ &= L_2\dot{\phi}\cos\theta\,\mathbf{j} - L_2\dot{\phi}\sin\theta\,\mathbf{i}\end{aligned}$$

Notice that θ and ϕ are related by

$$\sin\phi = \frac{L_1}{L_2}\sin\theta$$

Differentiating the above equation further yields

$$\dot{\phi} = \frac{L_1}{L_2}\frac{\cos\theta}{\cos\phi}\dot{\theta}$$

We can eliminate $\cos\phi$ from the denominator by expressing it in terms of θ, hence

$$\dot{\phi} = \frac{L_1}{L_2}\frac{\cos\theta}{\sqrt{1 - \left(\frac{L_1}{L_2}\right)^2 \sin^2\theta}}\dot{\theta}$$

Therefore, the velocity of point P in a fixed reference frame R is expressed as follows:

$$\bar{v}_P = \left[-L_1\dot{\theta}\sin\theta - L_2\dot{\theta}\frac{L_1}{L_2}\frac{\cos\theta}{\sqrt{1-\left(\frac{L_1}{L_2}\right)^2\sin^2\theta}}\frac{L_1}{L_2}\sin\theta \right]\mathbf{i}$$

There are no j components because P moves along x only. The virtual displacement is then found to be

$$\delta\bar{r}_P = -L_1\sin\theta\left[1 + \frac{L_1}{L_2}\frac{\cos\theta}{\sqrt{1-\left(\frac{L_1}{L_2}\right)^2\sin^2\theta}}\right]\delta\theta\,\mathbf{i}$$

The virtual work is then derived from the basic definition of

$$\delta W = \bar{F}\cdot\delta\bar{r}_P$$

where

$$\bar{F} = F\mathbf{i}$$

resulting into the following virtual work

$$\delta W = -FL_1\sin\theta\left[1 + \frac{L_1}{L_2}\frac{\sin\theta\cos\theta}{\sqrt{1-\left(\frac{L_1}{L_2}\right)^2\sin^2\theta}}\right]\delta\theta$$

Since the virtual work is zero, we can obtain a value for θ for which the system is in equilibrium.

Example 1.9.2
Consider another slider crank mechanism where the force is not constant but is defined by a linear spring of constant k. Find θ under which the potential energy is constant.

Solution
First note that

$$L\sin\theta = \frac{L}{2}\sin\phi$$

where

$$\sin\phi = 2\sin\theta$$

Let the potential energy be the energy required to produce work, hence

$$\int \bar{F}\cdot d\bar{r} = -\int dV$$

The potential energy can be divided into two parts

$$V = V_s + V_g$$

When V_S is the contribution of the spring, and V_G denote the contribution from the gravity force $m\bar{g}$, hence

$$V = \frac{1}{2}k\delta^2 + \left[-\frac{mgL\sin\theta}{2}\right] + \left[-\frac{mgL\sin\theta}{8}\right]$$

The spring deformation denoted by δ is given by

$$\delta = (L - L\cos\theta) + \left[\frac{L}{2} - \frac{L}{2}\cos\phi\right]$$

Which can be written as

$$\delta = (L - L\cos\theta) + \frac{L}{2}\left[1 - \sqrt{1 - 4\sin^2\theta}\right]$$

Let the deformation be small and neglect the potential energy due to the gravity forces, then the total potential energy becomes

$$V = \frac{1}{2}kL^2\left[(1-\cos\theta) + \frac{1}{2}\left(1 - \sqrt{1 - 4\sin^2\theta}\right)\right]^2$$

So,

$$\frac{\partial V}{\partial\theta} = 0 = kL^2\left[\frac{3}{2} - \cos\theta - \frac{\sqrt{1 - 4\sin^2\theta}}{2}\right]$$

$$\times\left(\sin\theta + \frac{1}{4}\left(1 - 4\sin^2\theta\right)^{1/2} 8\sin\theta\cos\theta\right)$$

from which we solve for θ.

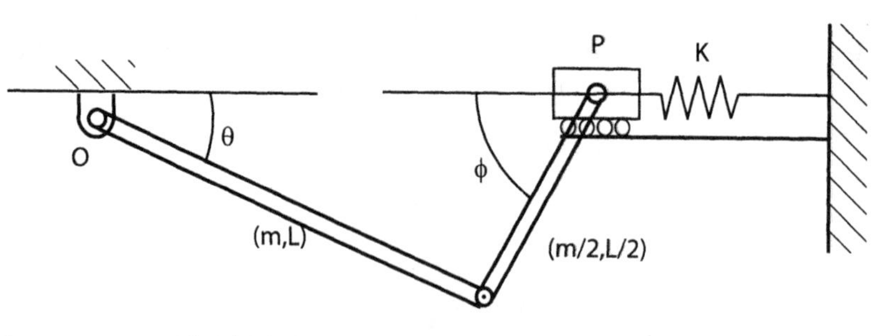

Fig. 1.16. A slider crank with spring mechanism

1.10 Problems

1.1. Show that $d\bar{v}/dt$ where $\bar{v} = \dot{r}\bar{e}_r + r\dot{\phi}\, e_\phi + r\dot{\theta}\sin\phi\,\bar{e}_\theta$ is given by equation (1.45).

1.2. Derive equation (1.51) from equation (1.50).

1.3. Obtain the acceleration of a point p in cylindrical coordinates using the projection of $\bar{e}_r$ and $\bar{e}_\theta$ along $\bar{i}$ and $\bar{j}$.

1.4. Express the velocity of a particle m in spherical coordinates using the concept of angular velocity.

1.5. A rod is constrained by two joints at point A and point B as shown in Figure 1.17. The rod slides without friction at a speed of 50 cm/s when the slider A reaches a height of 50 cm. Find the velocity of point B.

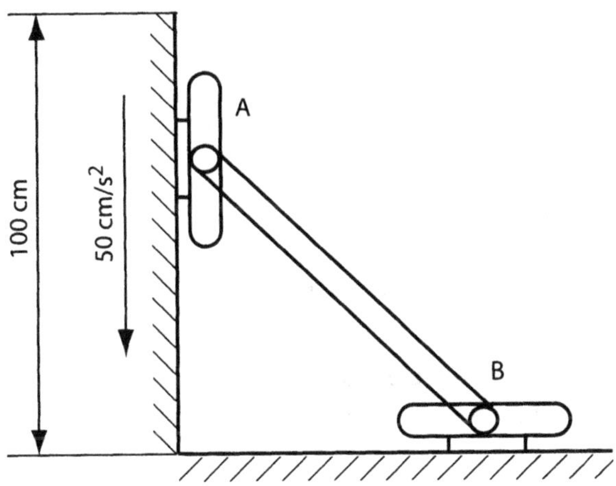

Fig. 1.17. A slider mechanism

1.6. In all three cases m moves along the path AB as shown in Figure 1.18. If the initial conditions are such that $v_A = 0$, find v_B.

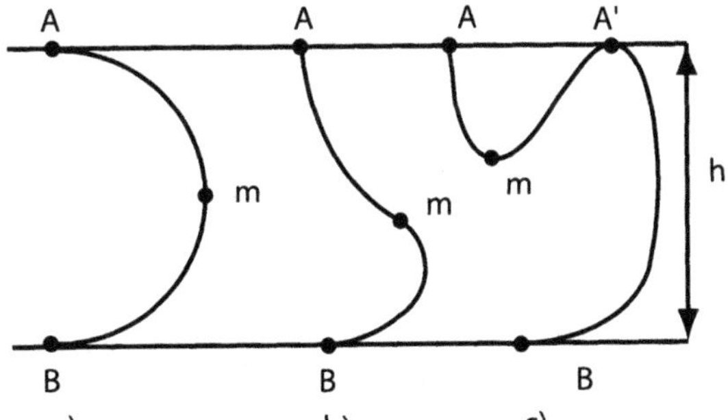

Fig. 1.18. 3 Cases Of M moving along path AB

1.7. Let F be a force acting on m as shown in Figure 1.19 given by

$$\bar{F} = \frac{F_0}{a^3}(xy^2\vec{i} + x^2y\vec{j})$$

and $v_A = v_0$ find v_B. Also check if the force F is conservative.

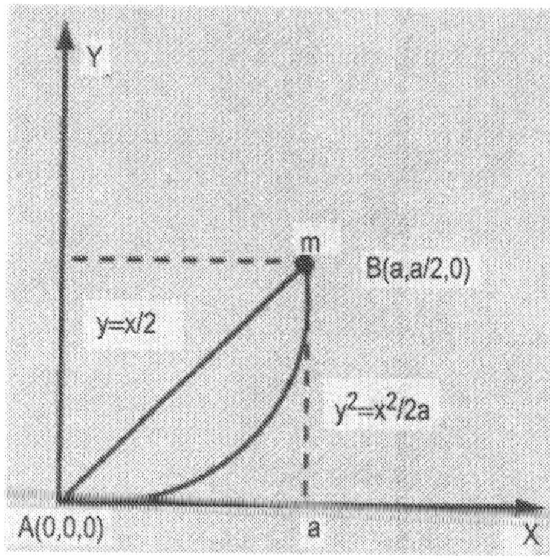

Fig. 1.19. Force acting on m

1.8. Find the angle θ at which the rod AB in Figure 1.20 rests in equilibrium using the principle of virtual work. (Assume smooth surfaces.)

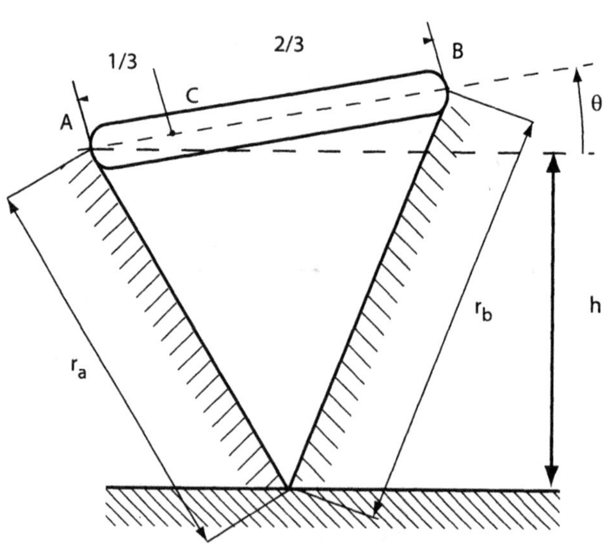

Fig. 1.20. Rod AB

1.9. Use the principle of virtual work to solve Example 1.6.4 completely.

1.10. A particle m moves along a smooth surface with a constant angular velocity as shown in Figure 1.21. Two particles m and m_k are connected through a cable that pass through a hole at O. Cable OB hangs vertically and OA is horizontal. Use the concept of angular momentum to derive the equations of motion of both masses.

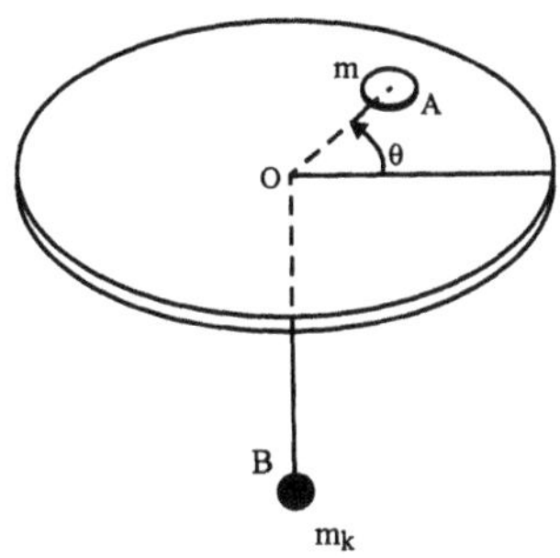

Fig. 1.21. Two particles m and m_k

2

Rigid-Body Kinematics

2.1 Introduction

Kinematics is the study of position, velocity and acceleration without the need of forces. Kinematics, however, for years used kinematical relations to design linkages, gears and a number of complex engineering problems. Although in principle we can analyze the motion of a system of N degrees of freedom, we often rely on simple models to obtain realistic answers. The concept of a rigid body is a good example where we assume that the body doesn't deform under the influence of external forces or simply the deformation is negligible. In this chapter, we set the foundation for the kinematical quantities needed to describe a rigid-body motion.

2.2 Vectors Differentiation

Vectors are used to represent kinematical quantities and expressions such as position, velocity and acceleration. They are also used to represent forces and dynamical quantities such as linear and angular momentum. A vector is usually used to locate a particular point in space with respect to a reference frame. The arrow from the origin of the reference frame to the point represents the vector. Each vector has a scalar quantity denoted as its magnitude and a direction. The scalar quantities of a vector are usually its Cartesian coordinates.

Example 2.2.1
Consider a point P in space as shown in Figure 2.1. Let the coordinates of P with respect to R be given by (x, y, z). Then the vector $\bar{r}$ could be written as

$$\bar{r} = x\mathbf{i} + y\mathbf{j} + z\mathbf{k} \tag{2.1}$$

where **i, j** and **k** are unit vectors fixed in R. The scalar quantities, x, y and z are essential to describe the concise position of point P in the reference frame R.

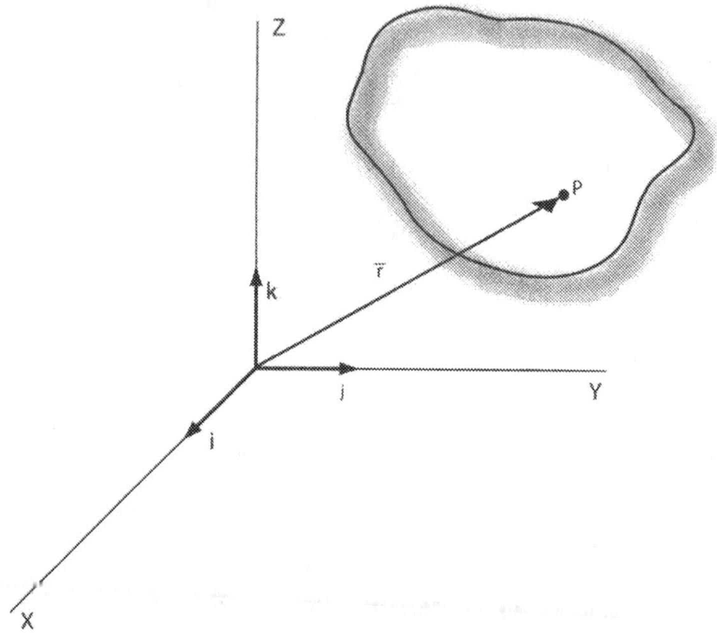

Fig. 2.1. Defining a point in space with respect to a fixed reference frame $R(x, y, z)$

Example 2.2.2
Let a point P' be located with respect to the frame R' denoted by (x', y', z') and (x, y, z) with respect to R (Figure 2.2). Then their corresponding vectors can be expressed as

$$^{R'}\bar{r}' = x'\mathbf{i}' + y'\mathbf{j}' + z'\mathbf{k}' \tag{2.2}$$

The superscript R' simply stands for $\bar{r}'$ with respect to R'. When more than one inertial frame is used it is best if the superscript on top of vectors are used to define which frame is being used, and

$$^{R'}\bar{r}' = x\mathbf{i} + y\mathbf{j} + z\mathbf{k} \tag{2.3}$$

On the other hand, if the orientation of the unit vectors **i'**, **j'** and **k'** in R' with respect to **i**, **j** and **k** in R are known, a relationship between them could be found. For example, if the axis x' of R' remains parallel to x in R and through a clockwise rotation of that axis y' and z' are obtained, it is apparent that

$$\begin{aligned} \mathbf{j}' &= \cos\theta\mathbf{j} - \sin\theta\mathbf{k} \\ \mathbf{k}' &= \sin\theta\mathbf{j} + \cos\theta\mathbf{k} \\ \mathbf{i}' &= \mathbf{i} \end{aligned} \tag{2.4}$$

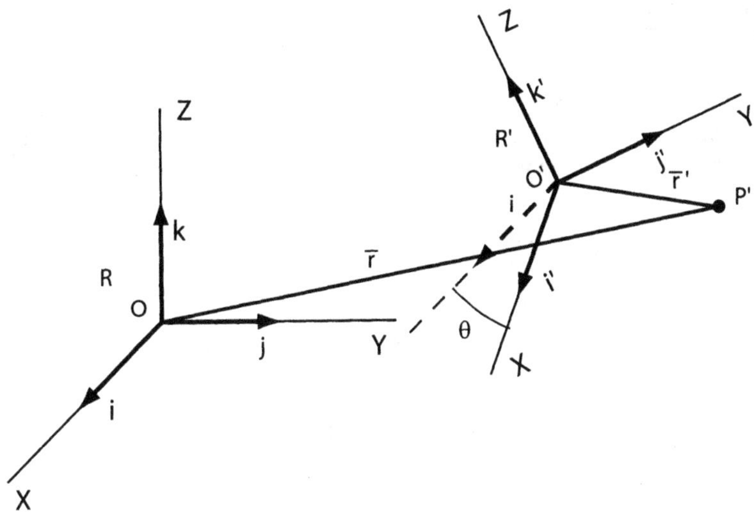

Fig. 2.2. Representation of a position vector locating an arbitrary point p' with respect to two reference frames R_1 and R'

We conclude by substitution of equation (2.4) into (2.2) that:

$$\begin{aligned} x &= x' \\ y &= y' \cos\theta + z' \sin\theta \\ z &= z' \cos\theta - y' \sin\theta \end{aligned} \tag{2.5}$$

2.3 First Derivatives and Partial Derivatives

2.3.1 Definition of a Derivative

The first derivative of a vector $\bar{v}$ is given by

$$\frac{d\bar{v}}{dt} = \lim_{\Delta t \to 0} \frac{\bar{v}(t + \Delta t) - \bar{v}(t)}{\Delta t} \tag{2.6}$$

Since $\bar{v}$ can be expressed with respect to a set of unit vectors **i**, **j** and **k** as

$$\bar{v} = v_x \mathbf{i} + v_y \mathbf{j} + v_z \mathbf{k} \tag{2.7}$$

then its derivative is

$$\frac{d\bar{v}}{dt} = \frac{dv_x}{dt}\mathbf{i} + \frac{dv_y}{dt}\mathbf{j} + \frac{dv_z}{dt}\mathbf{k} \tag{2.8}$$

which can be expressed further as

$$\dot{v} = \dot{v}_x \mathbf{i} + \dot{v}_y \mathbf{j} + \dot{v}_z \mathbf{k} \tag{2.9}$$

2.3.2 Definition of Partial Derivatives

Suppose that the vector $\bar{v}$ is a function of n scalar variables $q_1, q_2, \ldots, q_n$ in a reference frame R. The q's represent the degree of freedom of the system. This includes translation and rotation. For example q_1 is θ and q_2 is ϕ. Then $\partial^R \bar{v}/\partial q_i$ $(i = 1, \ldots, n)$ is called the *partial derivative* of $\bar{v}$ with respect to q_i in R. Notice how in this chapter we have introduced the superscript R to refer to the reference frame. This is useful when more than one frame is used in the dynamics of complex systems.

Example 2.3.1

Using Example 2.2.2, we can find the first time derivative of ${}^R\bar{v}$ and the partial velocity with respect to θ, that is, $\partial^R \bar{v}/\partial\theta$. Expressing ${}^R\bar{v}$ with respect to **i, j** and **k** we get

$$\bar{v} = x'\mathbf{i} + (y'\cos\theta + z'\sin\theta)\mathbf{j} + (z'\cos\theta - y'\sin\theta)\mathbf{k} \tag{2.10}$$

Then

$$\frac{d\bar{v}}{dt} = \frac{dx'}{dt}\mathbf{i} + \frac{d}{dt}(y'\cos\theta + z'\sin\theta)\mathbf{j} + \frac{d}{dt}(z'\cos\theta - y'\sin\theta)\mathbf{k} \tag{2.11}$$

or simply

$$\begin{aligned}\frac{d\bar{v}}{dt} &= \dot{x}\mathbf{i} + (\dot{y}'\cos\theta - \dot{\theta}\sin\theta y' + \dot{z}'\sin\theta + \dot{\theta}z'\cos\theta)\mathbf{j} \\ &\quad + (\dot{z}'\cos\theta - \dot{\theta}z'\sin\theta - \dot{y}'\sin\theta - \dot{\theta}y'\cos\theta)\mathbf{k}\end{aligned} \tag{2.12}$$

On the other hand,

$$\frac{\partial\bar{v}}{\partial\theta} = \frac{\partial x'}{\partial\theta}\mathbf{i} + \frac{\partial}{\partial\theta}(y'\cos\theta + z'\sin\theta)\mathbf{j} + \frac{\partial}{\partial\theta}(z'\cos\theta - y'\sin\theta)\mathbf{k} \tag{2.13}$$

which yields

$$\frac{\partial\bar{v}}{\partial\theta} = 0\mathbf{i} + (-y'\sin\theta + z'\cos\theta)\mathbf{j} + (-z'\sin\theta - y'\cos\theta)\mathbf{k} \tag{2.14}$$

2.3.3 Total Derivative of a Vector

There are two cases to be highlighted here.

1. First, suppose that the vector $\bar{v}$ is a function of n scalar variables $q_1, q_2, \ldots, q_n$ and that $q_1, \ldots, q_n$ are functions of time t. The total derivative of $\bar{v}$ in the fixed reference frame R is then obtained as follows:

$$\frac{d\bar{v}}{dt} = \frac{\partial\bar{v}}{\partial q_1}\frac{dq_1}{dt} + \frac{\partial\bar{v}}{\partial q_2}\frac{dq_2}{dt} + \cdots + \frac{\partial\bar{v}_n}{\partial q_n}\frac{dq_n}{dt} \tag{2.15}$$

or

$$\frac{d\bar{v}}{dt} = \sum_{i=1}^{n} \frac{\partial \bar{v}}{\partial q_i} \frac{dq_i}{dt} \tag{2.16}$$

or simply

$$\frac{d\bar{v}}{dt} = \sum_{i=1}^{n} \frac{\partial \bar{v}}{\partial q_i} \dot{q}_i \tag{2.17}$$

2. Suppose that the vector $\bar{v}$ is a function of both n scalar variables $q_1, \ldots, q_n$ and time t: This basically states that the time t appears explicitly in the expression of $\bar{v}$. If the n scalar variables $q_1, \ldots, q_n$ are also a function of time t, the total time derivative of $\bar{v}$ in R is

$$\frac{d\bar{v}}{dt} = \sum_{i=1}^{n} \frac{\partial \bar{v}}{\partial q_i} \dot{q}_i + \frac{\partial \bar{v}}{\partial t} \tag{2.18}$$

Example 2.3.2

Suppose that a position vector $\bar{v}$ is given by

$$\bar{v} = r\cos\theta\mathbf{i} + r\sin\phi\sin\theta\mathbf{j} + \sin\theta\cos\phi\ \mathbf{k} \tag{2.19}$$

where ϕ is a function given by a harmonic function

$$\phi = a\sin b\omega t$$

and the angular displacement θ is proportional to time t where

$$\theta = ct$$

a, b and c are constants that can be obtained through the initial condition. Define the total time derivative of $\bar{v}$.

Solution

Using equation (2.18), we write the expression for the derivative as

$$\frac{d\bar{v}}{dt} = \frac{\partial \bar{v}}{\partial \theta}\dot{\theta} + \frac{\partial \bar{v}}{\partial \phi}\dot{\phi} + \frac{\partial \bar{v}}{\partial t} \tag{2.20}$$

where the derivative components are found to be

$$\frac{\partial \bar{v}}{\partial \theta} = -r\sin\theta\mathbf{i} + \mathrm{r}\sin\phi\cos\theta\mathbf{j} + \cos\theta\cos\phi\ \mathbf{k} \tag{2.21}$$

$$\frac{\partial \bar{v}}{\partial \phi} = 0\mathbf{i} + \mathrm{r}\cos\phi\sin\theta\mathbf{j} - \sin\theta\sin\phi\ \mathbf{k} \tag{2.22}$$

$$\frac{\partial \bar{v}}{\partial t} = 0 \tag{2.23}$$

$$\dot{\theta} = c \tag{2.24}$$

$$\dot{\phi} = ab\omega\cos b\omega t \tag{2.25}$$

Therefore, the total derivative of $\bar{r}$ is

$$\frac{d\bar{v}}{dt} = (-r\sin\theta\mathbf{i} + r\sin\phi\cos\theta\mathbf{j} + \cos\theta\sin\phi\ \mathbf{k})\dot{\theta}$$

$$+ (0\mathbf{i} + r\cos\phi\ \ \sin\theta\mathbf{j} - \sin\theta\sin\phi\ \mathbf{k})\dot{\phi} \tag{2.26}$$

Obviously one can express the total derivative by expressing it in terms of $\bar{i}$, $\bar{j}$ and $\bar{k}$ components.

2.4 Generalized Coordinates

In the study of multibody dynamics, different sets of coordinates are often used to describe the rigid-body orientations with respect to a fixed reference frame R. Generalized coordinates are defined to be a set of convenient coordinates, usually independent of one another, used to describe the configuration of a particular system. If the system is subject to some additional constraints that will result in some dependency between the generalized coordinates, the number of independent generalized coordinates defines the number of degrees of freedom of the system. For instance, if n generalized coordinates are used to describe a particular configuration and there are m constraint equations ($m < n$), the difference $n - m$ is equal to the total degrees of freedom of the system. The advantage of finding the exact number of independent generalized coordinates to describe the configuration of a multibody system is the fact that the constraint forces need not be computed even if the system is subject to constraints.

Proper selection of the generalized coordinates allows one to obtain simpler expressions for velocities and accelerations and faster algorithms to derive the governing equations of motion. It is essential that the coordinates be selected to best describe the configuration of the system and ultimately lead to simpler expressions in the kinematics and dynamics of the system. In Chapter 3 we demonstrate how, for example, relative coordinates are suited for multibody systems with treelike topology.

2.4.1 Cartesian Coordinates

We define a set of rectangular Cartesian coordinates as three orthogonal axes (x, y, z) with unit vectors $(\mathbf{i}, \mathbf{j}, \mathbf{k})$. Every vector position in space can be expressed as

$$\bar{v} = v_x\mathbf{i} + v_y\mathbf{j} + v_z\mathbf{k} \tag{2.27}$$

where v_x, v_y and v_z represent the scalar components of the magnitude of $\bar{v}$ along the x, y and z axes, respectively.

For example, let the vector position $\bar{r}$ be as shown in Figure 2.3. Then,

$$^R\bar{r} = x\mathbf{i} + y\mathbf{j} + z\mathbf{k} \tag{2.28}$$

It is easily seen that the scalar variables x, y and z are independent of one another. To see this further, let us find the first derivative of $^R\bar{r}$, where

$$\frac{d^R\bar{r}}{dt} = \dot{x}\mathbf{i} + \dot{y}\mathbf{j} + \dot{z}\mathbf{k} \tag{2.29}$$

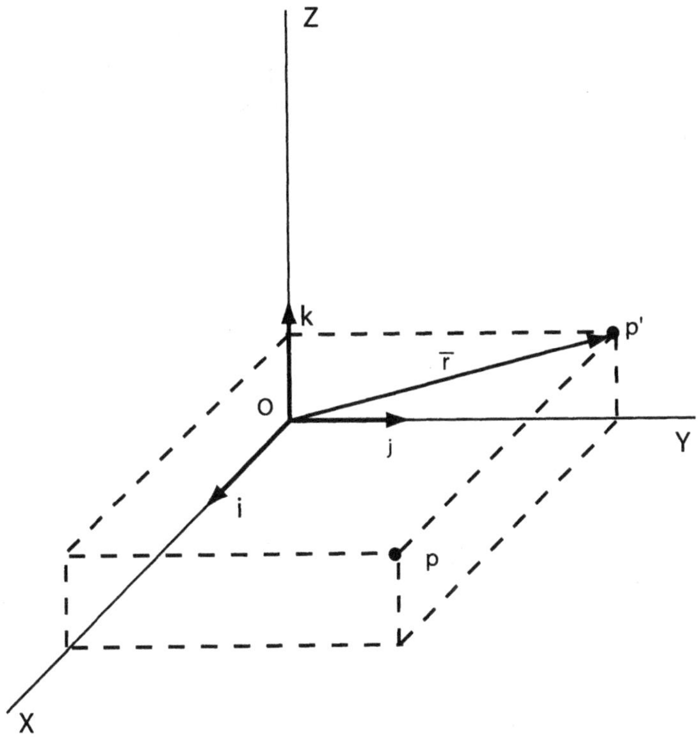

Fig. 2.3. A vector defined in a Cartesian reference frame $R(x, y, z)$

Clearly, the advantage of representing any position vector in space, with respect to a fixed reference frame R, defined by the Cartesian coordinates x, y and z is that their motion is uncoupled. However, in practical situations, the motion, given in terms of fixed directions, is not abundant.

Example 2.4.1
Consider Figure 2.3. Derive the first derivative of the vector $\overline{OP}'$.

Solution

$$\overline{OP}' = 0\mathbf{i} + y\mathbf{j} + z\mathbf{k}$$

Hence

$$\frac{d\overline{OP}'}{dt} = \dot{y}\mathbf{j} + \dot{z}\mathbf{k}$$

2.4.2 Euler Angles and Direction Cosines

In many dynamical problems, especially in multibody dynamics, it is useful to express the motion of rotating bodies, using a set of noninertial frames attached to the bodies and a fixed frame R. It is also known that any rotating frame fixed to a rigid body can be related to the fixed inertial frame by three successive rotations. If we denote the three successive rotations by α, β and γ, we define a matrix R called the direction cosine matrix that relates the base unit vectors $\bar{n}_i$ to the body unit vectors $\bar{n}_i^A$ $(i = 1, 2, 3)$. The direction cosine matrix can be found as follows.

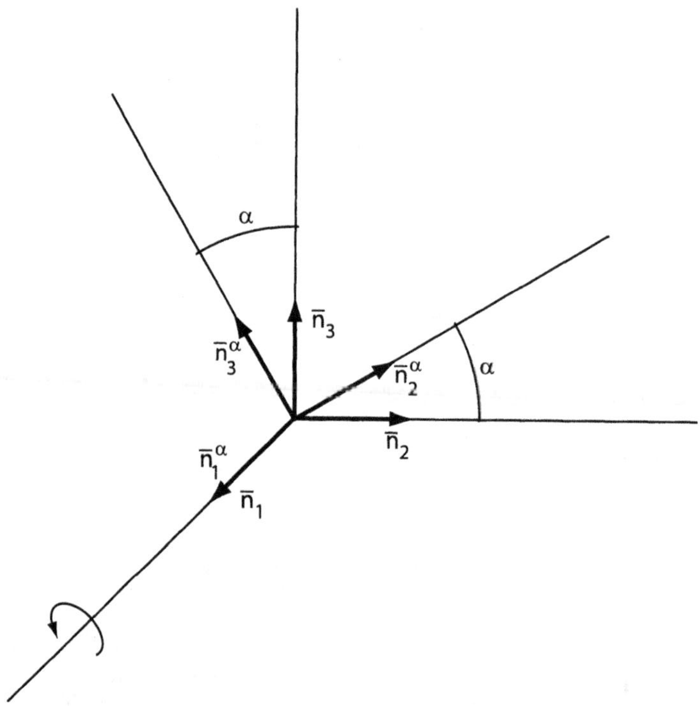

Fig. 2.4. Rotation of the orthogonal unit vectors $\bar{n}^\alpha$ about $\bar{n}_2^\alpha$ (ccw)

Let $\bar{n}_1$, $\bar{n}_2$ and $\bar{n}_3$ represent a set of orthogonal unit vectors as shown in Figure 2.4. Rotate the $\bar{n}_1$ axis counterclockwise by an angle α; the resulting axes $\bar{n}_1^\alpha$, $\bar{n}_2^\alpha$ and $\bar{n}_3^\alpha$ are related to the old axes $\bar{n}_1$, $\bar{n}_2$ and $\bar{n}_3$ by

$$\begin{aligned}\bar{n}_1^\alpha &= \bar{n}_1 \\ \bar{n}_2^\alpha &= \bar{n}_2 \cos\alpha + \bar{n}_3 \sin\alpha \\ \bar{n}_3^\alpha &= -\bar{n}_2 \sin\alpha + \bar{n}_3 \cos\alpha\end{aligned} \tag{2.30}$$

which can be written in matrix form as

$$\begin{bmatrix}\bar{n}_1^\alpha \\ \bar{n}_2^\alpha \\ \bar{n}_3^\alpha\end{bmatrix} = \begin{bmatrix}1 & 0 & 0 \\ 0 & \cos\alpha & \sin\alpha \\ 0 & -\sin\alpha & \cos\alpha\end{bmatrix}\begin{bmatrix}\bar{n}_1 \\ \bar{n}_2 \\ \bar{n}_3\end{bmatrix} \tag{2.31}$$

Expressing equation (2.31) in compact form yields

$$\{\bar{n}^{\alpha}\} = \{R_{\alpha}\}\{\bar{n}\} \tag{2.32}$$

where R_{α} denotes the 3×3 matrix given in equation (2.31) and relating $\{\bar{n}^{\alpha}\}$ to $\bar{n}$. Note that in order to have $\{\bar{n}^{\alpha}\}$ coincide with $\bar{n}$, you must rotate $\{\bar{n}_1^{\alpha}\}$ clockwise by angle α. As we will show later, it is based on this idea that any set of orthogonal vectors fixed in any rigid body could be brought to coincide with the axis fixed in an inertial frame through three successive rotations about each of its axes. Next, consider Figure 2.5, where $\{\bar{n}_2^{\alpha}\}$ is rotated by an angle β counterclockwise; the resulting axes $\{\bar{n}^{\beta}\}$ are related to $\bar{n}^{\alpha}$ by

$$\begin{bmatrix} \bar{n}_1^{\beta} \\ \bar{n}_2^{\beta} \\ \bar{n}_3^{\beta} \end{bmatrix} = \begin{bmatrix} \cos\beta & 0 & -\sin\beta \\ 0 & 1 & 0 \\ \sin\beta & 0 & \cos\beta \end{bmatrix} \begin{bmatrix} \bar{n}_1^{\alpha} \\ \bar{n}_2^{\alpha} \\ \bar{n}_3^{\alpha} \end{bmatrix} \tag{2.33}$$

or

$$\{\bar{n}^{\beta}\} = \{R_{\beta}\}\{\bar{n}^{\alpha}\} \tag{2.34}$$

where R_{β} is the 3×3 matrix given by equation (2.33).

Rotation of $\bar{n}_3^{\beta}$ counterclockwise by an angle γ as shown in Figure 2.6 results in

$$\begin{bmatrix} \bar{n}_1^{\gamma} \\ \bar{n}_2^{\gamma} \\ \bar{n}_3^{\gamma} \end{bmatrix} = \begin{bmatrix} \cos\gamma & \sin\gamma & 0 \\ -\sin\gamma & \cos\gamma & 0 \\ 0 & 0 & 1 \end{bmatrix} \begin{bmatrix} \bar{n}_1^{\beta} \\ \bar{n}_2^{\beta} \\ \bar{n}_3^{\beta} \end{bmatrix} \tag{2.35}$$

or

$$\{\bar{n}^{\gamma}\} = \{R_{\gamma}\}\{\bar{n}^{\beta}\} \tag{2.36}$$

It can easily be seen from equations (2.32), (2.34) and (2.36) that $\bar{n}^{\gamma}$ and $\bar{n}$ are related by

$$\{\bar{n}^{\gamma}\} = \left[S^{AO}\right]\{\bar{n}\} \tag{2.37}$$

where

$$\left[S^{AO}\right] = [R_{\gamma}][R_{\beta}][R_{\alpha}] \tag{2.38}$$

which is the matrix of direction cosines given by

$$\left[S^{AO}\right] = \begin{bmatrix} c\beta c\gamma & c\alpha s\gamma + s\alpha s\beta c\gamma & s\gamma s\alpha - c\alpha s\beta c\gamma \\ -c\beta s\gamma & c\alpha c\gamma - s\alpha s\beta s\gamma & s\alpha c\gamma + c\alpha s\beta s\gamma \\ s\beta & -s\alpha c\beta & c\alpha c\beta \end{bmatrix} \tag{2.39}$$

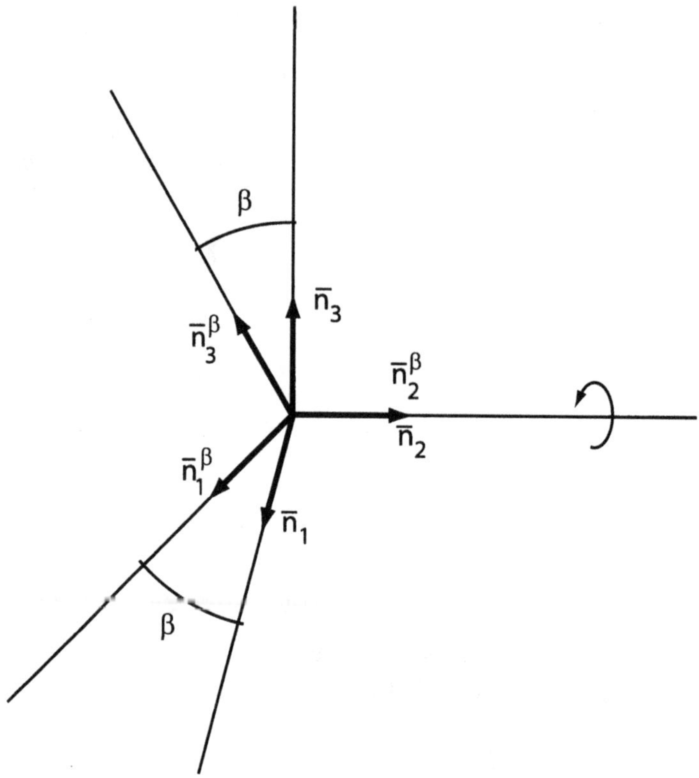

Fig. 2.5. Rotation of the orthogonal unit vectors $\bar{n}^{\beta}$ about $\bar{n}_2^{\beta}$ (ccw)

in which the symbols c = cosine and s = sine, the superscript A denotes an arbitrary body in space and O the center of the fixed inertial reference frame R. In the light of equation (2.38), we define S^{AO} as the transformation of A with respect to $R(0)$ and its direction cosines is given by

$$\left[S^{AO}\right] = \left\{\bar{n}^{\gamma}\right\} \cdot \{\bar{n}\}^{T}$$

or in general form we write the transformation between two frames s and r as

$$\left[S^{sr}\right] = \left\{\bar{n}^{s}\right\} \cdot \left\{\bar{n}^{r}\right\}^{T} \tag{2.40}$$

where S^{sr} is seen to be obtained by the dot product of two dextral sets of unit vectors $\bar{n}_1^r, \bar{n}_2^r, \bar{n}_3^r$ and $\bar{n}_1^s, \bar{n}_2^s, \bar{n}_3^s$. Hence, S^{sr} is a square matrix

$$S^{sr} = \begin{bmatrix} S_{11} & S_{12} & S_{13} \\ S_{21} & S_{22} & S_{23} \\ S_{31} & S_{32} & S_{33} \end{bmatrix} \tag{2.41}$$

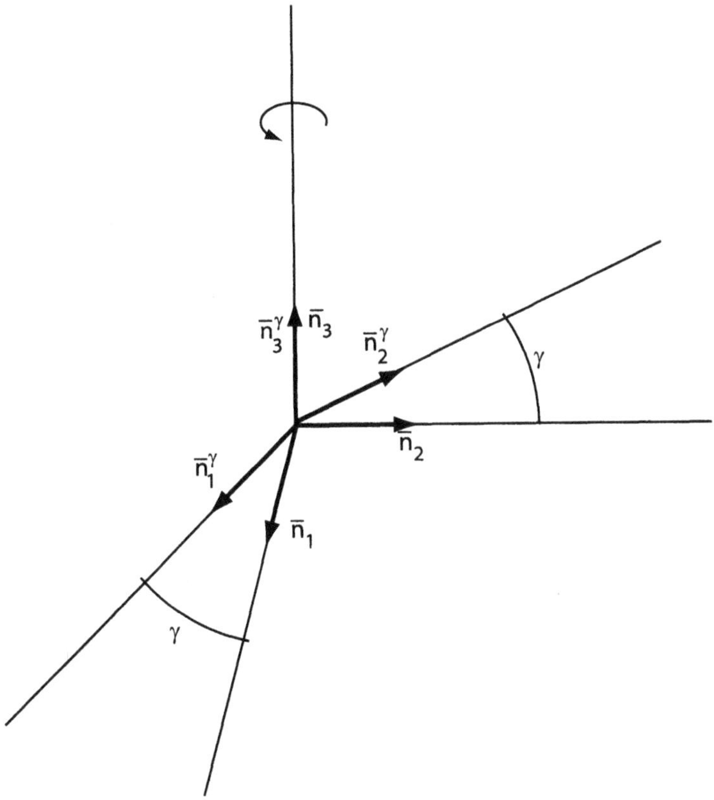

Fig. 2.6. Rotation of the orthogonal unit vectors $\bar{n}^\gamma$ about $\bar{n}_3^\gamma$ (ccw)

The matrix relation between the two vectors $\bar{n}_j^s$ and $\bar{n}_i^r$ $(i, j = 1, 2, 3)$ is given by equation (2.38). So the direction cosine matrix S^{sr} can be used to describe the relative orientation between two reference frames $\bar{n}_j^s$ and $\bar{n}_i^r$ of two rigid bodies, as shown in Figure 2.7.

Example 2.4.2

Consider an arbitrary vector $\bar{v}$ in body A (Figure 2.8). Let $\bar{n}_1^A$, $\bar{n}_2^A$ and $\bar{n}_3^A$ represent a set of dextral orthogonal unit vectors attached to the rigid body A and let $\bar{n}_1, \bar{n}_2$ and $\bar{n}_3$ be the dextral orthogonal unit vectors representing an inertial reference frame R. The vector $\bar{v}$ could be expressed with respect to either $\bar{n}_i^A$ or $\bar{n}_i$. In either case, we obtain

$$\bar{v} = \bar{v}_1^A \bar{n}_1^A + \bar{v}_2^A \bar{n}_2^A + \bar{v}_3^A \bar{n}_3^A \tag{2.42}$$

$$\bar{v} = v_1^R \bar{n}_1 + v_2^R \bar{n}_2 + v_3^R \bar{n}_3 \tag{2.43}$$

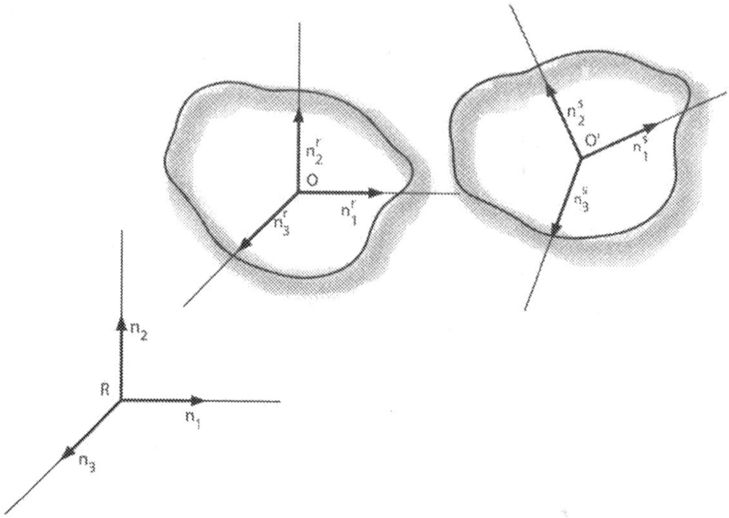

Fig. 2.7. Orientation of two arbitrary reference frames n^r and n^s and of two rigid bodies

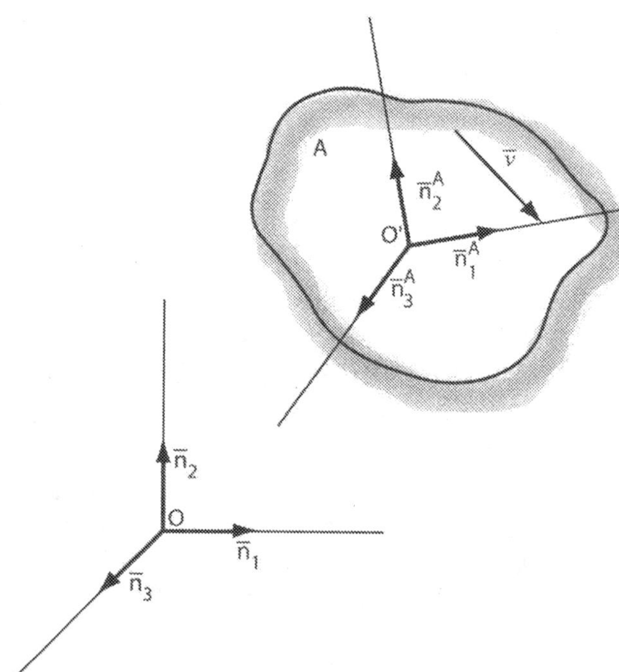

Fig. 2.8. A representation of a vector $\bar{v}$ attached to body A using direction cosines between $\bar{n}^A$ and $\bar{n}$

In matrix form, by equating (2.42) and (2.43) we obtain the following relation between the unit vectors $\bar{n}_i^A$ and $\bar{n}_i$

$$\begin{bmatrix} v_1^A & v_2^A & v_3^A \end{bmatrix} \begin{bmatrix} \bar{n}_1^A \\ \bar{n}_2^A \\ \bar{n}_3^A \end{bmatrix} = \begin{bmatrix} v_1^R & v_2^R & v_3^R \end{bmatrix} \begin{bmatrix} \bar{n}_1 \\ \bar{n}_2 \\ \bar{n}_3 \end{bmatrix} \tag{2.44}$$

We also know that the orthogonal unit vector $\bar{n}_i$ and $\bar{n}_j^A$ are related by the direction cosine matrix S^{AO}:

$$\begin{bmatrix} \bar{n}_1^A \\ \bar{n}_2^A \\ \bar{n}_3^A \end{bmatrix} = \begin{bmatrix} S^{AO} \end{bmatrix} \begin{bmatrix} \bar{n}_1 \\ \bar{n}_2 \\ \bar{n}_3 \end{bmatrix} \tag{2.45}$$

Substituting equation (2.45) into the left-hand side of equation (2.44), we obtain

$$\begin{bmatrix} v_1^A & v_2^A & v_3^A \end{bmatrix} \begin{bmatrix} S^{AO} \end{bmatrix} \begin{bmatrix} \bar{n}_1 \\ \bar{n}_2 \\ \bar{n}_3 \end{bmatrix} = \begin{bmatrix} v_1^R & v_2^R & v_3^R \end{bmatrix} \begin{bmatrix} \bar{n}_1 \\ \bar{n}_2 \\ \bar{n}_3 \end{bmatrix} \tag{2.46}$$

Hence we obtain the relation of the vector $\bar{v}$ expressed in R and A, respectively.

$$\begin{bmatrix} v_1^A & v_2^A & v_3^A \end{bmatrix} \begin{bmatrix} S^{AO} \end{bmatrix} = \begin{bmatrix} v_1^R & v_2^R & v_3^R \end{bmatrix} \tag{2.47}$$

Basically, equation (2.47) states that the components of the vector $\bar{v}$, with respect to the reference frame R, are obtained by multiplying its components with respect to the reference frame attached to body A by the direction cosine matrix S^{AO}. Equation (2.47) defines an important relationship that can be used in multibody systems when relative coordinates are involved.

Some properties of the direction cosine matrix, which is an orthogonal matrix, are

$$\begin{bmatrix} S^{AO} \end{bmatrix}^T = \begin{bmatrix} S^{AO} \end{bmatrix}^{-1} = \begin{bmatrix} S^{OA} \end{bmatrix} \tag{2.48}$$

where $[\,]^T$ denotes the transpose and $[\,]^{-1}$ denotes the inverse matrix. The transpose is simply written by altering the superscripts A and O:

$$\begin{bmatrix} S^{AO} \end{bmatrix} \begin{bmatrix} S^{OA} \end{bmatrix} = I$$

or

$$\begin{bmatrix} S^{OA} \end{bmatrix} \begin{bmatrix} S^{AO} \end{bmatrix} = I \tag{2.49}$$

where I is the identity matrix:

$$I = \begin{bmatrix} 1 & 0 & 0 \\ 0 & 1 & 0 \\ 0 & 0 & 1 \end{bmatrix} \tag{2.50}$$

and the determinant of S^{AO} is equal to 1.

In tensor notation, we have

$$S_{ik}^{AO} S_{jk}^{AO^T} = \delta_{ij} \quad (i, j = 1, 2, 3) \tag{2.51}$$

or equivalently,

$$S_{ki}^{AO} S_{kj}^{AO^T} = \delta_{ij} \quad (i, j = 1, 2, 3) \tag{2.52}$$

where δ_{ij}, the *Kronecker delta function*, takes the value of unity when the subscript i is equal to the subscript j and zero when the subscripts are different.

2.5 Euler and Rodriguez Formula

2.5.1 Euler Parameters

We learned in Section 2.3 that the dextral orthogonal unit vectors $\bar{n}_i^A$ $(i = 1, 2, 3)$ can be brought into alignment with $\bar{n}_i$ $(i = 1, 2, 3)$ representing the inertial reference frame R through three successive rotations R_α, R_β and R_γ resulting into the direction cosine matrix S. The Euler theorem on rotation states that any set of dextral orthogonal unit vectors such as $\bar{n}_i^A$ $(i = 1, 2, 3)$ fixed in body A can be brought into alignment with the unit vectors $\bar{n}_i$ $(i = 1, 2, 3)$ of a fixed reference by a simple rotation through an axis L by an angle θ (Figure 2.9). Let λ be a unit vector along the axis of rotation L. This is achieved by introducing the Euler vector $\bar{\varepsilon}$ and four scalar quantities $\varepsilon_1, \varepsilon_2, \varepsilon_3, \varepsilon_4$, called *Euler parameters*, such that

$$\bar{\varepsilon}_i = \bar{\lambda}_i \sin\frac{\theta}{2} \quad (i = 1, 2, 3) \tag{2.53}$$

or

$$\varepsilon_1 = \lambda_1 \sin\frac{\theta}{2} \tag{2.54}$$

$$\varepsilon_2 = \lambda_2 \sin\frac{\theta}{2} \tag{2.55}$$

$$\varepsilon_3 = \lambda_3 \sin\frac{\theta}{2} \tag{2.56}$$

and

$$\varepsilon_4 = \cos\frac{\theta}{2} \tag{2.57}$$

Unlike the orientation angles used in finding the direction cosine matrix, the Euler parameters are not independent. The sum of their squares is equal to unity:

$$\varepsilon_1^2 + \varepsilon_2^2 + \varepsilon_3^2 + \varepsilon_4^2 = 1 \tag{2.58}$$

$$\lambda_1^2 + \lambda_2^2 + \lambda_3^2 = 1 \tag{2.59}$$

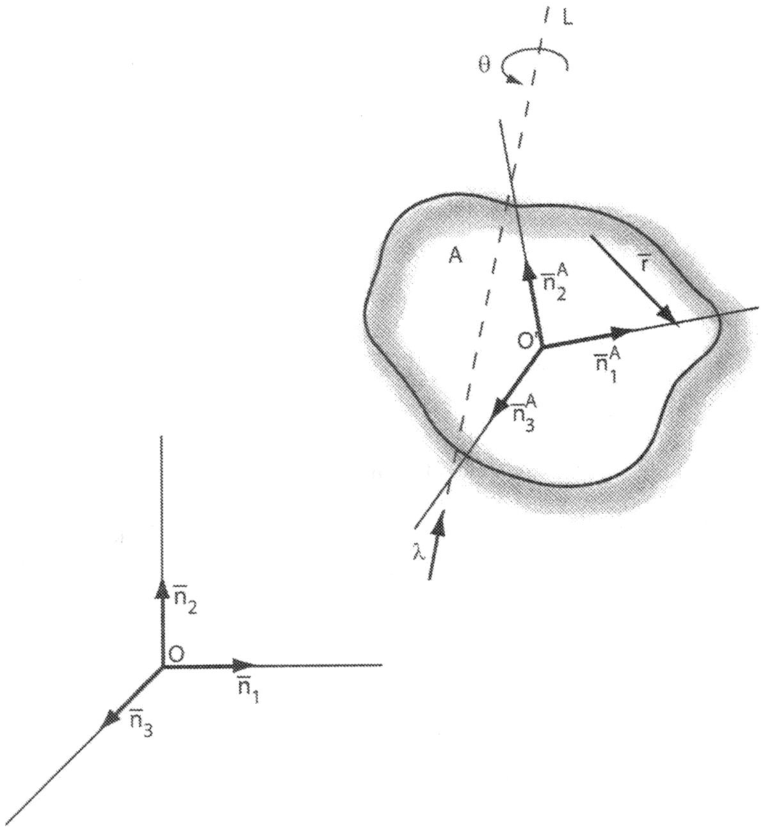

Fig. 2.9. A simple rotation of body A through an axis L

To obtain the axis of rotation: (1) obtain ε_1–ε_4 from matrix S; (2) obtain θ from $\varepsilon_4 = \cos\frac{\theta}{2}$; (3) obtain $\lambda_1, \lambda_2, \lambda_3$ from equations (2.54)–(2.56).

Otherwise if θ and axis $\lambda_1, \lambda_2, \lambda_3$ are known as ε_1. ε_4 can be obtained and the transformation matrix S can be calculated by formula 2.62 to 2.10.

The advantage to using Euler parameters stems from the simple form the matrix S takes when expressed in terms of Euler parameters, where

$$S_{ij} = \bar{n}_i \cdot \bar{n}_j^A \tag{2.60}$$

$$S_{ij} = \begin{bmatrix} S_{11} & S_{12} & S_{13} \\ S_{21} & S_{22} & S_{23} \\ S_{31} & S_{32} & S_{33} \end{bmatrix} \tag{2.61}$$

where the components of S_{ij} in terms of Euler parameters are

$$S_{11} = \varepsilon_1^2 - \varepsilon_2^2 - \varepsilon_3^2 + \varepsilon_4^2 \tag{2.62}$$

$$S_{12} = 2(\varepsilon_1\varepsilon_2 - \varepsilon_3\varepsilon_4) \tag{2.63}$$

$$S_{13} = 2(\varepsilon_1\varepsilon_3 + \varepsilon_2\varepsilon_4) \tag{2.64}$$

$$S_{21} = 2(\varepsilon_1\varepsilon_2 + \varepsilon_3\varepsilon_4) \tag{2.65}$$

$$S_{22} = -\varepsilon_1^2 + \varepsilon_2^2 - \varepsilon_3^2 + \varepsilon_4^2 \tag{2.66}$$

$$S_{23} = 2(\varepsilon_2\varepsilon_3 - \varepsilon_1\varepsilon_4) \tag{2.67}$$

$$S_{31} = 2(\varepsilon_1\varepsilon_3 - \varepsilon_2\varepsilon_4) \tag{2.68}$$

$$S_{32} = 2(\varepsilon_2\varepsilon_3 + \varepsilon_1\varepsilon_4) \tag{2.69}$$

$$S_{33} = -\varepsilon_1^2 - \varepsilon_2^2 + \varepsilon_3^2 + \varepsilon_4^2 \tag{2.70}$$

Derivation

Consider the rigid body A depicted in Figure 2.10. Let P define an arbitrary point on the body and L be the axis of rotation as defined by Euler rotation. Let λ be a unit vector along the line L. If the rigid body is rotated about L by an angle θ, the new position of P is represented by $P*$. Let O be a point on the line L; then we define the Position vectors $OP*$ and OP to be $\bar{P}*$ and $\bar{P}$. From Figure 2.10 we can write

$$\overline{OP}^* = \overline{OS} + \overline{SP}^* \tag{2.71}$$

where $\overline{OS}$ is the projection of $\bar{P}$ onto the axis of the line L such that

$$\overline{OS} = (\bar{P} \cdot \bar{\lambda})\bar{\lambda} \tag{2.72}$$

On the other hand,

$$\overline{SP}^* = \overline{SB} + \overline{BP}^* \tag{2.73}$$

Consider the vector $\overline{SP}^*$; it has two components—one along $\mathbf{m}$ and one along ℓ—hence

$$\overline{SP}^* = \left|\overline{SP}^*\right| \cos\theta \ell + \left|\overline{SP}^*\right| \sin\theta\, \mathbf{m} \tag{2.74}$$

We also know that $\left|\overline{SP}^*\right|$ is equal to $\left|\overline{SP}\right|$. Equation (2.74) can then be written as

$$\overline{SP}^* = \left|\overline{SP}\right| \cos\theta \ell + \left|\mathbf{SP}\right| \sin\theta\, \mathbf{m} \tag{2.75}$$

or

$$\overline{SP}^* = \overline{SP} \cos\theta + \bar{\lambda} \times \overline{SP} \sin\theta \tag{2.76}$$

Note that

$$\overline{SP} = \overline{OP} - \overline{OS} \tag{2.77}$$

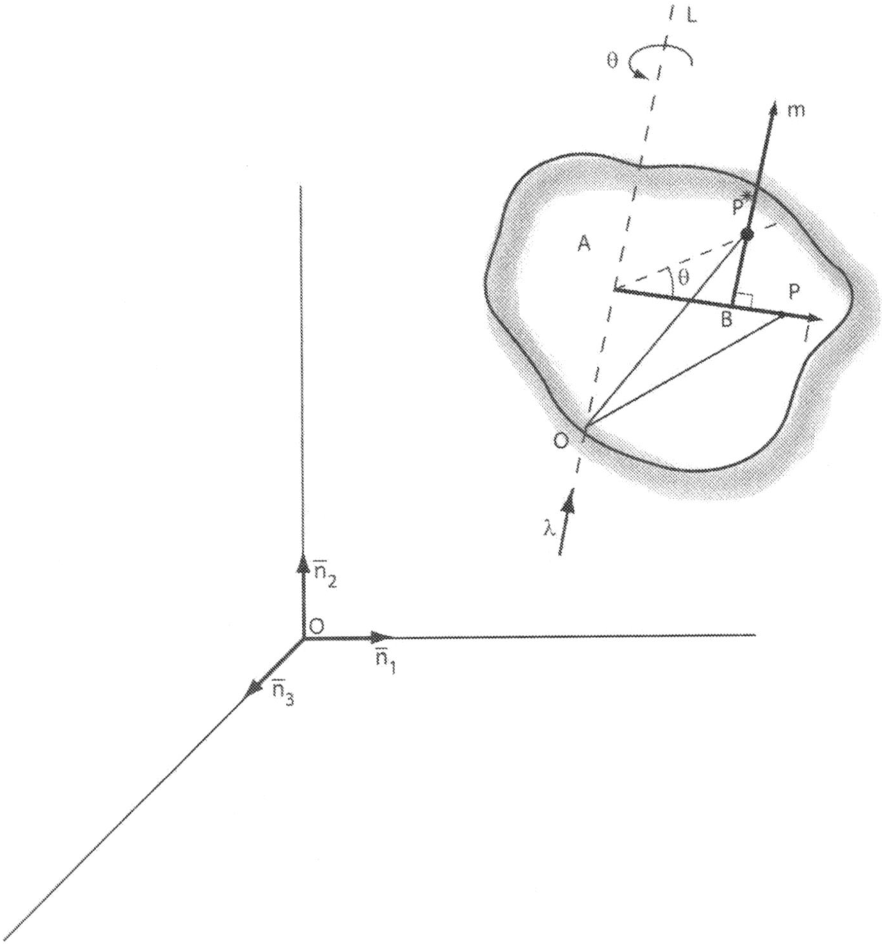

Fig. 2.10. Geometrical description of a point P rotation about LL' axis

Replacing $\overline{OP}$ simply by $\bar{P}$ and $\overline{OS}$ by the projection of $\bar{P}$ onto the axis of rotation L, equation (2.77) can be expressed as

$$\overline{SP} = \bar{P} - (\bar{P} \cdot \bar{\lambda})\bar{\lambda} \tag{2.78}$$

Substituting equation (2.78) into (2.76) and resubstituting the result and equation (2.72) into (2.71), we obtain

$$\overline{OP}^* = (\bar{P} \cdot \bar{\lambda})\bar{\lambda} + \left[\bar{P} - (\bar{P} \cdot \bar{\lambda})\bar{\lambda}\right]\cos\theta + \bar{\lambda} \times \bar{P} \sin\theta \tag{2.79}$$

or simply

$$\bar{P}^* = (\bar{P} \cdot \bar{\lambda})\bar{\lambda} + \left[\bar{P} - (\bar{P} \cdot \bar{\lambda})\bar{\lambda}\right]\cos\theta + \bar{\lambda} \times \bar{P} \sin\theta \tag{2.80}$$

Factorizing $\bar{P}$ from the right-hand side of equation (2.80) yields

$$\bar{P}^* = \bar{\bar{S}} \cdot \bar{P} \tag{2.81}$$

where

$$\bar{\bar{S}} = (1 - \cos\theta)\bar{\lambda}\bar{\lambda} + \cos\theta\bar{\bar{I}} + \sin\theta\lambda \times \bar{\bar{I}} \tag{2.82}$$

$\bar{\bar{I}}$ is the unit (or identity) dyadic. We can easily see that the dyadic $\bar{\bar{S}}$ is defined as

$$\bar{\bar{S}} = S_{ij}\bar{n}_i\bar{n}_j \tag{2.83}$$

where

$$S_{ij} = (1 - \cos\theta)\lambda_i\lambda_j + \cos\theta\delta_{ij} - \sin\theta e_{ijk}\lambda_k \tag{2.84}$$

where e_{ijk} is the permutation symbol and δ_{ij} is the Kronecker delta defined in equation (2.51). The matrix representation of the direction cosine matrix δ_{ij} is obtained as follows: Consider equation (2.80), which is commonly expressed as

$$\bar{P}^* = \bar{P} + (\bar{\lambda} \times \bar{P})\sin\theta + \bar{\lambda} \times (\bar{\lambda} \times \bar{P})(1 - \cos\theta) \tag{2.85}$$

Let Λ be defined as the skew matrix corresponding to the unit vector $\bar{\lambda} = [\lambda_1, \lambda_2, \lambda_3]^T$, where

$$\Lambda = \begin{bmatrix} 0 & -\lambda_3 & \lambda_2 \\ \lambda_3 & 0 & -\lambda_1 \\ -\lambda_2 & \lambda_1 & 0 \end{bmatrix} \tag{2.86}$$

The cross product $\bar{\lambda} \times \bar{P}$ can be expressed in terms of the skew matrix Λ as

$$\bar{\lambda} \times \bar{P} = \Lambda\bar{P} \tag{2.87}$$

Substituting equation (2.87) into (2.85), we obtain the following matrix representation of finite rotations:

$$\bar{P}^* = S\bar{P} \tag{2.88}$$

where

$$S = I + \Lambda\sin\theta + \Lambda^2(1 - \cos\theta) \tag{2.89}$$

where I is the identity matrix and S is the rotation matrix defined by equation (2.86).

Note that S_{ij} serves as the transformation needed to obtain a relationship between any arbitrary $\bar{P}$ fixed in body A, and is subjected to a simple rotation θ about an arbitrary axis such as the one given by line LL'. Hence S_{ij} forms a 3×3 square matrix whose elements are obtained from equation (2.84). What makes equation (2.84) even more interesting is the fact that it forms a square matrix (transformation) totally in terms of Euler parameters such that all the elements of S are given by equations (2.62) to (2.70). For example, consider S_{11}; from equation (2.84) we have

$$S_{11} = (1 - \cos\theta)\lambda_1^2 + \cos\theta - \sin\theta(0) \tag{2.90}$$

or

$$S_{11} = (1 - \cos\theta)\lambda_1^2 + \cos\theta \tag{2.91}$$

Using the definition of Euler parameters given by equations (2.54), (2.56) and (2.58), we can further express S_{11} as:

$$\begin{aligned} S_{11} &= 2\lambda_1^2 \sin^2\frac{\theta}{2} + \cos^2\frac{\theta}{2} - \sin^2\frac{\theta}{2} \\ &= \lambda_1^2 \sin^2\frac{\theta}{2} + (\lambda_1^2 - 1)\sin^2\frac{\theta}{2} + \cos^2\frac{\theta}{2} \\ &= \lambda_1^2 \sin^2\frac{\theta}{2} - \lambda_2^2 \sin^2\frac{\theta}{2} - \lambda_3^2 \sin^2\frac{\theta}{2} + \cos^2\frac{\theta}{2} \end{aligned} \tag{2.92}$$

or

$$S_{11} = \varepsilon_1^2 - \varepsilon_2^2 - \varepsilon_3^2 + \varepsilon_4^2 \tag{2.93}$$

Proceeding in the same manner, we can show that all the elements of S can be expressed in terms of Euler parameters. So

$$S = \begin{bmatrix} \varepsilon_1^2 - \varepsilon_2^2 - \varepsilon_3^2 + \varepsilon_4^2 & 2(\varepsilon_1\varepsilon_2 - \varepsilon_3\varepsilon_4) & 2(\varepsilon_1\varepsilon_3 + \varepsilon_2\varepsilon_4) \\ 2(\varepsilon_1\varepsilon_2 + \varepsilon_3\varepsilon_4) & -\varepsilon_1^2 + \varepsilon_2^2 - \varepsilon_3^2 + \varepsilon_4^2 & 2(\varepsilon_2\varepsilon_3 - \varepsilon_1\varepsilon_4) \\ 2(\varepsilon_1\varepsilon_3 - \varepsilon_2\varepsilon_4) & 2(\varepsilon_2\varepsilon_3 + \varepsilon_1\varepsilon_4) & -\varepsilon_1^2 - \varepsilon_2^2 + \varepsilon_3^2 + \varepsilon_4^2 \end{bmatrix} \tag{2.94}$$

Once more, the practical use of Euler parameters relies on the fact that, first, the transformation matrix S elements are linear in ε_i, and most of all they eliminate some singularities that occur with the use of Euler angles. To illustrate this further, let trS denote the trace of matrix S; then

$$trS = -1 + 4\varepsilon_4^2 \tag{2.95}$$

from which

$$\varepsilon_4 = \sqrt{\frac{trS + 1}{4}} \tag{2.96}$$

and using other elements of the diagonal matrix S, we obtain

$$\varepsilon_1 = \frac{S_{32} - S_{23}}{4\varepsilon_4} \tag{2.97}$$

$$\varepsilon_2 = \frac{S_{13} - S_{31}}{4\varepsilon_4} \tag{2.98}$$

$$\varepsilon_3 = \frac{S_{21} - S_{12}}{4\varepsilon_4} \tag{2.99}$$

In contrast to the singularities resulting from the transformation given by equation (2.66), where the evaluation of the Euler angles results in a division by a cosine or sine of an angle, which makes it undefined when the denominator is near or equal to zero. This will be further illustrated in the evaluation of the rate of changes of the Euler angles in Section 2.6.

Example 2.5.1

A set of mutually perpendicular unit vectors $\bar{n}_i^A$ $(i = 1, 2, 3)$ fixed in body A can be brought into alignment with the unit vectors $\bar{n}_j$ $(j = 1, 2, 3)$ fixed in an inertial frame R through three successive rotations $\alpha = 30°$, $\beta = 60°$ and $\gamma = 30°$ about $\bar{n}_1$, $\bar{n}_2$ and $\bar{n}_3$, respectively (Figure 2.11). An alternative to the alignment of $\bar{n}_i^A$ with $\bar{n}_j$ can be attained by rotating the rigid body A about the line axis LL' by an angle θ.

(a) Find the transformation matrix S_{ij} in terms of Euler's parameters.
(b) Determine the corresponding angle of rotation θ due to the rotation about LL'.

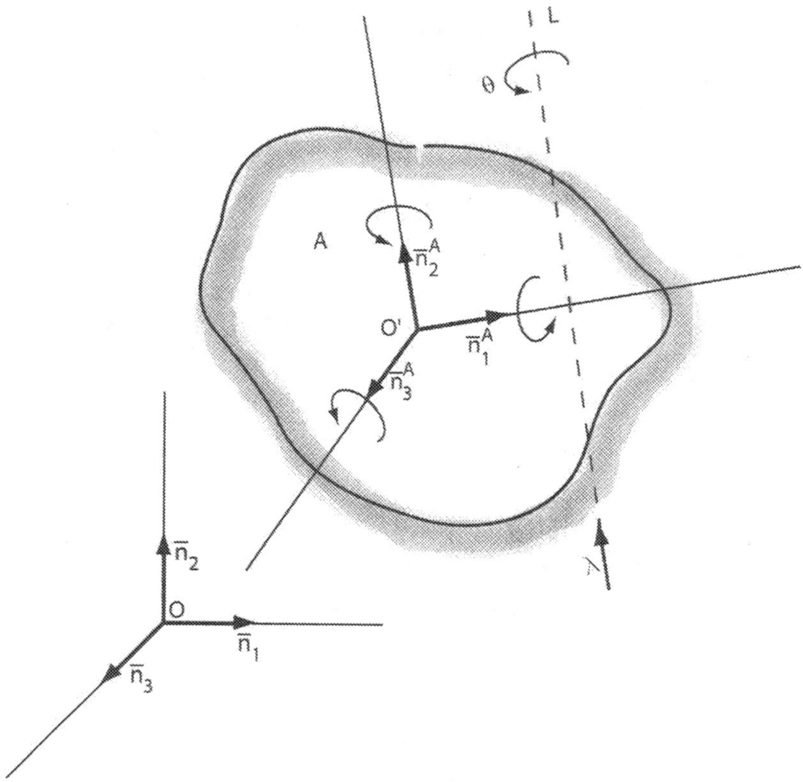

Fig. 2.11. Equivalence of the rotation of body A for alignment of the unit vectors $\bar{n}^A$ with $\bar{n}$

Solution
First we note that all three Euler angles are given; hence the transformation matrix of the direction cosines can be used to relate $\bar{n}_i^A$ $(i = 1, 2, 3)$ to $\bar{n}_j (j = 1, 2, 3)$. Using equation (2.39), S_{ij}^{AO} is given by

$$S_{ij}^{AO} = \begin{bmatrix} c\beta c\gamma & c\alpha s\gamma + s\alpha s\beta c\gamma & s\gamma s\alpha - c\alpha s\beta c\gamma \\ -c\beta s\gamma & c\alpha c\gamma - s\alpha s\beta s\gamma & s\alpha c\gamma + c\alpha s\beta s\gamma \\ s\beta & -s\alpha c\beta & c\alpha c\beta \end{bmatrix} \tag{2.100}$$

We also know that S_{ij} in terms of Euler parameters is given by equation (2.94). To solve for ε_1, ε_2, ε_3 and ε_4 we first find

$$trS = S_{11} + S_{22} + S_{33} \tag{2.101}$$

which yields

$$trS = c\beta c\gamma + c\alpha c\gamma - s\alpha s\beta s\gamma + c\alpha c\beta \tag{2.102}$$

Substituting the values of α, β, γ into equation (2.102), we obtain

$$\begin{aligned} trS = {} & \cos 60^\circ \cos 30^\circ + \cos 30^\circ \cos 30^\circ \\ & - \sin 30^\circ \sin 60^\circ \sin 30^\circ + \cos 30^\circ \cos 60^\circ \end{aligned} \tag{2.103}$$

or

$$trS = \frac{3\sqrt{3}}{8} + \frac{3}{4} = 1.3995 \tag{2.104}$$

Using equation (2.96), which states that

$$\varepsilon_4 = \sqrt{\frac{trS + 1}{4}} \tag{2.105}$$

which yields

$$\varepsilon_4 = \sqrt{\frac{3\sqrt{3}}{32} + \frac{7}{16}} = \frac{1}{4}\sqrt{\frac{3\sqrt{3}}{32} + 7} = 0.7745 \tag{2.106}$$

Similarly, using equations (2.97) to (2.99) we get

$$\varepsilon_3 = -0.3415 \tag{2.107}$$

$$\varepsilon_2 = -0.4085 \tag{2.108}$$

$$\varepsilon_1 = -0.2824 \tag{2.109}$$

(a) A simple check will be to add all the squares of the ε_i $(i = 1, 2, 3, 4)$, which must add up to 1. As expected, of course,

$$(\varepsilon_1)^2 + (\varepsilon_2)^2 + (\varepsilon_3)^2 + (\varepsilon_4)^2 = 1$$

(b) Using equation (2.57), θ is then found to be

$$\theta = 2\cos^{-1}\varepsilon_4 = 78.48^\circ \tag{2.110}$$

In this context, one can say that any change in the relative orientation of body A can be produced by simple rotation about an arbitrary axis such that equations (2.53) to (2.58) are satisfied.

Example 2.5.2

Consider the rigid body A given by the cube depicted in Figure 2.12. The cube is to be rotated about the line axis LL', which passes through its diagonal. Let the angle of rotation θ be equal to 60°. Find the new position of vector $\bar{P}$ after the rotation.

Solution

Let $\bar{n}_j^A (j = 1, 2, 3)$ be fixed at point O' in the lower left corner of the cube and $\bar{n}_i (i = 1, 2, 3)$ be fixed in an inertial frame R. Initially, both $\bar{n}_i^A$ and $\bar{n}_i$ are aligned. The new position $\bar{P}'$ of $\bar{P}$ can be found as

$$\bar{P}' = S_{ij}\bar{P} \tag{2.111}$$

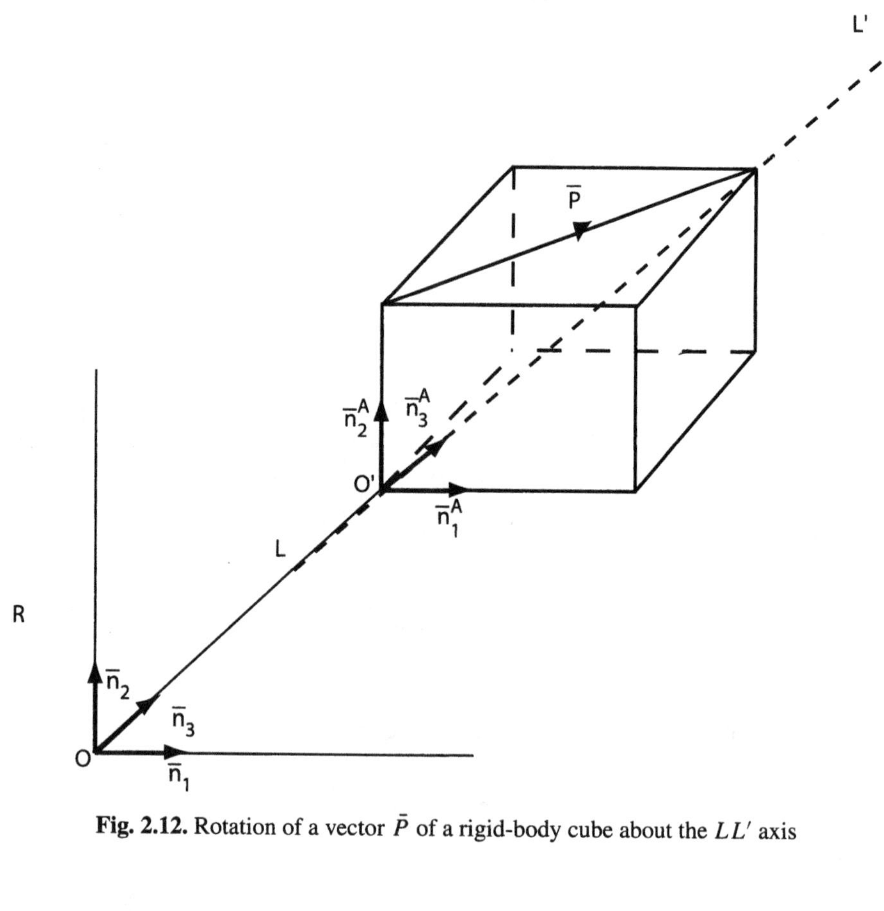

Fig. 2.12. Rotation of a vector $\bar{P}$ of a rigid-body cube about the LL' axis

where $\bar{P} = \begin{bmatrix} a \\ a \\ 0 \end{bmatrix}$ and S_{ij} is the transformation given by equation (2.84). To find the transformation matrix S in terms of Euler's parameters as given in equation (2.94), we need to solve for Euler's parameters. Using equations (2.53) to (2.57) together with equation (2.84), we obtain

$$\varepsilon_i = \lambda_i \sin\frac{\theta}{2} \quad (i = 1, 2, 3) \tag{2.112}$$

and

$$\varepsilon_4 = \cos\frac{\theta}{2} \tag{2.113}$$

where $\bar{\lambda}$ is a unit vector along LL' such that

$$\bar{\lambda} = \lambda_1 \bar{n}_1^A + \lambda_2 \bar{n}_2^A + \lambda_3 \bar{n}_3^A \tag{2.114}$$

From Figure 2.12 the scalar quantities $\lambda_i (i = 1, \ldots, 3)$ are seen to be

$$\lambda_1 = \lambda_2 = \lambda_3 = \frac{\sqrt{3}}{3} \tag{2.115}$$

Consequently,

$$\varepsilon_1 = \lambda_1 \sin\frac{\theta}{2} = \frac{\sqrt{3}}{6} \tag{2.116}$$

$$\varepsilon_2 = \varepsilon_3 = \varepsilon_1 = \frac{\sqrt{3}}{6} \tag{2.117}$$

and

$$\varepsilon_4 = \cos\frac{\theta}{2} = \frac{\sqrt{3}}{2} \tag{2.118}$$

Substituting the value of $\varepsilon_i (i = 1, \ldots, 4)$ into equation (2.94), the desired transformation matrix is found to be [using (2.62) to (2.70)]

$$S = \begin{bmatrix} \frac{2}{3} & -\frac{1}{3} & \frac{2}{3} \\ \frac{2}{3} & \frac{2}{3} & -\frac{1}{3} \\ -\frac{1}{3} & -\frac{2}{3} & \frac{2}{3} \end{bmatrix} \tag{2.119}$$

Therefore,

$$\bar{P}' = S\bar{P} = a\begin{bmatrix} \frac{1}{3} \\ \frac{4}{3} \\ \frac{1}{3} \end{bmatrix} \tag{2.120}$$

Example 2.5.3
Consider the position vector $\bar{P}$ on the cube in Figure 2.13. If the cube is to be rotated about LL' by 30°, find the new position of $\bar{P}'$.

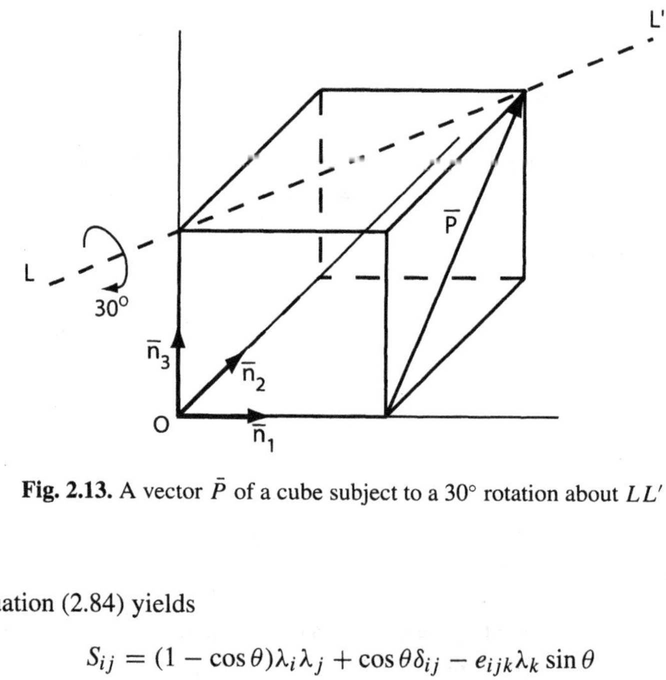

Fig. 2.13. A vector $\bar{P}$ of a cube subject to a 30° rotation about LL'

Solution
Using equation (2.84) yields

$$S_{ij} = (1 - \cos\theta)\lambda_i\lambda_j + \cos\theta\delta_{ij} - e_{ijk}\lambda_k \sin\theta$$

where

$$\lambda_1 = \frac{1}{\sqrt{2}} \quad \lambda_2 = \frac{1}{\sqrt{2}} \quad \lambda_3 = 0$$

and

$$\theta = 30^\circ$$

Thus the direction cosines can be found as

$$S_{11} = (1 - \cos 30^\circ)\lambda_1\lambda_2 + \cos 30^\circ = \frac{2 + \sqrt{3}}{4} = 0.933$$

$$S_{12} = (1 - \cos 30°)\lambda_1\lambda_2 - \sin 30°\lambda_3 = \frac{2-\sqrt{3}}{4} = 0.067$$

$$S_{13} = (1 - \cos 30°)\lambda_1\lambda_3 + \sin 30°\lambda_2 = \frac{\sqrt{2}}{4} = 0.354$$

$$S_{21} = (1 - \cos 30°)\lambda_2\lambda_1 + \sin 30°\lambda_3 = \frac{2-\sqrt{3}}{4} = 0.067$$

$$S_{22} = (1 - \cos 30°)\lambda_2\lambda_2 + \cos 30° = \frac{2+\sqrt{3}}{4} = 0.933$$

$$S_{23} = (1 - \cos 30°)\lambda_2\lambda_3 - \sin 30°\lambda_1 = -\frac{\sqrt{2}}{4} = -0.354$$

$$S_{31} = (1 - \cos 30°)\lambda_3\lambda_1 - \sin 30°\lambda_2 = -\frac{\sqrt{2}}{4} = -0.354$$

$$S_{32} = (1 - \cos 30°)\lambda_3\lambda_2 + \sin 30°\lambda_1 = \frac{\sqrt{2}}{4} = 0.354$$

$$S_{33} = (1 - \cos 30°)\lambda_3\lambda_3 + \cos 30° = \frac{\sqrt{3}}{2} = 0.866$$

The direction cosines matrix is

$$[S_{ij}] = \begin{bmatrix} 0.933 & 0.067 & 0.354 \\ 0.067 & 0.933 & -0.354 \\ -0.354 & 0.354 & 0.866 \end{bmatrix} \quad (i, j = 1, 3)$$

Position $\bar{P}$ is defined by the vector

$$\bar{P} = 0\mathbf{i} + a\mathbf{j} + a\mathbf{k}$$

or in matrix form,

$$\bar{P} = \begin{bmatrix} 0 \\ a \\ a \end{bmatrix}$$

Applying equation (2.111) to find the new position $\bar{P}'$, we get

$$\{\bar{P}'\} = [S][\bar{P}]$$

$$\bar{P}' = a \begin{bmatrix} 0.421 \\ 0.579 \\ 1.22 \end{bmatrix}$$

2.5.2 Rodriguez Parameters

The rotation of a rigid body with respect to an inertial fixed frame R can be obtained through a simple rotation θ about an axis LL' whose direction is given by λ, making use of the Rodriguez parameters. Similar to the Euler vector, the Rodriguez vector is defined as

$$\bar{\eta} = \bar{\lambda} \tan \frac{\theta}{2} \tag{2.121}$$

where the Rodriguez parameters are

$$\eta_1 = \lambda_1 \tan \frac{\theta}{2}$$

$$\eta_2 = \lambda_2 \tan \frac{\theta}{2}$$

and

$$\eta_3 = \lambda_3 \tan \frac{\theta}{2} \tag{2.122}$$

Recall that the Euler vector is given by

$$\bar{\varepsilon} = \bar{\lambda} \sin \frac{\theta}{2} \tag{2.123}$$

Hence a relationship between the Rodriguez and Euler parameters is given by

$$\bar{\eta} = \bar{\varepsilon} \frac{1}{\cos(\theta/2)} \tag{2.124}$$

or

$$\bar{\eta} = \frac{\bar{\varepsilon}}{\varepsilon_4} \tag{2.125}$$

The direction cosine matrix S in terms of Rodriguez parameters is given by

$$S = \begin{bmatrix} 1+\eta_1^2-\eta_2^2-\eta_3^2 & 2(\eta_1\eta_2-\eta_3) & 2(\eta_1\eta_3+\eta_2) \\ 2(\eta_1\eta_2+\eta_3) & 1-\eta_1^2+\eta_2^2-\eta_3^2 & 2(\eta_2\eta_3-\eta_1) \\ 2(\eta_1\eta_3-\eta_2) & 2(\eta_2\eta_3+\eta_1) & 1-\eta_1^2-\eta_2^2+\eta_3^2 \end{bmatrix} \frac{1}{1+\eta_1^2+\eta_2^2+\eta_3^2} \tag{2.126}$$

Proof
Making use of the direction cosines in terms of Euler parameters together with equation (2.125), the elements of S in terms of Rodriguez parameters can be obtained as follows:

$$S_{11} = \varepsilon_1^2 - \varepsilon_2^2 - \varepsilon_3^2 - \varepsilon_4^2 \tag{2.127}$$

Remembering the relationship

$$\eta_i = \frac{\varepsilon_i}{\varepsilon_4} \quad (i = 1, 2, 3) \tag{2.128}$$

$$\varepsilon_i = \eta_i \varepsilon_4 \quad (i = 1, 2, 3)$$

we get

$$S_{11} = \eta_1^2\varepsilon_4^2 - \eta_2^2\varepsilon_4^2 - \eta_3^2\varepsilon_4^2 + \varepsilon_4^2 \tag{2.129}$$

or factoring out ε_4^2,

$$S_{11} = (\eta_1^2 - \eta_2^2 - \eta_3^2 + 1)\varepsilon_4^2 \tag{2.130}$$

The sum of the square of Euler parameters is

$$\varepsilon_1^2 + \varepsilon_2^2 + \varepsilon_3^2 + \varepsilon_4^2 = 1 \tag{2.131}$$

which yields

$$\varepsilon_4^2 = 1 - (\varepsilon_1^2 + \varepsilon_2^2 + \varepsilon_3^2) \tag{2.132}$$

Dividing equation (2.130) by ε_4^2 and substituting equation (2.132), making use of equation (2.128), gives

$$S_{11} = \frac{1 + \eta_1^2 - \eta_2^2 - \eta_3^2}{1 + \eta_1^2 + \eta_2^2 + \eta_3^2} \tag{2.133}$$

Similarly, all the elements of S can be found.

Example 2.5.4
Find the corresponding transformation matrix S in terms of the Rodriguez parameters for Example 2.5.2.

Solution
From equations (2.115) and (2.122), the Rodriguez parameters are found to be

$$\eta_1 = \frac{\sqrt{3}}{3} \tan 30^\circ = \frac{1}{3} \tag{2.134}$$

Also,

$$\eta_2 = \eta_3 = \eta_1 = \frac{1}{3} \tag{2.135}$$

Substituting the values of η_1, η_2 and η_3 into equation (2.126), we obtain

$$S = 0.57 \begin{bmatrix} 0.75 & -0.5 & 1.5 \\ 1.56 & 0.75 & -0.5 \\ -0.55 & 1.5 & 0.75 \end{bmatrix} \tag{2.136}$$

Note: The advantage of the Rodriguez parameters over the Euler parameters is that there are only three components, whereas there are four Euler parameters. It should also be noted that the sum of Euler's parameters is always unity (they are usually small numbers); on the other hand, the sum of the Rodriguez parameters can be infinite for $\theta = \pi$.

2.6 Angular Velocity

Consider a vector $\bar{r}$ fixed in a rigid body A as shown in Figure 2.14. Let a set of mutually perpendicular unit vectors fixed in body A be defined by $\bar{n}_1^A$, $\bar{n}_2^A$ and $\bar{n}_3^A$and $\bar{n}_1$, $\bar{n}_2$ and $\bar{n}_3$ be a set of mutually perpendicular unit vectors fixed in an inertial reference frame R. Let body A be moving with respect to R; then the elements S_{ij} of the transformation matrix S given by equation (2.100) are a function of time t. The angular velocity of body A with respect to R and denoted by ${}^R\bar{\omega}^A$ is defined such that

$$\frac{d\bar{r}}{dt} = {}^R\bar{\omega}^A \times \bar{r} \tag{2.137}$$

where ${}^R\bar{\omega}^A$ is viewed as an operator that produces the total derivative of $\bar{r}$ by cross multiplication. The components of the angular velocity are given by a skew matrix Ω such that

$$\Omega = \dot{S}S^T = \begin{bmatrix} \dot{S}_{11} & \dot{S}_{12} & \dot{S}_{13} \\ \dot{S}_{21} & \dot{S}_{22} & \dot{S}_{23} \\ \dot{S}_{31} & \dot{S}_{32} & \dot{S}_{33} \end{bmatrix} \begin{bmatrix} S_{11} & S_{21} & S_{31} \\ S_{12} & S_{22} & S_{32} \\ S_{13} & S_{23} & S_{33} \end{bmatrix} \tag{2.138}$$

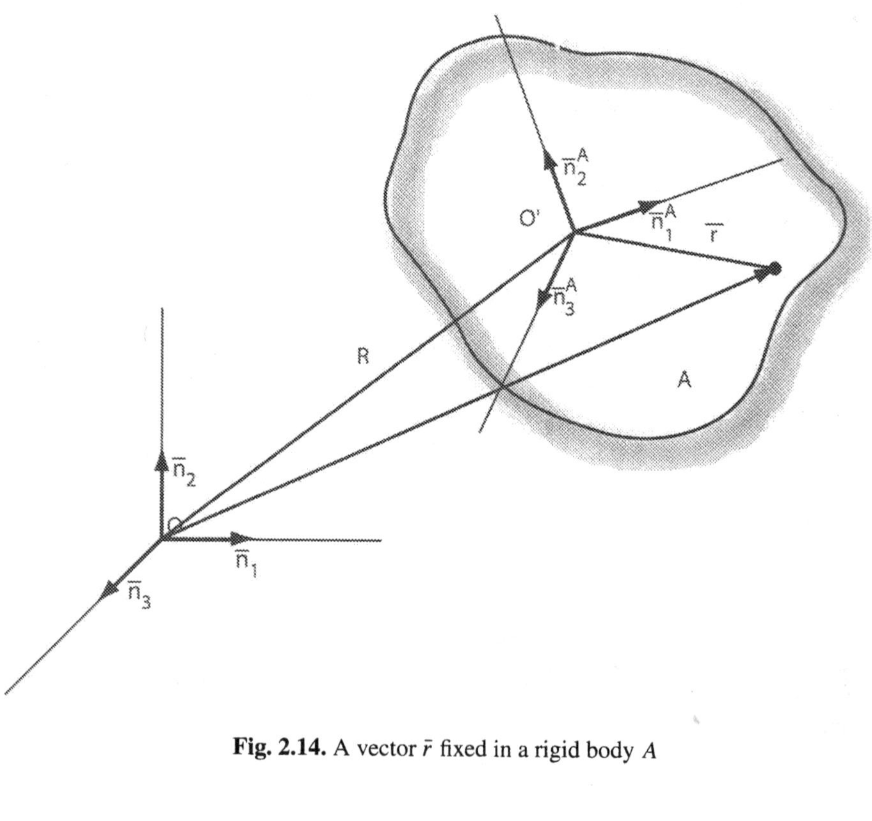

Fig. 2.14. A vector $\bar{r}$ fixed in a rigid body A

or

$$\dot{S} = \Omega S \tag{2.139}$$

Proof
The position vector $\bar{r}$ can be expressed as

$$\bar{r} = r_i \bar{n}_i^A \quad (i = 1, 2, 3) \tag{2.140}$$

where r_i $(i = 1, 2, 3)$ are the scalar components of $\bar{r}$ with respect to $\bar{n}_1^A$, $\bar{n}_2^A$ and $\bar{n}_3^A$, respectively. The rate of change of $\bar{r}$ with respect to the body reference frame denoted by $d\bar{r}/dt$ is given by

$$\frac{d\bar{r}}{dt} = \frac{d}{dt}(r_i \bar{n}_i^A) \tag{2.141}$$

or

$$\frac{d\bar{r}}{dt} = r_i \frac{d\bar{n}_i^A}{dt} + \frac{dr_i}{dt} \bar{n}_i^A \tag{2.142}$$

Since the r_i are not a function of time, $d\bar{r}_i/dt$ is equal to zero. This leads to the rate of change of $\bar{r}$ to be

$$\frac{d\bar{r}}{dt} = r_i \frac{d\bar{n}_i^A}{dt} \tag{2.143}$$

Finding an expression for $d\bar{n}_i^A/dt$ is the objective of the problem. The unit vectors $\bar{n}_i^A$ are rotating with body A relative to the fixed unit vectors $\bar{n}_j$ in the inertial reference frame R. Expressing $\bar{n}_i^A$ in terms of $\bar{n}_j$, $\bar{n}_i^A = S_{ij}\bar{n}_j$, equation (2.143) leads to

$$\frac{d\bar{r}}{dt} = r_i \dot{S}_{ij} \bar{n}_j \tag{2.144}$$

One can easily see that the rate of change of a fixed vector $\bar{r}$ in a rotating body A can be obtained by differentiation of the direction cosines matrix. This of course would lead to a tedious and complicated matrix. On the other hand, we assume [from equation (2.143)] that the rate of change of the unit vectors $\bar{n}_i$ could be expressed as a linear expression in terms of some components along $\bar{n}_j$, such that

$$\frac{d\bar{n}_i}{dt} = \Omega_{ij} \bar{n}_j \tag{2.145}$$

An important characteristic of the matrix Ω_{ij} can be seen by first considering the relationship between the unit vectors, where

$$\bar{n}_i \cdot \bar{n}_j = \delta_{ij} (i, j = 1, 2, 3) \tag{2.146}$$

where δ_{ij} is equal to 1 when i equals j, and zero otherwise. Differentiation of equation (2.146) yields

$$\frac{d\bar{n}_i}{dt} \cdot \bar{n}_j + \bar{n}_i \cdot \frac{d\bar{n}_j}{dt} = 0 \tag{2.147}$$

Substituting equation (2.145) into (2.147), we obtain

$$\Omega_{ij} + \Omega_{ji} = 0 \tag{2.148}$$

This implies that Ω_{ij} is a skew matrix of the form

$$\Omega_{ij} = \begin{bmatrix} 0 & -\omega_3 & \omega_2 \\ \omega_3 & 0 & -\omega_1 \\ -\omega_2 & \omega_1 & 0 \end{bmatrix} \tag{2.149}$$

Let us call the elements of this skew matrix ω_1, ω_2 and ω_3, where we will set

$$\Omega_{32} = -\Omega_{23} = \omega_1 \tag{2.150}$$

$$\Omega_{13} = -\Omega_{31} = \omega_2 \tag{2.151}$$

and

$$\Omega_{21} = -\Omega_{12} = \omega_3 \tag{2.152}$$

Then Ω_{ij} becomes

$$\Omega_{ij}^{OA} = e_{jin}\omega_n = \begin{bmatrix} 0 & -\omega_3 & \omega_2 \\ \omega_3 & 0 & -\omega_1 \\ -\omega_2 & \omega_1 & 0 \end{bmatrix} \tag{2.153}$$

Thus ω_1, ω_2 and ω_3 represent the components of the vector $\bar{\omega}$ called the *angular velocity*, where

$$\bar{\omega} = \omega_1\bar{n}_1 + \omega_2\bar{n}_2 + \omega_3\bar{n}_3 \tag{2.154}$$

To verify equation (2.137), we can easily see that

$$\bar{\omega} \times \bar{n} = \omega_{ij}\bar{n}_j \tag{2.155}$$

Hence, from equations (2.153) and (2.143) we deduce that

$$\bar{\omega} \times \bar{r} = \frac{d\bar{r}}{dt} \tag{2.156}$$

Now that we have established Ω_{ij} is a skew matrix, using equations (2.156) and (2.144), it follows that the skew matrix is given by equation (2.138). Some properties of the skew matrix are:

$$(a) \qquad \Omega + \Omega^T = \frac{d}{dt}(SS^T) = \frac{d}{dt}(I) = 0 \tag{2.157}$$

where I denotes the identity matrix and Ω^T the transpose of Ω.

$$(b) \qquad \Omega = -\Omega^T \tag{2.158}$$

The skew matrix Ω_{ij} $(i,\ j = 1, 2, 3)$, also called the *angular velocity matrix*, can be used together with the direction cosines matrix R to express $\dot{R}$. From equation (2.138) we can write

$$\dot{S}_{ij} = \Omega_{ik} S_{kj} \;\; (i,\ j, k = 1, 2, 3) \tag{2.159}$$

or

$$\left[\dot{S}^{AO}\right] = \left[\Omega^{AO}\right]\left[S^{AO}\right] \tag{2.160}$$

We can also prove that $\left[\dot{S}^{OA}\right] = \left[\Omega^{OA}\right]\left[S^{OA}\right]$, where $\left[S^{AO}\right] = \left[S^{OA}\right]^T$ and $\left[\Omega^{OA}\right] = \left[\Omega^{OA}\right]^T = -\left[\Omega^{AO}\right]$.

Equation (2.159) leads to Poisson's kinematical equations.

Notation is very important in defining kinematical equations and will make an understanding of the vectors involved clearer if a consistent well-defined notation is maintained. In the sequel, we refer to the angular velocity as ${}^R\bar{\omega}^A$, read as the angular velocity of body A in R.

Example 2.6.1
Suppose that

$$\left[S^{AO}\right] = \begin{bmatrix} \cos\theta & \sin\theta & 0 \\ -\sin\theta & \cos\theta & 0 \\ 0 & 0 & 1 \end{bmatrix}$$

denotes the shifter transformation matrix between body A and the inertia frame R with center O. Find $\left[\Omega^{AO}\right]$ and $\left[\Omega^{OA}\right]$, respectively.

Solution
From equation (2.160) we get

$$\begin{aligned}
\left[\Omega^{AO}\right] &= \left[\dot{S}^{AO}\right]\left[S^{AO}\right]^T = \left[\dot{S}^{AO}\right]\left[S^{OA}\right] \\
&= \dot{\theta}\begin{bmatrix} -\sin\theta & \cos\theta & 0 \\ -\cos\theta & -\sin\theta & 0 \\ 0 & 0 & 0 \end{bmatrix}\begin{bmatrix} \cos\theta & -\sin\theta & 0 \\ \sin\theta & \cos\theta & 0 \\ 0 & 0 & 1 \end{bmatrix} \\
&= \dot{\theta}\begin{bmatrix} 0 & 1 & 0 \\ -1 & 0 & 0 \\ 0 & 0 & 0 \end{bmatrix}
\end{aligned} \tag{2.161}$$

We can also get

$$\left[\Omega^{OA}\right]=\left[\dot{S}^{OA}\right]\left[S^{AO}\right]=\dot{\theta}\begin{bmatrix}0 & -1 & 0\\ 1 & 0 & 0\\ 0 & 0 & 0\end{bmatrix} \tag{2.162}$$

This proves that $\left[\Omega^{AO}\right]=-\left[\Omega^{OA}\right]$. Comparing with equation (2.153), we know that

$$\Omega_{ij}^{OA}=e_{jin}\omega_n=\begin{bmatrix}0 & -\dot{\theta} & 0\\ \dot{\theta} & 0 & 0\\ 0 & 0 & 0\end{bmatrix} \tag{2.163}$$

where $\omega_3=\dot{\theta}$ and $\omega_2=\omega_1=0$.

Example 2.6.2
Suppose that the transformation matrices between B and O, and C and O are given by

$$\left[S^{BO}\right]=\begin{bmatrix}1 & 0 & 0\\ 0 & \cos\alpha & \sin\alpha\\ 0 & -\sin\alpha & \cos\alpha\end{bmatrix}$$

and

$$\left[S^{CO}\right]=\begin{bmatrix}\cos\beta & 0 & -\sin\beta\\ 0 & 1 & 0\\ \sin\beta & 0 & \cos\beta\end{bmatrix}$$

Find their corresponding skew matrices $\left[\Omega^{BO}\right]$ and $\left[\Omega^{CO}\right]$.

Solution
Comparing with equation (2.153), we know that $\omega_1=\dot{\alpha}$ for $\left[S^{BO}\right]$ and $\omega_2=\dot{\beta}$ for $\left[S^{CO}\right]$. Hence

$$\left[\Omega^{BO}\right]=-\left[\Omega^{OB}\right]=\begin{bmatrix}0 & 0 & 0\\ 0 & 0 & \dot{\alpha}\\ 0 & -\dot{\alpha} & 0\end{bmatrix} \tag{2.164}$$

and

$$\left[\Omega^{CO}\right]=-\left[\Omega^{OC}\right]=\begin{bmatrix}0 & 0 & -\dot{\beta}\\ 0 & 0 & 0\\ \dot{\beta} & 0 & 0\end{bmatrix} \tag{2.165}$$

2.7 Further Derivation of the Angular Velocity

One of the most useful forms of the angular velocity is a form that involves strictly the unit vectors fixed in the rotating frame. As will be demonstrated, the angular velocity can be expressed as

$$^{R}\bar{\omega}^{A} = \left(\frac{d\bar{n}_2}{dt} \cdot \bar{n}_3\right) \bar{n}_1 + \left(\frac{d\bar{n}_3}{dt} \cdot \bar{n}_1\right) \bar{n}_2 + \left(\frac{d\bar{n}_1}{dt} \cdot \bar{n}_2\right) \bar{n}_3 \tag{2.166}$$

or

$$^{R}\bar{\omega}^{A} = \left(\dot{\bar{n}}_2 \cdot \bar{n}_3\right) \bar{n}_1 + \left(\dot{\bar{n}}_3 \cdot \bar{n}_1\right) \bar{n}_2 + \left(\dot{\bar{n}}_1 \cdot \bar{n}_2\right) \bar{n}_3 \tag{2.167}$$

Proof

Consider Figure 2.14 again. The angular velocity is defined as

$$^{R}\bar{\omega}^{A} = \omega_1\bar{n}_1 + \omega_2\bar{n}_2 + \omega_3\bar{n}_3 \tag{2.168}$$

where ω_1, ω_2 and ω_3 represent the components of the $^{R}\bar{\omega}^{A}$. An alternative representation of equation (2.168) is

$$^{R}\bar{\omega}^{A} = \left(^{R}\bar{\omega}^{A} \cdot \bar{n}_1\right) \bar{n}_1 + \left(^{R}\bar{\omega}^{A} \cdot \bar{n}_2\right) \bar{n}_2 + \left(^{R}\bar{\omega}^{A} \cdot \bar{n}_3\right) \bar{n}_3 \tag{2.169}$$

The unit vectors inside the parentheses can be replaced by their corresponding cross products:

$$^{R}\bar{\omega}^{A} = \left(^{R}\bar{\omega}^{A} \cdot \bar{n}_2 \times \bar{n}_3\right) \bar{n}_1 + \left(^{R}\bar{\omega}^{A} \cdot \bar{n}_3 \times \bar{n}_1\right) \bar{n}_2 + \left(^{R}\bar{\omega}^{A} \cdot \bar{n}_1 \times \bar{n}_2\right) \bar{n}_3 \tag{2.170}$$

Interchanging the dot and cross product in equation (2.170), we get

$$^{R}\bar{\omega}^{A} = \left(^{R}\bar{\omega}^{A} \times \bar{n}_2 \cdot \bar{n}_3\right) \bar{n}_1 + \left(^{R}\bar{\omega}^{A} \times \bar{n}_3 \cdot \bar{n}_1\right) \bar{n}_2 + \left(^{R}\bar{\omega}^{A} \times \bar{n}_1 \cdot \bar{n}_2\right) \bar{n}_3 \tag{2.171}$$

We also know from equation (2.137) that

$$\frac{d\bar{n}_i}{dt} = {}^{R}\bar{\omega}^{A} \times \bar{n}_i \quad (i = 1, 2, 3) \tag{2.172}$$

Hence equation (2.171) takes the following form:

$$^{R}\bar{\omega}^{A} = \left(\frac{d\bar{n}_2}{dt} \cdot \bar{n}_3\right) \bar{n}_1 + \left(\frac{d\bar{n}_3}{dt} \cdot \bar{n}_1\right) \bar{n}_2 + \left(\frac{d\bar{n}_1}{dt} \cdot \bar{n}_2\right) \bar{n}_3 \tag{2.173}$$

which proves equation (2.166). This equation is extremely useful in further derivation of *simple angular velocity*.

2.8 Angular Velocity and Euler Angles

Consider Figure 2.15. In the derivation of the direction cosine matrix S_{ij} $(i, j = 1, 2, 3)$, we associated the matrix with the relationship between a set of dextral unit vectors fixed in a rigid body A, and a set of dextral unit vectors fixed in an inertial reference frame R. The transformation matrix S_{ij} is the result of three successive rotations about $\bar{n}_1$, $\bar{n}_2$ and $\bar{n}_3$.

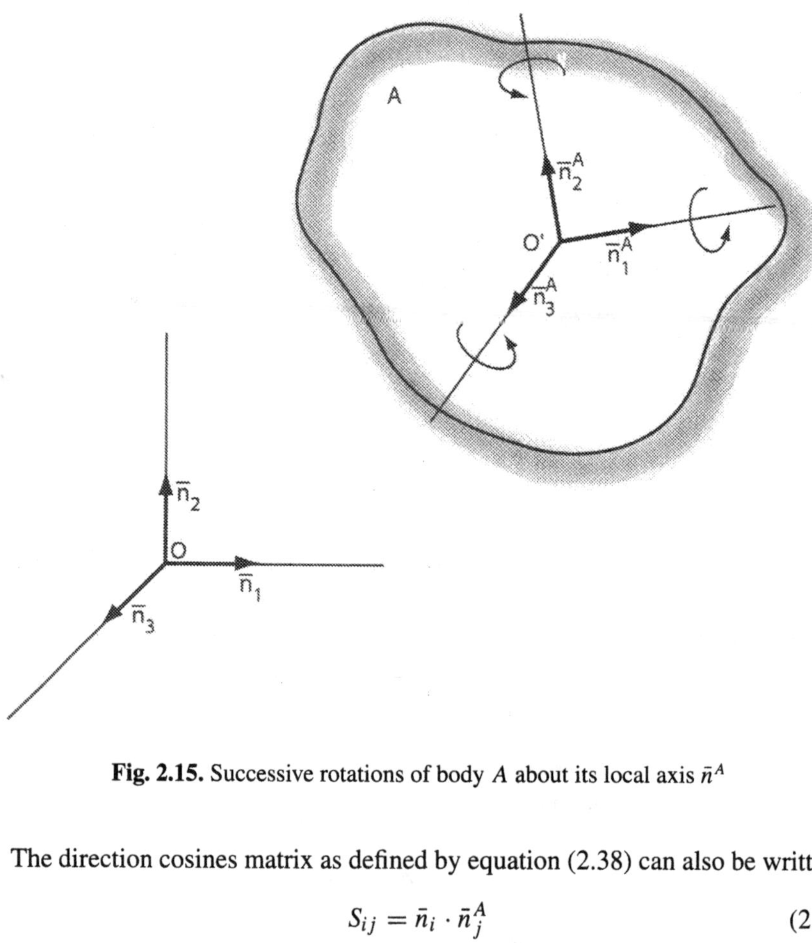

Fig. 2.15. Successive rotations of body A about its local axis $\bar{n}^A$

The direction cosines matrix as defined by equation (2.38) can also be written as

$$S_{ij} = \bar{n}_i \cdot \bar{n}_j^A \tag{2.174}$$

Using equation (2.138), the components of the angular velocity are found to be

$$\omega_1 = \dot{S}_{31}S_{21} + \dot{S}_{32}S_{22} + \dot{S}_{33}S_{23} \tag{2.175}$$

$$\omega_2 = \dot{S}_{11}S_{31} + \dot{S}_{12}S_{32} + \dot{S}_{13}S_{33} \tag{2.176}$$

$$\omega_3 = \dot{S}_{21}S_{11} + \dot{S}_{22}S_{12} + \dot{S}_{23}S_{13} \tag{2.177}$$

Differentiation of equation (2.174) leads to

$$\dot{S}_{ij} = \bar{n}_i \cdot \frac{d\bar{n}_j^A}{dt} + \frac{d\bar{n}_i}{dt} \cdot \bar{n}_j^A \tag{2.178}$$

Using equation (2.178) or differentiating the components of S given by equation (2.39), the rate of changes of the components of $S_{ij}(i, j = 1, 2, 3)$ will easily be found, and using equations (2.175) to (2.177) an explicit form of the angular velocity will be given by

$$\begin{aligned} {}^R\bar{\omega}^A &= (\dot{\alpha}\cos\beta\cos\gamma + \dot{\beta}\sin\gamma)\bar{n}_1 \\ &\quad + (-\dot{\alpha}\cos\beta\sin\gamma + \dot{\beta}\cos\gamma)\bar{n}_2 + (\dot{\gamma} + \dot{\alpha}\sin\beta)\bar{n}_3 \end{aligned} \tag{2.179}$$

Equation (2.179) defines an important form of the angular velocity when a particular rigid body undergoes three rotations, respectively. This will serve as a basis for defining the angular velocity when relative coordinates are used in the study of multibody dynamics. From equation (2.179) we deduce the components of the angular velocity to be

$$\omega_1 = \dot{\alpha}\cos\beta\cos\gamma + \dot{\beta}\sin\gamma \tag{2.180}$$

$$\omega_2 = -\dot{\alpha}\cos\beta\sin\gamma + \dot{\beta}\cos\gamma \tag{2.181}$$

$$\omega_3 = \dot{\gamma} + \dot{\alpha}\sin\beta \tag{2.182}$$

The representation of the angular velocity in terms of Euler's angle derivatives is not unique and depends on the assumed rotations of $\bar{n}_i^A$ $(i = 1, 2, 3)$ to make them aligned with $\bar{n}_i$ $(i = 1, 2, 3)$.

2.9 Angular Velocity and Euler Parameters

A simple form of the angular velocity of a rigid body A, rotating with respect to an inertial reference frame in terms of Euler parameters is given as follows:

$${}^R\bar{\omega}^A = \omega_1\bar{n}_1 + \omega_2\bar{n}_2 + \omega_3\bar{n}_3 \tag{2.183}$$

where

$$\omega_1 = 2(\varepsilon_4\dot{\varepsilon}_1 - \varepsilon_3\dot{\varepsilon}_2 + \varepsilon_2\dot{\varepsilon}_3 - \varepsilon_1\dot{\varepsilon}_4) \tag{2.184}$$

$$\omega_2 = 2(\varepsilon_3\dot{\varepsilon}_1 + \varepsilon_4\dot{\varepsilon}_2 - \varepsilon_1\dot{\varepsilon}_3 - \varepsilon_2\dot{\varepsilon}_4) \tag{2.185}$$

$$\omega_3 = 2(-\varepsilon_2\dot{\varepsilon}_1 + \varepsilon_1\dot{\varepsilon}_2 + \varepsilon_4\dot{\varepsilon}_3 - \varepsilon_3\dot{\varepsilon}_4) \tag{2.186}$$

Equations (2.184) to (2.186) reveal that we have three equations, ω_1, ω_2 and ω_3 in terms of four $\dot{\varepsilon}_1$ $(i = 1, \ldots, 4)$. Thus, we cannot uniquely determine $\dot{\varepsilon}_1$ $(1 = 1, \ldots, 4)$. However, we have established by means of equation (2.58) that Euler's parameters are dependent. Thus, differentiating equation (2.58) with respect to time yields

$$0 = 2(\varepsilon_1\dot{\varepsilon}_1 + \varepsilon_2\dot{\varepsilon}_2 + \varepsilon_3\dot{\varepsilon}_3 + \varepsilon_4\dot{\varepsilon}_4) \tag{2.187}$$

Combining equations (2.184) to (2.187), we obtain in matrix form

$$\begin{bmatrix} \omega_1 \\ \omega_2 \\ \omega_3 \\ \omega_4 \end{bmatrix} = 2 \begin{bmatrix} \varepsilon_4 & -\varepsilon_3 & \varepsilon_2 & -\varepsilon_1 \\ \varepsilon_3 & \varepsilon_4 & -\varepsilon_1 & -\varepsilon_2 \\ -\varepsilon_2 & \varepsilon_1 & \varepsilon_4 & -\varepsilon_3 \\ \varepsilon_1 & \varepsilon_2 & \varepsilon_3 & \varepsilon_4 \end{bmatrix} \begin{bmatrix} \dot{\varepsilon}_1 \\ \dot{\varepsilon}_2 \\ \dot{\varepsilon}_3 \\ \dot{\varepsilon}_4 \end{bmatrix} \tag{2.188}$$

or

$$\omega_i = 2E_{ij}\dot{\varepsilon}_j \quad (i = 1, 2, 3, 4;\ j = 1, 2, 3, 4) \tag{2.189}$$

with ω_4 equal to zero. Since E is orthogonal, we also get

$$\dot{\varepsilon}_j = \frac{1}{2}E_{ij}^T\omega_i \tag{2.190}$$

Proof

The components of the angular velocity given by equations (2.184) to (2.186) are found using equation (2.138), which simplifies to equations (2.175) to (2.177). Using the transformation matrix given in equation (2.61) and completing the necessary differentiation needed for equations (2.175) to (2.177), we obtain the components of the angular velocity as stated.

The rate of changes of $\dot{\varepsilon}_j$ as given by equation (2.190) are linear expressions of ε_i $(i = 1, \ldots, 4)$ and ω_i $(i = 1, \ldots, 3)$, and do not possess any singularities. This is an important observation since the solution to $\dot{\alpha}$, $\dot{\beta}$ and $\dot{\gamma}$ from equations (2.180) to (2.182) leads to singularities for certain values of α, β and γ as

$$\dot{\alpha} = (\omega_1 \cos\gamma - \omega_2 \sin\gamma)\frac{1}{\cos\beta} \tag{2.191}$$

$$\dot{\beta} = \omega_1 \sin\gamma + \omega_2 \cos\gamma \tag{2.192}$$

$$\dot{\gamma} = \omega_3 - \frac{\sin\beta}{\cos\beta}(\omega_1 \cos\gamma - \omega_2 \sin\gamma) \tag{2.193}$$

2.10 Simple Angular Velocity

Consider again a rigid body A as depicted in Figure 2.10. If the unit vector $\bar{\lambda}$ is independent of time, the angular velocity of A with respect to R is given by

$${}^R\bar{\omega}^A = \frac{d\theta}{dt}\bar{\lambda} \tag{2.194}$$

Proof

Let θ and $\bar{\lambda}$ have the meaning discussed in Section 2.4. Then the components of the angular velocity are obtained using equations (2.175) to (2.177), where S_{ij} $(i, j = 1, 2, 3)$ is given by equation (2.84). For instance,

$$\omega_1 = [\lambda_2 \sin\theta + \lambda_3\lambda_1(1-\cos\theta)](-\lambda_3\cos\theta + \lambda_1\lambda_2\sin\theta)\dot{\theta}$$
$$- [-\lambda_1 \sin\theta + \lambda_2\lambda_3(1-\cos\theta)](-\lambda_3^2 + \lambda_1^2)\sin\theta\dot{\theta}$$
$$+ [1 - (-\lambda_1^2 + \lambda_2^2)(1-\cos\theta)](\lambda_1\cos\theta + \lambda_2\lambda_3\sin\theta)\dot{\theta} \tag{2.195}$$

which reduces to

$$\omega_1 = \lambda_1\dot{\theta} \tag{2.196}$$

Similarly,

$$\omega_2 = \lambda_2\dot{\theta} \tag{2.197}$$

and

$$\omega_3 = \lambda_3\dot{\theta} \tag{2.198}$$

2.11 Angular Velocity and Intermediate Reference Frames

The angular velocity of a rigid body A with respect to an inertial frame R can be expressed using intermediate reference frames as

$$^{R}\bar{\omega}^{A} = {}^{R}\bar{\omega}^{R_1} + {}^{R_1}\bar{\omega}^{R_2} + {}^{R_2}\bar{\omega}^{R_3} + \cdots + {}^{R_n}\bar{\omega}^{A} \tag{2.199}$$

Let a vector $\bar{c}$ be fixed into body A; then

$$\frac{d\bar{c}}{dt} = {}^{R_n}\bar{\omega}^{A} \times \bar{c} \tag{2.200}$$

and

$$^{R_{n-1}}\frac{d\bar{c}}{dt} = {}^{R_{n-1}}\bar{\omega}^{A} \times \bar{c} \tag{2.201}$$

We also know that

$$^{R_{n-1}}\frac{d\bar{c}}{dt} = {}^{R_n}\frac{d\bar{c}}{dt} + {}^{R_{n-1}}\bar{\omega}^{R_n} \times \bar{c} \tag{2.202}$$

From equations (2.201) and (2.202), we obtain

$$^{R_{n-1}}\bar{\omega}^{A} \times \bar{c} = {}^{R_n}\bar{\omega}^{A} \times \bar{c} + {}^{R_{n-1}}\bar{\omega}^{R_n} \times \bar{c} \tag{2.203}$$

Since equation (2.203) is satisfied for every $\bar{c}$,

$$^{R_{n-1}}\bar{\omega}^{A} = {}^{R_n}\bar{\omega}^{A} + {}^{R_{n-1}}\bar{\omega}^{R_n} \tag{2.204}$$

It can easily be established that for n intermediary frames as depicted by Figure 2.16, the angular velocity of rigid body A, with respect to a fixed frame R, is given by the sum of the relative angular velocities as stated by equation (2.199).

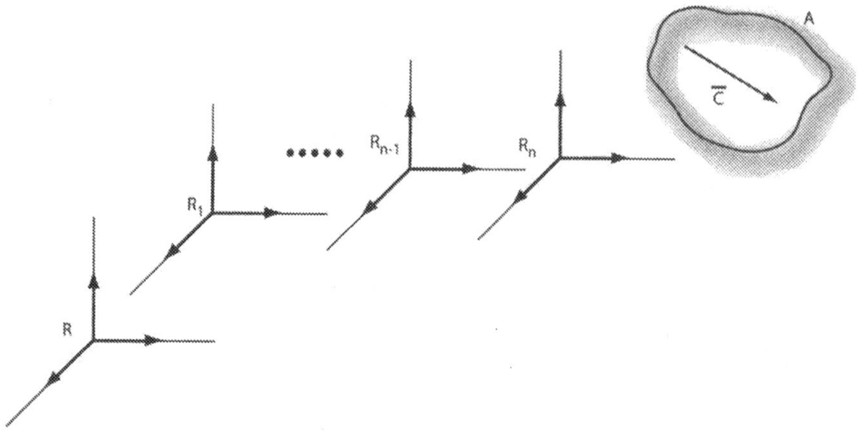

Fig. 2.16. A vector C fixed in body A and defined with respect to a set of auxiliary frames

Again, it is worth mentioning the importance of equation (2.199) in the study of multibody systems as the reference frames $R_1, R_2, \ldots, R_n$ can be fixed in other rigid bodies that undergo some rotations with respect to one another, and hence equation (2.199) can be used to find the angular velocity of a particular body with respect to a particular reference frame.

Example 2.11.1
A disk D attached to a shaft S_2 as shown in Figure 2.17 rotates with respect to the axis normal to the disk with an angular displacement $\psi(t)$. The shaft S_2 has a swing type of rotation $\phi(t)$ in the plane given by $\bar{n}_2^2$ and $\bar{n}_3^2$. In addition, the master shaft S_1 rotates about its center axis by $\theta(t)$. Find the angular velocity of the disk D in R.

Solution
Let us define the reference frames R_0, R_1 and R as shown, where their unit or orthogonal unit vectors are such that $\bar{n}_1^3$ and $\bar{n}_2^3$ are on the plane of the disk and $\bar{n}_3^2$ will be normal (coming out) to the disk and is along the shaft S_2 center axis. Let R_3 be at the tip of S_2. Second, we define the reference frame R_2 with its center at c such that $\bar{n}_3^2$ is along the S_2 center axis. Finally, let us choose the inertial fixed reference frame R such that initially it coincides with R_1, in which $\bar{n}_2^1$ is along the S_1 center axis.

Using the addition formula given by equation (2.199), we can write the angular velocity of the disk D in R to be

$$^{R}\bar{\omega}^{D} = {}^{R}\bar{\omega}^{R_1} + {}^{R_1}\bar{\omega}^{R_2} + {}^{R_2}\bar{\omega}^{D} \tag{2.205}$$

or

$$^{R}\bar{\omega}^{D} = \dot{\theta}\bar{n}_2^1 + \dot{\phi}\,\bar{n}_1^2 + \dot{\psi}\bar{n}_3^3 = \dot{\theta}\bar{n}_2^1 + \dot{\phi}\,\bar{n}_1^2 + \dot{\psi}\bar{n}_3^2 \tag{2.206}$$

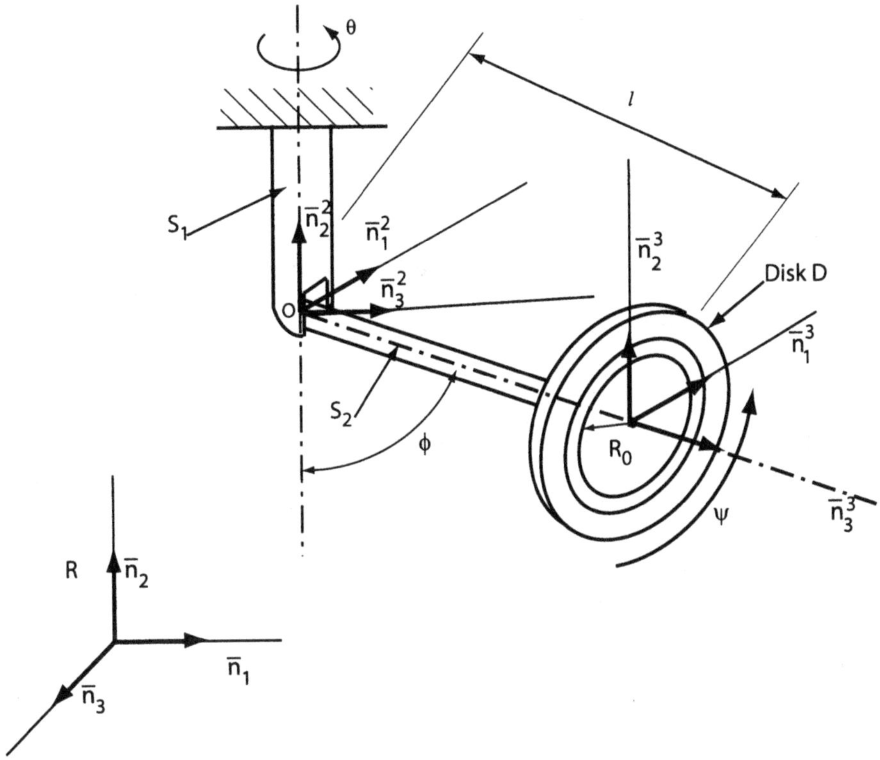

Fig. 2.17. A disk attached to a swinging shaft

The transformation matrix needed to express $\{\bar{n}^2\}$ with respect to $\{\bar{n}^1\}$ is

$$\{\bar{n}^2\} = \begin{bmatrix} 1 & 0 & 0 \\ 0 & \cos(\phi - 90°) & \sin(\phi - 90°) \\ 0 & -\sin(\phi - 90°) & \cos(\phi - 90°) \end{bmatrix} \{\bar{n}^1\}$$

$$= \begin{bmatrix} 1 & 0 & 0 \\ 0 & \sin\phi & -\cos\phi \\ 0 & \cos\phi & \sin\phi \end{bmatrix} \{\bar{n}^1\} \tag{2.207}$$

and the one relating $\bar{n}^1$ to $\bar{n}$ is

$$\{\bar{n}^1\} = \begin{bmatrix} \cos\theta & 0 & -\sin\theta \\ 0 & 1 & 0 \\ \sin\theta & 0 & \cos\theta \end{bmatrix} \{\bar{n}\} \tag{2.208}$$

Making use of the transformation matrices above, we obtain between $\bar{n}^2$ and $\bar{n}$

$$\{\bar{n}^2\} = \begin{bmatrix} 1 & 0 & 0 \\ 0 & \sin\phi & -\cos\phi \\ 0 & \cos\phi & \sin\phi \end{bmatrix} \begin{bmatrix} \cos\theta & 0 & -\sin\theta \\ 0 & 1 & 0 \\ \sin\theta & 0 & \cos\theta \end{bmatrix} \{\bar{n}\}$$

$$= \begin{bmatrix} \cos\theta & 0 & -\sin\theta \\ -\cos\phi\sin\theta & \sin\phi & -\cos\phi\sin\theta \\ \sin\phi\sin\theta & \cos\phi & \sin\phi\cos\theta \end{bmatrix} \{\bar{n}\} \tag{2.209}$$

The objective is to express ${}^{R}\bar{\omega}^{D}$ with respect to $\bar{n}_1$, $\bar{n}_2$ and $\bar{n}_3$. Expressing $\bar{n}_3^2$, $\bar{n}_1^2$ and $\bar{n}_2^1$ in terms of $\bar{n}_1$, $\bar{n}_2$ and $\bar{n}_3$, we get the final expression of the angular velocity of D with respect to 1

$$\begin{aligned} {}^{R}\bar{\omega}^{D} &= \dot{\psi}(\sin\phi\sin\theta\bar{n}_1 + \cos\phi\bar{n}_2 + \sin\phi\cos\theta\bar{n}_3) + \dot{\phi}(\cos\theta\bar{n}_1 - \sin\theta\bar{n}_3) + \dot{\theta}\bar{n}_2 \\ &= (\dot{\psi}\sin\phi\sin\theta + \dot{\phi}\cos\theta)\bar{n}_1 + (\dot{\psi}\cos\phi + \dot{\theta})\bar{n}_2 + (\dot{\psi}\sin\phi\cos\theta - \dot{\phi}\sin\theta)\bar{n}_3 \end{aligned} \tag{2.210}$$

2.12 Angular Acceleration

The angular acceleration of a rigid body A with respect to a reference frame R is defined as the time derivative in R of the angular velocity of A in R. That is,

$${}^{R}\bar{\alpha}^{A} = \frac{d}{dt}\left({}^{R}\bar{\omega}^{A}\right) \tag{2.211}$$

If ${}^{R}\bar{\omega}^{A}$ is expressed using intermediate reference frames, for instance,

$${}^{R}\bar{\omega}^{A} = {}^{R}\bar{\omega}^{B} + {}^{B}\bar{\omega}^{A}$$

then,

$${}^{R}\bar{\alpha}^{A} = {}^{R}\bar{\alpha}^{B} + {}^{B}\dot{\bar{\omega}}^{A} + \Omega \times {}^{B}\bar{\omega}^{A} \tag{2.212}$$

where Ω is viewed as the angular velocity of a frame c (where ω lies) in R. If B equals R, Ω equals zero.

Example 2.12.1
Consider the mechanism shown in Figure 2.18. The shaft rotates about its central axis by $\theta(t)$ while S_2 is rigidly attached to S_1. The shaft S_3 rotates freely about the axis normal to S_2 by angle $\phi\ (t)$. Find an expression for the angular acceleration of point A in R.

Solution
We know that

$$\bar{n}_m^A = S\,\bar{n}_m \tag{2.213}$$

and determining the direction cosine matrix to relate $\{\bar{n}^A\}$ unit vectors with respect to the (global) unit vector $\{\bar{n}\}$:

$$S = \begin{bmatrix} \cos\theta & \sin\theta & 0 \\ -\sin\theta & \cos\theta & 0 \\ 0 & 0 & 1 \end{bmatrix} \tag{2.214}$$

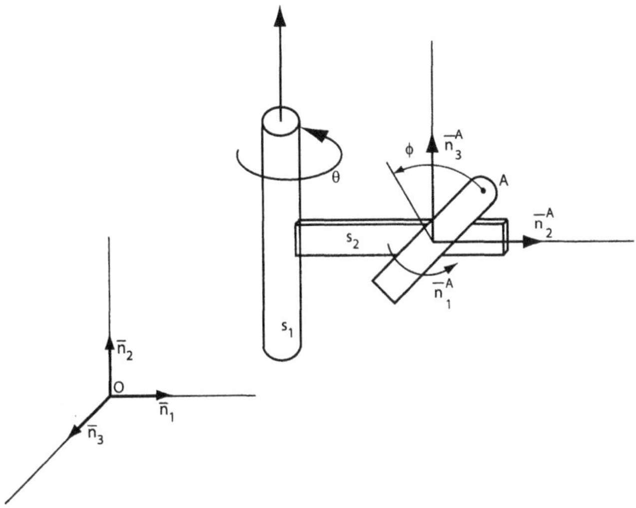

Fig. 2.18. Mechanism with shaft S_1, S_2 and S_3

The differentiation of the matrix $[S]$ yields

$$\dot{S} = \dot{\theta}\begin{bmatrix} -\sin\theta & \cos\theta & 0 \\ -\cos\theta & -\sin\theta & 0 \\ 0 & 0 & 0 \end{bmatrix} \tag{2.215}$$

Apply equation (2.199) and express it in terms of $(\bar{n})$ unit vectors:

$$\begin{aligned} {}^R\bar{\omega}^A &= \dot{\theta}\bar{n}_3 + \dot{\phi}\,\bar{n}_1^A = \dot{\theta}\bar{n}_3 + \dot{\phi}\,S_{1j}\bar{n}_j \\ &= \dot{\phi}\,\cos\theta\bar{n}_1 + \dot{\phi}\,\sin\theta\bar{n}_2 + \dot{\theta}\bar{n}_3 \end{aligned} \tag{2.216}$$

By using equation (2.211), we determine the angular acceleration of point A to be

$$\begin{aligned} {}^R\bar{\alpha}^A &= \frac{d}{dt}\left({}^R\bar{\omega}^A\right) = \ddot{\theta}\bar{n}_3 + \ddot{\phi}\,\bar{n}_1^A + \dot{\phi}\,\frac{d\bar{n}_1^A}{dt} \\ &= \ddot{\theta}\bar{n}_3 + \ddot{\phi}\,S_{1j}\bar{n}_j + \dot{\phi}\,\dot{S}_{1j}\bar{n}_j \\ &= \ddot{\theta}\bar{n}_3 + \ddot{\phi}\,(\cos\theta\bar{n}_1 + \sin\theta\bar{n}_2) + \dot{\phi}\,\dot{\theta}(-\sin\theta\bar{n}_1 + \cos\theta\bar{n}_2) \\ &= (\ddot{\phi}\cos\theta - \dot{\phi}\,\dot{\theta}\sin\theta)\bar{n}_1 + (\ddot{\phi}\sin\theta + \dot{\phi}\,\dot{\theta}\cos\theta)\bar{n}_2 + \ddot{\theta}\bar{n}_3 \end{aligned}$$

Of course, we can get ${}^R\bar{\alpha}^A$ from equation (2.216) directly by differentiation.

2.13 Velocity and Acceleration of a Fixed Point on a Rigid Body

Consider points P and Q fixed on a rigid body A that is moving relative to the fixed inertial frame R (Figure 2.19). Let ${}^R\bar{\omega}^A$ be the angular velocity of A in R. The velocity of point P relative to R is given by

$$ {}^R\bar{v}^P = {}^R\bar{v}^Q + {}^R\bar{\omega}^A \times \bar{r} \tag{2.217}$$

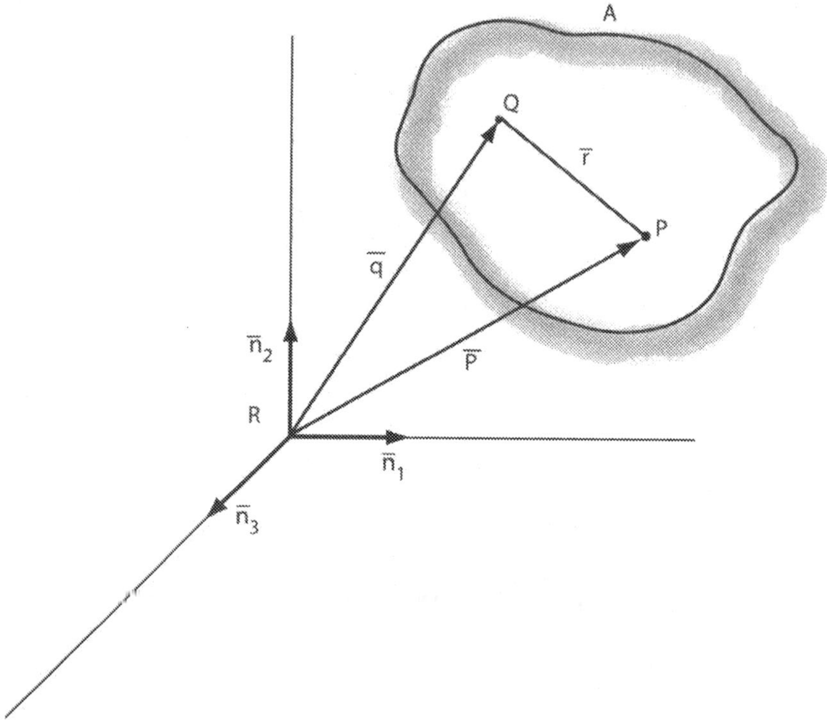

Fig. 2.19. Points P and Q are fixed to body A which is moving relative to the fixed frame R

Proof
Let

$$\bar{p} = \bar{q} + \bar{r} \tag{2.218}$$

where $\bar{p}$, $\bar{q}$ and $\bar{r}$ are all position vectors defined by Figure 2.19. The velocity of point P in R is given by differentiation in time of $\bar{p}$, that is,

$$\frac{d\bar{p}}{dt} = \frac{d\bar{q}}{dt} + \frac{d\bar{r}}{dt} \tag{2.219}$$

or

$${}^{R}\bar{v}^{P} = {}^{R}\bar{v}^{Q} + \frac{d}{dt}\bar{r} \tag{2.220}$$

Since $\bar{r}$ is fixed in A, its derivative is given by equation (2.137); hence

$${}^{R}\bar{v}^{P} = {}^{R}\bar{v}^{Q} + {}^{R}\bar{\omega}^{A} \times \bar{r} \tag{2.221}$$

The second term of equation (2.221) represents the velocity of point P relative to Q in R. The acceleration of point P in R, on the other hand, is given by differentiation of equation (2.217) as

$$ {}^{R}\bar{a}^{P} = \frac{d}{dt}\left({}^{R}\bar{v}^{Q}\right) + \frac{d}{dt}\left({}^{R}\bar{\omega}^{A} \times \bar{r}\right) \tag{2.222} $$

or

$$ {}^{R}\bar{a}^{P} = {}^{R}\bar{a}^{Q} + {}^{R}\bar{\alpha}^{A} \times \bar{r} + {}^{R}\bar{\omega}^{A} \times \left({}^{R}\bar{\omega}^{A} \times \bar{r}\right) \tag{2.223} $$

The second and third terms of equation (2.223) represent the acceleration of point P relative to Q in R.

Example 2.13.1
Find the velocity and acceleration of point P (on the disk) in R for Example 2.11.1, if $\dot{\theta} = 0$.

Solution
Let the position vector of point P (on the disk) in R be given by

$$ \bar{r}_p = \ell \bar{n}_3^2 + r\cos\psi \bar{n}_1^2 + r\sin\psi \bar{n}_2^2 \tag{2.224} $$

Equation (2.224) is similar to (2.218). Hence using the definition given by equation (2.217), the velocity of point p in R could be expressed as

$$ \bar{v}_p = {}^{R}\bar{v}^{S} + {}^{R}\bar{\omega}^{D} \times \bar{r} \tag{2.225} $$

where the components of $\bar{v}_p$ are

$$ {}^{R_2}\bar{\omega}^{D} = \dot{\psi}\bar{n}_3^3 = \dot{\psi}\bar{n}_3^2 $$

$$ {}^{R}\bar{v}^{S} = \ell\frac{d\bar{n}_3^2}{dt} = \ell\,{}^{R}\bar{\omega}^{R_2} \times \bar{n}_3^2 \tag{2.226} $$

$$ {}^{R}\bar{\omega}^{R_2} = \dot{\phi}\,\bar{n}_1^2,\ \bar{n}_3^2 = \cos\phi\,\bar{n}_2 + \sin\phi\,\bar{n}_3,\ \bar{n}_2^2 = \sin\phi\,\bar{n}_2 - \cos\phi\,\bar{n}_3 $$

$$ {}^{R}\bar{\omega}^{D} = {}^{R}\bar{\omega}^{R_2} + {}^{R}\bar{\omega}^{D} = \dot{\phi}\,\bar{n}_1^2 + \dot{\psi}\bar{n}_3^2 = \dot{\phi}\,\bar{n}_1 + \dot{\psi}(\cos\phi\,\bar{n}_2 + \sin\phi\,\bar{n}_3) \tag{2.227} $$

$$ \bar{r} = r(\cos\psi\bar{n}_1^1 + \sin\psi\bar{n}_1^2) = r[\cos\psi\bar{n}_1 + \sin\psi(\sin\phi\,\bar{n}_2 - \cos\phi\,\bar{n}_3)] \tag{2.228} $$

Therefore,

$$ \begin{aligned} \bar{v}_p &= \ell\dot{\phi}\,\bar{n}_1 \times (\cos\phi\,\bar{n}_2 + \sin\phi\,\bar{n}_3) + [\dot{\phi}\,\bar{n}_1 + \dot{\psi}(\cos\phi\,\bar{n}_2 + \sin\phi\,\bar{n}_3)] \\ &\quad \times r[\cos\psi\bar{n}_1 + \sin\psi(\sin\phi\,\bar{n}_2 - \cos\phi\,\bar{n}_3)] \\ &= -r\dot{\psi}\sin\psi\bar{n}_1 + (-\ell\dot{\phi}\sin\phi + r\dot{\psi}\cos\psi\sin\phi + r\dot{\phi}\sin\psi\cos\phi)\bar{n}_2 \\ &\quad + (\ell\dot{\phi}\cos\phi - r\dot{\psi}\cos\psi\cos\phi + r\dot{\phi}\sin\psi\sin\phi)\bar{n}_3 \end{aligned} \tag{2.229} $$

We can also obtain the solution by differentiation of equation (2.224). Let us express $\bar{r}$ with respect to $\bar{n}_1$, $\bar{n}_2$ and $\bar{n}_3$ as

$$\bar{r}_p = \ell \bar{n}_3^2 + r(\cos\psi \bar{n}_1^2 + \sin\psi \bar{n}_2^2) = \ell(\cos\phi\, \bar{n}_2 + \sin\phi\, \bar{n}_3)$$
$$+ r[\cos\psi \bar{n}_1 + \sin\psi(\sin\phi\, \bar{n}_2 - \cos\phi\, \bar{n}_3)] \qquad (2.230)$$

The velocity is then calculated in the same fashion of previous differentiation.

$$\bar{v}_p = \frac{d}{dt}\bar{r}_p = \ell\dot{\phi}\,(\sin\phi\, \bar{n}_2 - \cos\phi\, \bar{n}_3)$$
$$= +r[-\dot{\psi}\sin\psi \bar{n}_1 + \dot{\psi}\cos\psi(\sin\phi\, \bar{n}_2 - \cos\phi\, \bar{n}_3)$$
$$+ \sin\psi(\dot{\phi}\cos\phi\, \bar{n}_2 - \dot{\phi}\sin\phi\, \bar{n}_3)] \qquad (2.231)$$
$$= -r\dot{\psi}\sin\psi \bar{n}_1 + (-\ell\dot{\phi}\sin\phi + r\dot{\psi}\cos\psi\sin\phi + r\dot{\phi}\sin\psi\cos\phi)\bar{n}_2$$
$$+ (\ell\dot{\phi}\cos\phi - r\dot{\psi}\cos\psi\cos\phi + r\dot{\phi}\sin\psi\sin\phi)\bar{n}_3$$

which matches exactly with the answers given by equation (2.229).

The acceleration of point P in R can be obtained using equation (2.223) or by direct differentiation of equation (2.231), which yields

$$\bar{a}_p = -r(\ddot{\psi}\sin\psi + \dot{\psi}^2\cos\psi)\bar{n}_1 + (-\ell\dot{\phi}^2\cos\phi - \ell\ddot{\phi}\sin\dot{\phi} + r\ddot{\psi}\cos\psi\sin\phi$$
$$- r\dot{\psi}^2\sin\psi\sin\phi + r\dot{\psi}\dot{\phi}\cos\psi\cos\phi + r\ddot{\phi}\sin\psi\cos\psi + r\dot{\phi}\,\dot{\psi}\cos\psi\cos\phi$$
$$- r\dot{\phi}^2\sin\psi\sin\phi)\bar{n}_2 + (-\ell\dot{\phi}^2\cos\phi - \ell\ddot{\phi}\sin\dot{\phi} + r\ddot{\psi}\cos\psi\cos\phi$$
$$+ r\dot{\psi}^2\sin\psi\cos\phi + r\dot{\psi}\dot{\phi}\cos\psi\sin\phi + r\ddot{\phi}\sin\psi\sin\phi + r\dot{\phi}^2\sin\psi\cos\phi$$
$$+ r\dot{\phi}\,\dot{\psi}\cos\psi\sin\phi)\bar{n}_3 \qquad (2.232)$$

Example 2.13.2

Consider the 2-body mechanism where the complete FBD is drawn to show the rotation of each body. Find the mass center (point c as shown in Figure 2.20), velocity, and acceleration of body 2.

Solution

The direction cosine matrix of body 2 with respect to R is

$$\bar{n}_m^2 = \begin{bmatrix} \cos(\pi-\theta) & 0 & -\sin(\pi-\theta) \\ 0 & 1 & 0 \\ \sin(\pi-\theta) & 0 & \cos(\pi-\theta) \end{bmatrix} \begin{bmatrix} \cos\alpha & \sin\alpha & 0 \\ -\sin\alpha & \cos\alpha & 0 \\ 0 & 0 & 1 \end{bmatrix} \bar{n}_m \quad (m = 1, 3)$$
$$= \begin{bmatrix} -\cos\theta & 0 & -\sin\theta \\ 0 & 1 & 0 \\ \sin\theta & 0 & -\cos\theta \end{bmatrix} \begin{bmatrix} \cos\alpha & \sin\alpha & 0 \\ -\sin\alpha & \cos\alpha & 0 \\ 0 & 0 & 1 \end{bmatrix} \bar{n}_m \qquad (2.233)$$
$$= \begin{bmatrix} -\cos\theta\cos\alpha & -\cos\theta\sin\alpha & -\sin\theta \\ -\sin\alpha & \cos\alpha & 0 \\ \sin\theta\cos\alpha & \sin\theta\sin\alpha & -\cos\theta \end{bmatrix} \bar{n}_m$$

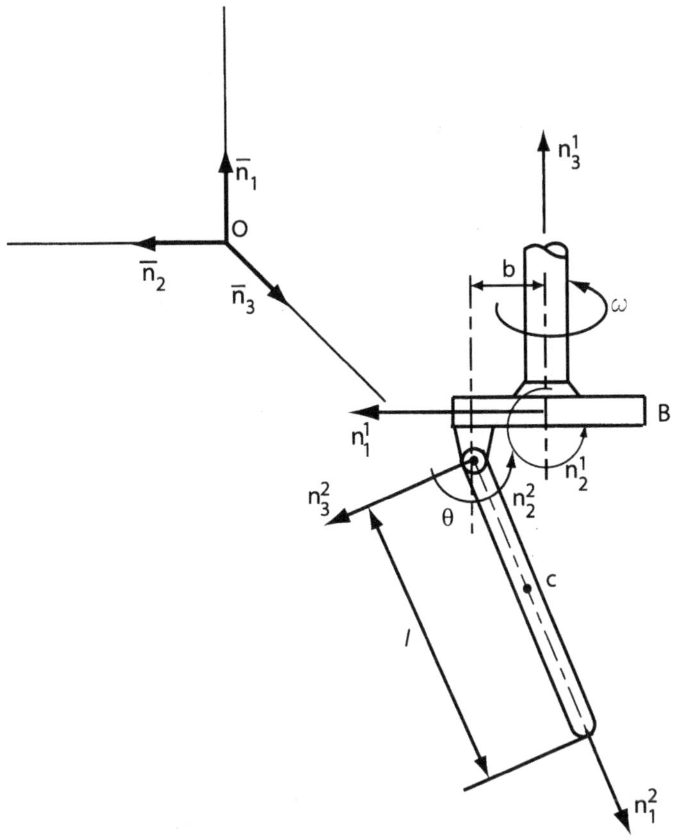

Fig. 2.20. Two-body mechanism

where $\dot{\alpha} = \omega$. The position vector of point c is given by

$$\begin{aligned}\bar{p}_c &= b\bar{n}_1^1 + \frac{1}{2}\ell\bar{n}_1^2 = b(\cos\alpha\bar{n}_1 + \sin\alpha\bar{n}_2) \\ &\quad + \frac{1}{2}\ell(-\cos\theta\cos\alpha\,\bar{n}_1 - \cos\theta\sin\alpha\,\bar{n}_2 - \sin\theta\,\bar{n}_3 \\ &= \left(b\cos\alpha - \frac{1}{2}\ell\cos\theta\cos\alpha\right)\bar{n}_1 \\ &\quad + \left(b\sin\alpha - \frac{1}{2}\ell\cos\theta\sin\alpha\right)\bar{n}_2 - \frac{1}{2}\ell\sin\theta\,\bar{n}_3 \end{aligned} \tag{2.234}$$

Differentiation of $\bar{p}_c$ leads to the velocity $\bar{v}_c$

$$\begin{aligned}\bar{v}_c = \frac{d}{dt}\bar{p}_c &= \left[-b\omega\sin\alpha + \frac{1}{2}\ell(\dot{\theta}\sin\theta\cos\alpha + \omega\cos\theta\sin\alpha)\right]\bar{n}_1 \\ &+ \left[b\omega\cos\alpha + \frac{1}{2}\ell(\dot{\theta}\sin\theta\sin\alpha - \omega\cos\theta\cos\alpha)\right]\bar{n}_2 - \frac{1}{2}\ell\dot{\theta}\cos\theta\bar{n}_3\end{aligned}$$

Differentiating the velocity further we get $\bar{a}_c$:

$$\bar{a}_c = \frac{d^2}{dt^2}\bar{p}_c = [-b(\dot{\omega}\sin\alpha + \omega^2\cos\alpha) + \frac{1}{2}\ell(\ddot{\theta}\sin\theta\cos\alpha$$
$$+ \dot{\theta}^2\cos\theta\cos\alpha - \dot{\theta}\omega\sin\theta\sin\alpha + \dot{\omega}\cos\theta\sin\alpha$$
$$- \omega\dot{\theta}\sin\theta\sin\alpha + \omega^2\cos\theta\cos\alpha]\bar{n}_1 + [b(\dot{\omega}\cos\alpha - \omega^2\sin\alpha)$$
$$+ \frac{1}{2}\ell(\ddot{\theta}\sin\theta\sin\alpha + \dot{\theta}^2\cos\theta\sin\alpha + \dot{\theta}\omega\sin\theta\cos\alpha - \dot{\omega}\cos\theta\cos\alpha$$
$$+ \omega\dot{\theta}\sin\theta\cos\alpha + \omega^2\cos\theta\sin\alpha)]\bar{n}_2 - \frac{1}{2}\ell(\ddot{\theta}\cos\theta - \dot{\theta}^2\sin\theta)\bar{n}_3$$

Example 2.13.3
A disc fixed to a cylinder is mounted on a rotating platform to position the disc in space. Find the velocity and acceleration of point A in body C as shown in Figure 2.21.

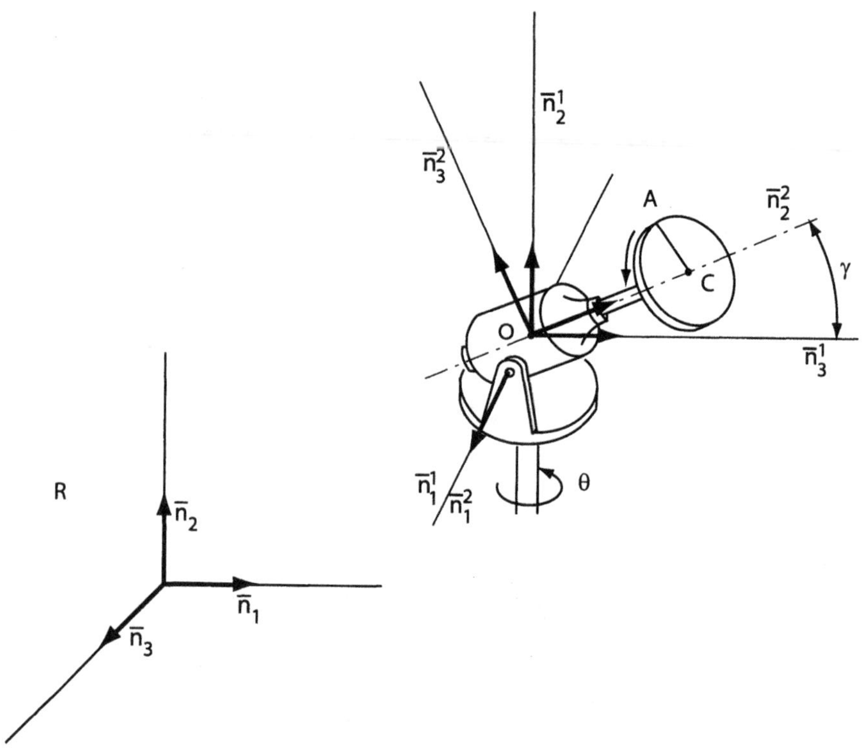

Fig. 2.21. Velocity and acceleration of point A in body C

Solution
The unit vectors $\bar{n}^2$ can be expressed in terms of $\bar{n}$ as

$$\bar{n}_m^2 = \begin{bmatrix} 1 & 0 & 0 \\ 0 & \cos\gamma & \sin\gamma \\ 0 & -\sin\gamma & \cos\gamma \end{bmatrix} \begin{bmatrix} \cos\theta & \sin\theta & 0 \\ -\sin\theta & \cos\theta & 0 \\ 0 & 0 & 1 \end{bmatrix} \bar{n}_m$$

$$= \begin{bmatrix} \cos\theta & \sin\theta & 0 \\ \cos\gamma\sin\theta & \cos\gamma\cos\theta & \sin\gamma \\ \sin\gamma\sin\theta & -\sin\gamma\cos\theta & \cos\gamma \end{bmatrix} \bar{n}_m \tag{2.235}$$

Let $|\overline{OC}| = \ell$, $|\overline{CA}| = R$, and $\dot{\alpha} = \omega$; the position of point A is then found to be

$$\begin{aligned} \bar{P}_A &= \overline{OC} + \overline{CA} = \ell\bar{n}_2^2 + R\sin\alpha\,\bar{n}_1^2 + R\cos\alpha\,\bar{n}_3^2 \\ &= \ell(-\cos\gamma\sin\theta\,\bar{n}_1 + \cos\gamma\cos\theta\,\bar{n}_2 + \sin\gamma\,\bar{n}_3) + R\sin\alpha\,(\cos\theta\,\bar{n}_1 + \sin\theta\,\bar{n}_2) \\ &\quad + R\cos\alpha\,(\sin\gamma\sin\theta\,\bar{n}_1 - \sin\gamma\cos\theta\,\bar{n}_2 + \cos\gamma\,\bar{n}_3) \qquad (2.236) \\ &= (-\ell\cos\gamma\sin\theta + R\sin\alpha\cos\theta + R\cos\alpha\sin\gamma\sin\theta)\bar{n}_1 \\ &\quad + (\ell\cos\gamma\cos\theta + R\sin\alpha\sin\theta - R\cos\alpha\sin\gamma\cos\theta)\bar{n}_2 \\ &\quad + (\ell\sin\gamma + R\cos\alpha\cos\gamma)\bar{n}_3 \end{aligned}$$

The velocity of point A in R is

$$\begin{aligned} \bar{v}_A &= \frac{d\bar{P}_A}{dt} \\ &= (+\ell\dot{\gamma}\sin\gamma\sin\theta - \ell\dot{\theta}\cos\gamma\cos\theta + R\omega\cos\alpha\cos\theta - R\dot{\theta}\sin\alpha\sin\theta \\ &\quad - R\omega\sin\alpha\sin\gamma\sin\theta + R\dot{\gamma}\cos\alpha\cos\gamma\sin\theta + R\dot{\theta}\cos\alpha\sin\gamma\cos\theta)\bar{n}_1 \\ &\quad + (-\ell\dot{\gamma}\sin\gamma\cos\theta - \ell\dot{\theta}\cos\gamma\sin\dot{\theta} + R\omega\cos\alpha\sin\theta \\ &\quad + R\dot{\theta}\sin\alpha\cos\theta + R\omega\sin\alpha\sin\gamma\cos\theta - R\dot{\gamma}\cos\alpha\cos\gamma\cos\theta \\ &\quad + R\dot{\theta}\cos\alpha\sin\gamma\sin\theta)\bar{n}_2 + (\ell\dot{\gamma}\cos\gamma - R\omega\sin\alpha\cos\gamma - R\dot{\gamma}\cos\alpha\sin\gamma)\bar{n}_3 \end{aligned} \tag{2.237}$$

Then by differentiation of equation (2.237), we can obtain the acceleration $\bar{a}_A$:

$$\bar{a}_A = \frac{d}{dt}\bar{v}_A \tag{2.238}$$

Example 2.13.4
Consider a cylinder rotating about a rotating shaft. Find the velocity and acceleration of point A as shown in Figure 2.22, if ω_1 and ω_2 are constant.

Solution
Let $|\overline{OA}| = \ell$ and the coordinate of point A in $\{\bar{n}^2\}$ (not shown) be $[\ell\cos\theta \;\; \ell\sin\theta \;\; 0]$, $\dot{\alpha}_1 = \omega_1$ and $\dot{\alpha}_2 = \omega_2$. The unit vectors $\bar{n}^1$, $\bar{n}^2$ and $\bar{n}$ are related as follows:

$$\bar{n}_m^1 = \begin{bmatrix} \cos\alpha_2 & \sin\alpha_2 & 0 \\ -\sin\alpha_2 & \cos\alpha_2 & 0 \\ 0 & 0 & 1 \end{bmatrix} \bar{n}_m \quad \bar{n}_m^2 = \begin{bmatrix} 1 & 0 & 0 \\ 0 & \cos\alpha_1 & \sin\alpha_1 \\ 0 & -\sin\alpha_1 & \cos\alpha_1 \end{bmatrix} \bar{n}_m^1 \tag{2.239}$$

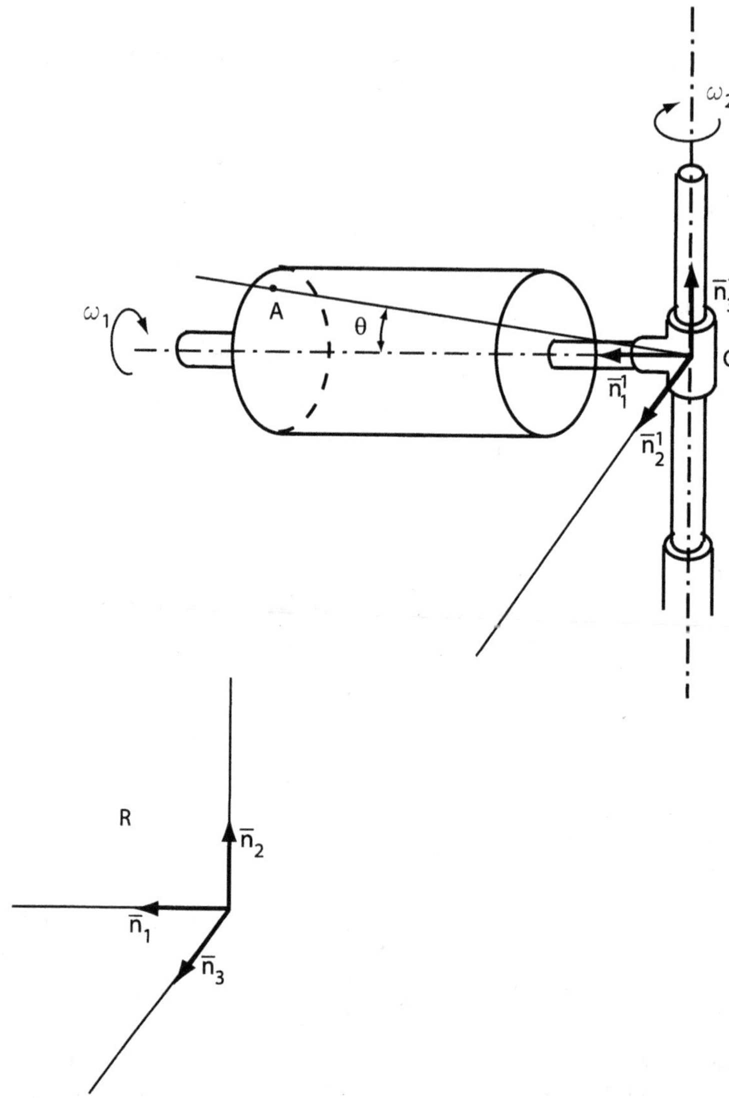

Fig. 2.22. Velocity and acceleration of point A

The position vector to point A can be expressed as

$$\begin{aligned}\bar{P}_A &= \ell(\cos\theta\bar{n}_1^1 + \sin\theta\cos\alpha_1\bar{n}_2^1 + \sin\theta\sin\alpha_1\bar{n}_3^1)\\ &= \ell(\cos\theta\cos\alpha_2\bar{n}_1 - \sin\theta\cos\alpha_1\sin\alpha_2\bar{n}_1 + \cos\theta\sin\alpha_2\bar{n}_2\\ &\quad + \sin\theta\cos\alpha_1\cos\alpha_2\bar{n}_2 + \sin\theta\sin\alpha_1\bar{n}_3)\end{aligned} \tag{2.240}$$

The differentiation in time of the above yields the velocity of point A:

$$
\begin{aligned}
\bar{v}_A &= \frac{d}{dt}\bar{P}_A \\
&= \ell[-\omega_2 \cos\theta \sin\alpha_2 \bar{n}_1 + \omega_2 \cos\theta \cos\alpha_2 \bar{n}_2 + \omega_1 \sin\theta \sin\alpha_2 \sin\alpha_1 \bar{n}_1 \\
&\quad - \omega_2 \sin\theta \cos\alpha_1 \cos\alpha_2 \bar{n}_1 - \omega_1 \sin\theta \sin\alpha_1 \cos\alpha_2 \bar{n}_2 \\
&\quad - \omega_2 \sin\theta \cos\alpha_1 \sin\alpha_2 \bar{n}_2 + \omega_1 \sin\theta \cos\alpha_1 \bar{n}_3] \\
&= \ell[(-\omega_2 \cos\theta \sin\alpha_2 + \omega_1 \sin\theta \sin\alpha_1 \sin\alpha_2 - \omega_2 \sin\theta \cos\alpha_1 \cos\alpha_2)\bar{n}_1 \\
&\quad + (\omega_2 \cos\theta \cos\alpha_2 - \omega_1 \sin\theta \sin\alpha_1 \cos\alpha_2 - \omega_2 \sin\theta \cos\alpha_1 \sin\alpha_2)\bar{n}_2 \\
&\quad + (\omega_1 \sin\theta \sin\alpha_1 \cos\alpha_1 \bar{n}_3))]
\end{aligned}
\tag{2.241}
$$

Finally, we obtain the acceleration by differentiating once more in time:

$$
\begin{aligned}
\bar{a}_A = \frac{d^2}{dt^2}\bar{P}_A = \ell[(-\omega_2^2 \cos\theta \cos\alpha_2 + \omega_1^2 \sin\theta \cos\alpha_1 \sin\alpha_2 \\
+ 2\omega_2\omega_1 \sin\theta \sin\alpha_1 \cos\alpha_2 + \omega_2^2 \sin\theta \cos\alpha_1 \sin\alpha_2)\bar{n}_1 \\
+ (-\omega_2^2 \cos\theta \sin\alpha_2 - \omega_1^2 \sin\theta \cos\alpha_1 \cos\alpha_2 \\
+ 2\omega_1\omega_2 \sin\theta \sin\alpha_1 \sin\alpha_2 - \omega_2^2 \sin\theta \cos\alpha_1 \cos\alpha_2)\bar{n}_2 \\
- \omega_1^2 \sin\theta \sin\alpha_1 \bar{n}_3]
\end{aligned}
\tag{2.242}
$$

Notice how the acceleration components have products of ω_1, ω_2, ω_1^2 and ω_2^2, which explains why their magnitudes are important. The acceleration can be simplified if $\omega_1 \ll \omega_2$ or other conditions are assumed. To gain insight into the problem of dynamics, one is often led to linearize the system and hence look at conditions where the Euler angles are assumed small.

2.14 Velocity and Acceleration of a Moving Point on a Rigid Body

The velocity and acceleration of moving parts on a rigid body, which in turn is rotating with respect to an inertial fixed frame R, are important aspects of kinematics in a number of engineering applications. For instance, a sliding particle on a rotating rod, and raindrops sliding on a windshield of a moving car, are just a few illustrative examples.

The velocity of a point P moving on a rigid body A that is moving in a reference frame is given by

$$
{}^R\bar{v}^P = {}^R\bar{v}^Q + {}^A\bar{v}^P + {}^R\bar{\omega}^A \times \bar{r} \tag{2.243}
$$

where Q is a point fixed on body A and $\bar{r}$ defines the position of P with respect to Q.

The acceleration of point P, on the other hand, is given by

$$ {}^R\bar{a}^P = {}^R\bar{a}^Q + {}^A\bar{a}^P + 2\,{}^R\bar{\omega}^A \times {}^A\bar{v}^P + {}^R\bar{\alpha}^A \times \bar{r} + {}^R\bar{\omega}^A \times ({}^R\bar{\omega}^A \times \bar{r}) \tag{2.244} $$

Proof

Consider Figure 2.19 again, where $\bar{p}$ defines the position of point P in R; then

$$ \bar{p} = \bar{q} + \bar{r} \tag{2.245} $$

Differentiation of (2.245) yields the velocity of point P in R; thus

$$ {}^R\bar{v}^P = \frac{d}{dt}(\bar{q} + \bar{r}) \tag{2.246} $$

or

$$ {}^R\bar{v}^P = \frac{d}{dt}\bar{q} + \frac{d}{dt}\bar{r} \tag{2.247} $$

which yields

$$ {}^R\bar{v}^P = {}^R\bar{v}^Q + {}^A\bar{v}^P + {}^R\bar{\omega}^A \times \bar{r} \tag{2.248} $$

Differentiating (2.248) once more, we obtain the acceleration of point P in R as

$$ {}^R\bar{a}^P = {}^R\bar{a}^A + \frac{d}{dt}{}^A\bar{v}^P + \frac{d}{dt}({}^R\bar{\omega}^A \times \bar{r}) \tag{2.249} $$

More explicitly,

$$ {}^R\bar{a}^P = {}^R\bar{a}^Q + {}^A\bar{a}^P + {}^R\bar{\omega}^A \times {}^A\bar{v}^P + {}^R\bar{\alpha}^A \times \bar{r} + {}^R\bar{\omega}^A \times \frac{d}{dt}\bar{r} \tag{2.250} $$

$$ = {}^R\bar{a}^P + {}^A\bar{a}^P + {}^R\bar{\omega}^A \times {}^A\bar{v}^P + {}^R\bar{\alpha}^A \times \bar{r} + {}^R\bar{\omega}^A \times ({}^A\bar{v}^P + {}^R\bar{\omega}^A \times \bar{r}) \tag{2.251} $$

Finally,

$$ {}^R\bar{a}^P = {}^R\bar{a}^Q + {}^A\bar{a}^P + 2\,{}^R\bar{\omega}^A \times {}^A\bar{v}^P + {}^R\bar{\alpha}^A \times \bar{r} + {}^R\bar{\omega}^A \times ({}^R\bar{\omega}^A \times \bar{r}) \tag{2.252} $$

If we choose point Q such that at any instant P and Q coincide, equation (2.244) reduces to

$$ {}^R\bar{a}^P = {}^R\bar{a}^Q + {}^A\bar{a}^P + 2\,{}^R\bar{\omega}^A \times {}^A\bar{v}^P \tag{2.253} $$

Example 2.14.1

Consider once more a disk D attached to a shaft S_2 that swings in the plane given by $\bar{n}_2$ and $\bar{n}_3$ in S_1 (Figure 2.23). The plane of the disk D is normal to the axis of S_2. A mass m is free to rotate in the slot of the disk as shown. Find the velocity and acceleration of point mass m in R.

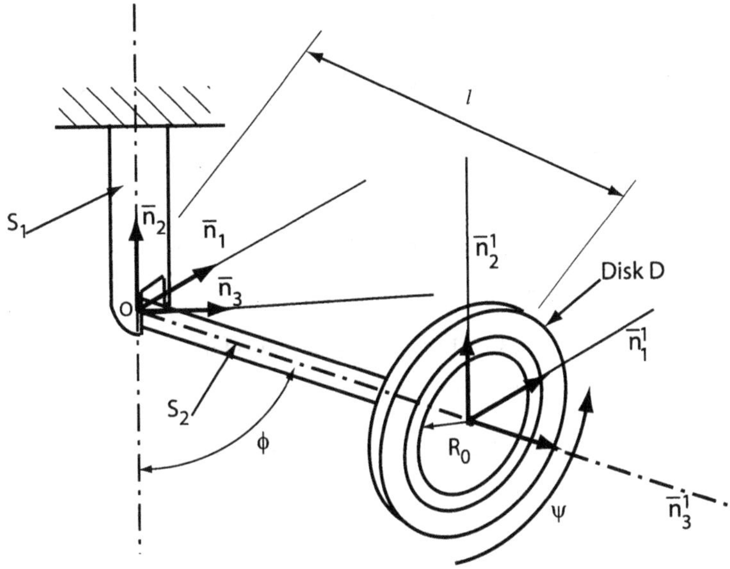

Fig. 2.23. Disk D attached to a shaft S_2

Solution

First, we write the relationship between the unit vectors $\bar{n}^1$ and $\bar{n}$:

$$\bar{n}_m^1 = \begin{bmatrix} 1 & 0 & 0 \\ 0 & \cos(\phi - 90^\circ) & \sin(\phi - 90^\circ) \\ 0 & -\sin(\phi - 90^\circ) & \cos(\phi - 90^\circ) \end{bmatrix} \bar{n}_m = \begin{bmatrix} 1 & 0 & 0 \\ 0 & \sin\phi & -\cos\phi \\ 0 & \cos\phi & \sin\ \phi \end{bmatrix} \bar{n}_m \tag{2.254}$$

Let r define the position of m with respect to $\bar{n}_1^1$ and $\bar{n}_2^1$:

$$\begin{aligned} \bar{r} &= r_1\bar{n}_1^1 + r_2\bar{n}_2^1 = r\cos\psi\bar{n}_1^1 + r\sin\psi\bar{n}_2^1 \\ &= r\cos\psi\bar{n}_1 + r\sin\psi(\sin\phi\,\bar{n}_2 - \cos\phi\,\bar{n}_3) \end{aligned} \tag{2.255}$$

The velocity of point mass m is given by

$$\bar{v}_m = {}^R\bar{v} + {}^{R_o}\bar{v}^m + {}^R\bar{\omega}^D \times \bar{r} \tag{2.256}$$

where

$$\begin{aligned} {}^R\bar{v}^o &= {}^R\bar{\omega}^{S_2} \times \ell\bar{n}_3^1 \\ &= \dot{\phi}\,\bar{n}_1 \times \ell(\cos\phi\,\bar{n}_2 - \sin\phi\,\bar{n}_3) \\ &= \ell\dot{\phi}\,(\cos\phi\,\bar{n}_3 - \sin\phi\,\bar{n}_2) \end{aligned} \tag{2.257}$$

$${}^{R_o}\bar{v}^m = r\dot{\psi}(-\sin\psi\bar{n}_1^1 + \cos\psi\bar{n}_2^1)$$

$${}^{R_o}\bar{v}^m = r\dot{\psi}[-\sin\psi\bar{n}_1 + \cos\psi(\sin\phi\,\bar{n}_2 - \cos\phi\,\bar{n}_3)] \tag{2.258}$$

and

$$
\begin{aligned}
{}^R\bar{\omega}^D \times \bar{r} &= \dot{\phi}\,\bar{n}_1 \times r[\cos\psi \bar{n}_1 + \sin\psi(\sin\phi\,\bar{n}_2 - \cos\phi\,\bar{n}_3)] \\
&= r\dot{\phi}\,(\sin\psi\,\sin\phi\,\bar{n}_3 + \sin\psi\,\cos\phi\,\bar{n}_2)
\end{aligned} \tag{2.259}
$$

Therefore,

$$
\begin{aligned}
\bar{v}_m &= \ell\dot{\phi}\,(\cos\phi\,\bar{n}_3 - \sin\phi\,\bar{n}_2) + r\dot{\psi}[-\sin\psi\bar{n}_1 + \cos\psi(\sin\phi\,\bar{n}_2 - \cos\phi\,\bar{n}_3)] \\
&\quad + r\dot{\phi}\,(\sin\psi\,\sin\phi\,\bar{n}_3 + \sin\psi\,\cos\phi\,\bar{n}_2) \\
&= -r\dot{\psi}\,\sin\psi\bar{n}_1 + (-\ell\dot{\phi}\,\sin\phi + r\dot{\psi}\,\cos\psi\,\sin\phi + r\dot{\psi}\,\cos\psi\,\cos\phi)\bar{n}_2 \\
&\quad + (-\ell\dot{\phi}\,\cos\phi - r\dot{\psi}\,\cos\psi\,\cos\phi + r\dot{\psi}\,\sin\psi\,\sin\phi)\bar{n}_3
\end{aligned} \tag{2.260}
$$

We can also obtain the solution by differentiation as follows:

$$
\begin{aligned}
\bar{v}_m = \frac{d}{dt}P_m = \frac{d}{dt}(\ell\bar{n}_3^1 + \bar{r}) = \frac{d}{dt}\{[(\ell\cos\phi\,\bar{n}_2 + \ell\sin\phi\,\bar{n}_3)] \\
+ r[\cos\psi\bar{n}_1 + \sin\psi(\sin\phi\,\bar{n}_2 - \cos\phi\,\bar{n}_3)]\}
\end{aligned} \tag{2.261}
$$

The solution is exactly the same as that given by equation (2.260).

Following the definition given by equation (2.252), the acceleration of m in R can be expressed as

$$
\bar{a}_m = {}^R\bar{a}^o + {}^{R_o}\bar{a}^m + 2\,{}^R\bar{\omega}^D \times {}^{R_o}\bar{v}^m + {}^R\bar{\alpha}^D \times \bar{r} + {}^R\bar{\omega}^D \times ({}^R\bar{\omega}^D \times \bar{r}) \tag{2.262}
$$

where

$$
{}^R\bar{a}^o = \frac{d\,{}^R\bar{v}^o}{dt} = \ell\ddot{\phi}\,(\cos\phi\,\bar{n}_3 - \sin\phi\,\bar{n}_2) + \ell\dot{\phi}^2(-\sin\phi\,\bar{n}_3 - \cos\phi\,\bar{n}_2) \tag{2.263}
$$

$$
{}^{R_o}\bar{a}^m = \frac{d\,{}^{R_o}\bar{v}^m}{dt} = r(-\cos\psi\bar{n}_1^1 + \sin\psi\bar{n}_2^1)\dot{\psi}^2 \tag{2.264}
$$

$$
{}^R\bar{\alpha}^D = \frac{d}{dt}\,{}^R\bar{\omega}^D = \ddot{\phi}\,\bar{n}_1 \tag{2.265}
$$

Substituting ${}^R\bar{a}^{Bo}$, ${}^{Ro}\bar{a}^m$ and ${}^R\bar{\alpha}^m$ into equation (2.262) will result in the final answer. We can also obtain $\bar{a}_m$ by differentiating equation (2.260) or (2.261).

$$
\begin{aligned}
\bar{a}_m &= -r(\ddot{\psi}\,\sin\psi + \dot{\psi}^2\cos\psi)\bar{n}_1 + (-\ell\ddot{\phi}\,\sin\phi - \ell\dot{\phi}^2\cos\phi + r\ddot{\psi}\,\cos\psi\,\sin\phi \\
&\quad - r\dot{\psi}^2\,\sin\psi\,\sin\phi + \dot{r}\dot{\psi}\dot{\phi}\,\cos\psi\,\cos\phi + r\ddot{\phi}\,\sin\psi\,\cos\phi \\
&\quad + r\dot{\phi}\,\dot{\psi}\,\cos\psi\,\cos\phi - r\dot{\psi}^2\,\sin\psi\,\sin\phi)\bar{n}_2 + (\ell\ddot{\phi}\,\cos\phi - \ell\dot{\phi}^2\,\sin\phi \\
&\quad - r\ddot{\psi}\,\cos\psi\,\cos\phi + r\dot{\psi}^2\,\sin\psi\,\cos\phi + r\dot{\phi}\,\dot{\psi}\,\cos\phi\,\sin\phi + r\ddot{\phi}\,\sin\psi\,\sin\phi \\
&\quad + r\dot{\phi}^2\,\sin\psi\,\cos\phi + r\dot{\phi}\,\dot{\psi}\,\cos\phi\,\sin\phi)\bar{n}_3
\end{aligned} \tag{2.266}
$$

2.15 Small Rotations

In the kinematic analysis of rigid bodies, if we assume small rotations, the previously developed quantities, such as the direction cosines S and $\dot{S}$, the skew matrix Ω, and the angular velocity ${}^R\bar{\omega}^A$, take on simpler forms. Consider the direction cosine matrix S given by equation (2.39). For small angles, we obtain

$$S^{OA} = \begin{bmatrix} 1 & -\gamma & \beta \\ \gamma & 1 & -\alpha \\ -\beta & \alpha & 1 \end{bmatrix} \quad \dot{S}^{OA} = \begin{bmatrix} 0 & -\dot{\gamma} & \dot{\beta} \\ \dot{\gamma} & 0 & -\dot{\alpha} \\ -\dot{\beta} & \dot{\alpha} & 0 \end{bmatrix} \tag{2.267}$$

where higher-order terms are dropped. It follows from equation (2.138) that

$$\Omega = \dot{S}S^T \tag{2.268}$$

or

$$\Omega = \begin{bmatrix} 0 & -\dot{\gamma} & \dot{\beta} \\ \dot{\gamma} & 0 & -\dot{\alpha} \\ -\dot{\beta} & \dot{\alpha} & 0 \end{bmatrix} \tag{2.269}$$

where the components of the angular velocity in R are

$$\begin{aligned} \omega_1 &= \dot{\alpha} \\ \omega_2 &= \dot{\beta} \\ \omega_3 &= \dot{\gamma} \end{aligned} \tag{2.270}$$

Then,

$${}^R\bar{\omega}^A = \dot{\alpha}\bar{n}_1 + \dot{\beta}\bar{n}_2 + \dot{\gamma}\bar{n}_3 \tag{2.271}$$

The representation of the angular velocity given by equation (2.271) is independent of the orientation. It should be pointed out that the skew matrix obtained in equation (2.269) agrees with the definition given by equation (2.149).

2.16 Problems

2.1. The angular velocity of body A in R is shown to be given by

$${}^R\bar{\omega}^A = \frac{d\bar{n}_2}{dt}\cdot\bar{n}_3\bar{n}_1 + \frac{d\bar{n}_3}{dt}\cdot\bar{n}_1\bar{n}_2 + \frac{d\bar{n}_1}{dt}\cdot\bar{n}_2\bar{n}_3 \tag{2.272}$$

where $\bar{n}_i$ $(i = 1, 2, 3)$ are a set of mutually perpendicular unit vectors fixed in body A. Use the relationship between the unit vectors fixed in A and R together with equation (2.173) to obtain equation (2.179).

2.2. The simple angular velocity is defined by equation (2.194) to be

$$^{R}\bar{\omega}^{A} = \frac{d\theta}{dt}\bar{\lambda}$$

Show that the angular velocity components are $\omega_1 = \lambda_1\dot{\theta}$, $\omega_2 = \lambda_2\dot{\theta}$ and $\omega_3 = \lambda_3\dot{\theta}$.

2.3. Consider the unit cube shown in Figure 2.24. If the cube is to be rotated about the axis LL' passing through its center by an angle $\theta = 30°$, find the new position of $\bar{P}$ after rotation. (Make use of Euler's parameters.)

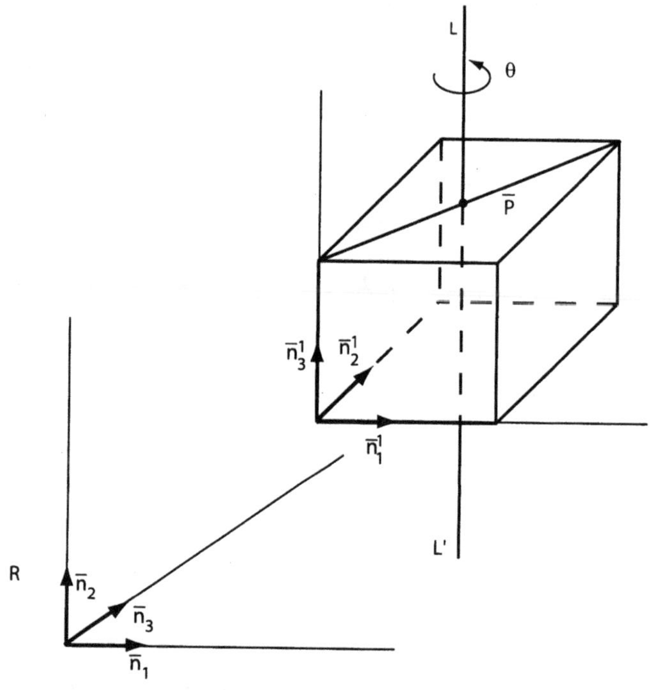

Fig. 2.24. Unit cube

2.4. Equation (2.94) defines the transformation matrix $\boldsymbol{S}$ between two arbitrary unit vectors $\bar{n}_i$ and $\bar{n}_j (i, j = 1, 2, 3)$ in terms of Euler's parameters. Use equation (2.84) to show that

$$S_{12} = 2(\varepsilon_1\varepsilon_2 - \varepsilon_3\varepsilon_4)$$

$$S_{13} = 2(\varepsilon_1\varepsilon_3 + \varepsilon_2\varepsilon_4)$$

$$S_{22} = (-\varepsilon_1^2 + \varepsilon_2^2 - \varepsilon_3^2 + \varepsilon_4^2)$$

2.5. Consider the disk of the gyroscope shown in Figure 2.25. The disk rotates about its own axis with a constant angular velocity $\dot{\gamma}$, and the outer gimbal support has an angular velocity $\dot{\alpha}$. Determine the angular velocity and angular acceleration of the disk with respect to the outer gimbal support G_0.

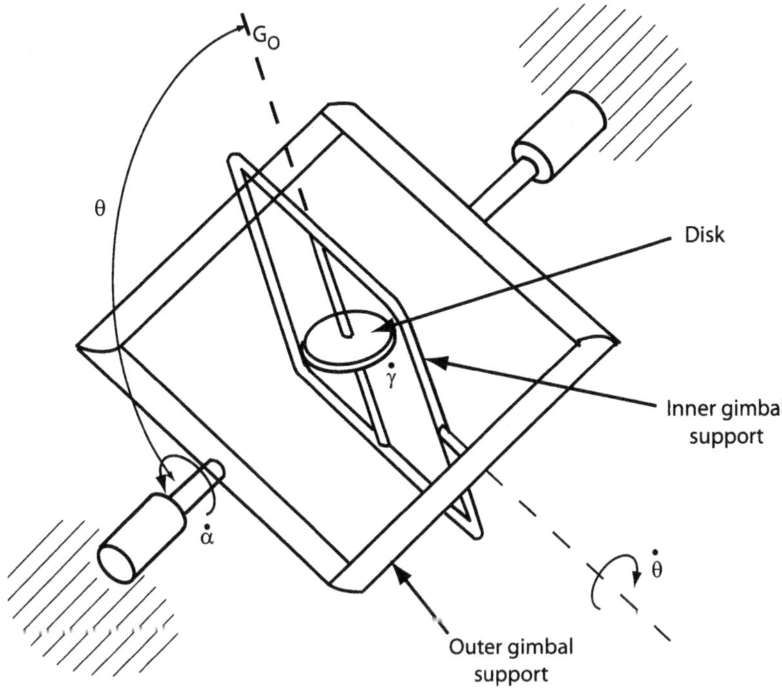

Fig. 2.25. Gyroscope

2.6. For the two-link mechanism shown in Figure 2.26, a slider s can move freely along link 2. Find the center of mass velocity and acceleration of S in R.

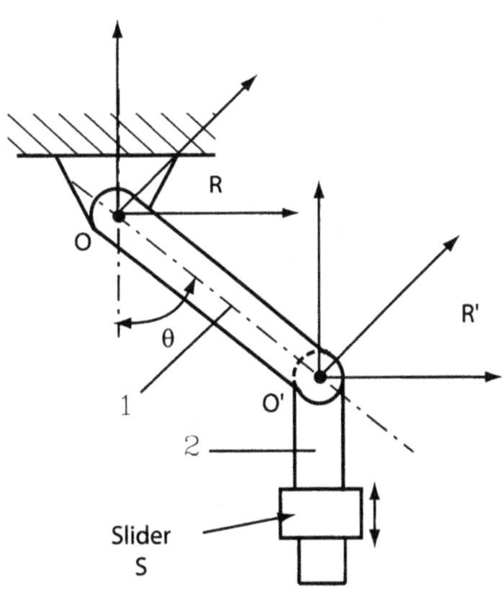

Fig. 2.26. Two-link mechanism

2.7. Prove the identities given by equations (2.184) to (2.186).

2.8. Given two arbitrary rigid bodies as shown in Figure 2.27. Body A is hinged at O and body B is hinged at a point in body A: namely, O'. Find the angular velocity of body B in R in terms of Euler's parameters. [*Hint:* Make use of equation (2.188).]

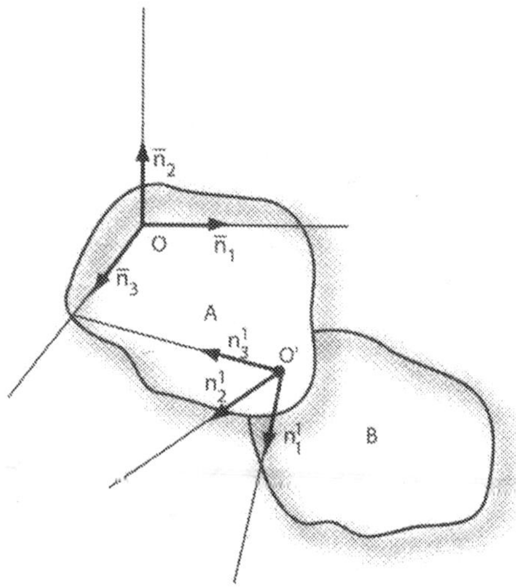

Fig. 2.27. Two arbitrary rigid bodies

2.9. Show the equivalence between the two representations of the direction cosines given by equations (2.84) and (2.89).

2.10. Find the relationship between Euler parameters and Euler angles for small rotations.

2.11. Find an expression for the angular velocity of a rigid body A with respect to a fixed reference frame R in terms of Rodriguez parameters.

2.12. The angular velocity components of body A in R are governed by the differential equations

$$\dot{\omega}_1 - k\omega_2\omega_3 = 0$$

$$\dot{\omega}_2 + k\omega_3\omega_1 = 0$$

$$\dot{\omega}_3 = 0$$

(a) Find the angular velocity components in time if $\omega_1(t = 0), \omega_2(t = 0) = 3$ rad/s and $\omega_3 = 0.5$ rad/s. **(b)** Determine the orientation of body A in R by solving for Euler's parameters in time. The initial conditions are $\varepsilon_1 = \varepsilon_2 = \varepsilon_3 = 0$ and $\varepsilon_4 = 1$ at time $t = 0$.

MATLAB Program to perform transformation matrices between vectors in different reference frames

MAIN PROGRAM

```
{\%}THE PROGRAM SHOULD HAVE THE OPTION OF USING
  EITHER{\%}

{\%}EULER ANGLES (DIRECTION COSINES) OR EULER
  PARAMETERS.{\%}


clear;

clc;

temp=0;

diary out.hwl

while temp-=3

disp('This program can be used to solve
  transformation matrices.')

disp('You can use two methods for the problem.')

disp('Please choose the method :')

disp('--------- Matrice transformtion
  program --------- ');

disp('1)Euler angles;')

disp('2)Euler parameter;')

disp('3)exit.')

disp('Please enter 1 or 2 or 3.');

numb=input('your input');

if numb==1

clc;
```

```
disp('Welcome to Euler angle method');

{\%} INPUT ANGLES FOR PROGRAM TO GET NEW POSITION {\%}

a=input('Input alfa angle $<$degree$>$:')

b=input('Input beta angle $<$degree$>$:')

c=input('Input gama angle $<$degree$>$:')

{\%}INPUT OLD POSITION TO TRANSFORM NEW POSITION {\%}

po=input('Input x component of the old point P:')

po=[po

input('Input y component of the old point P:')]

po=[po

input('Input z component of the old point P:')]

{\%} CALL THE FUNCTION OF EULER ANGLE TO CALCULATE
  NEW POSITION {\%}

pa=eang(a,b,c,po)

pause;

clc;

elseif numb==2

clc;

disp('welcome to Euler parameter method')

{\%} INPUT ANGLE FOR PROGRAM TO GET NEW POSITION {\%}

a=input('Input angle $<$deg.$>$:')

{\%}INPUT THE AXIS OF ROTATION L FOR EULER
  PARAMETER {\%}

disp(' You should input unit vecter at below:')
```

```
l=input('Input x component of the axis L:')

l:[l

input('Input y component of the axis L:')]

l=[l

input('Input z component of the axis L:')]

{\%} INPUT OLD POSITION TO TRANSFORM NEW POSITION {\%}

po=input('Input x component of the old point P:')

po= [po

input('Input y component of the old point P:')]

po=[po

input('Input z component of the old point P:')]

{\%} CALL THE FUNCTION OF EULER PAR$\sim {\%}
{\%} $TER TO CALCULATE NEW POSITION{\%}

pa=eparam(a,l,po)

pause;

clc;

elseif numb==3

disp(' ');

disp(' Good luck!By-by! ');

temp=numb;

end

end

diary off

{\%}{\%}{\%}{\%}{\%}{\%}{\%}{\%}{\%}{\%}{\%}
```

```
  {\%}{\%}{\%}

{\%}This function is used for the {\%}
transformation of Euler angle {\%}

{\%}{\%}{\%}{\%}{\%}{\%}{\%}{\%}{\%}{\%}{\%}
  {\%}{\%}{\%}

function pa=eang(a,b,c,p)

a=a*pi/180;

b=b*pi/180;

c=c*pi/180;

sll=cos(b)*cos(c);

sl2=cos(a)*sin(c)+sin(a)*sin(b)*cos(c);

sl3=sin(c)*sin(a)-cos(a)*sin(b)*cos(c);

s21=-cos(b)*sin(c);

s22=cos(a)*cos(c)+sln(a)*sin(b)*sin(c);

s23=sin(a)*cos(c)+cos(a)*sin(b)*sin(c);

s31=sin(b);

s32=-sin(a)*cos(b);

s33=cos(a)*cos(b);

disp(' ');

disp(' Transformation matrix and new position:');

s=[sll s12 s13;s21 s22 s23;s31 s32 s33]

pa=[sll s12 s13;s21 s22 s23;s31 s32 s33]*p;

{\%}{\%}{\%}{\%}{\%}{\%}{\%}{\%}{\%}{\%}{\%}
  {\%}{\%}{\%}

{\%} This function is used for the transformation
```

```
{\%} of Euler parameter {\%}

{\%}{\%}{\%}{\%}{\%}{\%}{\%}{\%}{\%}{\%}{\%}
  {\%}{\%}{\%}

function pa=eparam(a,l,p)

a=a*pi/180;

e1=[1 0 0]*l'sin(a/2);

e2=[0 1 0]*l'sin(a/2);

e3=[0 0 1]*l*sin(a/2);

e4=cos(a/2);

s11=e1*e1-e2*e2-e3*e3+e4*e4;

s12=2*(e1*e2-e3*e4);

s13=2*(e1*e3+e2*e4);

s21=2*(e1*e2+e3*e4);

s22=-e1*e1+e2*e2-e3*e3+e4*e4;

s23=2*(e2*e3-e1*e4);

s31=2*(e1*e3-e2*e4);

s32=2*(e2*e3+e1*e4);

s33=-e1*e1-e2*e2+e3*e3+e4*e4;

disp(' ');

disp(' Transformation matrix and new position:');

s=[s11 s12 s13;s21 s22 s23;s31 s32 s33]

pa=[s11 s12 s13;s21 s22 s23;s31 s32 s33]*p;

{\%}{\%}{\%}{\%}{\%}{\%}{\%}{\%}{\%}{\%}{\%}
  {\%}{\%}{\%}
{\%}OUTPUT Example 2.3{\%}
```

```
{\%}{\%}{\%}{\%}{\%}{\%}{\%}{\%}{\%}{\%}{\%}
  {\%}{\%}{\%}
This program can be used to solve
  transformation matrices.

You can use two methods for the problem.
Please choose the method :

---------- Matrice transformtion program ----------

1)Euler angles;

2)Euler parameter;

3)exit.

Please enter 1 or 2 or 3

your input Welcome to Euler angle method

Input alfa angle $<$degree$>$:

a =

30

Input beta angle $<$degree$>$:

b =

60

Input gamma angle $<$degree$>$:

c =

30

Input x component of the old point P:

po =

1

Input y component of the old point P:
```

```
po =

1

1

Input z component of the old point P:

po =

1

1

0

Transformation matrix and new position:

s =

0.4330 0.8080 -0.3995

-0.2500 0.9665 0.8080

0.8660 -0.2500 0.4330

pa =

1.2410

0.7165

0.6160

{\%}{\%}{\%}{\%}{\%}{\%}{\%}{\%}{\%}{\%}{\%}{\%}{\%}

{\%}OUTPUT Example 2.5.2{\%}

{\%}{\%}{\%}{\%}{\%}{\%}{\%}{\%}{\%}{\%}{\%}{\%}{\%}

This program can be used to solve
  transformation matrices.

You can use two methods for the problem.
```

```
Please choose the method :
----------- Matrice transformtion program -----------
1)Euler angles;
2)Euler parameter;
3)exit.
Please enter 1 or 2 or 3
your input Welcome to Euler angle method
Input angle $<$deg.$>$:
a = 60
You should input unit vecter at below
Input x component of the axis L:
l =
0.5770
Input y component of the axis L:
l =
0.5770
0.5770
Input z component of the axis L:
l =
0.5770
0.5770
0.5770
Input x component of the old point P:
```

```
po = 1

Input y component of the old point P:

po =

1

1

Input z component of the old point P:

po =

1

1

0

Transformation matrix and new position:

s =

0.6668 -0.3332 0.6662

0.6662 0.6668 -0.3332

-0.3332 0.6662 0.6668

pa =

0.3335

1.3329

0.3329
```

This program can be used to solve transformation matrices.
You can use two methods for the problem.
Please choose the method :
——— Matrice transformtion program ———
1) Euler angles;
2) Euler parameter;
3) exit.
Please enter 1 or 2 or 3
your input.

References

1. Coyley, A., On the Rotation of a Solid Body Round a Fixed Point. Cambridge Dublin Math J., Vol. **I**, 1846, pp. 167–173; also *Collect. Math. Pap.*, Vol. **I** (Paper 37), Cambridge University Press, 1889, pp. 237–252.
2. Gibbs, J. W., in *Vector Analysis* (E. B. Wilson, ed.), Scribners, New York, 1901, and Yale University Press, New Haven, Conn., 1931.
3. Hamel, G., *Theoretische Mechanik*, Springer-Verlag, Berlin, 1949.
4. Euler, L., Formulae Generates pro Translatione Quawnque Corporum Rigidorum, *Novi Commentari Acad. Imp. Petrop.* 20, 1775, pp. 189–207.
5. Euler, L., Nova Methodus Motum Corporum Rigidorum determinandi, *Novi Commentari Acad. Imp. Petrop.* 20, 1775, pp. 208–238.
6. Goldstein, H., *Classical Mechanics*, Addison-Wesley, Reading, MA, 1980.
7. Kane, T. R., Likins, P. W. and Levinson, D. A., *Spacecraft Dynamics*, McGraw-Hill, New York, 1983.
8. Rodriguez, O., Des lois géométriques qui régissent les déplacements d'un system solide dans l'espace, et la variation des coordonnées provenant des déplacements considéres indépendamment des causes qui peuvent les produire, *J. Math. Pures Appl.*, Vol. **5**, 1840, pp. 380–440.
9. Wittaker, E. T., *Analytical Dynamics of Particles and Rigid Bodies*, Cambridge University Press, Cambridge, 1937 (first edition, 1904).
10. Cheng, H. and Gupta, K. C., A Historical Note on Finite Rotations, *J. Appl. Mech.* Vol. **56**: No. 1, 1989, pp. 139–145.
11. Ginsberg, J. H., *Advanced Engineering Dynamics*, Harper & Row, New York, 1988.
12. Kane, T. R. and Levinson, D. A., *Dynamics: Theory and Applications*, McGraw-Hill, New York, 1985.
13. Paul, B., On the Composition of Finite Rotations, *Am. Math. Mon.*, Vol. **70**, 1963, pp. 949–954.
14. Hill, E. L., Rotation of a Rigid Body About a Fixed Point, *Am. J. Phys.*, Vol. **13**, 1945, 137–140.
15. Beatty, M. F., Vector Analysis of Finite Rigid Rotations, *Trans. ASME J. Appl. Mech.*, Vol. **44**, 1977, pp. 501–502.
16. Meriovitch, L., *Methods of Analytical Dynamics*, McGraw-Hill, New York, 1970.
17. Ronald L. Huston, *Multibody Dynamics*, Butterworth-Heinemann, 1990.

3

Kinematics for General Multibody Systems

3.1 Introduction

In this chapter, great emphasis is given to automatic generation of the matrices and arrays needed to express the kinematics of a general treelike multibody system. The coefficients that we develop will be in an explicit form ready for computer implementation. In the sequel, we introduce an array that contains the topological information about the structure interconnection and make use of the previously defined kinematics of rigid bodies and extend its definition to general open-structure problems. A more structured formulation for the partial angular velocity and partial velocity arrays and their derivatives is presented. At the end of the chapter the student and engineer are expected to be able to formulate the kinematics of any open mechanical system or chain following the procedures presented herein.

3.2 Configuration Graphs for Treelike Multibody Systems

The emphasis on the development of all-purpose dynamical programs capable of simulating the motion of complex systems relies on its effectiveness in the organization of the geometry, especially of its input and output. The bottom line is: How "user friendly" is the program? As the answer to this question is more of an art than science, in the following section we deal with the development of a structure-based array called a *configuration array, configuaration graph,* or *topology array* that describes the interconnection between bodies. This array alone should provide enough information to be able to reconstruct the structure without having to refer back to any figure.

Consider Figure 3.1, which describes a general multibody system without closed loops. This illustrates the typical open-chain system, or what we refer to as a *treelike system*. The first step is to select an arbitrary body; call it body 1. Next, number and label the bodies in ascending progression away from body 1, then number the bodies by moving from branch to branch until all bodies are numbered. A branch is defined to be any chain off the chain that contains body 1. Once the numbering is done, we define two arrays, which we call $\Gamma_0(k)$ and $\Gamma_1(k)$, such that

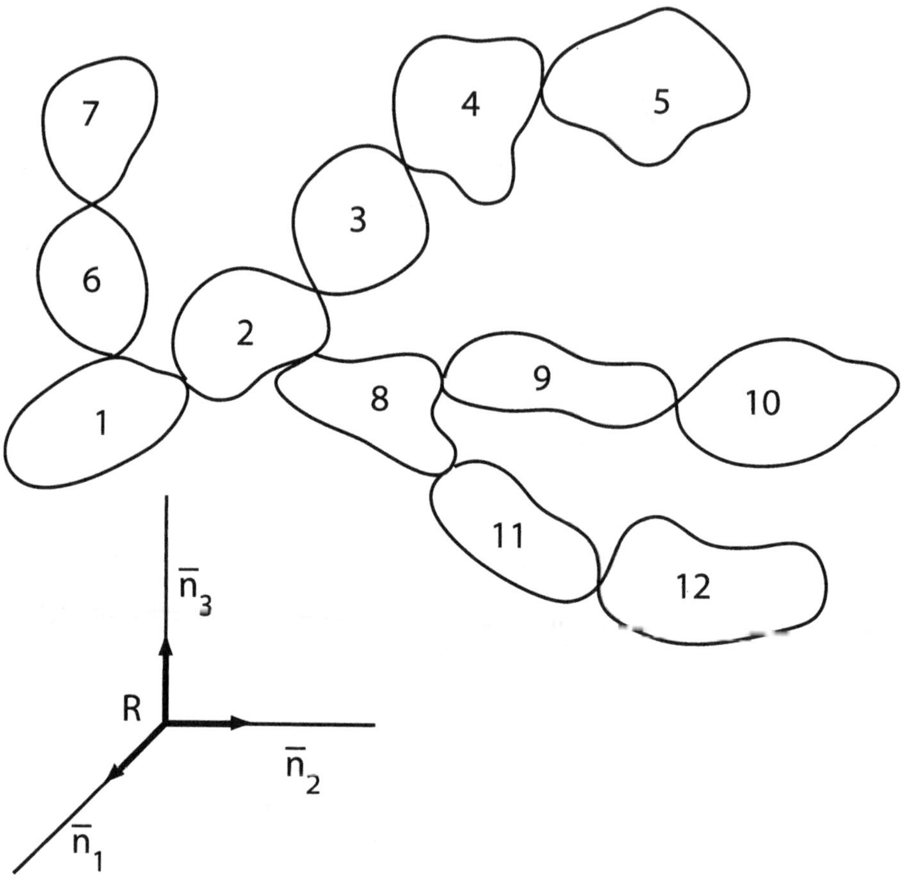

Fig. 3.1. Arbitrary treelike structure of a multibody system

$$\Gamma_0(k) = \{1\ \ 2\ \ 3\ \ 4\ \ 5\ \ 6\ \ 7\ \ 8\ \ 9\ \ 10\ \ 11\ \ 12\} \tag{3.1}$$

$$\Gamma_1(k) = \{0\ \ 1\ \ 2\ \ 3\ \ 4\ \ 5\ \ 6\ \ 7\ \ 8\ \ 9\ \ 10\ \ 11\} \tag{3.2}$$

where $\Gamma_0(k)$ defines the body array and is composed basically of all the body numbers. It is usually known simply by noting the number of bodies the structure has. On the other hand, $\Gamma_1(k)$ defines the *lower body array*, which describes the lower adjacent connecting body to each of the bodies defined by $\Gamma_0(k)$. For instance, from equations (3.1) and (3.2), we read that body 12 is connected to lower body 11. Note that body 1 is connected to body 0, which represents the fixed inertial frame. Interestingly, using the array $\Gamma_1(k)$, one can find its lower body array, say $\Gamma_2(k)$. Proceeding in this fashion and stopping when $\Gamma_m(k)$ has all the numbers equal to zero, we generate the configuaration graph of the structure, which we will call a *tree array*. For Figure 3.1, Table 3.1 illustrates the formation of the tree array by Γ.

The motivation behind the composition of Γ is to be able to express the kinematics of the multibody systems without having to refer to the figure anymore. We also

$\Gamma_0(k)$	1	2	3	4	5	6	7	8	9	10	11	12
$\Gamma_1(k)$	0	1	2	3	4	1	6	2	8	9	8	11
$\Gamma_2(k)$		0	1	2	3	0	1	1	2	8	2	8
$\Gamma_3(k)$			0	1	2		0	0	1	2	1	2
$\Gamma_4(k)$				0	1				0	1	0	1
$\Gamma_5(k)$					0					0		0

Table 3.1. $\Gamma(k)$ tree array for Figure 3.1 for the multibody system shown above

observe that the columns of Γ define the chain of interconnected bodies to go from any given body k to the reference frame R. Observe the utilization of the column corresponding to body 7. Suppose that we are interested in defining the angular velocity of body 7 with respect to R. Using the relative angular velocities and the addition formula defined in equation (2.199), we obtain

$$ {}^{R}\bar{\omega}^{7} = {}^{0}\bar{\omega}^{1} + {}^{1}\bar{\omega}^{6} + {}^{6}\bar{\omega}^{7} \tag{3.3} $$

Taking column 7 and working upward, we can derive equation (3.3) by considering all possible relative combinations between bodies that constitute the chain that goes from body B_7 to the reference frame R, which is zero. The column gives three sets of numbers—$(0, 1)$, $(1, 6)$ and $(6, 7)$—which exactly define the relative angular velocity components given in equation (3.3). Examining the tree array, we also observe that in each column, the zero appears at a different level of $\Gamma_i(k)(i = 1, 2, \ldots, P)$. This determines the degree of connectivity in a particular chain of bodies. To a certain extent, by counting the number of rows it takes to go from body $k(k = 1, \ldots, N)$ to R, we determine the number of bodies that are interconnected to one another. Let us denote by $H(k)$ the degree of connectivity for each body K. Then, for Figure 3.1 and using Table 3.1, we define $H(k)$ to be as shown in Table 3.2.

One final observation concerning the setup of Table 3.1 and Table 3.2: Only $\Gamma_0(k)$ and $\Gamma_1(k)$ are needed to generate the tree array and the exponent of $H(k)$.

k	1	2	3	4	5	6	7	8	9	10	11	12
H(k)	1	2	3	4	5	2	3	3	4	5	4	5

Table 3.2. Exponent array for $\Gamma(k)$ tree array of Figure 3.1

Example 3.2.1

Modeling of the human gait problem has long been the interest of researchers. Regardless of the complexity of the models employed, the equations of motion can be derived and the direct or inverse problem can then be solved. Walking could be looked upon as a function that could be triggered by two systems: the neurological and physiological systems. Consider the human model shown in Figure 3.2. It consists of 17 segments interconnected through revolute and spherical joints. Let us consider the model as a mechanical system with a treelike configuration where the lower torso,

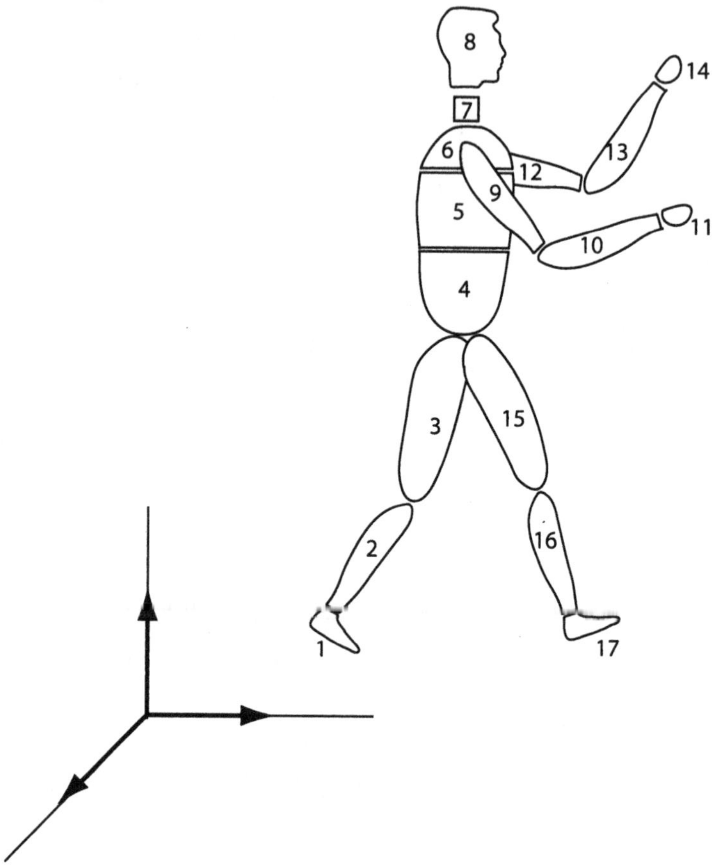

Fig. 3.2. Human model

middle torso, and upper torso constitute the trunk and the arms and legs constitute the branches. Given this assumption, we can adopt a numbering scheme to identify the configuration of the structure initially and at any time when it is undergoing certain motion. Table 3.3 gives an explicit tree array $\Gamma(k)$. We discuss its degrees of freedom, constraint and equations of motion in Chapter 6.

3.3 Generalized Coordinates Partitioning

The maximum degrees of freedom that a rigid body in a reference frame R can have is six: three for rotations and three for translations, where the arbitrary rotation in space is characterized by three successive rotations about the 3-axis (mutually perpendicular) fixed in the body. The three rotations as described through Euler's angles can be replaced with one rotation using Euler's parameters. In the sequel, we will use relative coordinates between bodies to describe the orientation of any rigid body B_k with respect to a fixed inertial frame R. Consider two rigid bodies as shown in Figure 3.3. Let $\bar{n}_i^1$ $(i = 1, 2, 3)$ be a set of mutually perpendicular axes fixed in body 1 (B_1) and $\bar{n}_i^2$ $(i = 1, 2, 3)$ be a set of similar vectors fixed in body 2 (B_2), as shown. Let $\bar{n}_i$ $(i = 1, 2, 3)$ be attached to the inertial fixed frame R. If we allow B_1 to have six degrees of freedom, it can rotate about $\bar{n}_1^1$, $\bar{n}_2^1$ and $\bar{n}_3^1$ by α_1, β_1 and γ_1 and translate along the same axis by ξ_{1x}, ξ_{1y} and ξ_{1z} (x, y and z are used to illustrate translation along $\bar{n}_1^1$, $\bar{n}_2^1$ and $\bar{n}_3^1$, respectively).

$\Gamma_0(k)$	1	2	3	4	5	6	7	8	9	10	11	12	13	14	15	16	17
$\Gamma_1(k)$	0	1	2	3	4	5	6	7	6	9	10	6	12	13	4	15	16
$\Gamma_2(k)$		0	1	2	3	4	5	6	5	6	9	5	6	12	3	4	15
$\Gamma_3(k)$			0	1	2	3	4	5	4	5	6	4	5	6	2	3	4
$\Gamma_4(k)$				0	1	2	3	4	3	4	5	3	4	5	1	2	3
$\Gamma_5(k)$					0	1	2	3	2	3	4	2	3	4	0	1	2
$\Gamma_6(k)$						0	1	2	1	2	3	1	2	3		0	1
$\Gamma_7(k)$							0	1	0	1	2	0	1	2			0
$\Gamma_8(k)$								0		0	1		0	1			
$\Gamma_9(k)$											0			0			

Table 3.3. Tree array for human model

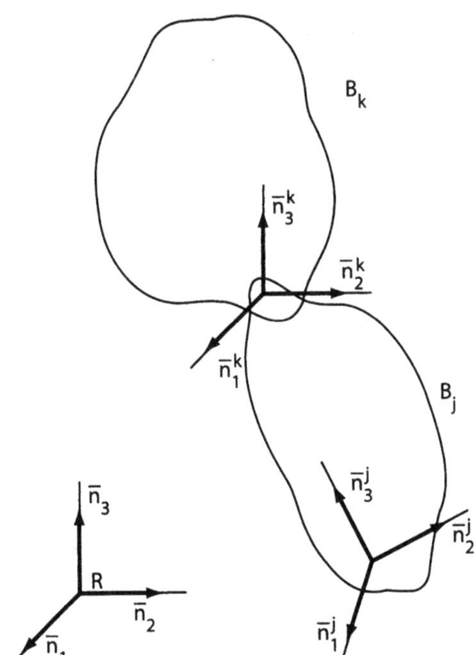

Fig. 3.3. Two arbitrary rigid bodies B_J and B_K

On the other hand, if body B_2 has all possible motions with respect to the fixed reference frame in body B_1, then α_2, β_2 and γ_2 and ξ_{2x}, ξ_{2y} and ξ_{2z} will correspond to the rotation- and translation-generalized coordinates. The total degrees of freedom for bodies B_1 and B_2 is 12. Now, let us introduce an array $x_\ell(\ell = 1, \ldots, 12)$ such that

$$\begin{bmatrix} x_1 \\ x_2 \\ x_3 \\ \hline x_4 \\ x_5 \\ x_6 \\ \hline x_7 \\ x_8 \\ x_9 \\ \hline x_{10} \\ x_{11} \\ x_{12} \end{bmatrix} = \begin{bmatrix} \alpha_1 \\ \beta_1 \\ \gamma_1 \\ \hline \alpha_2 \\ \beta_2 \\ \gamma_2 \\ \hline \xi_{1x} \\ \xi_{1y} \\ \xi_{1z} \\ \hline \xi_{2x} \\ \xi_{2y} \\ \xi_{2z} \end{bmatrix} \tag{3.4}$$

or

$$x_\ell = \begin{cases} x_\ell \quad \ell = 1, \ldots, 6 \\ \text{for rotation} \\ \hline x_\ell \quad \ell = 7, \ldots, 12 \\ \text{for translation} \end{cases} \tag{3.5}$$

Representing all generalized coordinates by x_ℓ, and grouping them in triplets as shown in equation (3.4) will allow us to build a strategy in establishing the kinematics, as will be shown in the next section. Based on the aforementioned organization of the generalized coordinates, one can extend equations (3.4) and (3.5) to the case where N bodies are interconnected to one another or form a chain of bodies. The generalized coordinates will then be given by $x_\ell(\ell = 1, \ldots, 6N)$, where

$$\left.\begin{matrix} x_1 \\ x_2 \\ x_3 \end{matrix}\right\} \rightarrow \text{ describe the relative orientation of } B_1 \text{ with respect to } R$$

$$\left.\begin{matrix} x_4 \\ x_5 \\ x_6 \end{matrix}\right\} \rightarrow \text{ describe the relative orientation of } B_2 \text{ with respect to } B_1$$

$$x_\ell = \left.\begin{matrix} x_{3N-2} \\ x_{3N-1} \\ x_{3N} \end{matrix}\right\} \rightarrow \text{ describe the relative orientation of } B_N \text{ with respect to } B_{N-1}$$

$$\left.\begin{matrix} x_{3N+1} \\ x_{3N+2} \\ x_{3N+3} \end{matrix}\right\} \rightarrow \text{ describe the relative translation of } B_1 \text{ with respect to } R$$

$$\vdots$$

$$\left.\begin{matrix} x_{6N-2} \\ x_{6N-1} \\ x_{6N} \end{matrix}\right\} \rightarrow \text{ describe the relative translation of } B_N \text{ with respect to } B_{N-1}$$

3.4 Transformation Matrices and Their Derivatives for N-Interconnected Rigid Bodies

3.4.1 Transformation Matrices

In actual practice, the bodies in an open treelike structure have one or two degrees of freedom. In some cases—in robotics, for example—a particular link or body can have three degrees of freedom. In view of the possible motion a particular body can undergo with respect to a fixed frame, three degrees of freedom are present when Euler's angles are used and one when Euler's parameters are used (this does not include translation). Since transformation matrices (direction cosines) and their derivatives are very important in kinematics, especially in the formulation of angular velocities, angular accelerations, mass center velocities, and accelerations, it is important to represent those kinematical expressions using block matrices. It should also be necessary to have algorithms that enable one to generate those kinematical expressions for any given multibody system with N interconnected bodies. The efficiency of the algorithm relies on its elegance to generate those expressions automatically without additional effort. Based on the generalized coordinate representation as stated in Section 3.4, and the description of the topology of the structure through the tree array, we will formulate a general expression for the kinematics applicable to any multibody system.

Before we get to the formulation for N interconnected bodies, let us consider the case of three interconnected rigid bodies, as shown in Figure 3.4. Let each body have three rotational and three translational degrees of freedom. Let R define a fixed inertial reference frame and R_1, R_2 and R_3 a set of reference frames attached to bodies B_1, B_2 and B_3, respectively, as shown. We denote by $\bar{n}_i, \bar{n}_i^1, \bar{n}_i^2$ and $\bar{n}_i^3 (i = 1, \ldots, 3)$ the corresponding mutually perpendicular unit vectors along the axis of R, R_1, R_2 and R_3, respectively. The orientation of each body with respect to its lower adjacent body is defined by α_i, β_j and γ_j (j is fixed and has values of 1, 2, or 3). Similarly, the relative translation of any body with respect to its lower adjacent body is given by

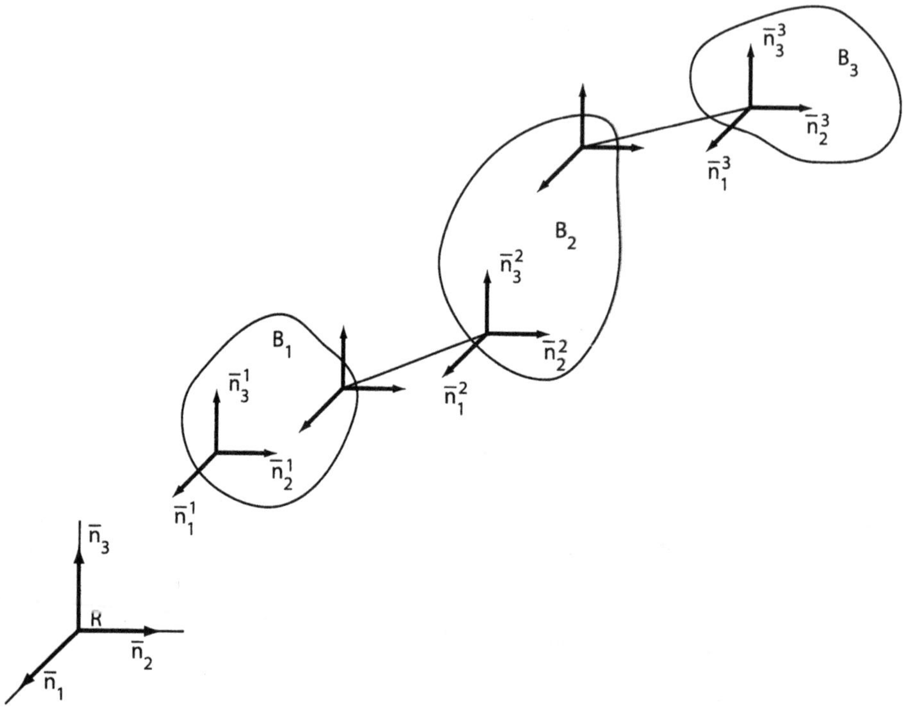

Fig. 3.4. Three arbitrary interconnected rigid bodies with three rotations and translations at each joint

$\xi_{jx}, \xi_{jy}, \xi_{jz}$. From equation (3.4), the generalized coordinates for the entire system could be expressed as

$$[X_\ell]^T = [\alpha_1, \beta_1, \gamma_1, \alpha_2, \beta_2, \gamma_2, \alpha_3, \beta_3, \gamma_3, \xi_{1x}, \xi_{1y}, \xi_{1z}, \xi_{2x}, \xi_{2y}, \xi_{2z}, \xi_{3x}, \xi_{3y}, \xi_{3z}]$$
$$(\ell = 1, \ldots, 18) \tag{3.6}$$

The unit vectors of any axis frame can be represented as a linear combination of the unit vectors of any other axis frame. Using equation (3.230), we can write

$$\bar{n}_i^3 = S_{ij}^{32} \bar{n}_j^2 \tag{3.7}$$

$$\bar{n}_i^2 = S_{ij}^{21} \bar{n}_j^1 \tag{3.8}$$

$$\bar{n}_i^1 = S_{ij}^{10} \bar{n}_j \tag{3.9}$$

Expressing all the unit vectors in bodies B_1, B_2 and B_3 with respect to the unit vectors fixed in the inertial fixed frame R, we obtain

$$\bar{n}_i^3 = S^{32} S^{21} S^{10} \bar{n}_j \tag{3.10}$$

$$\bar{n}_i^2 = S^{21} S^{10} \bar{n}_j \tag{3.11}$$

and

$$\bar{n}_i^1 = S^{10}\bar{n}_j \tag{3.12}$$

We can obtain equations (3.10) to (3.12) by making use of the tree array from Figure 3.4, given by Table 3.4. Taking the first column of $\Gamma(1)$, we see that body B_1 has lower body O (which denotes the reference frame R). Then the shifter matrix is deduced simply by seeing that there is only one direction cosine needed: S^{10}. So the shifter matrix S^{10} will be used to express the unit vectors in body B_1 with respect to R. This will lead automatically to equation (3.12). Column 2 gives an array with three numbers—2, 1 and 0—where only one zero is retained. To express unit vectors of body 2 with respect to R, two direction cosines need to be premultiplied: S^{10} and S^{21}. Those are obtained by reading two numbers at a time going upward in column $\Gamma(2)$ and using them as superscripts for S. Similarly, column 3 gives the array 3, 2, 1, 0, which gives three shifters: S^{10}, S^{21} and S^{32}. The product of those shifters gives a matrix that relates the unit vectors fixed in body B_3 to the unit vectors in R. The procedures outlined above can be used in a general multibody system with N interconnected bodies.

$\Gamma_0(k)$	1	2	3
$\Gamma_1(k)$	0	1	2
$\Gamma_2(k)$	0	0	1
$\Gamma_3(k)$	0	0	0

Table 3.4. Tree array for Figure 3.4

The shifter or transformation matrix between any set of unit vectors fixed in a reference frame of a rigid body, and those fixed in an inertia fixed frame R can be expressed as

$$\bar{n}_i^k = S_{ij}^{KO}\bar{n}_j \tag{3.13}$$

or

$$\begin{bmatrix} \bar{n}_1^k \\ \bar{n}_2^k \\ \bar{n}_3^k \end{bmatrix} = \left[S^{KO}\right] \begin{bmatrix} \bar{n}_1 \\ \bar{n}_2 \\ \bar{n}_3 \end{bmatrix} \tag{3.14}$$

where the $\bar{n}_i^k$ represent a set of mutually perpendicular unit vectors fixed in body k, and $\bar{n}_i$ the corresponding unit vectors fixed in R S^{KO} is a 3×3 matrix obtained by premultiplying all the transformation matrices between adjacent bodies in the chain interconnecting body k to R. The latter is obtained using the tree array and working the corresponding column upward. To make the analysis more general, S^{KO} can be defined as

$$S^{KO} = \prod_{r=1}^{H(k)} S^{sp} \tag{3.15}$$

with $s = \Gamma_{r-1}(k)$ and $p = \Gamma_r(k)$, where $\prod$ designates the successive products of S, from r equal to $H(k)$ to r equal to 1. Equation (3.15) defines the transformation matrix between any body B_k and R. This is very important in the sequel, as it will prove effective to express the kinematical expressions needed.

Example 3.4.1
Consider Figure 3.1, which forms a treelike structure of arbitrary interconnected rigid bodies. Find the transformation matrix between body B_9 and R.

Solution
Let the transformations be given by equation (3.15); hence

$$S^{90} = \prod_{r=1}^{H(9)} S^{sp} \tag{3.16}$$

where from Table 3.2, $H(k)$ is equal to 4 and the tree array is given by Table 3.1. Also, we note that the column vector of $\Gamma(9)$ is 0 1 2 8 9. So

$$S^{90} = S^{98}S^{82}S^{21}S^{10} \tag{3.17}$$

or

$$S^{09} = S^{01}S^{12}S^{28}S^{89} \tag{3.18}$$

The usefulness of equation (3.15) stems from the information it provides to generate the appropriate transformations, once the tree array and the exponent array are formed. Therefore, it will be rather a simple exercise for programming.

Example 3.4.2
Find the shifter matrix between body 3 and the inertial reference frame R S^{03} for Figure 3.4.

Solution
Once again, application of equation (3.15) gives

$$S^{30} = \prod_{r=1}^{H(3)} S^{sp} \tag{3.19}$$

We note from Table 3.2 that the corresponding exponent array $H(3)$ is equal to 3; therefore,

$$S^{30} = S^{32}S^{21}S^{10} \tag{3.20}$$

or

$$S^{03} = S^{01}S^{12}S^{23} \tag{3.21}$$

which could be checked with equation (3.10).

Example 3.4.3
Give the explicit form of S in Example 3.4.2, if body 1 has one degree of freedom, denoted by α_1 (rotation about the $\bar{n}_1^1$ axis), body 2 can rotate with respect to body 1 by β_2 (rotation about $\bar{n}_2^2$), and body 3 can only rotate about $\bar{n}_3^3$ by γ_3. (Note that a counterclockwise rotation is assumed for all rotations.)

Solution
The corresponding transformations are

$$S^{10} = \begin{bmatrix} 1 & 0 & 0 \\ 0 & \cos\alpha_1 & \sin\alpha_1 \\ 0 & -\sin\alpha_1 & \cos\alpha_1 \end{bmatrix} \tag{3.22}$$

$$S^{21} = \begin{bmatrix} \cos\beta_2 & 0 & -\sin\beta_2 \\ 0 & 1 & 0 \\ \sin\beta_2 & 0 & \cos\beta_2 \end{bmatrix} \tag{3.23}$$

$$S^{32} = \begin{bmatrix} \cos\gamma_3 & \sin\gamma_3 & 0 \\ -\sin\gamma_3 & \cos\gamma_3 & 0 \\ 0 & 0 & 1 \end{bmatrix} \tag{3.24}$$

Therefore,

$$\begin{aligned} S^{30} &= S^{32}S^{21}S^{10} \\ &= \begin{bmatrix} c\beta_2 c\gamma_3 & c\alpha_1 s\gamma_3 + s\alpha_1 s\beta_2 c\gamma_3 & s\gamma_3 s\alpha_1 - c\alpha_1 s\beta_2 c\gamma_3 \\ -s\gamma_3 c\beta_2 & c\alpha_1 c\gamma_3 - s\alpha_1 s\beta_2 s\gamma_3 & s\alpha_1 c\gamma_3 + c\alpha_1 s\beta_2 s\gamma_3 \\ s\beta_2 & -s\alpha_1 c\beta_2 & c\alpha_1 c\beta_2 \end{bmatrix} \end{aligned} \tag{3.25}$$

3.4.2 Time Derivatives of Transformation Matrices

In the course of the development of the angular acceleration of the first time derivative, transformation matrices will be required. Let the transformation matrix from frame K to reference frame J be given by S^{KJ}; hence

$$\frac{d[S^{KJ}]}{dt} = [\dot{S}^{KJ}] = \left[\omega^{KJ}\right]\left[S^{KJ}\right] \tag{3.26}$$

where ω is the skew matrix defined in Section 2.5. We note, however, that $\dot{S}^{KJ}$ is expressed in terms of a product of two known matrices. The skew matrix ω is composed of the components of the angular velocity of body K with respect to body J. From equation (3.26), we can express the skew matrix as

$$\left[\omega^{KJ}\right] = \left[\dot{S}^{KJ}\right]\left[S^{KJ}\right]^T \tag{3.27}$$

An explicit form of ω could be obtained by noting that

$$\left[S^{KJ}\right] = [R_\gamma][R_\beta][R_\alpha]$$
$$= \begin{bmatrix} c\gamma & -s\gamma & 0 \\ -s\gamma & c\gamma & 0 \\ 0 & 0 & 1 \end{bmatrix} \begin{bmatrix} c\beta & 0 & -s\beta \\ 0 & 1 & 0 \\ s\beta & 0 & c\beta \end{bmatrix} \begin{bmatrix} 1 & 0 & 0 \\ 0 & c\alpha & s\alpha \\ 0 & -s\alpha & c\alpha \end{bmatrix} \tag{3.28}$$

which is the same as equation (3.25), where R_γ, R_β and R_α are the corresponding rotation about $\bar{n}_3$, $\bar{n}_2$ and $\bar{n}_1$ axes in body K, respectively. Differentiation of equation (3.28) and substitution into equation (3.27) yields

$$\left[\omega^{KJ}\right] = [\dot{R}_\gamma][R_\gamma]^T + [R_\gamma][\dot{R}_\beta]\left[R_\beta^T\right]\left[R_\gamma^T\right] + [R_\gamma][R_\beta][\dot{R}_\alpha]\left[R_\alpha^T\right]\left[R_\beta^T\right]\left[R_\gamma^T\right]$$
$$= \begin{bmatrix} 0 & \dot{\alpha}s\beta + \dot{\gamma} & \dot{\alpha}s\gamma c\beta - \dot{\beta}c\gamma \\ -\dot{\alpha}s\beta - \dot{\gamma} & 0 & \dot{\alpha}c\beta c\gamma + \dot{\beta}s\gamma \\ -\dot{\alpha}s\gamma c\beta + \dot{\beta}c\gamma & -\dot{\alpha}c\beta c\gamma - \dot{\beta}s\gamma & 0 \end{bmatrix} \tag{3.29}$$

Consider Figure 3.4 again; the skew matrix relating the time derivative of the transformation matrix and the transformation matrix between body B_1 and the reference frame R denoted by ω^{10} is obtained using equation (3.29). Similarly, we can obtain ω^{20} and ω^{30}. From equation (3.20), we note that

$$\left[\dot{S}^{30}\right] = \left[\omega^{30}\right]\left[S^{30}\right] \tag{3.30}$$

and

$$\left[\dot{S}^{20}\right] = \left[\omega^{20}\right]\left[S^{20}\right]$$

where

$$[\omega^{20}] = [\dot{S}^{20}][S^{02}] = \left([\dot{S}^{21}][S^{10}] + [S^{21}][\dot{S}^{10}]\right)[S^{02}]$$
$$= \left[\omega^{21}\right]\left[S^{21}\right]\left[S^{10}\right] + \left[S^{21}\right]\left[\omega^{10}\right]\left[S^{10}\right]\Big)\left[S^{02}\right] \tag{3.31}$$
$$= \left[\omega^{21}\right] + \left[S^{21}\right]\left[\omega^{10}\right]\left[S^{12}\right]$$

and

$$\left[\omega^{30}\right] = \left[\dot{S}^{30}\right]\left[S^{03}\right]$$
$$= \left(\left[\dot{S}^{32}\right]\left[S^{21}\right]\left[S^{10}\right] + \left[S^{32}\right]\left[\dot{S}^{21}\right]\left[S^{10}\right] + \left[S^{32}\right]\left[S^{21}\right]\left[\dot{S}^{10}\right]\right)\left[S^{03}\right]$$
$$= \left(\left[\omega^{32}\right]\left[S^{32}\right]\left[S^{21}\right]\left[S^{10}\right] + \left[S^{32}\right]\left[\omega^{21}\right]\left[S^{21}\right]\left[S^{10}\right]\right.$$
$$\left. + \left[S^{32}\right]\left[S^{21}\right]\left[\omega^{10}\right]\left[S^{10}\right]\right)\left[S^{03}\right]$$
$$= \left[\omega^{32}\right] + \left[S^{32}\right]\left[\omega^{21}\right]\left[S^{23}\right] + \left[S^{31}\right]\left[\omega^{10}\right]\left[S^{13}\right] \tag{3.32}$$

ω^{10}, ω^{21} and ω^{32} are all obtained through equation (3.29) using the proper transformation matrices.

More important, we can express the time derivative of the transformation matrix between any body B_k corresponding to a multibody structure as

$$\left[\dot{S}^{k0}\right] = \left[\omega^{k0}\right]\left[S^{k0}\right] \tag{3.33}$$

where the elements of ω^{k0} are

$$\omega_{im}^{k0} = e_{\min} y_{kn} \tag{3.34}$$

where y_{kn} are the components of the angular velocity of body k and R, and e_{imm} is the permutation symbol.

Example 3.4.4
From equations (3.34) and (3.29), find the components of the angular velocity $\{y_k\}$.

Solution
The components of angular velocity of body k in R can be obtained from the skew matrix given in equation (3.29) and represented in matrix form as

$$\{y_k\} = \begin{bmatrix} \dot{\alpha} c\beta c\gamma + \dot{\beta} s\dot{\gamma} \\ -\dot{\alpha} s\gamma c\beta + \dot{\beta} c\gamma \\ \dot{\alpha} s\beta + \dot{\gamma} \end{bmatrix} \tag{3.35}$$

3.5 Angular Velocities and Partial Angular Velocities

Consider two rigid bodies B_k and B_j to be interconnected to one another by a free joint; the angular velocity of B_k reference frame $\bar{n}_i^k$ $(i = 1, 2, 3)$ (fixed in B_k) with respect to $\bar{n}_i^j$ $(i = 1, 2, 3)$ (fixed in B_j) can be expressed as

$${}^{j}\bar{\omega}^{k} = \dot{\alpha}_k \bar{n}_1^j + \dot{\beta}_k \bar{n}_2^j + \dot{\gamma}_k \bar{n}_3^j \tag{3.36}$$

We can write equation (3.36) in a matrix form as

$${}^{j}\bar{\omega}^{k} = \{\dot{x}_n\}^T \left\{\bar{n}^j\right\} = \{\dot{x}_k\}^T \left[S^{jo}\right]\{\bar{n}\} \tag{3.37}$$

where

$$\{\dot{x}_k\} = \left[\dot{\alpha}_k \dot{\beta}_k \dot{\gamma}_k\right]^T \tag{3.38}$$

Then,

$${}^{R}\bar{\omega}^{k} = {}^{R}\bar{\omega}^{1} + {}^{1}\bar{\omega}^{2} + \cdots + {}^{k-1}\bar{\omega}^{k} = \sum_{i=1}^{k} {}^{i-1}\bar{\omega}^{i} = \sum_{i-1}^{k} \{\dot{x}_i\}^T \left[s^{i-1,o}\right]\{\dot{n}\} \tag{3.39}$$

Using the representation of equation (3.37), we can write the angular velocity of any body B_k with respect to R in matrix form:

$$^{R}\bar{\omega}^{k} = \{\dot{x}\}^{T}\left[\omega^{k}\right]\{\bar{n}\} \tag{3.40}$$

The dimension of $\dot{x}$ and ω^k will depend on the number of bodies in the structure and the number of generalized coordinates used. For example, the partial angular velocities of B_1, B_2 and B_3 of Figure 3.4 are

$$\left[\omega^{1}\right] = \begin{bmatrix} I \\ 0 \\ 0 \end{bmatrix} \tag{3.41}$$

$$\left[\omega^{2}\right] = \begin{bmatrix} I \\ S^{10} \\ 0 \end{bmatrix} \tag{3.42}$$

$$\left[\omega^{3}\right] = \begin{bmatrix} I \\ S^{10} \\ S^{20} \end{bmatrix} \tag{3.43}$$

and $\{\dot{x}\} = [\dot{\alpha}_1, \dot{\beta}_1, \dot{\gamma}_1\ \dot{\alpha}_2, \dot{\beta}_2, \dot{\gamma}_2\ \dot{\alpha}_3, \dot{\beta}_3, \dot{\gamma}_3]^T$; I denotes a 3×3 identity matrix. For a system having n bodies, the partial angular velocity array of the k, the body is

$$\left[\omega^{k}\right] = \begin{bmatrix} I \\ S^{10} \\ S^{20} \\ \vdots \\ S^{k-1,0} \\ \vdots \\ 0 \end{bmatrix} \tag{3.44}$$

Note how for each k in equation (3.40) we can generate the corresponding array $\left[\omega^{k}\right]$. It is formed of block matrices suitable for computer implementation. Consider the multibody system given by Figure 3.1, where its tree array is given by Table 3.1. Then one can write the angular velocity of any body B_k of the structure with respect to the unit vectors fixed in R as

$$^{R}\bar{\omega}^{k} = \sum_{i=1}^{H(k)} {}^{s}\bar{\omega}^{p} \tag{3.45}$$

where

$$\begin{aligned} p &= \Gamma_{i=1}(k) \\ s &= \Gamma_{i}(k) \end{aligned} \tag{3.46}$$

For example, referring to Figure 3.1, ${}^{R}\bar{\omega}^{11}$ is found to be

$$\underbrace{{}^{R}\bar{\omega}^{11}}_{\substack{\text{absolute}\\ \text{angular}\\ \text{velocity}}} = \underbrace{{}^{8}\bar{\omega}^{11} + {}^{2}\bar{\omega}^{8} + {}^{1}\bar{\omega}^{2} + {}^{0}\bar{\omega}^{1}}_{\substack{\text{relative angular}\\ \text{velocities}}} \tag{3.47}$$

The s and p superscripts of ${}^{s}\bar{\omega}^{p}$ are found by making use of the array $\Gamma_i(11)$ and $\Gamma_{i-1}(11)$. Thus one can easily see that the partial angular velocity is simply given by the matrix S^{j0}. Indeed, if the relative angular velocity components are used as generalized speeds x_ℓ, the angular velocity expression leads to a simple form of equation (3.40). Once more $x_\ell(\ell = 1, \ldots, 6N)$ is partitioned such that, it will correlate with the representation of the generalized coordinates described in Section 3.5.1. What remains is to give an expression for the partial angular velocity for any body B_k in a multibody system, such as the one depicted by Figure 3.1. An illustration of equation (3.40) can be seen in the building of a block array $[\omega^k]$ for the configuration of the multibody system given by Figure 3.1. Table 3.5 gives an explicit form of $[\omega^k]$ for all the structures, where empty spaces denote zero matrices.

Example 3.5.1

Give the partial velocity array for body 3 of Figure 3.5, if the generalized speeds are chosen to be the components of the relative angular velocities and each joint is three-dimensional.

Solution

From equation (3.47), we write the angular velocity of body 3 with respect to body 2 as

$$ {}^{2}\bar{\omega}^{3} = \dot{\alpha}_3\bar{n}_1^2 + \dot{\beta}_3\bar{n}_2^2 + \dot{\gamma}_3\bar{n}_3^2 \tag{3.48}$$

or

$$ {}^{2}\bar{\omega}^{3} = \{\dot{x}_3\}^T\{\bar{n}^2\} = [\dot{\alpha}_3\dot{\beta}_3\dot{\gamma}_3][S^{20}]\bar{n} \tag{3.49}$$

Similarly, we can express the angular velocity of body 2 with respect to body 1:

$$ {}^{1}\bar{\omega}^{2} = \{\dot{x}_2\}^T\left\{\bar{n}^1\right\} = \left[\dot{\alpha}_2\dot{\beta}_2\,\dot{\gamma}_2\right]\left[S^{10}\right]\ \{\bar{n}\} \tag{3.50}$$

and the angular velocity of body 1 with respect to R:

$$ {}^{0}\bar{\omega}^{1} = \{\dot{x}_1\}^T\left\{\bar{n}^1\right\} = \left[\dot{\alpha}_1\dot{\beta}_1\dot{\gamma}_1\right]^T\ \{\bar{n}\} \tag{3.51}$$

Finally, we obtain the absolute angular velocity of body 3 in R:

$$ {}^{R}\bar{\omega}^{3} = {}^{o}\bar{\omega}^{1} + {}^{1}\bar{\omega}^{2} + {}^{2}\bar{\omega}^{3} \tag{3.52}$$

Also, from equation (3.44) the angular velocity of body 3 in R is given by

$$ {}^{R}\bar{\omega}^{3} = \{\dot{x}\}^T\left[\omega^3\right]\{\bar{n}\} \tag{3.53}$$

ℓ k	B_1 1,2,3	B_2 4,5,6	B_3 7,8,9	B_4 10,11,12	B_5 13,14,15	B_6 16,17,18	B_7 19,20,21	B_8 22,23,24	B_9 25,26,27	B_{10} 28,29,30	B_{11} 31,32,33	B_{12} 34,35,36
$[\omega^1_{\ell m}]$	I											
$[\omega^2_{\ell m}]$	I	S^{10}										
$[\omega^3_{\ell m}]$	I	S^{10}	S^{20}									
$[\omega^4_{\ell m}]$	I	S^{10}	S^{20}	S^{30}								
$[\omega^5_{\ell m}]$	I	S^{10}	S^{20}	S^{30}	S^{40}							
$[\omega^6_{\ell m}]$	I					S^{10}						
$[\omega^7_{\ell m}]$	I					S^{10}	S^{60}					
$[\omega^8_{\ell m}]$	I	S^{10}						S^{20}				
$[\omega^9_{\ell m}]$	I	S^{10}						S^{20}	S^{80}			
$[\omega^{10}_{\ell m}]$	I	S^{10}						S^{20}	S^{80}	S^{90}		
$[\omega^{11}_{\ell m}]$	I	S^{10}						S^{20}			S^{80}	
$[\omega^{12}_{\ell m}]$	I	S^{10}						S^{20}			S^{80}	S^{110}

Table 3.5. Table for three arbitrary rigid bodies with spherical joints

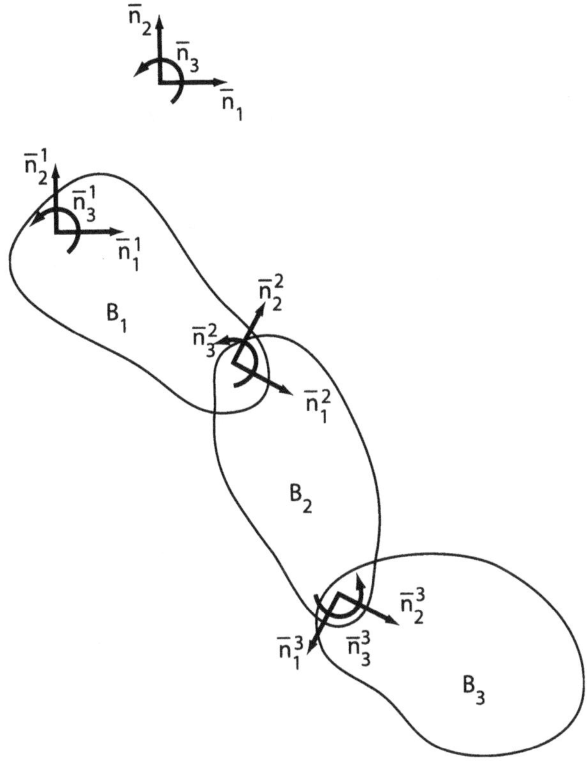

Fig. 3.5. Three arbitrary rigid bodies with spherical joints

where

$$\{\dot{x}\}^T = \begin{bmatrix} \dot{\alpha}_1 & \dot{\beta}_1 & \dot{\gamma}_1 & \dot{\alpha}_2 & \dot{\beta}_2 & \dot{\gamma}_2 & \dot{\alpha}_3 & \dot{\beta}_3 & \dot{\gamma}_3 \end{bmatrix} \tag{3.54}$$

and

$$\left[\omega^3\right] = \begin{bmatrix} I \\ S^{10} \\ S^{20} \end{bmatrix} \tag{3.55}$$

From equation (3.25) we know that

$$\left[S^{21}\right] = \begin{bmatrix} c\beta_2 c\gamma_2 & c\alpha_2 s\gamma_2 + s\alpha_2 s\beta_2 c\gamma_2 & s\alpha_2 s\gamma_2 - c\alpha_2 s\beta_2 c\gamma_2 \\ -c\beta_2 s\gamma_2 & c\alpha_2 c\gamma_2 - s\alpha_2 s\beta_2 s\gamma_2 & c\alpha_2 s\beta_2 s\gamma_2 + s\alpha_2 c\gamma_2 \\ s\beta_2 & -s\alpha_2 c\beta_2 & c\alpha_2 c\beta_2 \end{bmatrix} \tag{3.56}$$

$$\left[S^{10}\right] = \begin{bmatrix} c\beta_1 c\gamma_1 & c\alpha_1 s\gamma_1 + s\alpha_1 s\beta_1 c\gamma_1 & s\alpha_1 s\gamma_1 - c\alpha_1 s\beta_1 c\gamma_1 \\ -c\beta_1 s\gamma_1 & c\alpha_1 c\gamma_1 - s\alpha_1 s\beta_1 s\gamma_1 & c\alpha_1 s\beta_1 s\gamma_1 + s\alpha_1 c\gamma_1 \\ s\beta_2 & -s\alpha_1 c\beta_1 & c\alpha_1 c\beta_1 \end{bmatrix} \tag{3.57}$$

Then we get

$$\left[S^{20}\right] = \left[S^{21}\right]\left[S^{10}\right] \tag{3.58}$$

Table 3.6 gives an explicit form of $\left[\omega^k\right]$ for all three bodies of Figure 3.5

ℓ k	B_1 1,2,3	B_2 4,5,6	B_3 7,8,9
$[\omega^1_{\ell m}]$	I	0	0
$[\omega^2_{\ell m}]$	I	S^{10}	0
$[\omega^3_{\ell m}]$	I	S^{10}	S^{20}

Table 3.6. $\omega^k_{\ell m}$ for Figure 3.6

Example 3.5.2
Consider three rigid bodies interconnected to one another as shown in Figure 3.6. Body B_1 is pinned to the ground and has one degree of freedom γ_1 (rotation about $\bar{n}^1_3$). Bodies B_2 and B_3 have spherical joints and therefore have three degrees of freedom each: α_2, β_2, γ_2 and α_3, β_3, γ_3. Use the matrix notation to derive the corresponding expressions for ${}^R\bar{\omega}^1$, ${}^R\bar{\omega}^2$ and ${}^R\bar{\omega}^3$, where R is an inertial fixed frame.

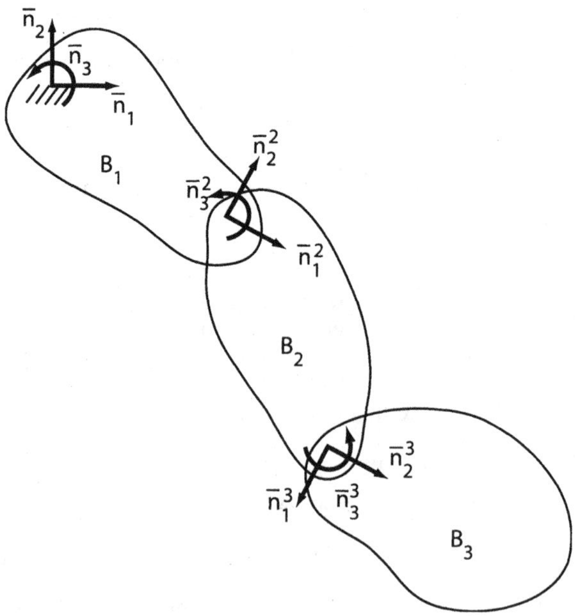

Fig. 3.6. Three arbitrary rigid bodies with body B_1 pinned to the ground

Solution
The generalized coordinates are

$$\{x\}^T = \left[0,\ 0,\ \gamma_1,\ \alpha_2,\ \beta_2,\ \gamma_2,\ \alpha_3,\ \beta_3,\ \gamma_3\right] \tag{3.59}$$

The partial angular velocity array for the three bodies of the system of Figure 3.5 is

$$[\omega] = \begin{bmatrix} I & I & I \\ 0 & S^{10} & S^{10} \\ \underbrace{0}_{[\omega^1]} & \underbrace{0}_{[\omega^2]} & \underbrace{S^{20}}_{[\omega^3]} \end{bmatrix} \tag{3.60}$$

where,

$$\left[S^{10}\right] = \begin{bmatrix} c\gamma_1 & s\gamma_1 & 0 \\ -s\gamma_1 & c\gamma_1 & 0 \\ 0 & 0 & 1 \end{bmatrix} \tag{3.61}$$

$$\left[S^{20}\right] = \left[S^{21}\right]\left[S^{10}\right]$$

$$= \begin{bmatrix} c\gamma_2 & s\gamma_2 & 0 \\ -s\gamma_2 & c\gamma_2 & 0 \\ 0 & 0 & 1 \end{bmatrix} \begin{bmatrix} c\beta_2 & 0 & -s\beta_2 \\ 0 & 1 & 0 \\ s\beta_2 & 0 & c\beta_2 \end{bmatrix} \begin{bmatrix} 1 & 0 & 0 \\ 0 & c\alpha_2 & s\alpha_2 \\ 0 & -s\alpha_2 & c\alpha_2 \end{bmatrix} \begin{bmatrix} c\gamma_1 & s\gamma_1 & 0 \\ -s\gamma_1 & c\gamma_1 & 0 \\ 0 & 0 & 1 \end{bmatrix}$$

$$= \left[\begin{matrix} c\beta_1 c\gamma_1 c\alpha_1 - s\gamma_1(c\alpha_2 s\gamma_2 + s\alpha_2 s\beta_2 s\gamma_2) \\ -c\beta_1 s\gamma_1 c\alpha_1 - s\gamma_1(c\alpha_2 c\gamma_2 - s\alpha_2 s\beta_2 c\gamma_2) \\ s\beta_2\gamma_1 + s\gamma_1 s\alpha_2 c\beta_2 \end{matrix} \right.$$

$$\left. \begin{matrix} c\beta_1 c\gamma_1 c\alpha_1 - s\gamma_1(c\alpha_2 s\gamma_2 + s\alpha_2 s\beta_2 c\gamma_2) & s\alpha_2 s\gamma_2 - c\alpha_2 s\beta_2 c\gamma \\ -c\beta_1 s\gamma_1 c\alpha_1 - s\gamma_1(c\alpha_2 c\gamma_2 - s\alpha_2 s\beta_2 c\gamma_2) & c\alpha_2 s\beta_2 s\gamma_2 + s\alpha_2 c\gamma \\ s\beta_2\gamma_1 + s\gamma_1 s\alpha_2 c\beta_2 & c\alpha_2 c\beta_2 \end{matrix} \right] \tag{3.62}$$

Substitution of $\left[S^{10}\right]$ and $\left[S^{20}\right]$ in the partial angular velocity array given by equation (3.60) gives the expression for partial angular velocity of bodies 1, 2 and 3. Subsequently, substitution of $\left[\omega^1\right]$, $\left[\omega^2\right]$ and $\left[\omega^3\right]$ in the equations given below leads to the angular velocities of bodies 1, 2 and 3 with respect to reference frame R.

$$^R\bar{\omega}^1 = \{\dot{x}\}^T \left[\omega^1\right] \{\bar{n}\} \tag{3.63}$$

$$^R\bar{\omega}^2 = \{\dot{x}\}^T \left[\omega^2\right] \{\bar{n}\} \tag{3.64}$$

$$^R\bar{\omega}^3 = \{\dot{x}\}^T \left[\omega^3\right] \{\bar{n}\} \tag{3.65}$$

3.6 Angular Accelerations

The angular acceleration of a particular rigid body k in a multibody system is given by simply differentiating in time equation (3.40). Hence denoting the angular acceleration of body k in R by ${}^R\bar{\alpha}^k$, we obtain

$$ {}^R\bar{\alpha}^k = \left(\{\ddot{x}\}^T \left[\omega^k \right] + \{\dot{x}\}^T \left[\dot{\omega}^k \right] \right) \{\bar{n}\} \tag{3.66} $$

where $\dot{\omega}^k$ is the rate of change (first time derivative) of ω^k. From equation (3.66), one can make the following observation. First there are two terms in the angular acceleration, where one requires the formulation of the partial angular velocity array ω^k and its time derivative $\dot{\omega}^k$. Second, we notice that ω^k has already been developed for ${}^R\bar{\omega}^k$; therefore, we only need $\dot{\omega}^k$. The development of $\dot{\omega}^k$ is shown using Table 3.6. One should not forget that the derivatives of the transformation matrices could be expressed as a product of two matrices, as shown in equation (3.33). Table 5.1 gives the corresponding ω^k for Table 3.6.

ℓ k	B_1 1,2,3	B_2 4,5,6	B_3 7,8,9
$[\dot{\omega}^1_{\ell m}]$	0	0	0
$[\dot{\omega}^2_{\ell m}]$	0	S^{10}	0
$[\dot{\omega}^3_{\ell m}]$	0	$\dot{S}^{10}$	$\dot{S}^{20}$

Table 3.7. $\dot{\omega}^k_{\ell m}$ for Figure 3.6

Example 3.6.1
Find an expression for the angular acceleration of body 3, shown in Figure 3.6.

Solution
The angular acceleration of body 3 is given by

$$ {}^R\dot{\bar{\omega}}^3 = \left(\{\ddot{x}\}^T \left[\omega^3 \right] + \{\dot{x}\}^T \left[\dot{\omega}^3 \right] \right) \{\bar{n}\} \tag{3.67} $$

or simply

$$ {}^R\bar{\alpha}^3 = \{\alpha\}^T \{\bar{n}\} \tag{3.68} $$

where

$$ \{\alpha\} = \{\ddot{x}\}^T \left[\omega^3 \right] + \{\dot{x}\}^T \left[\dot{\omega}^3 \right] \tag{3.69} $$

where $\left[\omega^3\right]$ is given by equation (3.43), and

$$ \left[\dot{\omega}^3 \right] = \begin{bmatrix} 0 \\ \omega^{10} S^{10} \\ \omega^{20} S^{20} \end{bmatrix} \tag{3.70} $$

where $\left[\omega^{10}\right]$ is given by equation (3.29) and $\left[\omega^{20}\right]$ by equation (3.31).

Example 3.6.2
Consider the robot arm shown in Figure 3.7. Find the angular velocity and angular acceleration of arm AB in R.

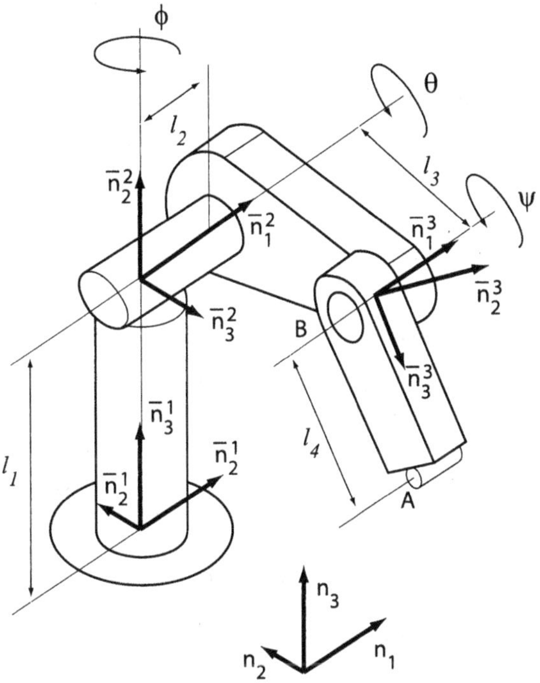

Fig. 3.7. Robot system and its axis of rotation

Solution
The transformation matrices of bodies 1 and 2 with respect to the reference frame R are given by S^{10} and S^{20}, respectively:

$$S^{10} = \begin{bmatrix} c\phi & s\phi & 0 \\ s\phi & c\phi & 0 \\ 0 & 0 & 1 \end{bmatrix} \tag{3.71}$$

$$S^{20} = S^{21}S^{10}$$

$$= \begin{bmatrix} 1 & 0 & 0 \\ 0 & c\theta & s\theta \\ 0 & -s\theta & c\theta \end{bmatrix} \cdot \begin{bmatrix} c\phi & s\phi & 0 \\ -s\phi & c\phi & 0 \\ 0 & 0 & 1 \end{bmatrix} = \begin{bmatrix} c\phi & s\phi & 0 \\ -c\theta s\phi & c\theta c\phi & s\theta \\ s\theta c\phi & -s\theta c\phi & c\theta \end{bmatrix} \tag{3.72}$$

For arm AB denoted as body 3, the angular velocity is

$$^{R}\bar{\omega}^{3} = \dot{\phi}\,\bar{n}_3^1 + \dot{\theta}\bar{n}_1^2 + \dot{\psi}\bar{n}_1^3 \tag{3.73}$$

where

$$\bar{n}_1^3 = \bar{n}_1^2 \text{ and } \bar{n}_3^1 = \bar{n}_3$$

Expressing the angular velocity with respect to the unit vectors in R, we obtain

$$\begin{aligned} {}^R\bar{\omega}^3 &= \dot{\phi}\bar{n}_3 + \dot{\theta}(c\phi\bar{n}_1 + s\phi\bar{n}_2) + \dot{\psi}(c\phi\bar{n}_1 + s\phi\bar{n}_2) \\ &= (\dot{\theta} + \dot{\psi})c\phi\,\bar{n}_1 + (\dot{\theta} + \dot{\psi})s\phi\,\bar{n}_2 + \dot{\phi}\,\bar{n}_3 \end{aligned}$$

The angular acceleration is then found to be

$$\begin{aligned} {}^R\bar{\alpha}^3 &= \frac{d}{dt}{}^R\bar{\omega}^3 \\ &= [(\ddot{\theta} + \ddot{\psi})c\phi - (\dot{\theta} + \dot{\psi})\dot{\phi}\,s\phi]\bar{n}_1 + [(\ddot{\theta} + \ddot{\psi})s\phi + (\dot{\theta} + \dot{\psi})\dot{\phi}\,c\phi]\bar{n}_2 + \ddot{\phi}\,\bar{n}_3 \end{aligned} \tag{3.74}$$

3.7 Generalized Speeds

The generalized speeds are a convenient form of representing the velocities in multibody systems. They allow the user to represent the velocity of an arbitrary point or center of mass of a rigid body as a function of new generalized speeds. These generalized speeds are related to the generalized coordinate derivatives through a transformation matrix [w] called the gimbal matrix. The generalized speeds are defined by a new set of generalized coordinate systems at the velocity level to make the representation of the dynamic equations simpler to follow. In essence, the generalized speeds are the result of a mapping or a transformation in a fixed reference frame R between the generalized coordinates and their derivatives.

$$\{y\} = [\mathrm{w}] \quad \{\dot{x}\} \tag{3.75}$$

where $\{y\}$ is the generalized speed vector and $\{\dot{x}\}$ is the generalized coordinate derivatives and [w] is the transformation matrix mapping the two vectors.

In the context of multibody dynamics, following the procedure outlined in this chapter and thereafter, a general form of the gimbal matrix [w] is defined, which can be used in the kinematics and dynamic formulation of multibody systems.

Let the relationship between the generalized coordinate derivatives and the generalized speeds be defined as follows

$$\{y\} = \{\dot{x}\}^T\,[\mathrm{w}] \tag{3.76}$$

Where the mapping function w is given by

$$[\mathrm{w}] = \begin{bmatrix} [S^{10}] & [S^{10}] & [S^{10}] & \dots & [S^{10}] \\ [0_{3\times3}] & [S^{20}] & [S^{20}] & \dots & [S^{20}] \\ [0_{3\times3}] & [0_{3\times3}] & [S^{30}] & \dots & [S^{30}] \\ & & & \ddots & \vdots \\ [0_{3\times3}] & [0_{3\times3}] & [0_{3\times3}] & \cdots & [S^{n0}] \end{bmatrix} \tag{3.77}$$

In the above matrix, S_{ij} denotes the transformation matrix between body j and body i, $0_{3\times3}$ is a zero matrix and I is the identity matrix.

Proof
In general multibody system, we define the angular velocity of body n with respect to R as follows:

$$^{R}\bar{\omega}^{n} = {}^{R}\bar{\omega}^{1}\left\{\bar{n}^{1}\right\} + {}^{1}\bar{\omega}^{2}\left\{\bar{n}^{2}\right\} + \cdots + {}^{(n-1)}\bar{\omega}^{n}\left\{\bar{n}^{n}\right\} \tag{3.78}$$

Expressing each unit body vector with respect to a fixed inertia frame R we obtain

$$^{R}\bar{\omega}^{n} = {}^{R}\bar{\omega}^{1}[s^{10}]\{\bar{n}\} + {}^{1}\bar{\omega}^{2}[s^{20}]\{\bar{n}\} + \cdots + {}^{(n-1)}\bar{\omega}^{n}[s^{n0}]\{\bar{n}\} \tag{3.79}$$

where $\overline{n}$ is the vector of the inertial reference frame R. Let α_i, β_i, γ_i be the angular rotation about each of the body i axis. These angular displacements represent the generalized coordinates of the ith body in the inertial reference frame R.

Equation (3.79) can be expressed further as

$$^{R}\bar{\omega}^{n} = \left\{[\dot{\alpha}_1\ \dot{\beta}_1\ \dot{\gamma}_1]\,[s^{10}] + [\dot{\alpha}_2\ \dot{\beta}_2\ \dot{\gamma}_2]\,[s^{20}] + \cdots + [\dot{\alpha}_n\ \dot{\beta}_n\ \dot{\gamma}_n]\,[s^{n0}]\right\}\ \{\overline{n}\}$$

$$^{R}\bar{\omega}^{n} = \left[\dot{\alpha}_1\ \dot{\beta}_1\ \dot{\gamma}_1\ \ \dot{\alpha}_2\ \dot{\beta}_2\ \dot{\gamma}_2 \cdots \dot{\alpha}_n\ \dot{\beta}_n\ \dot{\gamma}_n\right] \begin{bmatrix} [s^{10}] \\ [s^{20}] \\ \vdots \\ [s^{n0}] \end{bmatrix} \{\overline{n}\} \tag{3.80}$$

In a compact form, making use of the generalized coordinate, we write

$$\{y_n\}^T = \{\dot{x}\}^T \begin{bmatrix} [s^{10}] \\ [s^{20}] \\ \vdots \\ [s^{n0}] \end{bmatrix} \tag{3.81}$$

It follows that the generalized speeds for a multibody system composed of n bodies can be expressed as:

$$\{y\}^T = \{\dot{x}\}^T \begin{bmatrix} [s^{10}] & [s^{10}] & \cdots & [s^{10}] \\ [0_{3\times3}] & [s^{20}] & \cdots & [s^{20}] \\ \vdots & [0_{3\times3}] & \ddots & \vdots \\ [0_{3\times3}] & \cdots & \cdots & [s^{n0}] \end{bmatrix}_{3n\times3n} \tag{3.82}$$

Recall that $\{x\}$ denotes the generalized coordinates of the whole system, therefore the zero entries in the gimbal matrix allow for omission of the generalized coordinates that are not associated with a particular body. If any element in $\dot{x}$ is zero, then the corresponding row in [w] is useless and can be set to zero. Equation (3.82) can be expressed further in a general form as

$$\{y\}^T = \{\dot{x}\}^T\,[\mathrm{w}] \tag{3.83}$$

where [w] is defined by equation (3.77) and (3.82). Note that generalized speeds are represented in a matrix form ready for computer implementation.

Example 3.7.1

Consider Example 3.6.2 a 3-body robot. In this example, derive the angular velocity of body 3 with respect to R making use of equation (3.82). Essentially demonstrate the utility of equation (3.82).

Solution

The robot has 3 degree of freedom defined as follows

$$\{\dot{x}\} = \begin{bmatrix} 0 & 0 & \dot{\phi}, & \dot{\theta} & 0 & 0, & \dot{\psi} & 0 & 0 \end{bmatrix} \tag{3.84}$$

and the corresponding shifter matrices for each body with respect to R are

$$S^{10} = \begin{bmatrix} c\phi & s\phi & 0 \\ -s\phi & c\phi & 0 \\ 0 & 0 & 1 \end{bmatrix}, \quad S^{20} = \begin{bmatrix} c\phi & s\phi & 0 \\ -c\theta.s\phi & c\theta.c\phi & s\theta \\ s\theta.c\phi & -s\theta.c\phi & c\theta \end{bmatrix} \tag{3.85}$$

$$S^{30} = \begin{bmatrix} 1 & 0 & 0 \\ 0 & c\psi & s\psi \\ 0 & -s\psi & c\psi \end{bmatrix} \begin{bmatrix} c\phi & s\phi & 0 \\ -c\theta.s\phi & c\theta.c\phi & s\theta \\ s\theta.c\phi & -s\theta.c\phi & c\theta \end{bmatrix} = \begin{bmatrix} c\phi & s\phi & 0 \\ S^{30}_{21} & S^{30}_{22} & S^{30}_{23} \\ S^{30}_{31} & S^{30}_{32} & S^{30}_{33} \end{bmatrix} \tag{3.86}$$

We define the gimbal matrix making use of the above transformation matrices, where

$$[w] = \begin{bmatrix} [s^{10}] & [s^{10}] & [s^{10}] \\ [0] & [s^{20}] & [s^{20}] \\ [0] & [0] & [s^{30}] \end{bmatrix} \tag{3.87}$$

Explicitly the generalized speeds are defined by equation (3.82) as

$$[y] = \begin{bmatrix} \begin{bmatrix} c\phi & s\phi & 0 \\ -s\phi & c\phi & 0 \\ 0 & 0 & 1 \end{bmatrix} & \begin{bmatrix} c\phi & s\phi & 0 \\ -s\phi & c\phi & 0 \\ 0 & 0 & 1 \end{bmatrix} & \begin{bmatrix} c\phi & s\phi & 0 \\ -s\phi & c\phi & 0 \\ 0 & 0 & 1 \end{bmatrix} \\ \begin{bmatrix} 0 & 0 & 0 \\ 0 & 0 & 0 \\ 0 & 0 & 0 \end{bmatrix} & \begin{bmatrix} c\phi & s\phi & 0 \\ S^{20}_{21} & S^{20}_{22} & S^{20}_{23} \\ S^{20}_{31} & S^{20}_{32} & S^{20}_{33} \end{bmatrix} & \begin{bmatrix} c\phi & s\phi & 0 \\ S^{20}_{21} & S^{20}_{22} & S^{20}_{23} \\ S^{20}_{31} & S^{20}_{32} & S^{20}_{33} \end{bmatrix} \\ \begin{bmatrix} 0 & 0 & 0 \\ 0 & 0 & 0 \\ 0 & 0 & 0 \end{bmatrix} & \begin{bmatrix} 0 & 0 & 0 \\ 0 & 0 & 0 \\ 0 & 0 & 0 \end{bmatrix} & \begin{bmatrix} c\phi & s\phi & 0 \\ S^{20}_{21} & S^{20}_{22} & S^{20}_{23} \\ S^{20}_{31} & S^{20}_{32} & S^{20}_{33} \end{bmatrix} \end{bmatrix} \begin{bmatrix} 0 \\ 0 \\ \dot{\phi} \\ \dot{\theta} \\ 0 \\ 0 \\ \dot{\psi} \\ 0 \\ 0 \end{bmatrix} \tag{3.88}$$

Making use of the $\{\dot{x}\}$ vector we eliminate the elements from [w] that do not contribute to the product of $\{\dot{x}\}^T$ [w] and [w] becomes

$$[\mathrm{w}] = \begin{bmatrix} 0 & 0 & 0 & 0 & 0 & 0 & 0 & 0 & 0 \\ 0 & 0 & 0 & 0 & 0 & 0 & 0 & 0 & 0 \\ 0 & 0 & 1 & 0 & 0 & 1 & 0 & 0 & 1 \\ & & & c\phi & s\phi & 0 & c\phi & s\phi & 0 \\ & 0_{3\times 3} & & 0 & 0 & 0 & 0 & 0 & 0 \\ & & & 0 & 0 & 0 & 0 & 0 & 0 \\ & & & & & & c\phi & s\phi & 0 \\ & 0_{3\times 3} & & & 0_{3\times 3} & & 0 & 0 & 0 \\ & & & & & & 0 & 0 & 0 \end{bmatrix} \tag{3.89}$$

The above matrix is then used to obtain the generalized speeds of the system

$$\{y\}^T = \left\{\dot{x}\right\}^T [\mathrm{w}] = \begin{bmatrix} 0 \\ 0 \\ \dot{\phi} \\ \dot{\theta} \\ 0 \\ 0 \\ \dot{\psi} \\ 0 \\ 0 \end{bmatrix}^T \begin{bmatrix} 0 & 0 & 0 & 0 & 0 & 0 & 0 & 0 & 0 \\ 0 & 0 & 0 & 0 & 0 & 0 & 0 & 0 & 0 \\ 0 & 0 & 1 & 0 & 0 & 1 & 0 & 0 & 1 \\ & & & c\phi & s\phi & 0 & c\phi & s\phi & 0 \\ & 0_{3\times 3} & & 0 & 0 & 0 & 0 & 0 & 0 \\ & & & 0 & 0 & 0 & 0 & 0 & 0 \\ & & & & & & c\phi & s\phi & 0 \\ & 0_{3\times 3} & & & 0_{3\times 3} & & 0 & 0 & 0 \\ & & & & & & 0 & 0 & 0 \end{bmatrix}$$

$$= \begin{bmatrix} 0 \\ 0 \\ \dot{\phi} \\ \dot{\theta}\, c\phi \\ \dot{\theta}\, s\phi \\ \dot{\phi} \\ (\dot{\theta} + \dot{\psi})c\phi \\ (\dot{\theta} + \dot{\psi})s\phi \\ \dot{\phi} \end{bmatrix} \tag{3.90}$$

The last three components of $\{y\}$ represent the angular velocity components of body 3 with respect to R. Therefore we can write the angular velocity of body 3 as

$${}^{R}\bar{\omega}^{3} = (\dot{\theta} + \dot{\psi})c\phi\{\bar{n}_1\} + (\dot{\theta} + \dot{\psi})s\phi\{\bar{n}_2\} + \dot{\phi}\{\bar{n}_3\} \tag{3.91}$$

which is exactly the result obtained in Example 3.8.2 for the angular velocity of body 3 with respect to R.

Example 3.7.2

For further proof of the 2-body (2-D) system, consider Example 3.9.2, where we need to define the angular velocity of body 2 with respect to R making use of equation (3.82).

Solution

The transformation matrices of bodies 1 and 2 with respect to the reference frame R are given by S^{10} and S^{20} respectively

$$S^{10} = \begin{bmatrix} s\theta_1 & -c\theta_1 & 0 \\ c\theta_1 & s\theta_1 & 0 \\ 0 & 0 & 1 \end{bmatrix} \tag{3.92}$$

$$S^{20} = \begin{bmatrix} s(\theta_1+\theta_2) & -c(\theta_1+\theta_2) & 0 \\ c(\theta_1+\theta_2) & s(\theta_1+\theta_2) & 0 \\ 0 & 0 & 1 \end{bmatrix} \tag{3.93}$$

the generalized coordinate derivatives are

$$\{\dot{x}\}^T = [0 \ 0 \ \dot{\theta}_1 \ 0 \ 0 \ \dot{\theta}_2] \tag{3.94}$$

Following the derivation of the gimbal matrix [w], we express the matrix as follows

$$[\dot{\mathrm{w}}] = \begin{bmatrix} \begin{bmatrix} s_1 & c_1 & 0 \\ c_1 & s_1 & 0 \\ 0 & 0 & 1 \end{bmatrix} & \begin{bmatrix} s_1 & c_1 & 0 \\ c_1 & s_1 & 0 \\ 0 & 0 & 1 \end{bmatrix} \\ \begin{bmatrix} & & \\ & 0_{3\times 3} & \\ & & \end{bmatrix} & \begin{bmatrix} s_{12} & -c_{12} & 0 \\ c_{12} & s_{12} & 0 \\ 0 & 0 & 1 \end{bmatrix} \end{bmatrix} \tag{3.95}$$

The generalized speeds are then obtained

$$\{y\}^T = \{\dot{x}\}^T [\mathrm{w}] = \begin{bmatrix} 0 \\ 0 \\ \dot{\theta} \\ 0 \\ 0 \\ \dot{\theta}_2 \end{bmatrix} \begin{bmatrix} 0 & 0 & 0 & 0 & 0 & 0 \\ 0 & 0 & 0 & 0 & 0 & 0 \\ 0 & 0 & 1 & 0 & 0 & 1 \\ & & & 0 & 0 & 0 \\ & 0_{3\times 3} & & 0 & 0 & 0 \\ & & & 0 & 0 & 1 \end{bmatrix} \tag{3.96}$$

$$= \left[0 \ \ 0 \ \ \dot{\theta}, \ \ 0 \ \ 0 \ \ \dot{\theta}_1 + \dot{\theta}_2\right]$$

Again the results match exactly the results obtained in Example 3.9.2.

Note: The components of the angular velocities due to their representation through the shifters and generalized coordinate derivatives played a major role in defining the generalized speeds used in the derivation and representation of the velocities and subsequently the acceleration of the rigid bodies. Furthermore, the equations of motion will be shown to be much simpler when expressed in terms of $\{y\}$ and $\{\dot{y}\}$.

3.8 Mass Center Velocities of Arbitrary Interconnected Rigid Bodies in Treelike Systems

To develop the expressions for the mass center velocities of the rigid bodies of a given multibody system, we need to define their relative position vectors. To this end, consider Figure 3.8. Let each rigid body be connected to its adjoining body by a spherical joint. To define the position vectors, we locate the center of the reference frames fixed in each body and label them O_1, O_2 and O_3. Those points are located at the connecting joints. Let $\bar{q}_1$, $\bar{q}_2$ and $\bar{q}_3$ be the position vectors locating these reference points relative to the reference points of the adjoining lower connecting body. Ultimately, let $\bar{r}_1$, $\bar{r}_2$ and $\bar{r}_3$ locate the mass centers G_i $(i = 1, 2, 3)$ relative to their reference points. If we denote by $\bar{P}i$ $(i = 1, 2, 3,)$ the position vectors locating the mass centers G_i relative to the origin O of the inertial reference frame R, we obtain

$$\bar{P}_1 = \bar{\xi}_1 + \bar{r}_1 \tag{3.97}$$

$$\bar{P}_2 = \bar{\xi}_1 + \bar{q}_2 + \bar{r}_2 \tag{3.98}$$

$$\bar{P}_3 = \bar{\xi}_1 + \bar{q}_2 + \bar{q}_3 + \bar{r}_3 \tag{3.99}$$

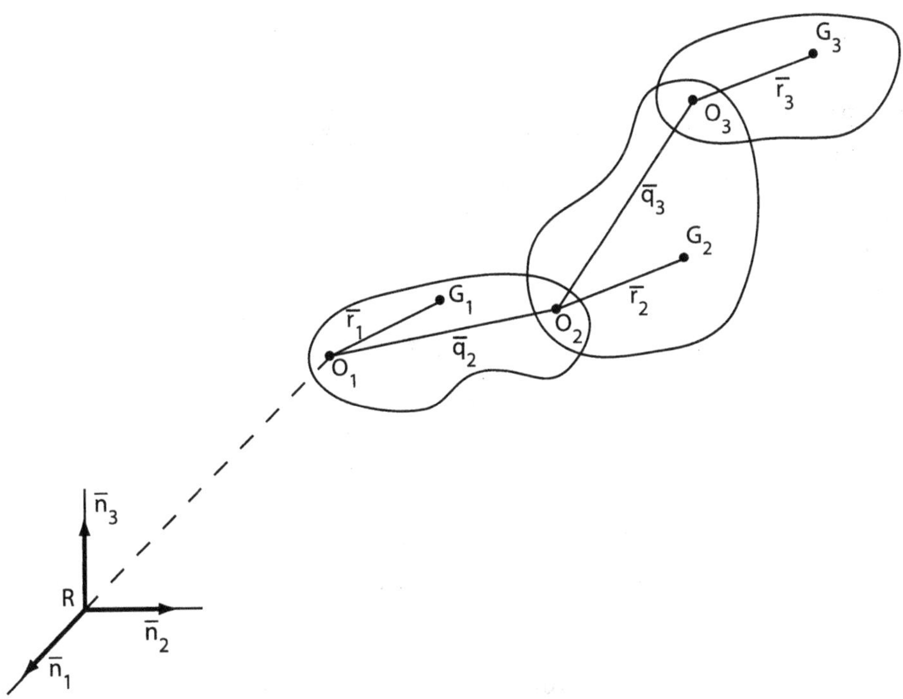

Fig. 3.8. Three arbitrary interconnected rigid bodies with no translation allowed at joints O_2 and O_3

Let $\bar{n}_i^1$, $\bar{n}_i^2$ and $\bar{n}_i^3 = (i = 1, 2, 3)$ be the mutually perpendicular unit vectors at points O_1, O_2 and O_3 respectively; then the relative position vectors used to describe $\bar{P}_i$ $(i = 1, 2, 3)$ can be expressed as

$$\bar{q}_1 = 0$$

$$\bar{q}_2 = q_{2m}\bar{n}_m^1 \tag{3.100}$$

$$\bar{q}_3 = q_{3m}\bar{n}_m^2 \tag{3.101}$$

$$\bar{r}_1 = r_{1m}\bar{n}_m^1 \tag{3.102}$$

$$\bar{r}_2 = r_{2m}\bar{n}_m^2 \tag{3.103}$$

$$\bar{r}_3 = r_{3m}\bar{n}_m^3 \tag{3.104}$$

and

$$\bar{\xi}_1 = \xi_{1m}\bar{n}_m \tag{3.105}$$

or in an index notation,

$$\bar{q}_j = q_{jm}\bar{n}_m^{j-1} \tag{3.106}$$

$$\bar{r}_j = r_{jm}\bar{n}_m^j \tag{3.107}$$

and

$$\bar{\xi}_1 = \xi_{1m}\bar{n}_m \quad (j = 1, 2, 3;\ m = 1, 2, 3) \tag{3.108}$$

The components q_{jm}, r_{jm} and ξ_{1m} are scalar quantities defining $\bar{q}$, $\bar{r}$ and $\bar{\xi}$ in their respective reference frames. Equations (3.97) to (3.99) can then be written in combined form as

$$\bar{P}_j = \bar{\xi}_1 + \sum_{i=1}^{j} \bar{q}_i + \bar{r}_j \tag{3.109}$$

where $q_1 = 0$. The position vectors $\bar{P}_j$ can be expressed with respect to the mutually perpendicular unit vectors $\bar{n}_m$ $(m = 1, 2, 3)$ fixed in R. Then it follows from the transformation matrices between unit vectors that

$$\bar{\xi}_i = \xi_{im}\bar{n}_m \tag{3.110}$$

$$\bar{r}_j = r_{jm}S_{nm}^{j0}\bar{n}_m \tag{3.111}$$

$$\bar{q}_j = q_{jn}S_{nm}^{j-1,0}\bar{n}_m \tag{3.112}$$

Substituting equations (3.110) to (3.112) into (3.109), we obtain

$$\bar{P}_1 = \xi_{1m}\bar{n}_m + \sum_{i=1}^{j} q_{in}S_{nm}^{i-1,0}\bar{n}_m + r_{jn}S_{nm}^{j0}\bar{n}_m \tag{3.113}$$

or use matrix notation,

$$\bar{P}_j = \left(\{\xi_1\}^T + \sum_{i=1}^{j} \{q_i\}^T \left[S^{i-1,0} \right] + \{r_j\}^T \left[S^{j0} \right] \right) \{\bar{n}\} \tag{3.114}$$

To define the mass center velocity of G_k in R, we take the first time derivative of $\bar{P}_j$; hence differentiation of equation (3.113) yields

$$\bar{v}_j = \frac{d\bar{P}_j}{dt} = \left[\dot{\xi}_{im} + \sum_{i=1}^{j} q_{in} \dot{S}_{nm}^{i-1,0} + r_{jn} \dot{S}_{nm}^{j0} \right] \bar{n}_m \tag{3.115}$$

Substituting the expression for $\dot{S}$ from equation (3.33) into (3.115), we obtain an explicit form of the mass center velocity of body J in R :

$$\bar{v}_j = \left[\dot{\xi}_{im} + \sum_{i=1}^{j} q_{in} \omega_{nt}^{i-1,0} S_{tm}^{i-1,0} + r_{jn} \omega_{nt}^{j0} S_{tm}^{j0} \right] \bar{n}_m \tag{3.116}$$

Recall the relationship between vector and skew matrix that was given in Section 3.2 and note that the sign is already considered in the skew matrix. We get:

$$\bar{v} = \left(\{\dot{\xi}_1\}^T + \sum_{i=1}^{j} \{y\}^T [Sq_i][S^{i-1,0}] + \{y\}^T [Sr_j][S^{j0}] \right) \{\bar{n}\} \tag{3.117}$$

where $\{y\}$ is the velocity coordinate associated with skew matrix ω, and $[Sq_i]$ and $[Sr_j]$ are the skew matrices of vectors q_i and r_j, which are given by

$$[Sq_i] = \begin{bmatrix} 0 & -q_{i3} & q_{i2} \\ q_{i3} & 0 & -q_{i1} \\ -q_{i2} & q_{i1} & 0 \end{bmatrix} \tag{3.118}$$

$$[Sr_j] = \begin{bmatrix} 0 & -r_{j3} & r_{j2} \\ r_{j3} & 0 & -r_{j1} \\ -r_{j2} & r_{j1} & 0 \end{bmatrix} \tag{3.119}$$

Except for $\dot{\xi}$, which takes care of any translation allowed between the inertial fixed frame R and body 1, the expression associated with the velocity of the center mass of body j can be expressed in compact form by factorizing the generalized coordinate derivates to yield

$$\bar{v}_j = (\{\dot{\xi}_1\}^T + \{y\}^T [V^j])\{\bar{n}\} \tag{3.120}$$

or

$$\bar{v}_j = (\{\dot{\xi}_1\}^T + \{\dot{x}\}^T [W][V^j])\{\bar{n}\} \tag{3.121}$$

where

$$[V^j] = \begin{bmatrix} [Sq_2][S^{10}] \\ [Sq_3][S^{20}] \\ \vdots \\ [Sq_j][S^{j-1,0}] \\ [Sr_j][S^{j0}] \\ 0 \end{bmatrix} \tag{3.122}$$

The matrix $[V^j]$ is known as the *partial velocity matrix* of $\bar{v}_j$ in R associated with the so-called generalized speeds $\{y\}$. This plays a major role in the formulation of the equations of motion, as we have the explicit form for its computation, and matrix $[W]$ plays a role in transforming $\{y\}$ to $\{\dot{x}\}$. Note that the equations above are suitable for any arbitrary point on body J when $\{r_j\}$ denotes the coordinate of the point $\{\bar{n}^j\}$.

For rigid bodies that have relative motion in between their joints in addition to their rotational degrees of freedom, their mass center velocities can be found similar to equation (3.117). For example, consider Figure 3.9 which is similar to Figure 3.8 except that we added translational vectors between relative bodies. The position vector to the mass center G_J $(J = 1, 2, 3)$ from the center O of R is given by modifying equation (3.109), where

$$\bar{P}_j = \sum_{i=1}^{j} \bar{\xi}_i + \sum_{i=1}^{j} \bar{q}_i + \bar{r}_j = \sum_{i=1}^{j} (\bar{\xi}_i + \bar{q}_i) + \bar{r}_j \tag{3.123}$$

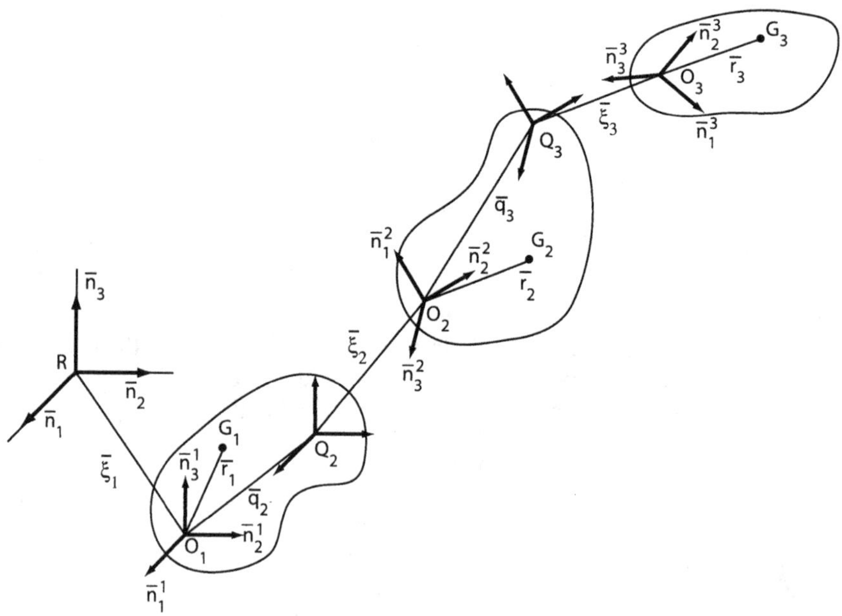

Fig. 3.9. Three arbitrary rigid bodies with all possible motion at each joint

Observe that equation (3.109) and (3.122) are similar except that there is a summation on the translation vectors as well. Following the procedures outlined for the derivation of the mass center velocity G_j, we write the mass center velocity G_j in R :

$$\bar{\upsilon}_j = (\{\dot{\xi}\}^T [V_s^j] + \{y\}^T [V^j])\{\bar{n}\} \tag{3.124}$$

where the first term in the equation is associated with the translation, and the second term defines the rotational contribution from the velocity

$$[V_s^j] = \begin{bmatrix} [I] \\ [S^{10}] \\ [S^{20}] \\ \vdots \\ [S^{j-1,0}] \\ 0 \end{bmatrix} \tag{3.125}$$

$$[V^j] = \begin{bmatrix} [Sq_2][S^{10}] \\ [Sq_3][S^{20}] \\ \vdots \\ [Sq_j][^{j-1,0}] \\ [Sr_j][S^{j0}] \\ 0 \end{bmatrix} \tag{3.126}$$

and

$$[S_{\xi_j}] = \begin{bmatrix} 0 & -\xi_{j3} & \xi_{j2} \\ \xi_{j3} & 0 & -\xi_{j1} \\ -\xi_{j2} & \xi_{j1} & 0 \end{bmatrix} \tag{3.127}$$

For a general multibody system, equation (3.126) represents the general definition of the velocity of the center mass of a rigid body with translation and rotational degrees of freedom. We can use the procedure developed herein plus the tree array and exponent array to change the notation on the summation subscript to make equation (3.126) read for arbitrary tree like structures. Take, for example, the general multibody system depicted by Figure 3.1. We can write the position vector from the center O of R to any mass center G_k of body B_k as

$$\bar{P}_k = \bar{\xi}_1 + \sum \bar{q}_n + \bar{r}_k \tag{3.128}$$

where $\bar{q}_1 = 0$ and $\bar{\xi}$, $\bar{q}$ and $\bar{r}$ are vectors with the same representation and meaning as the one used for the three-body problem.

Expressing the individual vectors in $\bar{n}_m$ fixed in R, the general expression for $\bar{P}_k$ becomes

$$\bar{P}_k = \left[\xi_{1m} + r_{ki} S_{im}^{k0} + \sum_{i=1}^{H(k)} q_{ui} S_{im}^{\upsilon 0} \right] \bar{n}_m \tag{3.129}$$

where

$$u = \Gamma_{h-1}(K) \quad \upsilon = T_h(k) \tag{3.130}$$

The position vector given in equation (3.129) is similar to the one given by equation (3.113), where for a general treelike structure the tree-array is used to define the direct path to the center masses. Similarly, the mass center velocity $\bar{\upsilon}_k$ is obtained by differentiation of equation (3.129):

$$\bar{\upsilon}_k = \{\dot{\xi}_1\}^T + \{r_k\}^T [\omega^{k0}][S^{k0}] + \sum_{i=1}^{H(k)} \{q_u\}^T [\omega^{\upsilon 0}]^T \, [S^{\upsilon 0}]\{\bar{n}\} \tag{3.131}$$

$$= (\{\dot{\xi}_1\}^T + \{y\}^T [V^j])\{\bar{n}\}$$

where $[V^j]$ is the same as equation (3.126), except that the order is according to the tree array. On the other hand, if the bodies were all to have relative translation denoted by ξ_i $(i = 1, \ldots N)$, where N is the number of bodies, then the mass center velocity of any rigid body of a multibody system could be expressed in a fashion similar to equation (3.124).

Example 3.8.1

For the robot mechanism described in Problem 3.6.2, find the angular velocity of the arm AB and derive the expressions for the mass center velocities of link 1, link 2, and the arm using the matrix formulation. Give their corresponding partial velocities and assume $\dot{\phi}$, $\dot{\theta}$ and ω to be constant.

Solution

A general expression for mass center velocity is

$$\overline{\upsilon_k} = \{y\}^T \, [V^k] \, \{\overline{n}\}$$

Thus various components needed for the expression above are given starting with the body vectors and their corresponding skew matrices.

$$\{q_2\} = [0 \quad 0 \quad l_1]$$

Hence

$$Sq_2 = \begin{bmatrix} 0 & -l_1 & 0 \\ l_1 & 0 & 0 \\ 0 & 0 & 0 \end{bmatrix}$$

Similarly

$$\{q_3\} = [l_2 \quad 0 \quad l_3]$$

and

$$Sq_3 = \begin{bmatrix} 0 & -l_3 & 0 \\ l_3 & 0 & -l_2 \\ 0 & l_2 & 0 \end{bmatrix}$$

The center mass vector in body 3 is:

$$\{r_3\} = [0 \quad 0 \quad l_4]$$

Hence

$$sr_3 = \begin{bmatrix} 0 & -l_4 & 0 \\ l_4 & 0 & 0 \\ 0 & 0 & 0 \end{bmatrix}$$

Shifter matrices for all bodies are defined as follows:

$$S^{10} = \begin{bmatrix} c\phi & s\phi & 0 \\ -s\phi & c\phi & 0 \\ 0 & 0 & 1 \end{bmatrix}$$

$$S^{20} = S^{21} \cdot S^{10} = \begin{bmatrix} 1 & 0 & 0 \\ 0 & c\theta & s\theta \\ 0 & -s\theta & c\theta \end{bmatrix} \begin{bmatrix} c\theta & s\theta & 0 \\ -s\theta & c\theta & 0 \\ 0 & 0 & 1 \end{bmatrix} = \begin{bmatrix} c\theta & s\theta & 0 \\ -c\theta s\phi & c\theta c\phi & s\theta \\ s\theta s\phi & -s\theta c\phi & c\theta \end{bmatrix}$$

$$S^{30} = S^{32} \cdot S^{20} = \begin{bmatrix} 1 & 0 & 0 \\ 0 & c\phi & s\psi \\ 0 & -s\psi & c\psi \end{bmatrix} \begin{bmatrix} c\phi & s\phi & 0 \\ -c\theta s\phi & c\theta c\phi & s\theta \\ s\theta s\phi & -s\theta c\phi & c\theta \end{bmatrix}$$

$$= \begin{bmatrix} c\phi & s\phi & 0 \\ (-c\psi.c\theta.s\phi + s\psi.s\theta.s\phi) & (c\psi.c\theta.c\phi - s\psi.s\theta.c\phi) & (c\psi.s\theta + s\psi.c\theta) \\ (s\psi.c\theta.s\phi + c\psi.s\theta.s\phi) & (-s\psi.c\theta.c\phi - c\psi.s\theta.c\phi) & (-s\psi.s\theta + c\psi.c\theta) \end{bmatrix}$$

Skew matrices associated with the angular velocities are then defined together with the generalized speeds. For body 1, we have

$$\omega^{10} = \dot{S}^{10} S^{01} = \begin{bmatrix} 0 & \dot{\phi} & 0 \\ -\dot{\phi} & 0 & 0 \\ 0 & 0 & 0 \end{bmatrix}$$

and the generalized speeds are

$$y_1 = [0 \quad 0 \quad \dot{\phi}]$$

Body 2 skew matrix is

$$\omega^{20} = \dot{S}^{20}\, S^{02} = \left(\dot{S}^{21} S^{10} + S^{21}\, \dot{S}^{10}\right) S^{02} = \left(\omega^{21} S^{21} S^{10} + S^{21}\omega^{10} S^{10}\right) S^{2}$$

$$\omega^{20} = \omega^{21} + S^{21}\omega^{10} S^{12}$$

$$\omega^{21} = \dot{S}^{21}\, S^{12} = \begin{bmatrix} 0 & 0 & 0 \\ 0 & 0 & \dot{\theta} \\ 0 & -\dot{\theta} & 0 \end{bmatrix}$$

$$\omega^{20} = \begin{bmatrix} 0 & 0 & 0 \\ 0 & 0 & \dot{\theta} \\ 0 & -\dot{\theta} & 0 \end{bmatrix} + \begin{bmatrix} 1 & 0 & 0 \\ 0 & c\theta & s\theta \\ 0 & -s\theta & c\theta \end{bmatrix} \begin{bmatrix} 0 & \dot{\phi} & 0 \\ -\dot{\phi} & 0 & 0 \\ 0 & 0 & 0 \end{bmatrix} \begin{bmatrix} 1 & 0 & 0 \\ 0 & c\theta & -s\theta \\ 0 & s\theta & c\theta \end{bmatrix}$$

$$\omega^{20} = \begin{bmatrix} 0 & \dot{\phi}\, c\theta & -\dot{\phi}\, s\theta \\ -\dot{\phi}\, c\theta & 0 & \dot{\theta} \\ \dot{\phi}\, s\theta & -\dot{\theta} & 0 \end{bmatrix}$$

Hence

$$y_2 = [\dot{\theta} \quad \dot{\phi}\, s\theta \quad \dot{\phi}\, c\theta]$$

Similarly we define the body 3 skew matrix as:

$$\omega^{30} = \dot{S}^{30}\, S^{03} = \left(\dot{S}^{32}\, S^{21} S^{10} + S^{32}\, \dot{S}^{21}\, S^{10} + S^{32} S^{21}\, \dot{S}^{10}\right) \cdot S^{03}$$

$$\omega^{30} = \omega^{32} + S^{32}\omega^{21} S^{23} + S^{32} S^{21}\omega^{10} S^{10} S^{03}$$

$$\omega^{30} = \begin{bmatrix} 0 & 0 & 0 \\ 0 & 0 & \dot{\psi} \\ 0 & -\dot{\psi} & 0 \end{bmatrix} + \begin{bmatrix} 1 & 0 & 0 \\ 0 & c\psi & s\psi \\ 0 & -s\psi & c\psi \end{bmatrix} \begin{bmatrix} 0 & 0 & 0 \\ 0 & 0 & \dot{\theta} \\ 0 & -\dot{\theta} & 0 \end{bmatrix} \begin{bmatrix} 1 & 0 & 0 \\ 0 & c\psi & -s\psi \\ 0 & s\psi & c\psi \end{bmatrix}$$

$$+ \begin{bmatrix} 1 & 0 & 0 \\ 0 & c\psi & s\psi \\ 0 & -s\psi & c\psi \end{bmatrix} \begin{bmatrix} 1 & 0 & 0 \\ 0 & c\theta & s\theta \\ 0 & -s\theta & c\theta \end{bmatrix} \begin{bmatrix} 0 & \dot{\phi} & 0 \\ -\dot{\phi} & 0 & 0 \\ 0 & 0 & 0 \end{bmatrix} + \begin{bmatrix} c\phi & s\phi & 0 \\ -s\phi & c\phi & 0 \\ 0 & 0 & 1 \end{bmatrix} S^{03}$$

$$\omega^{30} = \begin{bmatrix} 0 & \dot{\phi}\, c\,(\psi + \theta) & -\dot{\phi}\, s\,(\psi + \theta) \\ -\dot{\phi}\, c\,(\psi + \theta) & 0 & \left(\dot{\psi} + \dot{\theta}\right) \\ \dot{\phi}\, s\,(\psi + \theta) & -\left(\dot{\psi} + \dot{\theta}\right) & 0 \end{bmatrix}$$

From the above matrix, we obtain body 3 generalized speeds as:

$$y_3 = \begin{bmatrix} \dot{\psi} + \dot{\theta} & \dot{\phi}\, s\,(\psi + \theta) & \dot{\phi}\, c\,(\psi + \theta) \end{bmatrix}$$

Hence,

$$y = [0 \quad 0 \quad \dot{\phi} \quad \dot{\theta} \quad \dot{\phi}\, s\theta \quad \dot{\phi}\, c\theta \quad (\dot{\psi} + \dot{\theta}) \quad \dot{\phi}\, s(\psi + \theta) \quad \dot{\phi}\, c\,(\psi + \theta)]$$

The partial velocity associated with the mass center velocity of link 3 is

$$\left[v^3\right] = \begin{bmatrix} [sq_2] & [s^{10}] \\ [sq_3] & [s^{20}] \\ [sr_3] & [s^{30}] \end{bmatrix}$$

$$\left[v^3\right] = \begin{bmatrix} \begin{bmatrix} 0 & -l_1 & 0 \\ l_1 & 0 & 0 \\ 0 & 0 & 0 \end{bmatrix} & \begin{bmatrix} c\phi & s\phi & 0 \\ -s\phi & c\phi & 0 \\ 0 & 0 & 1 \end{bmatrix} \\ \begin{bmatrix} 0 & -l_3 & 0 \\ l_3 & 0 & -l_2 \\ 0 & l_2 & 0 \end{bmatrix} & \begin{bmatrix} c\phi & s\phi & 0 \\ -c\phi s\phi & c\theta\, c\phi & c\theta \\ s\theta s\phi & -s\theta c\phi & c\theta \end{bmatrix} \\ \begin{bmatrix} 0 & -l_4 & 0 \\ l_4 & 0 & 0 \\ 0 & 0 & 0 \end{bmatrix} & \begin{bmatrix} c\phi & s\phi & 0 \\ -s\phi c(\psi + \theta) & c\phi c(\psi + \theta) & s(\psi + \theta) \\ s\phi s(\psi + \theta) & -c\phi s(\psi + \theta) & c(\psi - \theta) \end{bmatrix} \end{bmatrix}$$

$$\left[v^3\right] = \begin{bmatrix} l_3 c\phi s\phi & -l_3 c\theta c\phi & -l_3 s\theta \\ l_3 c\phi - l_2 s\theta s\phi & l_3 s\phi + l_2 s\theta c\phi & -l_2 c\theta \\ l_2 c\phi s\phi & l_2 c\theta c\phi & l_2 s\theta \\ l_4 s\phi c(\psi + \theta) & -l_4 c\phi c(\psi + \theta) & -l_4 s(\psi + \theta) \\ l_4 c\phi & l_4 s\phi & 0 \end{bmatrix}$$

$$v = [0 \quad 0 \quad \dot{\phi} \quad \dot{\theta} \quad \dot{\phi}\, s\theta \quad \dot{\phi}\, c\theta \quad (\dot{\psi} + \dot{\theta}) \quad \dot{\phi}\, s(\psi + \theta) \quad \dot{\phi}\, c(\psi + \theta)][v^3] \begin{bmatrix} \bar{n}_1 \\ \bar{n}_2 \\ \bar{n}_3 \end{bmatrix}$$

$$\begin{aligned} v = {} & [\dot{\phi}(-l_3 s\theta c\phi + s^2\theta s\phi - c\theta l_2 + s(\psi + \theta) + \dot{\theta}(l_3 s\phi c\phi + l_4 s\phi c(\psi + \theta) \\ & + \dot{\psi}\, l_4 s\phi c(\psi + \theta)]\bar{n}_1 \; + [\dot{\phi}(l_2 c^2\theta c\phi - l_2 s^2\theta c\phi + l_3 s\theta s\phi - l_4 c\phi c(\psi + \theta)) \\ & - \dot{\theta}(l_3 c\theta c\phi + l_4 c\phi c(\psi + \theta)) - \dot{\psi}\, l_4 c\phi c(\psi + \theta)]\bar{n}_2 \\ & - [\dot{\phi}(-2 l_2 s\theta c\theta) - \dot{\theta}(l_3 s\theta + l_4 s(\psi + \theta)) - \dot{\psi}\, l_4 s(\psi + \theta)]\bar{n}_3 \end{aligned}$$

Example 3.8.2
Consider the planar mechanism shown in Figure 3.10, where link 1 is attached to a slider that moves horizontally. Find the velocity of point c attached to link 2 in R.

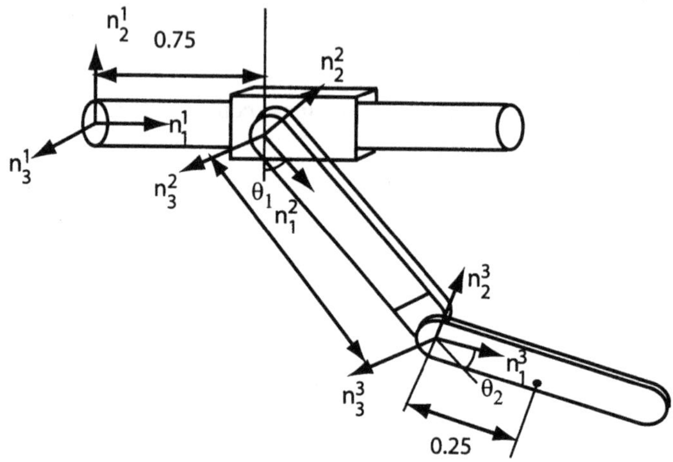

Fig. 3.10. Planar mechanism with two links and a slider

Solution
Let the position vector to the center mass of link 2 be given by

$$\bar{P} = \bar{\xi} + \bar{q}_2 + \bar{r}_2 \tag{3.132}$$

or

$$\bar{P} = \xi_1 \bar{n}_1 + \bar{q}_{21} \bar{n}_1^1 + r_{21} \bar{n}_1^2 \tag{3.133}$$

where ξ_1, q_{21} and r_{21} are the scalar components of the respective position vectors ξ, q and r. The velocity of point c in R is simply

$$\frac{d\bar{P}}{dt} = \bar{v} = \dot{\xi}_1 \bar{n}_1 + q_{21} \dot{\bar{n}}_1^1 + r_{21} \dot{\bar{n}}_1^2 \tag{3.134}$$

where the local body axis unit vectors $\bar{n}^1$ and $\bar{n}^2$ can be written as a function of the fixed frame inertia unit vectors $\bar{n}$. We have

$$\bar{n}_t^1 = S_{tm}^{10} \bar{n}_m \tag{3.135}$$

where

$$\bar{n}_1^1 = S_{1m}^{10} \bar{n}_m \tag{3.136}$$

with

$$S^{10} = \begin{bmatrix} \cos(\theta_1 - 90°) & -\sin(\theta_1 - 90°) & 0 \\ -\sin(\theta_1 - 90°) & \cos\ (\theta_1 - 90°) & 0 \\ 0 & 0 & 1 \end{bmatrix}$$

$$= \begin{bmatrix} \sin\theta_1 & -\cos\theta_1 & 0 \\ \cos\theta_1 & \sin\theta_1 & 0 \\ 0 & 0 & 1 \end{bmatrix} \tag{3.137}$$

and

$$\bar{n}_t^2 = S_{tm}^{20}\bar{n}_m \tag{3.138}$$

where

$$\bar{n}_1^2 = S_{1m}^{20}\bar{n}_m \tag{3.139}$$

with

$$S^{20} = S^{21}S^{10}$$

$$= \begin{bmatrix} \sin(\theta_1 + \theta_2) & -\cos\ (\theta_1 + \theta_2) & 0 \\ \cos(\theta_1 + \theta_2) & \sin\ (\theta_1 + \theta_2) & 0 \\ 0 & 0 & 1 \end{bmatrix} \tag{3.140}$$

We can easily see that

$$\bar{n}_1^1 = \sin\theta_1\bar{n}_1 - \cos\theta_1\bar{n}_2 \tag{3.141}$$

and

$$\bar{n}_1^2 = \sin(\theta_1 + \theta_2)\bar{n}_1 - \cos(\theta_1 + \theta_2)\bar{n}_2 \tag{3.142}$$

Equation (3.134) can be expressed further as

$$\frac{d\bar{P}}{dt} = \bar{v} = \dot{\xi}_1\bar{n}_1 + q_{21}\dot{S}_{1m}^{10}\bar{n}_m + r_{21}\dot{S}_{1m}^{20}\bar{n}_m \tag{3.143}$$

Recall that

$$\dot{S}^{10} = \omega^{10}S^{10} \tag{3.144}$$

$$\dot{S}^{20} = \omega^{20}S^{20} \tag{3.145}$$

where

$$\omega^{10} = \begin{bmatrix} 0 & \dot{\theta}_1 & 0 \\ -\dot{\theta}_1 & 0 & 0 \\ 0 & 0 & 0 \end{bmatrix} \tag{3.146}$$

$$\omega = \begin{bmatrix} 0 & \dot{\theta}_1 + \dot{\theta}_2 & 0 \\ -\left(\dot{\theta} + \dot{\theta}_2\right) & 0 & 0 \\ 0 & 0 & 0 \end{bmatrix} \tag{3.147}$$

Therefore, equation (3.132) becomes

$$\frac{d\bar{P}}{dt} = \bar{\upsilon} = \dot{\xi}_1\bar{n}_1 + q_{21}\omega_{1m}^{10}S_{tm}^{10}\bar{n}_m + r_{21}\omega_{1m}^{20}S_{tm}^{20}\bar{n}_m$$

$$= \dot{\xi}_1\bar{n}_1 + q_{21}\begin{bmatrix}0 & \dot{\theta}_1 & 0\end{bmatrix}\begin{bmatrix}\sin\theta_1 & -\cos\theta_1 & 0\\ \cos\theta_1 & \sin\theta_1 & 0\\ 0 & 0 & 1\end{bmatrix}\bar{n}_m$$

$$+ r_{21}\begin{bmatrix}0 & \dot{\theta}_1+\dot{\theta}_2 & 0\end{bmatrix}\begin{bmatrix}\sin(\theta_1+\theta_2) & -\cos(\theta_1+\theta_2) & 0\\ \cos(\theta_1+\theta_2) & \sin(\theta_1+\theta_2) & 0\\ 0 & 0 & 1\end{bmatrix}\bar{n}_m \quad (3.148)$$

$$\bar{\upsilon} = \dot{\xi}_1\bar{n}_1 + q_{21}[\dot{\theta}_1\cos\theta_1\bar{n}_1 + \dot{\theta}_1\sin\theta_1\bar{n}_2 + 0] + r_{21}[(\dot{\theta}_1+\dot{\theta}_2)\cos(\theta_1+\theta_2)\bar{n}_1$$
$$+ (\dot{\theta}_1+\dot{\theta}_2)\sin(\theta_1+\theta_2)\bar{n}_2 + 0]$$
$$= [\dot{\xi}_1 + q_{21}\dot{\theta}_1\cos\theta_1 + r_{21}(\dot{\theta}_1+\dot{\theta}_2)\cos(\theta_1+\theta_2)]\bar{n}_1$$
$$+ [q_{21}\dot{\theta}_1\sin\theta_1 + r_{21}(\dot{\theta}_1+\dot{\theta}_2)\sin(\theta_1+\theta_2)]\bar{n}_2 \quad (3.149)$$

Of course, we could have obtained (3.149) directly from equation (3.134) by substituting the derivative of equations (3.141) and (3.142). In any case, $\bar{\upsilon}$ could be written in matrix form as

$$\bar{\upsilon} = \{\dot{x}\}^T[\mathrm{w}][V]\{\bar{n}\} \quad (3.150)$$

where

$$\{\dot{x}\}^T = [\dot{\theta}_1 \;\; \dot{\theta}_2 \;\; \dot{\xi}_1] \quad (3.151)$$

and

$$[V] = \begin{bmatrix}q_{21}\cos\theta_1 + r_{21}\cos(\theta_1+\theta_2) & q_{21}\sin\theta_1 + r_{21}\sin(\theta_1+\theta_2) & 0\\ r_{21}\cos(\theta_1+\theta_2) & r_{21}\sin(\theta_1+\theta_2) & 0\\ 1 & 0 & 0\end{bmatrix} \quad (3.152)$$

Now, let us solve this problem using the matrix representation approach. The components of the position vectors in the local frame axes are given below.

$$\{\xi_1\} = [\xi \;\; 0 \;\; 0]^T \quad (3.153)$$

$$\{\dot{\xi}_1\} = [\dot{\xi} \;\; 0 \;\; 0]^T \quad (3.154)$$

$$\{q_2\} = [q_{21} \;\; 0 \;\; 0]^T \quad (3.155)$$

$$\{r_2\} = [r_{21} \;\; 0 \;\; 0]^T \quad (3.156)$$

The skew matrices associated with the position vectors q_2 and r_2 are given by

$$[Sq_2] = \begin{bmatrix} 0 & 0 & 0 \\ 0 & 0 & -q_{21} \\ 0 & q_{21} & 0 \end{bmatrix} \tag{3.157}$$

$$[Sr_2] = \begin{bmatrix} 0 & 0 & 0 \\ 0 & 0 & -r_{21} \\ 0 & r_{21} & 0 \end{bmatrix} \tag{3.158}$$

Let the generalized velocity components be given by

$$\{y\} = [0 \;\; 0 \;\; \dot{\theta}_1 \;\; 0 \;\; 0 \;\; \dot{\theta}_1 + \dot{\theta}_2]^T \tag{3.159}$$

Then, from equation (3.122), we get the partial velocity

$$[V^2] = \left[\frac{[Sq_2][S^{10}]}{[Sr_2][S^{20}]}\right]$$

$$= \left[\begin{array}{c} \begin{bmatrix} 0 & 0 & 0 \\ 0 & 0 & -q_{21} \\ 0 & q_{21} & 0 \end{bmatrix} \begin{bmatrix} \sin\theta_1 & -\cos\theta_1 & 0 \\ \cos\theta_1 & \sin\theta_1 & 0 \\ 0 & 0 & 1 \end{bmatrix} \\ \begin{bmatrix} 0 & 0 & 0 \\ 0 & 0 & -r_{21} \\ 0 & r_{21} & 0 \end{bmatrix} \begin{bmatrix} \sin(\theta_1+\theta_2) & -\cos(\theta_1+\theta_2) & 0 \\ \cos(\theta_1+\theta_2) & \sin(\theta_1+\theta_2) & 0 \\ 0 & 0 & 1 \end{bmatrix} \end{array}\right]$$

$$= \begin{bmatrix} 0 & 0 & 0 \\ 0 & 0 & -q_{21} \\ q_{21}\cos\theta_1 & q_{21}\sin\theta_1 & 0 \\ 0 & 0 & 0 \\ 0 & 0 & -r_{21} \\ r_{21}\cos(\theta_1+\theta_2) & r_{21}\sin(\theta_1+\theta_2) & 0 \end{bmatrix} \tag{3.160}$$

It follows that

$$\{y\}^T[V^2] = \begin{bmatrix} q_{21}\dot{\theta}_1\cos\theta_1 + r_{21}(\dot{\theta}_1+\dot{\theta}_2)\cos(\theta_1+\theta_2) \\ q_{21}\dot{\theta}_1\sin\theta_1 + r_{21}(\dot{\theta}_1+\dot{\theta}_2)\sin(\theta_1+\theta_2) \\ 0 \end{bmatrix}^T \tag{3.161}$$

The velocity of point c is given by equation (3.120):

$$\bar{V}_2 = (\{\dot{\xi}\}^T + \{y\}^T[V^2])\{\bar{n}\} \tag{3.162}$$

Substituting equations (3.153) to (3.161) in the equation (3.162), we obtain

$$\bar{v} = [\dot{\xi}^1 + q_{21}\dot{\theta}_1 \cos\theta_1 + r_{21}(\dot{\theta}_1 + \dot{\theta}_2)\cos(\theta_1 + \theta_2)]\bar{n}_1$$
$$+ [q_{21}\dot{\theta}_1 \sin\theta_1 + r_{21}(\dot{\theta}_1 + \dot{\theta}_2)\sin(\theta_1 + \theta_2)\bar{n}_2 \qquad (3.163)$$

The solution is, of course, the same as the one given by equation (3.149).

Example 3.8.3

As shown in Figure 3.11, the test chamber of a flight simulator consists of a drum that spins about axis $a - a$ at an angular rate $\dot{\theta}$ relative to the cylindrical housing A. The housing, in turn, is mounted on transverse horizontal bearings and rotates about axis $b - b$ with $\dot{\phi}$. The entire assembly is made to rotate about the fixed vertical z- axis at a constant rate ω.

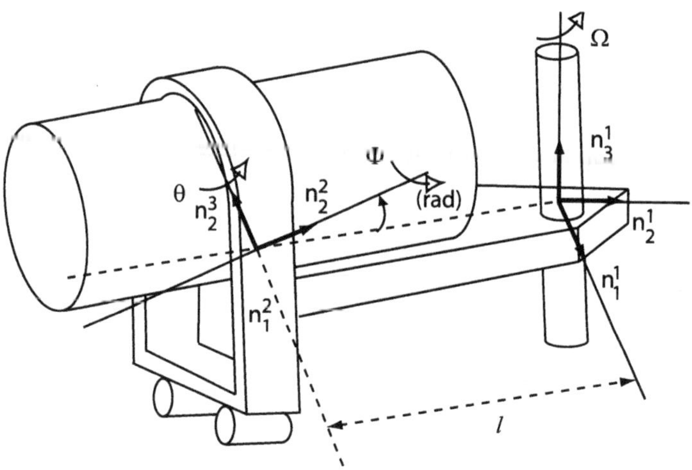

Fig. 3.11. Test chamber of a flight simulator

(a) Let the inertia fixed frame be given by x, y, z. Find the angular acceleration of the drum in R.

(b) Find the velocity of an arbitrary point on the drum in R.

Solution

(a) The various components of the expression for angular acceleration are obtained below. The shifter matrix between unit vectors of body 1 and R is

$$[S^{10}] = \begin{bmatrix} \cos\alpha & \sin\alpha & 0 \\ -\sin\alpha & \cos\alpha & 0 \\ 0 & 0 & 1 \end{bmatrix} \qquad (3.164)$$

Similarly, for body 2 unit vectors $\bar{n}^2$, we have

$$\left[S^{20}\right] = \begin{bmatrix} 1 & 0 & 0 \\ 0 & c\phi & s\phi \\ 0 & -s\phi & c\phi \end{bmatrix} \begin{bmatrix} c\alpha & s\alpha & 0 \\ -s\alpha & c\alpha & 0 \\ 0 & 0 & 1 \end{bmatrix} = \begin{bmatrix} c\alpha & s\alpha & 0 \\ -c\phi s\alpha & c\phi c\alpha & s\phi \\ s\phi s\alpha & -s\phi c\alpha & c\phi \end{bmatrix} \quad (3.165)$$

The partial angular velocity arrays are obtained following the definition as

$$[\omega^1] = \begin{bmatrix} [S^{10}] \\ --- \\ 0 \end{bmatrix} \quad (3.166)$$

$$[\omega^2] = \begin{bmatrix} [S^{10}] \\ --- \\ [S^{20}] \end{bmatrix} \quad (3.167)$$

Note how the components of $[\omega^1]$ are a function of the shifter matrix $[S^{10}]$ instead of the identity $[I]$ for the purpose of expressing the angular velocity of body 2 in R, which has two rotations.

Let the generalized coordinate derivative array be given by

$$\{\dot{x}\} = \begin{bmatrix} 0 & 0 & \omega & \dot{\phi} & \dot{\theta} & 0 \end{bmatrix}^T \quad (3.168)$$

We can easily see that the expression of the angular velocity of body 2 in R is given by premultiplication of equations (3.168) and (3.167), which yields

$$\begin{aligned} {}^R\bar{\omega}^2 &= \{\dot{x}\}^T[\omega^2]\{\bar{n}\} \\ &= (\dot{\phi}c\alpha - \dot{\theta}c\phi s\alpha)\bar{n}_1 + (\dot{\phi}s\alpha - \dot{\theta}c\phi c\alpha)\bar{n}_2 + (\omega + \dot{\theta}s\phi)\bar{n}_3 \end{aligned} \quad (3.169)$$

The convenience of matrix multiplication stems from the partitioning of the generalized coordinates and the partial angular velocity array, as shown above. The same is true for evaluating the velocity (linear) of any point in R.

The angular acceleration of body 2, denoted by $\bar{n}^2$, is found to be

$$ {}^R\bar{\alpha}^2 = (\{\ddot{x}\}[\omega^2] + \{\dot{x}\}[\dot{\omega}^2])\{\bar{n}\} \quad (3.170)$$

where

$$[\dot{\omega}] = \begin{bmatrix} [\dot{S}^{10}] \\ --- \\ [\dot{S}^{20}] \end{bmatrix} \quad (3.171)$$

Substituting equations (3.167) and (3.171) into (3.170), we obtain

$$\begin{aligned} {}^R\bar{\alpha}^2 &= (-\dot{\phi}\,\omega s\alpha - \ddot{\theta}c\phi\,s\alpha + \dot{\theta}\dot{\phi}\,s\phi\,s\alpha - \dot{\theta}\,\omega c\phi\,c\alpha)\bar{n}_i \\ &\quad + (\dot{\phi}\,\omega c\alpha + \ddot{\theta}c\phi\,c\alpha - \dot{\theta}\dot{\phi}\,s\phi\,c\alpha - \dot{\theta}\,\omega c\phi\,s\alpha)\bar{n}_2 + (\ddot{\theta}s\phi + \dot{\theta}\dot{\phi}c\phi)\bar{n}_3 \end{aligned} \quad (3.172)$$

Note that $\ddot{\phi} = \dot{\omega} = 0$. Since our objective is to show how matrix multiplication, together with the definition of the kinematical quantities, can be used, one can always obtain ${}^{R}\bar{\alpha}^{2}$ directly from equation (3.169) by differentiation. In the context of multibody systems, this would be long, tedious and prone to errors.

(b) The velocity of a point located at the centerline of the drum along the line of rotation for $\dot{\theta}$, shown in Figure 3.11, is given by

$$\bar{V}_2 = \{y\}^T \{V^2\}\{\bar{n}\} \tag{3.173}$$

where $\{y\}^T$ denotes the generalized speeds and $\{V^2\}$ the corresponding partial velocity array. The various components for equation (3.173) are obtained below.

$$\{q_2\} = [0 \;\; \ell_1 \;\; 0] \tag{3.174}$$

$$[Sq_2] = \begin{bmatrix} 0 & 0 & \ell_1 \\ 0 & 0 & 0 \\ -\ell_1 & 0 & 0 \end{bmatrix} \tag{3.175}$$

Similarly, we write the skew matrix for the position vector $\bar{r}_2$ as

$$[Sr_2] = \begin{bmatrix} 0 & 0 & \ell_2 \\ 0 & 0 & 0 \\ -\ell_2 & 0 & 0 \end{bmatrix} \tag{3.176}$$

The skew matrix associated with the angular velocity of body 2 is given by

$$[\omega^{20}] = \begin{bmatrix} 0 & \omega c\phi & -\omega s\phi \\ -\omega c\phi & 0 & \dot{\phi} \\ \omega s\phi & -\dot{\phi} & 0 \end{bmatrix} \tag{3.177}$$

from which we define the generalized speeds to be

$$\{y\} = \begin{bmatrix} 0 \\ 0 \\ \omega \\ \dot{\phi} \\ \omega s\phi \\ \omega c\phi \end{bmatrix} \tag{3.178}$$

From definition, we write the partial velocity array $[V^2]$ as

$$[V^2] = \begin{bmatrix} [Sq_2][S^{10}] \\ ----- \\ [Sr_2][S^{20}] \end{bmatrix} \tag{3.179}$$

Substituting the appropriate skew matrices and shifters in equation (3.173), we obtain

$$[V^2] = \begin{bmatrix} \begin{bmatrix} 0 & 0 & -\ell_1 \\ 0 & 0 & 0 \\ \ell_1 & 0 & 0 \end{bmatrix} \begin{bmatrix} c\alpha & s\alpha & 0 \\ s\alpha & c\alpha & 0 \\ 0 & 0 & 1 \end{bmatrix} \\ \begin{bmatrix} 0 & 0 & \ell_2 \\ 0 & 0 & 0 \\ -\ell_2 & 0 & 0 \end{bmatrix} \begin{bmatrix} c\alpha & s\alpha & 0 \\ -c\phi\, s\alpha & c\phi\, c\alpha & c\phi \\ s\phi\, s\alpha & -s\phi\, c\alpha & c\phi \end{bmatrix} \end{bmatrix}$$

$$= \begin{bmatrix} 0 & 0 & -\ell_1 \\ 0 & 0 & 0 \\ \ell_1 c\alpha & \ell_1 s\alpha & 0 \\ \ell_2 s\phi\, s\alpha & -\ell_2 s\phi\, c\alpha & \ell_2 c\phi \\ 0 & 0 & 0 \\ -\ell_2 c\alpha & -\ell_2 s\alpha & 0 \end{bmatrix} \tag{3.180}$$

Premultiplying the matrix above with the generalized speeds vector $\{y\}$ and the unit vectors fixed in R as shown in equation (3.173), we get the final expression for the velocity of point p in R as

$$\begin{aligned} {}^R\bar{V}^P &= (\ell_1 \omega c\alpha + \ell_2 \dot{\phi}\, s\phi\, s\alpha - \ell_2 \omega c\phi\, c\alpha)\bar{n}_1 \\ &\quad + (\ell_1 \omega s\alpha + \ell_2 \dot{\phi}\, s\phi\, c\alpha - \ell_2 \omega c\phi\, s\alpha)\bar{n}_2 + (\ell_2 \dot{\phi}\, c\phi)\bar{n}_3 \end{aligned} \tag{3.181}$$

Example 3.8.4

Assume that two arbitrary rigid bodies are interconnected by a spherical joint that allows three degrees of freedom for rotation: α, β and γ. For the skew matrix $[\omega^{kj}]$ given by equation (3.29), find the transformation matrix [w] that expresses the generalized speeds $\{y_k\}$ in terms of the generalized coordinate derivatives $\{\dot{x}_k\}$.

Solution

From equations (3.34) and (3.29), we know that

$$\{y_k\} = \begin{bmatrix} \dot{\alpha} c\beta c\gamma + \dot{\beta} s\gamma \\ -\dot{\alpha} c\beta s\gamma + \dot{\beta} c\gamma \\ \dot{\gamma} + \dot{\alpha} s\beta \end{bmatrix} \tag{3.182}$$

In addition, we have $\{\dot{x}_k\} = [\dot{\alpha} \quad \dot{\beta} \quad \dot{\gamma}]^T$; therefore,

$$[\mathrm{w}] = \begin{bmatrix} c\beta c\gamma & -c\beta s\gamma & s\beta \\ s\gamma & c\gamma & 0 \\ 0 & 0 & 1 \end{bmatrix} \tag{3.183}$$

which makes

$$\{y_k\}^T = \{\dot{x}_k\}^T [\mathrm{w}] \tag{3.184}$$

Note that the transformation matrix given by equation (3.183) is also referred to as the *gimbal matrix*. The latter is not unique and will depend on the order of rotations adapted in correlating the different body axes.

3.9 Mass Center Accelerations of Arbitrary Interconnected Rigid Bodies in Treelike Systems

Consider equation (3.209), which defines the mass center velocity of B_j. The mass center acceleration is then found by differentiating in time the velocity $\bar{\upsilon}_j$. Hence

$$\bar{a}_j = \frac{d}{dt}\bar{\upsilon}_j = (\{\ddot{\xi}_1\}^T + \{\dot{y}\}^T[V^j] + \{y\}^T[\dot{V}^j])\{\bar{n}\} \tag{3.185}$$

where

$$[\dot{V}^j] = \begin{bmatrix} [Sq_2][\omega^{10}][S^{10}] \\ [Sq_3][\omega^{20}][S^{20}] \\ \vdots \\ [Sq_j][\omega^{j-1,0}][S^{j-1,0}] \\ [Sr_j][\omega^{j0}][S^{j0}] \\ 0 \end{bmatrix} \tag{3.186}$$

On the other hand, if the bodies have relative translation denoted by $\xi_i (i = 1, \ldots, N$ where N is the number of bodies), the mass center acceleration of any rigid body of a multibody system can be expressed as equation (3.185), where $\dot{V}^k$ can be obtained by differentiating equation (3.186).

Example 3.9.1
Find the velocity and acceleration of point p as shown in Figure 3.12.

Solution
Since point G is fixed, we obtain

$$\bar{r}_p = \{r_p\}^T\{\bar{n}^2\} = \{r_p\}^T[S^{20}]\{\bar{n}\} \tag{3.187}$$

where $\{r_p\}^T = [r_1 \ r_2 \ r_3]$ denotes the coordinates of point p in $\{\bar{n}^2\}$. The transformation matrix of body 2 in R is given by

$$S^{20} = S^{21}S^{10}$$

$$= \begin{bmatrix} \cos\theta & \sin\theta & 0 \\ -\sin\theta & \cos\theta & 0 \\ 0 & 0 & 1 \end{bmatrix} \begin{bmatrix} 1 & 0 & 0 \\ 0 & \cos 120^\circ & \sin 120^\circ \\ 0 & -\sin 120^\circ & \cos 120^\circ \end{bmatrix} \begin{bmatrix} \cos\alpha & \sin\alpha & 0 \\ -\sin\alpha & \cos\alpha & 0 \\ 0 & 0 & 1 \end{bmatrix}$$

$$= \begin{bmatrix} \cos\alpha\cos\theta + \frac{1}{2}\sin\alpha\sin\alpha & \sin\alpha\cos\theta + \frac{1}{2}\cos\alpha\ \sin\theta & \frac{\sqrt{3}}{2}\sin\theta \\ -\cos\alpha\sin\theta + \frac{1}{2}\sin\alpha\cos\theta & -\sin\alpha\sin\theta + \frac{1}{2}\cos\alpha cos\theta & \frac{\sqrt{3}}{2}\cos\theta \\ \frac{\sqrt{3}}{2}\sin\alpha & -\frac{\sqrt{3}}{2}\sin\alpha & -\frac{1}{2} \end{bmatrix} \tag{3.188}$$

Velocity $\bar{v}_p$ can be obtained by differentiating the position vector $\bar{r}_p$:

$$\bar{v}_p = \frac{d}{dt}\bar{r}_p = \{r_p\}^T[\dot{S}^{20}]\{\bar{n}\} \tag{3.189}$$

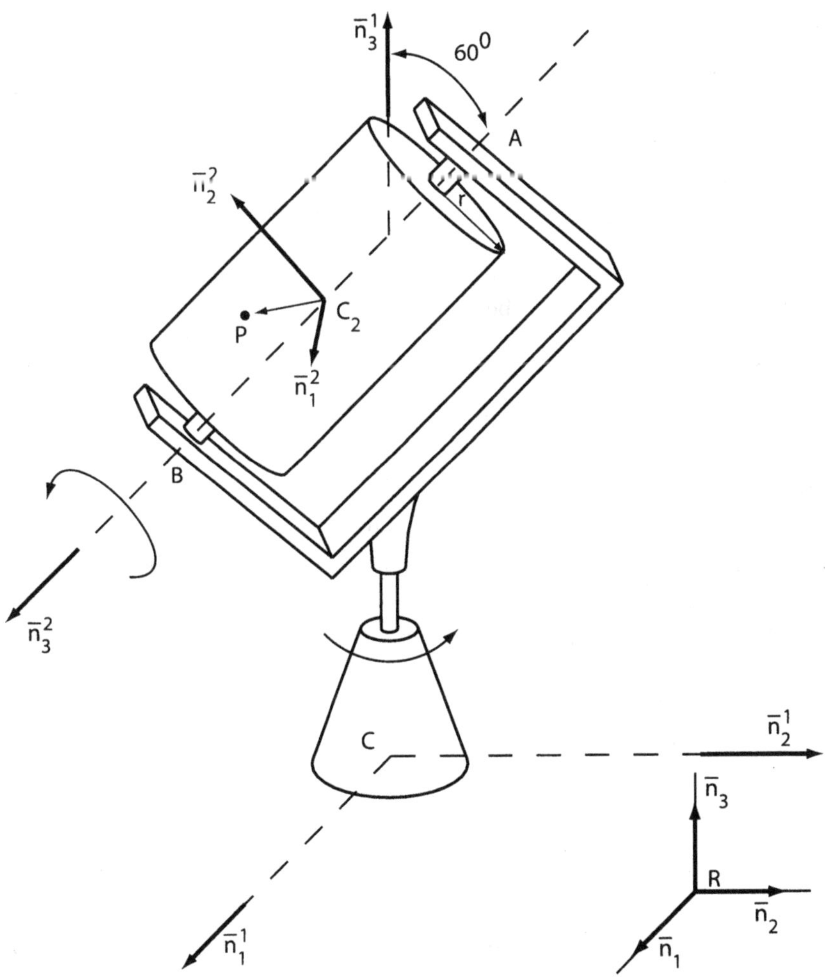

Fig. 3.12. Rotating disk mounted on a spinning shaft

$$\bar{v}_p = \left[r_1\left(-\omega s\alpha c\theta - \dot{\theta} c\alpha s\theta + \frac{1}{2}\omega c\alpha s\theta + \frac{1}{2}\theta s\alpha c\theta\right)\right.$$

$$\left. + r_2\left(\omega s\alpha s\theta - \dot{\theta} c\alpha c\theta + \frac{1}{2}\omega c\alpha c\theta - \frac{1}{2}\dot{\theta} s\alpha s\theta\right) + \frac{\sqrt{3}}{2} r_3 \omega c\alpha\right]\bar{n}_1$$

$$+ \left[r_1\left(\omega c\alpha\theta - \dot{\theta}\alpha\, s\theta + \frac{1}{2}\omega s\alpha s\theta + \frac{1}{2}\dot{\theta} c\alpha\, c\theta\right)\right.$$

$$\left. + r_2\left(-\omega c\alpha s\theta - \dot{\theta} s\alpha c\theta + \frac{1}{2}\omega s\alpha c\theta + \frac{1}{2}\dot{\theta} c\alpha s\theta\right) + \frac{\sqrt{3}}{2} r_3 \omega s\alpha\right]\bar{n}_2$$

$$+ \frac{\sqrt{3}}{2}\dot{\theta}(r_1 c\theta - r_2 s\theta)\bar{n}_3 \tag{3.190}$$

where $c\theta = \cos\theta$ and $s\theta = \sin\theta$; similarly, $c\alpha = \cos\alpha$ and $s\alpha = \sin\alpha$. We can obtain the acceleration of point p by further differentiation of equation (3.190)

$$\bar{a}_p = \frac{d^2}{dt^2}\bar{r}_p = \frac{d}{dt}\bar{v}_p \tag{3.191}$$

which can be expressed in compact matrix form as

$$\bar{a}_p = \{r_p\}^T([\dot{\omega}^{20}][S^{20}] + [\omega^{20}][\omega^{20}][S^{20}])\{\bar{n}\} \tag{3.192}$$

Observe how the matrices are expressions that can easily be obtained from the velocity equation, hence minimizing the burden of computation.

Example 3.9.2

For Example 2.13.2, find the mass center velocity and acceleration of body 2 in R following the matrix formulation.

Solution

The unit vectors in body 2 can be expressed in terms of those attached to R as

$$\{\bar{n}^2\} = [S^{21}][S^{10}]\{\bar{n}\} \tag{3.193}$$

where $\{\bar{n}\}$ denote unit vectors in R. From equation (3.202) we obtain the position vector to the center mass of body 2.

$$\bar{P}_2 = \left(\{\xi_1\}^T + \sum_{i=1}^{2}\{q_i\}^T[S^{i-1,0}] + \{r_2\}^T[S^{20}]\right)\{\bar{n}\}$$

$$= (0 + 0 + \{q_2\}^T[S^{10}] + \{r_2\}^T[S^{20}])\{\bar{n}\} \tag{3.194}$$

where

$$\{q_2\}^T = \begin{bmatrix} b & 0 & 0\end{bmatrix} \tag{3.195}$$

$$\{r_2\}^T = \begin{bmatrix}\frac{1}{2}\ell & 0 & 0\end{bmatrix} \tag{3.196}$$

From equation (3.204), we obtain

$$\bar{v}_2 = (\{q_2\}^T[\omega^{10}][S^{10}] + \{r_2\}^T[\omega^{20}][S^{20}])\{\bar{n}\} \tag{3.197}$$

After substitution we get

$$\bar{v}_2 = \begin{bmatrix} b & 0 & 0 \end{bmatrix} \begin{bmatrix} 0 & \omega & 0 \\ -\omega & 0 & 0 \\ 0 & 0 & 0 \end{bmatrix} \begin{bmatrix} \cos\alpha & \sin\alpha & 0 \\ -\sin\alpha & \cos\alpha & 0 \\ 0 & 0 & 1 \end{bmatrix} \bar{n}_m$$

$$+ \begin{bmatrix} \frac{1}{2}\ell & 0 & 0 \end{bmatrix} \begin{bmatrix} 0 & -\omega\cos\theta & \dot{\theta} \\ \omega\cos\theta & 0 & -\omega\sin\theta \\ -\dot{\theta} & \omega\sin\theta & 0 \end{bmatrix}$$

$$\cdot \begin{bmatrix} -\cos\theta\cos\alpha & -\cos\theta\sin\alpha & -\sin\theta \\ -\sin\alpha & -\sin\alpha & 0 \\ \sin\theta\cos\alpha & \sin\theta\sin\alpha & -\cos\theta \end{bmatrix} \bar{n}_m \tag{3.198}$$

$$= \left[-bw\sin\alpha + \frac{1}{2}\ell(\dot{\theta}\sin\theta\cos\alpha + \omega\cos\theta\sin\theta)\right]\bar{n}_1$$

$$+ \left[-bw\cos\alpha + \frac{1}{2}\ell(\dot{\theta}\sin\theta\sin\alpha - \omega\cos\theta\cos\alpha)\right]\bar{n}_2 - \frac{1}{2}\ell\dot{\theta}\cos\theta\bar{n}_3 \tag{3.199}$$

This result can be checked further by using equation (3.208). Let the position skew matrices be defined as usual by

$$[Sq_2] = \begin{bmatrix} 0 & 0 & 0 \\ 0 & 0 & -b \\ 0 & b & 0 \end{bmatrix} \tag{3.200}$$

$$[Sr_2] = \begin{bmatrix} 0 & 0 & 0 \\ 0 & 0 & -\frac{1}{2}\ell \\ 0 & \frac{1}{2}\ell & 0 \end{bmatrix} \tag{3.201}$$

The partial velocity array is given explicitly by

$$[V^2] = \begin{bmatrix} [Sq_2] & [S^{10}] \\ [Sr_2] & [S^{20}] \end{bmatrix}$$

$$= \begin{bmatrix} 0 & 0 & 0 \\ 0 & 0 & -b \\ -b\sin\alpha & b\cos\alpha & 0 \\ 0 & 0 & 0 \\ -\frac{1}{2}\ell\sin\theta\cos\alpha & -\frac{1}{2}\ell\sin\alpha\cos\alpha & \frac{1}{2}\ell\cos\theta \\ -\frac{1}{2}\ell\sin\alpha & \frac{1}{2}\ell\sin\alpha & 0 \end{bmatrix} \tag{3.202}$$

From equation (3.198) we write the generalized speeds

$$\{y\} = \begin{bmatrix} 0 & 0 & \omega & -\omega\sin\theta & -\theta & -\omega\cos\theta \end{bmatrix}^T \tag{3.203}$$

The velocity components of the center mass of body 2 in R are then obtained, where

$$\begin{aligned}
\{V_2\}^T &= \{y\}^T[V^2] \\
&= \begin{bmatrix} 0 \\ 0 \\ \omega \\ -\omega\sin\theta \\ -\dot{\theta} \\ -\omega\cos\theta \end{bmatrix} \begin{bmatrix} 0 & 0 & 0 \\ 0 & 0 & -b \\ -b\sin\alpha & b\cos\alpha & 0 \\ 0 & 0 & 0 \\ -\frac{1}{2}\ell\sin\theta\cos\alpha & -\frac{1}{2}\ell\sin\theta\sin\alpha & \frac{1}{2}\ell\cos\theta \\ -\frac{1}{2}\ell\sin\alpha & \frac{1}{2}\ell\cos\alpha & 0 \end{bmatrix} \\
&= \begin{bmatrix} -b\omega\sin\alpha + \frac{1}{2}\ell\dot{\theta}\sin\theta\cos\alpha + \frac{1}{2}\ell\omega\cos\theta\sin\alpha \\ b\omega\cos\alpha + \frac{1}{2}\ell\dot{\theta}\sin\theta\cos\alpha + \frac{1}{2}\ell\omega\cos\theta\cos\alpha \\ -\frac{1}{2}\ell\dot{\theta}\cos\theta \end{bmatrix}^T
\end{aligned} \tag{3.204}$$

Indeed, the solution is the same as the one given by equation (3.199). For the acceleration of point p, we can differentiate equation (3.204) once more, yielding

$$\begin{aligned}
\bar{a}_2 = \{&\{q_2\}^T \left([\dot{\omega}^{10}][S^{10}] + [\omega^{10}][\omega^{10}]\right) \\
&+ \{r_2\}^T \left([\dot{\omega}^{20}][S^{20}] + [\omega^{20}][S^{20}][S^{10}]\right)\}\{\bar{n}\}
\end{aligned} \tag{3.205}$$

We can also solve for $\bar{a}_2$ by equation (3.185). Differentiation of equation (3.202) gives the rate of change of the partial velocity array:

$$[\dot{V}^2] = \begin{bmatrix} 0 & 0 & 0 \\ 0 & 0 & 0 \\ -b\omega\cos\alpha & -b\omega\sin\alpha & 0 \\ 0 & 0 & 0 \\ -\frac{1}{2}\ell(\dot{\theta}\sin\theta\cos\alpha & -\frac{1}{2}\ell(\dot{\theta}\cos\theta\sin\alpha & -\frac{1}{2}\ell\dot{\theta}\sin\theta \\ -\omega\sin\theta\sin\alpha) & -\omega\sin\theta\cos\alpha) & \\ -\frac{1}{2}\ell\omega\cos\alpha & -\frac{1}{2}\ell\sin\alpha & 0 \end{bmatrix} \tag{3.206}$$

Differentiation of equation (3.203) yields

$$\{\dot{y}\} = \begin{bmatrix} 0 \\ 0 \\ \dot{\omega} \\ -\dot{\omega}\sin\theta - \omega\dot{\theta}\cos\theta\, A \\ -\ddot{\theta} \\ -\dot{\omega}\cos\theta + \omega\dot{\theta}\sin\theta \end{bmatrix} \tag{3.207}$$

From equation (3.185) we write the mass center acceleration of body 2 as

$$\{a_2\}^T = \{\dot{y}\}^T[V^2] + \{\dot{y}\}^T[V^2] \tag{3.208}$$

or

$$\{a_2\} = [V^2]^T\{\dot{y}\} + [\dot{V}^2]^T\{y\}$$

$$= \begin{bmatrix} 0 & 0 & -b\sin\alpha & 0 & -\frac{1}{2}\ell\sin\theta\cos\alpha & -\frac{1}{2}\ell\sin\alpha \\ 0 & 0 & b\cos\alpha & 0 & -\frac{1}{2}\ell\sin\theta\sin\alpha & \frac{1}{2}\ell\cos\alpha \\ 0 & -b & 0 & 0 & \frac{1}{2}\ell\cos\theta & 0 \end{bmatrix}$$

$$\begin{bmatrix} 0 \\ 0 \\ \dot{\omega} \\ -\dot{\omega}\sin\theta - \omega\dot{\theta}\cos\theta \\ -\ddot{\theta} \\ -\dot{\omega}\cos\theta + \omega\dot{\theta}\sin\theta \end{bmatrix}$$

$$+ \begin{bmatrix} 0 & 0 & -b\omega\cos\alpha & 0 & -\frac{1}{2}\ell(\dot{\theta}\cos\theta\cos\alpha - \omega\sin\theta\sin\alpha) & -\frac{1}{2}\ell\omega\cos\alpha \\ 0 & 0 & -b\omega\sin\alpha & 0 & -\frac{1}{2}\ell(\dot{\theta}\cos\theta\sin\alpha + \omega\sin\theta\cos\alpha) & \frac{1}{2}\ell\omega\cos\alpha \\ 0 & 0 & 0 & 0 & -\frac{1}{2}\ell\dot{\theta}\sin\theta & 0 \end{bmatrix}$$

$$\begin{bmatrix} 0 \\ 0 \\ \omega \\ -\dot{\omega}\sin\theta \\ -\dot{\theta} \\ -\omega\cos\theta \end{bmatrix}$$

$$= \begin{bmatrix} a_2^1 \\ a_2^2 \\ a_2^3 \end{bmatrix} \tag{3.209}$$

where

$$a_2^1 = -b\dot{\omega}\sin\alpha + \frac{1}{2}\ell\ddot{\theta}\sin\theta\cos\alpha + \frac{1}{2}\ell\sin\alpha(-\dot{\omega}\cos\theta + \omega\dot{\theta}\sin\theta)$$
$$- b\omega^2\cos\alpha + \frac{1}{2}\ell\dot{\theta}(\dot{\theta}\cos\theta\cos\alpha - \omega\sin\theta\sin\alpha) + \frac{1}{2}\ell\omega^2\cos\alpha\cos\theta \quad (3.210)$$

$$a_2^2 = -b\dot{\omega}\cos\alpha + \frac{1}{2}\ell\ddot{\theta}\sin\theta\sin\alpha + \frac{1}{2}\ell\cos\alpha(-\dot{\omega}\cos\theta + \omega\dot{\theta}\sin\theta)$$
$$- b\omega^2\sin\alpha + \frac{1}{2}\ell\dot{\theta}(\cos\theta\cos\alpha - \omega\sin\theta\sin\alpha) + \frac{1}{2}\ell\omega^2\sin\alpha\cos\theta \quad (3.211)$$

$$a_2^3 = -\frac{1}{2}\ell\ddot{\theta}\cos\theta + \frac{1}{2}\ell\dot{\theta}^2\sin\theta \quad (3.212)$$

3.10 Modeling of Joints

In the study of multibody dynamics, several types of joints illustrating the interconnection between bodies are encountered. The joint connections usually restrict the motion of a particular body with respect to another, and hence determine its possible degrees of freedom. An illustration of the possible joints used is given below.

3.10.1 Free Joint

A free joint is defined as one in which a particular body can undergo all possible motions, that is, three rotations and three translations. Consider Figure 3.13; let O and O' be the points on B_j and B_k, respectively. Let point O' be the center of the axis $\bar{n}_i^k (i = 1, 2, 3)$, which initially, is aligned with the body axis of B_j with center O. If body B_k can have six degrees of freedom, then point O' is allowed to have a translation along $\bar{n}_1^j$, $\bar{n}_2^j$ and $\bar{n}_1^j$, respectively. In addition, it can rotate about $\bar{n}_1^j$ by θ_1, $\bar{n}_2^j$ by θ_2 and $\bar{n}_3^j$ by θ_3. We can see that in the case where O coincides with O' during the motion of B_j, the joint reduces to a ball socket (spherical) type.

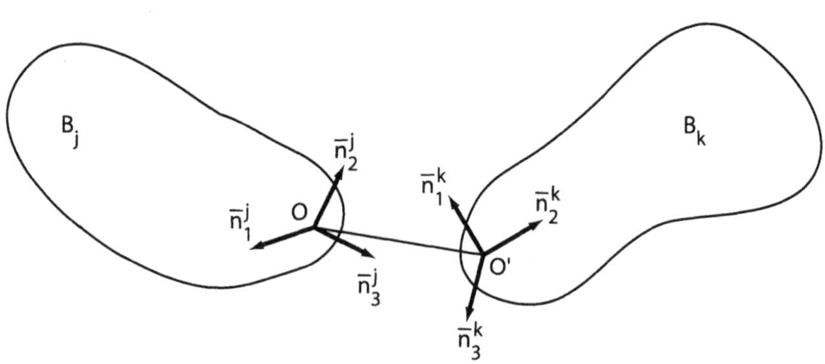

Fig. 3.13. Description of a free joint in multibody systems

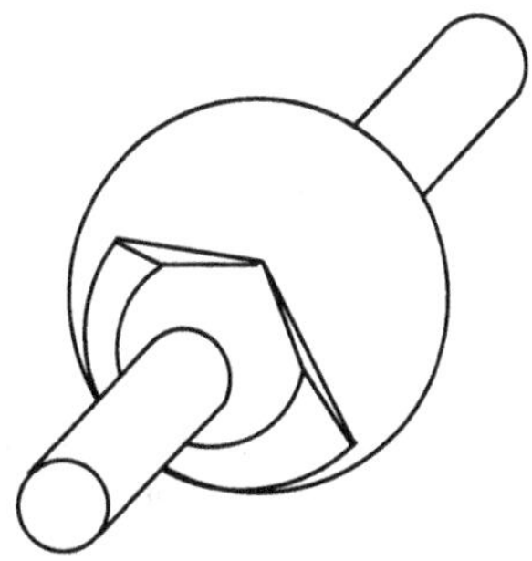

Fig. 3.14. Ball-and-socket type of joint interconnecting two rigid bodies

3.10.2 Ball-and-Socket Joint

The ball-and-socket joint is also known as a *spherical joint*. A typical example of such a connection is shown in Figure 3.14. The ball-and-socket joint allows B_k to have an arbitrary orientation with respect to B_j. This can be achieved better by body B_k capable of rotating about $\bar{n}_1^k, \bar{n}_2^k$ and $\bar{n}_3^k$. The orientation of $\bar{n}_i^k (i = 1, 2, 3)$ with respect to $\bar{n}_j^j (j = 1, 2, 3)$ defines the position of B_k with respect to B_j. The kinematical interpretation of a ball-and-socket joint is that the point O common to both B_k and B_j must be such that (see Figure 3.15)

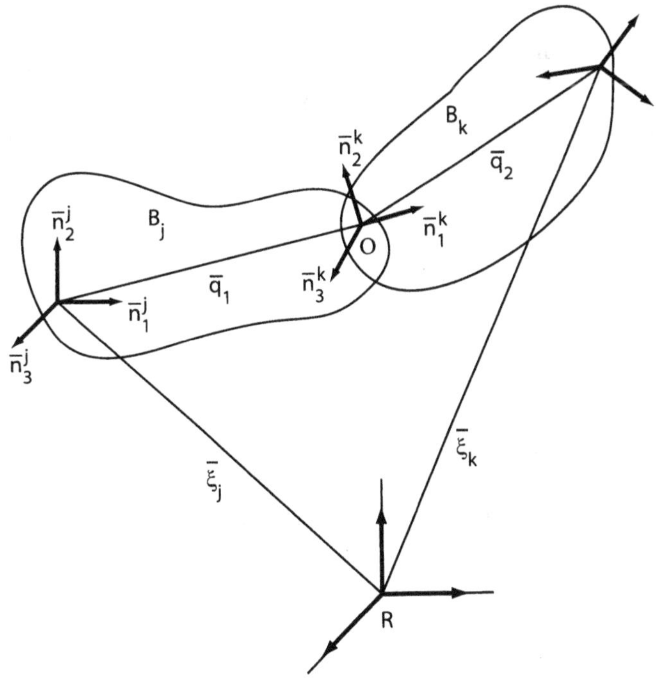

Fig. 3.15. Two interconnected rigid bodies with a ball-and-socket type of joint modeled by a common point O

$$\underbrace{\bar{\xi}_1 + \bar{q}_1}_{\bar{r}_1} = \underbrace{\bar{\xi}_2 + \bar{q}_2}_{\bar{r}_2} \tag{3.213}$$

or

$$\phi\,(x_1, x_2, \ldots) = \bar{r}_1 - \bar{r}_2 = 0 \tag{3.214}$$

where ϕ denotes the constraint equation that defines the kinematical condition for B_j and B_k such that their imposed joint connection is satisfied. The function ϕ is known as a holonomic constraint resulting from a geometrical condition imposed between bodies. A differentiation of equation (3.213) yields the velocity of the constraint representation as

$$^{R}\bar{\upsilon}_j^O = {}^{R}\,\bar{\upsilon}_k^O$$

$$\{y\}^T[V^j]\{\bar{n}\} - \{y\}^T[V^k]\{\bar{n}\} = 0 \tag{3.215}$$

that is,

$$[B]\{y\} = 0$$

where $^{R}\bar{\upsilon}_j^O$ is the velocity of point O in B_j with respect to R, whereas $^{R}\bar{\upsilon}_k^O$ represents the velocity of point O in B_k with respect to R, and $[B]$ is a constraint matrix defined by the partial velocity array of B_j and B_k, respectively

$$[B] = [V^j]^T - [V^k]^T \tag{3.216}$$

3.10.3 Revolute Joint

The revolute is a common connection between bodies in robotics (Figure 3.16). This particular connection allows one body to rotate with respect to the other about one particular axis.

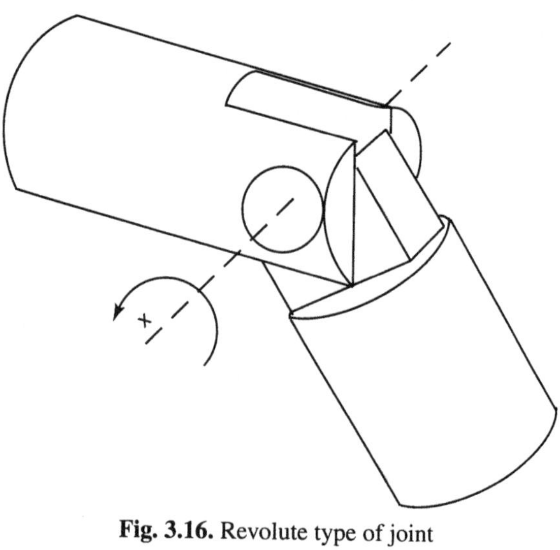

Fig. 3.16. Revolute type of joint

3.11 Problems

3.1. For the human model as shown in Figure 3.2, give (a) the exponent array $H(k)$ for tree array $\Gamma(k)$ given by Table 3.2, and (b) the partial velocity array table for $[\omega^k]$.

3.2. Consider Figure 3.17, where B_1 is pin connected to the reference frame R at point O and to body B_1 through a spherical joint. Use the matrix representation of angular velocities given by equation (3.6.5) to find the corresponding partial angular velocities of B_1 and B_2 in R.

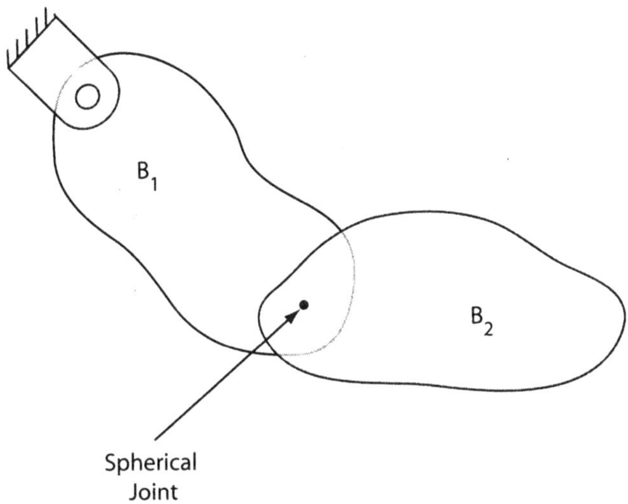

Fig. 3.17. Body B_1 is pin-connected to R, and B_2 is connected to B_1 through a spherical joint

3.3. Derive the corresponding array $[\omega^k]$ associated with the angular velocity of body k with respect to a fixed adjacent body R for the case where the orientation of k in R can be expressed in terms of a rotation $\bar{n}_1^k$ followed by a rotation about $\bar{n}_2^k$ and finally, a rotation about $\bar{n}_1^k$.

3.4. For the general multibody system depicted in Figure 3.18, derive (a) the tree array, and (b) the partial angular velocity matrix $[\omega^k]$ as shown in Table 3.5 (assuming that all bodies are connected by spherical joints).

3.5. Consider the slider on the shaft S_2 as shown in Figure 3.19. Give the explicit form of the partial velocity of the slider with respect to R.

3.6. Consider the three-link mechanism as shown in Figure 3.20, point O is hinged to the ground, whereas point C moves along the horizontal rail without sliding. The mechanism can be treated as an open system as shown, then adding the condition that $S(t)$ must be equal to $\bar{L}_1 + \bar{L}_2 + \bar{L}_3$. Then, we formulate the corresponding constraint equation needed to keep the mechanism in its configuration. Find the condition associated with the velocity of point C in R.

3.7. As shown in Figure 3.21, the wheel D is attached to the shaft BC and rotates about its center axis by an angular velocity $\dot{\psi}$ (const). The shaft BC is pinned

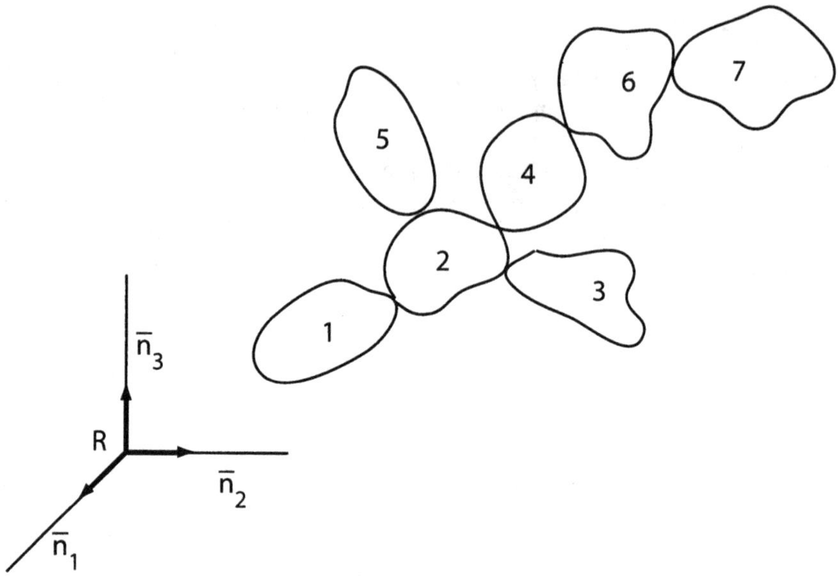

Fig. 3.18. Treelike multibody system

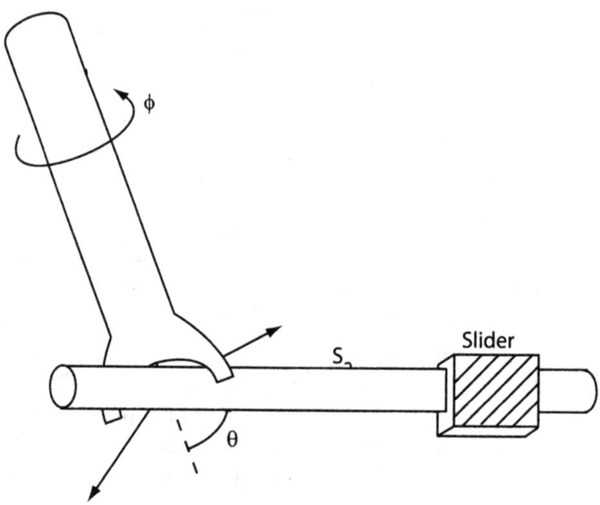

Fig. 3.19. Slider mounted on a rotating shaft S_2 which is connected to a spinning master shaft S_1

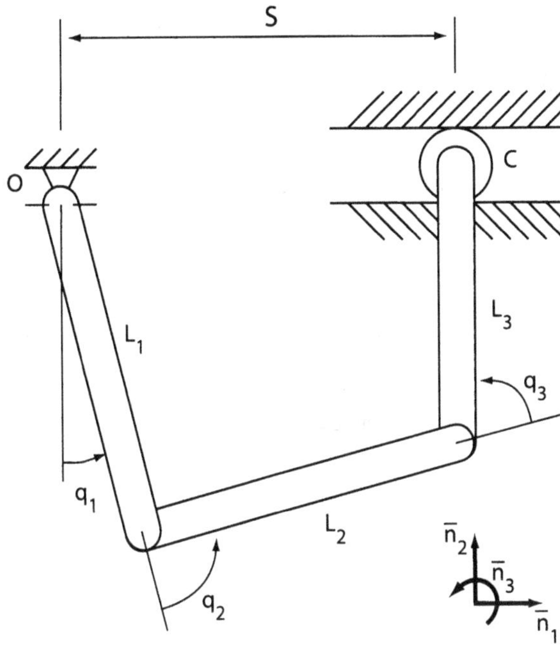

Fig. 3.20. Three-link mechanism with point C

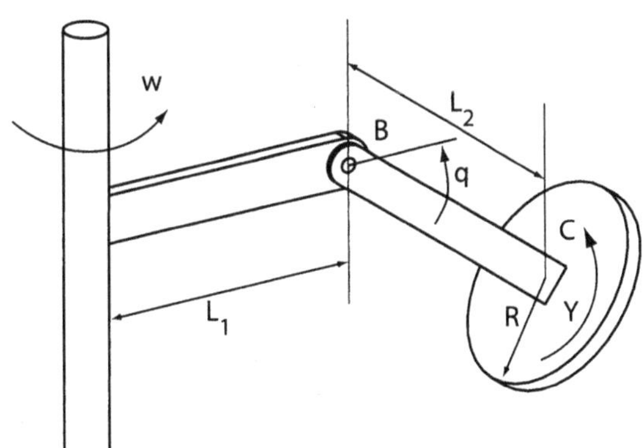

Fig. 3.21. Rotating T bar driving a shaft and a wheel

to the T bar at point B. The bar rotates with a constant angular velocity ω. The wheel does not slip during its rotation.

a) Find the angular velocity of the wheel with respect to R (fixed inertial frame).
b) Find the partial angular velocity of the wheel in R.
c) Give the expression for the velocity of an arbitrary point fixed on the wheel in R and derive its corresponding partial velocity array.

3.8. For the multibody as shown in Figure 3.18, give the expression for the velocity of the mass center of body 3 in R. (Assume that all bodies are connected by revolute joints.)

MATLAB Program to Compute the Kinematics of Open Chain Multibody System

MAIN PROGRAM

MATLAB program to compute the kinematics of open chain multibody system.

```
% clear all removes all variables from the workspace
and frees up system memory
clear all
%-----
%INPUT
%-----
% Reads inpu.m to get the input data
% SHIFTER OR TRANSFORMATION MATRIX
% SHIFTER MATRIX OF BODY 'K' WITH RESPECT TO
  ITS PARENT
%-------------------------
  -----------------------------
% S10,S21,S32,S43.....Sn(n-1)
% Function "shift" generates the shifter matrix using
% generalized coordinates
for i=1:n
b(i).si=shift(i,x)
end
% SHIFTER MATRIX OF BODY 'K' WITH RESPECT TO
% REFERENCE FRAME
%-----------------------------
  -----------------------------
% SHIFTERS S10,S20,S30,S40.....Sn0 are calculated
% using the below formula
% Sn0 = Sn(n-1) * S(n-1)(n-2) * S(n-2)(n-3)..*
  S(n-(n-1))0
for k=1:n
i3=eye(3)
for j=k:-1:1
i3=i3*b(j).si
end
b(k).s=i3
end
%%%%%%%%%%%%%%%%%%%%%%%%%%%%%%%%%%%%%%%%
% SKEW MATRIX
```

```
%%%%%%%%%%%%%%%%%%%%%%%%%%%%%%%%%%%%%%%%%
% SKEW MATRIX RELATING TIME DERIVATIVE OF
% THE SHIFTER AND
% SHIFTER MATRIX BETWEEN
% BODY 'K' AND ITS PARENT USING EQN: 3.5.22
for i=1:n
b(i).ski=skew(i,x,dx)
end
% SKEW MATRIX RELATING TIME DERIVATIVE OF THE SHIFTER
% AND SHIFTER MATRIX BETWEEN
% BODY 'K' AND THE REFERENCE FRAME USING
% EQN: 3.5.24 & 3.5.25
for i=1:n
b(i).skt=zeros(3)
for j=1:i-1
k=i
b(i).skt = b(i).skt + (b(k).si * b(k-1).ski *
  (b(k).si)')
k=k-1
end
b(i).sk = b(i).skt + b(i).ski
end
%---------------------------
% TIME DERIVATIVE OF SHIFTER
%---------------------------
% USING EQN: 3.5.19
for i=1:n
b(i).ds = b(i).sk * b(i).s
end
%-------------------------
% PARTIAL ANGULAR VELOCITY
%-------------------------
% USING EQN: 3.6.9
for i=1:n
b(i).pav=[1 0 0; 0 1 0; 0 0 1]
for k=1:(i-1)
b(i).pav=cat(1, b(i).pav, b(k).s)
end
for j=1:(n-i)
b(i).pav=cat(1, b(i).pav, zeros(3))
end
end
%-----------------------------------------
% ANGULAR VELOCITY OF THE BODY WITH RESPECT
% TO REFERENCE FRAME
%-----------------------------------------
```

```
% USING EQN: 3.6.5
for i=1:n
b(i).av = dx*b(i).pav
end
%-----------------------------------------
% TIME DERIVATIVE OF PARTIAL ANGULAR VELOCITIY
%-----------------------------------------
% USING EQN: 3.7.5
for i=1:n
b(i).dpav=[0 0 0; 0 0 0; 0 0 0]
for k=1:(i-1)
b(i).dpav=cat(1, b(i).dpav, b(k).ds)
end
for j=1:(n-i)
b(i).dpav=cat(1, b(i).dpav, zeros(3))
end
end
  %-----------------------
  % GENERALIZED SPEED VECTOR
  %-----------------------
  % USING EQN: 3.8.48
y=[]
for i=1:n
y = cat(2, y, b(i).av)
end
 %----------------------------------
 %SKEW MATRIX OF BODY VECTORS(Sq {\&} Sr)
 %----------------------------------
 % USING EQN: 3.8.22 {\&} 3.8.23
for i=2:(n+1)
b(i).sq=bskew(i,q)
end
for j=1:n
b(j).sr=bskew(j,r)
end
 %-----------------------
 % PARTIAL VELOCITY MATRIX
 %-----------------------
 % Partial velocity matrix of Body 'K' with respect
 % to reference frame
 % associated with generalized speed vector
 % {\{}y{\}}
 % USING EQN: 3.8.25
for i=1:n
b(i).pvel=[]
t=0
```

```
for k=1:i-1
j=i
b(i).pvel=cat(1, b(i).pvel, (b(j).sq * b(j-1).s))
t=t+1
end
b(i).pvel=cat(1,b(i).pvel, (b(i).sr * b(i).s))
t=t+1
if t$<$n
for k=(t+1):n
b(i).pvel=cat(1, b(i).pvel, zeros(3))
end
end
end
 %--------------------------------------------
 % TIME DERIVATIVE OF PARTIAL VELOCITY MATRIX
 %--------------------------------------------
 % USING EQN: 3.9.2
for i=1:n
b(i).dpvel=[]
t=0
for k=1:i-1
j=i
b(i).dpvel=cat(1, b(i).dpvel, (b(j).sq * b(j-1).sk *
  b(j-1).s))
t=t+1
end
b(i).dpvel=cat(1,b(i).dpvel, (b(i).sr * b(i).sk *
  b(i).s))
t=t+1
if t$<$n
for k=(t+1):n
b(i).dpvel=cat(1, b(i).dpvel, zeros(3))
end
end
end
```

FUNCTION M FILES

The following four M-files are called by the above main program while execution.

1).

```
 % % % % % % % % % % % % % % % %
 %----SHIFT.M----
 % % % % % % % % % % % % % % % %
 % SHIFTER MATRIX OF BODY 'K' WITH RESPECT
```

```
 % TO ITS PARENT
function s = shift(k,x)
al=x(3*k-2);
be=x(3*k-1);
ga=x(3*k);
sal=[1 0 0; 0 cos(al) sin(al); 0 -sin(al) cos(al)];
sbe=[cos(be) 0 -sin(be); 0 1 0; sin(be) 0 cos(be)];
sga=[cos(ga) sin(ga) 0; -sin(ga) cos(ga) 0; 0 0 1];
s=sga*sbe*sal;
2).
 % % % % % % % % % % % % % % % %
 %----SKEW.M-----
 % % % % % % % % % % % % % % % %
 % SKEW MATRIX RELATING TIME DERIVATIVE OF
 % THE SHIFTER
 % AND SHIFTER MATRIX BETWEEN
% BODY 'K' AND ITS PARENT USING EQN: 3.5.22
function sk = skew(k,x,dx)
al=x(3*k-2);
be=x(3*k-1);
ga=x(3*k);
dal=dx(3*k-2);
dbe=dx(3*k-1);
dga=dx(3*k);
sk=[0 ((dal*sin(be))+dga) ((dal*sin(ga)*cos(be))-dbe*
  cos(ga));
((-dal*sin(be))-dga) 0 ((dal*cos(be)*cos(ga))+(dbe*
  sin(ga)));
((-dal*sin(ga)*cos(be))+(dbe*cos(ga)))
((-dal*cos(be)*cos(ga))-(dbe*sin(ga))) 0];
3).
 % % % % % % % % % % % % % % % %
 %----BSKEW.M----
 % % % % % % % % % % % % % % % %
 % SKEW MATRIX OF BODY VECTORS(Sq {\&} Sr)
 % USING EQN: 3.8.22 {\&} 3.8.23
function sq = bskew(j,q)
sq=zeros(3)
for i=1:3
for k=1:3
if (i==1) {\&} (k==2)
sq(i,k) = -q(3*j)
sq(k,i) = -sq(i,k)
elseif (i==1) {\&} (k==3)
sq(i,k) = q(3*j-1)
sq(k,i) = -sq(i,k)
```

```
elseif (i==2) {\&} (k==3)
sq(i,k) = -q(3*j-2)
sq(k,i) = -sq(i,k)
end
end
end
4).
 % % % % % % % % % % % % % %
 %---INPU.m----
 % % % % % % % % % % % % % %
 %This file reads the input data from the file
 %''ex.dat'' and saves
 %them in vector form
n = csvread('ex.dat',1,0,[1,0,1,0])
constr = csvread('ex.dat',4,0,[4,0,4,(3*n-1)])
q = csvread('ex.dat',7,0,[7,0,7,(3*(n+1)-1)])
r = csvread('ex.dat',10,0,[10,0,10,(3*n-1)])
x = csvread('ex.dat',13,0,[13,0,13,(3*n-1)])
dx = csvread('ex.dat',16,0,[16,0,16,(3*n-1)])

INPUT FILE - (ex.dat)
NO OF BODIES
3

CONSTRAINTS
0,0,1,0,0,1,0,1,0

q VECTOR
0,0,0,935,0,0,960,0,0,980,0,0

r VECTOR
425,0,0,443,6,0,370,-3,0

INITIAL ANGLE
0,0,0.5,0,0,0.5,0,0.5,0

INITIAL ANGULAR VELOCITY
0,0,2,0,0,3,0,4,0
```

RESULTS

```
>> b(1).pav

ans =
1 0 0
0 1 0
0 0 1
0 0 0
0 0 0
0 0 0
0 0 0
0 0 0
0 0 0

>> b(1).dpav

ans =
0 0 0
0 0 0
0 0 0
0 0 0
0 0 0
0 0 0
0 0 0
0 0 0
0 0 0

>> b(1).pvel

ans =
0 0 0
0 0 -425.0
-203.7 372.9 0
0 0 0
0 0 0
0 0 0
0 0 0
0 0 0

0 0 0

>> b(1).dpvel

ans =
0 0 0
0 0 0
```

```
-745.9 -407.5 0
0 0 0
0 0 0
0 0 0
0 0 0
0 0 0
0 0 0

>> b(2).pav

ans =
1.00 0 0
0 1.00 0
0 0 1.00
0.87 0.47 0
-0.47 0.87 0
0 0 1.00
0 0 0
0 0 0
0 0 0

>> b(2).dpav

ans =
0 0 0
0 0 0
0 0 0
-0.96 1.75 0
-1.75 -0.96 0
0 0 0
0 0 0
0 0 0
0 0 0

>> b(2).pvel

ans =
0 0 0

0 0 -935.0000
-448.26 820.5397 0
0 0 6.0000
0 0 -443.0000
-376.0135 234.3051 0
0 0 0
0 0 0
0 0 0
```

```
>> b(2).dpvel

ans =
1.0e+003 *
0 0 0
0 0 0
-1.6411 -0.8965 0
0 0 0
0 0 0
-1.1715 -1.8801 0
0 0 0
0 0 0
0 0 0
```

Appendix: Kinematic Notation and Skew Matrices

Index Notation and Dyadics

Once they are mastered, the following notational conventions shorten and clarify the formulas. In a given expression, an index is called a *dummy* if it is used in a summation; otherwise, those indices are free. Consider the equation.

$$a_i = \sum_{j=1}^{n}\sum_{k=1}^{m} b_{ijk}c_j d_k + \sum_{j=1}^{n}\sum_{k=1}^{m} f_{ijk}g_{jk} + h_i \tag{3.217}$$

Indices j and k are dummy indices, whereas i is a free index. Equation (3.217) will be written as

$$a_i = b_{ijk}c_j d_k + f_{ijk}g_{jk} + h_i \quad (j = 1, \ldots, n; k = 1, \ldots, m) \tag{3.218}$$

Equation (3.218) reads as follows: For each i there is a sum on j from 1 to n and from 1 to m on k, respectively. For example.

$$a_1 = \underbrace{\sum_{j=1}^{n}\sum_{k=1}^{m} b_{1jk}}_{\text{term 1}} + \underbrace{\sum_{j=1}^{n}\sum_{k=1}^{m} f_{1jk}}_{\text{term 2}} + h_1 \tag{3.219}$$

The summation on terms 1 and 2 must be carried out first, and then adding to it the value of h_1 gives the value for a_1.

Dummy indices must appear *twice* or *more in a single term.* Then a summation is implied over the range of the repeated index. A free index must appear once in each term of the equation. The ranges of the indices should be specified.

Usually, subscript and superscript indices are treated the same way for the summation convention. However, throughout this book, since extensive labeling is necessary for the various tensors, the superscripts will be reserved for part of the labeling, and hence repeated superscripts will not imply summation unless explicitly stated otherwise.

Example 3.13.1
To illustrate how the index notation works, consider the following expression:

$$a_i = b_{ij} c_j \quad (i = 1, \ldots, 3; \; j = 1, \ldots, 4) \tag{3.220}$$

Note how the index j is repeated twice; that is, it appears once in b and once in c. Of course, j is a dummy index and i is a free index. Equation (3.220) reads as follows: For each i there is a sum on j from 1 to 4; for instance,

$$a_1 = b_{11}c_1 + b_{12}c_2 + b_{13}c_3 + b_{14}c_4 \tag{3.221}$$

$$a_2 = b_{21}c_1 + b_{22}c_2 + b_{23}c_3 + b_{24}c_4 \tag{3.222}$$

and

$$a_3 = b_{31}c_1 + b_{32}c_2 + b_{33}c_3 + b_{34}c_4 \tag{3.223}$$

In a matrix form we can write

$$\begin{bmatrix} a_1 \\ a_2 \\ a_3 \end{bmatrix} = \begin{bmatrix} b_{11} & b_{12} & b_{13} & b_{14} \\ b_{21} & b_{22} & b_{23} & b_{24} \\ b_{31} & b_{32} & b_{33} & b_{34} \end{bmatrix} = \begin{bmatrix} c_1 \\ c_2 \\ c_3 \\ c_4 \end{bmatrix} \tag{3.224}$$

or

$$[a] = [b][c] \tag{3.225}$$

Equation (3.220) is more convenient, as the summation is indicated over the range of the dummy indexes, and simple "do loops" in computation coding will usually handle such simple operations.

A (Cartesian) coordinate system is based on three mutually perpendicular coordinate planes. These intersect in three mutually perpendicular lines called axes. Let $\bar{n}_1, \bar{n}_2, \bar{n}_3$ denote unit vectors along these axes. The set of the three mutually perpendicular unit vectors characterizing the axis frame will be denoted as $\bar{n}$. In multibody systems it is usually necessary to deal with more than one axis frame. To designate unit vectors of different axis frames, superscripts will be used (e. g., $\bar{n}^1, \bar{n}^r$).

One can represent a vector $\bar{v}$ in different axis frames as

$$v = V_i^r \bar{n}_i^r = V_i^s \bar{n}_i^s \tag{3.226}$$

where the V_i^s, $i = 1, 2, 3$, represent the coordinates (components) of $\bar{v}$ in $\bar{n}^s$, and the V_i^r, $i = 1, 2, 3$, represent the coordinates $\bar{v}$ in $\bar{n}^s$. Denoting this direction cosine as S_{ij}^{rs}, which means the transformation from frame r to s,

$$\bar{n}_i^r = S_{ij}^{rs} \bar{n}_i^s$$

or in matrix form,

$$\{\bar{n}^r\} = \{S^{rs}\}\,\{\bar{n}^s\} \tag{3.227}$$

Since $\{\bar{n}^s\}\{\bar{n}^s\}^T = I$ (identity matrix),

$$[S^{rs}] = \{\bar{n}^r\}\{\bar{n}^s\}^T$$

or

$$S_{ij}^{rs} = \begin{bmatrix} \bar{n}_1^s \cdot \bar{n}_1^r & \bar{n}_2^s \cdot \bar{n}_1^r & \bar{n}_3^s \cdot \bar{n}_1^r \\ \bar{n}_1^s \cdot \bar{n}_2^r & \bar{n}_2^s \cdot \bar{n}_2^r & \bar{n}_3^s \cdot \bar{n}_2^r \\ \bar{n}_1^s \cdot \bar{n}_3^r & \bar{n}_2^s \cdot \bar{n}_3^r & \bar{n}_3^s \cdot \bar{n}_3^r \end{bmatrix} \tag{3.228}$$

where the components of S^{rs} are equivalent to the direction cosine matrix. For the same axis frame unit vectors, S_{ij} is equal to the Kronecker delta:

$$\bar{n}_i \cdot \bar{n}_j = \delta_{ij} = \begin{cases} 1 & \text{if } i = j \\ 0 & \text{if } i \neq j \end{cases} \quad (i, j = 1, 2, 3) \tag{3.229}$$

One can represent the unit vectors of one axis frame as linear combinations of the unit vectors of another axis frame,

$$\bar{n}_i^s = s_{ij}^{sr} \bar{n}_j^r \quad (i, j = 1, 2, 3) \tag{3.230}$$

which means the transformation of the frame s to r. Equations (3.229) and (3.230) lead to (3.227). Hence the 3×3 direction cosine matrix is the coordinate transformation matrix, which is equivalent to the transpose S_{ij} given by equation (2.40).

Substituting equation (3.230) into (3.227) for both $\bar{n}_i^s$ and $\bar{n}_j^s$, it is seen that the transformation matrix S^{sr} is orthogonal, that is,

$$S_{ij}^{sr^T} S_{jk}^{sr} = \delta_{ik} \tag{3.231}$$

The coordinates of a vector $\bar{v}$ in different axis frames are related through the transformation matrix. Substituting equation (3.230) into (3.226), we obtain

$$V_i^s = S_{ij}^{sr} V_j^r \quad (i, j = 1, 2, 3) \tag{3.232}$$

or

$$V_i^s = V_j^r S_{ji}^{rs} \tag{3.233}$$

Second-order tensors also play an important role in multibody dynamics. The tensor product $\bar{\bar{d}}$ of two vectors $\bar{a}$ and $\bar{b}$ yields and second-order tensor (called a *dyad*). The tensors product of two vectors is defined as

$$\bar{d} = \bar{a}\bar{b} = a_i \bar{n}_i b_j \bar{n}_j \tag{3.234}$$

Then, $\bar{\bar{d}}$ can be expressed as

$$\bar{\bar{d}} = d_{ij}\bar{n}_i\bar{n}_j \tag{3.235}$$

where d_{ij} is the component matrix of $\bar{\bar{d}}$:

$$d_{ij} = a_i b_j \tag{3.236}$$

Suppose that a_i^r and b_i^r are the components of $\bar{a}$ and $\bar{b}$ in $\bar{n}^r$, and a_i^s and b_i^s in $\bar{n}_i^s$. Then the components of $\bar{\bar{d}}$ in $\bar{n}_i^s$, d_{ij}^s can be obtained from d_{ij}^r and the coordinate transformation matrix S^{sr}. Using equations (3.230) and (3.235), we obtain

$$d_{ij}^s = S_{ik}^{sr} d_{k\ell}^r S_{\ell j}^{rs} \tag{3.237}$$

The scalar and cross products of a dyad $\bar{\bar{d}}$ from the left and right with a vector $\bar{v}$:

$$\bar{\bar{d}} \cdot \bar{v} = (\bar{a}\bar{b}) \cdot \bar{v} = \bar{a}(b \cdot \bar{v}) \tag{3.238}$$

$$\bar{v} \cdot \bar{\bar{d}} = \bar{v} \cdot (\bar{a}\bar{b}) = (\bar{v} \cdot \bar{a})b \tag{3.239}$$

$$\bar{v} \times \bar{\bar{d}} = (\bar{v} \times \bar{a})\bar{b} \tag{3.240}$$

$$\bar{\bar{d}} \times \bar{v} = \bar{a}(b \times \bar{v}) \tag{3.241}$$

$\bar{\bar{d}} \cdot \bar{v}$ and $\bar{v} \cdot \bar{\bar{d}}$ yield vectors that are equal only if the component matrix d_{ij} is symmetric. This is because if $\bar{c} = \bar{b}\bar{a}$ then $c_{ij} = d_{ji}$.

In general, an nth-order tensor T is defined such that:

1. In any axis frame there is a rule for associating $\bar{T}$ with a unique ordered set of 3^n quantities (called the components of $\bar{T}$).

2. If $T_{i_1,\cdots,i_n}^r$ and $T_{i_1,\cdots,i_n}^s$ are the components of $\bar{T}$ in $\bar{n}^r$ and $\bar{n}^s$, then

$$T_{i_1,\cdots,i_n}^r = S_{i_1 j_1}^{rs} \cdots S_{i_n j_n}^{rs} T_{j_1,\cdots,j_n}^s \tag{3.242}$$

For a second-order tensor $\bar{\bar{d}}$, equation (3.242) yields (3.237).

Let $\bar{A}$ and $\bar{B}$ be second-order tensors. Then $\bar{W} = \bar{A} \cdot \bar{B}$ is defined as

$$W_{ij} = A_{iq} B_{qj} \tag{3.243}$$

Let $\bar{A}$ be an nth-order tensor and $\bar{B}$ be an mth-order tensor. Then $\bar{A} \cdot \bar{B}$ is a tensor of order $n+m-2$. Let $\bar{W} = \bar{A}.\bar{B}$ where $\bar{A}$ is an nth-order tensor and $\bar{B}$ is an mth-order tensor; then $\bar{W}$ is an $(n+m)$th-order tensor.

In multibody dynamics, second-order tensors with symmetric and skew-symmetric component matrices are met. The inertia tensor and the unit tensor (a unit tensor $\bar{I}$ is a tensor whose component matrix is the identity matrix, $I_{ij} = \delta_{ij}$) have symmetric component matrices. Tensors with skew-symmetric component matrices are met in connection with vector cross products.

Other important properties of vectors that will be useful in the derivation of some equations are

$$\bar{a} \times (\bar{b} \times \bar{c}) = (\bar{a} \cdot \bar{c})\bar{b} - (\bar{a} \cdot \bar{b})\bar{c} \tag{3.244}$$

$$\bar{a} \cdot (\bar{b} \times \bar{c}) = \bar{a} \times \bar{b} \cdot \bar{c} \tag{3.245}$$

Example 3.13.2
Let $\bar{n}_1^s$, $\bar{n}_2^s$ and $\bar{n}_3^s$ be three mutually perpendicular unit vectors as shown in Figure 3.22. They are attached to a reference frame S. A reference frame R obtained by rotating $\bar{n}_1^s$ counterclockwise is, on the other hand, represented by $\bar{n}_1^r$, $\bar{n}_2^r$ and $\bar{n}_3^r$. Given $\bar{a}$ and $\bar{b}$ vectors as shown, find the dyadic $\bar{\bar{d}}$ given by $\bar{a}\bar{b}$. Give the coefficients of $\bar{\bar{d}}$ with respect to S and R, respectively, assume that $\theta = 30°$).

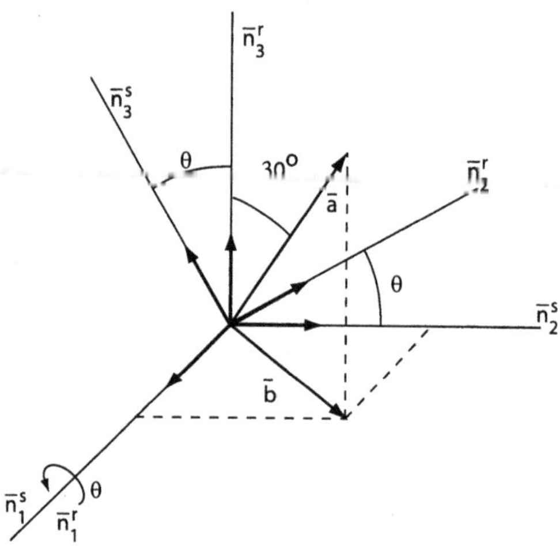

Fig. 3.22. Vectors and reference frames

Solution
The vectors $\bar{a}$ and $\bar{b}$ in the reference frame S can be represented in matrix form as

$$a^s = \begin{bmatrix} \frac{1}{\sqrt{3}} \\ \frac{1}{\sqrt{3}} \\ \frac{2}{3}\sqrt{2} \end{bmatrix} \qquad b^s = \begin{bmatrix} \frac{1}{\sqrt{3}} \\ \frac{1}{\sqrt{3}} \\ 0 \end{bmatrix}$$

$$S_{ij}^{rs} = \begin{bmatrix} 1 & 0 & 0 \\ 0 & \cos\theta & \sin\theta \\ 0 & -\sin\theta & \cos\theta \end{bmatrix} = \begin{bmatrix} 1 & 0 & 0 \\ 0 & \frac{\sqrt{3}}{2} & \frac{1}{2} \\ 0 & -\frac{1}{2} & \frac{\sqrt{3}}{2} \end{bmatrix} \tag{3.246}$$

The components of the dyadic in reference frame S are

$$d_{ij}^{s} = a^s b^s = \begin{bmatrix} \frac{1}{\sqrt{3}} \\ \frac{1}{\sqrt{3}} \\ \frac{2}{3}\sqrt{2} \end{bmatrix} \begin{bmatrix} \frac{1}{\sqrt{3}} & \frac{1}{\sqrt{3}} & 0 \end{bmatrix} = \frac{1}{3} \begin{bmatrix} 1 & 1 & 0 \\ 1 & 1 & 0 \\ 2\sqrt{\frac{2}{3}} & 2\sqrt{\frac{2}{3}} & 0 \end{bmatrix} \tag{3.247}$$

The components of the dyadic in the reference frame R are given by

$$d_{ij}^{r} = S_{ik}^{rs} d_{k}^{s} S_{\ell j}^{sr}$$

$$= \frac{1}{3} \begin{bmatrix} 1 & 0 & 0 \\ 0 & \frac{\sqrt{3}}{2} & \frac{1}{2} \\ 0 & -\frac{1}{2} & \frac{\sqrt{3}}{2} \end{bmatrix} \begin{bmatrix} 1 & 1 & 0 \\ 1 & 1 & 0 \\ 2\sqrt{\frac{2}{3}} & 2\sqrt{\frac{2}{3}} & 0 \end{bmatrix} \begin{bmatrix} 1 & 0 & 0 \\ 0 & \frac{\sqrt{3}}{2} & -\frac{1}{2} \\ 0 & \frac{1}{2} & \frac{\sqrt{3}}{2} \end{bmatrix}$$

$$= \frac{1}{3} \begin{bmatrix} 1 & \frac{\sqrt{3}}{2} & -\frac{1}{2} \\ \frac{\sqrt{3}}{2} + \sqrt{\frac{2}{3}} & \frac{3}{4} + \frac{\sqrt{2}}{2} & -\frac{\sqrt{3}}{4} - \frac{1}{\sqrt{6}} \\ \sqrt{2} - \frac{1}{2} & \sqrt{\frac{3}{2}} - \frac{\sqrt{3}}{4} & \frac{1}{4} - \frac{\sqrt{2}}{2} \end{bmatrix} \tag{3.248}$$

Example 3.13.3

For Example 3.2.2, find the relationship between the coefficients of $\overline{\overline{d}}$ in both R and S.

Solution

The components of the dyadic in reference frame R are given by

$$d_{ij}^{r} = S_{ik}^{rs} S_{\ell j}^{sr} d_{kl}^{s} \tag{3.249}$$

Then

$$d_{ij}^{r} = R d_{kl}^{s} \tag{3.250}$$

where

$$R = S_{ik}^{rs} S_{\ell j}^{sr}$$

Example 3.13.4
Consider the vectors $\bar{a}$, $\bar{b}$ and $\bar{c}$ shown in Figure 3.23.

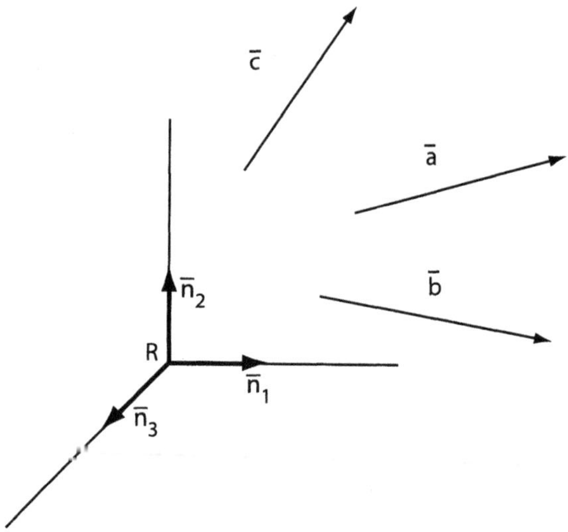

Fig. 3.23. Three arbitrary vectors in R

Let those vectors be given with respect to an inertial fixed frame R as

$$\bar{a} = a_i \bar{n}_i \quad (i = 1, 2, 3) \tag{3.251}$$

$$\bar{b} = b_j \bar{n}_j \quad (j = 1, 2, 3) \tag{3.252}$$

$$\bar{c} = c_k \bar{n}_k \quad (k = 1, 2, 3) \tag{3.253}$$

We define the dyadic $\bar{\bar{d}}$ to be

$$\bar{a}\bar{b} = a_i b_j \bar{n}_i \bar{n}_j \tag{3.254}$$

or simply

$$\bar{\bar{d}} = B_{ij} \bar{n}_i \bar{n}_j \tag{3.255}$$

Note that i and j are both dummy indexes in equation (3.255). Find an expression for

$$(\bar{a} \cdot \bar{\bar{d}}) \times \bar{c} \quad (\bar{a} \times \bar{c}) \cdot \bar{\bar{d}} \quad (\bar{a} \times \bar{b}) \times \bar{c} \tag{3.256}$$

Solution
(1) Substituting the expression for the dyadic $\bar{\bar{d}}$ in $(\bar{a} \cdot \bar{\bar{d}}) \times \bar{c}$ gives

$$\begin{aligned}
(\bar{a} \cdot \bar{\bar{d}}) \times \bar{c} &= [a_i \bar{n}_i \cdot B_{\ell k} \bar{n}_\ell \bar{n}_k] \times c_j \bar{n}_j \\
&= [a_i B_{\ell k} \delta_i \ell \bar{n}_k] \times c_j \bar{n}_j \\
&= a_i B_{\ell k} c_j \delta_i \ell e_{kjs} n_s
\end{aligned}$$

or

$$(\bar{a} \cdot \bar{\bar{d}}) \times \bar{c} = a_\ell B_{\ell k} c_j e_{kjs} \bar{n}_s \tag{3.257}$$

(2) Substituting the dyadic $\bar{\bar{b}}$ by equation (3.255) results in

$$\begin{aligned}(\bar{a} \times \bar{c}) \cdot \bar{\bar{d}} &= [a_i \bar{n}_i \times c_j \bar{n}_j] \cdot [B_{\ell k} \bar{n}_\ell \bar{n}_k] \\ &= \left[a_i c_j e_{ijt} \bar{n}_t\right] \cdot [B_{\ell k} \bar{n}_\ell \bar{n}_k] \\ &= a_i c_j B_{\ell k} e_{ijt} \delta_{t\ell} \bar{n}_k\end{aligned}$$

or

$$(\bar{a} \times \bar{c}) \cdot \bar{\bar{d}} = a_i c_j B_{tk} e_{ijt} \bar{n}_k \tag{3.258}$$

(3)

$$\begin{aligned}(\bar{a} \times \bar{b}) \times \bar{c} &= [a_i \bar{n}_i \times b_j \bar{n}_j] \times \bar{c}_k \bar{n}_k \\ &= [a_i b_j e_{ijt} \bar{n}_t] \times \bar{c}_k \bar{n}_k \\ &= a_i b_j c_k e_{ijt} e_{tks} \bar{n}_s\end{aligned} \tag{3.259}$$

In Example 3.13.4, we have illustrated the importance of the index notation in vectors. We should note that if we have a dot or cross product between two vectors, we must use a different index name for their labeling. In addition, we must keep in mind that the cross product between two unit vectors belonging to the same inertial frame can always be expressed as

$$\bar{n}_i \times \bar{n}_j = e_{ijt} \bar{n}_t \tag{3.260}$$

where the labeling of the permutation symbol comes from the two vectors $\bar{n}_i$ and $\bar{n}_j$ plus the resulting vector $\bar{n}_t$; therefore, we get i, j, t, in that order. Any interchange in the order of i, j and t associated with the permutation symbol will result in a change of sign. For example,

$$e_{ijt} = -e_{jit}$$

Meanwhile, we can show that

$$e_{ijk} e_{i\ell m} = \delta_{j\ell} \delta_{km} - \delta_{jm} \delta_{k\ell} \tag{3.261}$$

On the other hand, the dot product of two unit vectors yields the Kronecker delta, with its index notation resulting from the associated unit vectors. One final observation can be made on the omission of δ_{ij} from the resulting equation, where it can be achieved by replacing the index i by j in the same equation or j by i, since for $i \neq j$, $\delta_{ij} = 0$.

Skew Matrices

We can also use a skew matrix to express the cross product between two tensors. For example, let us assume that two vectors $\bar{a}$ and $\bar{b}$ are given by

$$\bar{a} = a_1\bar{n}_1 + a_2\bar{n}_2 + a_3\bar{n}_3 \tag{3.262}$$

$$\bar{b} = b_1\bar{n}_1 + b_2\bar{n}_2 + b_2\bar{n}_2 \tag{3.263}$$

where $\bar{n}_1, \bar{n}_2$ and $\bar{n}_3$ are unit vectors representing the inertial fixed frame R_0. Now let us define $\bar{c}$ such that

$$\bar{c} = \bar{a} \times \bar{b} \tag{3.264}$$

The determinant of the following matrix yields $\bar{c}$

$$\bar{c} = \begin{bmatrix} \bar{n}_1 & \bar{n}_2 & \bar{n}_3 \\ a_1 & a_2 & a_3 \\ b_1 & b_2 & b_3 \end{bmatrix} \tag{3.265}$$

in a matrix form the components of $\bar{c}$ can be expressed as:

$$\{c\} = [a]^{sk}\{b\} \tag{3.266}$$

where

$$\{c\} = [c_1 \;\; c_2 \;\; c_3]^T \tag{3.267}$$

$$\{b\} = [b_1 \;\; b_2 \;\; b_3]^T \tag{3.268}$$

where $[a]^{sk}$ denotes the skew matrix of tensor $\bar{a}$ and is given by

$$[a]^{sk} = \begin{bmatrix} 0 & -a_3 & a_2 \\ a_3 & 0 & -a_1 \\ -a_2 & a_1 & 0 \end{bmatrix} \tag{3.269}$$

Equation (3.266) can also be expressed as

$$\{c\} = -[b]^{sk}\{a\} \tag{3.270}$$

where

$$[b]^{sk} = \begin{bmatrix} 0 & -b_3 & b_2 \\ b_3 & 0 & -b_1 \\ -b_2 & b_1 & 0 \end{bmatrix} \tag{3.271}$$

For the scalar triple product an expression that is useful in dynamics we define the determinant d as

$$\begin{aligned} d &= \bar{a} \cdot (\bar{b} \times \bar{c}) = [a \;\; b \;\; c] = [b \;\; c \;\; a] = [c \;\; a \;\; b] \\ &= -[b \;\; a \;\; c] = -[a \;\; c \;\; b] = -[c \;\; b \;\; a] \end{aligned} \tag{3.272}$$

$$|d| = \begin{bmatrix} a_1 & a_2 & a_3 \\ b_1 & b_2 & b_3 \\ c_1 & c_2 & c_3 \end{bmatrix}$$

making use of the skew matrices associated with $\bar{a}$, $\bar{b}$ and $\bar{c}$ we can express the above scalar triple product as

$$d = \{a\}^T[b]^{sk}\{c\} = \{b\}^T[c]^{sk}\{a\} = \{c\}^T[a]^{sk}\{b\}$$
$$= -\{b\}^T[a]^{sk}\{c\} = -\{c\}^T[b]^{sk}\{a\} = -\{a\}^T[c]^{sk}\{b\} \quad (3.273)$$
$$d = a_1(b_2c_3 - b_3c_2) + a_2(b_3c_1 - b_1c_3) + a_3(b_1c_2 - b_2c_1) \quad (3.274)$$

Example 3.13.5
Find the matrix representation of $(\bar{a}\times\bar{b})\times\bar{c}$ making use of the skew matrices of $\bar{c}$ and $\bar{a}$.

Solution
Let $\bar{d} = (\bar{a} \times \bar{b}) \times \bar{c}$; then

$$\{d\} = ([A]^{sk}\{b\}) \times \{c\}$$
$$= -[c]^{sk}[A]^{sk}\{b\} \quad (3.275)$$

References

1. Huston, R. L., *Multibody Dynamics,* Butterworth-Heinemann, Stoneham, MA, 1990.
2. Amirouche, F. M. L., computational Methods in Multibody Dynamics, *Class Notes,* Department of Mechanical Engineering University of Illinois at Chicago 1987.
3. Kane, T. R., *Analytical Elements of Mechanics,* Vol. **2**, Academic Press, New York, 1961.
4. Sokolnikoff, I. S., *Tensor Analysis Theory and Applications to Geometry and Mechanics of Continua,* Wiley, New York, 1964.
5. Likins, P. W., *Elements of Engineering Mechanics,* McGraw-Hill, New York, 1973, pp. 522–526.
6. Robertson, R. E., and Schwertassek, R., *Dynamics of Multibody Systems,* Springer-Verlag, Berlin, 1988.

4

Modeling of Forces in Multibody Systems

4.1 Introduction

The modeling of forces in the context of multibody dynamics is undoubtedly challenging and important. Although the reader might be familiar with some of those concepts, their contribution in a general system of dynamics is the main focus of this chapter. We make use of concepts developed in Chapter 3 to extend the formalism of forces in the presence of a treelike configuration of interconnected rigid bodies. To this end, a brief review of the concepts of forces and their classification is presented; then a general form of the representation of those forces in the dynamics of multibody systems is developed.

4.2 Forces, Moments and Equivalence Force Systems

Definition

A force is characterized by a vector and is thus defined by three criteria: (1) a point of application, (2) the line of action and (3) a magnitude, orientation, and sense. Consider Figure 4.1, where a rigid body is acted upon by two forces, $\bar{F}_1$ and $\bar{F}_2$. $\bar{F}_1$ acts on the rigid body at point A defined as the point of application, and LL' denotes the line of action that is colinear and parallel to the vector $\bar{F}_1$. In this case $\bar{F}_1$ is referred to as a *fixed vector* or *bound vector*. Points of application are very important in determining the resulting effect of the force on the rigid body. Similarly, $\bar{F}_2$ has the point of application B and line of action PP'. So if a vector is associated with a point in space, it is called a bound vector; otherwise, it is a *free vector*.

A moment in itself is a vector not associated with a particular point of application. The moment of a force about a point is defined as the cross product of the position vector with the force, where the position vector $\bar{r}$ defines the position from the given point at which the moment is considered to any point on the line action of the force. Figure 4.2 depicts such a situation. Let $\bar{M}_A$ be such a moment; then

$$\bar{M}_A = \bar{r} \times \bar{F} \tag{4.1}$$

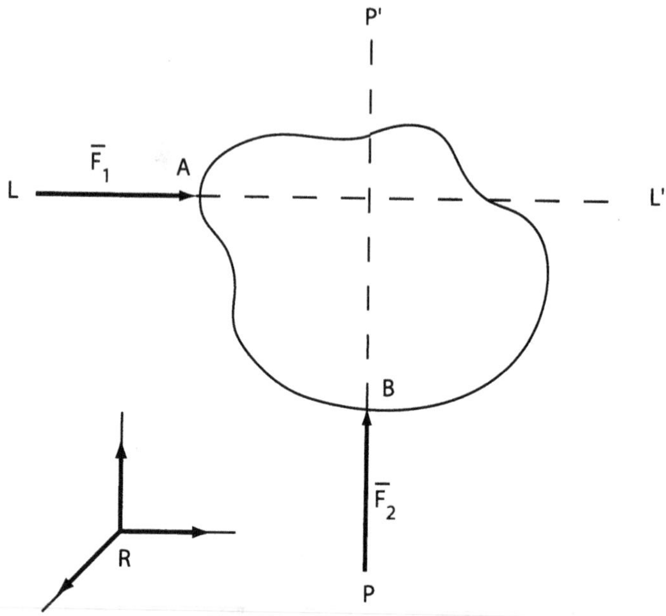

Fig. 4.1. Forces acting on a rigid body

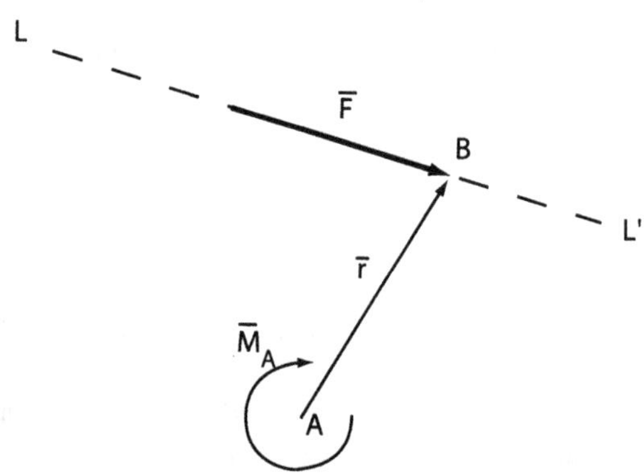

Fig. 4.2. Force acting along LL', position vector $\bar{r}$ from point A to B, and moment M_A about point A

Note that point B must be selected such that $\bar{r}$ is conveniently calculated or such that the cross product results in a simpler expression. On the other hand, the moment of a force about a given line is found by

$$M_{PP'} = [(\bar{r} \times \bar{F}) \cdot \bar{\lambda}]\lambda \tag{4.2}$$

where $\bar{\lambda}$ is a unit vector along line PP', and $\bar{r}$ is defined to be the position vector from a point on line PP' to the line of application of the force. Figure 4.3 depicts such a situation.

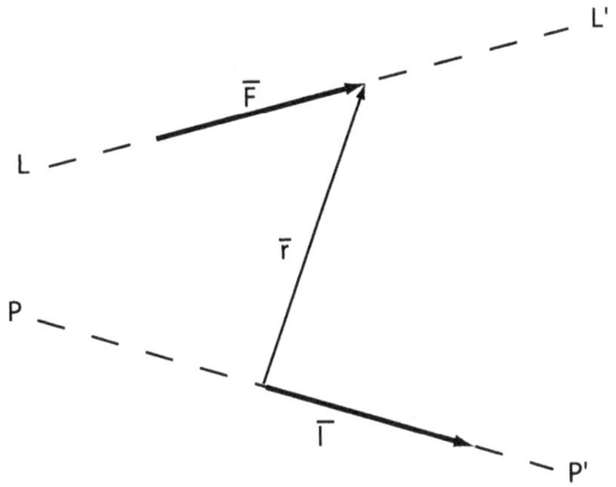

Fig. 4.3. Force acting on LL' and a unit vector $\bar{\lambda}$ defined along an arbitrary line PP'

Consider a rigid body A as shown in Figure 4.4, where it is subject to an arbitrary number of forces. Let the number of these forces be m; then we define $\bar{R}$ to be the resultant force of all the forces $\bar{F}_i$ $(i = 1, \dots, m)$ applied to the rigid body A.

$$\bar{R} = \sum_{i=1}^{m} \bar{F}_i \tag{4.3}$$

$$\bar{R} = R_1\bar{n}_1 + R_2\bar{n}_2 + R_3\bar{n}_3 \tag{4.4}$$

where R_1, R_2 and R_3 are the resulting components of $\bar{R}$ along $\bar{n}_1, \bar{n}_2$ and $\bar{n}_3$, respectively. Let $p_1, p_2, \dots, p_m$ denote the points of applications of the forces $\bar{F}_1, \dots, \bar{F}_m$ as shown in Figure 4.4. All the forces $\bar{F}_i(i = 1, \dots, m)$ are fixed vectors (bound); the resultant force, however, is not. This simply says that the resultant force is not usually associated with a particular point.

When a particular rigid body is acted upon by a set of forces as shown in Figure 4.4, it is usually convenient to replace all those forces by an equivalent force system consisting a resultant force $\bar{R}$ and a moment $\bar{M}_o$ defined as

$$\bar{M}_o = \sum_{i=1}^{m} (\bar{r}_i \times \bar{F}_i) = r_c \times \overline{R} \tag{4.5}$$

where $\bar{r}_i(i = 1, \dots, m)$ is the position vector from the point of application $P_i(i = 1, \dots, m)$ to point O fixed in a reference frame R, and r_c defines the position from the resultant force point application to the fixed reference frame.

4.3 Generalized Active Force

Consider a multibody system as depicted by Figure 4.5, where all bodies are interconnected by free joints, allowing each rigid body to undergo six degrees of freedom. Let

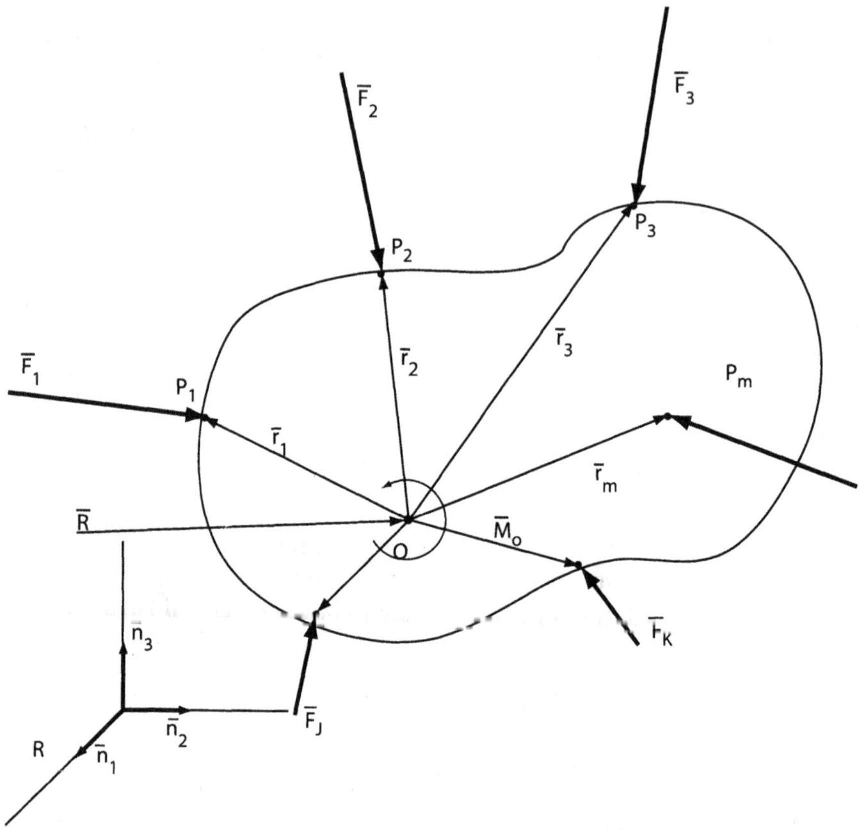

Fig. 4.4. Resultant force of m forces acting on a rigid body

body k be subject to external forces $\bar{F}(i = 1, \ldots, s)$ applied at points of application p_1 through p_s.

The equivalent force system of the forces acting on body k can be replaced by a resultant force $\bar{F}_R$ acting at a point Q and a couple torque $\bar{T}$. The generalized active forces are defined as

$$F = \bar{F}_R \cdot \frac{\partial \bar{\upsilon}_k}{\partial \dot{x}_\ell} + \bar{T} \cdot \frac{\partial \bar{\omega}^k}{\partial \dot{x}_\ell} \tag{4.6}$$

$x_\ell(\ell = 1, \ldots, n)$ are the generalized coordinates that describe the relative orientation and translation between bodies. Chapter 3 gives a full description of the organization of those coordinates. $\bar{\upsilon}_k$ and $\bar{\omega}^k$ are the mass center velocity and angular velocity of body k in the inertial reference frame R. Their respective partial derivatives are obtained from the procedures developed in Chapter 3. More explicitly, equation (4.6) becomes

$$F = \bar{F}_R \cdot V^k_{\ell m} \bar{n}_m + \bar{T} \cdot \bar{\omega}^k_{\ell m} \bar{n}_m \tag{4.7}$$

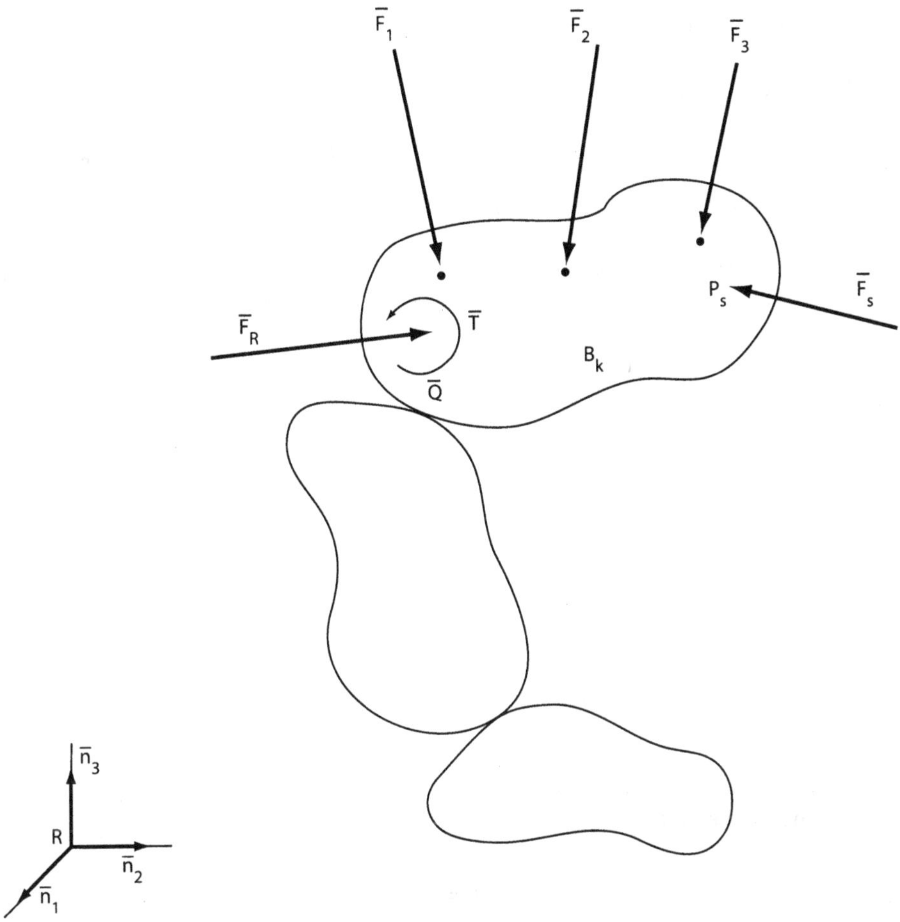

Fig. 4.5. Forces and torques acting on body B_k in a multibody system

Observe how the generalized active force is a scalar quantity, whereas $\bar{F}_R, \bar{T}$, and the partial derivatives are vectors. Both $\bar{F}_R$ and $\bar{T}$ can be expressed further with respect to the reference frame R along the unit vectors $\bar{n}_1$, $\bar{n}_2$ and $\bar{n}_3$ as

$$\bar{F}_R = F_{R_m}\bar{n}_m \tag{4.8}$$

and

$$\bar{T} = T_m\bar{n}_m \tag{4.9}$$

Hence equation (4.7) becomes

$$F = F_{R_m}V^k_{\ell m} + T_m\omega^k_{\ell m} \tag{4.10}$$

Remember that $\bar{F}_R$ and $\bar{T}$ are acting on body k and that in most cases there will be forces and torques applied simultaneously on different bodies of the system. To generalize equation (4.10), for the contribution of the generalized active forces to the system of N interconnected rigid bodies (open-chain multibody system), with respect to the reference frame R, we write equation (4.10) as

$$F_\ell = F_{R_{km}} V^k_{\ell m} + T_m \bar{\omega}^k_{\ell m} \tag{4.11}$$

where there is a sum from 1 to 3 on m for each ℓ and k is fixed. We should note that ℓ is associated with the generalized coordinates $x_\ell (\ell = 1, \ldots, n)$ and that equation (4.11) states that for each generalized coordinate x_ℓ there is one contribution to the generalized active forces F_ℓ. Or in a matrix notation we can write

$$\{f\} = \sum_k [V^k_w]\{F_k\} + \sum_k [\omega^k]\{T_k\} \tag{4.12}$$

where $[V^k_{\mathrm{w}}]$, a partial velocity matrix, is given by equation (3.122) but includes the terms $[V^k]$ and the transformation matrix $[W]$ needed in the generalized speeds, and $[\omega^k]$, a partial angular velocity matrix, is given by equation (3.44).

Example 4.3.1
Consider the planar two-link bar system shown in Figure 4.6. A constant torque $\bar{M}$ is applied at point B to drive link 2, and a force $\bar{P}$ acting at point c in the horizontal direction is used to move the system forward. Find the contribution of $\bar{P}$ and $\bar{M}$ to the generalized active forces.

Solution
The solution to this problem can proceed as follows. From equation (3.44) we get the partial angular velocity for links 1 and 2 as

$$[\omega^1] = \left[\frac{I}{0}\right] \tag{4.13}$$

$$[\omega^2] = \left[\frac{I}{S^{10}}\right] \tag{4.14}$$

where the transformation from body 1 to 0 is:

$$[S^{10}] = \begin{bmatrix} \cos(\theta 1 - 90°) & \sin(\theta 1 - 90°) & 0 \\ -\sin(\theta 1 - 90°) & \cos(\theta 1 - 90°) & 0 \\ 0 & 0 & 1 \end{bmatrix}$$

$$= \begin{bmatrix} \sin\theta 1 & -\cos\theta 1 & 0 \\ \cos\theta 1 & \sin\theta 1 & 0 \\ 0 & 0 & 1 \end{bmatrix} \tag{4.15}$$

Let the position to the center mass of link 1 be given by

$$\{r_1\} = \tfrac{1}{2}L\begin{bmatrix}1 & 0 & 0\end{bmatrix}^T \tag{4.16}$$

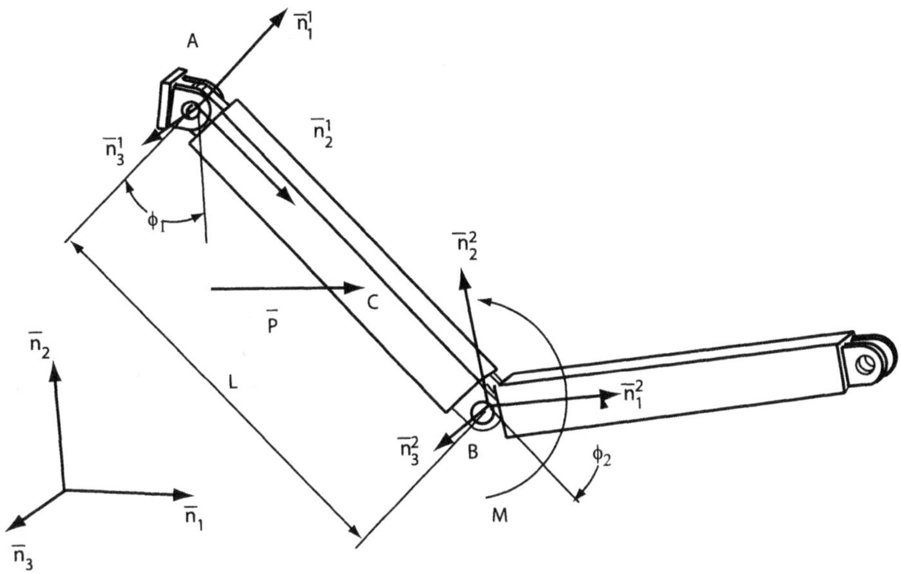

Fig. 4.6. Two-link manipulator with a force P acting on link 1 at the center mass

and its corresponding skew matrix

$$[Sr_1] = \frac{1}{2}L \begin{bmatrix} 0 & 0 & 0 \\ 0 & 0 & -1 \\ 0 & 1 & 0 \end{bmatrix} \tag{4.17}$$

From equation (3.122), we obtain the partial velocity

$$[V^1_w] = [W] \begin{bmatrix} [Sr1][S^{10}] \\ 0 \end{bmatrix} \tag{4.18}$$

where $[W]$ is found to be

$$[W] = \begin{bmatrix} \hat{\text{I}} & \hat{\text{I}} \\ 0 & \hat{\text{I}} \end{bmatrix} \quad \text{and} \quad \hat{\text{I}} = \begin{bmatrix} 0 & & \\ & 0 & \\ & & 1 \end{bmatrix}$$

Note that we have introduced $[W]$ to both eliminate the degrees of freedom that are not contributing to the system and make the following transformation hold:

$$\{y\}^T = \{\dot{x}\}^T [W] \tag{4.19}$$

In a more explicit form, we can see how the $\hat{\text{I}}$ is obtained through the generalized coordinates, derivatives, and generalized speeds.

$$\begin{bmatrix}0 & 0 & \dot{\theta}_1 & 0 & 0 & \dot{\theta}_1+\dot{\theta}_2\end{bmatrix} = \begin{bmatrix}0 & 0 & \dot{\theta}_1 & 0 & 0 & \dot{\theta}_2\end{bmatrix} \begin{bmatrix}\hat{\mathrm{I}} & \hat{\mathrm{I}} \\ 0 & \hat{\mathrm{I}}\end{bmatrix}$$

Substituting $[W]$, $[Sr_1]$ and $[S^{10}]$ in equation (4.18), $[V_{\mathrm{w}}^1]$ yields

$$[V_{\mathrm{w}}^1] = \begin{bmatrix} 0 & 0 & 0 \\ 0 & 0 & 0 \\ \frac{1}{2}L\cos\theta_1 & \frac{1}{2}L\sin\theta_1 & 0 \\ 0 & 0 & 0 \\ 0 & 0 & 0 \\ 0 & 0 & 0 \end{bmatrix} \tag{4.20}$$

From equation (4.12), we have the generalized active force expressed in matrix form as

$$\{f\} = \sum_k [V_{\mathrm{w}}^k]\{F\} + \sum_k [\omega^k]\{T\}$$

Expanding for two bodies and setting $\{F_2\} = 0$, we get

$$\{f\} = [V_{\mathrm{w}}^1]\{F_1\} + [\omega^1]\{T_1\} + [\omega^2]\{T_2\} \tag{4.21}$$

where

$$\{F_1\} = \begin{bmatrix}P & 0 & 0\end{bmatrix}^T \tag{4.22}$$

There is an opposing moment due to M in link 1 given by

$$\{T_1\} = \begin{bmatrix}0 & 0 & -M\end{bmatrix}^T \tag{4.23}$$

That same moment must act in the opposite direction in link 2, which yields

$$\{T_2\} = \begin{bmatrix}0 & 0 & M\end{bmatrix}^T \tag{4.24}$$

Therefore, the generalized active forces become

$$\begin{aligned}\{f\} &= \begin{bmatrix}0 & 0 & \frac{1}{2}PL\cos\theta_1 & 0 & 0 & 0\end{bmatrix}^T + \begin{bmatrix}0 & 0 & 0 & 0 & 0 & M\end{bmatrix}^T \\ &= \begin{bmatrix}0 & 0 & \frac{1}{2}PL\cos\theta_1 & 0 & 0 & M\end{bmatrix}^T\end{aligned} \tag{4.25}$$

Example 4.3.2

In Example 4.3.1 force P is acting in the horizontal direction. Now, suppose that P is acting on link 1 with a certain angle, say, α, shown in Figure 4.7. Find the new generalized active force expression.

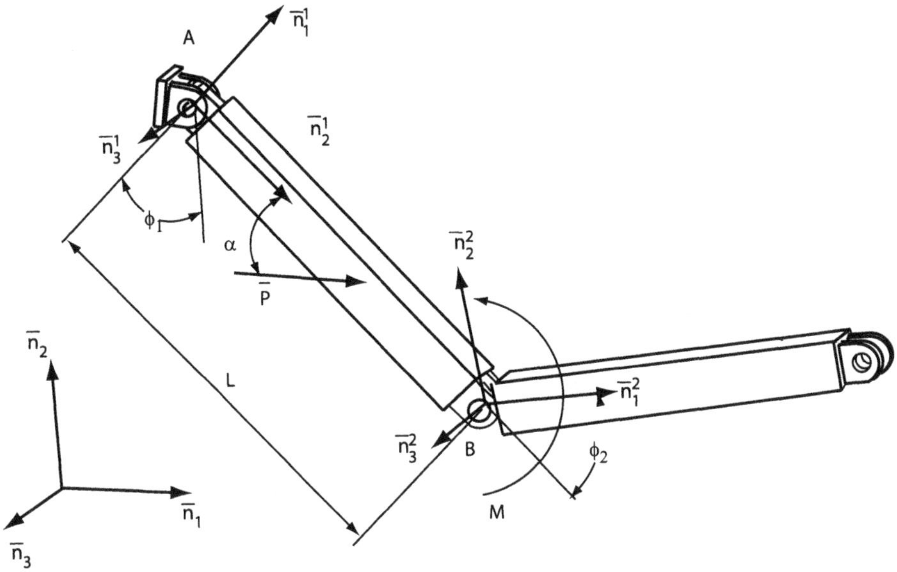

Fig. 4.7. Two-link manipulator with a force P acting on link 1 and a P moment acting at joint 2

Solution

The components of the force with respect to the reference frame R are

$$\{F1\}^T = [\,P\cos\alpha \qquad P\sin\alpha \qquad 0\,][S^{10}] \tag{4.26}$$

which can be factorized into the following form:

$$\{F1\} = P[S^{01}]\begin{bmatrix}\cos\alpha\\ \sin\alpha\\ 0\end{bmatrix}$$

$$= P\begin{bmatrix}\sin\theta1 & \cos\theta1 & 0\\ -\cos\theta1 & \sin\theta1 & 0\\ 0 & 0 & 1\end{bmatrix}\begin{bmatrix}\cos\alpha\\ \sin\alpha\\ 0\end{bmatrix}$$

$$= P\begin{bmatrix}\sin\theta1\cos\alpha + \cos\theta1\sin\alpha\\ -\cos\theta1\cos\alpha + \sin\theta1\sin\alpha\\ 0\end{bmatrix} = P\begin{bmatrix}\sin(\theta1+\alpha)\\ -\cos(\theta1+\alpha)\\ 0\end{bmatrix} \tag{4.27}$$

The generalized active force in matrix form is given by expanding equation (4.12)

$$\{f\} = [V_{\mathrm{w}}^1]\{F_1\} + [\omega^1]\{T_1\} + [\omega^2]\{T_2\} \tag{4.28}$$

where $[V_{\mathrm{w}}^1]$, $[\omega^1]$, $[\omega^2]$, $[T_1]$ and $[T_2]$ are given by equations (4.20), (4.13), (4.14), (4.23) and (4.24), respectively. Hence

$$\{f\} = \begin{bmatrix} 0 \\ 0 \\ PL[\cos\theta_1 \sin(\theta_1 + \alpha) - \sin\theta_1 \cos(\theta_1 + \alpha)] - M \\ 0 \\ 0 \\ M \end{bmatrix} = \begin{bmatrix} 0 \\ 0 \\ \frac{1}{2}PL\sin\alpha - M \\ 0 \\ 0 \\ M \end{bmatrix} \tag{4.29}$$

Compared with equation (4.25), we know that force P acting on the fixed coordinate system (Example 4.3.1) is different from that acting on the local coordinate system (for this example).

4.4 Modeling of Springs and Dampers at the Joints

The contribution of forces and torques to the generalized active forces resulting from axial and torsional springs and dampers acting at the joints can be expressed as follows. First, consider Figure 4.8(a), where springs and dampers connect two rigid bodies. Let $\bar{\delta}$ be the unit vector along which the spring and the damper do work. From the law of action and reaction, we can say that both the spring and damper can be replaced by forces equal in magnitude and opposite in direction acting on bodies j and k, respectively. Let the force acting on body j be $\bar{f}_j$ and the force acting on body k be $\bar{f}_k$ or $-\bar{f}_j$. In addition, let $\bar{\mu}$ measure the displacement along $\bar{\delta}$; then $\bar{f}$ can be expressed as

$$\bar{f} = f(\mu, \dot{\mu})\bar{\delta} \tag{4.30}$$

where $f(\mu, \dot{\mu})$ is a function of the physical properties of the springs and dampers employed.

The contribution of $\bar{f}_j$ and $\bar{f}_k$ to act as generalized active forces can be obtained using equation (4.6) as

$$F_\ell = \bar{f}_j \cdot \frac{\partial \bar{\upsilon}^{P1}}{\partial \dot{x}_\ell} + \bar{f}_k \cdot \frac{\partial \bar{\upsilon}^{P2}}{\partial \dot{x}_\ell} \tag{4.31}$$

Since $\bar{f}_k = -\bar{f}_j$, we get

$$F_\ell = \bar{f}_j \cdot \left[\frac{\partial \bar{\upsilon}^{P1}}{\partial \dot{x}_\ell} - \frac{\partial \bar{\upsilon}^{P2}}{\partial \dot{x}_\ell}\right]$$

In a matrix form the generalized active force vector becomes

$$\{F\} = ([V_w^1] - [V_w^2])\{f_j\}$$

where $\bar{\upsilon}^{P1}$ and $\bar{\upsilon}^{P2}$ are velocities of points $P1$ and $P2$, respectively, in R, and $[V_w^1]$ and $[V_w^2]$ are their partial velocity matrices. Note that no torsional springs and dampers are involved; hence there is no contribution from the second term of equation (4.6).

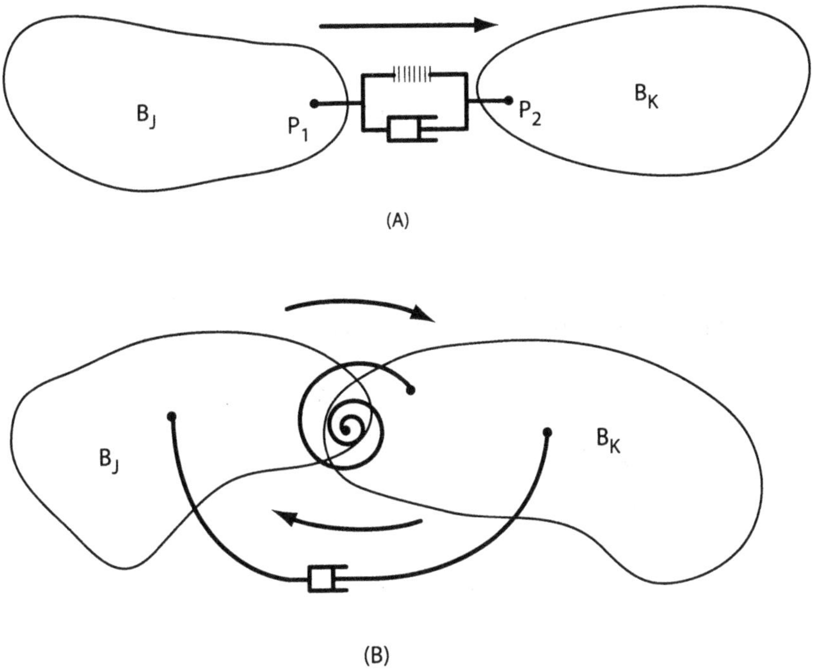

Fig. 4.8. (a) Linear spring and damper connecting two arbitrary rigid bodies; (b) Two arbitrary rigid bodies with torsional spring and torsional damper

Next, consider Figure 4.8(b), where the two bodies are hinged at point O. A torsion spring and a damper resist the relative motion of the bodies. Let the spring be undeformed when the two bodies are in some initial configuration. If we denote by θ the relative angle of rotation and $\bar{\eta}$ a unit vector directed as shown, the torques exerted on bodies j and k, respectively, are

$$\bar{T}_j = T(\theta, \dot{\theta})\bar{\eta} \tag{4.32}$$

and

$$\bar{T}_k = -T(\theta, \dot{\theta})\bar{\eta} \tag{4.33}$$

where $T(\theta, \dot{\theta})$ is a function depending on the properties of the torsional spring and damper. It follows from equation (4.6) that the contributions of $\bar{T}j$ and $\bar{T}k$ to the generalized active force are

$$F_\ell = \bar{T}_j \cdot \frac{\partial \bar{\omega}^j}{\partial \dot{x}_\ell} + \bar{T}_k \cdot \frac{\partial \bar{\omega}^k}{\partial \dot{x}_\ell} = \bar{T}_j \cdot \left[\frac{\partial \bar{\omega}^j}{\partial \dot{x}_\ell} - \frac{\partial \bar{\omega}^k}{\partial \dot{x}_\ell} \right] \tag{4.34}$$

Once again we note that the partial angular velocities of bodies k and j and their corresponding torques can be explicitly expressed with respect to the unit vectors fixed in the inertial reference frame R, and therefore equation (4.34) takes the form

$$\{F\} = [\omega^j][T_j] + [\omega^k][T_k] = ([\omega^j] - [\omega^k])\{T_j\} \tag{4.35}$$

where $[\omega^j]$ and $[\omega^k]$ are partial angular velocity matrices of bodies j and k, respectively.

Example 4.4.1
Consider Example 4.3.1, where a coil linear spring is placed at joint B as shown in Figure 4.9. Find the contribution of the spring to the generalized active forces.

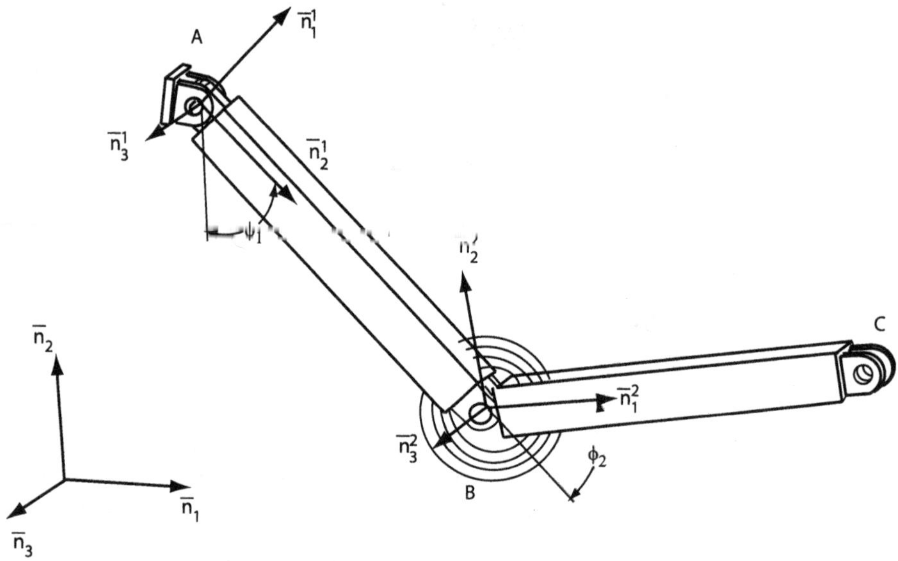

Fig. 4.9. Two-link manipulator with a torsional link at joint B

Solution
In Example 4.3.1, if we let $M = -k\theta_2$ and $P = 0$, we get the components of the torques

$$\begin{aligned}\{T_1\} &= \begin{bmatrix}0 & 0 & k\theta_2\end{bmatrix}^T \\ \{T_2\} &= \begin{bmatrix}0 & 0 & -k\theta_2\end{bmatrix}^T\end{aligned} \tag{4.36}$$

Then the contribution of the spring torque to the generalized active forces is found through the definition given by equation (4.35), that is

$$\begin{aligned}\{f\} &= [\omega^1]\{T1\} + [\omega^2]\{T_2\} \\ &= \begin{bmatrix}0 & 0 & 0 & k\theta_2 & 0 & 0 & -k\theta_2\end{bmatrix}^T\end{aligned} \tag{4.37}$$

Example 4.4.2

Linear springs and dampers acting at other points beside the joint are shown in Figure 4.10. Find the contribution of the spring to the generalized active force.

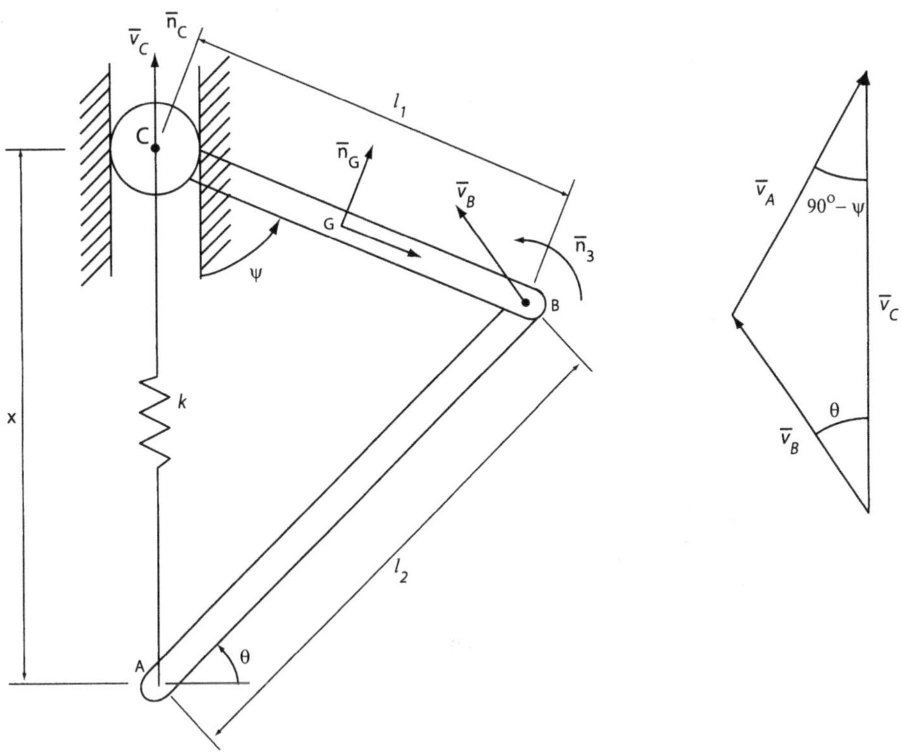

Fig. 4.10. Two-link mechanism constrained to move in the vertical direction

Solution

From the triangle, we can deduce the following relationship between the velocity vectors indicated:

$$\frac{|\bar{\upsilon}_{BC}|}{\sin\,\theta} = \frac{|\bar{\upsilon}_B|}{\sin(90^\circ - \psi)}$$

We can also see that

$$\ell_1 \sin\psi = \ell_2 \cos\theta \quad \sin\psi = \frac{\ell_2}{\ell_1}\cos\theta \tag{4.38}$$

Therefore, we write

$$\begin{aligned} |\bar{\upsilon}_{BC}| &= |\bar{\upsilon}_B|\,\frac{\sin\,\theta}{\cos\,\psi} \\ &= \ell_2\dot{\theta}\,\frac{\sin\theta}{\cos\psi} \end{aligned} \tag{4.39}$$

We also know that the velocity of a point at a distance l_1 on BC is

$$|\bar{\upsilon}_{BC}| = \ell_1 \, |\bar{\omega}_{BC}| \tag{4.40}$$

Combining equations (4.39) and (4.40), we obtain the angular velocity magnitude

$$|\bar{\omega}_{BC}| = \frac{\ell_2 \, \sin\theta}{\ell_1 \, \cos\psi}\dot{\theta} \quad \text{and} \quad |\bar{\omega}_{BC}| = -\frac{\ell_2 \, \sin\theta}{\ell_1 \, \cos\psi}\dot{\theta}\bar{n}_3 \tag{4.41}$$

The mass center velocity of point G as shown is therefore given by

$$\begin{aligned} \bar{\upsilon}_G &= \ell_2\dot{\theta}\bar{n}_B + \frac{1}{2}\ell_1 \cdot \frac{\ell_2 \, \sin\theta}{\ell_1 \, \cos\psi}\dot{\theta}\bar{n}_G \\ &= \ell_2\dot{\theta}\bar{n}_B + \frac{1}{2}\ell_2\frac{\sin\theta}{\cos\psi}\dot{\theta}\bar{n}_G \end{aligned} \tag{4.42}$$

The concentrated force and moment through G contributed by the spring are

$$\begin{aligned} \bar{F} &= -k(x - x_0)\bar{n}_c \\ &= -k[\ell_2(\sin\theta - \sin\theta_0) + \sqrt{\ell_1^2 - \ell_2^2\cos^2\theta} - \sqrt{\ell_1^2 - \ell_2^2\cos^2\theta_0}]\bar{n}_c \end{aligned} \tag{4.43}$$

and

$$\bar{M} = k(x - x_0) \cdot \frac{1}{2}\ell_1 \sin\psi\bar{n}_3 \tag{4.44}$$

The partial velocity of the center mass G is given by

$$\frac{\partial\bar{\upsilon}_G}{\partial\dot{\theta}} = \ell_2\bar{n}_B + \frac{1}{2}\ell_2\frac{\sin\theta}{\cos\psi}\bar{n}_G \tag{4.45}$$

Similarly the partial angular velocity of BC is

$$\frac{\partial\bar{\omega}_{BC}}{\partial\dot{\theta}} = -\frac{\ell_2 \, \sin\theta}{\ell_1 \, \cos\psi}\bar{n}_3 \tag{4.46}$$

The contribution of the spring to the generalized active force is then found from equation (4.6) to be

$$f = \bar{F} \cdot \frac{\partial\bar{\upsilon}_G}{\partial\dot{\theta}} + \bar{M} \cdot \frac{\partial\bar{\omega}_{BC}}{\partial\dot{\theta}} \tag{4.47}$$

which gives

$$f = -k(x - x_0)\ell_2 \left[\cos\theta + \sin\theta\frac{\ell_2\cos\theta}{x - \ell_2\sin\theta}\right] \tag{4.48}$$

Example 4.4.3
For the multibody system shown in Figure 4.11, the spring–damper set is acting on the arbitrary points of bodies i and j (2 and 4 in this case). Find the generalized force vector due to the spring–damper forces. It is assumed that each body has only one rigid body rotation, denoted by θ_i (revolute joints are assumed).

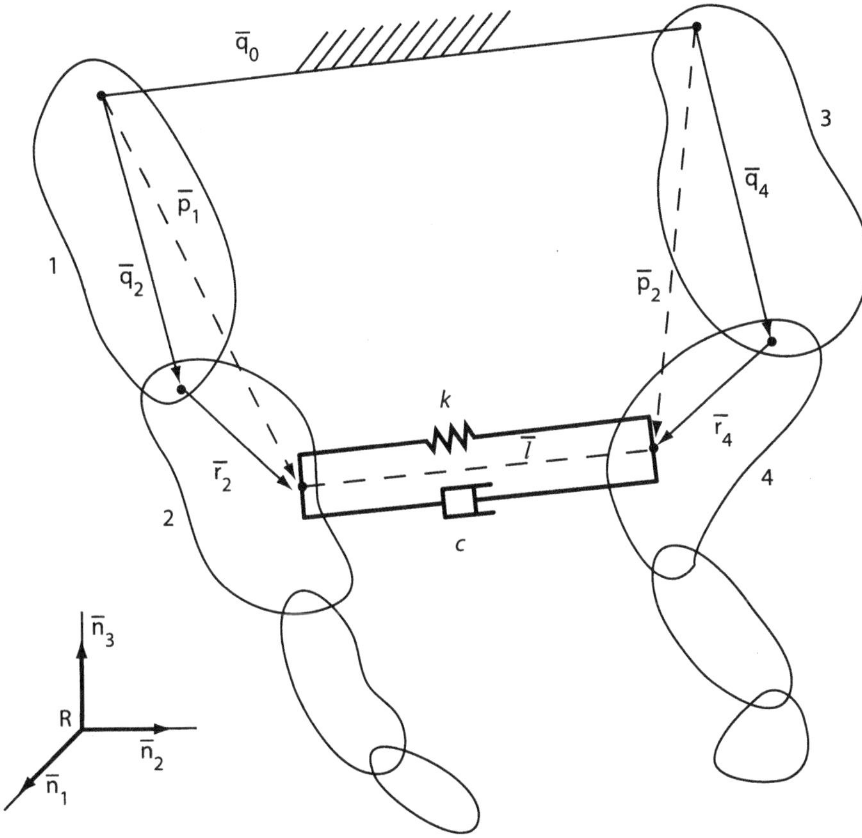

Fig. 4.11. Spring–damper force system acting between two arbitrary bodies in a multibody system

Solution
Suppose that position vectors of P_1 and P_2 in the local body coordinate systems are

$$\{P_1\}^T = \begin{bmatrix} P_{11} & P_{12} & P_{13} \end{bmatrix}$$
$$\{P_2\}^T = \begin{bmatrix} P_{21} & P_{22} & P_{23} \end{bmatrix}$$

In addition, let the local axes be chosen such that

$$\{r_2\}^T = \begin{bmatrix} r_2 & 0 & 0 \end{bmatrix}$$

and

$$\{r_4\}^T = \begin{bmatrix} r_4 & 0 & 0 \end{bmatrix} \tag{4.49}$$

where r_2 and r_4 denote the components of those vectors with respect to their local axis. The above is especially true for bodies composed of beams and plates. The partial velocity matrices for both points are

$$[V_{\mathrm{w}}^1] = [W] \begin{bmatrix} [Sq_2][S^{10}] \\ [Sr_2][S^{20}] \\ 0 \\ 0 \end{bmatrix} \tag{4.50}$$

$$[V_{\mathrm{w}}^2] = [W] \begin{bmatrix} 0 \\ 0 \\ [Sq_4][S^{30}] \\ [Sr_4][S^{40}] \end{bmatrix} \tag{4.51}$$

where

$$[W] = \begin{bmatrix} \hat{I} & \hat{I} & & 0 \\ 0 & \hat{I} & & \\ & & \hat{I} & \hat{I} \\ 0 & & 0 & \hat{I} \end{bmatrix} \tag{4.52}$$

and again

$$[\hat{I}] = \begin{bmatrix} 0 & & \\ & 0 & \\ & & 1 \end{bmatrix}$$

Recall, $[W]$ is the transformation matrix that describes the mapping $\{y\}^T = \{\dot{x}\}^T[W]$. For this problem, both $\{y\}$ and $\{\dot{x}\}$ can be found by inspection as

$$\{y\} = \begin{bmatrix} 0 & 0 & \dot{\theta}_1 & 0 & 0 & \dot{\theta}_1 + \dot{\theta}_2 & 0 & 0 & \dot{\theta}_3 & 0 & 0 & \dot{\theta}_3 + \dot{\theta}_4 \end{bmatrix}^T$$

and

$$\{\dot{x}\} = \begin{bmatrix} 0 & 0 & \dot{\theta}_1 & 0 & 0 & \dot{\theta}_2 & 0 & 0 & \dot{\theta}_3 & 0 & 0 & \dot{\theta}_4 \end{bmatrix}^T$$

The skew matrices of vectors q_i and r_i are

$$[Sq_2] = \ell_1 \begin{bmatrix} 0 & 0 & 0 \\ 0 & 0 & -1 \\ 0 & 1 & 0 \end{bmatrix} \tag{4.53}$$

$$[Sr_2] = r_2 \begin{bmatrix} 0 & 0 & 0 \\ 0 & 0 & -1 \\ 0 & 1 & 0 \end{bmatrix} \tag{4.54}$$

$$[Sr_4] = r_4 \begin{bmatrix} 0 & 0 & 0 \\ 0 & 0 & -1 \\ 0 & 1 & 0 \end{bmatrix} \tag{4.55}$$

$$[Sq_4] = \ell_3 \begin{bmatrix} 0 & 0 & 0 \\ 0 & 0 & -1 \\ 0 & 1 & 0 \end{bmatrix} \tag{4.56}$$

The transformation matrices of bodies 1, 2, 3 and 4 with respect to a reference frame R are

$$\left[S^{10}\right] = \begin{bmatrix} \cos\theta_1 & \sin\theta_1 & 0 \\ -\sin\theta_1 & \cos\theta_1 & 0 \\ 0 & 0 & 1 \end{bmatrix} \tag{4.57}$$

$$\left[S^{20}\right] = \begin{bmatrix} \cos(\theta_1 + \theta_2) & \sin(\theta_1 + \theta_2) & 0 \\ -\sin(\theta_1 + \theta_2) & \cos(\theta_1 + \theta_2) & 0 \\ 0 & 0 & 1 \end{bmatrix} \tag{4.58}$$

$$\left[S^{30}\right] = \begin{bmatrix} \cos\theta_3 & \sin\theta_3 & 0 \\ -\sin\theta_3 & \cos\theta_3 & 0 \\ 0 & 0 & 1 \end{bmatrix} \tag{4.59}$$

$$\left[S^{40}\right] = \begin{bmatrix} \cos(\theta_3 + \theta_4) & \sin(\theta_3 + \theta_4) & 0 \\ -\sin(\theta_3 + \theta_4) & \cos(\theta_3 + \theta_4) & 0 \\ 0 & 0 & 1 \end{bmatrix} \tag{4.60}$$

From the figure, we know that

$$\bar{q}_0 + \bar{p}_2 = \bar{\ell} + \bar{p}_1 \tag{4.61}$$

or

$$\ell = \bar{q}_0 + \bar{p}_2 - \bar{p}_1 \tag{4.62}$$

For body 1, the force due to the spring is

$$\bar{F}_{1k} = \left(\|\bar{\ell}\| - \ell_0\right) k \frac{\bar{\ell}}{\|\bar{\ell}\|} = \left(1 - \frac{\ell_0}{\|\bar{\ell}\|}\right) k\bar{\ell} \tag{4.63}$$

where k denotes the spring stiffness and ℓ_0 the initial length of the spring.

Expressing the position vectors with respect to unit vector $\bar{n}$ in R, we obtain

$$\bar{p}_1 = \bar{q}_2 + \bar{r}_2 = (\{q_2\}^T[S^{10}] + \{r_2\}^T[S^{20}])\{\bar{n}\} \tag{4.64}$$

$$\bar{p}_2 = \bar{q}_4 + \bar{r}_4 = (\{q_4\}^T[S^{30}] + \{r_4\}^T[S^{40}])\{\bar{n}\} \tag{4.65}$$

$$\bar{q}_0 = \{q_0\}^T\{\bar{n}\} \tag{4.66}$$

$$\begin{aligned}\bar{\ell} &= \bar{q}_0 + \bar{p}_2 - \bar{p}_1 \\ &= (\{q_0\}^T + \{q_4\}^T[S^{30}] + \{r_4\}^T[S^{40}] - \{q_2\}^T[S^{10}] - \{r_2\}^T[S^{20}])\{\bar{n}\}\end{aligned} \tag{4.67}$$

For body 2, the direction of the spring force is reversed; that is,

$$\{F_{2k}\} = -\{F_{1k}\} \tag{4.68}$$

where the components of $\{F_{1k}\}$ come from equation (4.63). Therefore, the generalized force is given by

$$\begin{aligned}\{F_k\} &= [V^1]\{F_{1k}\} + [V^2]\{F_{2k}\} \\ &= \left([V_w^1] - [V_w^2]\right)\{F_{1k}\} \\ &= [W]\begin{bmatrix}[Sq_2][S^{10}] \\ [Sr_2][S^{20}] \\ -[Sq_4][S^{30}] \\ -[Sr_4][S^{40}]\end{bmatrix}\{F_{1k}\}\end{aligned} \tag{4.69}$$

On the other hand, the force due to the damper is given by

$$\bar{F}_{1c} = c\left[(\bar{V}_2 - \bar{V}_1)\cdot\frac{\bar{\ell}}{\|\bar{\ell}\|}\right]\bar{\ell} \tag{4.70}$$

The term in brackets expresses the magnitude of the relative velocity in the direction of the damper (or $\bar{\ell}$), where the damping coefficient is c. The mass center velocity of bodies 1 and 2 is given by

$$\bar{V}_1 = \{\dot{x}\}^T[V_w^1]\{\bar{n}\} \tag{4.71}$$

$$\bar{V}_2 = \{\dot{x}\}^T[V_w^2]\{\bar{n}\} \tag{4.72}$$

From the law of action and reaction, the resulting damping forces on bodies 1 and 2 are

$$\bar{F}_{2c} = -\bar{F}_{1c} \tag{4.73}$$

The generalized active force due to the dampers is

$$\{f_c\} = [V_w^1]\{F_{1c}\} + [V_w^2]\{F_{2c}\}$$

$$= ([V_w^1] - [V_w^2])\{F_{1c}\}$$

$$= \begin{bmatrix} [Sq_2][S^{10}] \\ [Sr_2][S^{20}] \\ [-Sq_4][S^{30}] \\ [-Sr_4][S^{40}] \end{bmatrix} \{F_{1c}\} \tag{4.74}$$

where the components of $\{F_{1c}\}$ come from equation (4.70).

Combining equations (4.69) and (4.74), we get the generalized force due to the spring–damper set

$$\{f\} = \{f_k\} + \{f_c\}$$

$$= \begin{bmatrix} [Sq_2][S^{10}] \\ [Sr_2][S^{20}] \\ [-Sq_4][S^{30}] \\ [-Sr_4][S^{40}] \end{bmatrix} (\{F_{1k}\} + \{F_{1c}\}) \tag{4.75}$$

The explicit form of $\{f\}$ is left as an exercise.

Example 4.4.4

Consider the special case of Example 4.4.3 shown in Figure 4.12. Find the generalized force due to the spring–damper set.

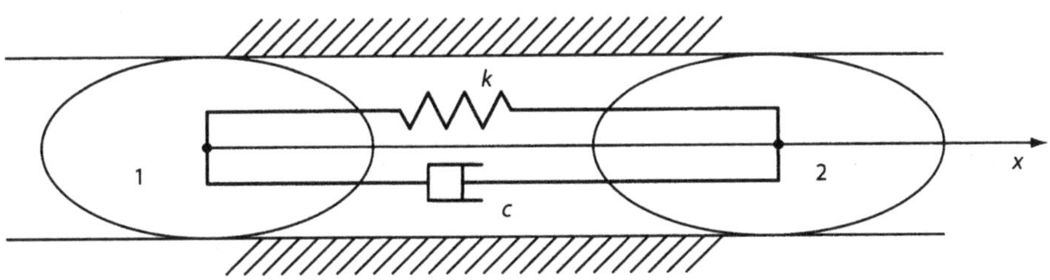

Fig. 4.12. Linear spring–damper system attached

Solution

The damper forces and spring forces of bodies 2 and 1 are

$$f_{2c} = c(\dot{x}_2 - \dot{x}_1) \tag{4.76}$$

$$f_{2k} = k(x_2 - x_1) \tag{4.77}$$

$$f_{1c} = -f_{2c} \tag{4.78}$$

$$f_{1k} = -f_{2k} \tag{4.79}$$

The velocities of the center masses of both bodies are

$$\{\bar{\upsilon}_1\} = \begin{bmatrix} \dot{x}_1 & 0 \end{bmatrix}^T \tag{4.80}$$

$$\{\bar{\upsilon}_2\} = \begin{bmatrix} 0 & \dot{x}_2 \end{bmatrix}^T \tag{4.81}$$

The generalized force f is explicitly given by

$$\begin{aligned} f &= \frac{\partial \bar{\upsilon}_1}{\partial \dot{x}_1} \cdot f_{1c} + \frac{\partial \bar{\upsilon}_2}{\partial \dot{x}_2} \cdot f_{2c} + \frac{\partial \bar{\upsilon}_1}{\partial \dot{x}_1} \cdot f_{1k} + \frac{\partial \bar{\upsilon}_2}{\partial \dot{x}_2} \cdot f_{2k} \\ &= \left[\frac{\partial \bar{\upsilon}_1}{\partial \dot{x}_1} - \frac{\partial \bar{\upsilon}_2}{\partial \dot{x}_2} \right] f_{1c} + \left[\frac{\partial \bar{\upsilon}_1}{\partial \dot{x}_1} - \frac{\partial \bar{\upsilon}_2}{\partial \dot{x}_2} \right] f_{1k} \\ &= \left[\frac{\partial \bar{\upsilon}_1}{\partial \dot{x}_1} - \frac{\partial \bar{\upsilon}_2}{\partial \dot{x}_2} \right] (f_{1c} + f_{1k}) \\ &= [k(x_1 - x_2) + c(\dot{x}_1 - \dot{x}_2)] \begin{bmatrix} 1 \\ -1 \end{bmatrix} \end{aligned} \tag{4.82}$$

4.5 Contact Forces

Forces resulting from contact of two rigid surfaces contribute nothing to the generalized forces f_ℓ. For instance, in a ball-and-socket joint, the contribution of the interaction forces is equal to zero. To prove that, let the point of contact be represented by point p of body j and p' of body k, as shown in Figure 4.13. Since the two bodies are in contact, the velocities of points p and p' in an inertial frame R must be equal, where

$$\bar{V}^p = \bar{V}^{p'} \tag{4.83}$$

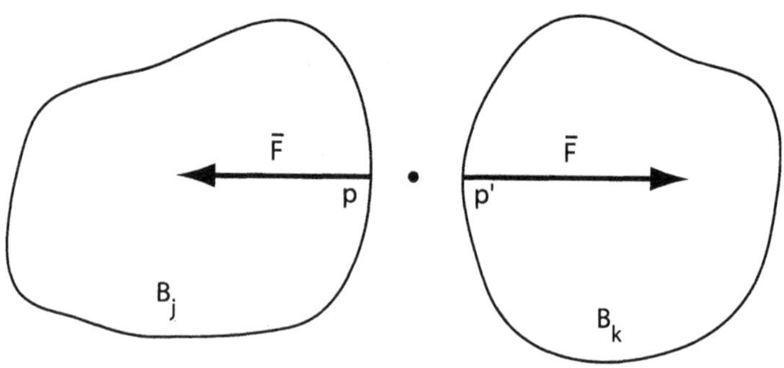

Fig. 4.13. Forces resulting from contact of two rigid surfaces

Let $\bar{F}_j$ and $\bar{F}_k$ be the resulting forces exerted on bodies j and k, respectively. Then, from equation (4.6), we write

$$\bar{F}_j \cdot \frac{\partial \bar{\upsilon}^p}{\partial \dot{x}_\ell} + (-\bar{F}_j) \cdot \frac{\partial \upsilon^{p'}}{\partial \dot{x}_\ell} = 0 \tag{4.84}$$

Contact forces are best illustrated by the rolling motion problem. If a rigid body B rolls on rigid body C, whose motion is prescribed as function of time, the contribution of the contact forces to the generalized active forces is zero. Contact forces fall into the category of workless forces or simply forces that do not contribute to the generalized active forces. One should also note from the definition of equation (4.6) that $\bar{F}_R$ can also be zero, if the partial velocities are zero and yet the forces are not, or when the forces are normal to the partial velocity vectors.

Example 4.5.1

For the two-link mechanism attached to the wheel shown in Figure 4.14, find the contribution of the contact force $\bar{N}_c$ to the generalized active force.

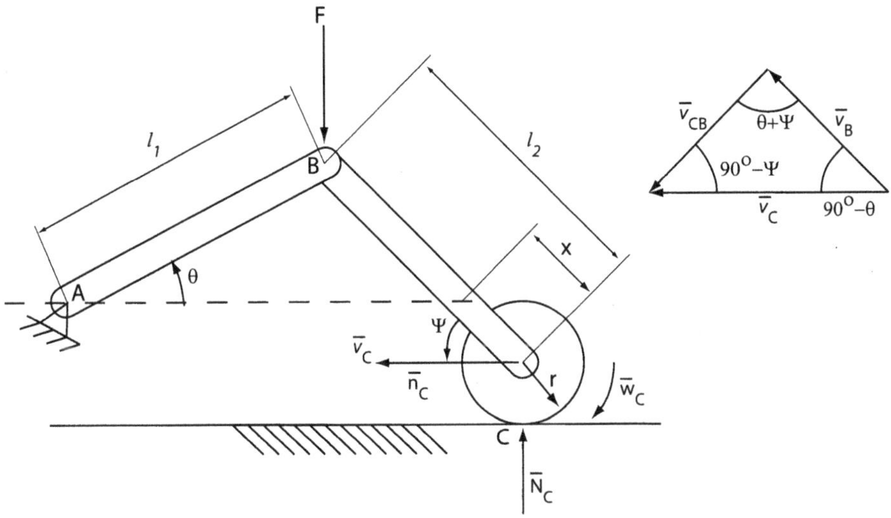

Fig. 4.14. Two-link mechanism attached to a rotating wheel

Solution

We must first determine the value of $\sin\psi$ using simple trigonometric identities:

$$\sin\psi = \frac{c-r}{x} \tag{4.85}$$

Using the "sine rule," the value of x is found through the relation:

$$\frac{\ell_2 - x}{\sin\theta} = \frac{\ell_1}{\sin\psi} \tag{4.86}$$

Hence,

$$x = \frac{(c-r)\ell_2}{\ell_1 \sin\theta + c - r} \tag{4.87}$$

Substituting the value of x into equation (4.85), we obtain

$$\sin\ \psi = \frac{\ell_1 \sin\theta + c - r}{\ell_2} \tag{4.88}$$

Furthermore, we establish the following relation

$$\frac{|\bar{V}_c|}{\sin(\theta + \psi)} = \frac{|\bar{V}_B|}{\sin(90^\circ - \psi)} \tag{4.89}$$

Hence we define the magnitude of $\bar{V}_c$ as

$$|\bar{V}_c| = \frac{\sin(\theta + \psi)}{\cos\psi}\ell_1\dot{\theta} \tag{4.90}$$

The vector $\bar{V}_c$ is then given by

$$\bar{V}_c = \frac{\sin(\theta + \psi)}{\cos\psi}\ell_1\dot{\theta}\bar{n}_c \tag{4.91}$$

Since $\bar{N}_c$ contributes only the concentrated force to wheel c through the mass center, the contribution of $\bar{N}_c$ to the generalized active force would be

$$f = \bar{N}_c \cdot \frac{\partial \bar{V}_c}{\partial \dot{\theta}} = 0 \tag{4.92}$$

The contribution of f is zero, due to the fact that $\bar{N}_c$ and $\bar{V}_c$ are mutually perpendicular.

4.6 Gravitational Forces

The gravitational forces exerted by the earth on a system of interconnected rigid bodies are usually given by

$$\bar{f}_i^G = m_i g \bar{n} \quad (i = 1, \ldots, N) \tag{4.93}$$

where N is the number of rigid bodies, m_i denotes the mass of body i, g is the gravity acceleration, and $\bar{n}$ is a unit vector directed vertically downward in the direction of g. Using the definition given by equation (4.6), the contribution of the gravitational forces to the generalized active forces are

$$F_\ell = \bar{f}_i^G \cdot \frac{\partial \bar{v}_i}{\partial \dot{x}_\ell} \tag{4.94}$$

where $\bar{v}_i$ denotes the velocity of the mass center in R of body i.

Example 4.6.1

Consider two pin-connected links, each of length L and mass m, shown in Figure 4.15. Find the contribution of $\bar{f}^G$ to the generalized active forces.

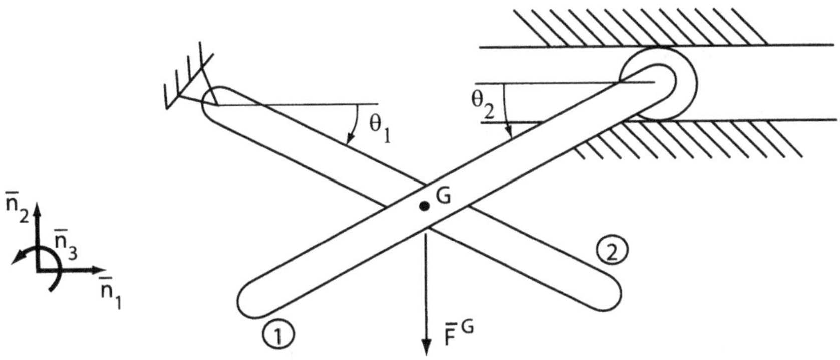

Fig. 4.15. Two pin-connected links

Solution

The general equation for the gravitational forces acting on each link is

$$\bar{f}_i = -mg\bar{n}_2 \quad (i = 1, 2) \tag{4.95}$$

The velocity of the mass center of each link is defined as

$$\bar{\upsilon}_1 = -\frac{1}{2}L\dot{\theta}_1(\sin\theta_1\bar{n}_1 + \cos\theta_1\bar{n}_2)$$

$$\bar{\upsilon}_2 = -\frac{1}{2}L\dot{\theta}_2(-\sin\theta_2\bar{n}_1 + \cos\theta_2\bar{n}_2)$$

Let

$$\theta_1 = \theta_2 = \theta$$

Taking the partial derivative with respect to $\dot{\theta}$, we get

$$\begin{aligned}\frac{\partial\bar{\upsilon}_1}{\partial\dot{\theta}} &= -\frac{1}{2}L(\sin\ \theta\bar{n}_1 + \cos\theta\bar{n}_2)\\ \frac{\partial\bar{\upsilon}_2}{\partial\dot{\theta}} &= -\frac{1}{2}L(-\sin\ \theta\bar{n}_1 + \cos\theta\bar{n}_2)\end{aligned} \tag{4.96}$$

Therefore, the contribution of $\bar{f}^G$ to the generalized active forces is

$$\begin{aligned}f &= \sum_{i=1}^{2} \bar{f}_i^G \cdot \frac{\partial\bar{\upsilon}_i}{\partial\dot{\theta}_i}\\ &= -mg\bar{n}_2 \cdot [-L\cos\ \theta\bar{n}_2]\\ &= mgL\cos\theta\end{aligned} \tag{4.97}$$

Example 4.6.2

Consider once more the multibody system shown in Figure 4.11. Find the generalized gravitational force vector. (Consider only the four bodies labeled 1 through 4.)

Solution
Suppose that the mass center coordinates of the four bodies are

$$\{r_1\} = \begin{bmatrix} r_1 & 0 & 0 \end{bmatrix}^T \tag{4.98}$$

$$\{r_2\} = \begin{bmatrix} r_2 & 0 & 0 \end{bmatrix}^T \tag{4.99}$$

$$\{r_3\} = \begin{bmatrix} r_3 & 0 & 0 \end{bmatrix}^T \tag{4.100}$$

$$\{r_4\} = \begin{bmatrix} r_4 & 0 & 0 \end{bmatrix}^T \tag{4.101}$$

Note that $r_i\,(i = 1, \ldots, 4)$ are different from those in Example 4.4.3, where r_2 and r_4 denote positions arbitrary points. Then we obtain

$$[Sr_i] = r_i \begin{bmatrix} 0 & 0 & 0 \\ 0 & 0 & -1 \\ 0 & 1 & 0 \end{bmatrix} \quad (i = 1, \ldots, 4) \tag{4.102}$$

where $[Sq_i]$ $(i = 2, 4)$ is given by equations (4.53) and (4.56). The generalized gravitational force is given by

$$\{f\} = \sum_{k=1}^{4} [V_w^k]\{F_k\} \tag{4.103}$$

where

$$\{F_k\} = \begin{bmatrix} 0 & -m_k g & 0 \end{bmatrix}^T \quad (k = 1, \ldots, 4) \tag{4.104}$$

The partial velocity matrices for all four mass center points of the bodies are

$$[V_w^1] = [W] \begin{bmatrix} [Sr_1][S^{10}] \\ 0 \\ 0 \\ 0 \end{bmatrix} = \begin{bmatrix} \hat{I} & \hat{I} & & 0 \\ 0 & \hat{I} & & \\ & & \hat{I} & \hat{I} \\ 0 & & 0 & \hat{I} \end{bmatrix} \begin{bmatrix} 0 & 0 & 0 \\ 0 & 0 & -r_1 \\ -r_1 s1 & r_1 c1 & 0 \\ & 0_{3\times 3} & \\ & 0_{3\times 3} & \\ & 0_{3\times 3} & \end{bmatrix} \tag{4.105}$$

$$= \begin{bmatrix} 0 & 0 & 0 \\ 0 & 0 & 0 \\ -r_1 s1 & r_1 c1 & 0 \\ & 0_{3\times 3} & \\ & 0_{3\times 3} & \\ & 0_{3\times 3} & \end{bmatrix}$$

where $c1 = \cos\theta_1$, $c2 = \cos\theta_2$, $s1 = \sin\theta_1$, and the partial velocities are explicitly given following the definitions outlined in Chapter 3.

$$
[V_{\mathrm{w}}^2] = [W] \begin{bmatrix} [Sq_1][S^{10}] \\ [Sr_2][S^{20}] \\ 0 \\ 0 \end{bmatrix}
$$

$$
= \begin{bmatrix} \hat{I} & \hat{I} & & 0 \\ 0 & \hat{I} & & \\ & & \hat{I} & \hat{I} \\ 0 & & 0 & \hat{I} \end{bmatrix}
\begin{bmatrix} 0 & 0 & 0 \\ 0 & 0 & -\ell_1 \\ -\ell_1 s1 & \ell_1 c1 & 0 \\ 0 & 0 & 0 \\ 0 & 0 & -r_2 \\ -r_2 s(1+2) & r_2 c(1+2) & 0 \\ & 0_{3\times 3} & \\ & 0_{3\times 3} & \end{bmatrix}
$$

$$
= \begin{bmatrix} 0 & 0 & 0 \\ 0 & 0 & 0 \\ -\ell_1 s1 - r_2 s(1+2) & --\ell_1 c1 - r_2 c(1+2) & 0 \\ 0 & 0 & 0 \\ 0 & 0 & 0 \\ -r_2 c(1+2) & r_2 c(1+2) & 0 \\ & 0_{3\times 3} & \\ & 0_{3\times 3} & \end{bmatrix} \tag{4.106}
$$

$$
[V_{\mathrm{w}}^3] = [W] \begin{bmatrix} 0 \\ 0 \\ [Sr_3][S^{30}] \\ 0 \end{bmatrix}
= \begin{bmatrix} \hat{I} & \hat{I} & & 0 \\ 0 & \hat{I} & & \\ & & \hat{I} & \hat{I} \\ 0 & & 0 & \hat{I} \end{bmatrix}
\begin{bmatrix} & 0_{3\times 3} & \\ & 0_{3\times 3} & \\ 0 & 0 & 0 \\ 0 & 0 & -r_3 \\ -r_3 s3 & r_3 c3 & 0 \\ & 0_{3\times 3} & \end{bmatrix}
$$

$$
= \begin{bmatrix} & 0_{3\times 3} & \\ & 0_{3\times 3} & \\ 0 & 0 & 0 \\ 0 & 0 & 0 \\ -r_3 s3 & r_3 c3 & 0 \\ & 0_{3\times 3} & \end{bmatrix} \tag{4.107}
$$

$$
[V_{\mathrm{w}}^4] = [W]\begin{bmatrix} 0 \\ 0 \\ [Sq_3][S^{30}] \\ [Sr_4][S^{40}] \end{bmatrix}
$$

$$
= \begin{bmatrix} \hat{I} & \hat{I} & & 0 \\ 0 & \hat{I} & & \\ & & \hat{I} & \hat{I} \\ 0 & & 0 & \hat{I} \end{bmatrix}
\begin{bmatrix} & 0_{3\times 3} & \\ & 0_{3\times 3} & \\ 0 & 0 & 0 \\ 0 & 0 & -\ell_3 \\ -\ell_3 s3 & \ell_3 c3 & 0 \\ 0 & 0 & 0 \\ 0 & 0 & -r_4 \\ -r_4 s(3+4) & r_4 c(3+4) & 0 \end{bmatrix}
$$

$$
= \begin{bmatrix} & 0_{3\times 3} & \\ & 0_{3\times 3} & \\ 0 & 0 & 0 \\ 0 & 0 & 0 \\ -\ell_3 s3 - r_4 s(3+4) & \ell_3 c3 + r_4 c(3+4) & 0 \\ 0 & 0 & 0 \\ 0 & 0 & 0 \\ -r_4 s(3+4) & r_4 c(3+4) & 0 \end{bmatrix} \tag{4.108}
$$

Hence

$$
\{f\} = \sum_{k=1}^{4} [V_{\mathrm{w}}^k]\{F_k\}
$$

$$
= -\begin{bmatrix} 0 \\ 0 \\ m_1 g r_1 c1 + m_2 g[\ell_1 c1 + r_2 c(1+2)] \\ 0 \\ 0 \\ m_2 g r_2 c(1+2) \\ 0 \\ 0 \\ m_3 g r_3 c3 + m_4 g[\ell_3 c3 + r_4 c(3+4)] \\ 0 \\ 0 \\ m_4 g r_4 c(3+4) \end{bmatrix} \tag{4.109}
$$

4.7 Generalized Inertia Forces

Let us recall the second law of motion stated by Sir Isaac Newton. If a particle of mass m is subject to a force $\bar{F}$, then viewing it in a fixed reference frame R, it is said that the particle acceleration is proportional to the force. This is usually written as

$$\bar{F} = m\bar{a} \tag{4.110}$$

where $\bar{F}$ is in the direction of the particle acceleration $\bar{a}$. In general, a particle can be subjected to n forces, in which case $\bar{F}$ is the resultant of all forces. Newton's second law forms the basis for most of the analysis in dynamics. The second law of motion can also be expressed in terms of the rate of change of the linear momentum, where equation (4.110) becomes

$$\bar{F} = \frac{d\bar{p}}{dt} = \frac{d}{dt}(m\bar{\upsilon}) \tag{4.111}$$

$\bar{P}$ denotes the linear momentum defined as the product of the mass m times the velocity $\bar{\upsilon}$ of the particle in the inertia reference frame R.

Equation (4.110) can be written as

$$\bar{F} - m\bar{a} = 0 \tag{4.112}$$

or

$$\bar{F} + \bar{F}^* = 0 \tag{4.113}$$

where

$$\bar{F}^* = -m\bar{a} \tag{4.114}$$

where $\bar{F}^*$ is known as the inertia force resulting directly from equation (4.113), which is known as the *d'Alembert principle.*

In the context of the study of multibody dynamics, we usually consider the bodies as rigid or elastic. Referring to Section 4.2, the total contribution to the generalized inertia forces in the presence of translational and rotational motion is defined as

$$F_\ell^* = \bar{F}^* \cdot \frac{\partial \bar{\upsilon}}{\partial \dot{x}_\ell} + \bar{T}^* \cdot \frac{\partial \bar{\omega}}{\partial \dot{x}_\ell} \tag{4.115}$$

where the $\dot{x}_\ell$ represent the time derivatives of the generalized coordinates, $\bar{\upsilon}$ is the mass center velocity of the particular rigid body, $\bar{\omega}$ its corresponding angular velocity, $\bar{F}^*$ the inertia force, and $\bar{T}^*$ the inertia torque. $\bar{T}^*$ is defined as

$$\bar{T}^* = \sum_{i=1}^{n} m_i \bar{r}_i \times \bar{a}_i$$

where the rigid body is assumed to be formed of m_i particles, $\bar{r}_i$ defines the position vector from the point mass to the mass center of the body, and $\bar{a}_i$ is the acceleration of point mass. In Chapter 5, we express $\bar{T}^*$ as a function of the angular acceleration, the inertia dyadic, and the angular velocity of the rigid body in R. Note that the contribution to the generalized inertia forces for the case of particles is given by only the first term to the right of the equal sign in equation (4.115).

4.8 Inertia Properties

In this section, we explain briefly the concepts of vectors and scalar quantities that characterize the inertia properties of both particles and rigid bodies. More important, those concepts are derived for the general case of multibody system dynamics. As we show in the sequel, the generalized inertia forces will be written explicitly to include the inertia properties of the interconnected bodies making use of the inertia vectors. In this section, we address only those concepts that are needed in the context of multibody kinematics formulation; hence the reader is assumed to have some knowledge of the topic.

A rigid body has an inherent characteristic about its composition that determines its inertia properties. This in most cases is independent of time, and therefore could be computed *a priori* aside from the kinematics and dynamics formulation. Let us begin by addressing the definition of the second moment of inertia, product of inertia, and radii of gyration of a continuous body A.

4.8.1 Second Moment

A vector $\bar{I}_a$ is called the second moment of a set S (comprised of N particles) relative to a fixed point O for a unit vector $\bar{n}_a$ if

$$\bar{I}_a = \sum_{i=1}^{n} m_i \bar{r}_i \times (\bar{n}_a \times \bar{r}_i) \tag{4.116}$$

where m_i denotes the mass of particle i, $\bar{r}_i$ the position vector from point O to the particle mass center, and $\bar{n}_a$ a unit vector parallel to some given line passing through O. Figure 4.16 illustrates such a description.

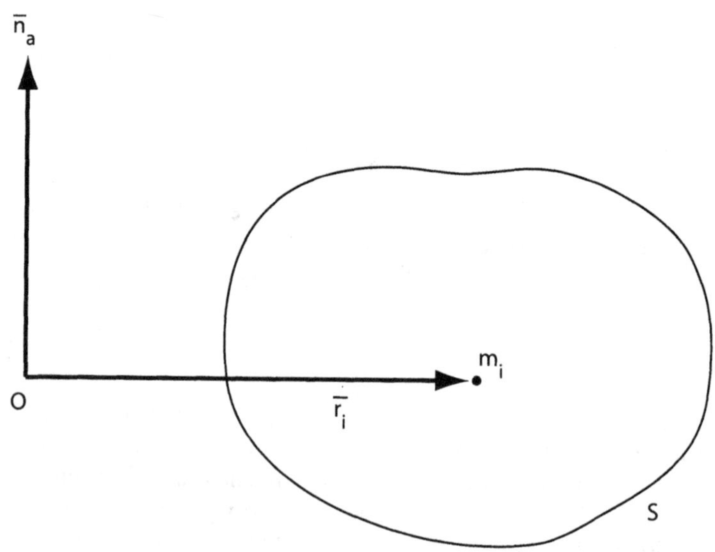

Fig. 4.16. A rigid body comprised of n mass particles m_i

Example 4.8.1
Find the second moment vector I_a for the mass particle m in body A as shown in Figure 4.17.

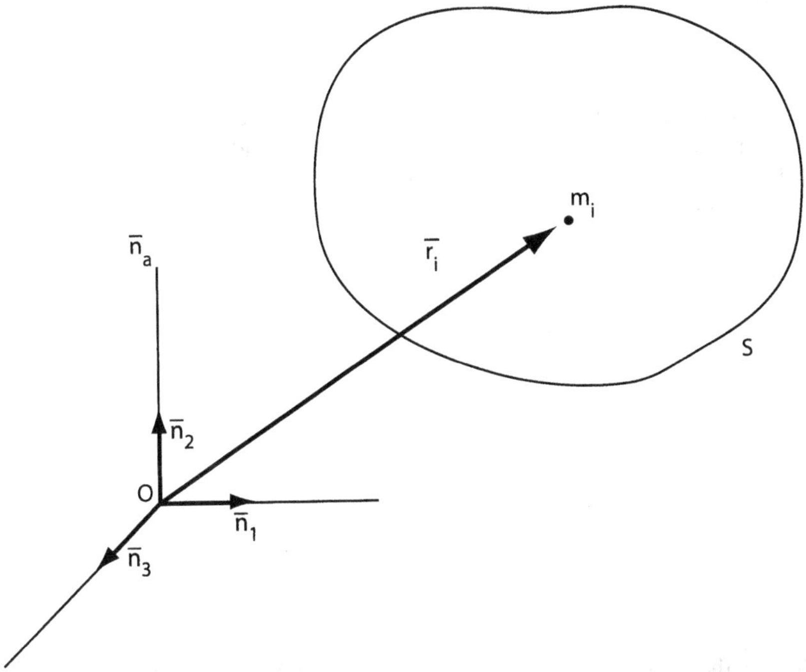

Fig. 4.17. A body with mass m_i

Solution
Let $\bar{n}_i$ $(i = 1, 2, 3)$ be a set of mutually perpendicular unit vectors. Let $\bar{n}_a$ be a unit vector parallel to $\bar{n}_3$. The second moment of the particle m relative to O for $\bar{n}_a$ is given as

$$\bar{I}_a = m\bar{r} \times (\bar{n}_a \times \bar{r}) \tag{4.117}$$

where

$$\bar{r} = r_1\bar{n}_1 + r_2\bar{n}_2 + r_3\bar{n}_3 \tag{4.118}$$

Substituting equation (4.118) into (4.117), we obtain

$$\bar{I}_a = \bar{I}_3 = -mr_1r_3\bar{n}_1 - mr_2r_3\bar{n}_2 + m(r_1^2 + r_2^2)\bar{n}_3 \tag{4.119}$$

Similarly, if $\bar{n}_a$ is chosen to be parallel to $\bar{n}_2$, we obtain

$$\bar{I}_a = \bar{I}_2 = -mr_1r_2\bar{n}_1 + m(r_1^2 + r_3^2)\bar{n}_2 - mr_2r_3\bar{n}_3 \tag{4.120}$$

and finally, taking $\bar{n}_a$ parallel to $\bar{n}_1$ yields

$$\bar{I}_a = \bar{I}_1 = m(r_2^2 + r_3^2)\bar{n}_1 - mr_1r_2\bar{n}_2 - mr_1r_3\bar{n}_3 \tag{4.121}$$

4.8.2 Product of Inertia

The vector $\bar{I}_a$ defining the second moment of S relative to O for $\bar{n}_a$, as defined by equation (4.116) is used to form a scalar quantity I_{ab}, called the product of inertia of S relative to O for $\bar{n}_a$ and $\bar{n}_b$, and is given by

$$I_{ab} = \bar{I}_a \cdot \bar{n}_b \tag{4.122}$$

Observe how the product of inertia is simply the projection of the second moment along a unit vector $\bar{n}_b$. It should also be noted that $\bar{n}_a$ and $\bar{n}_b$ do not necessarily have to differ from each other. As a matter of fact, if $\bar{n}_a$ is equal to $\bar{n}_b$, we obtain the scalar I_{aa} for the *product of inertia*, which is called the *moment of inertia*.

Example 4.8.2
For Example 4.8.1, let $\bar{n}_a$ be parallel to $\bar{n}_1$ and define the unit vector $\bar{n}_b$ to be parallel to $\bar{n}_3$. Find the product of inertia I_{ab}.

Solution
Using equation (4.122), we write

$$I_{ab} = I_{13} = \bar{I}_1 \cdot \bar{n}_3 \tag{4.123}$$

Using the results for $\bar{I}_1$ from equation (4.121), we obtain

$$I_{ab} = -mr_1r_3 \tag{4.124}$$

Following the same procedure, we can find the following product of inertia quantities:

$$I_{ij} = m\begin{bmatrix} r_2^2 + r_3^2 & -r_1r_2 & -r_1r_3 \\ -r_1r_2 & r_1^2 + r_3^2 & -r_2r_3 \\ r_3r_1 & -r_3r_2 & r_1^2 + r_2^2 \end{bmatrix} \quad (i, j = 1, \ldots, 3) \tag{4.125}$$

If $\bar{n}_a = \bar{n}_b$, we have the body uniform and symmetric; therefore, I_{ij} reduces to

$$I_{ij} = m\begin{bmatrix} r_2^2 + r_3^2 & 0 & 0 \\ 0 & r_1^2 + r_3^2 & 0 \\ 0 & 0 & r_2^2 + r_1^2 \end{bmatrix} \quad (i, j = 1, 2, 3) \tag{4.126}$$

In general, one writes the products of inertia in a matrix form as

$$I_{ij} = \begin{bmatrix} I_{11} & I_{12} & I_{13} \\ I_{21} & I_{22} & I_{23} \\ I_{31} & I_{32} & I_{33} \end{bmatrix} \quad (i, j = 1, 2, 3) \tag{4.127}$$

where the components are given by equation (4.125).

Consider a rigid body, this time given by a uniform beam as shown in Figure 4.18. Let O be the center of mass and the center of the mutually perpendicular unit vectors, $\bar{n}_x, \bar{n}_y$ and $\bar{n}_z$. Observe how we changed the notation of the unit vectors from $\bar{n}_1, \bar{n}_2, \bar{n}_3$ to $\bar{n}_x, \bar{n}_y, \bar{n}_z$ for convenience and assuming the familiarity of the readers with the conventional notation x, y, z.

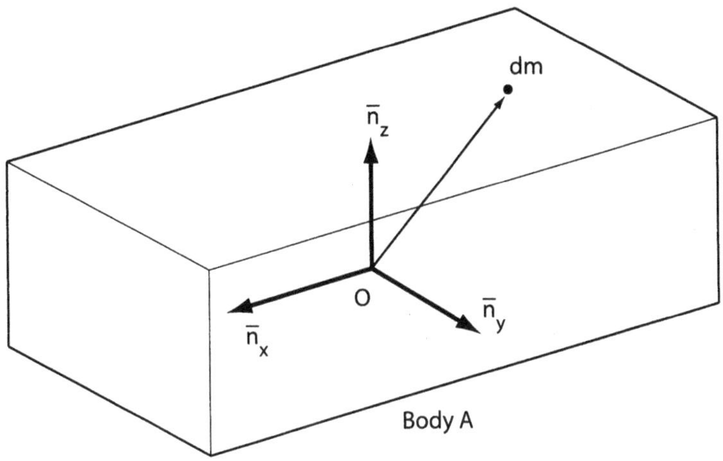

Fig. 4.18. A uniform beam rigid body

Let $\bar{r}$ denote the position vector to a point mass (dm). The moment of inertia of A relative to O—namely, I_{yy}, I_{zz}, I_{xx},—are found to be

$$I_{xx} = \int_m (r_y^2 + r_z^2)\, dm \tag{4.128}$$

$$I_{yy} = \int_m (r_x^2 + r_z^2)\, dm \tag{4.129}$$

and

$$I_{zz} = \int_m (r_x^2 + r_y^2)\, dm \tag{4.130}$$

where m denotes the total mass of body A. More explicitly, we can write the equation above by making use of the area and mass density as

$$I_{xx} = \rho \int_m (r_y^2 + r_z^2)\, dV \tag{4.131}$$

where ρ is the mass density and dV the unit volume.

4.8.3 Properties of the Product of Inertia

We denote by I_{ab} the inertia scalar of S relative to O for $\bar{n}_a$ and $\bar{n}_b$, and defined it as

$$I_{ab} = \bar{I}_a \cdot \bar{n}_b \tag{4.132}$$

One of its properties is in its symmetry, that is,

$$I_{ab} = I_{ba} \tag{4.133}$$

This can be proved as follows:

$$I_{ab} = \sum_{i=1}^{n} m_i[\bar{r}_i \times (\bar{n}_a \times \bar{r}_i)] \cdot \bar{n}_b$$

Applying equation (3.244) gives

$$\begin{aligned}
&= I_{ab} \sum_{i=1}^{n} m_i \left[\bar{r}_i^2(\bar{n}_a \cdot \bar{n}_b) - (\bar{r}_i \cdot \bar{n}_a)(\bar{r}_i \cdot \bar{n}_b)\right] \\
&= \sum_{i=1}^{n} m_i \left[\bar{r}_i^2 \bar{n}_b - (\bar{r}_i \cdot \bar{n}_b)\bar{r}_i\right] \cdot \bar{n}_a \qquad (4.134) \\
&= \sum_{i=1}^{n} m_i [\bar{r}_i \times (\bar{n}_b \times \bar{r}_i)] \cdot n_a
\end{aligned}$$

$$\bar{I}_b \cdot \bar{n}_a = I_{ba} \tag{4.135}$$

which proves equation (4.133). The product of inertia of S relative to O for $\bar{n}_a$ and $\bar{n}_b$ could be expressed in a convenient form as

$$I_{ab} = (\bar{n}_a \cdot \bar{n}_j) I_{jk} (\bar{n}_b \cdot \bar{n}_k) \tag{4.136}$$

where, as before, $\bar{n}_a$ and $\bar{n}_b$ are two chosen unit vectors. Equation (4.136) is a result of the definition of the product of inertia given by equation (4.122). If we select a set of three mutually unit vectors $\bar{n}_j (j = 1,\ 2,\ 3,)$, we can write the unit vector $\bar{n}_a$ as

$$\bar{n}_a = (\bar{n}_a \cdot \bar{n}_j)\bar{n}_j \quad (j = 1, 2, 3) \tag{4.137}$$

Substituting equation (4.137) into equation (4.117), we obtain

$$\bar{I}_a = \sum_{i=1}^{n} m_i \bar{r}_i \times [(\bar{n}_a \cdot \bar{n}_j)\bar{n}_j \times \bar{r}_i] \tag{4.138}$$

Factorizing out the dot product of $(\bar{n}_a \cdot \bar{n}_j)$ yields

$$\bar{I}_a = (\bar{n}_a \cdot \bar{n}_j) \sum_{i=1}^{n} m_i \bar{r}_i \times (\bar{n}_j \times \bar{r}_i) \tag{4.139}$$

or

$$\bar{I}_a = (\bar{n}_a \cdot \bar{n}_j)\bar{I}_j \tag{4.140}$$

From the definition of the product of inertia I_{ab}, equation (4.122) gives

$$I_{ab} = (\bar{n}_a \cdot \bar{n}_j)\bar{I}_j \cdot \bar{n}_b \quad (4.141)$$

Again $\bar{n}_b$ could be expressed as

$$\bar{n}_b = (\bar{n}_b \cdot \bar{n}_k)\bar{n}_k \ \ (k = 1, 2, 3) \quad (4.142)$$

where $\bar{n}_k$ $(k = 1, 2, 3)$ are three mutually perpendicular unit vectors. Substituting equation (4.142) into equation (4.141), we obtain equation (4.136). Observe how equation (4.136) can be interpreted as that in which I_{ab} is considered to be the product of inertia (inertia matrix) of S relative to the body reference frame, and I_{jk} is that of S relative to the fixed reference frame.

In addition, equation (4.136) can be expressed as

$$I_{ab} = I_{so} = S^{os} I_{jk} S^{so} \quad (4.143)$$

where S^{so} is a transformation matrix: $[S^{so}] = [S^{os}]^T$. Equation (4.143) is called the *similarity* of the inertia matrix. We should add that equation (4.143) will be useful for expressing the product of inertia, moment of inertia, and inertia Dyadic (discussed in the next section) in different frames.

4.8.4 Inertia Dyadic

The inertia dyadic is often useful to express the second moment and the product of inertia of S relative to O for the unit vectors $\bar{n}_a$ and $\bar{n}_b$. It is defined as

$$\bar{\bar{I}} = \bar{n}_j I_{jk} \bar{n}_k \ \ (j,\ k = 1, 2, 3) \quad (4.144)$$

There is a sum from 1 to 3 on j and k, using the definition of the second moment given by equation (4.122). The inertia dyadic can be expressed as

$$\bar{\bar{I}} = \sum_{i=1}^{N} m_i(\bar{\bar{u}} r_i^2 - \bar{r}_i \bar{r}_i) \quad (4.145)$$

where $\bar{\bar{u}}$ is a dyadic defined as

$$\bar{\bar{u}} = \bar{n}_i \bar{n}_j \quad (4.146)$$

where $\bar{n}_i \neq \bar{n}_j$ $(i, j = 1, 2, 3)$ are two three-mutually perpendicular unit vectors. It can easily be seen that the second moment of $\bar{I}_a$ of S relative to O is obtained as

$$\bar{I}_a = \bar{\bar{I}} \cdot \bar{n}_a = \bar{n}_a \cdot \bar{\bar{I}} \quad (4.147)$$

and that the product of inertia I_{ab} of S relative to O is

$$\bar{I}_a = \bar{n}_b \cdot I \cdot \bar{n}_a \quad (4.148)$$

A direct proof of equations (4.147) and (4.148) can be seen by noting that the second moment of S comprising of N particles of mass $m_i (i = 1, \ldots, N)$, relative to O for $\bar{n}_a$ can be expressed as

$$\bar{I}_a = \sum_{i=1}^{N} m_i[\bar{r}_i^2\bar{n}_a - (\bar{r}_i \cdot \bar{n}_a)\bar{r}_i] \tag{4.149}$$

Using equation (4.145), equation (4.149) takes on the form

$$\bar{I}_a = \bar{\bar{I}} \cdot \bar{n}_a \tag{4.150}$$

Equation (4.150) then yields the product of inertia as shown by equation (4.148).

The inertia dyadic is symmetric in which the dot product is interchangeable, as shown in equation (4.147). Observe the various ways the product of inertia can be written:

$$\bar{I}_{ab} = \bar{I}_a \cdot \bar{n}_b = \bar{I}_b \cdot \bar{n}_a \tag{4.151}$$

or

$$I_{ab} = \bar{\bar{I}} \cdot \bar{n}_a \cdot \bar{n}_b = \bar{n}_a \cdot \bar{\bar{I}} \cdot \bar{n}_b \tag{4.152}$$

Using the symmetry property, we can write it further as

$$\bar{I}_{ab} = \bar{\bar{I}} \cdot \bar{n}_b \cdot \bar{n}_a = \bar{n}_b \cdot \bar{\bar{I}} \cdot \bar{n}_a \tag{4.153}$$

We can express, the inertia dyadic further as

$$\bar{\bar{I}} = (\bar{\bar{I}} \cdot \bar{n}_j)\bar{n}_j \tag{4.154}$$

or simply

$$\bar{\bar{I}} = \bar{I}_j\bar{n}_j \quad (j = 1, 2, 3) \tag{4.155}$$

where

$$\bar{I}_j = (\bar{I}_j \cdot \bar{n}_k)\bar{n}_k \tag{4.156}$$

which can be written in a matrix notation as

$$\bar{I}_j = I_{jk}\bar{n}_k \tag{4.157}$$

for a rigid body k, the inertia dyadic of k relative to O is

$$\bar{\bar{I}} \triangleq \int_{m_k} (\bar{\bar{u}}r_i^2 - \bar{r}_i\bar{r}_i)\, dm \tag{4.158}$$

where m_k is the total mass of body k and $\bar{r}_i$ denotes the position vector from O to a particle dm.

4.8.5 Parallel Axes Theorem

Consider Figure 4.19; let $\bar{r}_i$ ($i = 1, \ldots, N$) denote the position vector from point O to the particle mass m_i and G be the center mass of the body. In body S, $\bar{r}_G$ is the position from point O to G and $\bar{P}_i$ is the relative position of m_i to G. Again, $\bar{n}_a$ and $\bar{n}_b$ are two arbitrarily chosen unit vectors. The parallel axis theorem for the second moment of S relative to O states that

$$ {}^{O}\bar{I}_a^S = {}^{O}\bar{I}_a^G + {}^{G}\bar{I}_a^S \tag{4.159} $$

where ${}^{O}\bar{I}^S$ represents the second moment of S relative to O, ${}^{O}\bar{I}^G$ is the second moment of G relative to O and ${}^{G}\bar{I}^S$ denotes the second moment of S relative to G. Note that G is conveniently chosen to be the center of mass of S, whereas in general cases it could be any arbitrary point on S. The parallel axes theorem is used to establish the relationship between the moments of inertia of S about two lines that are parallel to each other and pass through O and G, respectively (see Figure 4.19).

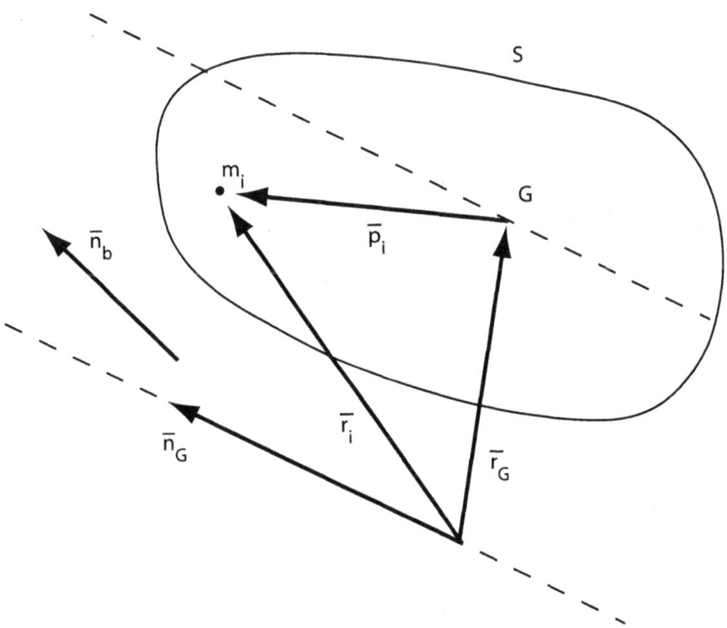

Fig. 4.19. Defining the position vectors from mass particles m_i to the center of mass of the rigid body

Example 4.8.3

Consider the rigid-body cube shown in Figure 4.20.

(a) Find the second moment of A relative to O for $\bar{n}_1$, $\bar{n}_2$ and $\bar{n}_3$, respectively.

(b) Find the inertia dyadic $\bar{\bar{I}}$ of A relative to O for $\bar{n}_1$ and $\bar{n}_3$.

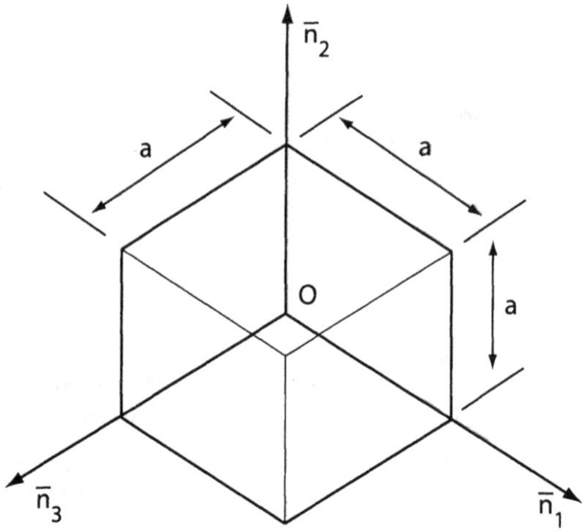

Fig. 4.20. Rigid-body cube

Solution
(**a**) Using equation (4.116), we find that

$$\bar{I}_1 = \int_m (\bar{r}_G \times (\bar{n}_1 \times \bar{r}_G))\, dm \tag{4.160}$$

where

$$\bar{r}_G = \frac{a}{2}(\bar{n}_1 + \bar{n}_2 + \bar{n}_3) \tag{4.161}$$

Substituting equation (4.161) into (4.160) yields

$$\begin{aligned}\bar{I}_1 &= \int_m \left[\bar{r}_G \times \left(\frac{a}{2}\bar{n}_3 - \frac{a}{2}\bar{n}_2\right)\right] dm \\ &= \int_m \left[\frac{a^2}{4}(2\bar{n}_1 - \bar{n}_3 - \bar{n}_2)\right] dm \\ &= \rho \int_V \frac{a^2}{4}(2\bar{n}_1 - \bar{n}_3 - \bar{n}_2)\, dV \\ &= \frac{a^2}{4}\rho \int_V (2\bar{n}_1 - \bar{n}_3 - \bar{n}_2)\, dV\end{aligned}$$

Therefore,

$$\bar{I}_1 = \frac{ma^2}{4}(2\bar{n}_1 - \bar{n}_3 - \bar{n}_2) \tag{4.162}$$

and similarly, using equation (4.160), we solve for $\bar{I}_2$ and $\bar{I}_3$ by substituting $\bar{n}_2$ and $\bar{n}_3$, respectively, into $\bar{n}_1$:

$$\bar{I}_2 = \frac{ma^2}{4}(2\bar{n}_2 - \bar{n}_1 - \bar{n}_3) \tag{4.163}$$

$$\bar{I}_3 = \frac{ma^2}{4}(2\bar{n}_3 - \bar{n}_1 - \bar{n}_2) \tag{4.164}$$

(b) The inertia dyadic $\bar{\bar{I}}$ of a relative to O is found using equations (4.154) and (4.157).

$$\begin{aligned}
\bar{\bar{I}} &= \bar{I}_j\bar{n}_j \quad (j = 1, 2, 3) \\
&= \bar{I}_1\bar{n}_1 + \bar{I}_2\bar{n}_2 + \bar{I}_3\bar{n}_3 \\
&= \frac{ma^2}{4}(2\bar{n}_1\bar{n}_1 - \bar{n}_3\bar{n}_1 - \bar{n}_2\bar{n}_1 + 2\bar{n}_2\bar{n}_2 - \bar{n}_2\bar{n}_3 - \bar{n}_2\bar{n}_1 \\
&\qquad + 2\bar{n}_3\bar{n}_3 - \bar{n}_1\bar{n}_3 - \bar{n}_2\bar{n}_3) \qquad (4.165) \\
&= \frac{ma^2}{4}(\bar{n}_1\bar{n}_1 + \bar{n}_2\bar{n}_2 + \bar{n}_3\bar{n}_3 - \bar{n}_1\bar{n}_2 - \bar{n}_1\bar{n}_3 - \bar{n}_2\bar{n}_3)
\end{aligned}$$

or in matrix form,

$$\bar{\bar{I}} = \frac{ma^2}{4}\begin{bmatrix} 2\bar{n}_1\bar{n}_1 & -\bar{n}_1\bar{n}_2 & -\bar{n}_1\bar{n}_3 \\ -\bar{n}_2\bar{n}_1 & 2\bar{n}_2\bar{n}_2 & -\bar{n}_2\bar{n}_3 \\ -\bar{n}_3\bar{n}_1 & -\bar{n}_3\bar{n}_2 & 2\bar{n}_3\bar{n}_3 \end{bmatrix} \tag{4.166}$$

For n_1 and $\bar{n}_3$,

$$\bar{\bar{I}}_{13} = -\frac{m}{4}a^2\bar{n}_1\bar{n}_3 \tag{4.167}$$

Example 4.8.4
Consider the rigid-body cube shown in Figure 4.21.

(a) Find the second moment of A relative to O for $\bar{n}_1$, $\bar{n}_2$ and $\bar{n}_3$, respectively.

(b) Find the inertia product I_{13} .

Solution
(a) Using equation (4.116), we find that

$$\bar{I}_1 = \int_m (\bar{r}_G \times (\bar{n}_1 \times \bar{r}_G))\, dm \tag{4.168}$$

where

$$\bar{r}_G = \frac{a}{2}(2\bar{n}_1 + \bar{n}_2 + \bar{n}_3) \tag{4.169}$$

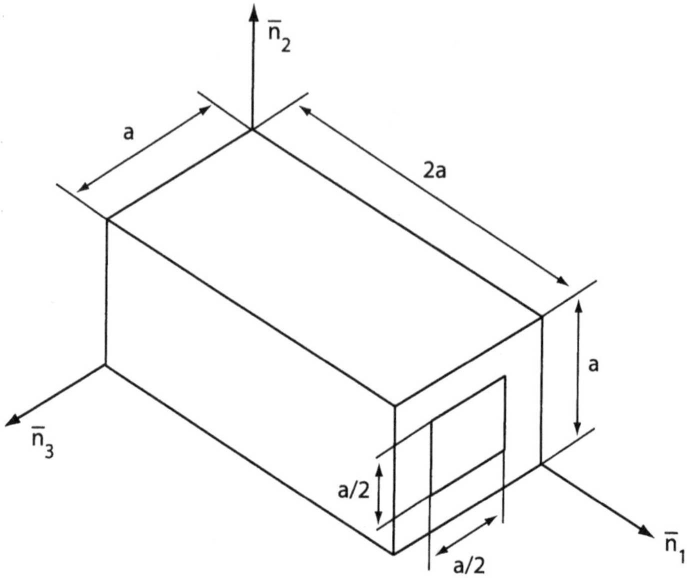

Fig. 4.21. Rigid hollow cube

Substituting equation (4.161) into (4.160) yields

$$\bar{I}_1 = \int_m \left[\frac{a}{2}(2\bar{n}_1 + \bar{n}_2 + \bar{n}_3) \times \left(\bar{n}_1 \times \frac{a}{2}(2\bar{n}_1 + \bar{n}_2 + \bar{n}_3)\right)\right] dm$$

$$= \int_m \left[\frac{a}{2}(2\bar{n}_1 + \bar{n}_2 + \bar{n}_3) \times \frac{a}{2}(\bar{n}_3 - \bar{n}_2)\right] dm$$

$$= \int_m \left[\frac{a^2}{2}(\bar{n}_1 - \bar{n}_3 - \bar{n}_2)\right] dm$$

$$= \frac{a^2}{2} m\,(\bar{n}_1 - \bar{n}_3 - \bar{n}_2)$$

Therefore,

$$\bar{I}_1 = \frac{3}{4}\rho \cdot a^5 \cdot (\bar{n}_1 - \bar{n}_3 - \bar{n}_2) \tag{4.170}$$

Similarly we solve for $\bar{I}_2$:

$$\bar{I}_2 = \int_m \left[\frac{a}{2}(2\bar{n}_1 + \bar{n}_2 + \bar{n}_3) \times \left(\bar{n}_2 \times \frac{a}{2}(2\bar{n}_1 + \bar{n}_2 + \bar{n}_3)\right)\right] dm$$

$$= \int_m \left[\frac{a^2}{4}(2\bar{n}_1 + \bar{n}_2 + \bar{n}_3) \times (\bar{n}_1 - 2\bar{n}_3)\right] dm$$

$$= \int_m \left[\frac{a^2}{4}(-2\bar{n}_1 + 5\bar{n}_2 - \bar{n}_3)\right] dm$$

$$= \frac{a^2}{4} m \cdot (-2\bar{n}_1 + 5\bar{n}_2 - \bar{n}_3)$$

Therefore,

$$\bar{I}_2 = \frac{3}{8}\rho \cdot a^5 \cdot (-2\bar{n}_1 + 5\bar{n}_2 - \bar{n}_3) \tag{4.171}$$

Solving for $\bar{I}_3$, we obtain:

$$\bar{I}_3 = \int_m \left[\frac{a}{2}(2\bar{n}_1 + \bar{n}_2 + \bar{n}_3) \times \left(\bar{n}_3 \times \frac{a}{2}(2\bar{n}_1 + \bar{n}_2 + \bar{n}_3)\right)\right] dm$$

$$= \int_m \left[\frac{a^2}{4}(2\bar{n}_1 + \bar{n}_2 + \bar{n}_3) \times (2\bar{n}_2 - \bar{n}_1)\right] dm$$

$$= \int_m \left[\frac{a^2}{4}(-2\bar{n}_1 - \bar{n}_2 + 5\bar{n}_3)\right] dm$$

$$= \frac{a^2}{4} m \cdot (-2\bar{n}_1 - \bar{n}_2 + 5\bar{n}_3)$$

which take the following form,

$$\bar{I}_3 = \frac{3}{8}\rho \cdot a^5 \cdot (-2\bar{n}_1 - \bar{n}_2 + 5\bar{n}_3) \tag{4.172}$$

(b) The product of inertia I_{13} is defined by projecting I_3 along n_1

$$I_{13} = \bar{I}_3 \cdot \bar{n}_1 = \frac{3}{8}\rho \cdot a^5 \cdot (-2\bar{n}_1 - \bar{n}_2 + 5\bar{n}_3) \cdot \bar{n}_1 \tag{4.173}$$

$$I_{13} = -\frac{3}{4}\rho \cdot a^5 \tag{4.174}$$

We can also obtain I_{13} as:

$$I_{13} = \bar{I}_1 \cdot \bar{n}_3 = \frac{3}{4}\rho \cdot a^5 \cdot (\bar{n}_1 - \bar{n}_2 - \bar{n}_3) \cdot \bar{n}_3 \tag{4.175}$$

$$I_{13} = -\frac{3}{4}\rho \cdot a^5 \tag{4.176}$$

The results are the same in both cases, equations (4.174) and (4.176). This is due to the symmetry of the inertia dyadic matrix.

4.9 Problems

4.1. A uniform bar with an attached mass M is made to rotate in an oscillating mode about point O (Figure 4.22). The bar is pinned to a rectangular frame that is given a translational motion along the horizontal axis. Attached to the mass M is a spring of stiffness K and a torsional spring of stiffness C.

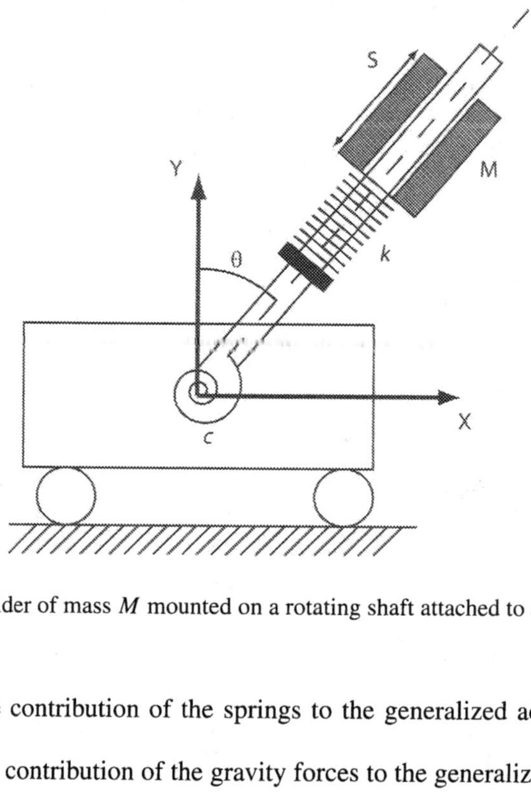

Fig. 4.22. Slider of mass M mounted on a rotating shaft attached to a moving cart

(a) Find the contribution of the springs to the generalized active force of the system.
(b) Find the contribution of the gravity forces to the generalized active force of the system.

4.2. The pendulum mechanism shown in Figure 4.23 has two bars, with bar 2 having a mass M to counterbalance the spring force. Bar 1 has an angular velocity $\omega(t)$ which drives the system about point O, while bar 2 with the attached mass can undergo small oscillations denoted by $\psi(t)$. Find the contribution of the spring forces to the generalized active force.

4.3. For the robot shown in Figure 4.24, give the explicit form of the contribution of the driving torques T_1, T_2, $T_3(t)$ to the generalized active forces.

4.4. For the robot in Problem 4.3, if the gravity forces are included, what orientation angles yield the maximum value of the generalized active forces for joint 1?

4.5. Consider a uniform bar of length ℓ and cross-sectional area ab shown in Figure 4.25. Find the moment of inertia with respect to point O and O', respectively, making use of the definitions.

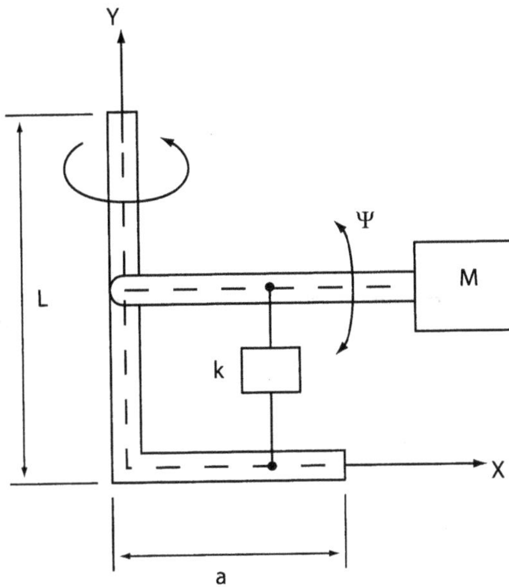

Fig. 4.23. Mass M fixed to an oscillating rod attached to a rotating base

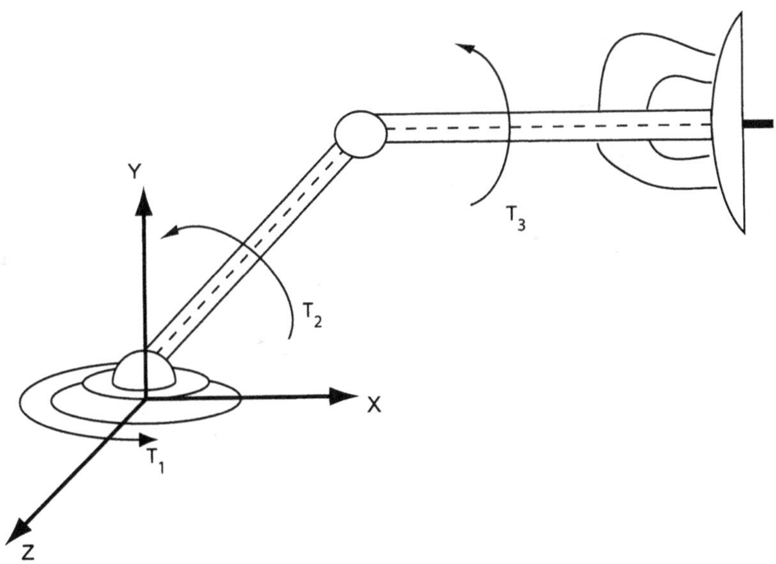

Fig. 4.24. Two-link robot manipulator

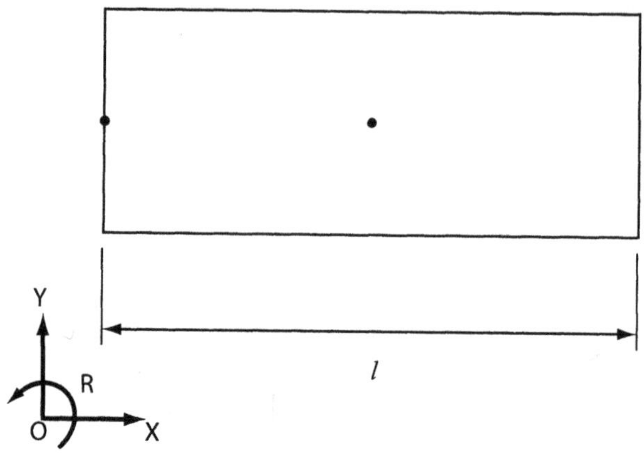

Fig. 4.25. Uniform bar of length L

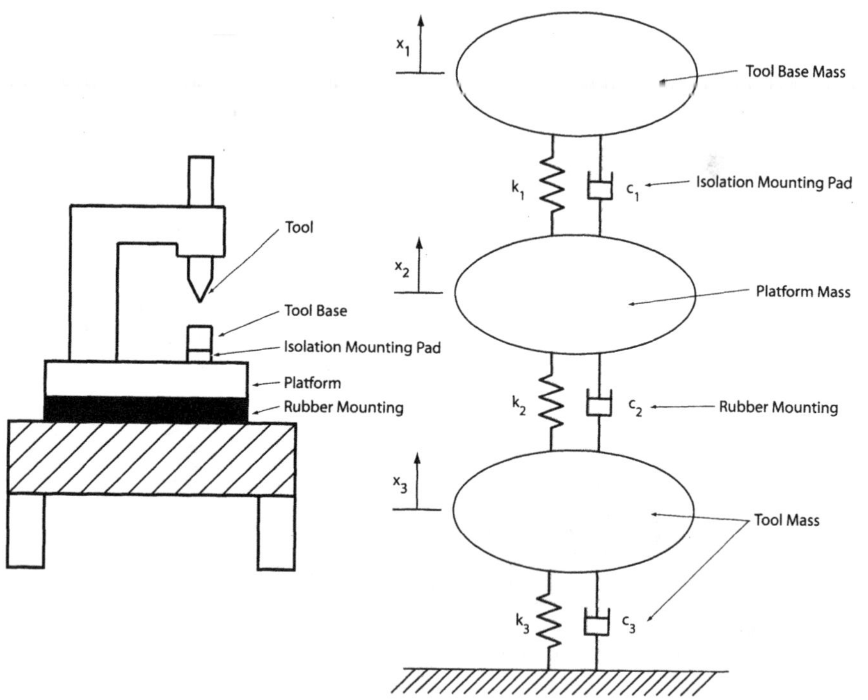

Fig. 4.26. (a) Schematic of a punch-press machine; (b) Vibration model of the punch-press

4.6. The punch-press machine shown in Figure 4.26(a) can be modeled as a multibody system of a lumped mass as shown in Figure 4.26(b). Given the mass spring and damper for each body mass, derive the generalized force contribution for the system. Write the generalized force as a sum of two forces, representing the springs and dampers separately.

4.7. The homogeneous rectangular plate weighs 50 lbs, with its end supported in the vertical plane by the light parallel links as shown in Figure 4.27. If a couple $M = 80$ lb-ft is applied to the end of link AB with the system initially at rest, show the contribution of the external forces to the generalized active forces.

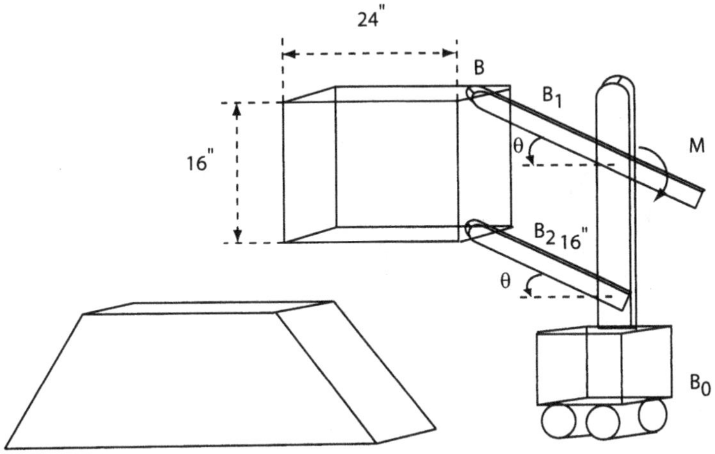

Fig. 4.27. A lift mechanism

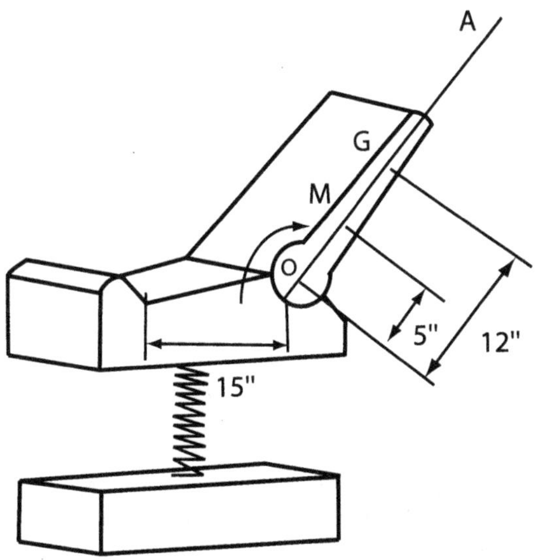

Fig. 4.28. A typical seat design

4.8. The 15-lb lever OA with 12-in. radius of gyration about O is initially at rest in the vertical position ($\theta = 90°$) as shown in Figure 4.28 where the attached spring of stiffness $= 10$ lb/in is unstretched. Calculate the contribution of moment M applied to the generalized active forces for an arbitrary angle θ, if the angular velocity of the lever is $\omega = 4$ rad/s. For $\theta = 120°$

4.9. Use Example 3.7.2 to find the contribution of the spring forces to the generalized active forces.

4.10. In Example 4.4.3 find the explicit form of the generalized active forces $\{f\}$.

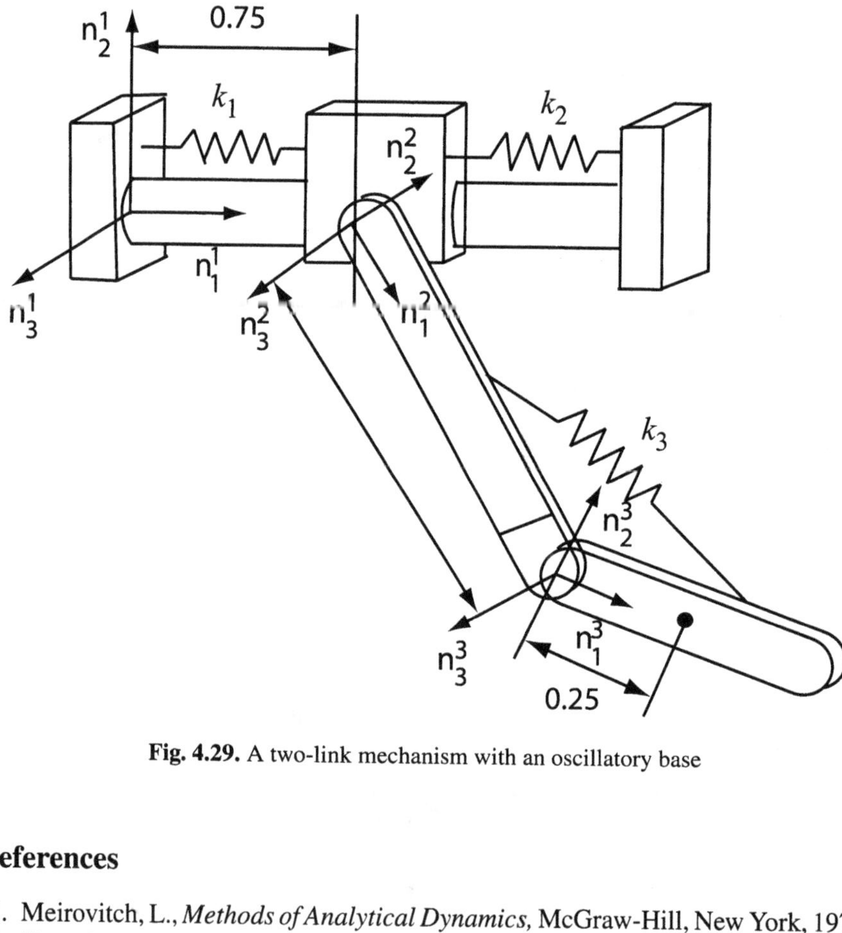

Fig. 4.29. A two-link mechanism with an oscillatory base

References

1. Meirovitch, L., *Methods of Analytical Dynamics,* McGraw-Hill, New York, 1970.
2. Kane, T. R., Likins, P. W. and Levinson, D. A., *Spacecraft Dynamics,* McGraw-Hill, New York, 1983.
3. Kane, T. R., *Dynamics,* Holt, Rinehart and Winston, New York, 1968.
4. Huston, R. L., *Multibody Dynamics,* Butterworth-Heinemann, Stoneham, MA, 1990.
5. Greenwood, D. T., *Principles of Dynamics,* Prentice Hall, Englewood Cliffs, NJ, 1988.

5

Equations of Motion of Multibody Systems

5.1 Introduction

The purpose of this chapter is to give the readers an overview of the task required in the formulation of the equations of motion, so that they can judge the utility of the methods as they apply to multibody dynamics. Most important, the equations are expressed in explicit forms, making use of the previously developed arrays in kinematics for computer application. The equations presented in this chapter are unique and present the explicit form of all the terms needed in the dynamics of MBS. The 'final' form of the equations can be achieved through most of the methods and should be viewed as a tool in getting the equations of motion without the use of any standard approach in dynamics.

5.2 Equations of Motion

The formulation of the governing equations of motion in multibody systems can be obtained by a number of methods. However, some methods are more efficient than others. It will become apparent to the reader that our effort in using Kane's equations, combined with the kinematic formalism and the partitioning of the generalized coordinates adapted in earlier chapters, lead to a simple way to derive the explicit form of equations of motion. This will be even more pronounced in linearization of the equations and formulation of the equations of motion for more complex systems with flexible bodies. Kane's first paper on his method appeared in 1961 but did not carry his name. In the 1970s, Huston et al. and Likins et al. used the equations in a number of applications and referred to them as the Lagrange form of D'Alembert's principle. This never conceptualized, and now these equations are best known as *Kane's equations* or *Kane's method.*

Kane's equation for a system S of N bodies modeled by n generalized coordinates X_ℓ in an inertial reference frame R is given by

$$f_\ell + f_\ell^* = 0 \quad (\ell = 1, \ldots, n) \tag{5.1}$$

where f_ℓ and f_ℓ^* denote the generalized active and inertia forces respectively. Those forces are defined in Chapter 4, more precisely by equations (4.6) and (4.115). Once the concept of the forces is understood, it is desirable to explore the benefits of the method and the information it provides in learning about the system dynamic behavior. What follows are proof of concept and derivation of Kane's equations.

5.3 Derivation of Kane's Equations Through the Principle of Virtual Work

Consider a general open-chain multibody system containing N interconnected bodies as depicted by Figure 5.1. Let the system be subjected to an externally applied force field and a constraint force field. The externally applied force field can be appropriately transformed into $\bar{F}_k$ and $\bar{M}_k (k = 1, 2, \ldots, N)$, where $\bar{F}_k$ are concentrated forces passing through G_k, the mass center of body k, and $\bar{M}_k$ are moments about G_k. The constraint force field can be transformed into $\bar{F}_k^c$ and $\bar{M}_k^c$, which represent a concentrated constraint force passing through G_k together with a moment about G_k. A typical body of the system will have forces as shown in Figure 5.2. Using D'Alembert's principle for the force equilibrium of body k, we obtain

$$\bar{F}_k + \bar{F}_k^* + \bar{F}_k^c = 0 \tag{5.2}$$

where $\bar{F}_k^* = -m_k a_k$ is the inertia force of body k.

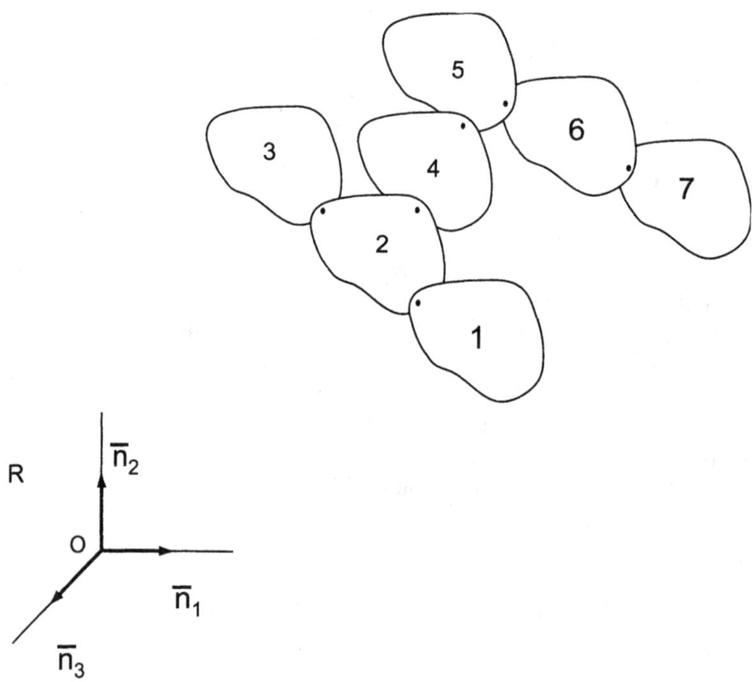

Fig. 5.1. A general treelike multibody system

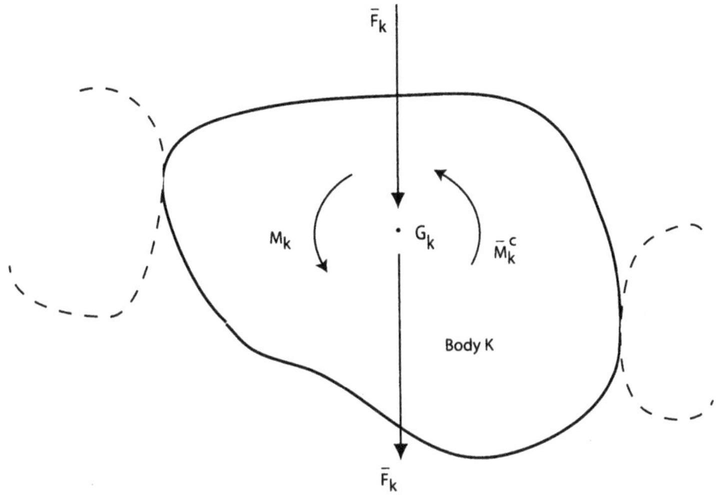

Fig. 5.2. Equivalent forces and moments acting at the center mass G_k of body K

5.3.1 Principle of Virtual Work

For a system S possessing $3n$ degrees of freedom in a fixed inertial reference frame and comprising n particles whose configuration is given by $x_1, x_2, \ldots, x_{3n}$ generalized coordinates. Suppose that the force components $F_1, F_2, \ldots, F_{3n}$ are applied at the particles along the corresponding coordinates. Then the virtual work, denoted by δW, of these forces in a virtual displacement δr is defined to be

$$\delta W = \sum_{i=1}^{3n} \bar{F}_i \,.\, \delta r_i \tag{5.3}$$

where $\bar{F}_i$ is the resultant force acting on the ith particle and r_i is the position vector of the particle in R. We should note that δr_i are virtual or imaginary in the sense that they are assumed to occur without the passage of time and do not necessarily conform to the constraints. In addition, a virtual displacement conforms to the instantaneous constraints; that is, any moving constraints are assumed to be stopped during the virtual displacement.

Extending the principle of virtual work to equation (5.2), we get

$$\delta W_f = (\bar{F}_k + \bar{F}_k^* + \bar{F}_k^c) \cdot \delta \bar{r}_k = 0 \quad (k = 1, 2, \ldots, 3n) \tag{5.4}$$

where δW_f is the virtual work of the system due to forces only. Many of the constraints that commonly occur are of a class known as *workless constraints.* Then it can be shown that the contribution of the constraint forces is

$$\bar{F}_k^c \cdot \delta \bar{r}_k = 0 \tag{5.5}$$

Equation (5.4) reduces to

$$\delta W_f = (\bar{F}_k + \bar{F}_k^*) \cdot \delta \bar{r}_k = 0 \quad (k = 1, 2, \ldots, 3n) \tag{5.6}$$

Substituting the $\bar{r}_k$ by its partial derivative, we obtain,

$$\delta W_f = (\bar{F}_k + \bar{F}_k^*) \cdot \frac{\partial \bar{r}_k}{\partial x_\ell} \delta x_\ell = 0 \quad (\ell = 1, \ldots, 3n) \tag{5.7}$$

We also know that the position vector can be written as a function of the generalized coordinates and time.

$$\bar{r}_k = \bar{r}_k(x_\ell,\ t) \tag{5.8}$$

Differentiation of the above equation results in

$$\dot{\bar{r}}_k = \frac{\partial \bar{r}_k}{\partial x_\ell}\frac{dx_\ell}{dt} + \frac{\partial \bar{r}_k}{\partial t} = \frac{\partial \bar{r}_k}{\partial x_\ell}\dot{x}_\ell + \frac{\partial \bar{r}_k}{\partial t} \tag{5.9}$$

The partial derivative of $\dot{\bar{r}}_k$ with respect to $\dot{x}_\ell$ from equation (5.9) is given by

$$\frac{\partial \dot{\bar{r}}_k}{\partial \dot{x}} = \frac{\partial \bar{r}_k}{\partial x_\ell} \tag{5.10}$$

or simply

$$\frac{\partial \bar{v}_k}{\partial \dot{x}_\ell} = \frac{\partial \bar{r}_k}{\partial x_\ell} \tag{5.11}$$

where $\bar{v}_k$ is the velocity of body k in R.

From equation (5.7), and since the virtual displacement δx_ℓ is arbitrary without violating the constraints, we obtain

$$F_\ell + F_\ell^* = 0 \tag{5.12}$$

where F_ℓ and F_ℓ^* are the generalized active and inertia forces, respectively, and are expressed in a unique form which makes use of the partial velocity vector. The generalized forces are defined as:

$$F_\ell = \bar{F}_k \cdot \frac{\partial \dot{\bar{r}}_k}{\partial \dot{x}_\ell} = \bar{F}_k \cdot \frac{\partial \bar{v}_k}{\partial \dot{x}_\ell} \tag{5.13}$$

and the generalized inertia forces are

$$F_\ell^* = \bar{F}_k^* \cdot \frac{\partial \dot{\bar{r}}_k}{\partial \dot{x}_\ell} = \bar{F}_k^* \cdot \frac{\partial \bar{v}_k}{\partial \dot{x}_\ell} \tag{5.14}$$

In a similar fashion, we can write the virtual work due to the moments as

$$\delta W_M = (\bar{M}_k + \bar{M}_k^* + \bar{M}_c) \cdot \delta\theta \tag{5.15}$$

where θ denotes the angular displacement. Similarly, assuming that the moment constraint performs zero work for $\delta\theta$, equation (5.15) reduces to

$$\delta W_M = (\bar{M}_k + \bar{M}_k^*) \cdot \delta\theta \tag{5.16}$$

In a similar fashion $\delta\theta$ can be expressed making use of partial angular velocity where

$$\delta W_M = (\bar{M}_k + \bar{M}_k^*) \cdot \frac{\partial \bar{\omega}^k}{\partial \dot{x}_\ell} \partial x_\ell \tag{5.17}$$

where $\bar{\omega}^k$ denotes the angular velocity of k in R. Equation (5.17) then becomes

$$\bar{M}_\ell + \bar{M}_\ell^* = 0 \tag{5.18}$$

where $\bar{M}_\ell$ and $\bar{M}_\ell^*$ represent the generalized active moments and generalized inertia moments respectively. By superposition of equations (5.12) and (5.18) for a system subjected to both external forces and moments, we obtain the equations of motion as

$$f_\ell + f_\ell^* = 0 \tag{5.19}$$

Equation (5.19) is known as *Kane's equations.*

For rigid-body dynamics, the inertia torque (moment) $\bar{M}_k^*$ is defined as follows:

$$\bar{M}_k^* = -\bar{\alpha}^k \cdot \bar{\bar{I}} - \bar{\omega}^k \times \left(\bar{\bar{I}} \cdot \bar{\omega}^k\right) \tag{5.20}$$

where $\bar{\alpha}^k$ is the angular acceleration of body k in R, and $\bar{\bar{I}}$ is the inertia dyadic of k relative to the mass center G_k. The concepts of inertia are presented in Chapter 4.

Proof

Employing the momentum principle of the mass center, the moment of inertia is given by

$$\bar{M}_k^* = -\frac{d}{dt} \int_{m_k} (\bar{\eta} \times \bar{v}^{dm})\, dm \tag{5.21}$$

where $\bar{\eta}$ is the position vector from the infinitesimal mass dm of body k to its center mass G_k, $\bar{v}^{dm}$ is the velocity of dm relative to G_k in R, and the integral is taken over the entire mass m_k of body k. Equation (5.21) takes the following form once the differentiation is taken inside the integral:

$$\bar{M}_k^* = -\int_{m_k} \left(\frac{d\bar{\eta}}{dt} \times \bar{v}^{dm} + \bar{\eta} \times \frac{d}{dt}\bar{v}^{dm}\right) dm \tag{5.22}$$

Substituting $\bar{v}^{dm}$ for $d\bar{\eta}/dt$, we obtain

$$\bar{M}_k^* = -\int_{m_k} \left[\bar{v}^{dm} \times \bar{v}^{dm} + \bar{\eta} \times \frac{d}{dt}(\bar{\omega}^k \times \eta)\right] dm \tag{5.23}$$

Note how the derivative of $\bar{v}^{dm}$ is replaced by its expression in terms of cross product of the angular velocity and the position vector. Dropping the cross product of the velocity of dm and carrying out the first derivative in the equation, we get

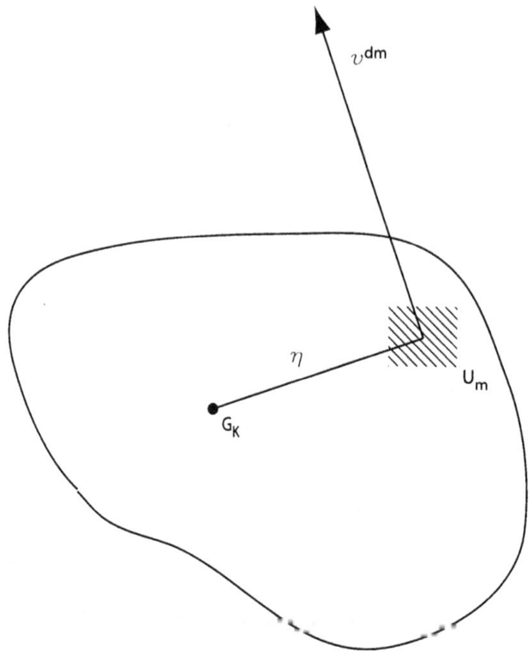

Fig. 5.3. Element mass dm and the center mass G_k of body K

$$\bar{M}_k^* = -\int_{m_k} \{(\bar{\upsilon}^{dm} \times \bar{\upsilon}^{dm}) + \bar{\eta} \times (\bar{\alpha}^k \times \bar{\eta}) + \bar{\eta} \times [\bar{\omega}^k \times (\bar{\omega}^k \times \bar{\eta})]\}dm \quad (5.24)$$

Equation (5.24) can be expressed further as

$$\bar{M}_k^* = -\int_{m_k} \{\bar{\upsilon}^{dm} \times \bar{\upsilon}^{dm} + (\bar{\eta}^2\bar{\alpha}^k - \bar{\alpha}^k \cdot \bar{\eta}\bar{\eta}) + \bar{\eta} \times [(\bar{\omega}^k \cdot \bar{\eta})\bar{\omega}^k - (\bar{\omega}^k)^2\bar{\eta}]\}\, dm \quad (5.25)$$

or

$$\begin{aligned} \bar{M}_k^* &= -\bar{\alpha}^k \cdot \int_{m_k} [\bar{\bar{u}}\bar{\eta}^2 - \bar{\eta}\bar{\eta}]\, dm - \bar{\eta} \times \int_{m} [(\bar{\omega}^k \cdot \bar{\eta})\bar{\omega}^k + (\bar{\omega}^k)^2\bar{\eta}]\, dm \\ &= -\bar{\alpha}^k \cdot \bar{\bar{I}} - \bar{\omega}^k \times \left\{ \int_{m_k} \left[\bar{\bar{u}}\bar{\eta}^2 - \bar{\eta}\bar{\eta}\right] dm \right\} \cdot \bar{\omega}^k \end{aligned} \quad (5.26)$$

Finally, identifying the inertia dyadic term, equation (5.26) becomes

$$\bar{M}_k^* = -\bar{\alpha}^k \cdot \bar{\bar{I}} - \bar{\omega}^k \times (\bar{\bar{I}} \cdot \bar{\omega}^k) \quad (5.27)$$

which is equivalent to equation (5.20). In equation (5.26), $\bar{\bar{u}}$ is the unit dyadic and $\bar{\bar{I}}$ is the inertia dyadic of body k relative to G_k for $\bar{n}_i$ $(i = 1, 2, 3)$ fixed in R [see equation (4.158)].

Further Proof of Equation (5.20)
The inertia torque (moment) $\bar{M}_k^*$ is given by

$$\bar{M}_k^* = -\sum_{i=1}^{n} m_i \bar{\eta}_i \times \bar{a}_i \tag{5.28}$$

where m_i denotes the particle mass, and $\bar{\eta}_i$ and $\bar{a}_i$ are the position from the center mass to the element and acceleration of m_i in R (body k is composed of n particle masses). We also know that the acceleration of particle i can be expressed as

$$\bar{a}_i = \bar{a}_k^* + \bar{\alpha}^k \times \bar{\eta}_i + \bar{\omega}^k \times (\bar{\omega}^k \times \bar{\eta}_i) \tag{5.29}$$

where $\bar{a}_k^*$ is the acceleration of body k in R (center of mass), and $\bar{\omega}^k$ and $\bar{\alpha}^k$ its corresponding angular velocity and acceleration; hence

$$\bar{M}_k^* = -\sum_{i=1}^{n} m_i \bar{\eta}_i \times \bar{a}_k^* - \sum_{i=1}^{n} m_i \bar{\eta}_i \times \bar{\alpha}^k \times \bar{\eta}_i - \sum_{i=1}^{n} m_i \bar{\eta}_i \times \bar{\omega}^k \times (\bar{\omega}^k \times \bar{\eta}_i) \tag{5.30}$$

Consider the three expressions given by equation (5.30). First,

$$\sum_{i=1}^{n} m_i \bar{\eta}_i \times \bar{a}_k^* = \left(\sum_{i=1}^{n} m_i \bar{\eta}_i\right) \times \bar{a}_k^* = 0 \tag{5.31}$$

Second,

$$\sum_{i=1}^{n} m_i \bar{\eta}_i \times \bar{\alpha}^k \times \bar{\eta}_i = \sum_{i=1}^{n} m_i \bar{\eta}_i \times (\alpha_a^k \bar{\eta}_a) \times \bar{\eta}_i \tag{5.32}$$

where α_a^k denotes the scalar components of the angular acceleration $\bar{\alpha}^k$ in R along a unit vector $\bar{n}_a$. Factorizing the scalar quantity out, we get

$$\alpha_a^k \sum_{i=1}^{n} m_i \bar{\eta}_i \times n_a \times \bar{\eta}_i = \alpha_a^k \bar{I}_a \tag{5.33}$$

where $\bar{I}_a$ is the second moment of body k with respect to R for $\bar{n}_a$. We also know that the inertia dyadic $\bar{\bar{I}}$ and $\bar{I}_a$ are related by

$$\bar{I}_a = \bar{\bar{I}} \cdot \bar{n}_a \tag{5.34}$$

Hence

$$\alpha_a^k \bar{I}_a = \alpha_a^k \bar{\bar{I}} \cdot \bar{n}_a = \bar{\bar{I}} \cdot \alpha_a^k \bar{n}_a = \bar{\bar{I}} \cdot \bar{\alpha}^k \tag{5.35}$$

The third term consequently yields

$$\begin{aligned} \sum_{i=1}^{n} m_i \bar{\eta}_i \times \bar{\omega}^k \times (\bar{\omega}^k \times \bar{\eta}_i) &= \bar{\omega}^k \times \sum_{i=1}^{n} m_i \bar{\eta}_i \times (\bar{\omega}^k \times \bar{\eta}_i) \\ &= \bar{\omega}^k \times (\bar{\bar{I}} \cdot \bar{\omega}^k) \end{aligned} \tag{5.36}$$

Substituting equations (5.31), (5.35) and (5.36) into (5.30), we get

$$\bar{M}_k^* = -\bar{\bar{I}} \cdot \bar{\alpha}^k - \bar{\omega}^k \times (\bar{\bar{I}} \cdot \bar{\omega}^k) \tag{5.37}$$

We should use equation (5.27) to define the inertia torque (moment) in our formulation of the equations of motion. Once the angular velocity of k in R is defined, the angular acceleration is then deduced automatically. Making use of the definition of the inertia dyadic, the explicit form of (5.18) can be found.

5.4 Automated Form of the Equations of Motion

Our interest is to develop an automated approach in representing the equations of motion for general multibody systems. To this end, we have considered the systems to be unconstrained, yet in most applications most of those mechanical systems are subjected to some form of constraints. This topic is addressed in Chapter 6 in more detail. The objective at this point is to use the kinematical expressions developed previously, together with Kane's equations to derive a numerical form of the governing equations ready for computer implementation and for use to explicitly obtain the equations of motion. Readers can use this form for further studies without having to rederive the governing equations. The direct representation of the equations of motion is by far the ultimate goal in dynamics. What follows is the matrix representation of the equation of motions which can be used directly and explicitly without the need of any mathematical tool as long as the kinematics in Chapter 3 and the generalized force definition in Chapter 4 are followed.

5.4.1 Matrix Representation of the Equations of Motion

In Chapter 3, the strategy was to derive the kinematic of a treelike multibody system using matrices, making use of a partitioning of the generalized coordinates. We followed our analysis of contribution of forces and moments in Chapter 4 making use of previously developed partial angular velocities and partial velocities. It will now become apparent that the work we have done in Chapter 3 and Chapter 4 will pay off at the end to obtain an explicit form of the equation of motions for any multibody system. Following equations (5.14 and 5.20), we define term 1 as

$$\begin{aligned}
\text{term 1} &= -m_k \bar{a}_k \cdot \frac{\partial \bar{v}_k}{\partial \dot{x}} \\
&= -m_k(\{\ddot{x}\}^T [V_{\mathrm{w}}^k] + \{\dot{x}\}^T [\dot{V}_{\mathrm{w}}^k])\{\bar{n}\} \cdot \{n\}^T \cdot [V_{\mathrm{w}}^k]^T \\
&= -m_k(\{\ddot{x}\}^T [V_{\mathrm{w}}^k][V_{\mathrm{w}}^k]^T + \{\dot{x}\}^T [\dot{V}_{\mathrm{w}}^k][V_{\mathrm{w}}^k]^T)
\end{aligned} \tag{5.38}$$

The contribution of moments is shown in term 2.

$$\text{term 2} = -[\bar{\bar{I}} \cdot \bar{\alpha}^k + \bar{\omega}^k \times (\bar{\bar{I}} \cdot \bar{\omega}^k)] \cdot \frac{\partial \bar{\omega}^k}{\partial \dot{x}}$$

$$= \underbrace{-(\bar{\bar{I}} \cdot \bar{\alpha}_k^k) \cdot \frac{\partial \bar{\omega}^k}{\partial \dot{x}}}_{\text{part 1}} \underbrace{-(\bar{\omega}^k \times (\bar{\bar{I}} \cdot \bar{\omega}^k)) \cdot \frac{\partial \bar{\omega}^k}{\partial \dot{x}}}_{\text{part 2}} \tag{5.39}$$

where,

$$\begin{aligned} \text{part 1} &= -\{(\{\ddot{x}\}^T[\omega^k] + \{\dot{x}\}^T[\dot{\omega}^k])\{\bar{n}\} \cdot \{\bar{n}\}^T[I_{ko}]\{\bar{n}\}\} \cdot \{\bar{n}\}^T[\omega^k]^T \\ &= -(\{\ddot{x}\}^T[\omega^k] + \{\dot{x}\}^T[\dot{\omega}^k])[I_{ko}][\omega^k]^T \\ &= -\{\ddot{x}\}^T[\omega^k][I_{ko}][\omega^k]^T - \{\dot{x}\}^T[\dot{\omega}^k][I_{ko}][\omega^k]^T \end{aligned} \tag{5.40}$$

and

$$\begin{aligned} \text{part 2} = &\ -\{\dot{x}\}^T[\omega^k]\{\bar{n}\} \times (\{\dot{x}\}^T[\omega^k]\{\bar{n}\} \cdot \{\bar{n}\}^T[I_{ko}]\{\bar{n}\}) \cdot \{\bar{n}\}^T[\omega^k]^T \\ = &\ -\{\dot{x}\}^T[\omega^k][\Omega_x^{ko}][I_{ko}][\omega^k]^T \end{aligned} \tag{5.41}$$

In the above equation, $[\Omega_x^{ko}]$ is the skew matrix of $\{\dot{x}\}^T[\omega^k]$, $[\omega^k]$ is given by equation (3.40), and

$$[V_{\mathrm{w}}^k]_{3n\times 3} = [W]_{3n\times 3}[V^k]_{3n\times 3} \tag{5.42}$$

where $[V^k]$ is given by equation (3.76) and $[W]$ is a transformation matrix used to map $\{y\}$ to $\{\dot{x}\}$. The components of $\{y\}$ represent the generalized speeds, which are a linear function of the generalized coordinate derivatives x.

$$\{y\}_{1\times 3n}^T = \{\dot{x}\}_{1\times 3n}^T[W]_{3n\times 3n} \tag{5.43}$$

Before we get to the evaluation of the coefficients needed to formulate the equations of motion, we should have the expressions for the inertia dyadic matrix. We know that the inertia matrices for the body in local coordinate (moment of inertia) can expressed as

$$[I_k] = \begin{bmatrix} I_{k11} & I_{k12} & I_{k13} \\ I_{k21} & I_{k22} & I_{k23} \\ I_{k31} & I_{k32} & I_{k33} \end{bmatrix} \tag{5.44}$$

If we locate the coordinate system at symmetric axes or principal centroid axes of body k properly, then

$$[I_k] = \begin{bmatrix} I_{k1} & 0 & 0 \\ 0 & I_{k2} & 0 \\ 0 & 0 & I_{k3} \end{bmatrix} \tag{5.45}$$

The coefficients I_{km} $(m = 1,\ 2,\ 3)$ are evaluated with respect to the unit vectors, as shown in Figure 4.18. Since $[I_k]$ is expressed in local coordinate $\{n\}^k$, then for the fixed coordinate $\{\bar{n}\}$ we have

$$[I_{ko}] = [S^{ok}][I_k][S^{ko}] \tag{5.46}$$

Combining the equations above, we arrive at the final representation of the equations of motion

$$[a]\{\ddot{x}\} + [b]\{\dot{x}\} + [c]\{\dot{x}\} = \{f\} \tag{5.47}$$

where

$$[a] = \sum_k m_k [V_w]^k [V_w^k]^T + \sum_k [\omega^k][I_{ok}][\omega^k]^T \tag{5.48}$$

$$[b] = \sum_k m_k [V_w^k][\dot{V}_w^k]^T + \sum_k [\omega^k][I_{ok}][\dot{\omega}^k]^T \tag{5.49}$$

$$[c] = \sum_k [\omega^k][I_{ok}][\Omega_x^{ok}][\omega^k]^T \tag{5.50}$$

and

$$\{x\} = \{x_1,\ x_2,\ x_3, \ldots, x_{3N},\ x_{3N+1},\ x_{6N}\}$$

where $\{f\}$ denotes the generalized external forces expressed in matrix form as

$$\{f\} = \sum_k [V_{\mathrm{W}}^k]\{F\} + \sum_k [\omega^k]\{M_k\} \tag{5.51}$$

where $\{F_x\}$ and $\{M_k\}$ are the resultant external force and moment acting on body k, respectively. A flowchart for computer coding that can be used to write general-purpose codes based on the equations above is presented in Appendix A. The uniqueness of equation (5.47) stems from the matrix representation of all the kinematical quantities already defined in Chapters 3 and 4. Once more, observe the dominant role of the matrices $[V_{\mathrm{W}}]$ and $[\omega^k]$, which constitutes the bulk of computation. The matrices $[a]$, $[b]$ and $[c]$ in their developed form identify for us the contribution from both the translational and rotational degrees of freedom; [b] and $[c]$ account for all the quadratic velocity terms, coriolis effects and essentially the nonlinear terms in the velocity contribution to the equations of motion. The generalized coordinates are partitioned in the same manner as described in Chapter 3.

Note: In the context of dynamics the a, b, c and f components of the equations of motion represent an alternative to Lagrange equations, Newton–Euler equations, D'Alembert principles, Hamilton's equations and Gibbs–Appel dynamical equations.

In the study of complex systems where the equations of motion are needed to simulate the system dynamics, a direct application of equation (5.47) will be beneficial.

Example 5.4.1
Using equation (5.47), derive the explicit kinematical expressions needed in the formulation of the equations of motion for a planar two-link manipulator shown in Figure 5.4. Use the concepts introduced in Chapters 3, 4 and 5. The final equations of motion should have the same form as equation (5.47). This exercise should assist readers in writing general codes based on equation (5.47).

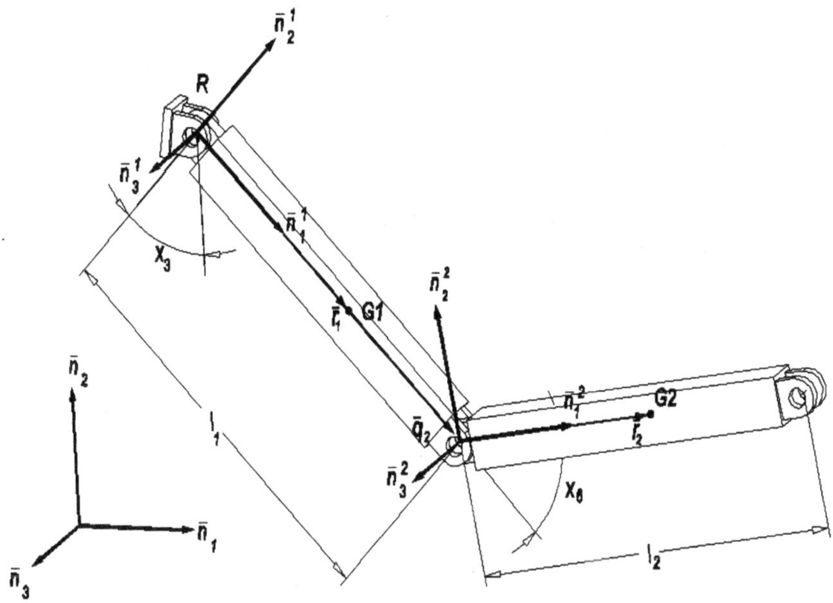

Fig. 5.4. Reference frames, body and center mass position vectors and joint angle for a planar two-link manipulator

Solution
Following are the concepts we have introduced that lead us to the simple form in representing the equations of motion of open-chain multibody system. We will use all the steps to demonstrate the utilities of equation (5.47), but one needs to keep in mind that the problem can be reduced if we adapt directly the exact number of degrees of freedom.

Number of Bodies (links)

$N = 2$
Our index k takes the values of 1 and 2, respectively:
$k = (1, 2)$

Degrees of Freedom

$n = 2$

Tree Array
Although we don't need the tree array because of the fact, this is a simple 2-body problem, the tree array is computed to get accustomed to generating all the necessary information.

$\Gamma_0(k)$	1	2
$\Gamma_1(k)$	0	1
$\Gamma_2(k)$	0	0

Table 5.1. Tree array for Figure 5.4

Transformation Matrices (Shifters)
Based on the reference frames $\bar{n}_i$, $\bar{n}_i^1$ and $\bar{n}_i^2 (i = 1,\ 2,\ 3)$ as shown, where $\bar{n}$ is the inertial frame and $\bar{n}^1$ and $\bar{n}^2$ are the local axis for links 1 and 2, respectively, we write the following transformation matrices:

$$[S^{10}] = \begin{bmatrix} c1 & s1 & 0 \\ -s1 & c1 & 0 \\ 0 & 0 & 1 \end{bmatrix}, \quad [S^{21}] = \begin{bmatrix} c2 & s2 & 0 \\ -s2 & c2 & 0 \\ 0 & 0 & 1 \end{bmatrix} \tag{5.52}$$

and

$$[S^{20}] = [S^{21}][S^{10}] = \begin{bmatrix} c(1+2) & s(1+2) & 0 \\ -s(1+2) & c(1+2) & 0 \\ 0 & 0 & 1 \end{bmatrix} \tag{5.53}$$

where

$$s1 = \sin\theta_1 \text{ and } c1 = \cos\theta_1$$

$$s2 = \sin\theta_2 \text{ and } c2 = \cos\theta_2$$

The expression $c\ (1 + 2)$ represents the cosine of the sum of the two angles, and $s\ (1 + 2)$ the sine of the sum of the two angles.

Generalized Coordinates
The values $\theta_1$and θ_2 are the only coordinates used to describe the motion of the two-link manipulator. In Chapter 3, we introduced a partition procedure for representing the rotation and translation coordinates. From Chapter 3, it follows that

$$\{x\}^T = [0\ \ 0\ \ x_3\ \ 0\ \ 0\ \ x_6] \tag{5.54}$$

where $x_3 = \theta_1$ and $x_6 = \theta_2$ and x_3 and x_6 indicate the rotation of link 1 about the $\bar{n}_3^1$ axis and $\bar{n}_3^2$ axis, respectively. The first two zeros show that no rotation is allowed along the other two axes, so one should keep in mind that in the transformation matrices we can substitute θ_1 with x_3 and θ_2 with x_6.

Derivatives (with Respect to Time) of the Transformation Matrices
Given the transformation matrix between body 1 and R, S^{10}, its derivative is found to be

$$[\dot{S}^{10}] = \begin{bmatrix} -s1 & c1 & 0 \\ -c1 & -s1 & 0 \\ 0 & 0 & 0 \end{bmatrix} \dot{x}_3 \tag{5.55}$$

The time derivative of S^{20} between body 2 and R is

$$[\dot{S}^{20}] = \begin{bmatrix} -s(1+2) & c(1+2) & 0 \\ -c(1+2) & -s(1+2) & 0 \\ 0 & 0 & 0 \end{bmatrix} (\dot{x}_3 + \dot{x}_6) \tag{5.56}$$

Note that $\dot{x}_3$ is the generalized coordinate derivative for body 1, and is also represented by $\dot{\theta}1$. Similarly, $\dot{x}_6$ is denoted by $\dot{\theta}2$.

Partial Angular Velocities
For body 1, the partial angular velocity is

$$[\omega^1] = \left[\frac{I_{3\times 3}}{0_{3\times 3}}\right] \tag{5.57}$$

Following the definition of the angular velocity for body 2, the partial angular velocity is

$$[\omega^2] = \left[\frac{I_{3\times 3}}{S^{10}}\right] \tag{5.58}$$

Angular Velocities
Body 1 angular velocity is defined by

$$\begin{aligned} {}^R\bar{\omega}^1 &= \{\dot{x}\}^T[\omega^1]\{\bar{n}\} \\ &= [0 \;\; 0 \;\; \dot{x}_3 \;\; 0 \;\; 0 \;\; \dot{x}_6]\left[\frac{I_{3\times 3}}{0_{3\times 3}}\right]\begin{bmatrix} \bar{n}_1 \\ \bar{n}_2 \\ \bar{n}_3 \end{bmatrix} \\ &= \dot{x}_3\bar{n}_3 \end{aligned} \tag{5.59}$$

Similarly, body 2 yields

$$\begin{aligned} {}^R\bar{\omega}^2 &= \{\dot{x}\}^T[\omega^2]\{\bar{n}\} \\ &= [0 \;\; 0 \;\; \dot{x}_3 \;\; 0 \;\; 0 \;\; \dot{x}_6]\left[\frac{I_{3\times 3}}{S^{10}}\right]\begin{bmatrix} \bar{n}_1 \\ \bar{n}_2 \\ \bar{n}_3 \end{bmatrix} \\ &= (\dot{x}_3 + \dot{x}_6)\bar{n}_3 \end{aligned} \tag{5.60}$$

As expected, the angular velocities are simply a function of $\dot{x}_3$ and $\dot{x}_6$. Observe how in the context of complex multibody systems, the angular velocities can easily be evaluated through the procedure above.

Time Derivatives of Partial Angular Velocities
For body 1, we have

$$[\dot{\omega}^1] = \left[\frac{0_{3\times 3}}{0_{3\times 3}}\right] \tag{5.61}$$

For body 2

$$[\dot{\omega}^2] = \left[\frac{I_{3\times 3}}{\dot{S}^{10}}\right] = \dot{x}_3 \begin{bmatrix} & 0_{3\times 3} & \\ -s1 & c1 & 0 \\ -c1 & -s1 & 0 \\ 0 & 0 & 0 \end{bmatrix} \tag{5.62}$$

where $I_{3\times 3}$ denotes the identity matrix, and $0_{3\times 3}$ the zero matrix.

Angular Accelerations
The angular acceleration of body 1 is defined as follows

$$\begin{aligned} {}^R\bar{\alpha}^1 &= (\{\ddot{x}\}^T[\omega^1] + \{\dot{x}\}^T[\dot{\omega}^1])\{\bar{n}\} \\ &= \left[[0\ 0\ \ddot{x}_3\ 0\ 0\ \ddot{x}_6]\left[\frac{I}{0}\right] + [0\ 0\ \dot{x}_3\ 0\ 0\ \dot{x}_6]\left[\frac{0}{0}\right]\right]\{\bar{n}\} \\ &= \ddot{x}_3\bar{n}_3 \end{aligned} \tag{5.63}$$

Similarly, we define body 2 angular velocity where

$$\begin{aligned} {}^R\bar{\alpha}^2 &= (\{\ddot{x}\}^T[\omega^2] + \{\dot{x}\}^T[\dot{\omega}^2])\{\bar{n}\} \\ &= \left[[0\ 0\ \ddot{x}_3\ 0\ 0\ \ddot{x}_6]\left[\frac{I}{S^{10}}\right] + [0\ 0\ \dot{x}_3\ 0\ 0\ \dot{x}_6]\left[\frac{0}{\dot{S}^{10}}\right]\right]\{\bar{n}\} \\ &= (\ddot{x}_3 + \ddot{x}_6)\bar{n}_3 \end{aligned} \tag{5.64}$$

Note that the angular velocities and angular accelerations could have been defined easily by simple inspection of the time derivatives of the angular displacements. This is just another reminder how the methods outlined in this chapter are effective tools when dealing with large systems.

Body Vectors
The vector components below follow from the representation of the q's and r's used in Chapter 3, modeling of the velocities in multibody systems. Note that for more complicated, asymmetric bodies, q and r may not have any components equal to zero. Following the notation proposed in Chapter 3 and examining the model presented in Figure 5.4, we collect the following geometrical information for body 1 and 2.

$$\{q_1\} = 0 \tag{5.65}$$

$$\{q_2\} = [\ell_1 \;\; 0 \;\; 0]^T \tag{5.66}$$

$$\{q_3\} = [\ell_2 \;\; 0 \;\; 0]^T \tag{5.67}$$

$$\{r_1\} = \begin{bmatrix} \frac{\ell_1}{2} & 0 & 0 \end{bmatrix}^T \tag{5.68}$$

$$\{r_2\} = \begin{bmatrix} \frac{\ell_1}{2} & 0 & 0 \end{bmatrix}^T \tag{5.69}$$

Their Skew Matrices
The above vectors can be expressed in a matrix form using the skew matrix.

$$[Sq_2] = \ell_1 \begin{bmatrix} 0 & 0 & 0 \\ 0 & 0 & -1 \\ 0 & 1 & 0 \end{bmatrix} \tag{5.70}$$

$$[Sq_3] = \ell_2 \begin{bmatrix} 0 & 0 & 0 \\ 0 & 0 & -1 \\ 0 & 1 & 0 \end{bmatrix} \tag{5.71}$$

$$[Sr_1] = \frac{\ell_1}{2} \begin{bmatrix} 0 & 0 & 0 \\ 0 & 0 & -1 \\ 0 & 1 & 0 \end{bmatrix} \tag{5.72}$$

$$[Sr_2] = \frac{\ell_2}{2} \begin{bmatrix} 0 & 0 & 0 \\ 0 & 0 & -1 \\ 0 & 1 & 0 \end{bmatrix} \tag{5.73}$$

Partial Velocity Matrices
Using equation (3.120) where $\{\dot{\xi}_1\} = \{\xi_1\} = 0$, for body 1 we get

$$[V^1] = \begin{bmatrix} [S_{r1}][S^{10}] \\ \hline 0 \end{bmatrix} = \frac{1}{2}\ell_1 \begin{bmatrix} 0 & 0 & 0 \\ 0 & 0 & -1 \\ -s1 & c1 & 0 \\ 0 & 0 & 0 \\ 0 & 0 & 0 \\ 0 & 0 & 0 \end{bmatrix} \tag{5.74}$$

and, for body 2

$$
[V^2] = \begin{bmatrix} [Sq_2][S^{10}] \\ ------ \\ [Sr_2][S^{20}] \end{bmatrix}
= \begin{bmatrix} 0 & 0 & 0 \\ 0 & 0 & -\ell_1 \\ -\ell_1 s1 & \ell_1 c1 & 0 \\ 0 & 0 & 0 \\ 0 & 0 & -\frac{1}{2}\ell_2 \\ -\frac{1}{2}\ell_2 s(1+2) & \frac{1}{2}\ell_2 c(1+2) & 0 \end{bmatrix} \tag{5.75}
$$

The differentiation of the partial velocity for body 1 is given by

$$
[\dot{V}^1] = \frac{1}{2}\ell_1 \dot{x}_3 \begin{bmatrix} 0 & 0 & 0 \\ 0 & 0 & 0 \\ -c1 & -s1 & 0 \\ 0 & 0 & 0 \\ 0 & 0 & 0 \\ 0 & 0 & 0 \end{bmatrix} \tag{5.76}
$$

In a similar fashion, we obtain the partial velocity for body 2

$$
[\dot{V}^2] = \begin{bmatrix} 0 & 0 & 0 \\ 0 & 0 & 0 \\ -\ell_1 \dot{x}_3 c1 & -\ell_1 \dot{x}_3 s1 & 0 \\ 0 & 0 & 0 \\ 0 & 0 & 0 \\ -\frac{1}{2}\ell_2(\dot{x}_3 + \dot{x}_6) \times c(1+2) & -\frac{1}{2}\ell_2(\dot{x}_3 + \dot{x}_6) \times s(1+2) & 0 \end{bmatrix} \tag{5.77}
$$

Angular Velocities Skew Matrices

From equation (3.27), we make use of the skew matrix related to the time derivative of the shifter, where

$$
[\Omega^{10}] = [\dot{S}^{10}][S^{01}] = \dot{x}_3 \begin{bmatrix} 0 & 1 & 0 \\ -1 & 0 & 0 \\ 0 & 0 & 0 \end{bmatrix} \tag{5.78}
$$

$$
[\Omega^{20}] = [\dot{S}^{20}][S^{02}] = (\dot{x}_3 + \dot{x}_6) \begin{bmatrix} 0 & 1 & 0 \\ -1 & 0 & 0 \\ 0 & 0 & 0 \end{bmatrix} \tag{5.79}
$$

The expression $[\Omega_x^{01}]$ is a skew matrix of angular velocity components defined by $\{\dot{x}\}^T[\omega^1]$.

$$\{\dot{x}\}^T[\omega^1] = [0 \;\; 0 \;\; \dot{x}_3 \;\; 0 \;\; 0 \;\; \dot{x}_6]\begin{bmatrix} I \\ \hline 0 \end{bmatrix} = \begin{bmatrix} 0 \\ 0 \\ \dot{x}_3 \end{bmatrix}^T \tag{5.80}$$

$$[\Omega_x^{01}] = \dot{x}_3\begin{bmatrix} 0 & -1 & 0 \\ 1 & 0 & 0 \\ 0 & 0 & 0 \end{bmatrix} = [\Omega^{10}]^T \tag{5.81}$$

In a similar fashion, $[\Omega_x^{02}]$ is a skew matrix defining the angular velocity components of $\{\dot{x}\}^T[\omega^2]$.

$$\{\dot{x}\}^T[\omega^2] = [0 \;\; 0 \;\; \dot{x}_3 \;\; 0 \;\; 0 \;\; \dot{x}_6]\begin{bmatrix} 1 & 0 & 0 \\ 0 & 1 & 0 \\ 0 & 0 & 1 \\ c1 & s1 & 0 \\ -s1 & c1 & 0 \\ 0 & 0 & 1 \end{bmatrix} \tag{5.82}$$

$$= \begin{bmatrix} 0 \\ 0 \\ \dot{x}_3 + \dot{x}_6 \end{bmatrix}$$

$$[\Omega_x^{02}] = (\dot{x}_3 + \dot{x}_6)\begin{bmatrix} 0 & -1 & 0 \\ 1 & 0 & 0 \\ 0 & 0 & 0 \end{bmatrix} = [\Omega^{20}]^T \tag{5.83}$$

From the equations above, we get

$$[W] = \begin{bmatrix} \hat{I} & \hat{I} \\ 0 & \hat{I} \end{bmatrix}_{6\times 6} \tag{5.84}$$

where

$$\hat{I} = \begin{bmatrix} 0 & & \\ & 0 & \\ & & 1 \end{bmatrix} \tag{5.85}$$

Note that $\hat{I}$ is just a form of an identity-zero matrix, with entries that correspond to the nonzero degrees of freedom.

Matrix $[W]$ is used to map $\{y\} = [0 \;\; 0 \;\; \dot{x}_3 \;\; 0 \;\; 0 \;\; \dot{x}_3 + \dot{x}_6]^T$ to

$$\{\dot{x}\} = [0 \quad 0 \quad \dot{x}_3 \quad 0 \quad 0 \quad \dot{x}_6]^T$$

Mass Center Velocities
The mass center velocities for body 1 and 2 are then obtained following the definitions and substituting the information derived above. Therefore, we obtain expressions which when premultiplied yield the velocities in their final form

$$\bar{V}_1 = \{y\}^T[V^1]\{\bar{n}\} = \{\dot{x}\}^T[W][V^1]\{\bar{n}\}$$

$$= \frac{1}{2}\ell_1\dot{x}_3(-s1\ \bar{n}_1 + c1\ \bar{n}_2) \tag{5.86}$$

$$\bar{V}_2 = \{y\}^T[V^2]\{\bar{n}\} = \{\dot{x}\}^T[W][V^2]\{\bar{n}\}$$

$$= \left[-\ell_1\dot{x}_3 s1 - \frac{1}{2}\ell_2(\dot{x}_3 + \dot{x}_6)s(1+2)\right]\bar{n}_1$$

$$+ \left[-\ell_1\dot{x}_3 c1 + \frac{1}{2}\ell_2(\dot{x}_3 + \dot{x}_6)c(1+2)\right]\bar{n}_2 \tag{5.87}$$

where $\bar{V}_1$ and $\bar{V}_2$ are the velocities of the center masses of links 1 and 2.

Mass Center Accelerations
In a similar fashion, we obtain the mass center accelerations, where

$$\bar{a}_1 = (\{\dot{y}\}^T[V^1] + \{y\}^T[\dot{V}^1])\{\bar{n}\}$$

$$= \frac{1}{2}\ell_1\ddot{x}_3(s1\bar{n}_1 + c1\bar{n}_2) - \frac{1}{2}\ell_1\dot{x}_3^2(c1\bar{n}_1 + s1\bar{n}_2) \tag{5.88}$$

$$\bar{a}_2 = (\{\dot{y}\}^T[V^2] + \{y\}^T[\dot{V}^2])\{\bar{n}\}$$

$$= \left[-\ell_1\ddot{x}_3 s1 - \frac{1}{2}\ell_2(\ddot{x}_3 + \ddot{x}_6)s(1+2)\right]\bar{n}_1$$

$$+ \left[-\ell_1\ddot{x}_3 c1 + s\frac{1}{2}\ell_2(\ddot{x}_3 + \ddot{x}_6)c(1+2)\right]\bar{n}_2$$

$$+ \left[-\ell_1\ddot{x}_3 c1 - \frac{1}{2}\ell_2(\ddot{x}_3 + \ddot{x}_6)^2 c(1+2)\right]\bar{n}_1$$

$$- \left[\ell_1\ddot{x}_3^2 s1 + \tfrac{1}{2}\ell_2(\ddot{x}_3 + \ddot{x}_6)^2 s(1+2)\right]\bar{n}_2 \tag{5.89}$$

where $\bar{a}_1$ and $\bar{a}_2$ are the mass center accelerations of links 1 and 2, respectively.

Equations of Motion
Once the kinematics have been computed the equations of motion become a simple mathematical task. From equation (5.48) the generalized mass matrix is defined by

$$[a] = \sum_{k=1}^{2} m_k [V_w^k][V_w^k]^T + \sum_{k=1}^{2} [\omega^k][I_{ok}][\omega^k]^T \tag{5.90}$$

or

$$[a] = m_1[V_w^1][V_w^1]^T + m_2[V_w^2][V_w^2]^T + [\omega^1][I_{10}][\omega^1]^T + [\omega^2][I_{20}][\omega^2]^T \tag{5.91}$$

where the explicit form for the above matrices are

$$[V_w^1] = [W][V^1] \tag{5.92}$$

$$[V_w^2] = [W][V^2] \tag{5.93}$$

$$[I_{10}] = [S^{01}][I_1][S^{10}] \tag{5.94}$$

$$[I_{20}] = [S^{02}][I_2][S^{20}] \tag{5.95}$$

$$[I_1] = \begin{bmatrix} 0 & & \\ & 0 & \\ & & I_{13} \end{bmatrix} \tag{5.96}$$

$$[I_2] = \begin{bmatrix} 0 & & \\ & 0 & \\ & & I_{23} \end{bmatrix} \tag{5.97}$$

Let the moments of inertia components be of a slender beam. Then

$$I_{13} = \tfrac{1}{12} m_1 \ell_1^2 \tag{5.98}$$

and

$$I_{23} = \tfrac{1}{12} m_2 \ell_2^2 \tag{5.99}$$

The generalized mass matrix then becomes

$$[a] = m_1 \begin{bmatrix} \hat{I} & \hat{I} \\ 0 & \hat{I} \end{bmatrix} \left[\frac{[S_{r1}][S^{10}]}{0}\right] [[S^{01}][S_{r1}^T]|0] \begin{bmatrix} \hat{I} & 0 \\ \hat{I} & \hat{I} \end{bmatrix}$$

$$+ m_2 \begin{bmatrix} \hat{I} & \hat{I} \\ 0 & \hat{I} \end{bmatrix} \left[\frac{[S_{q2}][S^{10}]}{[S_{r2}][S^{20}]}\right] \left[[S^{01}][S_{q2}^T][S^{02}]|[S_{r2}^T]\right] \begin{bmatrix} \hat{I} & 0 \\ \hat{I} & \hat{I} \end{bmatrix}$$

$$+ \left[\frac{I}{0}\right] [S^{01}][I_1][S^{01}][I|0] + \left[\frac{I}{S^{10}}\right] [S^{02}][I_2][S^{20}][I|S^{01}]$$

Simplifying the above equation further yields

$$[a] = m_1 \begin{bmatrix} \hat{I} & \hat{I} \\ 0 & \hat{I} \end{bmatrix} \begin{bmatrix} [S_{r1}][S_{r1}^T] & 0 \\ 0 & 0 \end{bmatrix} \begin{bmatrix} \hat{I} & 0 \\ \hat{I} & \hat{I} \end{bmatrix}$$

$$+ m_2 \begin{bmatrix} \hat{I} & \hat{I} \\ 0 & \hat{I} \end{bmatrix} \begin{bmatrix} [S_{q2}][S_{q2}^T] & [S_{q2}][S^{12}][S_{r2}^T] \\ [S_{r2}][S^{10}][S_{q2}^T] & [S_{r2}][S_{r2}^T] \end{bmatrix} \begin{bmatrix} \hat{I} & 0 \\ \hat{I} & \hat{I} \end{bmatrix}$$

$$+ \left[\frac{I}{0}\right] \begin{bmatrix} 0 & & \\ & 0 & \\ & & I_{13} \end{bmatrix} [I|0] + \left[\frac{I}{[S^{10}]}\right] \begin{bmatrix} 0 & & \\ & 0 & \\ & & I_{23} \end{bmatrix} [I|S^{01}]$$

Substitution of the submatrices yields

$$[a] = \frac{1}{4} m_1 \ell_1^2 \begin{bmatrix} \hat{I} & 0 \\ 0 & 0 \end{bmatrix} + m_2 \begin{bmatrix} (\ell_1^2 + \ell_1 \ell_2 c2 + \frac{1}{4}\ell_2^2)\hat{I} & (\frac{1}{2}\ell_1\ell_2 c2 + \frac{1}{4}\ell_2^2)\hat{I} \\ (\frac{1}{2}\ell_1\ell_2 c2 + \frac{1}{4}\ell_2^2)\hat{I} & \frac{1}{4}\ell_2^2 \hat{I} \end{bmatrix}$$

$$+ \begin{bmatrix} \hat{I}_{13}\hat{I} & 0 \\ 0 & 0 \end{bmatrix} + \begin{bmatrix} I_{23}\hat{I} & I_{23}\hat{I} \\ I_{23}\hat{I} & I_{23} I \end{bmatrix}$$

Expanding the matrix $\hat{I}$, we obtain the generalized mass matrix as

$$[a] = \begin{bmatrix} 0 & 0 & 0 & 0 & 0 & 0 \\ 0 & 0 & 0 & 0 & 0 & 0 \\ 0 & 0 & \begin{matrix} \frac{1}{4}m_1\ell_1^2 + m_2 \\ \times(\ell_1^2 + \ell_1\ell_2 c2 \\ +\frac{1}{4}\ell_2^2) + I_{13} + I_{23} \end{matrix} & 0 & 0 & \begin{matrix} m_2(\frac{1}{2}\ell_1\ell_2 c2 \\ +\frac{1}{4}\ell_2^2) + I_{23} \end{matrix} \\ 0 & 0 & 0 & 0 & 0 & 0 \\ 0 & 0 & 0 & 0 & 0 & 0 \\ 0 & 0 & \begin{matrix} m_2(\frac{1}{2}\ell_1\ell_2 c2 \\ +\frac{1}{4}\ell_2^2) + I_{23} \end{matrix} & 0 & 0 & \frac{1}{4}m_2\ell_2^2 + I_{23} \end{bmatrix} \tag{5.100}$$

From equation (5.49), we derive the remaining dynamic terms

$$[b] = \sum_k m_k [V_w^k][\dot{V}_w^k]^T + \sum_k [\omega^k][I_{ko}][\dot{\omega}^k]^T$$

$$= m_1[V_w^1][\dot{V}_w^1]^T + m_2[V_w^2][\dot{V}_w^2]^T + [\omega^1][I_{10}][\dot{\omega}^1]^T + [\omega^2][I_{20}][\dot{\omega}^2]^T$$

$$= m_1 \begin{bmatrix} \hat{I} & \hat{I} \\ 0 & \hat{I} \end{bmatrix} \left[\frac{[S_{r1}][S^{10}]}{0}\right] [[\dot{S}^{01}][S_{r1}^T] + [S^{01}][\dot{S}_{r1}^T] \mid 0] \begin{bmatrix} \hat{I} & 0 \\ \hat{I} & \hat{I} \end{bmatrix} \tag{5.101}$$

$$+ m_2 \begin{bmatrix} \hat{I} & \hat{I} \\ 0 & \hat{I} \end{bmatrix} \left[\frac{[S_{r1}][S^{10}]}{[S_{r2}][S^{20}]} \right] [[\dot{S}^{01}][S_{q2}^T]] + [S^{01}][\dot{S}_{q2}^T] | [\dot{S}^{02}][S_{r2}^T]$$

$$+ [S^{02}][\dot{S}_{r2}^T] \begin{bmatrix} \hat{I} & 0 \\ \hat{I} & \hat{I} \end{bmatrix} + \left[\frac{I}{0} \right] [S^{01}][I_1][S^{10}][0|0]$$

$$+ \left[\frac{I}{[S^{10}]} \right] [S^{02}][I_2][S^{20}][I|[\dot{S}^{01}]]$$

Note that

$$[\dot{S}_{r1}] = [\dot{S}_{q2}] = 0 \tag{5.102}$$

$$[S^{10}][\dot{S}^{01}] = [\Omega^{01}] \tag{5.103}$$

and

$$[S^{20}][\dot{S}^{02}] = [\Omega^{02}] \tag{5.104}$$

Therefore,

$$[b] = m_1 \begin{bmatrix} \hat{I} S_{r1} \Omega^{01} S_{r1}^T \hat{I} & 0 \\ 0 & 0 \end{bmatrix} + m_2 \begin{bmatrix} \hat{I} & \hat{I} \\ 0 & \hat{I} \end{bmatrix}$$

$$\times \begin{bmatrix} [S_{q2}][\Omega^{01}][S_{q2}^T] & [S_{q2}][S^{12}][\Omega^{02}][S_{r2}^T] \\ [S_{r2}][S^{21}][\Omega^{01}][S_{q2}^T] & [S_{r2}][\Omega^{02}][S_{r2}^T] \end{bmatrix} \begin{bmatrix} \hat{I} & 0 \\ \hat{I} & \hat{I} \end{bmatrix} + 0 + 0$$

$$= m_2 \begin{bmatrix} \frac{1}{2}\ell_1\ell_2\dot{x}_3 s2\hat{I} & -\frac{1}{2}\ell_1\ell_2(\dot{x}_3 + \dot{x}_6)s2\hat{I} \\ \frac{1}{2}\ell_1\ell_2\dot{x}_3 s2\hat{I} & 0 \end{bmatrix} \begin{bmatrix} \hat{I} & 0 \\ \hat{I} & \hat{I} \end{bmatrix} \tag{5.105}$$

$$= m_2 \begin{bmatrix} -\frac{1}{2}\ell_1\ell_2\dot{x}_6 s2\hat{I} & -\frac{1}{2}\ell_1\ell_2(\dot{x}_3 + \dot{x}_6)s2\hat{I} \\ \frac{1}{2}\ell_1\ell_2\dot{x}_3 s2\hat{I} & 0 \end{bmatrix}$$

Expanding the above we obtain,

$$= \begin{bmatrix} 0 & 0 & 0 & 0 & 0 & 0 \\ 0 & 0 & 0 & 0 & 0 & 0 \\ 0 & 0 & -\frac{1}{2}m_2\ell_1\ell_2\dot{x}_6 s2 & 0 & 0 & -\frac{1}{2}m_2\ell_1\ell_2(\dot{x}_3 + \dot{x}_6)s2 \\ 0 & 0 & 0 & 0 & 0 & 0 \\ 0 & 0 & 0 & 0 & 0 & 0 \\ 0 & 0 & \frac{1}{2}m_2\ell_1\ell_2\dot{x}_3 s2 & 0 & 0 & 0 \end{bmatrix}$$

From equation (5.50) we get matrix $[c]$:

$$[c] = \sum_k [\omega^k][\Omega_x^{ok}][I_{ko}][\omega^k]^T$$

$$= [\omega^1][\Omega_x^{01}][I_{10}][\omega^1]^T + [\omega^2][\Omega_x^{02}][I_{20}][\omega^2]^T \tag{5.106}$$

$$= \left[\frac{I}{0}\right][\Omega_x^{01}][S^{01}][I_1][S^{10}][I|0] + \left[\frac{I}{S^{10}}\right][\Omega_x^{02}][S^{02}][I_2][S^{20}][I|S^{01}] = 0$$

From equation (5.51), we have the generalized active force

$$\{f\} = \sum_k [V_w^k]\{f_k\} + \sum_k [\omega^k]\{M_k\} \tag{5.107}$$

Since there is no external moment applied to any link, the generalized active force vector reduces to

$$\{f\} = [V_w^1]\{f_1\} + [V_w^2]\{f_2\} \tag{5.108}$$

where $\{f_1\}$ and $\{f_2\}$ are the external force of links 1 and 2 due to the gravity force alone.

$$\{f_1\} = [0 \quad -m_1 g \quad 0]^T \tag{5.109}$$

$$\{f_2\} = [0 \quad -m_2 g \quad 0]^T \tag{5.110}$$

Substituting the two equations above into equation (5.108), we obtain

$$\{f\} = \begin{bmatrix} \hat{I} & \hat{I} \\ 0 & \hat{I} \end{bmatrix} \begin{bmatrix} 0 & 0 & 0 \\ 0 & 0 & -\frac{1}{2}\ell_1 \\ -\frac{1}{2}\ell_1 s1 & -\frac{1}{2}\ell_1 c1 & 0 \\ 0 & 0 & 0 \\ 0 & 0 & 0 \\ 0 & 0 & 0 \end{bmatrix} \begin{bmatrix} 0 \\ -m_1 g \\ 0 \end{bmatrix}$$

$$+ \begin{bmatrix} \hat{I} & \hat{I} \\ 0 & \hat{I} \end{bmatrix} \begin{bmatrix} 0 & 0 & 0 \\ 0 & 0 & -\ell_1 \\ -\ell_1 s1 & \ell_1 c1 & 0 \\ 0 & 0 & 0 \\ 0 & 0 & -\frac{1}{2}\ell_2 \\ -\frac{1}{2}\ell_2 s(1+2) & \frac{1}{2}\ell_2 c(1+2) & 0 \end{bmatrix} \begin{bmatrix} 0 \\ -m_2 g \\ 0 \end{bmatrix}$$

$$= \begin{bmatrix} 0 \\ 0 \\ -(\frac{1}{2}m_1 + m_2)g\ell_1 c1 - \frac{1}{2}m_2 g\ell_2 c(1+2) \\ 0 \\ 0 \\ -\frac{1}{2}m_2 g\ell_2 c(1+2) \end{bmatrix} \tag{5.111}$$

Canceling the zero rows and columns in matrix $[a]$ corresponding to the zero generalized coordinates, we obtain

$$
\begin{aligned}
a_{33} &= \tfrac{1}{4}m_1\ell_1^2 + m_2(\ell_1^2 + \ell_1\ell_2 c2 + \tfrac{1}{4}\ell_2^2) + I_{13} + I_{23} \\
a_{36} &= m_2(\tfrac{1}{2}\ell_1\ell_2 c2 + \tfrac{1}{4}\ell_2^2) + I_{23} \\
a_{63} &= a_{36} = m_2(\tfrac{1}{2}\ell_1\ell_2 c2 + \tfrac{1}{4}\ell_2^2) + I_{23} \\
a_{66} &= \tfrac{1}{4}m_2\ell_2^2 + I_{23}
\end{aligned}
\tag{5.112}
$$

Similarly, we can obtain the coefficients of the remaining $[b]$ and $\{f_\ell\}$ matrices as

$$
\begin{aligned}
b_{33} &= -\tfrac{1}{2}m_2\ell_1\ell_2\dot{X}_6 s2 \\
b_{36} &= -\tfrac{1}{2}m_2\ell_1\ell_2(\dot{X}_3 + \dot{X}_6)s2 \\
b_{63} &= \tfrac{1}{2}m_2\ell_1\ell_2\dot{X}_3 s2 \\
b_{66} &= 0 \\
f_3 &= -(\tfrac{1}{2}m_1 + m_2)g\ell_1 c1 - \tfrac{1}{2}m_2 g\ell_2 c(1+2) \\
f_6 &= -\tfrac{1}{2}m_2 g\ell_2 c(1+2)
\end{aligned}
\tag{5.113}
$$

Since the coefficients of $[c]$ are zero, the final equations of motion after substituting the X's with the θ's take the following form:

$$
\begin{bmatrix} a_{33} & a_{36} \\ a_{63} & a_{66} \end{bmatrix}
\begin{bmatrix} \ddot{\theta}_1 \\ \ddot{\theta}_2 \end{bmatrix}
+
\begin{bmatrix} b_{33} & b_{36} \\ b_{63} & b_{66} \end{bmatrix}
\begin{bmatrix} \dot{\theta}_1 \\ \dot{\theta}_2 \end{bmatrix}
=
\begin{bmatrix} f_3 \\ f_6 \end{bmatrix}
\tag{5.114}
$$

The intention of this exercise is to assist the reader interested in the development of general-purpose computer programs to see all the intermediate steps needed in the evaluation of the proper coefficients in the development of the governing equations of motion. All the arrays are expressed in a matrix form suited for computer coding.

The advantage of the foregoing method of developing the equations of motion relies on the fact that all the coefficient matrices are written for the general case, when the bodies can have all possible degrees of freedom. The resulting matrices given by equations (5.100), (5.105) and (5.111) show the zero coefficients associated with the omitted (suppressed) degrees of freedom automatically. If we choose to derive the equations of motion for the same two-link robot problem by adding more degrees of freedom to the links, one would update the transformation matrices and the generalized coordinates, then proceed with the method to obtain the final equations.

Example 5.4.2

A painting robot with three articulated links is shown in the Figure 5.5. To test the design of its workspace conditions, we need to derive its equations of motion. A

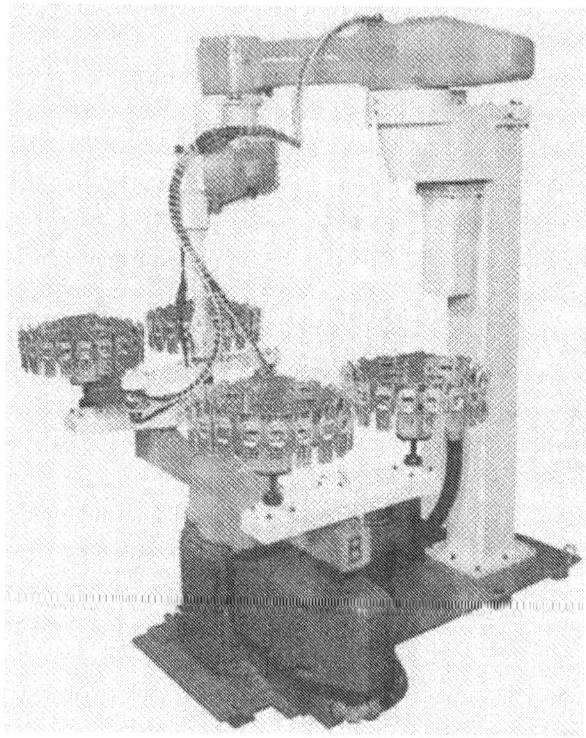

Fig. 5.5. Mitsubishi MRP-S70 painting robot

constant torque of 300 Kg-mm, 200 Kg-mm and 100 Kg-mm is applied at joints A, B and C, respectively. In addition to torque, gravity forces are also considered.

(a) For the initial conditions given in the above table, derive the equations of motion coefficients a, b, c and f matrices and their corresponding kinematical expressions V, $\dot{V}$, ω and $\dot{\omega}$. Use MATLAB or any other programming tool making use of the appendix in Chapter 3 and the flow chart at the end of Chapter 5 to compute the necessary coefficients.

(b) Simulate the motion of the robot for $t = 2$ seconds. Plot the time history of the joint angles.

Solution

The above painting robot is modeled as a simple 3-body system. Each body is considered rigid and assumed to be connected with the other by a revolute joint (one degree of freedom). So the system has three degrees of freedom. 3D model of each body was developed in Pro Engineer. Mass, moment of inertia and position of center of gravity of each body were obtained from the model.

Based on the derivations of equation of motion as described in this section of the matrix representation of the equations of motion, a MATLAB program was written for simulation purposes. The dynamical equations of motion for the given initial

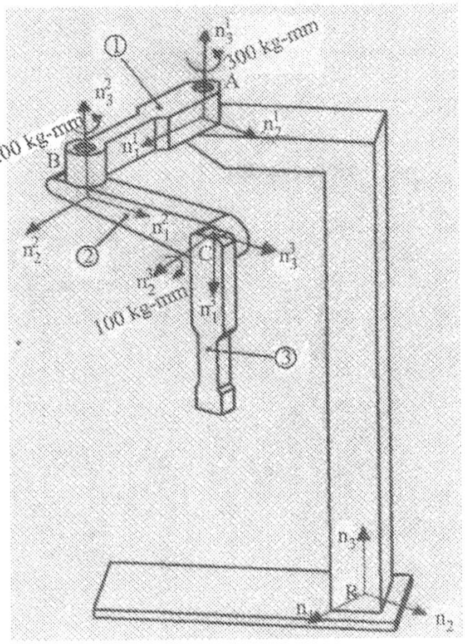

Fig. 5.6. Multibody dynamics model

conditions are obtained. Essentially the equations are derived and then the initial conditions at time equal zero are substituted yielding the coefficients of

$$a\ddot{x} + b\dot{x} + c\dot{x} = f$$

where each matrix is found to be:

$$[\mathbf{a}] = \begin{bmatrix} 1.65e+9 & 0.83e+9 & -0.01e+9 \\ 0.83e+9 & 0.62e+9 & 0.0 \\ -0.01e+9 & 0.0 & 0.15e+9 \end{bmatrix}$$

$$[\mathbf{b}] = \begin{bmatrix} -5.42e+8 & -9.03e+8 & 1.8e+8 \\ 3.61e+8 & 0.0 & 2.59e+8 \\ 1.08e+8 & 0.89e+8 & 0.0 \end{bmatrix}$$

$$[\mathbf{c}] = \begin{bmatrix} -2.33e+8 & -2.33e+8 & 1.42e+8 \\ -2.33e+8 & -2.33e+8 & 1.42e+8 \\ 1.91e+8 & 1.91e+8 & 0.0 \end{bmatrix}$$

$$\{f\} = \begin{Bmatrix} -5.11e+6 \\ -1.06e+6 \\ 0.12e+6 \end{Bmatrix}$$

		BODY 1			BODY 2			BODY 3		
1	Mass (×10 kg)	22.9			24.7			14.2		
	Direction	$n1$	$n2$	$n3$	$n1$	$n2$	$n3$	$n1$	$n2$	$n3$
2	Moment of inertia (×10^4N-mm^3)	1364	22768	22931	24912	1544	24390	14470	14698	524
3	Length (mm)	935	0	0	960	0	0	980	0	0
4	Distance between center of mass and origin of body coordinate. system (mm)	425	0	0	443	6	0	370	−3	0
5	External Torque (×10 N-mm)	0	300	0	0	200	0	0	100	0
6	External Force (×10 N)	0	−229	0	0	−247	0	0	−142	0
7	Initial angle (rad)	0	0	0.5	0	0	0.5	0	0.5	0
8	Initial angular velocity (rad/sec)	0	0	2.0	0	0	3.0	0	4.0	0

Table 5.2. Geometrical data and Initial conditions for MRP-S70. Material: steel (density - 7400 kg/cub. mm)

Observe how the generalized mass matrix $[a]$ is symmetric and $[b]$ and $[c]$ are nonsymmetric.

The kinematical expression of each link for the initial conditions given are provided for checking the procedures and the programming performed.

BODY 1

$$[\omega^1] = \begin{bmatrix} 1 & 0 & 0 \\ 0 & 1 & 0 \\ 0 & 0 & 1 \\ 0 & 0 & 0 \\ 0 & 0 & 0 \\ 0 & 0 & 0 \\ 0 & 0 & 0 \\ 0 & 0 & 0 \\ 0 & 0 & 0 \end{bmatrix}$$

$$[\dot{\omega}^1] = \begin{bmatrix} 0 & 0 & 0 \\ 0 & 0 & 0 \\ 0 & 0 & 0 \\ 0 & 0 & 0 \\ 0 & 0 & 0 \\ 0 & 0 & 0 \\ 0 & 0 & 0 \\ 0 & 0 & 0 \\ 0 & 0 & 0 \end{bmatrix}$$

$$[V^1] = \begin{bmatrix} 0 & 0 & 0 \\ 0 & 0 & -425 \\ -203 & 373 & 0 \\ 0 & 0 & 0 \\ 0 & 0 & 0 \\ 0 & 0 & 0 \\ 0 & 0 & 0 \\ 0 & 0 & 0 \\ 0 & 0 & 0 \end{bmatrix}$$

$$[\dot{V}^1] = \begin{bmatrix} 0 & 0 & 0 \\ 0 & 0 & 0 \\ -746 & 407 & 0 \\ 0 & 0 & 0 \\ 0 & 0 & 0 \\ 0 & 0 & 0 \\ 0 & 0 & 0 \\ 0 & 0 & 0 \\ 0 & 0 & 0 \end{bmatrix}$$

BODY 2

$$[\omega^2] = \begin{bmatrix} 1 & 0 & 0 \\ 0 & 1 & 0 \\ 0 & 0 & 1 \\ 0.87 & 0.48 & 0 \\ -0.48 & 0.87 & 0 \\ 0 & 0 & 1 \\ 0 & 0 & 0 \\ 0 & 0 & 0 \\ 0 & 0 & 0 \end{bmatrix}$$

$$[\dot{\omega}^2] = \begin{bmatrix} 0 & 0 & 0 \\ 0 & 0 & 0 \\ 0 & 0 & 0 \\ 0.96 & 1.75 & 0 \\ -1.75 & 0.96 & 0 \\ 0 & 0 & 0 \\ 0 & 0 & 0 \\ 0 & 0 & 0 \\ 0 & 0 & 0 \end{bmatrix}$$

$$[V^2] = \begin{bmatrix} 0 & 0 & 0 \\ 0 & 0 & -935 \\ -448 & 820 & 0 \\ 0 & 0 & 6 \\ 0 & 0 & -443 \\ -415 & 155 & 0 \\ 0 & 0 & 0 \\ 0 & 0 & 0 \\ 0 & 0 & 0 \end{bmatrix}$$

$$[V^3] = \begin{bmatrix} 0 & 0 & 0 \\ 0 & 0 & -935 \\ -448 & 820 & 0 \\ 0 & 0 & 0 \\ 0 & 0 & -960 \\ -895 & 348 & 0 \\ -0.78 & -2 & -2 \\ -96 & -247 & -257 \\ -344 & 136 & -2.15 \end{bmatrix}$$

$$[\dot{V}^2] = \begin{bmatrix} 0 & 0 & 0 \\ 0 & 0 & 0 \\ -1641 & -896 & 0 \\ 0 & 0 & 0 \\ 00 & 0 & \\ -775 & -2075 & 0 \\ 0 & 0 & 0 \\ 0 & 0 & 0 \\ 0 & 0 & 0 \end{bmatrix}$$

BODY 3

$$[\omega^3] = \begin{bmatrix} 1 & 0 & 0 \\ 0 & 1 & 0 \\ 0 & 0 & 1 \\ 0.87 & 0.48 & 0 \\ -0.48 & 0.87 & 0 \\ 0 & 0 & 1 \\ 0.36 & 0.93 & 0 \\ -0.93 & 0.36 & 0 \\ 0 & 0 & 1 \end{bmatrix}$$

$$[\dot{\omega}^3] = \begin{bmatrix} 0 & 0 & 0 \\ 0 & 0 & 0 \\ 0 & 0 & 0 \\ -0.96 & 1.75 & 0 \\ -1.75 & -0.96 & 0 \\ 0 & 0 & 0 \\ -5.59 & 2.17 & 0 \\ -2.17 & -5.59 & 0 \\ 0 & 0 & 0 \end{bmatrix}$$

$$[\dot{V}^3] = \begin{bmatrix} 0 & 0 & 0 \\ 0 & 0 & 0 \\ -1641 & -896 & 0 \\ 0 & 0 & 0 \\ 0 & 0 & 0 \\ -1740 & -4473 & 0 \\ 7 & -11.7 & 8.6 \\ 863 & -1442 & 1062 \\ -683 & -1728 & -8.4 \end{bmatrix}$$

(b) In addition to the derivation of the equations of motion, we can integrate to obtain the time history of each joint. What follows are the results of our MATLAB program. For the time history of the joint angles and the angular velocity of each link, see Figures 5.7 and 5.8.

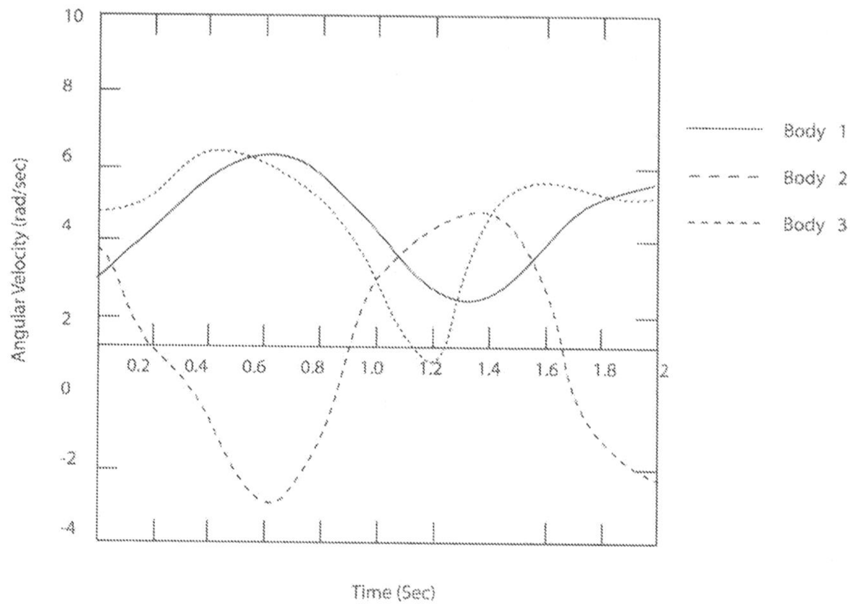

Fig. 5.7. Angular velocity (rad/sec) vs. time (sec)

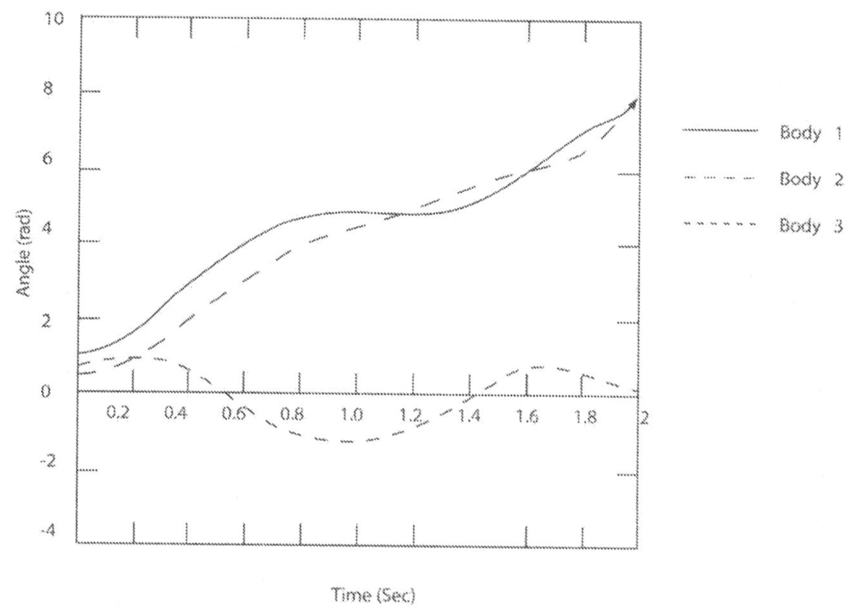

Fig. 5.8. Angle (rad) vs. time (sec)

5.5 Problems

5.1. Derive the equations of motion for the spacecraft given in Figure 5.9 using Kane's equations through the matrix representation given by equation (5.47). The spacecraft has a constant rotational rate Ω about its z-axis. The x, y, and z axes are attached to the body of the craft, and the solar panels rotate about the y axis at an angular velocity $\dot{\theta}(t) = a_1 \sin \omega t$ with respect to the aircraft.

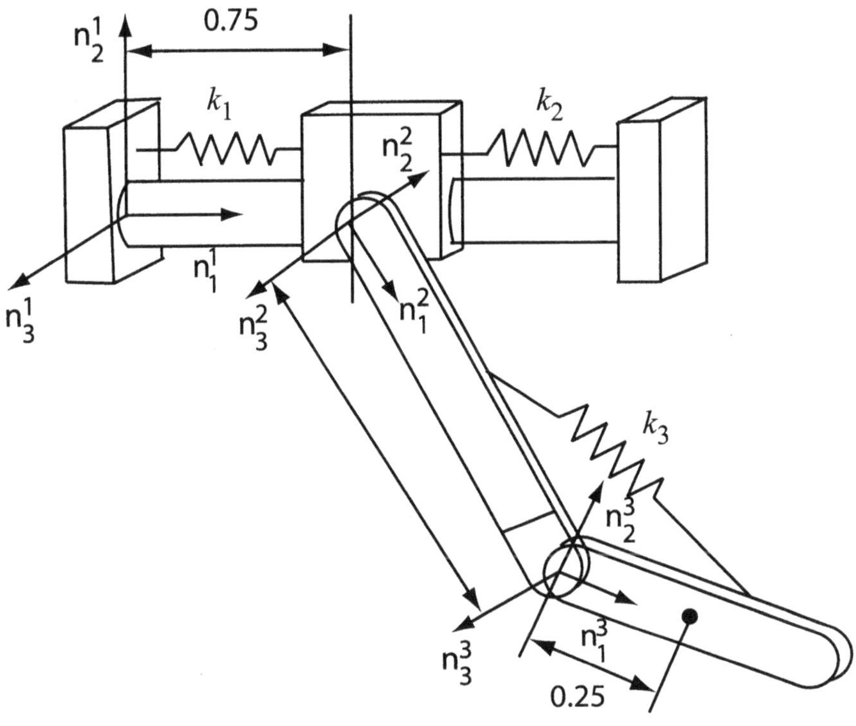

Fig. 5.9. A two-link mechanism with an oscillatory base

5.2. Consider the mechanism given in Example 2.13.2, which consists of a shaft rotating at an angular velocity $\omega = \omega(t)$ and a rod attached to the shaft base with a rotation along the $\bar{n}_2^2$ unit vector as shown. Develop the explicit governing equations of motion using equation (5.47). (*Note:* The kinematics of the problem have been worked out in Chapter 2).

5.3. A mechanism with a sliding mass is constrained by two springs k_1 and k_2 where k_2 maximum displacement is 0.2 in. A two-arm flexion extension is mounted on the mass as shown. If the mass initial position is at 0.2 in. along n_1^1, find the dynamical response as a function of time after it is released.

5.4. For the pendulum given in Problem 4.2, find the contribution of the generalized inertia forces. Give the matrix form of the equations of motion as stated in equation (5.47).

5.5. The robotic system given in Problem 4.3 is to be studied for possible use in General Motors' welding operations at the assembly. To simulate its motion, we need to derive the governing equations of motion. Find the first-order differential equations that govern its motion.

5.6. For the drive shaft for a belt machine shown in Figure 5.10 and Figure 5.11 write the equations of motion using multibody dynamics formulation as described in Chapter 5. Let $J_1 = J_2 = 2J_3 = 10\,\text{kgm}^2/\text{rad}$, $k_1 = k_2 = k_3 = 10^3\,\text{Nm / rad}$ and $c = 2\,\text{Nms / rad}$.

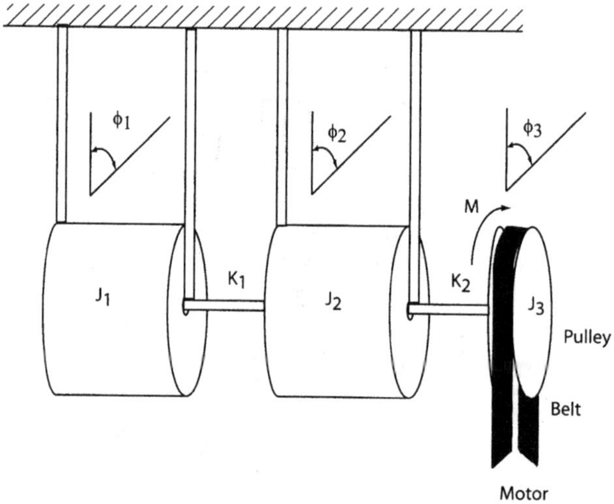

Fig. 5.10. Schematic of the belt-driven shaft

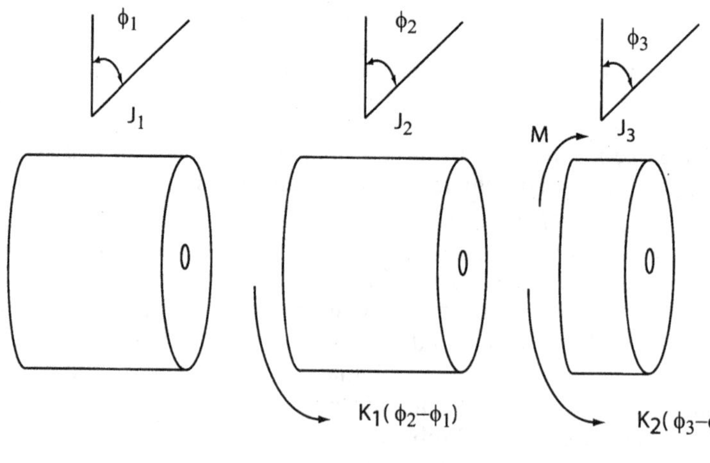

Fig. 5.11. Free-body diagram of the three inertias in the rotating system of the belt-driven shaft

5.7. In Problem 4.6 suppose that the tool base mass (m_1) is subjected to an impact force of $1000\delta(t)$, where δ(t) is the Kronecher delta function. Find the dynamic response of the system analytically. Let $m_1 = 400$ kg, $m_2 = 2000$ kg, $m_3 = 8000$ kg, $k_1 = 400,000$ N/m, $k_3 = 2k_1$, $k_2 = 800,000$ N / m. *Note*: First derive the equations of motion, and then solve the dynamical system.

5.8. For Example 3.8.1, derive the equations of motion coefficients a, b, c using MATLAB. Use the following geometrical data for the robot given in Table 5.3 and Table 5.4.

	Link 1	Link 2	Link 3
Mass	1	2	0.5
Center mass position in local axis	(0, 0, 0.5)	(0.2, 0, 0.53)	(0.2, −0.1, 0)
Body vector (q)	(0, 0, 0)	(0, 0, 1.15)	(0.5, 0, 0.3)
Moment of inertia	$I_1 = \begin{bmatrix} 2.4 & 0 & 0 \\ 0 & 0.08 & 0 \\ 0 & 0 & 2.4 \end{bmatrix} \times 10^{10}$	$I_2 = \begin{bmatrix} 2.34 & 0 & -2.2 \\ 0 & 8.2 & 0 \\ -2.2 & 0 & 6.28 \end{bmatrix} \times 10^{10}$	$I_2 = \begin{bmatrix} 0.76 & 0 & 0.76 \\ 0 & 3 & 0 \\ 0.76 & 0 & 2.8 \end{bmatrix} \times 10^{10}$
Initial configuration	0°	90°	45°

Table 5.3. Geometric data

	Link 1	Link 2	Link 3
Mass	2	1	0.5
Center mass position in local axis	(0, 0, 0)	(0.5, 0, 0)	(0.25, 0, 0)
Body vector (q)	(0, 0, 0)	(1, 0,0)	(0, 0,0)
Moment of inertia	$I_1 = \begin{bmatrix} 2 & 0 & 0 \\ 0 & 8 & 0 \\ 0 & 0 & 6 \end{bmatrix} \times 10^{10}$	$I_2 = \begin{bmatrix} 1 & 0 & 0 \\ 0 & 2 & 0 \\ 0 & 0 & 3 \end{bmatrix} \times 10^{10}$	$I_2 = \begin{bmatrix} 0.5 & 0 & 0 \\ 0 & 1 & 0 \\ 0 & 0 & 1.5 \end{bmatrix} \times 10^{10}$
Forces	Gravity	Gravity	Gravity
Initial configuration	0°	$\theta_1 = -60°$	$\theta_2 = 30°$

Table 5.4. Geometric properties

5.9. For Example 3.8.2, given by Figure 5.12, derive the equations of motion coefficients a, b, c using MATLAB. Use the geometrical data given in Table (5.5).

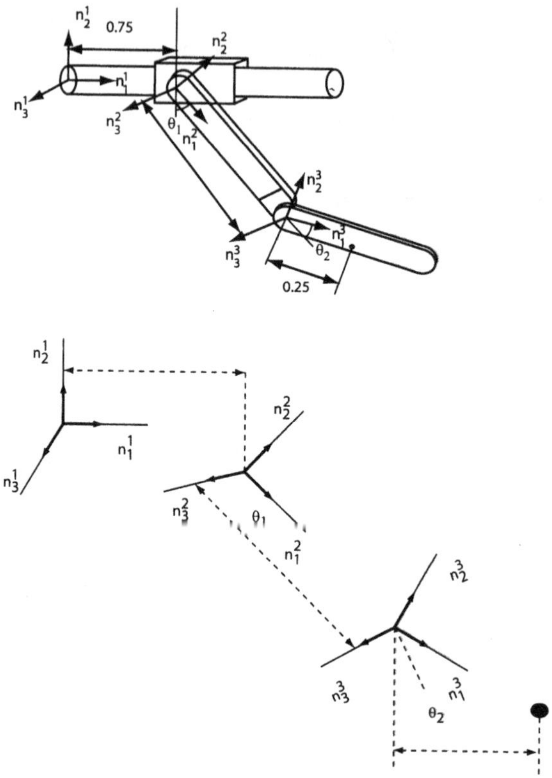

Fig. 5.12. Schematic of a two-body system

	Link 1	Link 2
Mass	1	2
Center mass position in local axis	(0, −0.5, 0.5)	(0, 0, 0)
Body vector (q)	(0, 0, 0)	(0, −1.5,0)
Moment of inertia	$I_1 = \begin{bmatrix} 2.4 & 1 & 0.5 \\ 0.5 & 8 & 4 \\ 2 & 3.5 & 2.4 \end{bmatrix} \times 10^{10}$	$I_2 = \begin{bmatrix} 2.34 & 0 & 0 \\ 0 & 1.2 & 0 \\ 0 & 0 & 6.28 \end{bmatrix} \times 10^{10}$
Initial configuration	$\Omega = 0^\circ$	$\theta = 30^\circ \quad \phi = 15^\circ$

Table 5.5. Geometric properties of the robot

Solution:

$$a = \begin{bmatrix} 4.5 & 1.5 \\ 1.5 & 1.5 \end{bmatrix} \times 10^{12} \quad b = \begin{bmatrix} 0 & 0 \\ 0 & 0 \end{bmatrix} \quad c = \begin{bmatrix} 0 & 0 \\ 0 & 0 \end{bmatrix}$$

5.10. The test chamber of a flight simulator described in Example 3.83 shown in Figure 5.13 is used to simulate the housing which rotates at a constant angular velocity ϕ. If the entire assembly rotates at 20 rpm, find the equations of motion time history of the simulator.

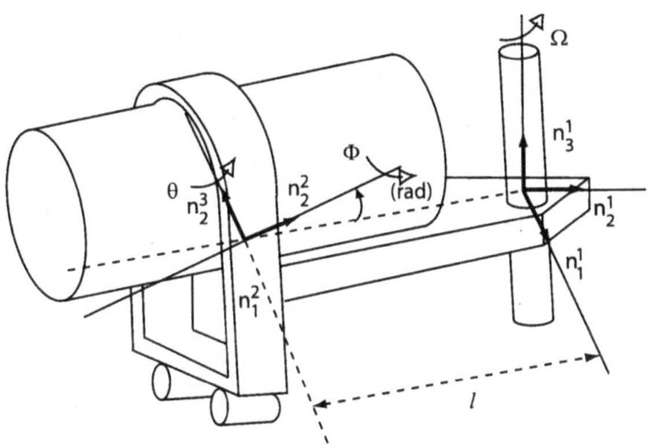

Fig. 5.13. Schematic of a two-link robot

5.11. The 5-kg disk and hub A have a radius of gyration of 85 mm about the z_0 axis and spin at the rate $p = 1250$ rev/min (Figure 5.14). Simultaneously, the assembly rotates about the vertical z-axis at the rate $\Omega = 400$ rev/min. Calculate the gyroscopic moment M exerted on the shaft at C by the disk and the bending moment Mo in the shaft at O. Neglect the mass of the shaft but otherwise account for all forces acting on it.

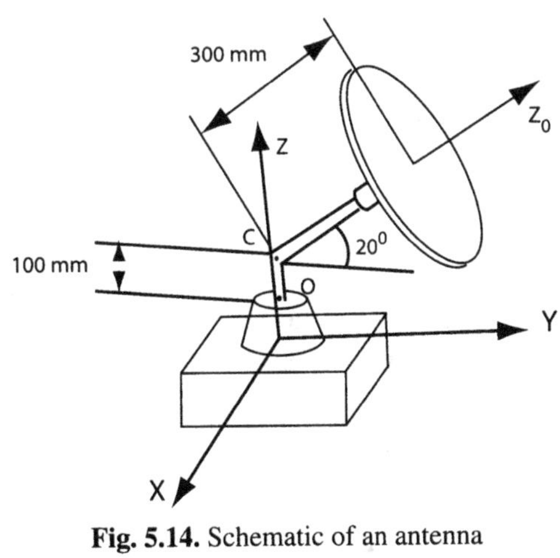

Fig. 5.14. Schematic of an antenna

5.12. The uniform slender bar of mass m and length l is centrally mounted on the shaft A-A as shown in Figure 5.15, about which it rotates with a constant speed $\dot{\phi}$ = constant. Simultaneously, the yoke is forced to rotate about the z-axis with a constant speed ω_0. As a function of ϕ, determine the magnitude of the torque M required to maintain the constant speed ω_0.

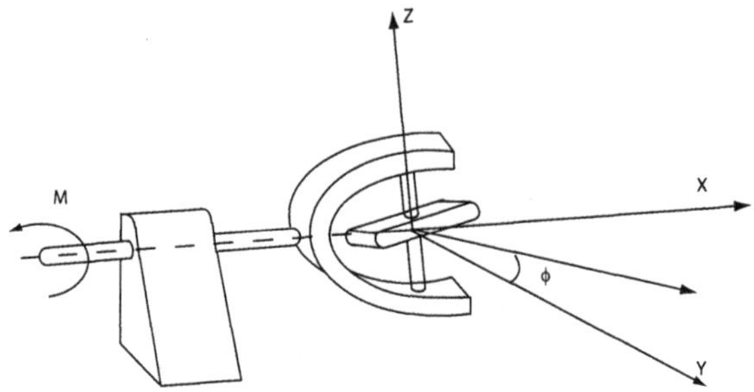

Fig. 5.15. Schematic of Yoke mechanism

5.13. Compute the generalized mass matrix a and the coefficient matrices b, c of the lift mechanism system shown in Figure 5.16 when the lifts are at 30° and body B_0 travels 6 meters at a constant speed of 1.2 m/s in the direction of n_1. Use MATLAB to solve this problem. (See 5.6 Project.)

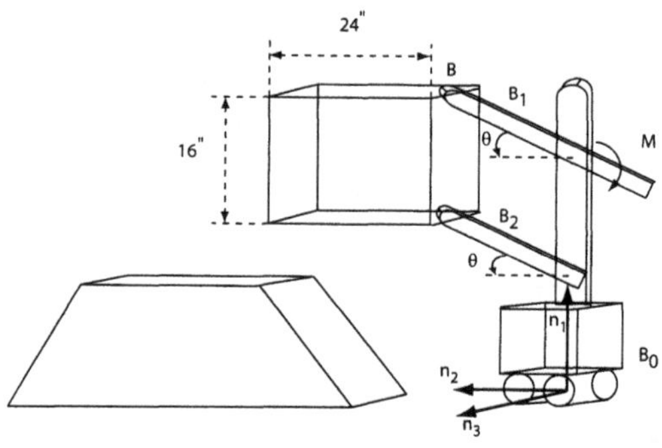

Fig. 5.16. Schematic of a lift

5.6 Project

Purpose

The purpose of this project is to show how dynamics can be used to solve a real-life problem. Our example shows the step-by-step process used to create the equations of motion for the 4-body system of a common construction machine.

Description of the excavator and why the dynamics equations are important

A hydraulic excavator is used to remove large amounts of soil and debris in a fast and efficient manner. A hydraulic excavator is a heavy and powerful machine that can be dangerous without the proper control. Understanding the equations of motion is essential in order to simulate the movements of this machine. With the equations developed in this book, a computer program can be created to simulate the motion of the hydraulic excavator. A manufacturer of such a device can save a tremendous amount of development time and money by using a computer simulation.

Certain assumptions have been made in this example. The equations have been simplified to aid in the reader's understanding of the process. Design engineers have hundreds of important considerations to take into account when designing a machine such as this. The handwritten calculations needed to describe our excavator could fill a book by itself. Hence we have simplified our excavator into a 4-body system.

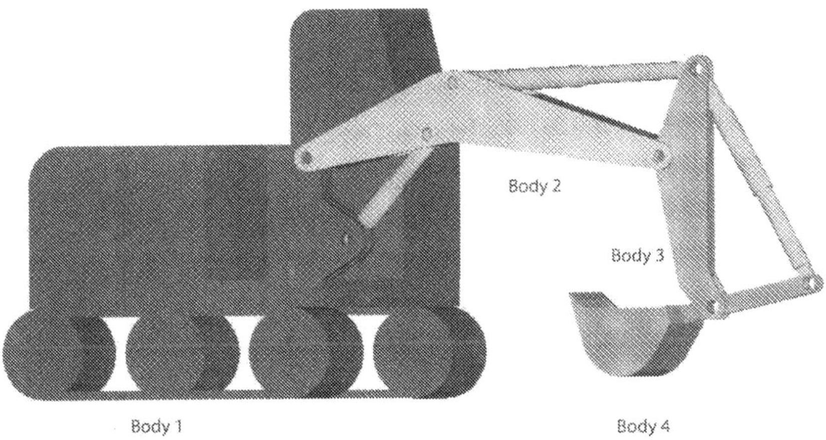

Fig. 5.17. An overview of the excavator

The following is the data collected from the actual excavator and used for analysis. We begin with the tree array as it is always a way of reminding oneself that the method and procedures developed are for open treelike structure.

Number of bodies (links) $N = 4$

Degrees of freedom $n = 4$

Table 5.6 gives an explicit form of $[\omega^k]$ for all the structure, where empty spaces denote zero matrices.

The Transformation Matrices

When solving equations in a multibody dynamics problem, it is crucial that all calculations are performed with respect to the origin. Therefore our first step is to generate the transformation matrices that shift the body coordinates to the global coordinate system (origin).

$\Gamma_0(k)$	1	2	3	4
$\Gamma_1(k)$	0	1	2	3
$\Gamma_2(k)$	0	0	1	2
$\Gamma_3(k)$			0	1
$\Gamma_4(k)$				0

Table 5.6. $\Gamma(k)$ tree array initial angle data

	Rotation (rad)		
	n_1	n_2	n_3
Body 1	0	0	0.3
Body 2	0.35	0	0
Body 3	4.8	0	0
Body 4	4.6	0	0

Table 5.7. Initial angle data for excavator configuration shown

	Velocity (rad/s)		
	n_1	n_2	n_3
Body 1	0	0	0
Body 2	0.1	0	0
Body 3	0.2	0	0
Body 4	0.1	0	0

Table 5.8. Initial angular velocity for excavator

	Mass (kg)
Body 1	5000
Body 2	680
Body 3	450
Body 4	180

Table 5.9. Mass data for excavator

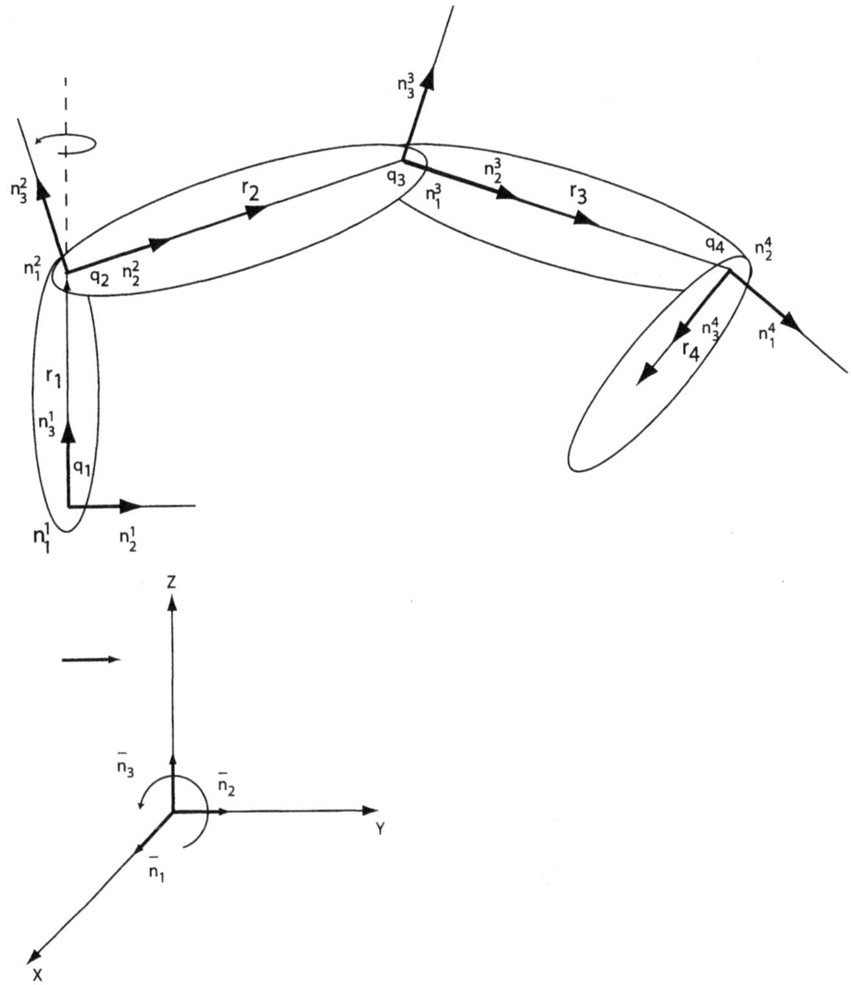

Fig. 5.18. A multibody model of the excavator

The matrix $[S^{10}]$ translates body 1 into the general coordinate system

$$\left[S^{10}\right] = \begin{bmatrix} c\alpha & s\alpha & 0 \\ -s\alpha & c\alpha & 0 \\ 0 & 0 & 1 \end{bmatrix}$$

Since there is no direct shifter between body 2 and the global coordinate system, we must first shift between body 2 and body 1. Matrix $[S^{21}]$ shifts the coordinates of body 2 into body 1. Shifter $[S^{20}]$ uses $[S^{21}]$ and $[S^{10}]$ to translate body 2 into the general coordinate system.

	Magnitude (m)		
	n_1	n_2	n_3
q_2	0	0	0
q_3	0	2.5	−0.9
q_4	0	1.25	0.06
r_1	0	0	0.5
r_2	0	1.5	0
r_3	0	0.65	0
r_4	0	0.6	0
$r_{d1\ 1}$	0	0.5	0
$r_{d2\ 1}$	0	0.06	−0.25
$r_{d2\ 2}$	0	1.4	−0.15
$r_{d3\ 1}$	0	1.6	0.13
$r_{d3\ 2}$	0	−0.2	0.18
$r_{d4\ 1}$	0	−0.2	0.18
$r_{d4\ 2}$	0	0.1	0

Table 5.10. Body vectors and mass center coordinates in local coordinates

$\ell\ k$	B_1 123	B_2 456	B_3 789	B_4 10,11,12
$[\omega^1_{\ell m}]$	I			
$[\omega^2_{\ell m}]$	I	S^{10}		
$[\omega^3_{\ell m}]$	I	S^{10}	S^{20}	
$[\omega^4_{\ell m}]$	I	S^{10}	S^{20}	S^{30}

Table 5.11. Partial angular matrix array

$$\left[S^{20}\right] = \left[S^{21}\right]\left[S^{10}\right] = \begin{bmatrix} 1 & 0 & 0 \\ 0 & c\beta & s\beta \\ 0 & -s\beta & c\beta \end{bmatrix} \begin{bmatrix} c\alpha & s\alpha & 0 \\ -s\alpha & c\alpha & 0 \\ 0 & 0 & 1 \end{bmatrix}$$

$$= \begin{bmatrix} c\alpha & s\alpha & 0 \\ -c\beta s\alpha & c\alpha c\beta & s\beta \\ -s\beta s\alpha & -s\beta c\alpha & c\beta \end{bmatrix}$$

Similarly $[S^{30}]$ converts body 3 into the global coordinate system.

$$\left[S^{30}\right]=\left[S^{32}\right]\left[S^{20}\right]=\begin{bmatrix}1 & 0 & 0\\ 0 & c\theta & s\theta\\ 0 & -s\theta & c\theta\end{bmatrix}\begin{bmatrix}c\alpha & s\alpha & 0\\ -c\beta s\alpha & c\alpha c\beta & s\beta\\ -s\beta s\alpha & -s\beta c\alpha & c\beta\end{bmatrix}$$

$$=\begin{bmatrix}c\alpha & s\alpha & 0\\ -s\alpha c\beta c\theta+s\alpha s\beta s\theta & c\alpha c\beta c\theta-c\alpha s\beta s\theta & s\beta c\theta+c\beta s\theta\\ s\alpha c\beta s\theta+s\alpha s\beta c\theta & -c\alpha c\beta s\theta-c\alpha s\beta c\theta & -s\beta s\theta+c\beta c\theta\end{bmatrix}$$

$$=\begin{bmatrix}c\alpha & s\alpha & 0\\ -s\alpha c(\beta+\theta) & c\alpha c(\beta+\theta) & s(\beta+\theta)\\ s\alpha s(\beta+\theta) & -c\alpha s(\beta+\theta) & c(\beta+\theta)\end{bmatrix}$$

Finally, $[S^{40}]$ converts body 4 into the global coordinate system.

$$\left[S^{40}\right]=\left[S^{43}\right]\left[S^{30}\right]$$

$$=\begin{bmatrix}1 & 0 & 0\\ 0 & c\phi & s\phi\\ 0 & -s\phi & c\phi\end{bmatrix}\begin{bmatrix}c\alpha & s\alpha & 0\\ -s\alpha c(\beta+\theta) & c\alpha c(\beta+\theta) & s(\beta+\theta)\\ s\alpha s(\beta+\theta) & -c\alpha s(\beta+\theta) & c(\beta+\theta)\end{bmatrix}$$

$$=\begin{bmatrix}c\alpha & s\alpha & 0\\ -s\alpha c\phi c(\beta+\theta) & c\alpha c\phi c(\beta+\theta) & c\phi s(\beta+\theta)\\ +s\alpha s\phi s(\beta+\theta) & -c\alpha s\phi s(\beta+\theta) & +s\phi c(\beta+\theta)\\ s\alpha s\phi c(\beta+\theta) & -c\alpha s\phi c(\beta+\theta) & -s\phi s(\beta+\theta)\\ +s\alpha c\phi s(\beta+\theta) & -c\alpha c\phi s(\beta+\theta) & +c\phi c(\beta+\theta)\end{bmatrix}$$

$$=\begin{bmatrix}c\alpha & s\alpha & 0\\ -s\alpha c(\beta+\theta+\phi) & c\alpha c(\beta+\theta+\phi) & s(\beta+\theta+\phi)\\ s\alpha s(\beta+\theta+\phi) & -c\alpha s(\beta+\theta+\phi) & c(\beta+\theta+\phi)\end{bmatrix}$$

The Time Derivatives of the Shifter Matrices

Later in our example, we will need the time derivatives of each transformation matrix to determine the skew matrix $[\Omega]$. The time derivatives of each transformation matrix are as follows:

For $[S^{10}]$ the time derivative is

$$\left[\dot{S}^{10}\right]=\dot{\alpha}\begin{bmatrix}-s\alpha & c\alpha & 0\\ -c\alpha & -s\alpha & 0\\ 0 & 0 & 1\end{bmatrix}$$

for $[S^{20}]$

$$\left[\dot{S}^{20}\right] = \begin{bmatrix} -\dot{\alpha}s\alpha & \dot{\alpha}c\alpha & 0 \\ -\dot{\alpha}c\alpha c\beta + \dot{\beta}s\alpha s\beta & -\dot{\alpha}s\alpha c\beta - \dot{\beta}c\alpha s\beta & \dot{\beta}c\beta \\ \dot{\alpha}c\alpha s\beta + \dot{\beta}s\alpha c\beta & \dot{\alpha}s\alpha s\beta + \dot{\beta}c\alpha c\beta & -\dot{\beta}s\beta \end{bmatrix}$$

and for $[S^{30}]$

$$\left[\dot{S}^{30}\right] = \begin{bmatrix} -\dot{\alpha}s\alpha & \dot{\alpha}c\alpha & 0 \\ -\dot{\alpha}c\alpha c(\beta+\theta) + (\dot{\beta}+\dot{\theta})s\alpha s(\dot{\beta}+\dot{\theta}) & -\dot{\alpha}s\alpha c(\beta+\theta) - (\dot{\beta}+\dot{\theta})c\alpha s(\beta+\theta) & (\dot{\beta}+\dot{\theta})c(\beta+\theta) \\ \dot{\alpha}c\alpha s(\beta+\theta) + (\dot{\beta}+\dot{\theta})s\alpha c(\dot{\beta}+\dot{\theta}) & \dot{\alpha}s\alpha s(\beta+\theta) - (\dot{\beta}+\dot{\theta})c\alpha c(\beta+\theta) & -(\dot{\beta}+\dot{\theta})s(\beta+\theta) \end{bmatrix}$$

and finally for $[S^{40}]$

$$[\dot{S}^{40}] = \begin{bmatrix} -\dot{\alpha}s\alpha & \dot{\alpha}c\alpha & 0 \\ -\dot{\alpha}c\alpha c(\beta+\theta+\phi) + (\dot{\beta}+\dot{\theta}+\dot{\phi}) \times s\alpha s(\beta+\theta+\phi) & -\dot{\alpha}s\alpha c(\beta+\theta+\phi) - (\dot{\beta}+\dot{\theta}+\dot{\phi}) \times c\alpha s(\beta+\theta+\phi) & (\dot{\beta}+\dot{\theta}+\dot{\phi})\, c(\beta+\theta+\phi) \\ \dot{\alpha}c\alpha s(\beta+\theta+\phi) + (\dot{\beta}+\dot{\theta}+\dot{\phi}) \times s\alpha c(\beta+\theta+\phi) & \dot{\alpha}s\alpha s(\beta+\theta+\phi) - (\dot{\beta}+\dot{\theta}+\dot{\phi}) \times c\alpha c(\beta+\theta+\phi) & -(\dot{\beta}+\dot{\theta}+\dot{\phi})\, s(\beta+\theta+\phi) \end{bmatrix}$$

The partial angular velocities are:

$$\left[\omega^1\right] = \begin{bmatrix} I_{3\times3} \\ 0_{3\times3} \end{bmatrix} \quad \left[\omega^2\right] = \begin{bmatrix} I_{3\times3} \\ S^{10} \\ 0_{3\times3} \end{bmatrix} \quad \left[\omega^3\right] = \begin{bmatrix} I_{3\times3} \\ S^{10} \\ S^{20} \\ 0_{3\times3} \end{bmatrix} \quad \left[\omega^4\right] = \begin{bmatrix} I_{3\times3} \\ S^{10} \\ S^{20} \\ S^{30} \\ 0_{3\times3} \end{bmatrix}$$

The Generalized Coordinates

The generalized coordinates denote the angle of rotation for each body and are given by α, β, θ and ϕ.

$$x = \begin{bmatrix} 0 & 0 & \alpha & \beta & 0 & 0 & \theta & 0 & 0 & \phi & 0 & 0 \end{bmatrix}$$

Angular Velocities

Following the definition given in Chapter 3, for an N multibody system we write the angular velocity for body 1 in this fashion:

$$^{R}\varpi^{1} = [\dot{x}]\left\{\omega^{1}\right\}[\bar{n}] = \begin{bmatrix} 0 & 0 & \dot{\alpha} \end{bmatrix} \begin{bmatrix} 1 & 0 & 0 \\ 0 & 1 & 0 \\ 0 & 0 & 1 \end{bmatrix} \begin{bmatrix} n^1 \\ n^2 \\ n^3 \end{bmatrix} = \alpha \bar{n_3}$$

Likewise for body 2

$$^{R}\varpi^{2} = \begin{bmatrix} \dot{\alpha} & \dot{\beta} \end{bmatrix} \begin{bmatrix} 0 & 0 & 1 \\ c\alpha & s\alpha & 0 \end{bmatrix} \begin{bmatrix} n^1 \\ n^2 \\ n^3 \end{bmatrix} = \begin{bmatrix} \dot{\beta} c\alpha & \dot{\beta} s\alpha & \dot{\alpha} \end{bmatrix} \begin{bmatrix} n^1 \\ n^2 \\ n^3 \end{bmatrix}$$

$$= \dot{\beta} c\alpha \bar{n_1} + \dot{\beta} s\alpha \bar{n_2} + \dot{\alpha} \bar{n_3}$$

for body 3

$$^{R}\varpi^{3} = \begin{bmatrix} \dot{\alpha} & \dot{\beta} & \dot{\theta} \end{bmatrix} \begin{bmatrix} 0 & 0 & 1 \\ c\alpha & s\alpha & 0 \\ c\alpha & s\alpha & 0 \end{bmatrix} \begin{bmatrix} n^1 \\ n^2 \\ n^3 \end{bmatrix}$$

$$= \begin{bmatrix} \dot{\beta} c\alpha + \dot{\theta} c\alpha & \dot{\beta} s\alpha + \dot{\theta} s\alpha & \dot{\alpha} \end{bmatrix} \begin{bmatrix} n^1 \\ n^2 \\ n^3 \end{bmatrix}$$

$$= c\alpha(\dot{\beta} + \dot{\theta})\bar{n_1} + s\alpha(\dot{\beta} + \dot{\theta})\bar{n_2} + \dot{\alpha}\bar{n_3}$$

and lastly for body 4

$$^{R}\varpi^{4} = \begin{bmatrix} \dot{\alpha} & \dot{\beta} & \dot{\theta} & \dot{\phi} \end{bmatrix} \begin{bmatrix} 0 & 0 & 1 \\ c\alpha & s\alpha & 0 \\ c\alpha & s\alpha & 0 \\ c\alpha & s\alpha & 0 \end{bmatrix} \begin{bmatrix} n^1 \\ n^2 \\ n^3 \end{bmatrix}$$

$$= \begin{bmatrix} c\alpha(\dot{\beta} + \dot{\theta} + \dot{\phi}) & s\alpha(\dot{\beta} + \dot{\theta} + \dot{\phi}) & \dot{\alpha} \end{bmatrix} \begin{bmatrix} n^1 \\ n^2 \\ n^3 \end{bmatrix}$$

$$= c\alpha(\dot{\beta} + \dot{\theta} + \dot{\phi})\bar{n_1} + s\alpha(\dot{\beta} + \dot{\theta} + \dot{\phi})\bar{n_2} + \dot{\alpha}\bar{n_3}$$

Time Derivatives of the Partial Angular Velocities

To compute the angular accelerations it is necessary to calculate the partial angular velocities.

For body 1 and 2, the time derivatives are

$$\left[\dot{\omega}^{1}\right] = \begin{bmatrix} 0_{3\times3} \\ 0_{3\times3} \end{bmatrix} \qquad \left[\dot{\omega}^{2}\right] = \begin{bmatrix} 0_{3\times3} \\ \dot{S}^{10} \\ 0_{3\times3} \end{bmatrix}$$

and for body 3 and 4

$$\left[\dot{\omega}^3\right] = \begin{bmatrix} 0_{3\times 3} \\ \dot{S}^{10} \\ \dot{S}^{20} \\ 0_{3\times 3} \end{bmatrix} \qquad \left[\dot{\omega}^4\right] = \begin{bmatrix} 0_{3\times 3} \\ \dot{S}^{10} \\ \dot{S}^{20} \\ \dot{S}^{30} \\ 0_{3\times 3} \end{bmatrix}$$

Angular Accelerations

A fundamental part of motion analysis is finding the angular acceleration of each body. The angular acceleration is also a vital element used to compute the mass center acceleration for each body.

The angular acceleration for body 1 is

$$\begin{aligned} {}^R\bar{\alpha}^1 &= \left(\{\ddot{x}\}^T\left[\omega^1\right] + \{\dot{x}\}^T\left[\dot{\omega}^1\right]\right)\{\bar{n}\} \\ &= \left([\ddot{\alpha}]\begin{bmatrix}0 & 0 & 1\end{bmatrix} + [\dot{\alpha}]\begin{bmatrix}0 & 0 & 0\end{bmatrix}\right)\{\bar{n}\} \\ &= \ddot{\alpha}\bar{n}_3 \end{aligned}$$

for body 2 the angular acceleration is

$$\begin{aligned} {}^R\bar{\alpha}^2 &= \left(\begin{bmatrix}\ddot{\alpha} & \ddot{\beta}\end{bmatrix}\begin{bmatrix} 0 & 0 & 1 \\ c\alpha & s\alpha & 0 \end{bmatrix} + \begin{bmatrix}\dot{\alpha} & \dot{\beta}\end{bmatrix}\begin{bmatrix} 0 & 0 & 1 \\ -\dot{\alpha}s\alpha & -\dot{\alpha}c\alpha & 0 \end{bmatrix}\right)\{\bar{n}\} \\ &= \left(\begin{bmatrix}\ddot{\beta}c\alpha & \ddot{\beta}s\alpha & \ddot{\alpha}\end{bmatrix} + \begin{bmatrix}-\dot{\beta}\dot{\alpha}s\alpha & \dot{\beta}\dot{\alpha}c\alpha & 0\end{bmatrix}\right)\{\bar{n}\} \\ &= \left(\ddot{\beta}c\alpha - \dot{\alpha}\dot{\beta}s\alpha\right)\bar{n}_1 + \left(\ddot{\beta}s\alpha + \dot{\alpha}\dot{\beta}c\alpha\right)\bar{n}_2 + \ddot{\alpha}\bar{n}_3 \end{aligned}$$

The angular acceleration for body 3 is

$$\begin{aligned} {}^R\bar{\alpha}^3 &= \left(\begin{aligned} &\begin{bmatrix}\ddot{\alpha} & \ddot{\beta} & \ddot{\theta}\end{bmatrix}\begin{bmatrix} 0 & 0 & 1 \\ c\alpha & s\alpha & 0 \\ c\alpha & s\alpha & 0 \end{bmatrix} \\ &+ \begin{bmatrix}\dot{\alpha} & \dot{\beta} & \dot{\theta}\end{bmatrix}\begin{bmatrix} 0 & 0 & 1 \\ -\dot{\alpha}s\alpha & \dot{\alpha}c\alpha & 0 \\ -\dot{\alpha}s\alpha & \dot{\alpha}c\alpha & 0 \end{bmatrix} \end{aligned}\right)\{\bar{n}\} \\ &= \left(\begin{aligned} &\begin{bmatrix}c\alpha(\ddot{\beta}+\ddot{\theta}) & s\alpha(\ddot{\beta}+\ddot{\theta}) & \ddot{\alpha}\end{bmatrix} \\ &+ \begin{bmatrix}-\dot{\alpha}s\alpha(\dot{\beta}+\dot{\theta}) & \dot{\alpha}c\alpha(\dot{\beta}+\dot{\theta}) & 0\end{bmatrix} \end{aligned}\right)\{\bar{n}\} \\ &= \left(c\alpha(\ddot{\beta}+\ddot{\theta}) - \dot{\alpha}s\alpha(\dot{\beta}+\dot{\theta})\right)\bar{n}_1 \\ &\quad + \left(s\alpha(\ddot{\beta}+\ddot{\theta}) + \dot{\alpha}c\alpha(\dot{\beta}+\dot{\theta})\right)\bar{n}_2 + \ddot{\alpha}\bar{n}_3 \end{aligned}$$

and for body 4

$$
{}^{R}\bar{\alpha}^{4} = \left(\begin{bmatrix} \ddot{\alpha} & \ddot{\beta} & \ddot{\theta} & \ddot{\phi} \end{bmatrix} \begin{bmatrix} 0 & 0 & 1 \\ c\alpha & s\alpha & 0 \\ c\alpha & s\alpha & 0 \\ c\alpha & s\alpha & 0 \end{bmatrix} + \begin{bmatrix} \dot{\alpha} & \dot{\beta} & \dot{\theta} \end{bmatrix} \begin{bmatrix} 0 & 0 & 0 \\ -\dot{\alpha}s\alpha & \dot{\alpha}c\alpha & 0 \\ -\dot{\alpha}s\alpha & \dot{\alpha}c\alpha & 0 \\ -\dot{\alpha}s\alpha & \dot{\alpha}c\alpha & 0 \end{bmatrix} \right) \{\bar{n}\}
$$

$$
= \left(\begin{array}{l} \begin{bmatrix} c\alpha(\ddot{\beta}+\ddot{\theta}+\ddot{\phi}) & s\alpha(\ddot{\beta}+\ddot{\theta}+\ddot{\phi}) & \ddot{\alpha} \end{bmatrix} \\ + \begin{bmatrix} -\dot{\alpha}s\alpha(\dot{\beta}+\dot{\theta}+\dot{\phi}) & \dot{\alpha}c\alpha(\dot{\beta}+\dot{\theta}+\dot{\phi}) & 0 \end{bmatrix} \end{array} \right) \{\bar{n}\}
$$

$$
= (c\alpha(\ddot{\beta}+\ddot{\theta}+\ddot{\phi}) - \dot{\alpha}s\alpha(\dot{\beta}+\dot{\theta}+\dot{\phi})\bar{n_1}
$$

$$
+ (s\alpha(\ddot{\beta}+\ddot{\theta}+\ddot{\phi}) + \dot{\alpha}c\alpha(\dot{\beta}+\dot{\theta}+\dot{\phi}))\bar{n_2} + \ddot{\alpha}\bar{n_3}
$$

Body Vectors
The vectors below describe the dimensions of the system. The column vector $\{q_n\}$ gives the dimensions between body coordinate systems, while $\{r_n\}$ gives the dimensions (in body coordinates) to the bodies center of gravity.

In our example, the body 1 coordinate system lies on the origin. Hence

$$
\{q_1\} = 0
$$

The vector between body 1 and body 2's coordinate system is

$$
\{q_2\} = \begin{bmatrix} 0 & 0 & l_1 \end{bmatrix}^T
$$

Between body 2 and body 3 we have

$$
\{q_3\} = \begin{bmatrix} 0 & l_2 & 0 \end{bmatrix}^T
$$

and finally

$$
\{q_4\} = \begin{bmatrix} 0 & l_3 & 0 \end{bmatrix}^T
$$

Using the body coordinate system for body 1, the center of gravity for body 1 is located at

$$
\{r_1\} = \begin{bmatrix} 0 & 0 & \frac{l_1}{2} \end{bmatrix}^T
$$

In the body 2 coordinate system the center of gravity for body 2

$$
\{r_2\} = \begin{bmatrix} 0 & \frac{l_2}{2} & 0 \end{bmatrix}^T
$$

likewise for body 3

$$\{r_3\} = \begin{bmatrix} 0 & \frac{l_3}{2} & 0 \end{bmatrix}^T$$

and for body 4

$$\{r_4\} = \begin{bmatrix} 0 & \frac{l_4}{2} & 0 \end{bmatrix}^T$$

The Skew Matrices

$$[Sq_2] = l_1 \begin{bmatrix} 0 & -1 & 0 \\ 1 & 0 & 0 \\ 0 & 0 & 0 \end{bmatrix} \quad [Sq_3] = l_2 \begin{bmatrix} 0 & 0 & 1 \\ 0 & 0 & 0 \\ -1 & 0 & 0 \end{bmatrix}$$

$$[Sq_4] = l_3 \begin{bmatrix} 0 & 0 & 1 \\ 0 & 0 & 0 \\ -1 & 0 & 0 \end{bmatrix}$$

$$[Sr_1] = \frac{l_1}{2} \begin{bmatrix} 0 & -1 & 0 \\ 1 & 0 & 0 \\ 0 & 0 & 0 \end{bmatrix} \quad [Sr_2] = \frac{l_2}{2} \begin{bmatrix} 0 & 0 & 1 \\ 0 & 0 & 0 \\ -1 & 0 & 0 \end{bmatrix}$$

$$[Sr_3] = \frac{l_3}{2} \begin{bmatrix} 0 & 0 & 1 \\ 0 & 0 & 0 \\ -1 & 0 & 0 \end{bmatrix} \quad [Sr_4] = \frac{l_4}{2} \begin{bmatrix} 0 & 0 & 1 \\ 0 & 0 & 0 \\ -1 & 0 & 0 \end{bmatrix}$$

Partial Velocity Matrices
With $\dot{\xi} = \xi = 0$ we use equation (3.120) to obtain the partial velocity matrices.

$$\left[V^1\right] = \begin{bmatrix} [Sr_1][S^{10}] \\ 0_{9\times 3} \end{bmatrix} = \begin{bmatrix} \frac{l_1}{2}s\alpha & -\frac{l_1}{2}c\alpha & 0 \\ \frac{l_1}{2}c\alpha & \frac{l_1}{2}s\alpha & 0 \\ 0 & 0 & 0 \\ & 0_{9\times 3} & \end{bmatrix}$$

and

$$\left[V^2\right] = \begin{bmatrix} [Sq_1][S^{10}] \\ [Sr_2][S^{20}] \\ 0_{6\times 3} \end{bmatrix} = \begin{bmatrix} l_1 s\alpha & -l_1 c\alpha & 0 \\ l_1 c\alpha & l_1 s\alpha & 0 \\ 0 & 0 & 0 \\ \frac{l_2}{2} s\alpha s\beta & -\frac{l_2}{2} c\alpha s\beta & \frac{l_2}{2} c\beta \\ 0 & 0 & 0 \\ -\frac{l_2}{2} c\alpha & -\frac{l_2}{2} s\alpha & 0 \\ & 0_{6\times 3} & \end{bmatrix}$$

and

$$\left[V^3\right] = \begin{bmatrix} [Sq_2][S^{10}] \\ [Sq_3][S^{20}] \\ [Sr_3][S^{30}] \\ 0_{3\times 3} \end{bmatrix} = \begin{bmatrix} l_1 s\alpha & -l_1 c\alpha & 0 \\ l_1 c\alpha & l_1 s\alpha & 0 \\ 0 & 0 & 0 \\ l_2 s\alpha s\beta & -l_2 c\alpha s\beta & l_2 c\beta \\ 0 & 0 & 0 \\ -l_2 c\alpha & -l_2 s\alpha & 0 \\ \frac{l_3}{2} s\alpha s \times(\beta+\theta) & -\frac{l_3}{2} c\alpha s \times(\beta+\theta) & \frac{l_3}{2} c (\beta+\theta) \\ 0 & 0 & 0 \\ -\frac{l_3}{2} c\alpha & -\frac{l_3}{2} s\alpha & 0 \\ & 0_{3\times 3} & \end{bmatrix}$$

and finally

$$\left[V^4\right] = \begin{bmatrix} [Sq_2][S^{10}] \\ [Sq_3][S^{20}] \\ [Sq_4][S^{30}] \\ [Sr_4][S^{40}] \end{bmatrix} = \begin{bmatrix} l_1 s\alpha & -l_1 c\alpha & 0 \\ l_1 c\alpha & l_1 s\alpha & 0 \\ 0 & 0 & 0 \\ l_2 s\alpha s\beta & -l_2 c\alpha s\beta & l_2 c\beta \\ 0 & 0 & 0 \\ -l_2 c\alpha & -l_2 s\alpha & 0 \\ l_3 s\alpha s(\beta+\theta) & -l_3 c\alpha s(\beta+\theta) & l_3 c(\beta+\theta) \\ 0 & 0 & 0 \\ -l_3 c\alpha & -l_3 s\alpha & 0 \\ \frac{l_4}{2} s\alpha s (\beta+\theta+\phi) & -\frac{l_4}{2} c\alpha s (\beta+\theta+\phi) & \frac{l_4}{2} c (\beta+\theta+\phi) \\ 0 & 0 & 0 \\ -\frac{l_4}{2} c\alpha & -\frac{l_4}{2} s\alpha & 0 \end{bmatrix}$$

Differentiation of Partial Velocity Matrices

$$
\left[\dot{V}^1\right] = \begin{bmatrix} \tfrac{1}{2}l_1\dot{\alpha}c\alpha & \tfrac{1}{2}l_1\dot{\alpha}s\alpha & 0 \\ -\tfrac{1}{2}l_1\dot{\alpha}s\alpha & \tfrac{1}{2}l_1\dot{\alpha}c\alpha & 0 \\ 0 & 0 & 0 \\ & 0_{3x3} & \\ & 0_{3x3} & \\ & 0_{3x3} & \end{bmatrix}
$$

$$
\left[\dot{V}^2\right] = \begin{bmatrix} l_1\dot{\alpha}c\alpha & l_1\dot{\alpha}s\alpha & 0 \\ -l_1\dot{\alpha}s\alpha & l_1\dot{\alpha}c\alpha & 0 \\ 0 & 0 & 0 \\ \tfrac{1}{2}l_2(\dot{\alpha}c\alpha s\beta + \dot{\beta}s\alpha c\beta) & -\tfrac{1}{2}l_2(-\dot{\alpha}s\alpha s\beta + \dot{\beta}c\alpha c\beta) & -\dot{\beta}\tfrac{1}{2}l_2 s\beta \\ 0 & 0 & 0 \\ \dot{\alpha}\tfrac{1}{2}l_2 s\alpha & -\alpha\tfrac{1}{2}l_2 c\alpha & 0 \\ & 0_{3x3} & \\ & 0_{3x3} & \end{bmatrix}
$$

$$
\left[\dot{V}^3\right] = \begin{bmatrix} l_1\dot{\alpha}c\alpha & l_1\dot{\alpha}s\alpha & 0 \\ -l_1\dot{\alpha}s\alpha & l_1\dot{\alpha}c\alpha & 0 \\ 0 & 0 & 0 \\ l_2(\dot{\alpha}c\alpha s\beta + \dot{\beta}s\alpha c\beta) & -l_2(-\dot{\alpha}s\alpha s\beta) + \dot{\beta}c\alpha c\beta & -l_2\dot{\beta}s\beta \\ 0 & 0 & 0 \\ l_2\dot{\alpha}s\alpha & -l_2\dot{\alpha}c\alpha & 0 \\ \tfrac{1}{2}l_3(\dot{\alpha}c\alpha s(\beta+\theta) + (\dot{\beta}+\dot{\theta}) \times s\alpha c(\beta+\theta)) & -\tfrac{1}{2}l_3(-\dot{\alpha}s\alpha s(\beta+\theta) + (\dot{\beta}+\dot{\theta}) \times c\alpha c(\beta+\theta)) & -\tfrac{1}{2}l_3(\dot{\beta}+\dot{\theta}) \times s(\beta+\theta)) \\ 0 & 0 & 0 \\ \tfrac{1}{2}l_3\dot{\alpha}s\alpha & -\tfrac{1}{2}l_3\dot{\alpha}c\alpha & 0 \\ & 0_{3x3} & \end{bmatrix}
$$

$$
\left[\dot{V}^4\right] = \begin{bmatrix}
l_1\dot{\alpha}c\alpha & l_1\dot{\alpha}s\alpha & 0 \\
-l_1\dot{\alpha}s\alpha & l_1\dot{\alpha}c\alpha & 0 \\
0 & 0 & 0 \\
\begin{array}{c} l_2(\dot{\alpha}c\alpha s\beta \\ +\dot{\beta}s\alpha c\beta) \end{array} & \begin{array}{c} -l_2(-\dot{\alpha}s\alpha s\beta \\ +\dot{\beta}c\alpha c\beta) \end{array} & -l_2\dot{\beta}s\beta \\
0 & 0 & 0 \\
l_2\dot{\alpha}s\alpha & -l_2\dot{\alpha}c\alpha & 0 \\
\begin{array}{l} l_3(\dot{\alpha}c\alpha s(\beta+\theta) \\ +(\dot{\beta}+\dot{\theta}) \\ \quad\times s\alpha c(\beta+\theta)) \end{array} & \begin{array}{l} -l_3(-\dot{\alpha}s\alpha s(\beta+\theta) \\ +(\dot{\beta}+\dot{\theta}) \\ \quad\times c\alpha c(\beta+\theta)) \end{array} & \begin{array}{l} -l_3(\dot{\beta}+\dot{\theta}) \\ \times s(\beta+\theta)) \end{array} \\
0 & 0 & 0 \\
l_3\dot{\alpha}s\alpha & -l_3\dot{\alpha}c\alpha & 0 \\
\begin{array}{l} 1/2\,l_4(\dot{\alpha}c\alpha s \\ \times(\beta+\theta+\phi) \\ +(\dot{\beta}+\dot{\theta}+\dot{\phi}) \\ \quad\times s\alpha c(\beta+\theta+\phi)) \end{array} & \begin{array}{l} -1/2\,l_4(-\dot{\alpha}s\alpha s \\ \times(\beta+\theta+\phi) \\ +(\dot{\beta}+\dot{\theta}+\dot{\phi}) \\ \quad\times c\alpha c(\beta+\theta+\phi)) \end{array} & \begin{array}{l} -1/2\,l_4(\dot{\beta}+\dot{\theta}+\dot{\phi}) \\ \times s(\beta+\theta+\phi) \end{array} \\
0 & 0 & 0 \\
1/2\,l_4\dot{\alpha}s\alpha & -1/2\,l_4\dot{\alpha}c\alpha & 0
\end{bmatrix}
$$

Angular Velocities Skew Matrices

A tool required for calculating the generalized speeds $\{y\}$ is the skew matrix $[\Omega^{KJ}]$ of the angular velocity components of body K with respect to body J.

For body 1 we calculate the skew matrix as follows:

$$
\Omega^{10} = \dot{S^{10}}S^{01} = \begin{bmatrix} -\dot{\alpha}s\alpha & \dot{\alpha}c\alpha & 0 \\ -\dot{\alpha}c\alpha & -\dot{\alpha}s\alpha & 0 \\ 0 & 0 & 0 \end{bmatrix} \begin{bmatrix} c\alpha & -s\alpha & 0 \\ s\alpha & c\alpha & 0 \\ 0 & 0 & 1 \end{bmatrix}
$$

$$
\Omega^{10} = \begin{bmatrix} 0 & \dot{\alpha} & 0 \\ -\dot{\alpha} & 0 & 0 \\ 0 & 0 & 0 \end{bmatrix}
$$

for body 2

$$
\Omega^{20} = \dot{S^{20}}S^{02} = \begin{bmatrix} -\dot{\alpha}s\alpha & \dot{\alpha}c\alpha & 0 \\ -\dot{\alpha}c\alpha c\beta + \dot{\beta}s\alpha s\beta & -\dot{\alpha}s\alpha c\beta - \dot{\beta}c\alpha s\beta & \dot{\beta}c\beta \\ \dot{\alpha}c\alpha s\beta + \dot{\beta}s\alpha c\beta & \dot{\alpha}s\alpha s\beta - \dot{\beta}c\alpha c\beta & -\dot{\beta}s\beta \end{bmatrix}
$$

$$
\times \begin{bmatrix} c\alpha & -s\alpha c\beta & s\alpha s\beta \\ s\alpha & c\alpha c\beta & -c\alpha s\beta \\ 0 & s\beta & c\beta \end{bmatrix}
$$

$$\Omega^{20} = \begin{bmatrix} 0 & \dot\alpha c\beta & -\dot\alpha s\beta \\ -\dot\alpha c\beta & 0 & \dot\beta \\ \dot\alpha s\beta & -\dot\beta & 0 \end{bmatrix}$$

for body 3 we have

$$\Omega^{30} = \dot{S}^{30} S^{03}$$

$$\times \begin{bmatrix} -\dot\alpha s\alpha & \dot\alpha c\alpha & 0 \\ \begin{matrix} -\dot\alpha c\alpha c(\beta+\theta) \\ +(\dot\beta+\dot\theta) s\alpha s(\dot\beta+\dot\theta) \end{matrix} & \begin{matrix} -\dot\alpha s\alpha c(\beta+\theta) \\ -(\dot\beta+\dot\theta) c\alpha s(\beta+\theta) \end{matrix} & \begin{matrix} (\dot\beta+\dot\theta) \\ \times c(\beta+\theta) \end{matrix} \\ \begin{matrix} \dot\alpha c\alpha s(\beta+\theta) \\ +(\dot\beta+\dot\theta) s\alpha c(\dot\beta+\dot\theta) \end{matrix} & \begin{matrix} \dot\alpha s\alpha s(\beta+\theta) \\ -(\dot\beta+\dot\theta) c\alpha c(\beta+\theta) \end{matrix} & \begin{matrix} -(\dot\beta+\dot\theta) \\ \times s(\beta+\theta) \end{matrix} \end{bmatrix}$$

$$\times \begin{bmatrix} c\alpha & -s\alpha c(\beta+\theta) & s\alpha s(\beta+\theta) \\ s\alpha & c\alpha c(\beta+\theta) & -c\alpha s(\beta+\theta) \\ 0 & s(\beta+\theta) & c(\beta+\theta) \end{bmatrix}$$

$$\Omega^{30} = \begin{bmatrix} 0 & \dot\alpha c(\beta+\theta) & -\dot\alpha s(\beta+\theta) \\ -\dot\alpha c(\beta+\theta) & 0 & \dot\beta+\dot\theta \\ \dot\alpha s(\beta+\theta) & -(\dot\beta+\dot\theta) & 0 \end{bmatrix}$$

and for body 4 we have

$$\Omega^{40} = \dot{S}^{40} S^{04}$$

$$= \begin{bmatrix} -\dot\alpha s\alpha & \dot\alpha c\alpha & 0 \\ \begin{matrix} -\dot\alpha c\alpha c(\beta+\theta+\phi) \\ +(\dot\beta+\dot\theta+\dot\phi) \\ \times s\alpha s(\beta+\theta+\phi) \end{matrix} & \begin{matrix} -\dot\alpha s\alpha c(\beta+\theta+\phi) \\ -(\dot\beta+\dot\theta+\dot\phi) \\ \times c\alpha s(\beta+\theta+\phi) \end{matrix} & \begin{matrix} (\dot\beta+\dot\theta+\dot\phi) \\ c(\beta+\theta+\phi) \end{matrix} \\ \begin{matrix} \dot\alpha c\alpha s(\beta+\theta+\phi) \\ +(\dot\beta+\dot\theta+\dot\phi) \\ \times s\alpha c(\beta+\theta+\phi) \end{matrix} & \begin{matrix} \dot\alpha s\alpha s(\beta+\theta+\phi) \\ -(\dot\beta+\dot\theta+\dot\phi) \\ \times c\alpha c(\beta+\theta+\phi) \end{matrix} & \begin{matrix} -(\dot\beta+\dot\theta+\dot\phi) \\ \times s(\beta+\theta+\phi) \end{matrix} \end{bmatrix}$$

$$\times \begin{bmatrix} c\alpha & -s\alpha c(\beta+\theta+\phi) & s\alpha s(\beta+\theta+\phi) \\ s\alpha & c\alpha c(\beta+\theta+\phi) & -c\alpha s(\beta+\theta+\phi) \\ 0 & s(\beta+\theta+\phi) & c(\beta+\theta+\phi) \end{bmatrix}$$

$$\Omega^{40} = \begin{bmatrix} 0 & \dot\alpha c(\beta+\theta+\phi) & -\dot\alpha s(\beta+\theta+\phi) \\ -\dot\alpha c(\beta+\theta+\phi) & 0 & (\dot\beta+\dot\theta+\dot\phi) \\ \dot\alpha s(\beta+\theta+\phi) & -(\dot\beta+\dot\theta+\dot\phi) & 0 \end{bmatrix}$$

The matrix of generalized speeds is found to be

$$y = \begin{bmatrix} 0 \\ 0 \\ \dot{\alpha} \\ \dot{\beta} \\ -\dot{\alpha} s\beta \\ \dot{\alpha} c\beta \\ (\dot{\beta} + \dot{\theta}) \\ -\dot{\alpha} s(\beta + \theta) \\ \dot{\alpha} c(\beta + \theta) \\ (\dot{\beta} + \dot{\theta} + \dot{\phi}) \\ -\dot{\alpha} s(\beta + \theta + \phi) \\ \dot{\alpha} c(\beta + \theta + \phi) \end{bmatrix}^T$$

The time derivative of the generalized coordinates is

$$\dot{x} = \begin{bmatrix} 0 & 0 & \dot{\alpha} & \dot{\beta} & 0 & 0 & \dot{\theta} & 0 & 0 & \dot{\phi} & 0 & 0 \end{bmatrix}$$

From

$$[y] = [\dot{x}][W]$$

we get

$$[W] = \begin{bmatrix}
0 & 0 & 0 & 0 & 0 & 0 & 0 & 0 & 0 & 0 & 0 & 0 \\
0 & 0 & 0 & 0 & 0 & 0 & 0 & 0 & 0 & 0 & 0 & 0 \\
0 & 0 & 1 & 0 & -s\beta & c\beta & 0 & \begin{matrix} -s(\beta \\ +\theta) \end{matrix} & \begin{matrix} c(\beta \\ +\theta) \end{matrix} & 0 & \begin{matrix} -s(\beta \\ +\theta + \phi) \end{matrix} & \begin{matrix} c(\beta \\ +\theta + \phi) \end{matrix} \\
0 & 0 & 0 & 1 & 0 & 0 & 1 & 0 & 0 & 1 & 0 & 0 \\
0 & 0 & 0 & 0 & 0 & 0 & 0 & 0 & 0 & 0 & 0 & 0 \\
0 & 0 & 0 & 0 & 0 & 0 & 0 & 0 & 0 & 0 & 0 & 0 \\
0 & 0 & 0 & 0 & 0 & 0 & 1 & 0 & 0 & 1 & 0 & 0 \\
0 & 0 & 0 & 0 & 0 & 0 & 0 & 0 & 0 & 0 & 0 & 0 \\
0 & 0 & 0 & 0 & 0 & 0 & 0 & 0 & 0 & 0 & 0 & 0 \\
0 & 0 & 0 & 0 & 0 & 0 & 0 & 0 & 0 & 1 & 0 & 0 \\
0 & 0 & 0 & 0 & 0 & 0 & 0 & 0 & 0 & 0 & 0 & 0 \\
0 & 0 & 0 & 0 & 0 & 0 & 0 & 0 & 0 & 0 & 0 & 0
\end{bmatrix}$$

Mass Center Velocities

The velocity at the center of mass for body 1 is

$$\overline{V_1} = \{y\}^T \left[V^1\right] \{\overline{n}\}$$

$$\overline{V_1} = 0$$

and for body 2

$$\overline{V_2} = [\dot{x}]^T \, [W] \left[V^2\right] \{\bar{n}\}$$

Let

$$\hat{I} = \begin{bmatrix} 1 & 0 & 0 \\ 0 & 0 & 0 \\ 0 & 0 & 0 \end{bmatrix}$$

$$[W]\left[V^2\right] = \begin{bmatrix} 0 & 0 & 0 & 0 & 0 & 0 & 0 & 0 & 0 & 0 & 0 & 0 \\ 0 & 0 & 0 & 0 & 0 & 0 & 0 & 0 & 0 & 0 & 0 & 0 \\ 0 & 0 & 1 & 0 & -s\beta & c\beta & 0 & \begin{matrix}-s(\beta \\ +\theta)\end{matrix} & \begin{matrix}c(\beta \\ +\theta)\end{matrix} & 0 & \begin{matrix}-s(\beta \\ +\theta+\phi)\end{matrix} & \begin{matrix}c(\beta \\ +\theta+\phi)\end{matrix} \\ \multicolumn{3}{c}{0_{3x3}} & \multicolumn{3}{c}{\hat{I}} & \multicolumn{3}{c}{\hat{I}} & \multicolumn{3}{c}{\hat{I}} \\ \multicolumn{3}{c}{0_{3x3}} & \multicolumn{3}{c}{0_{3x3}} & \multicolumn{3}{c}{\hat{I}} & \multicolumn{3}{c}{\hat{I}} \\ \multicolumn{3}{c}{0_{3x3}} & \multicolumn{3}{c}{0_{3x3}} & \multicolumn{3}{c}{0_{3x3}} & \multicolumn{3}{c}{\hat{I}} \end{bmatrix}$$

$$\times \begin{bmatrix} l_1 s\alpha & -l_1 c\alpha & 0 \\ l_1 c\alpha & l_1 s\alpha & 0 \\ 0 & 0 & 0 \\ \frac{l^2}{2} s\alpha s\beta & -\frac{l^2}{2} c\alpha s\beta & \frac{l^2}{2} c\beta \\ 0 & 0 & 0 \\ -\frac{l^2}{2} c\alpha & -\frac{l^2}{2} s\alpha & 0 \\ \multicolumn{3}{c}{0_{3x3}} \\ \multicolumn{3}{c}{0_{3x3}} \end{bmatrix}$$

$$\overline{V_2} = \begin{bmatrix} \dot{\alpha} & \dot{\beta} \end{bmatrix} \begin{bmatrix} -\frac{l^2}{2} c\alpha s\beta & -\frac{l^2}{2} s\alpha c\beta & 0 \\ \frac{l^2}{2} s\alpha s\beta & -\frac{l^2}{2} c\alpha s\beta & \frac{l^2}{2} c\beta \end{bmatrix} \begin{bmatrix} n_1 \\ n_2 \\ n_3 \end{bmatrix}$$

$$= \left[{}^1\!/\!{}_2 l_2 \left(-\dot{\alpha} c\alpha c\beta + \dot{\beta} s\alpha s\beta\right)\right] \overline{n_1} + \left[-\frac{l^2}{2}\left(\dot{\alpha} s\alpha c\beta + \dot{\beta} c\alpha s\beta\right)\right] \overline{n_2}$$

$$+ \frac{l^2}{2} \dot{\beta} c\beta \overline{n_3}$$

$$\overline{V_3} = [\dot{x}]^T [W] \left[V^3\right] \{\bar{n}\}$$

$$[W]\left[V^3\right] = \begin{bmatrix} 0 & 0 & 0 \\ 0 & 0 & 0 \\ \begin{matrix} -l_2 c\alpha c\beta \\ -\frac{l^3}{2} c\alpha c(\beta+\theta) \end{matrix} & \begin{matrix} -l_2 s\alpha c\beta \\ -\frac{l^3}{2} s\alpha c(\beta+\theta) \end{matrix} & 0 \\ \begin{matrix} l_2 s\alpha s\beta \\ +\frac{l^3}{2} s\alpha s(\beta+\theta) \end{matrix} & \begin{matrix} -l_2 c\alpha s\beta \\ -\frac{l^3}{2} c\alpha s(\beta+\theta) \end{matrix} & \begin{matrix} l_2 c\beta \\ +\frac{l^3}{2} c(\beta+\theta) \end{matrix} \\ 0 & 0 & 0 \\ 0 & 0 & 0 \\ \frac{l^3}{2} s\alpha s(\beta+\theta) & -\frac{l^3}{2} c\alpha s(\beta+\theta) & \frac{l^3}{2} c(\beta+\theta) \\ 0 & 0 & 0 \\ 0 & 0 & 0 \\ & 0_{3x3} & \end{bmatrix}$$

$$\overline{V_3} = \begin{bmatrix} \dot{\alpha} & \dot{\beta} & \dot{\theta} \end{bmatrix} \begin{bmatrix} \begin{matrix} -l_2 c\alpha c\beta - {}^1\!/\!{}_2 l_3 \\ \times c\alpha c(\beta+\theta) \end{matrix} & \begin{matrix} -l_2 s\alpha c\beta - {}^1\!/\!{}_2 l_3 \\ \times s\alpha c(\beta+\theta) \end{matrix} & 0 \\ \begin{matrix} l_2 s\alpha s\beta + \frac{l^3}{2} \\ \times s\alpha s(\beta+\theta) \end{matrix} & \begin{matrix} -l_2 c\alpha s\beta - \frac{l^3}{2} \\ \times c\alpha s(\beta+\theta) \end{matrix} & \begin{matrix} l_2 c\beta + \frac{l^3}{2} \\ \times c(\beta+\theta) \end{matrix} \\ \frac{l^3}{2} s\alpha s(\beta+\theta) & -\frac{l^3}{2} c\alpha s(\beta+\theta) & \frac{l^3}{2} c(\beta+\theta) \end{bmatrix}$$

$$\times \begin{bmatrix} n_1 \\ n_2 \\ n_3 \end{bmatrix}$$

$$
\begin{aligned}
&= \left[\dot{\alpha}\left(-l_2 c\alpha c\beta - \frac{l^3}{2} c\alpha c(\beta+\theta)\right) + \dot{\beta}\left(l_2 s\alpha s\beta + \frac{l^3}{2} s\alpha s(\beta+\theta)\right)\right.\\
&\quad \left. + \dot{\theta}\frac{l^3}{2} s\alpha s(\beta+\theta)\right]\overline{n_1} + \left[\dot{\alpha}\left(-l_2 s\alpha c\beta - \frac{l^3}{2} s\alpha c(\beta+\theta)\right)\right.\\
&\quad \left. + \dot{\beta}\left(-l_2 c\alpha s\beta - \frac{l^3}{2} c\alpha s(\beta+\theta)\right) - \dot{\theta}\frac{l^3}{2} c\alpha s(\beta+\theta)\right]\overline{n_2}\\
&\quad + \left[\dot{\beta}\left(l_2 c\beta + \frac{l^3}{2} c(\beta+\theta)\right) + \dot{\theta}\frac{l^3}{2} c(\beta+\theta)\right]\overline{n_3}
\end{aligned}
$$

$$
\overline{V_4} = [\dot{x}]^T\,[W]\left[V^4\right]\{\bar{n}\}
$$

$$
[W]\left[V^4\right] = \begin{bmatrix}
0 & 0 & 0\\
0 & 0 & 0\\
-l_2 c\alpha c\beta - l_3 c\alpha c \times(\beta+\theta) - \frac{l^4}{2} c\alpha c \times(\beta+\theta+\phi) & -l_2 s\alpha c\beta - l_3 s\alpha c \times(\beta+\theta) - \frac{l^4}{2} s\alpha c \times(\beta+\theta+\phi) & 0\\
l_2 s\alpha s\beta + l_3 s\alpha s \times(\beta+\theta) + \frac{l^4}{2} s\alpha s \times(\beta+\theta+\phi) & -l_2 c\alpha s\beta - l_3 c\alpha s \times(\beta+\theta) - \frac{l^4}{2} c\alpha s \times(\beta+\theta+\phi) & l_2 c\beta + l_3 c \times(\beta+\theta) + \frac{l^4}{2} c \times(\beta+\theta+\phi)\\
0 & 0 & 0\\
0 & 0 & 0\\
l_3 s\alpha s(\beta+\theta) + \frac{l^4}{2} \times s\alpha s(\beta+\theta+\phi) & -l_3 c\alpha s(\beta+\theta) - \frac{l^4}{2} c\alpha s(\beta+\theta+\phi) & l_3 c(\beta+\theta) + \frac{l^4}{2} \times c(\beta+\theta+\phi)\\
0 & 0 & 0\\
0 & 0 & 0\\
\frac{l^4}{2} s\alpha s(\beta+\theta+\phi) & -\frac{l^4}{2} c\alpha s(\beta+\theta+\phi) & l_4 c(\beta+\theta+\phi)\\
0 & 0 & 0\\
0 & 0 & 0
\end{bmatrix}
$$

$$\overline{V_4} = \begin{bmatrix} \dot{\alpha}\left(-l_2 c\alpha c\beta - l_3 c\alpha c(\beta+\theta) - \frac{l^4}{2}{}_4 c\alpha c(\beta+\theta+\phi)\right) \\ +\dot{\beta}\left(l_2 s\alpha s\beta + l_3 s\alpha s(\beta+\theta) + \frac{l^4}{2} s\alpha s(\beta+\theta+\phi)\right) \\ +\dot{\theta}\left(l_3 s\alpha s(\beta+\theta) + \frac{l^4}{2} s\alpha s(\beta+\theta+\phi)\right) + \dot{\phi}\frac{l^4}{2} s\alpha s(\beta+\theta+\phi) \end{bmatrix} \overline{n_1}$$

$$+ \begin{bmatrix} \dot{\alpha}\left(-l_2 s\alpha c\beta - l_3 s\alpha c(\beta+\theta) - \frac{l^4}{2} s\alpha c(\beta+\theta+\phi)\right) \\ +\dot{\beta}\left(-l_2 c\alpha s\beta - l_3 c\alpha s(\beta+\theta) - \frac{l^4}{2} c\alpha s(\beta+\theta+\phi)\right) \\ +\dot{\theta}\left(-l_3 c\alpha s(\beta+\theta) - \frac{l^4}{2} c\alpha s(\beta+\theta+\phi)\right) - \dot{\phi}\frac{l^4}{2} c\alpha s(\beta+\theta+\phi) \end{bmatrix}$$

$$\times \overline{n_2} + \left[\dot{\beta}\left(l_2 c\beta + l_3 c(\beta+\theta) + \frac{l^4}{2} c(\beta+\theta+\phi)\right)\right.$$

$$\left. +\dot{\theta}\left(l_3 c(\beta+\theta) + \frac{l^4}{2} c(\beta+\theta+\phi)\right) + \dot{\phi}\frac{l^4}{2} c(\beta+\theta+\phi)\right] \overline{n_3}$$

Mass Center Accelerations

The acceleration at the center of mass for body 1 is

$$\overline{a_1} = \left(\{\dot{Y}\}\left[V^1\right] + \{Y\}\left[\dot{V^1}\right]\right)\{\overline{n}\} = 0$$

$$\overline{a_2} = \left(\{\dot{Y}\}\left[V^2\right] + \{Y\}\left[\dot{V^2}\right]\right)\{\overline{n}\}$$

$$\dot{Y} = \begin{bmatrix} 0 \\ 0 \\ \ddot{\alpha} \\ \ddot{\beta} \\ -\ddot{\alpha} s\beta - \dot{\alpha}\dot{\beta} c\beta \\ \ddot{\alpha} c\beta - \dot{\alpha}\dot{\beta} s\beta \\ (\ddot{\beta}+\ddot{\theta}) \\ -\ddot{\alpha} s(\beta+\theta) - \dot{\alpha}(\dot{\beta}+\dot{\theta}) c(\beta+\theta) \\ \ddot{\alpha} c(\beta+\theta) - \dot{\alpha}(\dot{\beta}+\dot{\theta}) s(\beta+\theta) \\ (\ddot{\beta}+\ddot{\theta}+\ddot{\phi}) \\ -\ddot{\alpha} s(\beta+\theta+\phi) - \dot{\alpha}(\dot{\beta}+\dot{\theta}+\dot{\phi}) c(\beta+\theta+\phi) \\ \ddot{\alpha} c(\beta+\theta+\phi) - \dot{\alpha}(\dot{\beta}+\dot{\theta}+\dot{\phi}) s(\beta+\theta+\phi) \end{bmatrix}^T$$

and for body 2

$$\overline{a_2} = \begin{bmatrix} \ddot{\beta} & \ddot{\alpha}c\beta - \dot{\alpha}\dot{\beta}s\beta \end{bmatrix} \begin{bmatrix} \frac{l^2}{2}s\alpha s\beta & -\frac{l^2}{2}c\alpha c\beta & \frac{l^2}{2}c\beta \\ -\frac{l^2}{2}c\alpha & -\frac{l^2}{2}s\alpha & 0 \end{bmatrix} \begin{bmatrix} \overline{n_1} \\ \overline{n_2} \\ \overline{n_3} \end{bmatrix}$$

$$+ \begin{bmatrix} \dot{\beta} & \dot{\alpha}c\beta \end{bmatrix} \begin{bmatrix} \frac{l^2}{2}(\dot{\alpha}c\alpha s\beta \\ +\dot{\beta}s\alpha c\beta) & -\frac{l^2}{2}(-\dot{\alpha}s\alpha s\beta \\ +\dot{\beta}c\alpha c\beta) & -\dot{\beta}\frac{l^2}{2}s\beta \\ \dot{\alpha}\frac{l^2}{2}s\alpha & -\dot{\alpha}\frac{l^2}{2}c\alpha & 0 \end{bmatrix} \begin{bmatrix} \overline{n_1} \\ \overline{n_2} \\ \overline{n_3} \end{bmatrix}$$

The acceleration for body 3 is

$$\overline{a_3} = \left(\{\dot{Y}\}\left[V^3\right] + \{Y\}\left[\dot{V}^3\right]\right)\{\overline{n}\}$$

$$\overline{a_3} = \begin{bmatrix} \ddot{\beta} \; \ddot{\alpha}c\beta - \dot{\alpha}\dot{\beta}s\beta \; (\ddot{\beta} + \ddot{\theta}) \; \ddot{\alpha}c(\beta + \theta) - \dot{\alpha}(\dot{\beta} + \dot{\theta})s(\beta + \theta) \end{bmatrix}$$

$$\times \begin{bmatrix} l_2 s\alpha s\beta & -l_2 c\alpha s\beta & l_2 c\beta \\ -l_2 c\alpha & -l_2 s\alpha & 0 \\ \frac{l^3}{2}s\alpha s(\beta + \theta) & -\frac{l^3}{2}c\alpha s(\beta + \theta) & \frac{l^3}{2}c(\beta + \theta) \\ -1/2 l_3 c\alpha & -1/2 l_3 s\alpha & 0 \end{bmatrix} \begin{bmatrix} \overline{n_1} \\ \overline{n_2} \\ \overline{n_3} \end{bmatrix}$$

$$+ \begin{bmatrix} \dot{\beta} \; \dot{\alpha}c\beta \; (\dot{\beta} + \dot{\theta}) \; \dot{\alpha}c(\beta + \theta) \end{bmatrix}$$

$$\times \begin{bmatrix} l_2(\dot{\alpha}c\alpha s\beta + \dot{\beta}s\alpha c\beta) & -l_2(-\dot{\alpha}s\alpha s\beta + \dot{\beta}c\alpha c\beta) & -l_2\dot{\beta}s\beta \\ l_2\dot{\alpha}s\alpha & -l_2\dot{\alpha}c\alpha & 0 \\ \frac{l^3}{2}(\dot{\alpha}c\alpha s(\beta + \theta) \\ +(\dot{\beta} + \dot{\theta})s\alpha c(\beta + \theta)) & -\frac{l^3}{2}(-\dot{\alpha}s\alpha s(\beta + \theta) \\ +(\dot{\beta} + \dot{\theta})c\alpha c(\beta + \theta)) & -\frac{l^3}{2}(\dot{\beta} + \dot{\theta}) \\ s(\beta + \theta)) \\ \frac{l^3}{2}\dot{\alpha}s\alpha & -\frac{l^3}{2}\dot{\alpha}c\alpha & 0 \end{bmatrix}$$

$$\times \begin{bmatrix} \overline{n_1} \\ \overline{n_2} \\ \overline{n_3} \end{bmatrix}$$

lastly

$$\overline{a_4} = \left(\{\dot{Y}\}\left[V^4\right] + \{Y\}\left[\dot{V}^4\right]\right)\{\overline{n}\}$$

$$\overline{a_4} = \begin{bmatrix} 0 \\ 0 \\ \ddot{\alpha} \\ \ddot{\beta} \\ -\ddot{\alpha}s\beta - \dot{\alpha}\dot{\beta}c\beta \\ \ddot{\alpha}c\beta - \dot{\alpha}\dot{\beta}s\beta \\ (\ddot{\beta} + \ddot{\theta}) \\ -\ddot{\alpha}s(\beta+\theta) - \dot{\alpha}(\dot{\beta}+\dot{\theta})c(\beta+\theta) \\ \ddot{\alpha}c(\beta+\theta) - \dot{\alpha}(\dot{\beta}+\dot{\theta})s(\beta+\theta) \\ (\ddot{\beta}+\ddot{\theta}+\ddot{\phi}) \\ -\ddot{\alpha}s(\beta+\theta+\phi) - \dot{\alpha}(\dot{\beta}+\dot{\theta}+\dot{\phi})c(\beta+\theta+\phi) \\ \ddot{\alpha}c(\beta+\theta+\phi) - \dot{\alpha}(\dot{\beta}+\dot{\theta}+\dot{\phi})s(\beta+\theta+\phi) \end{bmatrix}^T$$

$$\times \begin{bmatrix} l_1 s\alpha & -l_1 c\alpha & 0 \\ l_1 c\alpha & l_1 s\alpha & 0 \\ 0 & 0 & 0 \\ l_2 s\alpha s\beta & -l_2 c\alpha s\beta & l_2 c\beta \\ 0 & 0 & 0 \\ -l_2 c\alpha & -l_2 s\alpha & 0 \\ l_3 s\alpha s(\beta+\theta) & -l_3 c\alpha s(\beta+\theta) & l_3 c(\beta+\theta) \\ 0 & 0 & 0 \\ -l_3 c\alpha & -l_3 s\alpha & 0 \\ \frac{l_4}{2} s\alpha s(\beta+\theta+\phi) & -\frac{l_4}{2} c\alpha s(\beta+\theta+\phi) & \frac{l_4}{2} c(\beta+\theta+\phi) \\ 0 & 0 & 0 \\ -\frac{l_4}{2} c\alpha & -\frac{l_4}{2} s\alpha & 0 \end{bmatrix} \begin{bmatrix} \overline{n_1} \\ \overline{n_2} \\ \overline{n_3} \end{bmatrix}$$

$$
+\begin{bmatrix} 0 \\ 0 \\ \dot{\alpha} \\ \dot{\beta} \\ -\dot{\alpha}s\beta \\ \dot{\alpha}c\beta \\ (\dot{\beta}+\dot{\theta}) \\ -\dot{\alpha}s(\beta+\theta) \\ \dot{\alpha}c(\beta+\theta) \\ (\dot{\beta}+\dot{\theta}+\dot{\phi}) \\ -\dot{\alpha}s(\beta+\theta+\phi) \\ \dot{\alpha}c(\beta+\theta+\phi) \end{bmatrix}^T X
$$

$$
\begin{bmatrix}
l_1\dot{\alpha}c\alpha & l_1\dot{\alpha}s\alpha & 0 \\
-l_1\dot{\alpha}s\alpha & l_1\dot{\alpha}c\alpha & 0 \\
0 & 0 & 0 \\
l_2(\dot{\alpha}c\alpha s\beta + \dot{\beta}s\alpha c\beta) & -l_2(-\dot{\alpha}s\alpha s\beta + \dot{\beta}c\alpha c\beta) & -l_2\dot{\beta}s\beta \\
0 & 0 & 0 \\
l_2\dot{\alpha}s\alpha & -l_2\dot{\alpha}c\alpha & 0 \\
\begin{array}{l} l_3(\dot{\alpha}c\alpha s(\beta+\theta) \\ +(\dot{\beta}+\dot{\theta})s\alpha c(\beta+\theta)) \end{array} & \begin{array}{l} -l_3(-\dot{\alpha}s\alpha s(\beta+\theta) \\ +(\dot{\beta}+\dot{\theta})c\alpha c(\beta+\theta)) \end{array} & \begin{array}{l} -l_3(\dot{\beta}+\dot{\theta}) \\ \times s(\beta+\theta)) \end{array} \\
0 & 0 & 0 \\
l_3\dot{\alpha}s\alpha & -l_3\dot{\alpha}c\alpha & 0 \\
\begin{array}{l} \frac{l^4}{2}(\dot{\alpha}c\alpha s(\beta+\theta+\phi) \\ +(\dot{\beta}+\dot{\theta}+\dot{\phi}) \\ \times s\alpha c(\beta+\theta+\phi)) \end{array} & \begin{array}{l} -\frac{l^4}{2}(-\dot{\alpha}s\alpha s(\beta+\theta+\phi) \\ +(\dot{\beta}+\dot{\theta}+\dot{\phi}) \\ \times c\alpha c(\beta+\theta+\phi)) \end{array} & \begin{array}{l} -\frac{l^4}{2}(\dot{\beta}+\dot{\theta}+\dot{\phi}) \\ s(\beta+\theta+\phi)) \end{array} \\
0 & 0 & 0 \\
\frac{l^4}{2}\dot{\alpha}s\alpha & -\frac{l^4}{2}\dot{\alpha}c\alpha & 0
\end{bmatrix}
\begin{bmatrix} \overline{n_1} \\ \overline{n_2} \\ \overline{n_3} \end{bmatrix}
$$

A general-purpose program was written in MATLAB to generate the equations. If motions are as derived in this chapter and for the initial conditions given, the following matrices are obtained:

```
a =

1.0e+004 *

0.0007 0.0002 0 0 -0.0001 0 0 0 0 0 0 0

0.0002 3.8796 0 -0.0001 0.0001 0 0 0 0 0 0 0

0 0 0.0012 0 0 0.0001 0 0 0.0001 0 0 0.0001

0 -0.0001 0 0 -0.0001 0 0 0 0 0 0 0

-0.0001 0.0001 0 -0.0001 0.0001 0 0 0 0 0 0 0

0 0 0.0001 0 0 0.5633 0 0 0.0385 0 0 -0.0099

0 0 0 0 0 0 0 0 0 0 0 0

0 0 0 0 0 0 0 0.0001 0 -0.0001 0 0

0 0 0.0001 0 0 0.0385 0  0  0.0495  0  0  0.0118

0  0  0  0  0  0  0  -0.0001  0  0.0001  0  0

0  0  0  0  0  0  0  0  0  0  0  0

0  0  0.0001  0  0  -0.0099  0  0  0.0118  0  0
  0.0065


b =

0 0 0 0 0 0 -0.0395 -0.0180 0 0.0427 -0.1188 0

0 0 0 0 0 0 0.0422 0.0179 0 -0.0419 0.1251 0

0 0 0 0 0 0 0 0 0 0 0 0

0 0 0 0 0 0 -0.0395 -0.0180 0 0.0427 -0.1188 0

0 0 0 0 0 0 0.0422 0.0179 0 -0.0419 0.1251 0

0 0 76.2430 0 0 76.2430 0 0 201.9657 0 0 133.8160
```

```
0 0 0 0 0 0 -0.0229 -0.0108 0 0.0260 -0.0695 0

0 0 0 0 0 0 0.0531 0.0229 0 -0.0539 0.1579 0

0 0 -76.2430 0 0 -76.2430 0 0 -13.4461 0 0 67.2305

0 0 0 0 0 0 -0.0526 -0.0232 0 0.0559 -0.1632 0

0 0 0 0 0 0 -0.0180 -0.0087 0 0.0213 -0.0559 0

0 0 -40.2093 0 0 -40.2093 0 0 -40.3383 0 0 0

C =

-0.1214 -0.1186 0 -0.1214 -0.1186 0 -0.1543 -0.0705 0
  0.0569 -0.1584 0

0.1299 0.1214 0 0.1299 0.1214 0 0.1633 0.0703 0
   0.0559 0.1668 0

0 0 0 0 0 0 0 0 0 0 0 0

-0.1214 -0.1186 0 -0.1214 -0.1186 0 -0.1543 -0.0705
  0 0.0569 -0.1584 0

0.1299 0.1214 0 0.1299 0.1214 0 0.1633 0.0703 0
  -0.0559 0.1668 0

0 0 0 0 0 0 0 0 0 0 0 0

-0.0703 -0.0705 0 -0.0703 -0.0705 0 -0.0900 -0.0426 0
  0.0347 -0.0926 0

0.1633 0.1543 0 0.1633 0.1543 0 0.2059 0.0900 0
  -0.0718 0.2106 0

0 0 0 0 0 0 0 0 0 0 0 0

-0.1668 -0.1584 0 -0.1668 -0.1584 0 -0.2106 -0.0926
  0 0.0745 -0.2176 0

-0.0559 -0.0569 0 -0.0559 -0.0569 0 -0.0718 -0.0347
  0 0.0285 -0.0745 0

0 0 0 0 0 0 0 0 0 0 0 0
```

and the generalized active forces are:

```
f =

1.0e+004 *

0

0.2500

-7.3536

0

0

0.8221

0

0

2.3931

0

0

0.4381
```

The time history of the dynamic system require that the equations of motion be integrated to yield the angular displacement, angular velocities and angular accelerations of each body.

References

1. Hamilton, W. R., Second Essay on a General Method of Dynamics, *Philos. Trans. R. Soc. London,* 1835, pp. 95–144.
2. Gibbs, J. W., On the Fundamental Formulae of Dynamics, *Am. J. Math.,* Vol. **II**, 1879, pp. 49–64.
3. Appel, P., Sur une forme générale des équations de dynamique, *Z. Reine Angew. Math.,* Vol. **121**, 1900, pp. 310–319.
4. Witaker, E. T., *Analytical Dynamics,* 4th ed., Cambridge University Press, London, 1937.
5. Rund, H., *The Hamilton–Jacobi Theory in the Calculus of Variations,* D. Van Nostrand, Princeton, NJ, 1966.

6. Goldstein, H., *Classical Mechanics,* Addison-Wesley, Reading, MA, 1950.
7. Lanczos, C., *The Variational Principle of Mechanics,* University of Toronto Press, Toronto, 1949.
8. Meirovitch, L., *Methods of Analytical Dynamics,* McGraw-Hill, New York, 1970.
9. Cajori, F., *Sir Isaac Newton's Mathematical Principles of Natural Philosophy and His System of the World,* University of California Press, Berkeley, CA, 1934.
10. Kane, T. R. and Wang, C. F., On the Derivation of Equations of Motion, *J. Soc. Ind. Appl. Math.,* Vol. **13**, 1965.
11. Huston, R. L., Passerello, C. E. and Harlow, M. W., Dynamics of Multi-Rigid-Body Systems, *Appl. Mech.,* Vol. **45**, Dec. 1978.
12. Gupta, V. K., Dynamic Analysis of Multi-Rigid-Body Systems, *J. Eng. Ind.,* Vol. **9**, 1979, pp. 886–892.
13. Kane, T. R., Likins, P. W. and Levinson, D. A., *Spacecraft Dynamics,* McGraw-Hill, New York, 1983.
14. Huston, R. L., *Multibody Dynamics,* Butterworth-Heinemann, Stoneham, MA, 1990.
15. Kane, T. R. and Levinson, D. A. *Dynamics: Theory and Applications*, McGraw-Hill, New York, 1985.
16. Greenwood, D. T., *Principles of Dynamics*, Prentice Hall, Englewood Cliffs, NJ, 1988.
17. Amirouche, F., *Computational Methods in Multibody Dynamics*, Prentice Hall, Englewood Cliffs, NJ, 1992.

6

Hamilton–Lagrange and Gibbs–Appel Equations

6.1 Introduction

Dynamics of particles stems from theoretical physics where most of the pioneering work started with Isaac Newton and Leibniz's discovery of calculus. Calculus of variation was then introduced by Leonard Euler which became an important tool in physics. Joseph-Louis Lagrange introduced the Lagrange function $L(q, \dot{q})$ which depends on a set of generalized coordinates and velocities of a system. The general work of Euler and Lagrange also led to the principle of least action, where differential equations are obtained by minimizing the action over an interval of time. The Hamiltonian formalism of mechanics was then introduced in 1830, where the Hamiltonian function represents the total energy of the system. The Hamilton principle is related to the Lagrange through a transformation called the Legendre transformation. Equations of motion can be derived from the Hamilton principle. While the history of mechanics was based on particle dynamics, Lagrange functions are extended to continuous system- or rigid-body dynamics. Gibbs–Appell introduced a function that deals with acceleration, velocity and position from which the equations of motion are derived. This chapter introduces the formulation for the various ways of deriving the equations of motion and highlights the benefits of each.

6.2 Energy Equations

6.2.1 Kinetic Energy

The kinetic energy of system S composed of N particles in reference frame R is defined as

$$T = \sum_{i=1}^{N} \frac{1}{2} m_i \bar{v}_i \cdot \bar{v}_i \tag{6.1}$$

where m_i is the mass of the particle and $\bar{v}_i$ is the velocity of the ith particle in R (Figure 6.1). The kinetic energy of S can be expressed as

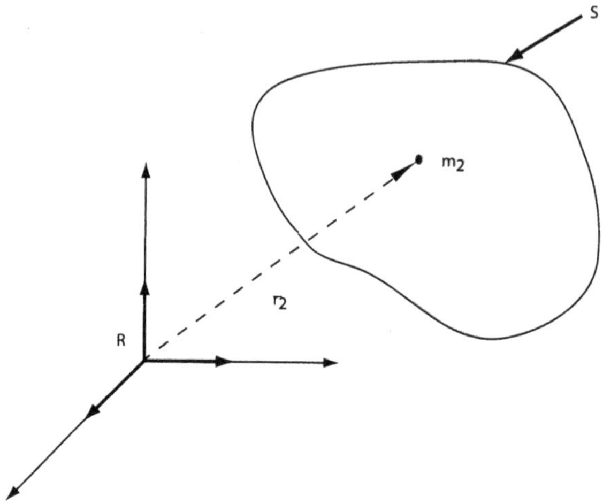

Fig. 6.1. Position of a point mass (element) m_i of body S in R

$$T = T_2 + T_1 + T_0 \tag{6.2}$$

If we denote the generalized coordinates by $x_p(\ell = 1, \ldots, n)$, then T_2 represents the quadratic function of $\dot{x}$'s, T_1 the linear function of the $\dot{x}$'s, and T_0 includes the remaining terms in x's and t. To show the derivation of equation (6.2), let the position of the ith particle of S in R be given by

$$\bar{r}_i = \bar{r}_i(x_1, x_2, x_3, \ldots, x_n, t) \tag{6.3}$$

where n is the number of degrees of freedom of the system. The velocity vector $\bar{\upsilon}_i$ can be written as

$$\bar{\upsilon}_i = \frac{d\bar{r}_i}{dt} = \frac{\partial \bar{r}_i}{\partial x_1}\dot{x}_1 + \frac{\partial \bar{r}_1}{\partial x_2}\dot{x}_2 + \cdots + \frac{\partial \bar{r}_i}{\partial x_n}\dot{x}_n + \frac{\partial \bar{r}_i}{\partial t} \tag{6.4}$$

Expressing the above in a compact form yields,

$$\bar{\upsilon}_i = \sum_{j=1}^{N} \frac{\partial \bar{r}_i}{\partial x_j}\dot{x}_j + \frac{\partial \bar{r}_i}{\partial t} \tag{6.5}$$

Substituting the velocity expressed given by equation (6.5) into equation (6.1), we obtain

$$T = \frac{1}{2}\sum_{i=1}^{N} m_i \left(\sum_{j=1}^{N} \frac{\partial \bar{r}_i}{\partial x_j}\dot{x}_j + \frac{\partial \bar{r}_i}{\partial t}\right)^2 \tag{6.6}$$

Expanding (6.6) yields the following; a kinetic energy function with several terms:

$$T = \frac{1}{2}\sum_{i=1}^{N} m_i \left(\sum_{j=1}^{N}\sum_{k=1}^{N} \frac{\partial \bar{r}_i}{\partial x_j} \cdot \frac{\partial \bar{r}_i}{\partial x_k}\dot{x}_j\dot{x}_k + 2\frac{\partial \bar{r}_i}{\partial t} \cdot \sum_{j=1}^{N} \frac{\partial \bar{r}_i}{\partial x_j}\dot{x}_j + \frac{\partial \bar{r}_i}{\partial t} \cdot \frac{\partial \bar{r}_i}{\partial t}\right) \quad (6.7)$$

Equation (6.7) is equivalent to equation (6.2), where one can see that there are terms with second-order velocity components and the product of velocities defined by

$$T_2 = \frac{1}{2}\sum_{i=1}^{N}\sum_{j=1}^{N}\sum_{k=1}^{N} m_i \frac{\partial \bar{r}_i}{\partial x_j} \cdot \frac{\partial \bar{r}_i}{\partial x_k}\dot{x}_j\dot{x}_k \quad (6.8)$$

or from which we can deduce the following expression,

$$T_2 = \frac{1}{2}\sum_{j=1}^{N}\sum_{k=1}^{N} C_{jk}\dot{x}_j\dot{x}_k \quad (6.9)$$

where

$$C_{jk} = \sum_{i=1}^{N} m_i \frac{\partial \bar{r}_i}{\partial x_j} \cdot \frac{\partial \bar{r}_i}{\partial x_k} \quad (6.10)$$

Similarly, equation (6.7) gives a term where the kinetic energy is a function of a single velocity component defined by,

$$T_1 = \sum_{j=1}^{N} d_j\dot{x}_j \quad (6.11)$$

where

$$d_j = \sum_{i=1}^{N} m_i \frac{\partial \bar{r}_i}{\partial t} \cdot \frac{\partial \bar{r}_i}{\partial x_j} \quad (6.12)$$

and finally, we see that equation (6.7) has a term that doesn't involve $\dot{x}$, given by

$$T_0 = \frac{1}{2}\sum_{i=1}^{N} m_i \frac{\partial \bar{r}_i}{\partial t} \cdot \frac{\partial \bar{r}_i}{\partial t} \quad (6.13)$$

where C_{jk}, d_j and T_0 are coefficients that are functions of the x's and time t. In summary we can say that the total kinetic energy is given by $T = T_2 + T_1 + T_0$.

6.2.2 Energy of a Rigid Body

The total kinetic energy of a rigid body B moving in a reference frame R is defined by its kinetic energy contribution for translation and rotation,

$$T = T_{\text{tr}} + T_{\text{rot}} \tag{6.14}$$

where T_{tr} denotes the kinetic energy due to the translation and is given by

$$T_{\text{tr}} = \frac{1}{2} m \bar{v}^2 \tag{6.15}$$

where m is the mass of the rigid body and $\bar{v}$ denotes the velocity of center mass of B in R.

The kinetic energy due to the rotation given by T_{rot} is defined as

$$T_{\text{rot}} = \frac{1}{2} \bar{\omega} \cdot \overline{\overline{I}} \cdot \bar{\omega} \tag{6.16}$$

where $\bar{\omega}$ is the angular velocity of B in R, and $\overline{\overline{I}}$ is the inertia dyadic of B for G (center mass of B); alternatively, we can write T_{rot} as

$$T_{\text{rot}} = \frac{1}{2} I \bar{\omega}^2 \tag{6.17}$$

where I is the moment of inertia of B relative to G. More simply, T_{rot} can be expressed as

$$T_{\text{rot}} = \frac{1}{2} \sum_{j=1}^{3} \sum_{k=1}^{3} \omega_j I_{jk} \omega_k \tag{6.18}$$

where I_{jk} is the product of inertia of B.

Equations (6.16), (6.17) and (6.18) represent the contribution of the kinetic energy to rotation, making use of the inertia dyadic, moment of inertia and product of inertia, respectively.

6.2.3 Work, Potential Energy and Generalized Forces

In Chapter 1, we introduced the concept of potential energy through work. We also derived relationships between potential energy and the external forces. It is evident that in rigid-body dynamics and multibody systems in particular, that we can derive functions from which the generalized forces can be derived. The generalized active forces f_ℓ are for the generalized coordinates $x_\ell (\ell = 1, \ldots, n)$. A potential function V_1 is such that

$$f_\ell = -\frac{\partial V}{\partial x_\ell} \tag{6.19}$$

where V is a function of x_ℓ and t. We can also note that there are forces that contribute to the generalized active forces but which are not derivable from the potential function V. Such forces include friction forces and time-varying forces, in which case equation (6.19) is extended to include those additional forces as

$$f_\ell = -\frac{\partial V}{\partial x_\ell} + f_\ell^c \tag{6.20}$$

Example 6.2.1
A rectangular plate of mass m is mounted on a shaft AB as shown in Figure 6.2. The shaft is given a constant angular speed Ω about a fixed axis. If the plate is to rotate without sliding on the shaft, find the kinetic and potential energy of the plate.

Solution
Consider the unit vectors $\bar{n}_i (i = 1, 2, 3)$ used to represent the plate local principal centroid axis of the plate. We have $\bar{n}_1$ along the shaft; $\bar{n}_2$ and $\bar{n}_3$ are normal to the plate surface. The moment of inertia matrix for a uniform plate is given by

$$I_{ij} = \begin{bmatrix} I_{11} & 0 & 0 \\ 0 & I_{22} & 0 \\ 0 & 0 & I_{33} \end{bmatrix} \tag{6.21}$$

where the coefficients of I_{ij} are

$$I_{11} = \frac{1}{4} m(a^2 + b^2) \tag{6.22}$$

$$I_{22} = \frac{1}{4} mb^2 \tag{6.23}$$

and

$$I_{33} = \frac{1}{4} ma^2 \tag{6.24}$$

The kinetic energy of the plate can be obtained from the definition of the kinetic energy

$$T = \frac{1}{2} m\bar{v} \cdot \bar{v} + \frac{1}{2} \bar{\omega} \cdot \bar{\bar{I}} \cdot \bar{\omega} \tag{6.25}$$

where $\bar{v}$ represents the mass center velocity of the plate in R, $\bar{\omega}$ its corresponding angular velocity, and $\bar{\bar{I}}$ the inertia dyadic. If R is chosen to be given by the mutually perpendicular unit vectors $\bar{n}_1 (i = 1, 2, 3)$, the I_{ij} represent the dyadic matrix of $\bar{\bar{I}}$.

The angular velocity of the plate in the reference frame R is

$$\bar{\omega} = \Omega \bar{n}_z + \dot{\theta} \bar{n}_1 \tag{6.26}$$

We can express n_z in terms of fixed inertia frame coordinates

$$\bar{n}_z = \cos\theta \bar{n}_3 + \sin\theta \bar{n}_2 \tag{6.27}$$

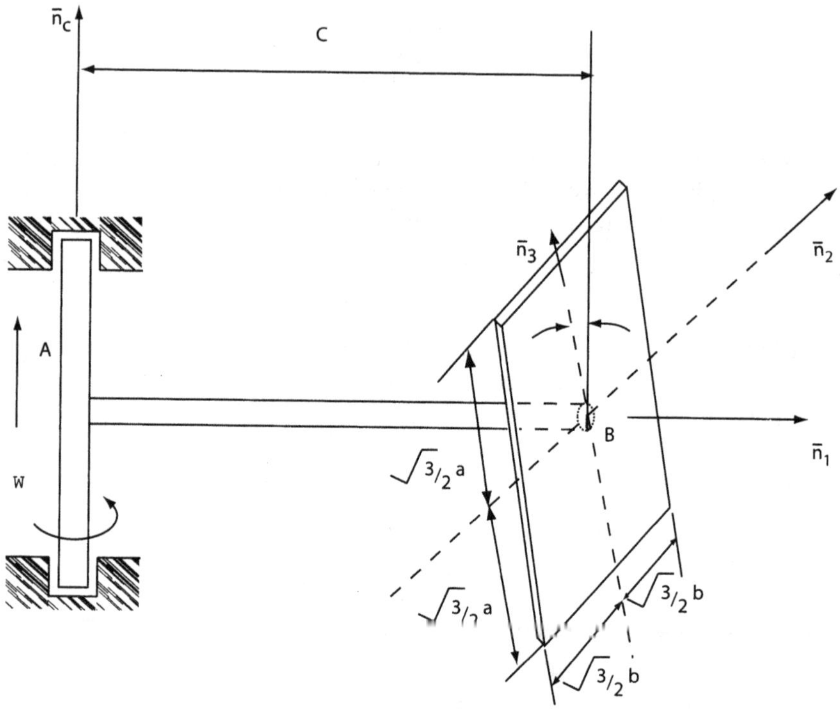

Fig. 6.2. A rotating plate attached to a T bar

Substituting equation (6.27) into (6.26), we obtain the angular velocity in terms of $\bar{n}_1$, $\bar{n}_2$ and $\bar{n}_3$

$$\bar{\omega} = \dot{\theta}\bar{n}_1 + \Omega \sin\theta \bar{n}_2 + \Omega \cos\theta \bar{n}_3 \tag{6.28}$$

and the center mass velocity of the plate in R is then found by

$$\bar{\upsilon} = c\Omega(\bar{n}_3 \times \bar{n}_1) = c\Omega\bar{n}_2 \tag{6.29}$$

The kinetic energy is then found by means of (6.25)

$$T = \frac{1}{2}m(c\Omega)^2 + \frac{1}{2}\left[\dot{\theta}\Omega\sin\theta\Omega\cos\theta\right] \begin{bmatrix} \frac{1}{4}m(a^2+b^2) & 0 & 0 \\ 0 & \frac{1}{4}mb^2 & 0 \\ 0 & 0 & \frac{1}{4}ma^2 \end{bmatrix} \begin{bmatrix} \dot{\theta} \\ \Omega\sin\theta \\ \Omega\cos\theta \end{bmatrix} \tag{6.30}$$

Expanding and simplifying the above equation we obtain the desired kinetic energy contribution.

$$T = \frac{m}{8}[4c^2\Omega^2 + (a^2+b^2)\dot{\theta}^2 + b^2\Omega^2\sin^2\theta + a^2\Omega^2\cos^2\theta] \tag{6.31}$$

The potential energy is obviously

$$V = 0 \tag{6.32}$$

This can be shown by denoting that the contribution of the gravity force to the generalized forces is zero.

$$(mg\bar{n}_3) \cdot \frac{\partial \bar{v}}{\partial \dot{\theta}} = 0 \tag{6.33}$$

Hence the generalized force is also zero.

6.3 Lagrange's Equations

The standard Lagrange function L is defined as

$$L = T - V \tag{6.34}$$

where T is the total kinetic energy of the system and V its corresponding potential energy. Lagrange equations of motion are then obtained by

$$\frac{d}{dt}\left[\frac{\partial L}{\partial \dot{x}_\ell}\right] - \frac{\partial L}{\partial x_\ell} = 0 \tag{6.35}$$

for the case when the generalized active forces are contributed through the potential function V,

$$\frac{d}{dt}\left[\frac{\partial L}{\partial \dot{x}_\ell}\right] - \frac{\partial L}{\partial x} = f^c \tag{6.36}$$

for the case when the generalized active forces are given by equation (6.54). The value f^c represents the generalized constraint forces associated with the constraints added to the system. The derivation of the expressions in equations (6.34) to (6.36) can be proved through the principal of virtual work applied to D'Alembert's principle (see reference [15]), the variation principle, or Hamilton's principle. In the next section, the relationship between the generalized inertia forces, generalized active forces, and Lagrange equations will be shown. The Lagrange equations represent a set of nonlinear n second-order differential equations known as equations of motion.

Example 6.3.1
The center O of a spacecraft is moving through space with a constant velocity as shown in Figure 6.3. The spacecraft has a constant rotational rate Ω about its z-axis. The x, y and z axes are attached to the body of the craft, and the solar panels rotate about the y-axis at an angular velocity $\dot{\theta}(t) = a_1 \sin \omega t$ with respect to the aircraft.

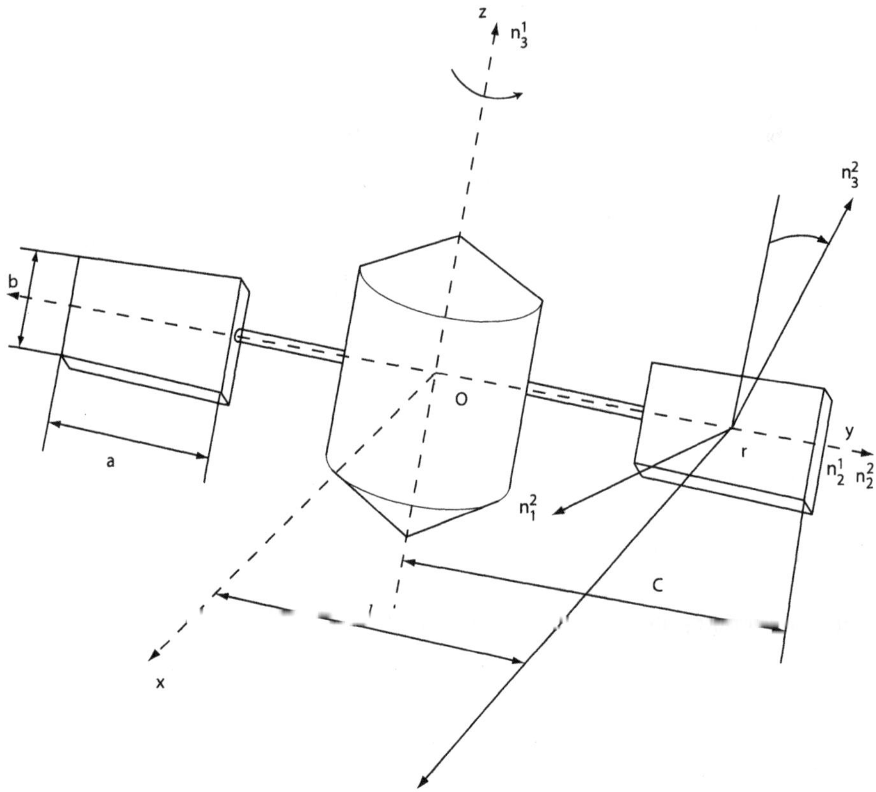

Fig. 6.3. Two solar panels attached to a spacecraft

(a) Find the total kinetic energy of the system.
(b) Derive the equations of motion using the Lagrangian method.

Solution

The solution to this problem is mainly to show how Lagrange equations are formulated. Therefore, we opt not to give the general approach used to solve multibody systems, as depicted in Section 6.4. That is left as an exercise.

(a) The moment of inertia components are taken to be

$$I_{11} = I_{22} + I_{33} = \frac{1}{12}ma^2 + \frac{1}{12}mb^2 \tag{6.37}$$

The transformation between unit vectors in body 1 and 2 as shown in Figure 6.3 is

$$\bar{n}^1_m = \begin{bmatrix} c\theta & 0 & -s\theta \\ 0 & 1 & 0 \\ s\theta & 0 & c\theta \end{bmatrix} \bar{n}^2_m \tag{6.38}$$

The angular velocity of the solar panel can be expressed as

$$\begin{aligned}\bar{\omega}^2 &= \Omega \bar{n}_3^1 + \dot{\theta}\bar{n}_2^2 \\ &= \Omega(s\theta\bar{n}_1^2 + c\theta\bar{n}_3^2) + \dot{\theta}\bar{n}_2^2 \\ &= \begin{bmatrix}\Omega s\theta \\ \dot{\theta} \\ \Omega c\theta\end{bmatrix}\bar{n}_m^2\end{aligned} \tag{6.39}$$

The mass center velocity of the solar panel is $-\ell\Omega\bar{n}_1^2$ and the kinetic energy of the plate panel is

$$\begin{aligned}T_p = T_t + T_{\text{rot}} &= \frac{1}{2}m_2\ell^2\Omega^2 + \frac{1}{2}\bar{\omega}^2 \cdot \bar{\bar{I}} \cdot \bar{\omega}^2 \\ &= \frac{1}{2}m_2\ell^2\Omega^2 + \frac{1}{2}[\Omega s\theta \quad \dot{\theta} \quad \Omega c\theta]\begin{bmatrix}\frac{1}{12}m_2(a^2+b^2) & 0 & 0 \\ 0 & \frac{1}{12}m_2a^2 & 0 \\ 0 & 0 & \frac{1}{12}m_2b^2\end{bmatrix}\begin{bmatrix}\Omega s\theta \\ \dot{\theta} \\ \Omega c\theta\end{bmatrix} \\ &= \frac{1}{2}m_2\ell^2\Omega^2 + \frac{1}{24}m_2[\Omega^2 s^2\theta(a^2+b^2) + a^2\dot{\theta}^2 + \Omega^2c^2\theta b^2]\end{aligned} \tag{6.40}$$

or

$$T_p = \frac{1}{2}m_2\left[\Omega^2\left(\ell^2 + \frac{1}{12}b^2 + \frac{1}{12}a^2s^2\theta\right) + \frac{1}{12}a^2\dot{\theta}^2\right] \tag{6.41}$$

The total kinetic energy of the system is given by

$$T = 2T_p + T_s + T_T \tag{6.42}$$

where $T_s = \frac{1}{2}I_s\Omega^2$, the rotation kinetic energy of the body of spacecraft; $T_T = \frac{1}{2}(m_1 + m_2)\upsilon^2$, the translation kinetic energy of total system; and υ is the velocity of spacecraft. Therefore,

$$T = m_2\left[\Omega^2\left(\ell^2 + \frac{1}{12}b^2 + \frac{1}{12}a^2s^2\theta\right) + \frac{1}{12}a^2\dot{\theta}^2\right] + \frac{1}{2}I_s\Omega^2 + \frac{1}{2}(m_1 + m_2)\upsilon^2 \tag{6.43}$$

(b) Using Lagrange equations, we derive the equations of motion for the generalized coordinates α and θ.

$$\begin{aligned}\frac{d}{dt}\left[\frac{\partial L}{\partial \Omega}\right] - \frac{\partial L}{\partial \alpha} &= 0 \\ \frac{d}{dt}\left[\frac{\partial L}{\partial \dot{\theta}}\right] - \frac{\partial L}{\partial \theta} &= 0\end{aligned} \tag{6.44}$$

where $L = T - V$, $V = 0$, and $\Omega = \dot{\alpha}$. The equations above yield the following two equations

$$2m_2\left[\dot{\Omega}\left(\ell^2+\frac{1}{12}b^2+\frac{1}{12}a^2s^2\theta\right)+\frac{1}{6}a^2\Omega\dot{\theta}s\theta c\theta\right]+I_s\dot{\Omega}=0$$
$$\frac{1}{6}m_2a^2(\ddot{\theta}-\Omega^2 s\theta c\theta)=0 \tag{6.45}$$

Simplifying further, we arrive at

$$2m_2\left[\dot{\Omega}\left(\ell^2+\frac{1}{12}b^2+\frac{1}{12}a^2s^2\theta\right)\right]+I_s\dot{\Omega}+\frac{1}{3}a^2s\theta c\theta\Omega\dot{\theta}=0$$
$$\ddot{\theta}-\Omega^2 s\theta c\theta=0 \tag{6.46}$$

where $\dot{\theta}=a_1\sin\omega t$.

Note: How we have used Lagrange equations is to derive the equations without making use of the general principle adapted in Chapter 5. This direct approach is one where we need to derive the equations for both the kinematics and Lagrange equations and perform all the derivations to obtain the final solution. For simple problems it is difficult to see the benefits of the automated approaches that are suited to large systems.

Example 6.3.2
Suppose a circular disk D with a mass m is rolling with slipping along a smooth surface S. The disc resembles a rolling coin, which has three rotations. Its angular velocity along the y-axis due to the coin rotation is denoted by $\omega_y=\dot{\theta}$, whereas the yaw and pitch are represented by ψ and ϕ, respectively. Derive the equation of motion, using the Lagrangian method and simulate its response for different initial conditions.

Solution
The interesting problem is a classical one and found in several dynamics books. To derive the equation of motion we need to formulate the Lagrangian function, which require the kinetic and potential energy of the disk.

The angular velocities of the disk with respect to R is found to be

$$\omega_x=\dot{\phi}-\dot{\psi}\sin\theta,\quad \omega_y=\dot{\theta},\quad \omega_z=\dot{\psi}\cos\theta \tag{6.47}$$

Let the moment of inertia be such that

$$I_a=I_{xx},\quad I_t=I_{yy}=I_{zz}\quad \text{(assuming symmetry)} \tag{6.48}$$

The disk has both translation and rotation; therefore we have to compute both the kinetic energy due to rotation as well as the kinetic energy due to translation. Kinetic energy due to rotation is given by

$$T_{\text{rot}}=\frac{1}{2}(I_{xx}\omega_x^2+I_{yy}\omega_y^2+I_{zz}\omega_z^2) \tag{6.49}$$

After substitution we obtain

$$T_{\text{rot}}=\frac{1}{2}I_a(\dot{\phi}-\dot{\psi}\sin\theta)^2+\frac{1}{2}I_t(\dot{\theta}^2+\dot{\psi}^2\cos^2\theta) \tag{6.50}$$

Kinetic energy due to translation is given by

$$T_{\text{tran}} = \frac{1}{2}m(v_x^2 + v_y^2) = \frac{1}{2}m(r^2\dot{\theta}^2 + R^2\dot{\psi}^2) \tag{6.51}$$

The velocity of the center mass is

$$v_0 = R\,\dot{\psi} = r\omega_x = r(\dot{\phi} - \dot{\psi}\sin\theta) \tag{6.52}$$

from which we deduce

$$R = (\dot{\phi} - \dot{\psi}\sin\theta)r/\,\dot{\psi} \tag{6.53}$$

This results in

$$T_{\text{tran}} = \frac{1}{2}mr^2(\dot{\theta}^2 + (\dot{\phi} - \dot{\psi}\sin\theta)^2) \tag{6.54}$$

Total kinetic energy is

$$T = \frac{1}{2}I\alpha(\dot{\phi} - \dot{\psi}\sin\theta)^2 + \frac{1}{2}I_t(\dot{\theta}^2 + \dot{\psi}^2\cos^2\theta) + \frac{1}{2}mr^2[\dot{\theta}^2 + (\dot{\phi} - \dot{\psi}\sin\theta)^2] \tag{6.55}$$

Potential energy of the coin is due to the gravity force mg where

$$V = mgr\cos\theta \tag{6.56}$$

Therefore, the net change in energy is given by the Lagrange function L

$$L = T - V \tag{6.57}$$

Partial derivation of the total energy with respect to the angular coordinate $\dot{\theta}, \dot{\phi}, \dot{\psi}$ about the three axes is given by the following expressions

$$\frac{\partial L}{\partial\,\dot{\theta}} = (I_t + mr^2)\,\dot{\theta} \tag{6.58}$$

$$\frac{\partial L}{\partial\,\dot{\phi}} = (I_a + mr^2)(\dot{\phi} - \dot{\psi}\sin\theta) \tag{6.59}$$

$$\frac{\partial L}{\partial\,\dot{\psi}} = I_t\,\dot{\psi}\cos^2\theta - (I_a + mr^2)(\dot{\phi} - \dot{\psi}\sin\theta)\sin\theta \tag{6.60}$$

And the time derivative of the above equations leads to the following equations

$$\frac{d}{dt}\left(\frac{\partial L}{\partial\,\dot{\theta}}\right) = (I + mr^2)\,\ddot{\theta} \tag{6.61}$$

$$\frac{d}{dt}\left(\frac{\partial L}{\partial\,\dot{\phi}}\right) = (I_a + mr^2)(\ddot{\phi} - \ddot{\psi}\sin\theta - \dot{\psi}\,\dot{\theta}\cos\theta) \tag{6.62}$$

$$\frac{d}{dt}\left(\frac{\partial L}{\partial \dot\psi}\right) = I_t(\ddot\psi\cos^2\theta - 2\dot\psi\dot\theta\cos\theta\sin\theta)$$
$$- (I_a + mr^2)(\ddot\phi\sin\theta + \dot\phi\,\dot\theta\cos\theta - \ddot\psi\sin^2\theta - 2\,\dot\psi\,\dot\theta\sin\theta\cos\theta) \tag{6.63}$$

To complete the Lagrangian formulation, we need to evaluate the partial derivatives with respect to the generalized coordinates. Note that only θ contributes to the potential energy, hence

$$\frac{\partial L}{\partial\theta} = -I_a(\dot\phi - \dot\psi\sin\theta)\,\dot\psi\cos\theta - I_t\,\dot\psi^2\cos\theta\sin\theta$$
$$- mr^2(\dot\phi - \dot\psi\sin\theta)\,\dot\psi\cos\theta + mgr\sin\theta \tag{6.64}$$

The equation of motion is obtained by combining the two terms we derived from L, that is

$$\frac{d}{dt}\left(\frac{\partial L}{\partial \dot\theta}\right) - \frac{\partial L}{\partial\theta} = 0 \tag{6.65}$$

Hence,

$$(I_t + mr^2)\,\ddot\theta = -(I_a + mr^2)(\dot\phi - \dot\psi\sin\theta)\,\dot\psi\cos\theta$$
$$-I_t\dot\psi^2\cos\theta\sin\theta + mgr\sin\theta$$

$$\frac{d}{dt}\left(\frac{\partial L}{\partial \dot\phi}\right) - \frac{\partial L}{\partial\phi} = 0 \tag{6.66}$$

$$(I_a + mr^2)(\ddot\phi - \ddot\psi\sin\theta - \dot\psi\,\dot\theta\cos\theta) = 0$$

$$\ddot\phi - \dot\psi\,\dot\theta\cos\theta = \ddot\psi\sin\theta$$

$$\frac{d}{dt}\left(\frac{\partial L}{\partial \dot\psi}\right) - \frac{\partial L}{\partial\psi} = 0 \tag{6.67}$$

$$[I_t\cos^2\theta + (I_a + mr^2)\sin^2\theta]\,\ddot\psi + 2(-I_t + I_a + mr^2)\,\dot\psi\,\dot\theta\cos\theta\sin\theta$$
$$= (I_a + mr^2)(\ddot\phi\sin\theta + \dot\phi\,\dot\theta\cos\theta)$$

The above equations (6.65), (6.3), (6.3) represent the equation of motion of the rolling coin/disk.

To simulate the dynamic response of the coin, we decide to numerically integrate the equation of motion. It is obvious that the behavior of the coin will depend on the initial conditions assumed at time zero. Let us explore some practical conditions and find out how the three-dimensional angles will behave. First, let us rewrite the equation of motion in a form suitable for integration.

Equation of motion (6.65) can be expressed as

$$\ddot{\theta} = \frac{(I_a + mr^2)}{(I_t + mr^2)} \cos\theta\, \dot{\psi}\, \dot{\phi} + \frac{I_a + mr^2 - I_t}{I_t + mr^2} \sin\theta \cos\theta \dot{\psi}^2 + \frac{mgr \sin\theta}{I_t + mr^2} \tag{6.68}$$

for equation (6.3).

Let the geometric parameters be represented by

$$a = I_t \cos^2\theta + (I_a + mr^2)\sin^2\theta$$

$$b = -(I_a + mr^2)\sin\theta$$

$$c = [-2(-I_t + I_a + mr^2)\sin\theta + (I_a + mr^2)]\cos\theta$$

which results in an equation of the form

$$a\,\ddot{\psi} + b\,\ddot{\phi} = c\,\dot{\phi}\,\dot{\theta} \tag{6.69}$$

Solving for $a\,\ddot{\psi} + b\,\ddot{\phi} = c\,\dot{\phi}\,\dot{\theta}$ from the above equation and the equation of motion, we get

$$\ddot{\phi} = \frac{c\sin\theta + a\cos\theta}{a + b\sin\theta}\,\dot{\phi}\,\dot{\theta} \tag{6.70}$$

and

$$\ddot{\psi} = \frac{(c - b\cos\theta)\,\dot{\phi}\,\dot{\theta}}{a + b\sin\theta} \tag{6.71}$$

Let y and $\dot{y}$ be matrices of the angular coordinates and their derivatives as given below

$$y = \begin{bmatrix} \dot{\theta} \\ \dot{\phi} \\ \dot{\psi} \\ \theta \\ \phi \\ \psi \end{bmatrix} \qquad \dot{y} = \begin{bmatrix} \ddot{\theta} \\ \ddot{\phi} \\ \ddot{\psi} \\ \theta \\ \phi \\ \psi \end{bmatrix} \tag{6.72}$$

Equations (6.68), (6.70) and (6.71) can be written in matrix form as

$$\dot{y} = Ry + E \tag{6.73}$$

where

$$R = \begin{bmatrix} A & B \\ C & D \end{bmatrix}$$

and the coefficient A, B, C and D are submatrices given by

$$A = \begin{bmatrix} 0 & \dfrac{I_a + mr^2}{I_t + mr^2} \cos\theta\,\dot{\psi} & -\dfrac{I_a + mr^2 - I_t}{I_t + mr^2} \cos\theta \sin\theta\,\dot{\psi} \\ \dfrac{c\sin\theta + a\cos\theta}{a + b\sin\theta}\dot{\phi} & 0 & 0 \\ 0 & \dfrac{c - b\cos\theta}{a + b\sin\theta}\dot{\theta} & 0 \end{bmatrix} \tag{6.74}$$

$$B = \begin{bmatrix} 0\,0\,0 \\ 0\,0\,0 \\ 0\,0\,0 \end{bmatrix} \quad C = \begin{bmatrix} 1\,0\,0 \\ 0\,1\,0 \\ 0\,0\,1 \end{bmatrix} \quad D = \begin{bmatrix} 0\,0\,0 \\ 0\,0\,0 \\ 0\,0\,0 \end{bmatrix} \quad E = \begin{bmatrix} \dfrac{mgr\sin\theta}{I_t + mr^2} \\ 0 \\ 0 \\ 0 \\ 0 \\ 0 \end{bmatrix}$$

We run three cases where we want to know what happens to the coin when the initial conditions are such that

(a) $y_{(t=0)} = [\dot{\theta}\ \dot{\phi}\ \dot{\psi}\ \theta\ \phi\ \psi] = [1\ 0\ 0\ 30^o\ 0\ 0]^T$

(b) $y_{(t=0)} = [\dot{\theta}\ \dot{\phi}\ \dot{\psi}\ \theta\ \phi\ \psi] = [0\ 1\ 0\ 0\ 0\ 0]^T$

(c) $y_{(t=0)} = [\dot{\theta}\ \dot{\phi}\ \dot{\psi}\ \theta\ \phi\ \psi] = [0.2\ 0.5\ 0\ 0\ 0\ 0]^T$

Solution

In case (a), the coin is seen to oscillate about the y-axis and there is no rolling involved. In case (b), there is only one angular component in the x-axis; the coin is rolling in a straight line. Even if the initial conditions are changed we will obtain the same dynamic response. In case (c), both the angular velocity $\dot{\theta}$ and $\dot{\phi}$ are given initial values, which result in the coin rotating about the x-axis and in the mean time we see its center of mass moving in a circular path. Furthermore, the coin is oscillating about the y-axis.

Source Code

```
% matlab program for simulation of coin motion
% This is a function to define the equation of motion
% m-mass, r-radius of the coin,
%g-gravitional acceleration
% y=[dot{\_}theta;dot{\_}fi;dot{\_}psi;theta;fi;psi]
function ydot=msd(t,y)
global r m g
dot{\_}theta=y(1);
dot{\_}fi=y(2);
dot{\_}psi=y(3);
theta=y(4);
```

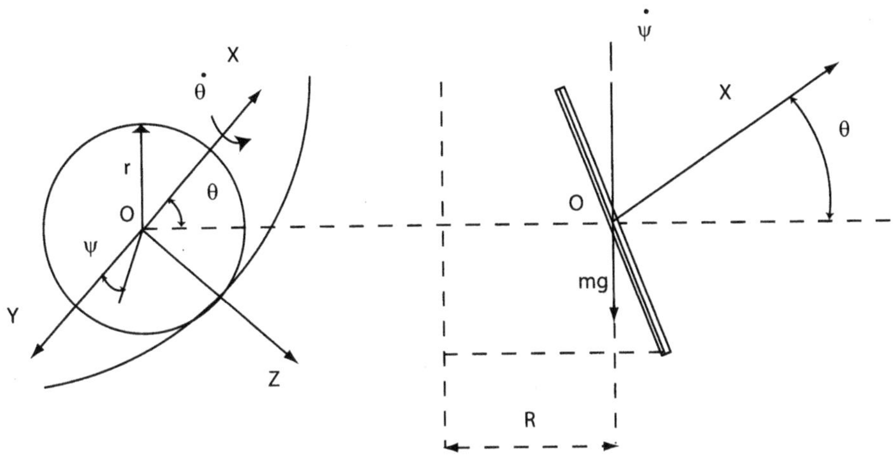

Fig. 6.4. Coin

```
psi=y(5);
psi=y(6);
Ia=m*r\^{}2/2;
It=m*r\^{}2/4;
a=It*cos(theta)\^{}2+(Ia+m*r\^{}2)*sin(theta)\^{}2;
b=-(Ia+m*r\^{}2)*sin(theta);
c=(-2*(-It+Ia+m*r\^{}2)*sin(theta)
                    +(Ia+m*r\^{}2))*cos(theta);
a12=-( Ia+m*r\^{}2)*cos(theta)*dot{\_}psi/( It
  +m*r\^{}2);
a13=( Ia+m*r\^{}2-It)*cos(theta)*sin(theta)
                    *dot{\_}psi/( It+m*r\^{}2);
a21=(c*sin(theta)
        +a*cos(theta))*dot{\_}fi/(a+b*sin(theta));
a32=(c-b*cos(theta))*dot{\_}theta/(a+b*sin(theta));
A=[0 ,a12, a13;a21, 0, 0;0 ,a32, 0];
B=zeros(3,3);
C=eye(3,3);
D=zeros(3,3);
R=[A B;C D];
E=zeros(6,1);
E(1,1)=m*g*sin(theta)/( It+m*r\^{}2);
Ydot=R*y+E;

% matlab program for simulation of coin motion
```

```
% main program
% A combination of fourth --
% and fifth -- order Runge -- kutta methods
% is used to solve the equations of motion
% m-mass, r-radius of the coin,
%g-gravitational acceleration
clear
global r m g
m=10;
r=0.1;
g=9.8;
% define the initial conditions
% y and
%y0=[dot{\_}theta;dot{\_}fi;dot{\_}psi;theta;fi;psi];
y0=[0.01;0.1;0.01;30*pi/180;0;0];
% solve equations by calling function msd
[t,y]=ode45('msd',[0,5],y0);
% calculate angular accelerations
for I=1:length(t)
ydot(:,I)=msd(t,y(I, : )');
end
% plot angular displacement
plot(t,y(:,4)*pi/180,'---',
          t,y(:,5)*pi/180,'-',t,y(:,6)*pi/180,'-');
legend('theta','fi','psi');
xlabel('time (sec)');
ylabel('Angular Displacement (deg)');
title('Angular Displacement ');
figure;
% plot angular velocities
plot(t,y(:,1),'---',t,y(:,2),'-',t,y(:,3),'-');
legend('dot-theta','dot-fi','dot-psi');
xlabel('Time (sec)');
ylabel('Angular velocity (rad/sec)');
title('Angular velocity');
figure;
% plot angular accelerations
plot(t,ydot(1,: ),'---',
         t,ydot(2,: ),'-',t,ydot(3,: ),'-');
legend('double-dot-theta',
         'double-dot-fi','double-dot-psi');
xlabel('Time (sec)');
ylabel('Angular Acceleration (rad/sec)');
title('Angular Acceleration');
```

6.4 Application of Lagrange Equations to Multibody Systems

Consider a general multibody system as depicted by Figure 6.5. Let $x_\ell(\ell = 1, \ldots, n)$ represent the generalized coordinates, and n denotes the total degrees of freedom of the system. If we allow each rigid body to undergo six degrees of freedom, then for N bodies, n is equal to $6N$. Let R be the inertial reference frame where the system motion is defined. To obtain the equations of motion of the multibody system in R using Lagrange equations, let us first compute the kinetic energy of the system, where

$$T = \frac{1}{2}\sum_{i=1}^{N} m_i \bar{\upsilon}_i \cdot \bar{\upsilon}_i + \frac{1}{2}\sum_{i=1}^{N} \bar{\omega}^i \cdot \bar{\bar{I}} \cdot \bar{\omega}^i \tag{6.75}$$

where $\bar{\upsilon}_i$ is the velocity of the mass center of the ith body, m_i its corresponding mass, $\bar{\omega}^i$ the angular velocity of body i in R, and I the inertia dyadic of body i for G_i, its mass center. Let the generalized active forces be given by

$$f_\ell = \frac{\partial V}{\partial x_\ell} \tag{6.76}$$

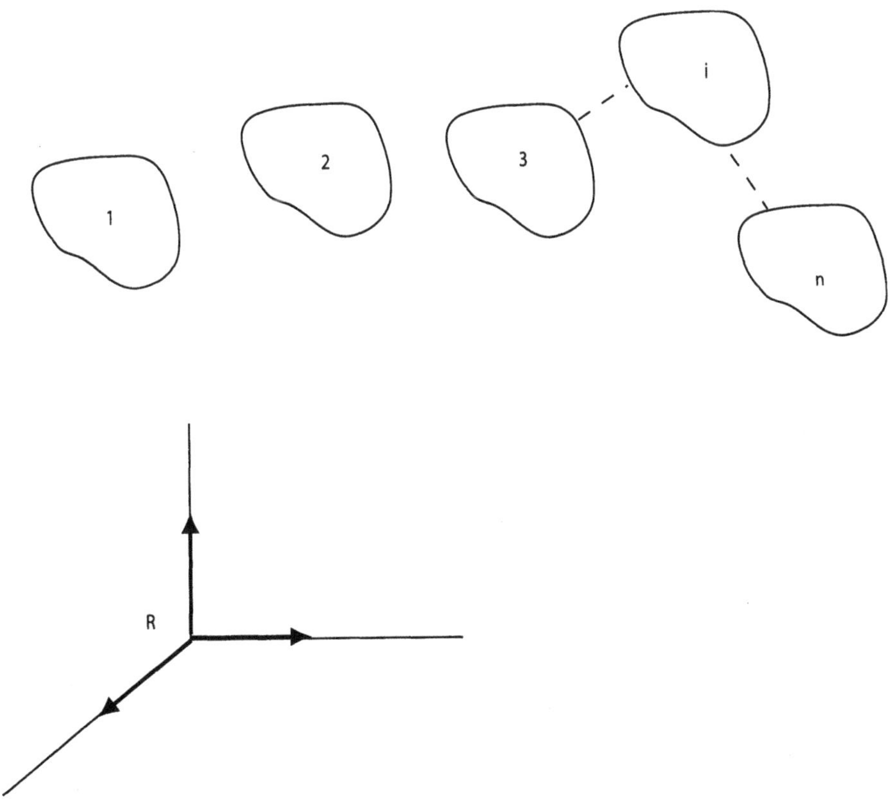

Fig. 6.5. Labeling of an arbitrary chain of interconnected rigid bodies

Then, Lagrange equations become

$$\frac{d}{dt}\left[\frac{\partial T}{\partial \dot{x}_\ell}\right] - \frac{\partial T}{\partial x_\ell} = f_\ell \tag{6.77}$$

Using expressions developed for the mass center velocity in Chapter 3, we write the velocity of body i center mass in R in an index notation. (See index representation at the end of the chapter.)

$$\bar{v}_i = V^i_{\ell m}\dot{x}_\ell \bar{n}_m \qquad (\ell = 1, \ldots, 6N; \quad m = 1, \ldots, 3) \tag{6.78}$$

where $V^i_{\ell m}$ is the partial velocity array associated with $\dot{x}_\ell$ (including the W term). Similarly, the angular velocity of body i in R is expressed as

$$\bar{\omega}^i = \omega^i_{\ell m}\dot{x}_\ell \bar{n}_m \tag{6.79}$$

where $\omega^i_{\ell m}$ is the partial angular velocity array associated with $\dot{x}_\ell$. Substituting equations (6.78) and (6.79) together with the definition of inertia dyadic $\bar{\bar{I}}$, the kinetic energy equation becomes

$$T = \frac{1}{2}m_i[(V^i_{\ell m}\dot{x}_\ell \bar{n}_m) \cdot (V^i_{pn}\dot{x}_p \bar{n}_n)] + \frac{1}{2}[(\omega^i_{\ell m}\dot{x}_\ell \bar{n}_m) \cdot I_{kj}\bar{n}_k\bar{n}_j \cdot (\omega^i_{pn}\dot{x}_p\bar{n}_n)] \tag{6.80}$$

or

$$T = \frac{1}{2}m_i(V^i_{\ell m}V^i_{pn}\dot{x}_p\dot{x}_\ell + I_{mn}\omega^i_{\ell m}\omega^i_{pn}\dot{x}_\ell\dot{x}_p) \tag{6.81}$$

Carrying out the differentiation of equation (6.77), we obtain

$$\begin{aligned}\frac{d}{dt}\left[\frac{\partial T}{\partial \dot{x}_\ell}\right] &= m_i V^i_{\ell m}\dot{V}^i_{pn}\dot{x}_p + I_{mn}\omega^i_{\ell m}\dot{\omega}^i_{pn}\dot{x}_p \\ &\quad + m_i V^i_{\ell m}V^i_{pn}\ddot{x}_p + I_{mn}\omega^i_{\ell m}\omega^i_{pn}\ddot{x}_p + e_{nmb}I_{mr}\omega^i_{pr}\omega^i_{sn}\dot{\omega}^i_{\ell h}\dot{x}_p\dot{x}_s\end{aligned} \tag{6.82}$$

and

$$\frac{\partial T}{\partial x_\ell} = 0 \tag{6.83}$$

Hence, the equations of motion given by equation (6.77) become

$$a_{\ell p}\ddot{x}_p + b_{\ell p}\dot{x}_p + c_{\ell ps}\dot{x}_p\dot{x}_s = f_\ell \tag{6.84}$$

where, as expected, the coefficients $a_{\ell p}$, $b_{\ell p}$ and $c_{\ell ps}$ are exactly the same as the one derived using the matrix approach in Chapter 5.

6.5 Relationship Between Kane's and Lagrange Equations

The equations of motion of a multibody system are unique; therefore irrelevant of the method they should be the same. In what follows is a proof that in essence Lagrange equations must be equivalent to Kane's method.

It follows from equation (6.77) that the generalized inertia forces are

$$f_\ell = \frac{d}{dt}\left[\frac{\partial T}{\partial \dot{x}_\ell}\right] + \frac{\partial T}{\partial x_\ell} \quad (\ell = 1, \ldots, n) \tag{6.85}$$

where n denotes the number of degrees of freedom of the system. The generalized inertia forces are also given through Kane's equations [see equation (6.42)] as

$$f_\ell^* = F_k^* \cdot \frac{\partial \bar{\upsilon}_k}{\partial \dot{x}_\ell} + T_k^* \cdot \frac{\partial \bar{\omega}^k}{\partial \dot{x}_\ell} \tag{6.86}$$

where F_k^* is the inertia force and T_k^* the inertia torque of body k and R. Let us denote by F_ℓ^* the contribution of the inertia forces and by T_ℓ^* the corresponding inertia torques to the generalized inertia forces. Then, equation (6.86) becomes

$$f_\ell^* = F_\ell^* + T_\ell^* \tag{6.87}$$

By definition,

$$\begin{aligned}
F_\ell^* &= F_k^* \cdot \frac{\partial \bar{\upsilon}_k}{\partial \dot{x}_\ell} \\
&= -m_k a_k \cdot \frac{\partial \bar{\upsilon}_k}{\partial \dot{x}_\ell} \\
&= -\left[\frac{d}{dt}\left(m_k \bar{\upsilon}_k \cdot \frac{\partial \bar{\upsilon}_k}{\partial \dot{x}_\ell}\right) - m_k \bar{\upsilon}_k \cdot \frac{d}{dt}\left(\frac{\partial \bar{\upsilon}_k}{\partial \dot{x}_\ell}\right)\right] \\
&= -\left[\frac{d}{dt}\frac{\partial}{\partial \dot{x}_\ell}\left(\tfrac{1}{2} m_k \bar{\upsilon}_k \cdot \bar{\upsilon}_k\right) - m_k \bar{\upsilon}_k \cdot \frac{d}{dt}\frac{\partial \bar{r}_k}{\partial x_\ell}\right]
\end{aligned} \tag{6.88}$$

note that $\partial \bar{r}_k / \partial x_\ell$ is equal to $\partial \bar{\upsilon}_k / \partial \dot{x}_\ell$ as stated in equation (6.44). Expressing equation (6.88) further, we get

$$\begin{aligned}
F_\ell^* &= -\left[\frac{d}{dt}\left(\frac{\partial T_f}{\partial \dot{x}_\ell}\right) - m_k \bar{\upsilon}_k \frac{\partial \bar{\upsilon}_k}{\partial \dot{x}_\ell}\right] \\
&= -\left[\frac{d}{dt}\left(\frac{\partial T_f}{\partial \dot{x}_\ell}\right) - \frac{\partial}{\partial x_\ell}\left(\frac{1}{2} m_k \bar{\upsilon}_k \cdot \bar{\upsilon}_k\right)\right]
\end{aligned} \tag{6.89}$$

$$F_\ell^* = -\left[\frac{d}{dt}\left(\frac{\partial T_f}{\partial \dot{x}_\ell}\right) - \frac{\partial T_f}{\partial x_\ell}\right] \tag{6.90}$$

where T_f is the kinetic energy due to the inertia forces only. Similarly, for the generalized inertia torque, we have

$$T_\ell^* = -\left[M_k^* \cdot \frac{\partial \bar{\omega}^k}{\partial \dot{x}_\ell} \right] \tag{6.91}$$

Substituting for the inertia torque, its expression as given by equation (6.60), we obtain

$$\begin{aligned} T_\ell^* &= -[\bar{\alpha}_k \cdot \bar{\bar{I}} + \bar{\omega}^k \times \bar{\bar{I}} \cdot \bar{\omega}^k] \cdot \frac{\partial \bar{\omega}^k}{\partial \dot{x}_\ell} \\ &= -\left[\frac{d}{dt} \left(\bar{\omega}^k \cdot \bar{\bar{I}} \cdot \frac{\partial \bar{\omega}^k}{\partial \dot{x}_\ell} \right) - \bar{\omega}^k \cdot \bar{\bar{I}} \cdot \frac{d}{dt} \frac{\partial \bar{\omega}^k}{\partial \dot{x}_\ell} \right] \end{aligned} \tag{6.92}$$

where

$$\frac{d}{dt}\left(\frac{\partial \bar{\omega}^k}{\partial \dot{x}_\ell} \right) = \frac{\partial \bar{\omega}^k}{\partial x_\ell}$$

Then equation (6.92) becomes

$$\begin{aligned} T_\ell^* &= -\left[\frac{d}{dt} \frac{\partial}{\partial \dot{x}_\ell} \left(\frac{1}{2} \omega^k \cdot \bar{\bar{I}} \cdot \bar{\omega}^k \right) - \bar{\omega}^k \cdot \bar{\bar{I}} \cdot \frac{\partial \bar{\omega}^k}{\partial x_\ell} \right] \\ &= -\left[\frac{d}{dt} \frac{\partial T_m}{\partial \dot{x}_\ell} - \frac{\partial}{\partial \dot{x}_\ell} \left(\frac{1}{2} \bar{\omega}^k \cdot \bar{\bar{I}} \cdot \omega^k \right) \right] \end{aligned} \tag{6.93}$$

which gives

$$T_\ell^* = -\left[\frac{d}{dt} \left(\frac{\partial T_m}{\partial \dot{x}_\ell} \right) - \frac{\partial T_m}{\partial \dot{x}_\ell} \right] \tag{6.94}$$

where T_m is the kinetic energy due to the rotation only. It follows from the summation of (6.90) and (6.94) that the generalized inertia forces obtained from Kane's and Lagrange's method are equivalent. Similarly, it can be shown that the generalized active forces from both methods are the same. Hence the equations of motion of a system S in R can be found through either method.

Example 6.5.1

Derive the equations of motion for the two-arm planar manipulator as shown in Figure 6.6 using Lagrange's method. Verify your results against the solution obtained by the method derived in Chapter 5.

Solution

The angular velocity components for links 1 and 2 are

$$\{\omega^1\} = \begin{bmatrix} 0 \\ 0 \\ \dot{x}_3 \end{bmatrix}, \qquad \{\omega^2\} = \begin{bmatrix} 0 \\ 0 \\ \dot{x}_3 + \dot{x}_6 \end{bmatrix} \tag{6.95}$$

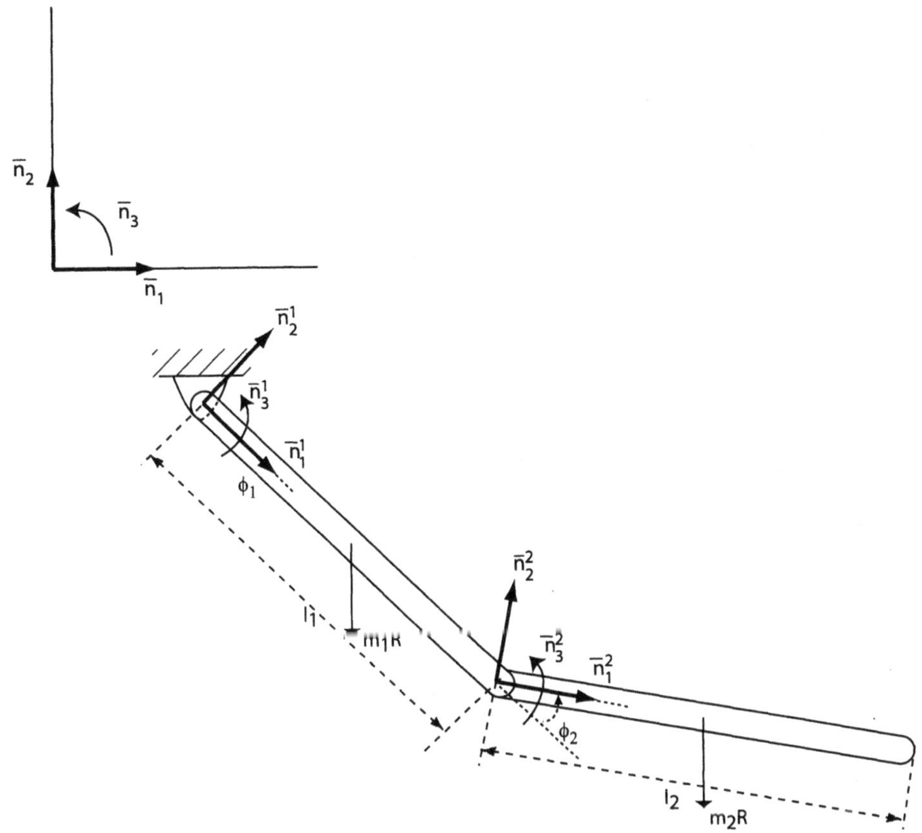

Fig. 6.6. External forces and local axis frames of a planar two-link system

The mass center velocity of each link is given by

$$\{v_1\} = \begin{bmatrix} \frac{1}{2}\ell_1 s1 \dot{x}_3 \\ \frac{1}{2}\ell_1 c1 \dot{x}_3 \\ 0 \end{bmatrix} \tag{6.96}$$

$$\{v_2\} = \begin{bmatrix} \left[\ell_1 s1 + \frac{1}{2}\ell_2 s(1+2)\right]\dot{x}_3 + \frac{1}{2}\ell_2 \dot{x}_6 s(1+2) \\ \dot{x}_3[\ell_1 c1 + \frac{1}{2}\ell_2 c(1+2)] + \frac{1}{2}\ell_2 \dot{x}_6 c(1+2) \\ 0 \end{bmatrix} \tag{6.97}$$

where $s1 = \sin\theta_1$ and $c1 = \cos\theta_1$; similarly, $s2 = \sin\theta_2$; $c2 = \cos\theta_2$, and in this problem $\theta_1 = x_3$ and $\theta_2 = x_6$, to follow the notation from the previous solution using the matrix approach. The kinetic energy of the system is defined as

$$T = \frac{1}{2}m_1(\bar{v}_1 \cdot \bar{v}_1) + \frac{1}{2}m_2(\bar{v}_2 \cdot \bar{v}_2) + \frac{1}{2}\bar{\omega}^1 \cdot \bar{\bar{I}}_1 \cdot \bar{\omega}^1 + \frac{1}{2}\bar{\omega}^2 \cdot \bar{\bar{I}}_2 \cdot \bar{\omega}^2 \tag{6.98}$$

where $\overline{\overline{I}}_1$ and $\overline{\overline{I}}_2$ are the inertia dyadics corresponding to links 1 and 2, respectively. Substituting the kinematical quantities needed to express the kinetic energy, we obtain

$$T = \frac{1}{6}m_1\ell_1^2\dot{x}_3^2 + \frac{1}{24}m_1\ell_2^2(\dot{x}_3+\dot{x}_6)^2 + \frac{1}{2}m_2\left[\left(\ell_1^2+\frac{1}{4}\ell_2^2\right)\dot{x}_3^2 + \frac{1}{2}\ell_2^2\dot{x}_6\left(\dot{x}_3+\frac{1}{2}\dot{x}_6\right) + \ell_1\ell_2\dot{x}_3(\dot{x}_3+\dot{x}_6)c2\right] \tag{6.99}$$

The potential energy function is given by

$$V = (m_1g)\delta r_1 + (m_2g)\delta r_2 = m_1g\frac{1}{2}\ell_1c1 + m_2g\left[\ell_1c1+\frac{1}{2}\ell_2c(1+2)\right] \tag{6.100}$$

Using the Lagrange function

$$L = T - V \tag{6.101}$$

The equations of motion are derived from Lagrange equations

$$\frac{d}{dt}\left[\frac{\partial L}{\partial \dot{x}}\right] - \frac{\partial L}{\partial x} = 0 \tag{6.102}$$

For x_3 we have

$$\frac{d}{dt}\left[\frac{\partial L}{\partial \dot{x}_3}\right] - \frac{\partial L}{\partial x_3} = 0 \tag{6.103}$$

Carrying out the differentiation in equation (6.103), we get

$$\ddot{x}_3\left[\frac{1}{3}m_1\ell_1^2 + m_2\left(\ell_1^2+\frac{1}{4}\ell_2^2+\ell_1\ell_2c2+\frac{1}{12}\ell_2^2\right)\right] + \ddot{x}_6\left[m_2\left(\frac{1}{12}\ell_2^2+\frac{1}{4}\ell_2^2+\frac{1}{2}\ell_1\ell_2c2\right)\right] - \frac{1}{2}m_2\ell_1\ell_2s2(2\dot{x}_3\dot{x}_6+\dot{x}_6^2) = -\frac{1}{2}\ell_1mgs1 - m_2g\left[\ell_1s1+\frac{1}{2}\ell_1s(1+2)\right] \tag{6.104}$$

Similarly, the Lagrange equation for x_6 yields

$$\ddot{x}_3\left(\frac{1}{3}m_2\ell_2^2+\frac{1}{2}m_2\ell_1\ell_2c2\right) + \ddot{x}_6\left(\frac{1}{3}m_2\ell_2^2\right) + \frac{1}{2}m_2\ell_1\ell_2\dot{x}_3^2s2 = -\frac{1}{2}m_2g\ell_2s(1+2) \tag{6.105}$$

Equations (6.104) and (6.105) check similarly with those obtained by equations (6.90), except for the external forces, which are different due to the selection of the coordinates.

6.6 Gibbs–Appel Equations

Gibbs–Appel equations were introduced by Gibbs in 1879 and were subsequently studied and formalized by Appel in 1900 [3]. It is our intention in this section, to introduce this method and correlate its formulation to the methods developed previously by Kane and Lagrange. Let us introduce the *Gibbs–Appel function.*

Consider a system S constituted of P particles with mass m_i $(i = 1, \ldots, p)$, and let R represent the inertial fixed frame with center O (Figure 6.7). The Gibbs–Appel function is defined as

$$G = \frac{1}{2}\sum_{i=1}^{p} m_i \bar{a}_i \cdot \bar{a}_i \tag{6.106}$$

and the equations of motion are obtained by

$$\frac{\partial G}{\partial \ddot{x}_\ell} = f_\ell \tag{6.107}$$

where f_ℓ denotes the generalized active forces and x_ℓ the generalized coordinates. Equation (6.106) for a rigid body can be expressed as

$$G = \frac{1}{2}M_k\bar{a}_k^2 + \frac{1}{2}\bar{\alpha}_k \cdot \bar{\bar{I}} \cdot \bar{\alpha}_k + \bar{\alpha}_k \cdot \bar{\omega}^k \times (\bar{\bar{I}} \cdot \bar{\omega}^k) \tag{6.108}$$

where M_k is the mass of body k, $\bar{a}_k$ its center of mass acceleration, $\bar{\alpha}_k$ the angular velocity, and $\bar{\bar{I}}$ the dyadic of body k. For a multibody system, the function G can easily be formulated by substituting into it the explicit form of $\bar{a}_k$, $\bar{\omega}^k$, and $\bar{\alpha}_k$, as developed in Chapter 3. From a simple observation of (6.107) one can see that the generalized inertia forces are determined by $\partial G/\partial \ddot{x}_\ell$. Making use of Gibbs equations, we can show that this method also yields the numerical representation of the equations of motion given by equation (6.75).

Proof

Let us use the index notation used in arriving at (6.75). Substituting the kinematical expressions developed in Chapter 3 into equation (6.108), we get

$$\begin{aligned} G = {} & \frac{1}{2}m_k(V_{pm}^k\ddot{x}_p + \dot{V}_{pm}^k\dot{x}_p)\bar{n}_m \cdot (V_{\ell n}^k\ddot{x}_\ell + \dot{V}_{\ell n}^k\dot{x}_\ell)\bar{n}_n \\ & + \frac{1}{2}(\dot{\omega}_{pm}^k\dot{x}_p + \omega_{pm}^k\ddot{x}_p)\bar{n}_m \cdot I_{ij}\bar{n}_i\bar{n}_j \cdot (\dot{\omega}_{\ell n}^k\dot{x}_\ell + \omega_{\ell n}^k\ddot{x}_\ell)\bar{n}_n \\ & + (\dot{\omega}_{pm}^k\dot{x}_p + \omega_{pm}^k\ddot{x}_p)\bar{n}_m \cdot \omega_{st}^k\dot{x}_s\bar{n}_t \times (I_{ij}\bar{n}_i\bar{n}_j \cdot \omega_{\ell r}^k\dot{x}_\ell\bar{n}_r) \end{aligned} \tag{6.109}$$

Equation (6.109) is divided into three parts, as stated in each line and calculated below.

$$\begin{aligned} \text{part 1} &= \frac{1}{2}m_k(V_{pm}^k V_{pm}^k\ddot{x}_p\ddot{x}_\ell + V_{pm}^k\ddot{x}_p\dot{V}_{\ell n}^k\dot{x}_\ell + \dot{V}_{pm}^k\dot{x}_p V_{\ell n}^k\ddot{x}_\ell + \dot{V}_{pm}^k\dot{x}_p\dot{V}_{\ell n}\dot{x}_\ell)\delta_{mn} \\ &= \frac{1}{2}m_k(V_{pm}^k V_{\ell m}^k\ddot{x}_p\ddot{x}_\ell + V_{pm}^k\dot{V}_{\ell m}^k\dot{x}_\ell\ddot{x}_p + \dot{V}_{pm}^k V_{\ell m}^k\dot{x}_p\ddot{x}_\ell + \dot{V}_{pm}^k\dot{V}_{pm}\dot{V}_{\ell m}\dot{x}_p\dot{x}_\ell) \end{aligned} \tag{6.110}$$

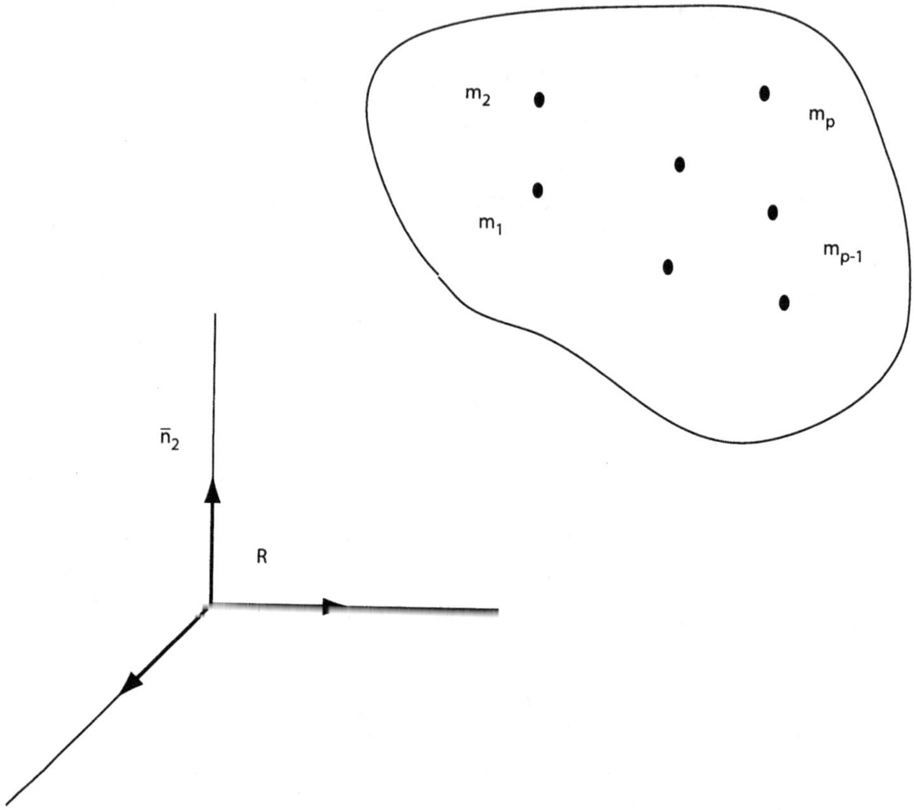

Fig. 6.7. A discretization of a rigid body into p element masses

$$\text{part 2} = \frac{1}{2}(\dot{\omega}^k_{pm}\dot{x}_p I_{mn} + \omega^k_{pm}\ddot{x}_p I_{mn})(\omega^k_{\ell n}\dot{x}_\ell + \omega^k_{\ell n}\dot{x}_\ell)$$

$$= \frac{1}{2} I_{mn}(\dot{\omega}^k_{pm}\omega^k_{\ell n}\dot{x}_p\dot{x}_\ell + \dot{\omega}^k_{pm}\omega^k_{\ell n}\dot{x}_p\ddot{x}_\ell + \omega^k_{pm}\omega^k_{\ell n}\ddot{x}_p\dot{x}_\ell + \omega^k_{pm}\omega^k_{\ell n}\ddot{x}_p\ddot{x}_\ell) \tag{6.111}$$

$$\text{part 3} = (\dot{\omega}^k_{pm}\dot{x}_p + \omega^k_{pm}\ddot{x}_p)\bar{n}_m \cdot \omega^k_{st}\dot{x}_s\bar{n}_t \times (I_{ir}\omega^k_{\ell r}\dot{x}_\ell\bar{n}_i)$$

$$= (\dot{\omega}^k_{pm}\dot{x}_p + \omega^k_{pm}\ddot{x}_p)\bar{n}_m \cdot \omega^k_{st} I_{ir}\omega^k_{\ell r}\dot{x}_\ell e_{tiu}\bar{n}_u)$$

$$= (\dot{\omega}^k_{pm}\dot{x}_p + \omega^k_{pm}\ddot{x}_p)\omega^k_{st}\omega^k_{\ell r}\dot{x}_\ell\dot{x}_s I_{ir} e_{tiu}\delta_{mu} \tag{6.112}$$

$$= (\dot{\omega}^k_{pm}\dot{x}_p + \omega^k_{pm}\ddot{x}_p) I_{ir}\omega^k_{st}\omega^k_{\ell r}\dot{x}_\ell\dot{x}_s e_{tim}$$

In part 3, substituting i by m, m by h and t by n, we obtain

$$\text{part 3} = (\dot{\omega}^k_{pm}\dot{x}_p + \omega^k_{pm}\ddot{x}_p) I_{mr}\omega^k_{sn}\omega^k_{\ell r}\dot{x}_\ell\dot{x}_s e_{nmh} \tag{6.113}$$

From equation (6.107), we obtain the governing equations of motion:

$$\begin{aligned}\frac{\partial G}{\partial \ddot{x}_\ell} &= m_k\left(V^k_{pm}V^k_{\ell m}\ddot{x}_p + \frac{1}{2}V^k_{\ell m}\dot{V}^k_{pm}\dot{x}_p + \frac{1}{2}\dot{V}^k_{pm}V^k_{\ell m}\dot{x}_p\right)\\ &\quad + I_{mn}\left(\frac{1}{2}\dot{\omega}^k_{pm}\omega^k_{\ell n}\dot{x}_p + \frac{1}{2}\omega^k_{\ell n}\dot{\omega}^k_{pm}\dot{x}_p + \dot{\omega}^k_{pm}\omega^k_{\ell n}\ddot{x}_p\right) + I_{mr}\omega^k_{sn}\omega^k_{pr}\omega^k_{\ell h}\dot{x}_p\dot{x}_s e_{nmh}\\ &= (m_k V^k_{pm}V^k_{\ell m} + I_{mn}\omega^k_{pm}\omega^k_{\ell n})\ddot{x}_p + (m_k \dot{V}^k_{pm}V^k_{\ell m} + I_{mn}\dot{\omega}^k_{pm}\omega^k_{\ell m})\dot{x}_p \qquad (6.114)\\ &\quad + I_{mr}\omega^k_{sn}\omega^k_{pr}\dot{\omega}^k_{\ell h}e_{nmh}\dot{x}_p\dot{x}_s\\ &= a_{\ell p}\ddot{x}_p + b_{\ell p}\dot{x}_p + c_{\ell ps}\dot{x}_p\dot{x}_s\end{aligned}$$

Therefore, the resulting equations of motion could be written in a tensor form as

$$a_{\ell p}\ddot{x}_p + b_{\ell p}\dot{x}_p + c_{\ell ps}\dot{x}_p\dot{x}_s = f_\ell \tag{6.115}$$

where $a_{\ell p}$, $b_{\ell p}$ and $c_{\ell ps}$ are the same as those given by equation (5.47). In a similar fashion, we can derive from Gibbs equations, the matrix form of the governing equations of motion given by equation (6.84).

Example 6.6.1
Consider Example 6.5.1 again, and solve the problem by using Gibbs–Appel equations.

Solution
From Equations (6.96) and (6.97), we obtain

$$\bar{a}_1 = \frac{d\bar{v}_1}{dt} = \begin{bmatrix} -\frac{1}{2}\ell_1 c1\dot{x}_3^2 + \frac{1}{2}\ell_1 s1\ddot{x}_3 \\ -\frac{1}{2}\ell_1 s1\dot{x}_3^2 + \frac{1}{2}\ell_1 c1\ddot{x}_3 \\ 0 \end{bmatrix}\bar{n}_m \tag{6.116}$$

and

$$\bar{a}_2 = \frac{d\bar{v}_2}{dt} = \begin{bmatrix} \left[\ell_1 c1\dot{x}_3 + \frac{1}{2}\ell_2(\dot{x}_3+\dot{x}_6)c(1+2)\right]\dot{x}_3 + [\ell_1 s1 + \ell_2 s(1+2)]\ddot{x}_3 \\ +\frac{1}{2}\ell_2\dot{x}_6(\dot{x}_3+\dot{x}_6)c(1+2) + \frac{1}{2}\ell_2\ddot{x}_6 s(1+2) \\ \ddot{x}_3\left[\ell_1 c1 + \frac{1}{2}\ell_2 c(1+2)\right] + \dot{x}_3\left[-\ell_1\dot{x}_3 s1 - \frac{1}{2}\ell_2(\dot{x}_3+\dot{x}_6)s(1+2)\right] \\ +\frac{1}{2}\ell_2\ddot{x}_6 c(1+2) - \frac{1}{2}\ell_2\dot{x}_6(\dot{x}_3+\dot{x}_6)s(1+2) \\ 0 \end{bmatrix} \tag{6.117}$$

In addition, we have

$$I_1 = \frac{1}{12}m_1\ell_1^2 \tag{6.118}$$

$$I_2 = \frac{1}{12} m_2 \ell_2^2 \tag{6.119}$$

$$\bar{\omega}^1 = \dot{x}_3 \bar{n}_3 \tag{6.120}$$

$$\bar{\omega}^2 = \dot{x}_6 \bar{n}_3 \tag{6.121}$$

$$\bar{\alpha}_1 = \ddot{x}_3 \bar{n}_3 \tag{6.122}$$

$$\bar{\alpha}_2 = \ddot{x}_6 \bar{n}_3 \tag{6.123}$$

$$f_1 = -\frac{1}{12} \ell_1 mgs1 - m_2 g \left[\ell_1^{s1} + \frac{1}{2} \ell_2 s(1+2) \right] \tag{6.124}$$

$$f_2 = -\frac{1}{2} m_2 g \ell_2 s(1+2) \tag{6.125}$$

Note that part 3 of equation (6.108) is equal to zero. Therefore, the Gibbs–Appel function reduces to

$$G = \frac{1}{2} m_1 \bar{a}_1^2 + \frac{1}{2} I_1 \ddot{x}_3^2 + \frac{1}{2} m_2 \bar{a}_2^2 + \frac{1}{2} I_2 \ddot{x}_6^2 \tag{6.126}$$

From equation (6.107), we obtain

$$\frac{\partial G}{\partial x_3} = f_1 \tag{6.127}$$

$$\frac{\partial G}{\partial \ddot{x}_6} = f_2$$

The solution is the same as equation (6.104) and (6.105).

6.7 Hamilton's Equations

As we know, Lagrange's equations consist of n second-order differential equations in the nx's and time. Once the Lagrangian function $L(x,\ \dot{x},\ t)$ is known for a given system, the equations of motion are obtained by a straightforward procedure. The n equations have a form that sometimes permits their complete integration, resulting in analytical solutions for the generalized coordinate x's as functions of time.

Hamilton's principle has not provided us with any new physical theories, but it is a convenient approach to solving problems in dynamics. From a theoretical viewpoint, it is often preferable to describe a system in terms of $2n$ first-order differential equations, instead of n second-order differential equations. As a matter of fact, because of the nonlinearity of the differential equations governing multibody systems, we can only solve those equations numerically; hence they need to be in the form of first-order equations. There are many ways in which this can be accomplished, and all are not equally suited to our needs. For example, we can define n new variables y_i, where $y_i = \dot{x}_i$, and thereby we can rewrite the second-order forms of the differential equations into first-order types. This procedure can be overcome by using the

Hamilton's canonical equations. Hamilton defines a new function H(x, p, t) known as the Hamiltonian function, which can be used to generate a set of first-order equations that are particularly symmetrical in form. These constitute the canonical equations of Hamilton.

Let us consider a holonomic system that can be described by the standard form of Lagrange's equations, that is,

$$\frac{d}{dt}\left[\frac{\partial L}{\partial \dot{x}_i}\right] - \frac{\partial L}{\partial x_i} = 0 \quad (i = 1, \dots, n) \tag{6.128}$$

where the x_i define the generalized coordinates of the holonomic system.

We define the generalized momentum conjugate to x_i to be

$$p_i = \frac{\partial L}{\partial \dot{x}_i} \quad (i = 1, \dots, n) \tag{6.129}$$

or simply, we can write p_i as a function of the kinetic energy function, as is the case when the potential energy does not contain any time derivatives in x_i.

$$p_i = \frac{\partial T}{\partial \dot{x}} \quad (i = 1, \dots, n) \tag{6.130}$$

The Hamiltonian function $H(x, p, t)$ for the system S in R is defined as follows:

$$H = \sum_{i=1}^{n} p_i \dot{x}_i - L(x, \dot{x}, t) \tag{6.131}$$

For the general case, H is an explicit function of the x's, p's and t. Since the right-hand side of equation (6.131) contains $\dot{x}$'s, we must eliminate these quantities by expressing them in terms of the $\dot{x}$'s. The total differential of H is

$$dH = \sum_{i=1}^{n}\left[\frac{\partial H}{\partial x_i}dx_i + \frac{\partial H}{\partial p_i}dp_i\right] - \frac{\partial H}{\partial t}dt \tag{6.132}$$

According to equation (6.131), we can also write equation (6.132) as

$$dH = \sum_{i=1}^{n}\left[\dot{x}_i dp_i + p_i d\dot{x}_i - \frac{\partial L}{\partial x_i}dx_i - \frac{\partial L}{\partial \dot{x}_i}d\dot{x}_i\right] - \frac{\partial L}{\partial t}dt \tag{6.133}$$

Using equations (6.129) and (6.130) to substitute for $\partial L/\partial x_i$ and $\partial L/\partial \dot{x}_i$, the second and fourth terms within parentheses in equation (6.133) cancel and the equation becomes

$$dH = \sum_{i=1}^{n}[\dot{x}_i dp_i + p_i d\dot{x}_i] - \frac{\partial L}{\partial t}dt \tag{6.134}$$

If we identify the coefficients of dx_i, dp_i and dt between equations (6.132) and (6.134), we find that

$$\frac{\partial H}{\partial p_i} = \dot{x}_i \quad (i = 1, \ldots, n) \tag{6.135}$$

$$\frac{\partial H}{\partial x_i} = -\dot{p}_i \quad (i = 1, \ldots, n) \tag{6.136}$$

and

$$\frac{\partial H}{\partial t} = -\frac{\partial L}{\partial t} \tag{6.137}$$

Equations (6.135) and (6.136) are Hamilton's canonical equations. Lagrange first obtained this set of equations in 1809; Poisson derived similar equations in the same year. But they did not recognize the equations as a basic set of equations of motion. Cauchy first realized this point in 1831. Hamilton first derived the equations in 1834 from a fundamental variational principle and made them the basis for a far-reaching theory of dynamics. Thus the designation *Hamilton's equations* is fully deserved.

From equation (6.137), we know that

$$\frac{\partial L}{\partial t} = 0 \tag{6.138}$$

because the Lagrangian describes a closed system (i.e., a system that does not intersect with anything outside the system) and cannot depend explicitly on the time. This implies that

$$H = \text{const.} \tag{6.139}$$

or

$$H = \sum_{i=1}^{n} p_i \dot{x}_i - L = \text{const.} \tag{6.140}$$

Using equation (6.129), the function above becomes

$$H = \sum_{i=1}^{n} \frac{\partial L}{\partial \dot{x}_i} \dot{x}_i - L = \text{const.} \tag{6.141}$$

If the potential energy V does not depend explicitly on the velocities $\dot{x}_i$ $(i = 1, \ldots, n)$, then

$$\frac{\partial L}{\partial \dot{x}_i} = \frac{\partial (T - V)}{\partial \dot{x}_i} = \frac{\partial T}{\partial \dot{x}_i} \tag{6.142}$$

On the other hand, according to Euler's theorem, which states that if $f(y_i)$ is a homogeneous function of y_i which is of degree N, then

$$\sum_{i=1}^{n} y_i \frac{\partial f}{\partial y_i} = Nf \tag{6.143}$$

in our case $y_i = \dot{x}_i$, $f = T$, and $N = 2$; therefore, equation (6.141) yields

$$H = \sum_{i=1}^{n} \frac{\partial L}{\partial \dot{x}_i}\dot{x}_i - L = \sum_{i=1}^{n} \frac{\partial L}{\partial \dot{x}_i}\dot{x}_i - (T - V) = 2T - T + V = T + V = \text{const.} \tag{6.144}$$

This simply means that the Hamilton function represents the total energy of the system.

Example 6.7.1
Using Hamilton's equation, solve for the equations of motion of the problem for Figure 6.6.

Solution
From equations (6.99) to (6.101), we obtain the Lagrange function

$$\begin{aligned} L &= T - V \\ &= \frac{1}{6}m_1\ell_1^2\dot{x}_3^2 + \frac{1}{24}m_1\ell_1^2(\dot{x}_3 + \dot{x}_6)^2 + \frac{1}{2}m_2\left[\left(\ell_1^2 + \frac{1}{4}\ell_2^2\right)\dot{x}_3^2\right. \\ &\quad \left. + \frac{1}{2}\ell_2^2\dot{x}_6\left(\dot{x}_3 + \frac{1}{2}\dot{x}_6\right) + \ell_1\ell_2\dot{x}_3(\dot{x}_3 + \dot{x}_6)\cos x_6\right] - \frac{1}{2}m_1 g\ell_1\cos x_3 \\ &\quad - m_2 g\left[\ell_1\cos x_3 + \frac{1}{2}\ell_2\cos(x_3 + x_6)\right] \end{aligned} \tag{6.145}$$

The linear momentum is then found to be

$$\begin{aligned} p_3 &= \frac{\partial L}{\partial \dot{x}_3} = \frac{1}{3}m_1\ell_1^2\dot{x}_3 + \frac{1}{12}m_2\ell_2^2(\dot{x}_3 + \dot{x}_6) + m_2\left(\ell_1^2 + \frac{1}{4}\ell_2^2\right)\dot{x}_3 + \frac{1}{4}m_2\ell_2^2\dot{x}_6 \\ &\quad + \frac{1}{2}m_2\ell_1\ell_2(2\dot{x}_3 + \dot{x}_6)\cos x_6 \end{aligned} \tag{6.146}$$

$$p_6 = \frac{\partial L}{\partial \dot{x}_6} = \frac{1}{12}m_2\ell_2^2(\dot{x}_3 + \dot{x}_6) + \frac{1}{4}m_2\ell_2^2(\dot{x}_3 + \dot{x}_6) + \frac{1}{2}m_2\ell_1\ell_2\dot{x}_3\cos x_6 \tag{6.147}$$

The Hamiltonian function is then obtained:

$$H = T + V \tag{6.148}$$

and from equation $\dot{p}_3 = \partial H/\partial x_i$ we get

$$\dot{p}_3 = \frac{-\partial H}{\partial x_3} = -\frac{1}{2}m_1 g\ell_1\sin x_3 - m_2 g\left[\ell_1\sin x_3 + \frac{1}{2}\ell_2\sin(x_3 + x_6)\right] \tag{6.149}$$

$$\dot{p}_3 = \frac{-\partial H}{\partial x_6} = -\frac{1}{2}m_2 g\ell_1\ell_2\dot{x}_3(\dot{x}_3 + \dot{x}_6)\sin x_6 - \frac{1}{2}m_2 g\ell_2\sin(x_3 + x_6) \tag{6.150}$$

From equations (6.146) to (6.150), we obtain the desired equations of motion:

$$\begin{aligned}
&\frac{1}{3}m_1\ell_1^2\ddot{x}_3 + \frac{1}{12}m_2\ell_2^2(\ddot{x}_3+\ddot{x}_6) + m_2\left(\ell_1^2+\frac{1}{4}\ell_2^2\right)\ddot{x}_3 + \frac{1}{4}m_2\ell_2^2\ddot{x}_6 \\
&\quad + \frac{1}{2}m_2\ell_1\ell_2(2\ddot{x}_3+\ddot{x}_6)\cos x_6 - \frac{1}{2}m_2\ell_1\ell_2\dot{x}_6(2\dot{x}_3+\dot{x}_6)\sin x_6 \\
&\quad = \frac{1}{2}m_1 g\ell_1\sin x_3 + m_2 g\left[\ell_1\sin x_3 + \frac{1}{2}\ell_2\sin(x_3+x_6)\right]
\end{aligned} \tag{6.151}$$

$$\begin{aligned}
&\frac{1}{12}m_2\ell_2^2(\ddot{x}_3+\ddot{x}_6) + \frac{1}{4}m_2\ell_2^2(\ddot{x}_3+\ddot{x}_6) + \frac{1}{2}m_2\ell_1\ell_2\ddot{x}_3\cos x_6 \\
&\quad + \frac{1}{2}m_2\ell_1\ell_2\dot{x}_3\sin x_6 = \frac{1}{2}m_2 g\ell_2\sin(x_3+x_6)
\end{aligned} \tag{6.152}$$

They are the same as those equations given by (6.104) and (6.105). Note that equations (6.146), (6.147), (6.149) and (6.150) constitute a set of four first-order differential equations that can be integrated separately to yield the time history of the system. Of course, this is equivalent to seeking solution to the two second-order differential equations given by (6.151) and (6.152).

6.8 Problems

6.1. Consider the mechanism given in Example 2.13.2, which consists of a shaft rotating at an angular velocity $\omega = \omega(t)$ and a rod attached to the shaft base with a rotation along the $\bar{n}_2^2$ unit vector as shown. Develop the explicit governing equations of motion using both Lagrange and Gibbs–Appel equations. (*Note:* The kinematics of the problem has been worked out in Chapter 2.)

6.2. A drum fixed to a rotating shaft has an angular velocity ω_2 about the unit vector along the master shaft that is given by

$$\omega_2 = A\sin\omega t$$

where A is constant and ω denotes the angular speed of the shaft set at 50 rpm. Give the Hamilton canonical equations for the system. (See Figure 2.22.)

6.3. The robotic system given in Problem 4.3 is to be studied for possible use in General Motors' welding operations at the assembly. To simulate its motion, we need to derive the governing equations of motion. Find the first-order differential equations that govern its motion using Lagrange equations.

6.4. Making use of the matrix representation of the kinematics as shown in Chapters 3, 4 and 5, derive the canonical equations of motion of Hamilton for a general chain of N interconnected bodies, if each body is allowed three possible degrees of freedom for rotation.

6.5. In Example 6.3.1 a plate is made to rotate with respect to shaft AB with a constant angular speed Ω. Show that the Hamiltonian function for such a system is constant.

6.6. A mass m_0 and length l is suspended from a cart of mass M as shown in Figure 6.8. The cart is constrained by a set of forces defined by a linear spring of stiffness k and a linear dash pot with a coefficient c. Furthermore, the pendulum rotation is subject to a torsional spring used to control the pendulum motion. Harmonic force acts on m_0 and is given by $F_0 \sin \omega\theta$. Use the principle of work potential energy to define the generalized forces of the system.

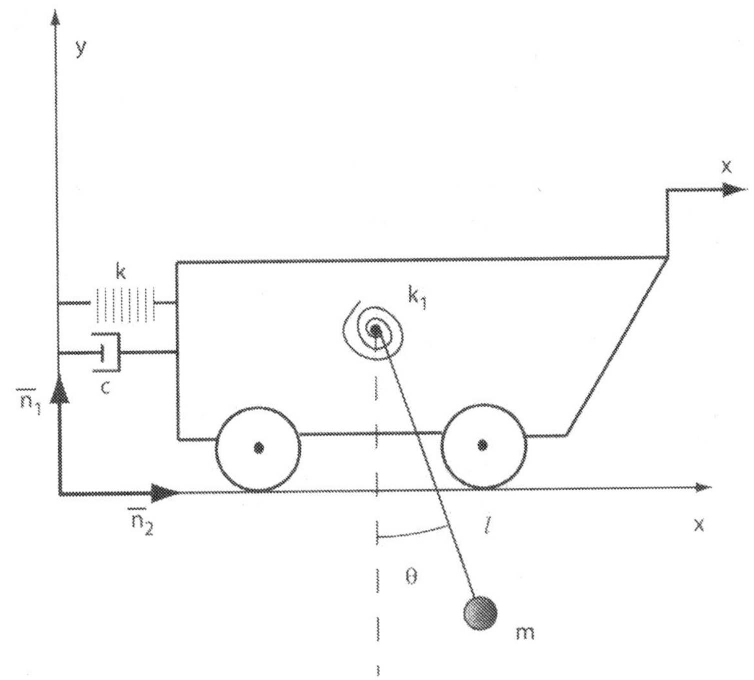

Fig. 6.8. Cart–pendulum system

6.7. Making use of the matrix representation of the kinematics as shown in Chapters 3, 4 and 5, derive the canonical equations of motion of Hamilton for a general chain of N interconnected bodies, if each body is allowed three possible degrees of freedom for rotation.

References

1. Hamilton, W. R., Second Essay on a General Method of Dynamics, *Philos. Trans. R. Soc. London,* 1835, pp. 95–144.
2. Gibbs, J. W., On the Fundamental Formulae of Dynamics, *Am. J. Math.,* Vol. **II**, 1879, pp. 49–64.
3. Appel, P., Sur une forme générale des équations de dynamique, *Z. Reine Angew. Math.,* Vol. **121**, 1900, pp. 310–319.

4. Witaker, E. T., *Analytical Dynamics,* 4th ed., Cambridge University Press, London, 1937.
5. Rund, H., *The Hamilton–Jacobi Theory in the Calculus of Variations,* D. Van Nostrand, Princeton, NJ, 1966.
6. Goldstein, H., *Classical Mechanics,* Addison-Wesley, Reading, MA, 1950.
7. Lanczos, C., *The Variational Principle of Mechanics,* University of Toronto Press, Toronto, 1949.
8. Meirovitch, L., *Methods of Analytical Dynamics,* McGraw-Hill, New York, 1970.
9. Cajori, F., *Sir Isaac Newton's Mathematical Principles of Natural Philosophy and His System of the World,* University of California Press, Berkeley, CA, 1934.
10. Kane, T. R. and Wang, C. F., On the Derivation of Equations of Motion, *J. Soc. Ind. Appl. Math.,* Vol. **13**, 1966.
11. Huston, R. L., Passerello, C. E. and Harlow, M. W., Dynamics of Multi-Rigid-Body Systems, *Appl. Mech.,* Vol. **45**, Dec. 1978.
12. Gupta, V. K., Dynamic Analysis of Multi-Rigid-Body Systems, *J. Eng. Ind.,* Vol. **9**, 1979, pp. 886–892.
13. Kane, T. R., Likins, P. W. and Levinson, D. A., *Spacecraft Dynamics,* McGraw-Hill, New York, 1983.
14. Huston, R. L., *Multibody Dynamics,* Butterworth-Heinemann, Stoneham, MA, 1990.
15. Kane, T. R. and Levinson, D. A. *Dynamics: Theory and Applications*, McGraw-Hill, New York, 1986.
16. Greenwood, D. T., *Principles of Dynamics*, Prentice Hall, Englewood Cliffs, NJ, 1988.

7

Handling of Constraints in Multibody Systems Dynamics

7.1 Introduction

In the preceding chapters, treatment of the generalization of kinematics and the equations of motion was done in the presence of no constraints. In general, we assumed that the system is not subjected to any external condition or forces restricting its global motion, in which case the generalized coordinates are looked upon as the independent coordinates of the system. If a multibody system becomes subjected to a kinematic or geometric condition during the cycle of its motion, the multibody system is said to be *constrained.* If the dynamical system has n independent generalized coordinates and m constraint equations, the total number of degrees of freedom is given by $n - m$. In this chapter, we study the dynamics of multibody systems in the presence of constraints and explore various techniques used to develop the proper governing equations of motion. Dynamics of multibody system requires some intuition where one needs to be able to differentiate between an open treelike system and identify the constraints. Conditions that must be imposed on rigid bodies are usually treated through the generalized forces and closed loops, prescribed motions and any geometrical or kinematical constraints are treated through the use of additional constraint equations to be solved together with the equations of motion.

7.2 Holonomic Constraints

Let us assume that a multibody system is specified by n generalized coordinates, and furthermore, that there are m independent equations of constraint of the form

$$\phi_i(x_1, x_2, \dots, x_n, t) = 0 \quad (i = 1, 2, \dots, m) \tag{7.1}$$

The constraints that can be expressed in this form are said to be holonomic.

Example 7.2.1
Consider the motion of a point mass attached by a rod to a fixed point O in R as shown in Figure 7.1. If the rod is subjected to an angular motion in the plane P, the

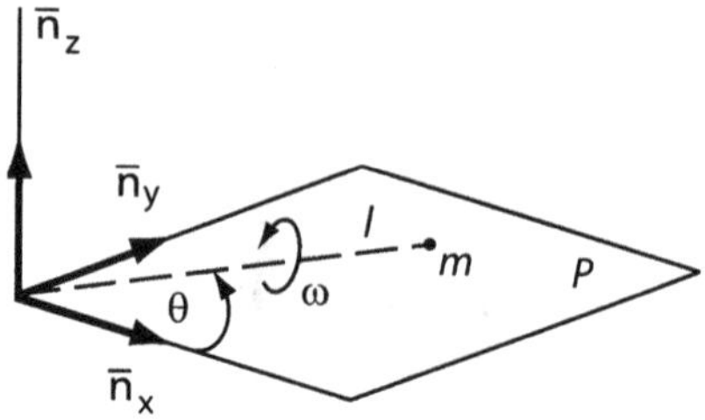

Fig. 7.1. Motion of a point mass attached by a rod

holonomic constraint is the condition that the mass m remains at a constant distance from O, or

$$x_1^2 + y_1^2 = \ell^2 \tag{7.2}$$

which can be expressed in the form of equation (7.1) as

$$x_1^2 + y_1^2 - \ell^2 = 0 \tag{7.3}$$

Note that if the particle is to move with respect to R, it has two degrees of freedom, x_1 and y_1, imposing on it one constraint equation to maintain its distance to R constant; then its degrees of freedom are reduced from two to one. This is expected since only one degree of freedom is needed to describe the point mass, and θ can be used as the independent coordinate. The holonomic constraints are further classified into two types: scleronomic and rheonomic. Constraints of the type of equation (7.1) that have time to appear explicitly in the equation are known as *rheonomic;* otherwise, they are *scleronomic.*

Equation (7.1) is often referred to as a representation of the constraint at the position level. Since it is an algebraic equation and must be integrated with the equations of motion, which are nonlinear differential equations, special numerical schemes must be adapted. This can make analysis of multibody systems a little bit complicated. As will be shown in subsequent sections of this chapter, a clever way of representing the constraints is to use the differentiated form of equation (7.1). This is usually referred to as constraints at the velocity level, and further differentiation yields the constraint representation at the acceleration level. Those forms are most suitable to the solution of the dynamical equations, since only differential equations are used to represent them.

The constraint equations usually represent a set of geometric or kinematic conditions. Geometric constraints are basically derived from a position type of condition, whereas kinematic constraints are defined by imposing a kinematic condition on the system, such as that the mass center velocity or angular velocity must be known.

7.3 Nonholonomic Constraints

Nonholonomic constraints are special constraints that are not integrable. Their differentiated form is usually handled numerically as is the case in this chapter. Let us define what nonholonomic constraints are. Differentiation of equation (7.1) with respect to time gives

$$\frac{\partial \phi_i}{\partial x_1}\dot{x}_1 + \frac{\partial \phi_i}{\partial x_2}\dot{x}_2 + \cdots + \frac{\partial \phi_i}{\partial t} = 0 \tag{7.4}$$

which must be satisfied at any point during the course of motion. The displacements $dx_1, dx_2, \ldots, dx_n$ must satisfy

$$\frac{\partial \phi_i}{\partial x_1}dx_1 + \frac{\partial \phi_i}{\partial x_2}dx_2 + \cdots + \frac{\partial \phi_i}{\partial t}dt = 0 \quad (i = 1, \ldots, m) \tag{7.5}$$

Equation (7.5) is said to be in *pfaffian form*. For a holonomic system, equation (7.5) is said to be integrable and yields a function ϕ_i of the form of equation (7.1). On the other hand, if equation (7.5) is not integrable, it is said to represent a nonholonomic constraint. In compact form, equation (7.5) can be written as

$$\sum_{i=1}^{n} B_{ji} dx_i + \phi_{jt} dt = 0 \tag{7.6}$$

where B_{ji}'s are function of x's and t, yet the time t may appear explicitly in the array. The matrix B_{ji} is of order $m \times n$ and has a rank of m. It is usually referred to as the *constraint Jacobian matrix*. All the m rows of B are assumed to be linearly independent. ϕ_{jt} defines the partial derivative of ϕ_j with respect to time.

Nonholonomic constraints are, then, differential equations of the form of equation (7.6) that are not integrable.

When a constraint condition is imposed on the system, there must be, besides the given forms, an additional force acting on the system to confine its motion to the constraint surface. This force, called the *constraint force*, is assumed to be normal to the surface at any point, which implies a perfectly smooth surface.

7.4 Constrained Multibody Systems

Consider the multibody system shown in Figure7.2. Let n denote the total number of generalized coordinates associated with x_ℓ ($\ell = 1, \ldots, n$). Let m represent the number of constraint equations resulting from the closed loop and the prescribed motion to body k. Let those constraints be holonomic of the form of an exact differential

$$d\phi_i = 0 \quad (i = 1, \ldots, m) \tag{7.7}$$

ϕ_i is function of x_1, $x_2, \ldots, x_n$ and time t. As a result, a set of n constraint forces are produced. The constraint forces, together with the other system forces, determine the actual motion of the multibody system. The motion must also satisfy equation (7.6) at all times. The constraint forces are such that they do not perform any work in any virtual displacement. The components of the constraint forces are proportional to the coefficient of the Jacobian constraint matrix B.

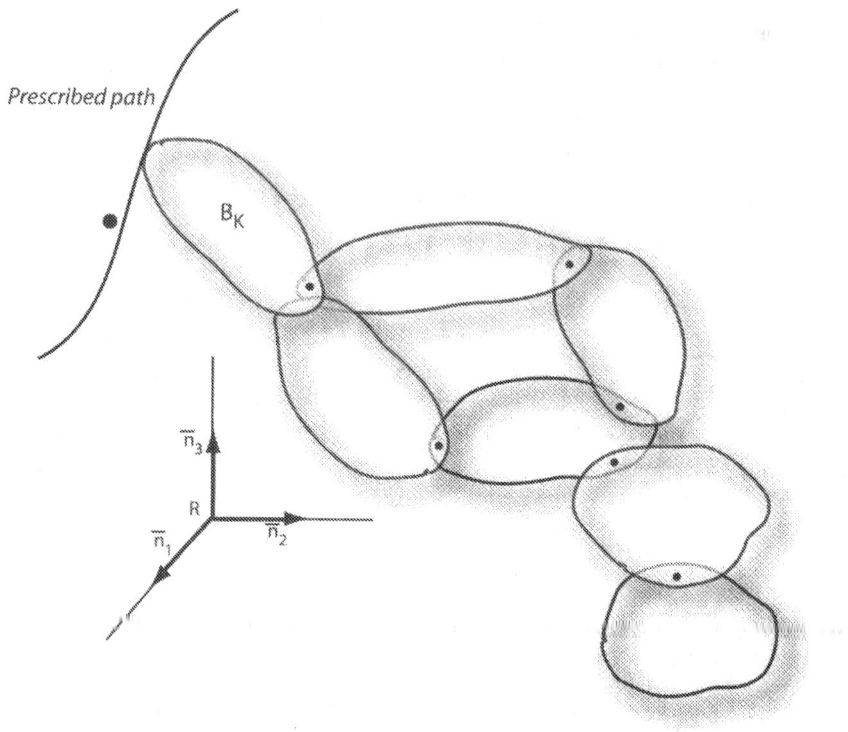

Fig. 7.2. General multibody system with a closed loop and a prescribed motion to body B_k

Those forces are expressed as

$$f_\ell^c = \sum_{i=1}^{m} \lambda_i \frac{\partial \phi_i}{\partial x_\ell} \tag{7.8}$$

or

$$f_\ell^c = \lambda_i B_{\ell i}^T \tag{7.9}$$

where

$$B_{\ell i}^T = \frac{\partial \phi_i}{\partial x_\ell} \tag{7.10}$$

and where $\lambda_i (i = 1, \ldots, m)$ are known as Lagrange undetermined multipliers. There is a sum on i from 1 to m in equation (7.9). Kane's equation can then be extended to include the generalized constraint forces as

$$f_\ell + f_\ell^* = f_\ell^c \tag{7.11}$$

or

$$f_\ell + f_\ell^* - \lambda_i B_{\ell i}^T = 0 \tag{7.12}$$

There are $n - m$ degrees of freedom; however, in the case of a system subject to nonholonomic constraints, we might be forced to work with a number of generalized coordinates exceeding the degrees of freedom by the number of constraints. For a multibody system, the computational burden of keeping the generalized constraint forces in the equation of motion makes the analysis complicated, since the undetermined multipliers have to be computed at each step of the integration. It is desirable, however, to find a set of independent coordinates to express the equations of motion or simply to eliminate the undetermined multipliers. The latter is achieved by premultiplying equation (7.12) by a matrix orthogonal to B. Numerical algorithms to handle such problems will be presented in subsequent sections.

7.5 The Augmented Method

The augmented method utilizes the equations of motion, including the undetermined multipliers, together with the differentiated form of the constraint equations. The two are combined to form a set of $n + m$ second-order differential equations used to describe the dynamics of the system. An illustration of the above is as follows: Let the equations of motion of a general multibody system (open chain) in R be given by

$$[a_{\ell p}]\dot{y}_p + [b_{\ell p}]y_p + [c_{\ell ps}]y_p = f_\ell \tag{7.13}$$

where $a_{\ell p}$, $b_{\ell p}$ and $c_{\ell ps}$ are defined by equations (5.3) to (5.5), and f_ℓ denotes the generalized active forces. Let this system be subject to m holonomic constraints of the form

$$\phi_i(x_1, x_2, \ldots, x_n, t) = 0 \quad (i = 1, \ldots, m) \tag{7.14}$$

We also define the generalized speeds to be of either two forms:

$$y = \dot{x} \quad \text{or} \quad y = W\dot{x} \tag{7.15}$$

where W is a matrix relating the generalized speeds y to the generalized coordinates derivatives $\dot{x}$, and is a function of the generalized coordinates x and probably t. Differentiation of equation (7.14) gives an expression, terms of the generalized speeds as

$$B_{i\ell}y_\ell = g_i \quad \text{or} \quad [B]\{y\} = \{g\} \quad (\ell = 1, \ldots, \; n; i = 1, \ldots, m) \tag{7.16}$$

where $B_{i\ell}$ is the constraint Jacobi matrix and g_i the time-dependent function. We also know that if the multibody system is constrained, there must be additional constraint forces to be added to equation (7.13). In matrix notation the equation could be expressed as

$$[a]\{\dot{y}\} + [b]\{y\} + [c]\{y\} = \{f\} - [B]^T\{\lambda\} \tag{7.17}$$

where λ_i $(i = 1, \ldots, m)$ are the undetermined multipliers. $[a]$, $[b]$ and $[c]$ are matrices given by equation (5.47). To form a set of $n + m$ differential equations, we differentiate equation (7.16) once more, which gives a constraint at the acceleration level in a matrix form as

$$[B]\{\dot{y}\} = -[\dot{B}]\{y\} + \{\dot{g}\} \tag{7.18}$$

Combining equations (7.17) and (7.18), we obtain the governing equations of motion:

$$\begin{bmatrix} a_{\ell p} & B_{p\ell} \\ B_{\ell p} & 0 \end{bmatrix} \begin{bmatrix} \dot{y}_p \\ \lambda_p \end{bmatrix} = \begin{bmatrix} H \\ G \end{bmatrix} + \begin{bmatrix} f_\ell \\ \dot{g}_i \end{bmatrix} \tag{7.19}$$

or

$$\begin{bmatrix} a & B^T \\ B & 0 \end{bmatrix} \begin{Bmatrix} \dot{y} \\ \lambda \end{Bmatrix} = \begin{Bmatrix} H \\ G \end{Bmatrix} + \begin{Bmatrix} f \\ \dot{g} \end{Bmatrix} \tag{7.20}$$

where

$$H = -\{[b] + [c]\}\{y\} \tag{7.21}$$

and

$$G = -[\dot{B}]\{y\} \tag{7.22}$$

Equation (7.19) can easily be integrated through the standard integration routines to yield the time history of the multibody system.

7.6 Coordinate Reduction

We learned that if a multibody system has n independent generalized coordinates and it becomes subjected to m constraint equations (holonomic or nonholonomic), the degrees of freedom reduces to $n - m$. We also know in the study of multibody dynamics that the governing equation of motion is often complex, and it is a very difficult task to identify the independent coordinates of the system. This applies to both holonomic and certainly to nonholonomic equations, since they are not integrable. Subsequently, one can adapt the augmented method procedure outlined in the preceding section and solve the differential equations augmented with m unknown undetermined multipliers. It is apparent that if the undetermined multipliers are not needed in the analysis, their computation is useless. The question we now ask is: Can we eliminate those undetermined multipliers from the equations of motion? Coordinate reduction techniques have long been used to answer this question. In coordinate reduction the objective is to determine a matrix T such that

$$[T][B]^T = 0 \tag{7.23}$$

where T is a $(n - m) \times n$ orthogonal complement of B, which is a $m \times n$ matrix and determines the null space of B.

Premultiplying equation (7.17) by T gives

$$[T]([a]\{\dot{y}\} + [b]\{y\} + [c]\{y\}) - [T]\{f\} = 0 \tag{7.24}$$

where the generalized constraint forces are eliminated as a result of equation (7.23). The matrix T is rectangular of order $m \times n$ with a full rank m. From equation (7.24) we make an important observation: The undetermined multipliers are eliminated and the equations of motion are ultimately reduced to $n - m$, which is the minimum dimension of the equations. Combining equations (7.24) and (7.18), we obtain the governing equations of motion in the form

$$A_{\ell p} \dot{y}_p = D_\ell \tag{7.25}$$

or

$$[A]\{\dot{y}\} = \{D\} \tag{7.26}$$

where A and D are two block matrices given by

$$A_{\ell p} = \left[\frac{T_{\ell i} a_{ip}}{B_{\ell p}} \right] \tag{7.27}$$

or

$$[A] = \left[\frac{[T][a]}{[B]} \right]_{n \times m} \tag{7.28}$$

and

$$D_\ell = \left[\frac{T_{\ell i}(f_i - b_{ip} y_p - c_{ips} y_p y_s)}{-\dot{B}_{\ell i} y_i + \dot{g}_\ell} \right] \tag{7.29}$$

or

$$\{D\} = \left[\frac{[T](\{f\} - [b]\{y\} - [c]\{y\})}{-[\dot{B}]\{y\} + \{\dot{g}\}} \right] \tag{7.30}$$

It can be shown that A is positive definite and that equation (7.26) can be expressed as

$$\dot{y}_p = A_{p\ell}^{-1} D_\ell \tag{7.31}$$

or

$$\{\dot{y}\} = [A]^{-1}\{D\} \tag{7.32}$$

where equation (7.32) represents a set of n second-order differential equations. Using the proper initial conditions, the time history of the dynamical system can be obtained. The solution of constrained multibody systems dynamics require the computation of T, the orthogonal complement to B. What follows are different techniques used to generate T.

7.7 Pseudo Upper Triangular Decomposition Method

The pseudo upper triangular decomposition (PUTD) method developed by Amirouche et al. [13,16] is used to generate the orthogonal complement array T of B. The method's objective is first to upper-triangularize the matrix B, then identify a second matrix that makes its product with the upper-triangularized B go to zero. The PUTD method utilizes the Householder transformation for the upper triangularization and identifies, through the Gramm–Schmidt process, the corresponding matrix that makes it vanish, thus determining the *null space* of the Jacobian matrix B. To illustrate this procedure, we are going to consider holonomic constraints of the form

$$\phi_i(x_1, x_2, \ldots, x_n) = 0 \tag{7.33}$$

which, once differentiated, yields a set of differential equations expressed as

$$By = 0 \tag{7.34}$$

where B is a matrix with any number of columns and any number of rows. Let n be the number of columns and m the number of rows, and y is equal to $\dot{x}$. First let us introduce the well-known Householder transformation, which states that for any two given vectors $\bar{X}$ and $\bar{Y} \in R^n$, where R^n denotes an n-dimensional real space, we have a unit vector $\bar{U}$ such that

$$H\bar{X} = \sigma\bar{Y} \tag{7.35}$$

where

$$H = I - 2\bar{U}\bar{U}^T \tag{7.36}$$

and

$$\sigma = \sqrt{\frac{\bar{X}^T\bar{X}}{\bar{Y}^T\bar{Y}}} \tag{7.37}$$

where $\bar{X}^T$ and $\bar{Y}^T$ denote the transpose of $\bar{X}$ and $\bar{Y}$, respectively, and I is the identity matrix. The Householder transformation also states that:

1. If $\bar{X} \neq \bar{O}$ and $\sigma\,\bar{Y} \neq \bar{X}$, then $\bar{U}$ is uniquely defined as

$$\bar{U} = \frac{\sigma\,\bar{Y} - \bar{X}}{\sqrt{(\sigma\,\bar{Y} - \bar{X})^T(\sigma\,\bar{Y} - \bar{X})}} \tag{7.38}$$

or

$$\bar{U} = \frac{\sigma\,\bar{Y} - \bar{X}}{\left\|\sigma\,\bar{Y} - \bar{X}\right\|}$$

2. If $\bar{X} \neq 0$ and $\sigma\,\bar{Y} = \bar{X}$, then

$$\bar{U} = \bar{O} \tag{7.39}$$

3. If $\bar{X} = \bar{O}$, U can be chosen by any vector that is not necessarily unity. For the case when the vectors are linearly independent, such as the case for the vectors given by the Jacobian matrix B, we can write

$$H\bar{X} = \bar{W} \tag{7.40}$$

where $\bar{W}$ is given by the right-hand side of equation (7.35). Equation (7.38) can then be rewritten such as

$$\bar{U} = \frac{\sigma\,\bar{Y} - \bar{X}}{\sqrt{(\bar{W} - \bar{X})^T(\bar{W} - \bar{X})}} \tag{7.41}$$

It can also be shown that

$$H = H^T \tag{7.42}$$

and

$$HH^T = I \tag{7.43}$$

Therefore

$$H^{-1} = H^T = H \tag{7.44}$$

which implies that the Householder transformation matrix is an orthogonal matrix that gives a full rank transformation.

The constraint Jacobian $B_{i\ell}(i = 1, 2, \ldots, m)$ has m linearly independent column vectors. Our objective is to use the Householder transformation technique to formulate an upper triangular matrix that consists of m linearly independent vectors, and then find $n - m$ orthonormal vectors in the n-dimensional space which are orthogonal to these vectors. Consider the first column vector of B, where

$$B_{\ell i}^{(1)} = \left[\bar{B}_1^{(1)}, \bar{B}_2^{(1)}, \ldots, \bar{B}_m^{(1)}\right]_{n\times m} \tag{7.45}$$

Let

$$\bar{X}^{(1)} = \bar{B}_1^{(1)} \tag{7.46}$$

and

$$\bar{Y}^{(1)} = \left[\left\|\bar{B}_1^{(1)}\right\|, 0, \ldots, 0\right]_{n\times 1}^T \tag{7.47}$$

where $\left\|B_1^{(1)}\right\|$ denotes the norm $\bar{B}_1^{(1)}$. Then

$$\bar{W}^{(1)} = \bar{Y}^{(1)} \tag{7.48}$$

From equation (7.40), we can write

$$H^{(1)}\bar{B}_1^{(1)} = \bar{W}^{(1)} \tag{7.49}$$

Using equations (7.36) and (7.41), we can easily obtain $H^{(1)}$. The Householder transformation matrix is then used to define a new set of linearly independent vectors as

$$B_{\ell i}^{(2)} = H_{\ell j}^{(1)} B_{ji}^{(1)} = \left\{ \bar{W}^{(1)}, \bar{B}_2^{(2)}, \bar{B}_3^{(2)}, \ldots, \bar{B}_m^{(2)} \right\} \tag{7.50}$$

Similarly, consider $\bar{B}_2^{(2)}$:

$$\bar{X}^{(2)} = \bar{B}_2^{(2)} \tag{7.51}$$

with

$$\bar{W}^{(2)} = \begin{bmatrix} B_{12}^{(2)} \\ \left[\left\| \bar{B}_2^{(2)} \right\|^2 - (B_{12}^{(2)})^2 \right]^{1/2} \\ 0 \\ 0 \\ \vdots \\ 0 \end{bmatrix}_{n \times 1} \tag{7.52}$$

where $\left\| \bar{B}_2^{(2)} \right\| \neq \left(B_{12}^{(2)} \right)^2$; otherwise, $\bar{W}^{(1)}$ and $\bar{B}_2^{(2)}$ become dependent. Proceeding with the foregoing procedure up to $q-1$ steps, with $2 \leq q \leq m$ we obtain

$$H_{\ell j}^{q-1} B_{ji}^{q-1} = B_{\ell i}^{(q)} = \left[\bar{W}^{(1)}, \bar{W}^{(2)}, \ldots, \bar{W}^{(q-1)}, \bar{B}_q^{(q)}, \bar{B}_m^{(q)} \right]_{n \times m} \tag{7.53}$$

Again let

$$\bar{X}^{(q)} = \bar{B}^{(q)} \tag{7.54}$$

and

$$\bar{W}^{(q)} = \begin{bmatrix} B_{1_q}^{(q)} \\ B_{2_q}^{(q)} \\ \vdots \\ B_{q-1,q}^{(q)} \\ \left\| \bar{B}_1^{(q)} \right\|^2 - \sum_{r=1}^{q-1} \left(B_{rq}^{(q)} \right)^2 \\ \vdots \\ 0 \end{bmatrix}_{n \times 1} \tag{7.55}$$

Solving for $H^{(q)}$, then $B_{\ell i}^{(q+1)}$ can be obtained and the procedure continues up to $q = m$. At the final step, the following result is obtained:

$$H_{\ell j}^{m} B_{ji}^{(m)} = \left[\bar{W}^{(1)},\ \bar{W}^{(2)}, \ldots, \bar{W}^{(m)}\right]_{n \times m} \tag{7.56}$$

It can easily be seen that

$$\begin{aligned} H_{\ell j}^{m} B_{ji}^{(m)} &= H_{\ell i}^{(m)} H_{js}^{(m-1)} \cdots H_{tp}^{(1)} B_{pi} \\ &= B_{\ell i}^{(m+1)} \end{aligned} \tag{7.57}$$

where $B_{\ell i}^{(m+1)}$ gives an upper triangular matrix of the form

$$B_{\ell i}^{(m+1)} = \begin{bmatrix} * & * & * & \ldots & * \\ 0 & * & * & \ldots & * \\ 0 & 0 & * & \ldots & * \\ \vdots & \vdots & \vdots & & \vdots \\ 0 & 0 & 0 & \ldots & * \\ 0 & 0 & 0 & & 0 \\ \vdots & \vdots & \vdots & & \vdots \\ 0 & 0 & 0 & & 0 \end{bmatrix}_{n \times m} \tag{7.58}$$

From equation (7.57), we define a transformation matrix H such that

$$\begin{aligned} H &\triangleq H^{(m)} H^{(m-1)} \cdots H^{(1)} \\ &= \prod_{i=1}^{m} \left(I - 2U^{(i)} \cdot U^{(i)^T}\right)_{n \times m} \end{aligned} \tag{7.59}$$

Equation (7.57) shows that the columns of B^{m+1} are linearly independent, and hence an orthogonal basis for the column vectors can be obtained through the Gram–Schmidt process.

Introducing the Gram–Schmidt process, an orthonormal set of vectors $\bar{d}_i \quad (i = 1, 2, \ldots, m)$ can be expressed as

$$\bar{d}_i = \frac{\bar{W}^{(1)}}{\left\| \bar{W}^{(1)} \right\|} \tag{7.60}$$

and

$$\bar{d}_i = \frac{\bar{W}^{(i)} - \sum\limits_{j=1}^{i-1} \left(\bar{d}_j \cdot \bar{W}^{(i)}\right) \bar{d}_j}{\left\| \bar{W}^{(i)} - \sum\limits_{j=1}^{i-1} \left(\bar{d}_j \cdot \bar{W}^{(i)}\right) \bar{d}_j \right\|} \qquad (i = 2, 3, \ldots, m) \tag{7.61}$$

From equations (7.61) and (7.60), we get

$$\begin{aligned} d_1 &= \{1, 0, 0, \ldots, 0, \ldots, 0\}^T_{n\times 1} \\ d_2 &= \{0, 1, 0, \ldots, 0, \ldots, 0\}^T_{n\times 1} \\ d_m &= \{0, 0, 0, \ldots, 1, 0, \ldots, 0\}^T_{n\times 1} \end{aligned} \tag{7.62}$$

Thus, we can define an identity matrix such as

$$\begin{aligned} D_{n\times n} &\triangleq \left[D_{1n\times m} : D_{2n\times(n-m)}\right] = \left[\bar{d}_1, \bar{d}_2, \ldots, \bar{d}_m : D_2\right] \\ &\triangleq I_{n\times n} \end{aligned} \tag{7.63}$$

where

$$D_2 = \left[\bar{d}_{m+1}, \bar{d}_{m+2}, \ldots, \bar{d}_n\right] \tag{7.64}$$

or

$$D_2^T = \begin{bmatrix} 0 \ldots 0\,1\,0 \ldots 0 \\ 0 \ldots 0\,0\,1 \ldots 0 \\ \vdots \quad\quad \vdots\ \vdots \\ 0 \ldots 0\,0\,0 \ldots 1 \end{bmatrix}_{(n\times m)\times n} \tag{7.65}$$

Since $n - m$ columns of D_2 are orthogonal to the m columns of $B_{\ell i}^{(m+1)}$, we have

$$D_2^T H B^T = 0 \tag{7.66}$$

that is,

$$[D_2]^T_{(n-m)\times n}\,[H]_{n\times m}\,[B]^T_{n\times m} = 0 \tag{7.67}$$

or

$$T B^T = 0 \tag{7.68}$$

where $T = D_2^T H$, the orthogonal complement of B, or

$$B T^T = 0 \tag{7.69}$$

A linear combination of the $n - m$ solution vectors is also a solution. Therefore, in general,

$$B H^T D_2 q = 0 \tag{7.70}$$

or

$$BT^T q = 0 \tag{7.71}$$

where q is an arbitrary column vector of order $n - m$. Hence the following theorem has been proved.

Theorem
Consider any set of m independent linear homogeneous equations in n unknowns, where $n > m$:

$$BY = 0 \tag{7.72}$$

There exists an orthogonal matrix $T^T = H^T D_2$ such that $BH^T D_2 = 0$, where H is defined by recursive multiplication of Householder's transformation, such that HB^T is an upper triangular matrix, and D_2 is a matrix whose columns are the last $n - m$ columns of the identity matrix of order n. Also, the general solution of the equations are

$$Y = H^T D_2 q = T^T q \tag{7.73}$$

where q is any arbitrary column vector of order $n - m$.

Example 7.7.1
The main intent is to illustrate an application of the method. This example consists of five point masses connected by linear massless springs, as shown in Figure 7.3. The masses move only in the horizontal position. Let F_ℓ ($\ell = 1, \ldots, 5$) denote the external forces acting on the masses. Those forces act horizontally and are a function of time only. If we let $x_n (n = 1, \ldots, 5)$ be the generalized coordinates and the springs constants be $k_{\ell}n$ ($\ell, n = 1, \ldots, 5$), the equation of motion can be expressed as

$$M_{\ell n}\ddot{x}_n + K_{\ell n}x_n = F_\ell \quad (n = 1, \ldots, 5;\ \ell = 1, \ldots, 5) \tag{7.74}$$
$$M = mI$$

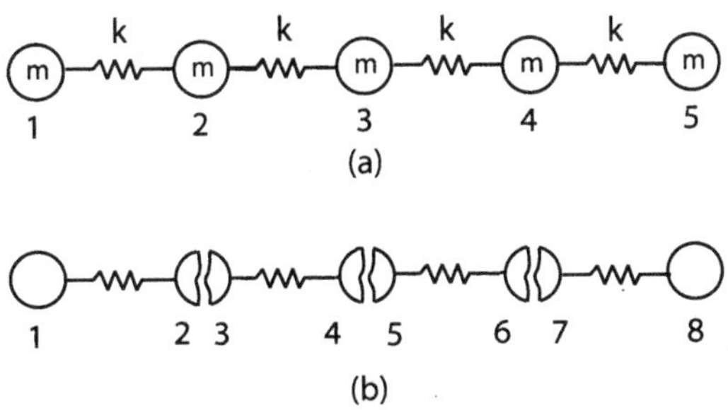

Fig. 7.3. Linear spring-mass system

where I is the identity matrix, $F_\ell = 0$ and

$$K = k\begin{bmatrix} 1 & -1 & 0 & 0 & 0 \\ -1 & 2 & -1 & 0 & 0 \\ 0 & -1 & 2 & -1 & 0 \\ 0 & 0 & -1 & 2 & -1 \\ 0 & 0 & 0 & -1 & 1 \end{bmatrix} \tag{7.75}$$

Equation (7.74) represents the equations of motion of the system of Figure 7.3(a) in terms of the independent coordinates. To illustrate the use of the PUTD method, let the system of Figure 7.3(a) be viewed as Figure 7.3(b), where three inner masses are cut equally. This in turn produces eight masses. The equations of motion become

$$M_{\ell n}\ddot{x}_n + K_{\ell n}x_n = F_\ell \quad (n = 1, \dots, 8;\ \ell = 1, \dots, 8) \tag{7.76}$$

where $F_\ell = 0$ and

$$[M] = \begin{bmatrix} 1&0&0&0&0&0&0&0 \\ 0&\frac{1}{2}&0&0&0&0&0&0 \\ 0&0&\frac{1}{2}&0&0&0&0&0 \\ 0&0&0&\frac{1}{2}&0&0&0&0 \\ 0&0&0&0&\frac{1}{2}&0&0&0 \\ 0&0&0&0&0&\frac{1}{2}&0&0 \\ 0&0&0&0&0&0&\frac{1}{2}&0 \\ 0&0&0&0&0&0&0&1 \end{bmatrix} \tag{7.77}$$

$$[M] = \begin{bmatrix} 1&-1&0&0&0&0&0&0 \\ -1&1&0&0&0&0&0&0 \\ 0&0&1&-1&0&0&0&0 \\ 0&0&-1&1&0&0&0&0 \\ 0&0&0&0&1&-1&0&0 \\ 0&0&0&0&-1&1&0&0 \\ 0&0&0&0&0&0&1&-1 \\ 0&0&0&0&0&0&-1&1 \end{bmatrix} \tag{7.78}$$

Explicit forms of M and K can easily be found by taking a free-body diagram for each mass and applying Newton's second law. In any case, the important problem is to impose a set of constraints on the system and then apply the PUTD method. Let us assume that three equations of constraints are imposed:

$$\begin{aligned} x_2 &= x_3 \\ x_4 &= x_5 \\ x_6 &= x_7 \end{aligned} \tag{7.79}$$

Adding the additional constraint forces resulting from the constraint equations, equation (7.76) becomes

$$[M]\{\ddot{x}\} + [K]\{x\} + [B]^T\{\lambda\} = \{F\} \tag{7.80}$$

Equations (7.80) and (7.77) to (7.79) constitute the governing equations of motion of the system. Applying the PUTD method, we can obtain a matrix T such that

$$[T][M]\{\ddot{x}\} + [T][K]\{x\} = [T]\{F\} \tag{7.81}$$

Equation (7.81) represents the reduced form of the equations of motion to minimum dimension. The governing equations are then given by equations (7.81) and (7.79). To find T, let the constraint equations be represented by their differentiated form at the velocity level as

$$[B]\dot{r} = 0 \quad (n - 1, \ldots, 8); \quad (m - 1, \ldots, 3) \tag{7.82}$$

where the B matrix is as follows:

$$B = \begin{bmatrix} 0 & 1 & -1 & 0 & 0 & 0 & 0 & 0 \\ 0 & 0 & 0 & 1 & -1 & 0 & 0 & 0 \\ 0 & 0 & 0 & 0 & 0 & 1 & -1 & 0 \end{bmatrix} \tag{7.83}$$

The equations of motion for the system shown in Figure 7.3 lead to eight second-order differentiation equations in terms of the coordinates $x_1, x_2, \ldots, x_8$. In solving for the system dynamic response, those equations, along with the constraint equations given by equations (7.77) to (7.79), must be solved together. It is obvious that the number of degrees of freedom for the system should be 5 = (8 – 3). To reduce the equations of motion automatically, the PUTD method presented herein can be employed. Let's find the orthogonal matrix $H^T D_2 = T$ which will eliminate the undetermined multipliers, and hence reduce the equations of motion to a set of five equations.

Consider the first row of the constraint matrix B given by equation (7.83):

$$\bar{B}_1^{(1)} = \{0\ 1\ -1\ 0\ 0\ 0\ 0\ 0\}^T \tag{7.84}$$

Let

$$X^{(1)} = B^T$$

where

$$\bar{X}_1^{(1)} = \bar{B}_1^{(1)} \tag{7.85}$$

and

$$\begin{aligned} \bar{W}^{(1)} &= \left\{ \left\| \bar{X}_1^{(1)} \right\| \; 0\,0\,0\,0\,0\,0\,0 \right\}^T \\ &= \left[\sqrt{2}\; 0\,0\,0\,0\,0\,0\,0 \right]^T \end{aligned} \tag{7.86}$$

From equation (7.41) we get

$$\begin{aligned} \bar{U}^1 &= \frac{\bar{W}^{(1)} - \bar{X}^{(1)}}{\left\| \bar{W}^{(1)} - \bar{X}_1^{(1)} \right\|} \\ &= \frac{1}{2}\left[\sqrt{2}\; -1\; 1\; 0\,0\,0\,0\,0 \right]^T \end{aligned}$$

From equation (7.36), we obtain the Householder transformation matrix:

$$H^{(1)} = I - 2\bar{U}^{(1)}\bar{U}^{(1)^T}$$

$$= I - 2\begin{bmatrix} \frac{\sqrt{2}}{2} \\ -\frac{1}{2} \\ \frac{1}{2} \\ 0 \\ 0 \\ 0 \\ 0 \\ 0 \end{bmatrix} \begin{bmatrix} \frac{\sqrt{2}}{2} & -\frac{1}{2} & \frac{1}{2} & 0\,0\,0\,0\,0 \end{bmatrix}$$

$$= \begin{bmatrix} 0.0000 & 0.7071 & -0.7071 & 0.0000 & 0.0000 & 0.0000 & 0.0000 & 0.0000 \\ 0.7071 & 0.5000 & 0.5000 & 0.0000 & 0.0000 & 0.0000 & 0.0000 & 0.0000 \\ -0.7071 & 0.5000 & 0.5000 & 0.0000 & 0.0000 & 0.0000 & 0.0000 & 0.0000 \\ 0.0000 & 0.0000 & 0.0000 & 1.0000 & 0.0000 & 0.0000 & 0.0000 & 0.0000 \\ 0.0000 & 0.0000 & 0.0000 & 0.0000 & 0.0000 & 0.0000 & 0.0000 & 0.0000 \\ 0.0000 & 0.0000 & 0.0000 & 0.0000 & 0.0000 & 1.0000 & 0.0000 & 0.0000 \\ 0.0000 & 0.0000 & 0.0000 & 0.0000 & 0.0000 & 0.0000 & 1.0000 & 0.0000 \\ 0.0000 & 0.0000 & 0.0000 & 0.0000 & 0.0000 & 0.0000 & 0.0000 & 1.0000 \end{bmatrix} \tag{7.87}$$

The value $H^{(1)}$ is then used to obtain the vector $\bar{X}^{(2)}$ through the householder transformation as

$$\bar{X}^{(2)} = H^{(1)} B^T$$

$$= \begin{bmatrix} 1.414 & 0 & 0 \\ 0 & 0 & 0 \\ 0 & 0 & 0 \\ 0 & 1 & 0 \\ 0 & -1 & 0 \\ 0 & 0 & 1 \\ 0 & 0 & -1 \\ 0 & 0 & 0 \end{bmatrix} \quad (7.88)$$

From equation (7.52), we obtain the vector $\bar{W}^{(2)}$, where we see that

$$B_{21}^{(2)} = 0$$

and

$$\left[\left\| \bar{B}_2^{(2)} \right\|^2 - (B_{21}^{(2)})^2 \right]^{1/2} = (2-0)^{1/2} = \sqrt{2}$$

Therefore,

$$\bar{W}^{(2)} = \left[0 \ \sqrt{2} \ 0\ 0\ 0\ 0\ 0\ 0 \right]^T$$

Then from equation (7.41) we obtain the corresponding unit vector:

$$\bar{U}^2 = \frac{\bar{W}^{(2)} - X_2^{(2)}}{\left\| \bar{W}^{(2)} - \bar{X}_2^{(2)} \right\|}$$

$$= \left[0 \ \frac{\sqrt{2}}{2} \ 0 \ -\frac{1}{2} \ \frac{1}{2} \ 0\ 0\ 0 \right]^T$$

Ultimately, we evaluate the transformation matrix, from the householder definition

$$H^{(2)} = I - 2\bar{U}^{(2)} \bar{U}^{(2)T}$$

$$= \begin{bmatrix} 1.0000 & 0.0000 & 0.0000 & 0.0000 & 0.0000 & 0.0000 & 0.0000 & 0.0000 \\ 0.0000 & 0.0000 & 0.0000 & 0.7071 & -0.7071 & 0.0000 & 0.0000 & 0.0000 \\ 0.0000 & 0.0000 & 1.0000 & 0.0000 & 0.0000 & 0.0000 & 0.0000 & 0.0000 \\ 0.0000 & 0.7071 & 0.0000 & 0.5000 & 0.5000 & 0.0000 & 0.0000 & 0.0000 \\ 0.0000 & -0.7071 & 0.0000 & 0.5000 & 0.5000 & 0.0000 & 0.0000 & 0.0000 \\ 0.0000 & 0.0000 & 0.0000 & 0.0000 & 0.0000 & 1.0000 & 0.0000 & 0.0000 \\ 0.0000 & 0.0000 & 0.0000 & 0.0000 & 0.0000 & 0.0000 & 1.0000 & 0.0000 \\ 0.0000 & 0.0000 & 0.0000 & 0.0000 & 0.0000 & 0.0000 & 0.0000 & 1.0000 \end{bmatrix} \quad (7.89)$$

Systematically, we update our matrix $X^{(2)}$:

$$X^{(3)} = H^{(2)} X^{(2)} = \begin{bmatrix} 1.414 & 0 & 0 \\ 0 & 1.414 & 0 \\ 0 & 0 & 0 \\ 0 & 0 & 0 \\ 0 & 0 & 0 \\ 0 & 0 & 1 \\ 0 & 0 & -1 \\ 0 & 0 & 0 \end{bmatrix} \tag{7.90}$$

where $X^{(3)}$ corresponds to our new B^T. Proceeding in a similar fashion by taking the third vector of B^T, we see that our $\bar{W}^{(3)}$ components are

$$B_{31}^{(3)} = 0, \qquad B_{32}^{(3)} = 0$$

$$\left[\left\| B_3^{(3)} \right\|^2 - \sum_{r=1}^{2} (B_{3r}^{(3)})^2 \right]^{1/2} = (2-0)^{1/2} = 1.414$$

which gives

$$\bar{W}^{(3)} = \left\{ 0.0\ 0.0\ 1.414\ 0.0\ 0.0\ 0.0\ 0.0 \right\}^T$$

Hence

$$\bar{U}^{(3)} = \frac{\bar{W}^{(3)} - X_2^{(3)}}{\left\| \bar{W}^{(3)} - \bar{X}_2^{(3)} \right\|} = \left[0\ 0\ \frac{\sqrt{2}}{2}\ 0\ 0\ -\frac{1}{2}\ \frac{1}{2}\ 0 \right]^T \tag{7.91}$$

This leads to the Householder transformation

$$H^{(3)} = I - 2\bar{U}^{(3)} \bar{U}^{(3)T}$$

$$= \begin{bmatrix} 1.0000 & 0.0000 & 0.0000 & 0.0000 & 0.0000 & 0.0000 & 0.0000 & 0.0000 \\ 0.0000 & 1.0000 & 0.0000 & 0.0000 & 0.0000 & 0.0000 & 0.0000 & 0.0000 \\ 0.0000 & 0.0000 & 0.0000 & 0.0000 & 0.0000 & 0.7071 & -0.7071 & 0.0000 \\ 0.0000 & 0.0000 & 0.0000 & 1.0000 & 0.0000 & 0.0000 & 0.0000 & 0.0000 \\ 0.0000 & 0.0000 & 0.0000 & 0.0000 & 1.0000 & 0.0000 & 0.0000 & 0.0000 \\ 0.0000 & 0.0000 & 0.7071 & 0.0000 & 0.0000 & 0.5000 & 0.5000 & 0.0000 \\ 0.0000 & 0.0000 & -0.7071 & 0.0000 & 0.0000 & 0.5000 & 0.5000 & 0.0000 \\ 0.0000 & 0.0000 & 0.0000 & 0.0000 & 0.0000 & 0.0000 & 0.0000 & 1.0000 \end{bmatrix} \tag{7.92}$$

Finally, we obtain the upper-triangularized B^T, which corresponds to $X^{(4)}$.

$$X^{(4)} = H^{(3)}X^{(3)} = \begin{bmatrix} 1.414 & 0 & 0 \\ 0 & 1.414 & 0 \\ 0 & 0 & 1.414 \\ 0 & 1 & 0 \\ 0 & 0 & 0 \\ 0 & 0 & 0 \\ 0 & 0 & 0 \\ 0 & 0 & 0 \end{bmatrix}$$

This is, of course, shown to be the result of

$$H^{(3)}H^{(2)}H^{(1)}B^T = X^{(4)} \tag{7.93}$$

In equation (7.93), B^T denotes the original constraint matrix. From equation (7.65) we get

$$D_2^T = \begin{bmatrix} 0\,0\,0\,1\,0\,0\,0\,0 \\ 0\,0\,0\,0\,1\,0\,0\,0 \\ 0\,0\,0\,0\,0\,1\,0\,0 \\ 0\,0\,0\,0\,0\,0\,1\,0 \\ 0\,0\,0\,0\,0\,0\,0\,1 \end{bmatrix}$$

Then we deduce the orthogonal complement to B:

$$T = D_2^T H = D_2^T H^{(3)}H^{(2)}H^{(1)}$$

which is found to be

$$T^T = H^T D_2 = \begin{bmatrix} 0.5000 & -0.5000 & -0.5000 & 0.0000 & 0.0000 \\ 0.3536 & -0.3536 & 0.3536 & -0.3536 & 0.0000 \\ 0.3536 & -0.3536 & 0.3536 & -0.3536 & 0.0000 \\ 0.5000 & 0.5000 & 0.0000 & 0.0000 & 0.0000 \\ 0.5000 & 0.5000 & 0.0000 & 0.0000 & 0.0000 \\ 0.0000 & 0.0000 & 0.5000 & 0.5000 & 0.0000 \\ 0.0000 & 0.0000 & 0.5000 & 0.5000 & 0.0000 \\ 0.0000 & 0.0000 & 0.0000 & 0.0000 & 1.0000 \end{bmatrix} \tag{7.94}$$

We can easily see that $TB^T = 0$, making use of equation (7.81). We obtain the reduced equation of motion. The latter, combined with equation (7.82), gives the solution to the dynamics of the system.

Example 7.7.2

To illustrate the PUTD method further, we study another example. Consider the mechanism shown in Figure 7.4. The system consists of two subsystems, the first of which is composed solely of a particle P_1 of mass m_0 which slides on a frictionless track T so that its position may be described by the single generalized coordinates x_1. The second subsystem consists of two rigid links, 1 and 2, each of length L with mass m.

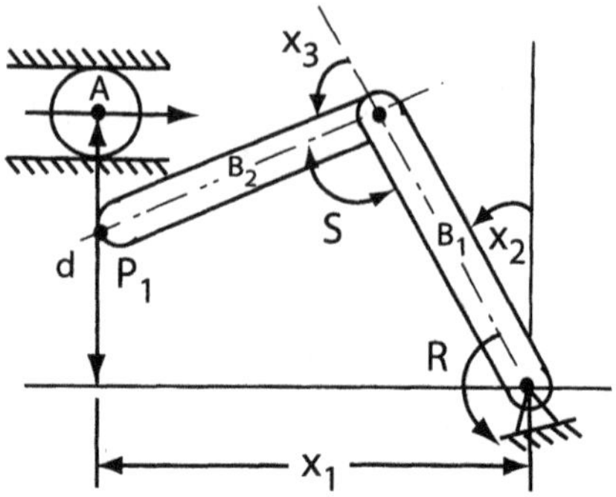

Fig. 7.4. Unconstrained system consisting of a two-link mechanism and a particle P sliding on a frictionless tract

Motors (not shown) exert a torque R on link 1 and a torque S between links 1 and 2. The equations of motion can easily be obtained using equation (5.47), where the generalized coordinates are x_1, x_2 and x_3.

$$m_0 \ddot{x}_1 = 0 \tag{7.95}$$

$$\left(\frac{5}{3}+c3\right) mL^2\ddot{x}_2 + \left(\frac{1}{3}+\frac{1}{2}c3\right) mL^2\ddot{x}_3 - \frac{1}{2}mL^2 s3\left(2\dot{x}_2\dot{x}_3+\dot{x}_3^2\right)$$

$$= R - \frac{3}{2}mgLs2 - \frac{1}{2}mgLs(2+3)$$

$$S - \frac{1}{2}mgLs(2+3) = \left(\frac{1}{3}+\frac{1}{2}c3\right) mL^2\ddot{x}_2 + \frac{1}{3}mL^2\ddot{x}_3 + \frac{1}{2}mL^2 s3\dot{x}_2^2$$

In matrix form the equations above can be written as

$$\begin{bmatrix} m_0 & 0 & 0 \\ 0 & \left(\frac{5}{3}+c3\right)mL^2 & \left(\frac{1}{3}+\frac{1}{2}c3\right)mL^2 \\ 0 & \left(\frac{1}{3}+\frac{1}{2}c3\right)mL^2 & \frac{1}{3}mL^2 \end{bmatrix} \ddot{x} + \begin{bmatrix} 0 & 0 & 0 \\ 0 & -mL^2 s3\dot{x}_3 & -\frac{1}{2}mL^2 s3\dot{x}_3 \\ 0 & \frac{1}{2}mL^2 s3\dot{x}_2 & 0 \end{bmatrix} \dot{x}$$

$$= \begin{bmatrix} 0 \\ R - \frac{1}{2}mgL\left[3s2 + s\left(2+3\right)\right] \\ S - \frac{1}{2}mgLs\left(2+3\right) \end{bmatrix} \tag{7.96}$$

where $si = \sin\theta_i$, $ci = \cos\theta_i$, $si+j = \sin\left(\theta_i+\theta_j\right)$, $ci+j = \cos\left(\theta_i+\theta_j\right)$ and $x = \{x_1 \quad x_2 \quad x_3\}$. Consider next the system given by Figure 7.5, where a constraint point A of B_2 is imposed such that it coincides with P_1. The constraints are obtained by equating the velocity of P_1 and point A in the horizontal and vertical directions, respectively, which yields

$$[B]\{\dot{x}\} = 0 \tag{7.97}$$

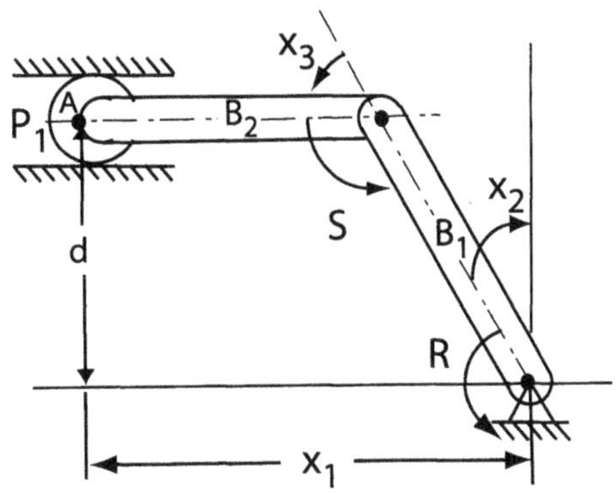

Fig. 7.5. Constrained system composed of subsystem shown in Figure 7.4

where

$$[B] = \begin{bmatrix} 1 & -L\,[c2 + c\,(2+3)] & -Lc\,(2+3) \\ 0 & L\,[s2 + s\,(2+3)] & Ls\,(2+3) \end{bmatrix} \tag{7.98}$$

Since the system now possesses one degree of freedom utilizing the PUTD method, we should be able to reduce the equations of motion given by (7.96) by embedding into them the constraints given by equation (7.97). It will be better to use numerical values for equation (7.98) so that we can illustrate all the steps of the pseudo upper triangular decomposition. Let us assume that $x_2 = 30^o$, $x_3 = 60^o$, and $L = 1$. Hence the matrix B becomes

$$B = \begin{bmatrix} 1 & -0.866 & 0 \\ 0 & 1.5 & 1 \end{bmatrix} \tag{7.99}$$

In step 1 of the transformation, consider the first row of B, then

$$\bar{X}_1^{(1)} = \left[1 \; -0.866 \; 0 \right]^T \tag{7.100}$$

Using equation (7.47), we obtain

$$\bar{W}^{(1)} = \left(1.3229 \; 0.0 \; 0.0 \right)^T \tag{7.101}$$

Equation (7.36) then gives the Householder transformation $H^{(1)}$ as

$$H^{(1)} = \begin{bmatrix} 0.3780 & 0.6547 & 0.6547 \\ 0.6547 & 0.3110 & -0.6940 \\ 0.6547 & -0.6890 & 0.3110 \end{bmatrix} \tag{7.102}$$

Using equation (7.51) together with the results of equation (7.102), we obtain $\bar{B}_2^{(2)}$. Performing step 2 yields the transformation matrix $H^{(2)}$:

$$H^{(2)} = \begin{bmatrix} 1.000 & 0.0 & 0.0 \\ 0.0 & 0.2943 & 0.9557 \\ 0.0 & 0.9557 & -0.2943 \end{bmatrix} \tag{7.103}$$

From equation (7.59), we obtain the desired transformation matrix $T^T = H^T D_2$, which will be used to reduce the governing equations of motion. The latter is found to be

$$T^T = H^T D_2 = \begin{bmatrix} 0.4330 \\ 0.5000 \\ -0.7500 \end{bmatrix} \tag{7.104}$$

It can easily be shown that the product of BT^T is equal to zero after substitution. Multiplying T by both sides of equation (7.99), we get

$$[T]\ [a]\{\ddot{x}\} + [T][b]\{\dot{x}\} = [T]\{f\} \tag{7.105}$$

that is,

$$0.433m_0\ddot{x}_1 + \left[0.5\left(\frac{5}{3} + c3\right)mL^2 - 0.75\left(\frac{1}{3} + \frac{1}{2}c3\right)mL^2\right]\ddot{x}_2$$

$$+\left[0.5\left(\frac{5}{3} + c3\right)mL^2 - 0.75\left(\frac{1}{3} + \frac{1}{2}c3\right)mL^2\right]\ddot{x}_3$$

$$-\left[0.5mL^2 s3\dot{x}_3 + 0.75\cdot\frac{1}{2}mL^2 s3\dot{x}_2\right]\dot{x}_2 - 0.5\cdot\frac{1}{2}mL^2 s3\dot{x}_3^2$$

$$= 0.5\left\{R - \frac{1}{2}mgL\left[3s2 + s\left(2+3\right)\right] - 0.75\left[S - \frac{1}{2}mgLs\left(2+3\right)\right]\right\}$$

We can write the equations above in the form of equation (7.35) with $c3 = \cos 60^o$ and $L = 1$:

$$[A] = \begin{bmatrix} [T][a] \\ \hline [B] \end{bmatrix} = \begin{bmatrix} 0.433m_0 & 0.645m & 0.417m \\ 1 & -0.866 & 0 \\ 0 & 1.5 & 1 \end{bmatrix} \tag{7.106}$$

$$\{D\} = \begin{bmatrix} [T]\{\{f\} - [b]\{\dot{x}\} - [c]\{\dot{x}\}\} \\ \hline -\left[\dot{B}\right]\{\dot{x}\} + \{\dot{g}\} \end{bmatrix} \tag{7.107}$$

where $\{\dot{g}\} = 0$ and $[c] = 0$ and

$$\left[\dot{B}\right] = \begin{bmatrix} 0 & L\left[\dot{x}_2 s2 + (\dot{x}_2 + \dot{x}_3)\,s\left(2+3\right)\right] & L\left(\dot{x}_2 + \dot{x}_3\right)s\left(2+3\right) \\ 0 & L\left[\dot{x}_2 c2 + (\dot{x}_2 + \dot{x}_3)\,c\left(2+3\right)\right] & L\left(\dot{x}_2 + \dot{x}_3\right)c\left(2+3\right) \end{bmatrix} \tag{7.108}$$

Substituting $x_2 = 30°$, $x_3 = 60°$, and $L = 1$ into equation (7.108), we get

$$[\dot{B}] = \begin{bmatrix} 0 & 1.5\dot{x}_2 + \dot{x}_3 & \dot{x}_2 + \dot{x}_3 \\ 0 & 0.866\dot{x}_2 & .0 \end{bmatrix} \tag{7.109}$$

Therefore,

$$\{D\} = \begin{bmatrix} 0.5R - 0.75S - 0.25mg + 0.188m\dot{x}_2^2 + 0.25m\dot{x}_2\dot{x}_3 + 0.188m\dot{x}_3^2 \\ -1.5\dot{x}_2^2 - 2\dot{x}_2\dot{x}_3 - \dot{x}_3^2 \\ -0.866\dot{x}_2^2 \end{bmatrix} \tag{7.110}$$

The equations of motion of the system for the given data are then expressed as

$$[A]\{\ddot{x}\} = \{D\} \tag{7.111}$$

or

$$\{\ddot{x}\} = [A]^{-1}\{D\} \tag{7.112}$$

Solving for the differential equations above, we obtain the time history of the system.

7.8 The Zero-Eigenvalue Theorem and Embedding Method

7.8.1 The Zero-Eigenvalue Theorem

The zero-eigenvalue theorem provides an alternative to computing the orthogonal complement of the Jacobi constraint matrix B. This method was introduced by Walton and Steeves [10] and later extended by Kamman and Huston [4] to the solution of constrained multibody systems. A brief outline of the method is as follows: Let equation (7.113) be written as

$$By = 0 \tag{7.113}$$

where y is a column vector equal to $\dot{X}$ and denotes the generalized speeds. B is a matrix of order $m \times n$. Next, a matrix S is formed as

$$S = B^T B \tag{7.114}$$

where B^T is the transpose of B. Since S is symmetric, there exists an orthogonal matrix T such that

$$T^T S T = \Lambda \tag{7.115}$$

where Λ is an $n \times n$ diagonal matrix with real eigenvalue elements λ_i. Those eigenvalues are nonnegative, with at least $n - m$ of them being zero. Next, let T be rearranged so that the zero eigenvalues will appear at the end of the diagonal matrix Λ. Define a matrix D such that

$$D = BT \tag{7.116}$$

Equation (7.116) then shows that $D^T D$ is equal to Λ. Therefore, D can be partitioned into two matrices,

$$D = \begin{bmatrix} \hat{D} & 0 \\ 0 & 0 \end{bmatrix} \tag{7.117}$$

The order $\hat{D}$ is $p \times p$, where $n - m < p < n$. From the orthogonality of T, and noting that TT^T is the identity matrix $(n \times n)$, the constraint equations may be expressed in the form

$$By = BTT^T y \triangleq DZ = 0 \tag{7.118}$$

where Z is $T^T y$. By inverting T, y may be expressed as

$$y = TZ \quad (Z \text{ is an } n \times 1 \text{ vector}) \tag{7.119}$$

Then T may be partitioned as

$$T = \left| \hat{R}_{n \times p} \middle| \hat{O}_{n \times (n-p)} \right|_{n \times n} \tag{7.120}$$

where the matrix $\hat{O}$ is on the order of $n - p$, whose columns are those associated with the zero eigenvalues of S and $\hat{R}$ corresponds to the column vectors with real values. The solution to equation (7.113) can be expressed in terms of p independent elements of Z as

$$y_\ell = \hat{T}_{\ell r} \hat{Z}_r \quad (\ell = 1, 2, \ldots, n;\ r = 1, 2, \ldots, p) \tag{7.121}$$

or

$$y_{n\times 1} = \hat{T}_{n\times p} \hat{Z}_{p\times 1} \tag{7.122}$$

where $\hat{T}_{\ell r}$ are elements of the matrix associated with $\hat{O}$. The reduced equations of motion can be obtained by premultiplying the equations of motion by $\hat{T}_{\ell r}$.

Example 7.8.1
Consider Example 7.7.1, where the constraint Jacobian matrix is given by equation (7.7.46). Let us apply the zero-eigenvalue theorem to generate the corresponding $\hat{T}$ to reduce the equations of motion. Following the procedures outlined above, we first compute the matrix S given by equation (7.114), then perform an eigenvalue problem on S to determine A.

The eigenvectors corresponding to the zero eigenvalues of $S = B^T B$ produce the orthogonal complement to B, which is found to be

$$\hat{T} = \begin{bmatrix} 1 & 0 & 0 & 0 & 0 \\ 0 & 1/\sqrt{2} & 0 & 0 & 0 \\ 0 & 1/\sqrt{2} & 0 & 0 & 0 \\ 0 & 0 & 1/\sqrt{2} & 0 & 0 \\ 0 & 0 & 1/\sqrt{2} & 0 & 0 \\ 0 & 0 & 0 & 1/\sqrt{2} & 0 \\ 0 & 0 & 0 & 0 & 0 \\ 0 & 0 & 0 & 0 & 1 \end{bmatrix} \tag{7.123}$$

where $\hat{T}$ is the transformation matrix relating the independent coordinates (which are five in this case) and the displacement used to describe the degrees of freedom of the system given by Figure 7.3.

7.8.2 Embedding Method

Essentially, if we have n equations of motion and m constraints, we should have m dependent coordinates and $n - m$ independent ones. The embedding method is used to eliminate the undetermined multipliers and reduce the equations of motion to minimum dimension, hence representing the equations of motion only in terms of the independent coordinates. Recall the acceleration representation of the m constraints equations given by (7.18)

$$[B]\{\dot{y}\} = -[\dot{B}]\{y\} + \{g\} \tag{7.124}$$

Rearrange the ordinate $\{y\}$ such that

$$\{y\} = \left[\frac{y_i}{y_d}\right] \tag{7.125}$$

where y_i and y_d are, respectively, the velocity arrays or general speeds of independent and dependent coordinates. Then the constraint equation above becomes

$$[B_i]\{\dot{y}_i\} + [\dot{B}_d]\{\dot{y}_d\} = \{g_c\} \tag{7.126}$$

where the Jacobian matrix is partitioned as $\left[B_i \middle| B_d\right] = [B]$ and the right-hand side of equation (7.126) is given by

$$\{g_c\} = -[\dot{B}]\{y\} + \{g\} \tag{7.127}$$

Premultiplying equation (7.126) by $[B_d]^{-1}$ yields

$$\{\dot{y}_d\} = [B_d]^{-1}\{g_c\} - [B_d]^{-1}[B_i]\{\dot{y}_i\} \tag{7.128}$$

Hence

$$\{\dot{y}\} = \begin{bmatrix} y_i \\ y_d \end{bmatrix} = \begin{bmatrix} I \\ -B_d^{-1} B_i \end{bmatrix} \{\dot{y}\} + \begin{bmatrix} O \\ B_d^{-1} g_c \end{bmatrix} \tag{7.129}$$

or in a compact form as

$$\{\dot{y}\} = [B_{di}] \{\dot{y}_i\} + \{g_{dc}\} \tag{7.130}$$

where

$$[B_{di}] = \begin{bmatrix} I \\ -B_d^{-1} B_i \end{bmatrix} \quad \text{and} \quad \{g_{dc}\} = \begin{bmatrix} O \\ B_d^{-1} g_c \end{bmatrix} \tag{7.131}$$

Substituting equation (7.130) into equation (7.17), we get

$$[a] ([B_{di}] \{\dot{y}_i\} + \{g_{dc}\}) + ([b] + [c]) \{y\} = \{f\} - [B]^T \{\lambda\} \tag{7.132}$$

Premultiplying this equation by $[B_{di}]^T$ and rearranging the terms yields

$$\begin{aligned} &[B_{di}]^T [a] [B_{di}] \{\dot{y}_i\} + [B_{di}]^T ([b] + [c]) \{y\} \\ &= [B_{di}]^T \{f\} - [B_{di}]^T [a] \{g_{dc}\} - [B_{di}]^T [B]^T \{\lambda\} \end{aligned} \tag{7.133}$$

Now let us prove that $[B_{di}]^T [B]^T = 0$. From equations (7.126) and (7.130), we can write

$$\begin{aligned} [B_{di}]^T [B]^T &= \begin{bmatrix} I - B_i^T B_d^{-T} \end{bmatrix} \begin{bmatrix} B_i^T \\ B_d^T \end{bmatrix} \\ &= \begin{bmatrix} B_i^T \end{bmatrix} - [B_i]^T [B_d]^{-T} [B_d]^T = 0 \end{aligned} \tag{7.134}$$

Thus we can delete the undetermined multipliers λ term and equation (7.133) becomes

$$[A] \{\dot{y}_i\} = [Q] \tag{7.135}$$

where

$$[A] = [B_{di}]^T [a] \; [B_{di}] \tag{7.136}$$

$$[Q] = [B_{di}]^T (\{f\} - ([b] + [c]) \{y\} - [a] \{g_{dc}\}) \tag{7.137}$$

This procedure of eliminating the constraint forces is called the *embedding method.* Observe how equation (7.135) is a function of the independent coordinates only and represents the minimum form of the equations of motion. The governing linearly dependent coordinates equations of motion given by equation (7.135) represent an ideal form where the dependent coordinates are eliminated in the process through a simple partitioning and resubstitution procedure.

Example 7.8.2

Consider Example 7.7.2 and let $m_0 = m = 1$ kg, $L = 1$ m, $S = R = 1$ Nm, $x_2 = 30°$, $x_3 = 60°$, and $\dot{x}_2 = 1$ rad/s. Find the acceleration of the slide using the embedding method.

Solution

From equation (7.99) the constraint equations are such that

$$B\dot{x} = g$$

since $g = 0$ and $\dot{B} = 0$. Then we write the constraints at the velocity and acceleration levels as

$$B\dot{x} = \begin{bmatrix} 1 & -0.866 & 0 \\ 0 & 1.5 & 1 \end{bmatrix} \begin{bmatrix} \dot{x}_1 \\ \dot{x}_2 \\ \dot{x}_3 \end{bmatrix} = 0 \tag{7.138}$$

and

$$B\ddot{x} = \begin{bmatrix} 1 & -0.866 & 0 \\ 0 & 1.5 & 1 \end{bmatrix} \begin{bmatrix} \ddot{x}_1 \\ \ddot{x}_2 \\ \ddot{x}_3 \end{bmatrix} = 0$$

Since the system has only one degree of freedom (3 – 2), we can select any of the three generalized coordinates to be the independent one. Let us choose x_1; then from equation (7.126) the partitioning of the Jacobian is done as follows:

$$[B_i] = \begin{bmatrix} 1 \\ 0 \end{bmatrix} \tag{7.139}$$

$$[B_d] = \begin{bmatrix} -0.866 & 0 \\ 1.5 & 1 \end{bmatrix} \tag{7.140}$$

and

$$\{g_c\} = 0 \tag{7.141}$$

that is,

$$\begin{bmatrix} 1 \\ 0 \end{bmatrix} \ddot{x}_1 + \begin{bmatrix} -0.866 & 0 \\ 1.5 & 1 \end{bmatrix} \begin{bmatrix} \ddot{x}_2 \\ \ddot{x}_3 \end{bmatrix} = 0 \tag{7.142}$$

Using equation (7.130), we write

$$[B_{di}] = \begin{bmatrix} I \\ -B_d^{-1} B_i \end{bmatrix} = \begin{bmatrix} 1 \\ -\begin{bmatrix} -0.154 & 0 \\ -1.732 & 1 \end{bmatrix} = \begin{bmatrix} 1 \\ 0 \end{bmatrix} \end{bmatrix}$$

$$= \begin{bmatrix} 1 \\ 1.154 \\ -1.732 \end{bmatrix} \tag{7.143}$$

and

$$\{g_{dc}\} = 0 \tag{7.144}$$

Obviously, we should accept that

$$[B_{di}]^T [B]^T = \begin{bmatrix} 1 & 1.154 & -1.732 \end{bmatrix} \begin{bmatrix} 1 & 0 \\ -0.866 & 1.5 \\ 0 & 1 \end{bmatrix} = \begin{bmatrix} 0 & 0 \end{bmatrix} \tag{7.145}$$

Hence from equation (7.135), the equations of motion of the system in terms of x_1 are

$$A\ddot{x}_1 = Q \tag{7.146}$$

where

$$A = [B_{di}]^T [a] [B_{di}]$$

and $[a]$ comes from equation (7.96) by letting $L = 1$, $c3 = \cos 60^o$:

$$\begin{aligned} A &= \begin{bmatrix} 1 & 1.154 & -1.732 \end{bmatrix} \begin{bmatrix} m_0 & 0 & 0 \\ 0 & 2.166m & 0.583m \\ 0 & 0.583m & 0.333m \end{bmatrix} \begin{bmatrix} 1 \\ 1.154 \\ -1.732 \end{bmatrix} \\ &= m_0 + 1.551m = 2.551 \end{aligned} \tag{7.147}$$

From equations (7.96) and (7.137), we get

$$\begin{aligned} Q &= [B_{di}]^T (\{f\} - ([b] + [c]) \{\dot{x}\} - [a] \{g_{dc}\}) \\ &= \begin{bmatrix} 1 & 1.154 & -1.732 \end{bmatrix} \left(\begin{bmatrix} 0 \\ 1 - 4.9\,(1.5 + 0) \\ 1 - 0 \end{bmatrix} \right. \\ &\quad \left. - \begin{bmatrix} 0 & 0 & 0 \\ 0 & -0.866\dot{x}_3 & -0.433\dot{x}_3 \\ 0 & 0.433\dot{x}_2 & 0 \end{bmatrix} \begin{bmatrix} \dot{x}_1 \\ \dot{x}_2 \\ \dot{x}_3 \end{bmatrix} - 0 \right) \\ &= \begin{bmatrix} 1 & 1.154 & -1.732 \end{bmatrix} \begin{bmatrix} 0 \\ -6.35 + 0.866\dot{x}_2\dot{x}_3 + 0.433\dot{x}_3^2 \\ 1 - 0.433x_2^2 \end{bmatrix} \\ &= -9.059 + 0.75\dot{x}_2^2 + \dot{x}_2\dot{x}_3 + 0.5\dot{x}_3^2 \end{aligned} \tag{7.148}$$

Using the constraint equation given by the second row of equation (7.138), we write

$$1.5\dot{x}_2 + \dot{x}_3 = 0 \tag{7.149}$$

Substituting the relation (7.149) into equation (7.148), we obtain

$$Q = -9.059 + 0.75\dot{x}_2^2 - 1.5\dot{x}_2^2 + 1.125\dot{x}_2^2 = 0.375\dot{x}_2^2 - 9.059 \\ = -8.684 \tag{7.150}$$

Therefore,

$$\ddot{x}_1 = \frac{Q}{A} = -\frac{8.684}{2.551} = -3.404\,\text{m/s}^2 \tag{7.151}$$

The method shows clearly how the embedding method works in isolating the equations of motion through single substitutions. The selection of the independent coordinate is, however, arbitrary.

7.9 Relation Between the PUTD and the Zero-Eigenvalue Theorem

In the analysis of constrained multibody systems, efficient computer algorithms are sought to compute the orthogonal complement matrix T corresponding to the constraint Jacobian matrix B. Both the PUTD method and the zero-eigenvalue theorem provide an elegant approach that is suitable for large computer codes for multibody dynamics. One advantage is that the PUTD method relies on the multiplication type of operation suited for computers. It does not involve any eigenvalue solving. In addition, the PUTD method is tailored for analysis of the stabilization of the constraint matrix B when it changes rank. This topic is discussed in Chapter 10.

Mathematically, we can deduce a relationship between the zero-eigenvalue theorem and the PUTD method [15]. Consider the holonomic constraint equations given by

$$B_y = 0 \tag{7.152}$$

Let T be the orthogonal complement of B computed through the PUTD and $\hat{T}$ the one computed using the zero-eigenvalue theorem. A solution to equation (7.152) is given by

$$y = Tq \tag{7.153}$$

or

$$y = \hat{T}\bar{q} \tag{7.154}$$

Since the vector of new generalized speeds, q is arbitrary in equation (7.153), the solution can be written in the form

$$y = H^T D_2 G\bar{q} \tag{7.155}$$

where G is any nonsingular square matrix of order $(n - m)$.

In general, the coordinate transformation matrix in each of the methods above corresponds to different generalized speeds. In fact, the new generalized speeds obtained by the transformation of equations (7.73) and (7.122) are abstract in character and do not lend themselves to physical interpretation. In view of equation (7.155), the relation between the transformation matrices is

$$H^T D_2 G = P \tag{7.156}$$

where P is either T^T or $\hat{T}^T$, with corresponding nonsingular matrix G. The columns of P are linear combination of the columns $D_2^T H = T$, and T forms a different basis for the same $(n-m)$-dimensional subspace of R^n, which is the space composed of the linear combinations of the solution vectors of equation (7.152).

In the special case that P is orthogonal, G is an orthogonal matrix. This is seen by writing the ith and jth columns of P, $P_i = \left(H^T D_2\right) g_i$ and $P_j = \left(H^T D_2\right) g_j$, where g_i is the ith column of G. Then since $T = D_2^T H$, $P_i \cdot P_j = 0$ if and only if $g_i \cdot g_j = 0\ (i \neq j)$ and $T_i \cdot T_i = 1$ if and only if $g_i \cdot g_i = 1$.

At this point it is a matter of computational efficiency as to which method is best suited for the determination of T. To obtain a matrix T to automatically eliminate the underdetermined multipliers in the equations of motion and hence reduce the governing equations from n and $n-m$ for m given constraints, we have shown the algorithmic procedures involved for both the PUTD method and the zero-eigenvalue theorem. Although the two methods lead to the same answer, their computational procedures differ.

Example 7.9.1

In Examples 7.7.1 and 7.8.1 we have computed the orthogonal complement T of B using the PUTD method and the zero-eigenvalue theorem, respectively. To find the relationship between the two methods, we observe the following: The columns of $H^T D_2$ constitute another basis of the vector space composed of the linear combinations of the eigenvectors of S [in equation (7.114)] corresponding to the zero eigenvalues.

The relation between the two sets [see equation (7.94) for T and equation (7.123) for $\hat{T}$] is the orthogonal G matrix defined in equation (7.155), which is calculated to be

$$T^T G = \hat{T} \tag{7.157}$$

or

$$G = T\hat{T} \tag{7.158}$$

$$G = \begin{bmatrix} 0.5 & 0.5 & 0.707 & 0 & 0 \\ -0.5 & -0.5 & 0.707 & 0 & 0 \\ -0.5 & 0.5 & 0 & 0.707 & 0 \\ 0.5 & -0.5 & 0 & 0.707 & 0 \\ 0 & 0 & 0 & 0 & 0 \end{bmatrix} \tag{7.159}$$

7.10 Derivation of the Constraint Equation for Closed Loops

Mechanisms and spacecraft structures are a good example of multibody systems with closed loops. The analysis of such systems can be broken into two parts: (1) the dynamical equations of motion are formulated for an open-chain system; and (2) a set of constraint conditions resulting from the loop formation will be imposed on the system. In this regard it becomes apparent that the question anyone would ask is: Where would we cut the loop to consider the multibody system as an open chain? Consider Figure 7.6, where a general multibody system with a closed loop composed of N rigid bodies is shown. Let $x_\ell(\ell = 1, \ldots, p)$ denote the generalized coordinates of the system. To derive the equations of motion of the given system, we must first make the structure an open one without closed loops. Figure 7.7 gives such a structure.

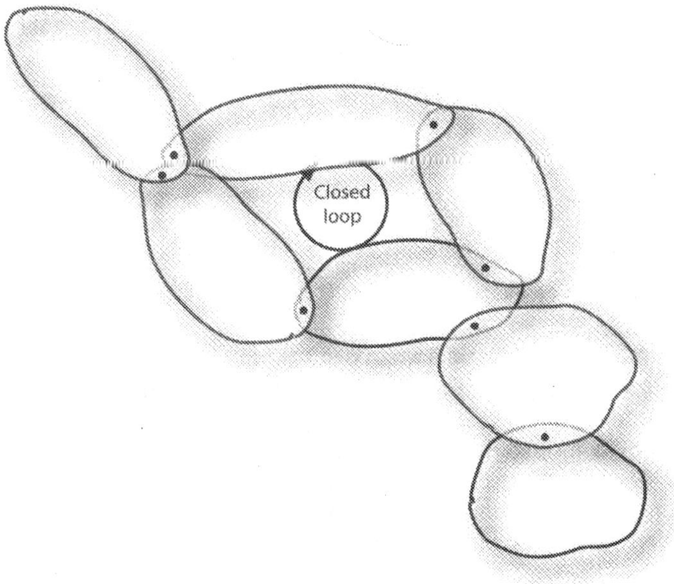

Fig. 7.6. General multibody system with a closed loop

The possible cuts could be anywhere at the joints of the loop. The selection is then arbitrary and should not affect the equations in their developed form. Making the multibody system an open chain results in additional forces and torques at the joint of interest (see Figure 7.7). Those forces and torques constitute the corresponding constraint forces and torques resulting from the constraint condition of maintaining the loop formation, as indicated in Figure 7.6.

The derivation of the equations of motion for the open system as in Figure 7.7 can be obtained by using Kane's equations with the undetermined multipliers as

$$f_\ell^* + f_\ell + \lambda_i B_{i\ell} = 0 \tag{7.160}$$

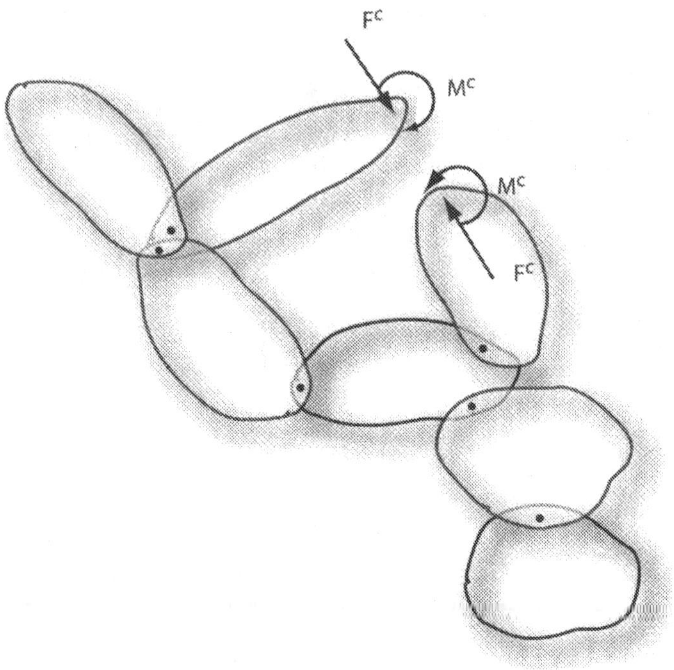

Fig. 7.7. Modeling of a loop constraint by cutting through an arbitrary joint of the loop

where B is the Jacobian constraint matrix and λ_i the undetermined multipliers, which correspond to the constraint forces and torques resulting from opening the closed loop. In addition to equation (7.160), we must impose a set of constraints that would maintain the loop formation at all times. This can be derived by considering the position vectors from the origins of the reference frame R to points p and q of body B_p and B_q, respectively. If the points p and q coincide, as is the condition of the loop formation, their position vectors must be equal, that is,

$$\bar{P}_p = \bar{P}_q \tag{7.161}$$

where $\bar{P}_p$ corresponds to the position vector of point p in R going through branch 1 and $\bar{P}_q$ the position vector of q in R going through branch 2. Expressing these vectors in terms of the unit vectors $\bar{n}_m (m = 1, 2, 3)$ attached to R and differentiating, we get

$$\frac{d\bar{P}_p}{dt} - \frac{d\bar{P}_q}{dt} = 0 \tag{7.162}$$

that is,

$$\bar{V}_p - \bar{V}_q = 0 \tag{7.163}$$

or in matrix form,

$$\{\dot{x}\}^T\left(\left[V_w^p\right]-\left[V_w^q\right]\right)\{\bar{n}\}=0 \tag{7.164}$$

where $\left[V_w^p\right]=[W]\left[V^p\right]$ and $\left[V_w^q\right]=[W]\left[V^q\right]$, the partial velocity matrices of points p and q [see equation (3.120)]. Equation (7.164) can be expressed in more recognizable form as

$$[B]\{\dot{x}\}=0 \tag{7.165}$$

where

$$[B]=\left[V_w^p\right]^T-\left[V_w^q\right]^T \tag{7.166}$$

In summary, we can state that the treatment of closed loops in multibody systems can be done as follows:

Step 1. Make the structure an open-treelike system by opening the loops at some arbitrary connecting joint within the loop.
Step 2. Formulate the equations of motion based on the procedures developed in previous chapters, especially using equation (5.47) through (5.51).
Step 3. Derive equation (7.165) making use of equation (7.166), where every component is computed previously in step 1. So this does not require any additional efforts.
Step 4. Solve the dynamics of multibody systems with closed loops combining the equations of motion from step 1 and the constraint equations from step 3.

Use either the augmented or PUTD method to obtain the explicit algorithmic governing equations given by equation (7.20) or (7.35).

Use either the augmented method or the PUTD to obtain the explicit algorithmic governing equations given by equation (7.20) or (7.35).

Example 7.10.1
For simply illustrating some of these ideas, consider the planar triple pendulum shown in Figure 7.8. There are frictionless pins at the joints: O_1, O_2 and O_3. The system has three degrees of freedom, which may be described by the generalized coordinates θ_1, θ_2 and θ_3.

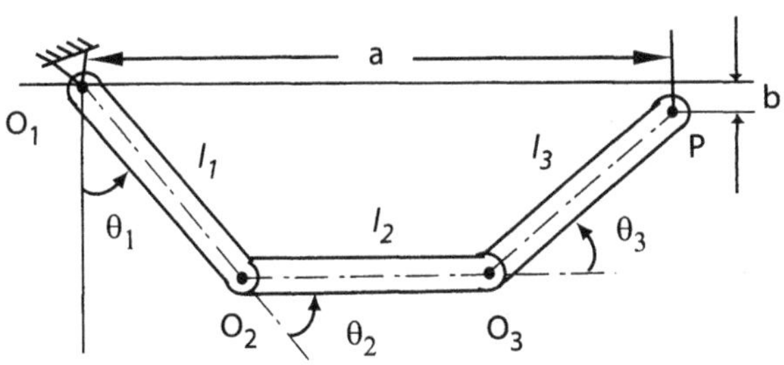

Fig. 7.8. Planar triple pendulum

Solution
First the equations of motion for the planar pendulum given by Figure 7.8 are found. The constraint for the "closed loop" may be formed by fixing the endpoint P of the pendulum. Hence let P be fixed at point P_0 having coordinates (a, b) relative to the $X - Y$ coordinate system shown in Figure 7.9. This constrained system has only one degree of freedom. Two scalar constraint equations relating the coordinates θ_1, θ_2 and θ_3 may be obtained from the position vector equation:

$$\overline{O_1 O_2} + \overline{O_2 O_3} + \overline{O_3 P_0} + \overline{P_0 O_1} = 0 \tag{7.167}$$

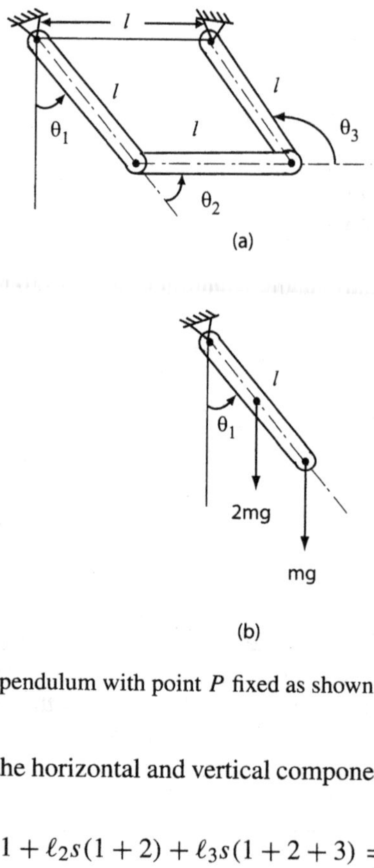

Fig. 7.9. (a) Planar triple pendulum with point P fixed as shown. (b) Single pendulum bar

That is, considering the horizontal and vertical components of this equation leads to the equations

$$\begin{aligned} \ell_1 s1 + \ell_2 s(1+2) + \ell_3 s(1+2+3) &= a \\ \ell_1 c1 + \ell_2 c(1+2) + \ell_3 c(1+2+3) &= b \end{aligned} \tag{7.168}$$

which, upon differentiation, becomes

$$\begin{aligned} &[\ell_1 c1 + \ell_2 c(1+2) + \ell_3 c(1+2+3)]\,\dot{\theta}_1 + [c\,(1+2)\ell_2 \\ &\quad + \ell_3 c(1+2+3)]\,\dot{\theta}_2 + \ell_3 c(1+2+3)\dot{\theta}_3 = 0 \\ &[\ell_1 s1 + \ell_2 s(1+2) + \ell_3 s(1+2+3)]\,\dot{\theta}_1 + [s(1+2)\ell_2 \\ &\quad + \ell_3 s(1+2+3)]\,\dot{\theta}_2 + \ell_3 s(1+2+3)\dot{\theta}_3 = 0 \end{aligned} \tag{7.169}$$

In equations (7.168) and (7.169), $s1 = \sin\theta_1$, $s(1+2) = \sin(\theta_1 + \theta_2)$, and so on. Similarly, $c1 = \cos\theta_1, c(1+2) = \cos(\theta_1 + \theta_2)$ and so on. Equations (7.169) represent the constraint equations as described by equation (7.165).

To simplify this analysis, let $\ell_1 = \ell_2 = \ell_3 = \ell$, $m_1 = m_2 = m_3$, and let P be fixed on the x axis at a point P_0, a distance ℓ from O_1. The system then takes the form of a rhombic linkage, as shown in Fig. 7.9a. In this case, $a = \ell$, $b = 0$, and the constraints are seen to be satisfied by the relations $\theta_2 = \pi/2 - \theta_1$ and $\theta_3 = \pi/2 + \theta_1$. The coefficient matrix B of equation (7.169) may then be expressed as

$$B = \begin{bmatrix} 0 & -c1 & -c1 \\ 1 & 1 - s1 & -s1 \end{bmatrix} \tag{7.170}$$

This problem was solved numerically making use of the PUTD method for the reduction of the equations of motion. The results are shown in Table 7.1.

Time (s)	θ_1 (rad)	$\dot{\theta}_1$ (m/s)	$\ddot{\theta}_1$ (m/s^2)
0.1	0.3095	−0.07758	−7.172
0.2	0.1997	−1.380	−4.670
0.3	0.0440	−1.672	−1.037
0.4	−0.1217	−1.578	2.858
0.5	−0.2595	−1.124	7.041
0.6	−0.3381	−0.4178	7.809
0.7	−0.3400	0.3799	7.851
0.8	−0.2649	1.094	7.163
0.9	−0.1293	1.564	3.036
1.0	0.0360	1.676	−0.8461

Table 7.1. Time history of θ_1, $\dot{\theta}_1$ and $\ddot{\theta}_1$

7.11 Prescribed Motion

In this section, we focus on the constraints resulting from certain bodies of the general multibody system being subjected to some prescribed motions. We are interested in developing suitable formalisms for defining and describing the desired motions of the multibody system between the path endpoints. The objective is to define the proper constraint equations in order for a certain point belonging to a certain rigid body in the multibody system to follow the prescribed path. This will be done in the context of nonredundant systems.

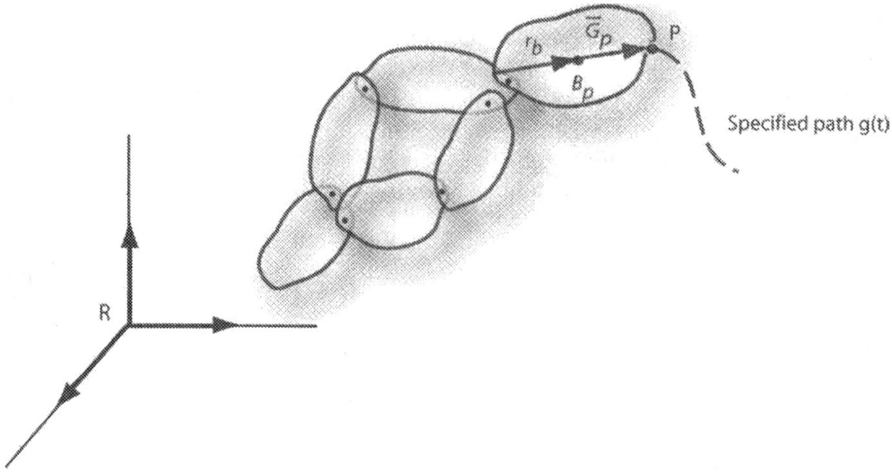

Fig. 7.10. General multibody system

Consider the multibody system as depicted in Figure 7.10, where point P on body B_p is to follow a certain specified path. If the path can be represented by a sequence of points in space R, there exists a function g, a function of time, such that the velocity of point P in R is

$$\bar{V}^p = \bar{g} \tag{7.171}$$

Making use of the matrix representation of a point belonging to a rigid body in multibody systems, equation (7.171) can then be expressed as

$$\{\dot{x}\}^T \left[V^p\right] = \{g(t)\}^T \tag{7.172}$$

where $g(t)$ represents the prescribed velocity of point p in body B_p. For a prescribed acceleration we write

$$\bar{a}^p = \overline{\dot{g}} \tag{7.173}$$

where the acceleration $\bar{a}^p$ is obtained by further differentiation of the left-hand side of equation (7.172).

Example 7.11.1
A two-link planar robotic system will be investigated (Figure 7.11). The system will be subjected to a one constraint equation. Two different types of constraints are considered. In the first type, one coordinate of the end effector is given a time-dependent motion, while the other coordinate is determined by the force acting on the system. As a matter of fact, both coordinates can be prescribed as a function of time. This constraint can then be written as

$$y_c = f(t) \tag{7.174}$$

where we are assuming that $f(t) = a_1 t^2 + a_2$, higher-order terms can be added for precision path control. The parameters a_1 and a_2 are two arbitrary constants. In the second case, a path will be specified for the end effector. We can also make further assumptions about how the end effector is going to follow the path by saying that the motion along this path will depend on the forces to which the system is subjected. This path will be of the form

$$y_c = f(x) \tag{7.175}$$

Here $f(x) = b_1 x^2 + b_2$, where b_1 and b_2 are two arbitrary constants.

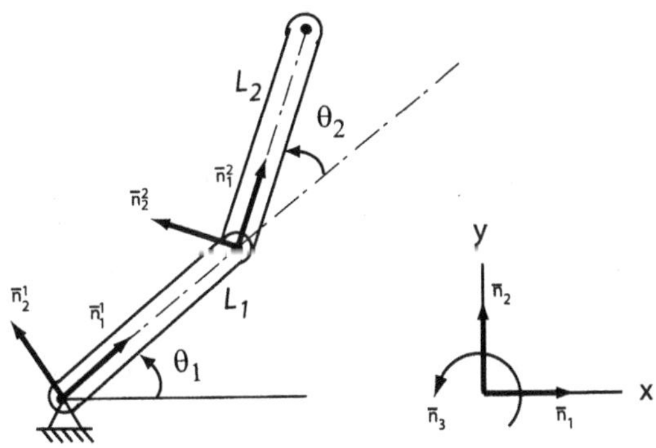

Fig. 7.11. Planar two-link manipulator

Equations of Motion
From Chapter 5, we know that the equations of motion for this particular example are given by

$$[a]\left\{\ddot{\theta}\right\} = \{b\} + \{f\} \tag{7.176}$$

where

$$\{\theta\} = \begin{Bmatrix} \theta_1 \\ \theta_2 \end{Bmatrix}$$

$$[a] = \begin{bmatrix} \frac{1}{4}m_1L_1^2 + m_2\left(L_1^2 + \frac{1}{4}L_2^2 + L_1L_2c2\right) & m_3\left(\frac{1}{4}L_2^2 + \frac{1}{2}L_1L_2c2\right) \\ m_3\left(\frac{1}{4}L_2^2 + \frac{1}{2}L_1L_2c2\right) & \frac{1}{4}m_2L_2^2 \end{bmatrix} \tag{7.177}$$

$$\{b\} = \begin{bmatrix} \frac{1}{2}m_1L_1L_2s2(2\dot{\theta}_1\dot{\theta}_2 + \dot{\theta}_2^2) \\ -\frac{1}{2}m_1L_1L_2s2\dot{\theta}_1^2 \end{bmatrix} \tag{7.178}$$

and

$$\{f\} = \begin{bmatrix} -\frac{1}{2}m_1gL_1c1 - m_2g\left[L_1c1 + \frac{1}{2}L_2c(1+2)\right] \\ -\frac{1}{2}m_2gL_2c(1+2) \end{bmatrix} \tag{7.179}$$

Note that $c1 = \cos\theta_1$, $s1 = \sin\theta_1$, and $c(1+2) = \cos(\theta_1 + \theta_2)$.

Constraining the System
The equations for a constrained multibody system are given by

$$f^* + f + f^c = 0 \tag{7.180}$$

or

$$f^* + f + B^T\lambda = 0 \tag{7.181}$$

where λ are the undetermined multipliers and B the constraint Jacobian matrix. Let T be the orthogonal complement to B. Then the equations above reduce to

$$Tf^* + Tf = 0 \tag{7.182}$$

Premultiplying our developed equations of motion by T for the two-link manipulator, we obtain

$$[T]\,[a]\left\{\ddot{\theta}\right\} = [T]\{b\} + [T]\{f\} \tag{7.183}$$

Notice how the generalized constraint forces are eliminated by virtue of TB^T. We also know that the constraint equations at the velocity level can be written as

$$[B]\left\{\dot{\theta}\right\} = \{g\} \tag{7.184}$$

The time derivative of equation (7.184) gives the acceleration representation of the equations of motion:

$$[B]\left\{\ddot{\theta}\right\} = \{\dot{g}\} - \left[\dot{B}\right]\left\{\dot{\theta}\right\} \tag{7.185}$$

Augmenting equation (7.185) and (7.183) we obtain the desired governing equations of motion for the constrained system as [see equations (7.35) to (7.37)]

$$[A]\left\{\ddot{\theta}\right\} = \{D\} \tag{7.186}$$

where

$$[A] = \left[\frac{[T]\;[a]}{[B]}\right] \tag{7.187}$$

and

$$\{D\} = \begin{bmatrix} [T](\{f\} - \{b\}) \\ \hline \{\dot{g}\} - [\dot{B}]\ \{y\} \end{bmatrix} \tag{7.188}$$

Equation (7.186) is an exact form of equation (7.35) developed for general constrained multibody systems. In equations (7.187) and (7.188), a, b and f are given by (7.177) to (7.179). B, $\dot{B}$ and $\dot{g}$ will be formulated in the next sections for the prescribed motion and the prescribed path type of the constraints that will be used in the simulations.

Prescribed Motion Type of Constraint

The position equation for this constraint as stated earlier is given by

$$y_c = a_1 t^2 + a_2 \tag{7.189}$$

Making use of the position of the tip of link 2 in the x, y frame, we write

$$L_1 s1 + L_2 s(1+2) = a_1 t^2 + a_2 \tag{7.190}$$

Differentiation of the above yields

$$\dot{\theta}_1\ [L_1 c1 + L_2 c(1+2)] = \dot{\theta}_2\, [L_2 c(1+2)] = 2a_1 t \tag{7.191}$$

or

$$[B]\left\{\dot{\theta}\right\} = \{g\} \tag{7.192}$$

where

$$B = [L_1 c1 + L_2 c(1+2) L_2 c(1+2)], \qquad g = 2a_1 t \tag{7.193}$$

Further differentiation gives the expression at the acceleration level:

$$\begin{aligned} &\ddot{\theta}_1\, [L_1 \mathrm{c}1 + L_2 \mathrm{c}(1+2)] + \ddot{\theta}_2 L_2 c(1+2) \\ &\qquad = 2a_1 + L_1 \dot{s} 1 \dot{\theta}_1^2 + L_2 s(1+2)(\dot{\theta}_1 + \dot{\theta}_2)^2 \end{aligned} \tag{7.194}$$

which can be put in the following form:

$$[B]\left\{\ddot{\theta}\right\} = \{\dot{g}\} - \left[\dot{B}\right]\{\theta\} \tag{7.195}$$

Simulation Results

Combining equations (7.183) and (7.195), the equations were put in the form of equation (7.186) and a standard IMSL routine, the Runge–Kutta, was used for the integration of the equations. For the simulation of this system, the following data were used:

$$L_1 = L_2 = 1\,\text{m} \quad m_1 = m_2 = 36\,\text{kg} \quad a_1 = 1\,\text{m/s}^2 \; a_2 = 0$$

That is, $f(t) = t^2$.

The initial position is given by Figure 7.12, where $\theta_1 = 70°$, $\theta_2 = -140°$. The initial velocities are assumed zero, $\dot{\theta}_1 = \dot{\theta}_2 = 0$. The system is simulated for 0.8 s. During this period θ_1 varies from 70° to 41°, and θ_2 varies from −140° to −41.6°. The path of the end effector is shown in Figure 7.12, and where the final position at the location of the end effector at different times are shown. θ_1 and θ_2 are given in Table 7.2.

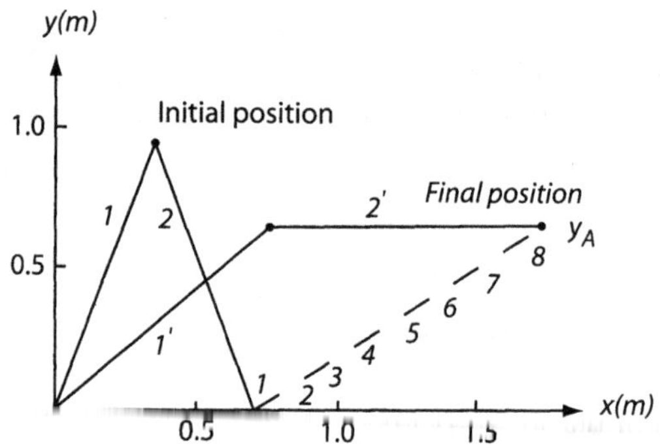

Fig. 7.12. Path of the end effector for a two-link arm robot

Position number	Time (s)	$\theta_1 - 70°$ (deg)	$\theta_2 + 140°$ (deg)
0	0	0	0
1	0.1	0.08	1.47
2	0.2	0.12	5.73
3	0.3	−0.36	12.55
4	0.4	−1.83	21.90
5	0.5	−4.73	33.80
6	0.6	−9.57	49.24
7	0.7	−17.13	69.57
8	0.8	−29.00	98.43

Table 7.2. Prescribed motion constraint

Since the specified motion for the y coordinate of the end effector is in the positive y direction, the end effector moves upward, while it also moves to the right by the action of the gravitational forces.

Prescribed Path Constraint

Let us assume that the path for the end effector (endpoint of line 2) is specified by the relationship

$$y_c = b_1 x^2 + b_2 \tag{7.196}$$

where b_1 and b_2 are selected constants. Expressing y and x in terms of θ_1 and θ_2 and differentiating in time, the constraint equations above become

$$\begin{aligned}
&\dot{\theta}_1\{L_1c1 + L_2c(1+2) + 2b_1[L_1^2c1s1 + L_2^2c(1+2)s(1+2) \\
&\quad + L_1L_2s1c(1+2) + L_1L_2c1s(1+2)]\} \\
&\quad + \dot{\theta}_2\{L_2c(1+2) + 2b_1[L_2^2c(1+2)s(1+2) + L_1L_2c1s(1+2)]\} = 0 \qquad (7.197)
\end{aligned}$$

and the differentiated form of the above equation yields

$$\begin{aligned}
&\ddot{\theta}_1\{L_1c1 + L_2c(1+2) + 2b_1[L_1^2c1s1 + L_2^2c(1+2)s(1+2) \\
&\quad + L_1L_2s1c(1+2) + L_1L_2cls(1+2)]\} \\
&\quad + \ddot{\theta}_2\{L_2c(1+2) + 2b_1[L_2^2c(1+2)s(1+2) + L_1L_2cls(1+2)]\} \\
&\quad + \dot{\theta}_1\{L_1s1\dot{\theta}_1 + L_2s(1+2)(\dot{\theta}_1+\dot{\theta}_2) + 2b_1(L_1^2s_1^2\dot{\theta}_1 - L_1^2{}_1^2\dot{\theta}_1) \\
&\quad + l_1l_2(2\dot{\theta}_1+\dot{\theta}_2)c(2\theta_1+\theta_2) + L_2^2s^2(1+2)(\dot{\theta}_1+\dot{\theta}_2) - L_2^2c^2(1+2)(\dot{\theta}_1+\dot{\theta}_2)]\} \\
&\quad + \dot{\theta}_2\{L_2s(1+2)(\dot{\theta}_1+\dot{\theta}_2) + 2b_1[L_2^2s^2(1+2)(\dot{\theta}_1+\dot{\theta}_2) - L_2^2c^2(1+2)(\dot{\theta}_1+\dot{\theta}_2) \\
&\quad - L_1L_2clc(1+2)(\dot{\theta}_1+\dot{\theta}_2)] + L_1L_2s1s(1+2)\dot{\theta}_1\} \qquad (7.198)
\end{aligned}$$

The variables B, $\dot{g}$, $\dot{B}\dot{\theta}$ in equations (7.185) and (7.188) are obtained from the left- and right-hand sides of equations (7.197) and (7.198), respectively. Essentially the constraints as represented in the automated form shown in equation (7.187) are easily obtained from factorizing the above equations.

Simulation Results

The data used for this system are $L = L_2 = 1\,\text{m}$, $m_1 = m_2 = 36\,\text{kg}$, $b_1 = 1\,\text{m}^{-1}$; and b_2 is determined by the specified initial configuration. The initial position is the same as in the previous simulation. Initially $\dot{\theta}_1 + \dot{\theta}_2 = 0$ again.

The system is simulated for 0.8 s and the simulation is shown in Figure 7.13. At this period θ_1 varies from 70° to −113°. The values $\theta 1$ and θ_2 are given in Table 7.3. In Figure 7.13, the initial and final positions and the position of the end effector at different times are shown. In this system, the end effector first moves downward, following the prescribed path by the effect of the gravitational forces. Then (since the parabolic path turns upwards when x becomes zero), the end effector moves upward with the speed gained.

7.12 Evaluation of the Constraint Forces

In most of the dynamic analysis problems, the evaluation of the constraint forces and torques is often of no concern or may be overlooked and therefore not computed. When the generalized constraint forces become important in the analysis of constrained multibody systems, it is essential to compute the corresponding restraining forces and torques of the system. The following procedure can be used to evaluate the constraint forces and torques.

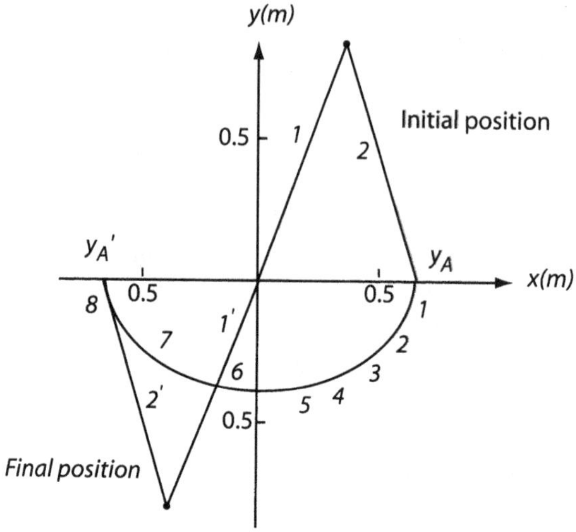

Fig. 7.13. Simulation of a two-link manipulator subject to a prescribed path

Position number	Time (s)	$\theta_1 - 70$ (deg)	$\theta_2 + 140$ (deg)
0	0	0	0
1	0.1	−2.29	−1.55
2	0.2	−9.57	−5.33
3	0.3	−22.75	9.45
4	0.4	−42.63	−12.38
5	0.5	−69.96	−13.58
6	0.6	−104.68	−13.52
7	0.7	−144.38	−10.71
8	0.8	−182.89	−3.50

Table 7.3. Prescribed motion constraint

Let point $P_j(j = 1, \ldots, N_1)$ and bodies $B_j(j = 1, \ldots, N_2)$ in the multibody structure shown in Figure 7.14 have specified motions described by

$$[B]\{\dot{x}\} = \{g\} \tag{7.199}$$

The number of the constraint equations is then $m = 3(N_1 + N)$.

Let $\overline{F}_j(j = 1, \ldots, N_1)$ represent the associated constraint forces at P_j and $\overline{M}_j \quad (j = 1, \ldots, N_2)$ be the constraint moments at B_j. Let the first $3N_1$ of the constraint equation (7.199) which are corresponding to P_j be designated as

$$\overline{v}_j = \{\ddot{x}\}^T [V_w^i]\{\bar{n}\} \quad (j = 1, \ldots, N_1) \tag{7.200}$$

Note that $\overline{v}_j$ is the specified velocity of P_j, and $[V_w^j]$ is the associated partial velocity array.

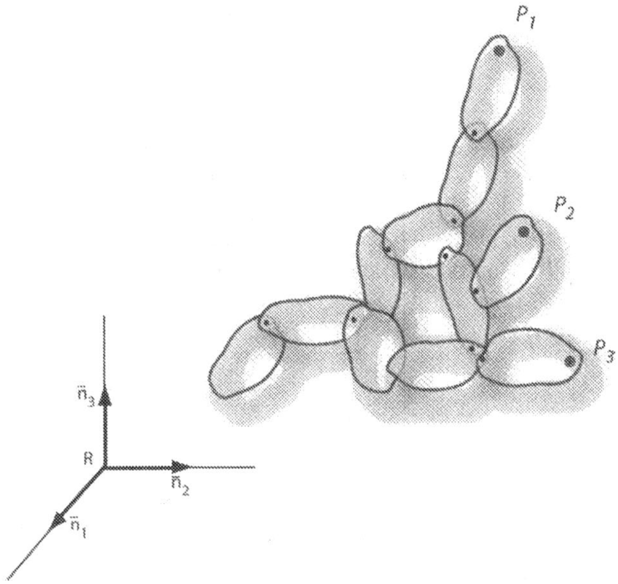

Fig. 7.14. Constrained multibody system with several points subject to specified motions

Similarly, the last three N_2 of the constraint equations (7.199) corresponding to B_i are

$$\overline{\omega}^j = \{\dot{x}\}^T[\omega^j]\{\bar{n}\} \quad (j = 1, \ldots, N_2) \tag{7.201}$$

Similarly, $\overline{\omega}^j$ is specified angular velocity of B_j and $[\omega^j]$ is the partial angular velocity array.

Then the generalized forces due to the constraint forces and moments are

$$\{f^c\} = \frac{\partial \bar{v}_j}{\partial \dot{x}} \cdot \bar{F}_j + \frac{\partial \bar{\omega}^j}{\partial \dot{x}} \cdot \bar{M}_j \tag{7.202}$$

or

$$\{f^c\} = [V_w^j]\{F_j\} + [\omega^j]\{M_j\} \tag{7.203}$$

Rewriting equation (7.203) as one array $B_{i\ell}$, $i = 1, \ldots, 3(N_1 + N_2)$, and letting λ be a $3(N_1 + N_2) \times 1$ array, one is left with

$$f^c = B^T \lambda \tag{7.204}$$

The equations of motion for the configuration system given by Figure 7.14 is then given by

$$f + f^* + f^c = 0 \tag{7.205}$$

Using the PUTD method, we can find a matrix T such that equation (7.205) becomes

$$Tf + Tf^* = 0 \tag{7.206}$$

Equations (7.206) and (7.199) can then be used to determine the kinematics of the system, especially the generalized speeds $\{\dot{x}\}$. Equation (7.205) can then be utilized to obtain the generalized constraint forces f^c. Then equation (7.204) can be utilized for determining the individual constraint force components $\lambda_i (i = 1, \ldots, 3(N_1 + N_2)$. Premultiplying by B, in equation (7.204) we obtain

$$E\lambda = Bf^c \tag{7.207}$$

where E is an $m \times m$ symmetric matrix given by

$$E = BB^T \tag{7.208}$$

Then

$$\lambda = E^{-1}Bf^c \tag{7.209}$$

provides an expression for calculating the individual constraint force components.

To avoid calculations involving matrix inversion, one can make use of the upper triangular matrix HB^T, which is available from the PUTD method used in obtaining the transformation matrix T. Premultiplying equation (7.204) by $D_2^T H$, we get

$$U\lambda = Q \tag{7.210}$$

where U is an $m \times m$ upper triangular matrix and Q is an $m \times 1$ array, given by

$$U = D_2^T HB^T \tag{7.211}$$

and

$$Q = D_2^T Hf^c = -D_2^T H(f + f^*) \tag{7.212}$$

Then λ_i can be calculated by back substitution.

Example 7.12.1
Determine the constraint forces and torques for Example 7.11.1 following the procedures outlined in this section.

Solution
The joint constraint torques for both constraint paths are computed through a general-purpose program. Results are displayed in Tables 7.4 and 7.5, and their corresponding graphs are given by Figures 7.15 and 7.16.

M_1 (N – m)	M_2 (N – m)
241.0	123.4
235.0	129.7
238.2	141.5
251.6	157.5
271.2	170.9
287.8	177.5
238.3	163.7
222.5	109.4

Table 7.4.

M_1 (N – m)	M_2 (N – m)
−3.80	9.91
−0.55	9.63
13.01	−37.83
15.96	−137.41
1.48	−295.80
−24.77	−438.51
−91.99	−412.15
−161.09	−324.10

Table 7.5.

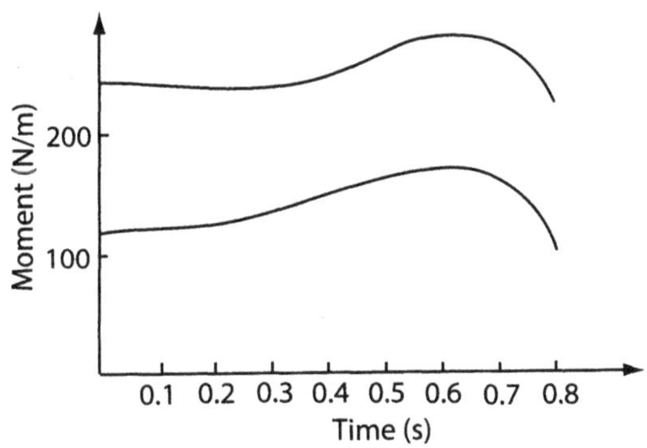

Fig. 7.15. Joint moments: (1) M_1 (2) M_2

Example 7.12.2
Consider the planetary gear system shown in Figure 7.17. Let I_0 be the equivalent moment of inertia of the motor and I_3 the equivalent moment of inertia of the load. The moments of inertia of gears 1 and 2 are I_1 and I_2, respectively, and the moments of inertia of shafts 1 and 2 are assumed to be small, hence negligible.

(a) Use multibody dynamics to model the system.
(b) Derive the constraints equations.
(c) Find the contact force between gears (constraint forces).

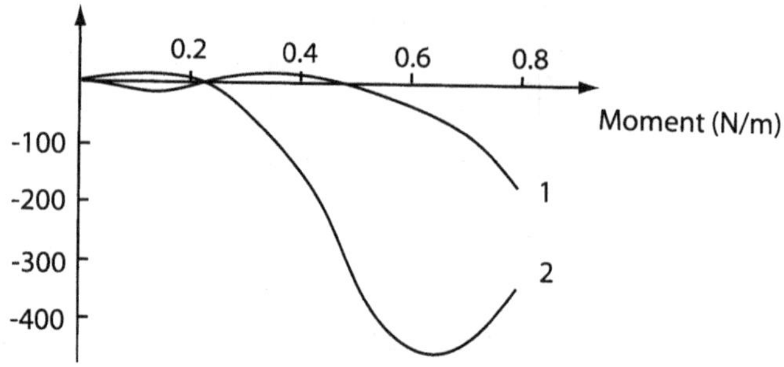

Fig. 7.16. Joint moments: (1) M_1 (2) M_2

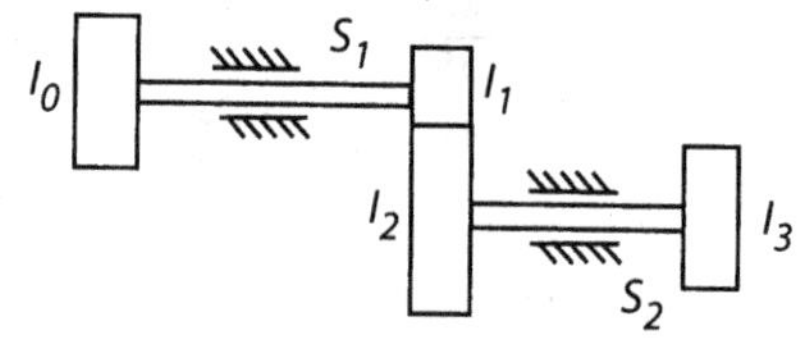

Fig. 7.17. Planetary gear system

Solution

(a) The system is modeled by four bodies: the motor, two gears, and the load shafts 1 and 2 are modeled as torsional springs. Assume that $\theta_i (i = 0, \ldots, 3)$ are the generalized coordinates of the motor, gear 1, gear 2, and the load respectively. Then the kinetic energy of the system is given by

$$T = \frac{1}{2} \sum_{i=0}^{3} I_i \theta_i^2 \tag{7.213}$$

The potential energy of the system is

$$V = \frac{1}{2} [k_i (\theta_1 - \theta_0)^2 + k_2 (\theta_3 - \theta_2)^2] \tag{7.214}$$

where k_1 and k_2 are stiffness coefficients of shafts 1 and 2, respectively. Deriving the equations of motion we obtain

$$[M]\{\ddot{\theta}\} + [K]\{\theta\} = \{f\} \tag{7.215}$$

where the mass matrix is

$$[M] = \begin{bmatrix} I_0 & & & \\ & I_1 & & \\ & & I_2 & \\ & & & I_3 \end{bmatrix} \tag{7.216}$$

and the stiffness matrix

$$[K] = \begin{bmatrix} k_1 & -k_1 & & \\ -k_1 & k_1 & & 0 \\ & 0 & k_2 & -k_2 \\ & & -k_2 & k_2 \end{bmatrix} \tag{7.217}$$

The external force vector

$$\{f\} = \begin{bmatrix} f_0 \\ 0 \\ 0 \\ f_3 \end{bmatrix} \tag{7.218}$$

(b) *Case* 1 : The constraint equations for the multibody system of the gear train are evaluated by imposing the conditions between the angular displacements of each body. Those constraints are given as follows:

$$\begin{aligned} \theta_0 &= \theta_1 \\ r_1\theta_1 &= r_2\theta_2 \\ \theta_3 &= \theta_2 \end{aligned} \tag{7.219}$$

where r_1 and r_2 are the radius of gears 1 and 2, respectively. In the equations above, the stiffness coefficients k_1 and k_2 are assumed to be infinite. Then the equations of motion of the system including the constraint forces become

$$[M]\{\ddot{\theta}\} + [K]\{\theta\} + [B]^T\{\lambda\} = \{f\} \tag{7.220}$$

where the Jacobi constraint matrix is found from (7.219) by factorizing θ_1, θ_2 and θ_3

$$[B] = \begin{bmatrix} 1 & -1 & 0 & 0 \\ 0 & r_1 & -r_2 & 0 \\ 0 & 0 & 1 & -1 \end{bmatrix} \tag{7.221}$$

Lagrange undetermined multiplier array can be expressed as

$$\{\lambda\} = \begin{bmatrix} \lambda_1 \\ \lambda_2 \\ \lambda_3 \end{bmatrix} \tag{7.222}$$

From equation (7.221), we can find a matrix $[T]$ such that

$$[T][B]^T\{\lambda\} = 0 \tag{7.223}$$

After some laborious work, using the PUTD method, we derive T and subsequently arrive at a simplified form of the equations of motion as

$$\left[I_0 + I_1 + \frac{r_1^2}{r_2^2}(I_2 + I_3) \right] \ddot{\theta}_0 = f_0 + \frac{r_1}{r_2} f_3 \tag{7.224}$$

It is obvious that our system is driven by the motor and possesses only one degree of freedom θ_0. From equation (7.224) we solve for the angular acceleration $\ddot{\theta}_0$:

$$\ddot{\theta}_0 = \frac{f_0 + (r_1/r_2) f_3}{I_0 + I_1 + (r_1^2/r_2^2)(I_2 + I_3)} \tag{7.225}$$

Observe if $f_0 + (r_1/r_2) f_3 = 0$; then $\ddot{\theta}_0 = 0$, $\dot{\theta}_0 = \text{const}$. The system will then rotate with a constant angular velocity ω.

Case 2 : Let us assume that the equations of the constraints for the system are simply given by

$$\begin{aligned} r_1\theta_1 &= r_2\theta_2 \\ \theta_3 &= \theta_2 \end{aligned} \tag{7.226}$$

The equations above mean that the stiffness coefficient k_1 is finite while k_2 is infinite. Then the Jacobi constraint matrix becomes

$$[B] = \begin{bmatrix} 0 & r_1 & -r_2 & 0 \\ 0 & 0 & 1 & -1 \end{bmatrix} \tag{7.227}$$

and Lagrange undetermined multipliers

$$\{\lambda\} = \begin{Bmatrix} \lambda_1 \\ \lambda_2 \end{Bmatrix} \tag{7.228}$$

We can also find a matrix $[T]$ such that

$$[T][B]^T\{\lambda\} = 0 \tag{7.229}$$

Similarly, after some work we obtain the desired equations of motion

$$\begin{bmatrix} I_0 & 0 \\ 0 & I_1 + \dfrac{r_1^2}{r_2^2}(I_2 + I_3) \end{bmatrix} \begin{Bmatrix} \ddot{\theta}_0 \\ \ddot{\theta}_1 \end{Bmatrix} + \begin{bmatrix} k_1 & -k_1 \\ -k_1 & k_1 \end{bmatrix} \begin{Bmatrix} \theta_0 \\ \theta_1 \end{Bmatrix} = \begin{Bmatrix} f_0 \\ \dfrac{r_1}{r_2} f_3 \end{Bmatrix} \tag{7.230}$$

Let the solution of the above equation be carried out,

$$\ddot{\theta} = -\omega^2\theta \quad \text{and} \quad \left| -\omega^2[M] + [K] \right| = 0 \tag{7.231}$$

that is,

$$\begin{vmatrix} -\omega^2 I_0 + k_1 & -k_1 \\ -k_1 & k_1 - \omega^2 \left[I_1 + \dfrac{r_1^2}{r_2^2}(I_2 + I_3) \right] \end{vmatrix} = 0 \tag{7.232}$$

The solution to equation (7.232) yields the modal frequencies of the system:

$$\omega_1 = 0 \tag{7.233}$$

$$\omega_2 = \sqrt{\frac{k_1[I_0 + I_1 + (r_1^2/r_2^2)(I_2 + I_3)]}{I_0[I_1 + (r_1^2/r_3^2)(I_2 + I_3)]}} \tag{7.234}$$

and their corresponding modal shapes:

$$[\phi] = \begin{bmatrix} \dfrac{1}{\sqrt{I + I_0}} - \sqrt{\dfrac{I}{I_0^2 + I I_0}} \\ \dfrac{1}{\sqrt{I + I_0}} \ \sqrt{\dfrac{I_0}{I_0^2 + I I_0}} \end{bmatrix} \tag{7.235}$$

where

$$I = I_1 + \frac{r_1^2}{r_2^2}(I_2 + I_3) \tag{7.236}$$

Let $[\theta] = [\phi][\Theta]$ and premultiply both sides of equation (7.231) by $[\phi]^T$:

$$\{\ddot{\Theta}\} + \begin{bmatrix} 0 & 0 \\ 0 & \omega_2^2 \end{bmatrix} \{\ddot{\Theta}\} = \frac{1}{\sqrt{I + I_0}} \left(f_0 + \frac{r_1}{r_2} f_3 \right)_3 \frac{1}{\sqrt{I_0 + I I_0}} f_0 + \frac{r_1}{r_2} - \sqrt{\frac{I_0}{I^2 + I I_0}} f_3 \tag{7.237}$$

The solution to the differential equations above is given by

$$\Theta_1 = \frac{1}{\sqrt{I + I_0}} \int_0^t \int_0^t \left[f_0(\xi) + \frac{r_1}{r_2} f_3(\xi) \right] d\xi \, d\tau + c_1 t + c_2$$

and

$$\Theta_2 = c_3 \cos \omega_2 t + c_4 \sin \omega_2 t + \frac{1}{\omega_2} \int_0^t - \left[\sqrt{\frac{I_0}{I_0^2 + I I_0}} f_0 \right.$$
$$\left. + \frac{r_1}{r_2} - \sqrt{\frac{I_0}{I^2 + I I_0}} f_3(\tau) \right] \sin \omega_2 (t - \tau) d\tau \tag{7.238}$$

where $c_i (i = 1, \ldots, 4)$ are constants depending on the initial conditions. Finally, by back substitution, we obtain

$$\{\theta\} = [\phi]\{\Theta\} \tag{7.239}$$

Obviously, the solution of the system consists of two parts. One is rigid-body movement and the other is the torsional vibration of shaft 1.

(c) Now let us discuss the constraint forces defined by

$$\{f^c\} = [B]^T\{\lambda\} = \{f\} - [M]\{\ddot{\theta}\} - [K]\{\theta\} \tag{7.240}$$

For case 1

$$\{f^c\} = \begin{bmatrix} f_0 - I_0\ddot{\theta}_0 \\ -I_1\ddot{\theta}_0 \\ -I_2\dfrac{r_1}{r_2}\ddot{\theta}_0 \\ f_3 - I_3\dfrac{r_1}{r_2}\ddot{\theta}_0 \end{bmatrix} \tag{7.241}$$

and for case 2

$$\{f^c\} = \begin{bmatrix} f_0 - I_0\ddot{\theta}_0 \\ -I_1\ddot{\theta}_1 - k_1(\theta_1 - \theta_0) \\ -I_2\dfrac{r_1}{r_2}\ddot{\theta}_1 \\ f_3 - I_3\dfrac{r_1}{r_2}\ddot{\theta}_1 \end{bmatrix} \tag{7.242}$$

where θ_0 and θ_1 are described by equation (7.239).

Example 7.12.3
For the pantograph shown in Figure 7.18, find the constraint equations and corresponding equations of motion of the system and the locus of point P.

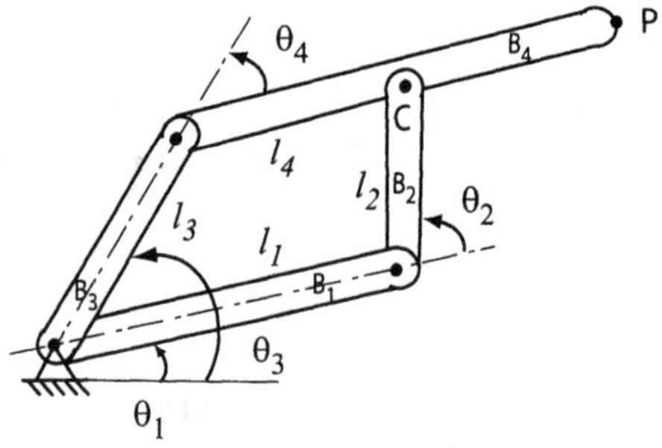

Fig. 7.18. Planar pantograph system with four links

Solution
From Sections 5.4 and 5.10 we get the equations of motion of the system, which have the following form:

$$[a]\{\ddot{\theta}\} + [b]\{\dot{\theta}\} = \{F\} \tag{7.243}$$

or

$$\begin{bmatrix} a_{11} & a_{12} & 0 & 0 \\ a_{21} & a_{22} & 0 & 0 \\ 0 & 0 & a_{33} & a_{34} \\ 0 & 0 & a_{43} & a_{44} \end{bmatrix} \begin{Bmatrix} \ddot{\theta}_1 \\ \ddot{\theta}_2 \\ \ddot{\theta}_3 \\ \ddot{\theta}_4 \end{Bmatrix} + \begin{bmatrix} b_{11} & b_{12} & 0 & 0 \\ b_{21} & b_{22} & 0 & 0 \\ 0 & 0 & b_{33} & b_{34} \\ 0 & 0 & b_{43} & b_{44} \end{bmatrix} \begin{Bmatrix} \dot{\theta}_1 \\ \dot{\theta}_2 \\ \dot{\theta}_3 \\ \dot{\theta}_4 \end{Bmatrix} = \begin{Bmatrix} f_1 \\ f_2 \\ f_3 \\ f_4 \end{Bmatrix} \tag{7.244}$$

Note that the equations above are derived by assuming that point c of link body B_2 is free. This allows us to open the loop formed by B_1, $B_2 B_3$ and B_4. Due to symmetry of branch 1 ($B_1 - B_2$) and branch 2 ($B_3 - B_4$), we get the equations for an open treelike system.

$$a_{11} = \frac{1}{3} m_1 \ell_1^2 + m_2 \left(\ell_1^2 + \ell_1 \ell_2 c2 + \frac{1}{3} \ell_2^2 \right) \tag{7.245}$$

$$a_{33} = \frac{1}{3} m_3 \ell_3^2 + m_4 \left[\ell_3^2 + \ell_3 (\ell_4 + \ell_5) c4 + \frac{1}{3} (\ell_4 + \ell_5)^2 \right] \tag{7.246}$$

$$a_{12} = a_{21} = m_2 \left(\frac{1}{2} \ell_1 \ell_2 c2 + \frac{1}{3} \ell_2^2 \right) \tag{7.247}$$

$$a_{34} = a_{43} = m_4 \left[\frac{1}{2} \ell_3 (\ell_4 + \ell_5) c4 + \frac{1}{3} (\ell_4 + \ell_5)^2 \right] \tag{7.248}$$

$$a_{22} = \frac{1}{3} m_2 \ell_2^2 \tag{7.249}$$

$$a_{44} = \frac{1}{3} m_4 (\ell_4 + \ell_5)^2 \tag{7.250}$$

$$b_{11} = -\frac{1}{2} m_2 \ell_1 \ell_2 s2 \dot{\theta}_2 \tag{7.251}$$

$$b_{33} = -\frac{1}{2} m_4 \ell_3 (\ell_4 + \ell_5) s4 \dot{\theta}_4 \tag{7.252}$$

$$b_{12} = -\frac{1}{2} m_2 \ell_1 \ell_2 s2 (\dot{\theta}_1 + \dot{\theta}_2) \tag{7.253}$$

$$b_{34} = -\frac{1}{2} m_4 \ell_3 (\ell_4 + \ell_5) s4 (\dot{\theta}_3 + \dot{\theta}_4) \tag{7.254}$$

$$b_{21} = \frac{1}{2} m_2 \ell_1 \ell_2 s2 \dot{\theta}_1 \tag{7.255}$$

$$b_{43} = \frac{1}{2} m_4 \ell_3 (\ell_4 + \ell_5) s4 \dot{\theta}_3 \tag{7.256}$$

$$b_{22} = b_{44} = 0 \tag{7.257}$$

$$f_1 = -\frac{1}{2}m_1 g \ell_1 s1 - m_2 g\left[\ell_1 s1 + \frac{1}{2}\ell_2 s(1+2)\right] \tag{7.258}$$

$$f_3 = -\frac{1}{2}m_3 g \ell_3 s3 - m_4 g\left[\ell_3 s3 + \frac{1}{2}(\ell_4 + \ell_5) s(3+4)\right] \tag{7.259}$$

$$f_2 = -\frac{1}{2}m_2 g \ell_2 s(1+2) \tag{7.260}$$

$$f_4 = -\frac{1}{2}m_4 g(\ell_4 + \ell_5) s(3+4) \tag{7.261}$$

where m_1 to m_4 are the masses of bodies 1 to 4, respectively.

There are four degrees of freedom associated with equations (7.243). Now if we consider the constraint resulting from point C being fixed to B_4, the system will have only two degrees of freedom. The constraint conditions, which are geometrical in nature, can be described by

$$x_c(\text{branch 1}) = x_c(\text{branch 2}) \tag{7.262}$$

$$y_c(\text{branch 1}) = y_c(\text{branch 2}) \tag{7.263}$$

that is

$$\bar{\ell}_1 + \bar{\ell}_2 = \bar{\ell}_3 + \bar{\ell}_4 \tag{7.264}$$

where

$$\bar{\ell}_1 = \left[\ell_1\ 0\ 0\right][S^{10}]\{\bar{n}\} \tag{7.265}$$

$$\bar{\ell}_2 = \left[\ell_2\ 0\ 0\right][S^{20}]\{\bar{n}\} \tag{7.266}$$

$$\bar{\ell}_3 = \left[\ell_3\ 0\ 0\right][S^{30}]\{\bar{n}\} \tag{7.267}$$

$$\bar{\ell}_4 = \left[\ell_4\ 0\ 0\right][S^{40}]\{\bar{n}\} \tag{7.268}$$

$$[S^{10}] = \begin{bmatrix} c1 & s1 & 0 \\ -s1 & c1 & 0 \\ 0 & 0 & 1 \end{bmatrix} \tag{7.269}$$

$$\left[S^{20}\right] = \begin{bmatrix} c(1+2) & s(1+2) & 0 \\ -s(1+2) & c(1+2) & 0 \\ 0 & 0 & 1 \end{bmatrix} \tag{7.270}$$

$$[S^{30}] = \begin{bmatrix} c3 & s3 & 0 \\ -s3 & c3 & 0 \\ 0 & 0 & 1 \end{bmatrix} \tag{7.271}$$

$$\left[S^{40}\right] = \begin{bmatrix} c(3+4) & s(3+4) & 0 \\ -s(3+4) & c(3+4) & 0 \\ 0 & 0 & 1 \end{bmatrix} \tag{7.272}$$

where $c1 = \cos\theta_1$, $c2 = \cos\theta_2$, $c(1+2) = \cos(\theta_1 + \theta_2)$, and so on. Differentiating both sides of equation (7.264), we get

$$\begin{aligned}&[\ell_1\ 0\ 0]\left[\Omega^{10}\right]\left[S^{10}\right]+[\ell_2\ 0\ 0]\left[\Omega^{10}\right]\left[S^{20}\right]\\&=[\ell_3\ 0\ 0]\left[\Omega^{30}\right]\left[S^{30}\right]-[\ell_4\ 0\ 0]\left[\Omega^{40}\right]\left[S^{40}\right]\end{aligned} \tag{7.273}$$

where

$$\left[\Omega^{10}\right]=\dot{\theta}_1\begin{bmatrix}0&1&0\\-1&0&0\\0&0&0\end{bmatrix} \tag{7.274}$$

$$\left[\Omega^{20}\right]=(\dot{\theta}_1+\dot{\theta}_2)\begin{bmatrix}0&1&0\\-1&0&0\\0&0&0\end{bmatrix} \tag{7.275}$$

$$\left[\Omega^{30}\right]=\dot{\theta}_3\begin{bmatrix}0&1&0\\-1&0&0\\0&0&0\end{bmatrix} \tag{7.276}$$

$$\left[\Omega^{40}\right]=(\dot{\theta}_3+\dot{\theta}_4)\begin{bmatrix}0&1&0\\-1&0&0\\0&0&0\end{bmatrix} \tag{7.277}$$

Simplifying and rearranging, we arrive at

$$\begin{aligned}&\ell_1\dot{\theta}_1\left[-s1\ c1\ 0\right]+\ell_2(\dot{\theta}_1+\dot{\theta}_2)\left[-s(1+2)\ c(1+2)\ 0\right]\cdot\ell_3\dot{\theta}_3\left[-s3\ c3\ 0\right]\\&\quad-\ell_4(\dot{\theta}_3+\dot{\theta}_4)\left[-s(3+4)\ c(3+4)\ 0\right]=0\end{aligned} \tag{7.278}$$

or

$$\ell_1 s1\dot{\theta}_1-\ell_2 s(1+2)(\dot{\theta}_1+\dot{\theta}_2)+\ell_3 c3\dot{\theta}_3+l_4 s(3+4)(\dot{\theta}_3+\dot{\theta}_4)=0 \tag{7.279}$$

$$\ell_1 c1\dot{\theta}_1-\ell_2 c(1+2)(\dot{\theta}_1+\dot{\theta}_2)-\ell_3 c3\dot{\theta}_3-\ell_4 c(3+4)(\dot{\theta}_3+\dot{\theta}_4)=0 \tag{7.280}$$

that is,

$$[B]\{\dot{\theta}\}=0 \tag{7.281}$$

where

$$\{\dot{\theta}\}=\left[\dot{\theta}_1\ \dot{\theta}_2\ \dot{\theta}_3\ \dot{\theta}_4\right]^T \tag{7.282}$$

and

$$[B]=\begin{bmatrix}-\ell_1 s1-\ell_2 s(1+2) & -\ell_2 s(1+2) & \ell_3 s3+\ell_4 s(3+4) & \ell_4 s(3+4)\\ \ell_1 c1+\ell_2 c(1+2) & \ell_2 c(1+2) & -\ell_3 c3-l_4 c(3+4) & \ell_4 c(3+4)\end{bmatrix} \tag{7.283}$$

The generalized constraint force is given by

$$\left\{f^c\right\}=[B]^T\ [\lambda] \tag{7.284}$$

where

$$\{\lambda\} = \left[\lambda_1 \; \lambda_2\right]^T \tag{7.285}$$

Expressing the equations of motion given by (7.243) by adding the generalized constraint forces, we obtain

$$[a]\{\ddot{\theta}\} + [b]\{\dot{\theta}\} + [B]^T\{\lambda\} = \{f\} \tag{7.286}$$

We also know that if we differentiate equation (7.281) once more, we get

$$[B]\{\ddot{\theta}\} + [B]\{\dot{\theta}\} = 0 \tag{7.287}$$

According to equation (7.20), equation (7.243) becomes

$$\begin{bmatrix} [a] & [B]^T \\ [B] & 0 \end{bmatrix} \begin{Bmatrix} \ddot{\theta} \\ \lambda \end{Bmatrix} = \begin{bmatrix} [H] \\ [G] \end{bmatrix} + \begin{bmatrix} f \\ 0 \end{bmatrix} \tag{7.288}$$

where $[a]$ is given by equation (7.243),

$$\{\ddot{\theta}\} = \left[\ddot{\theta}_1 \; \ddot{\theta}_2 \; \ddot{\theta}_3 \; \ddot{\theta}_4\right]^T \quad \{\lambda\} = \begin{Bmatrix} \lambda_1 \\ \lambda_2 \end{Bmatrix}$$

and $\{f\} = \left[f_1 \; f_2 \; f_3 \; f_4\right]^T$ is given by equation (7.244), the remaining coefficients are

$$[H] = -[b]\{\dot{\theta}\}$$

and

$$[G] = -[\dot{B}]\{\dot{\theta}\}$$

We can numerically solve equation (7.286) or (7.288) and find the constraint force given by equation (7.284). An alternative to the numerical solution will be to write equation (7.288) in the form of equation (7.35):

$$[A]\{\ddot{\theta}\} = \{D\} \tag{7.289}$$

where the coefficients A and D are found as

$$[A] = \begin{bmatrix} [T][a] \\ \hline [B] \end{bmatrix}_{4\times 4} \tag{7.290}$$

and

$$\{D\} = \begin{bmatrix} [T]\{f\} - [b]\{\dot{\theta}\} \\ \hline -[\dot{B}]\{\dot{\theta}\} \end{bmatrix}_{4\times 1} \tag{7.291}$$

$[T]_{2\times 4}$ is an orthogonal complement matrix that can be found numerically. Then the position vector of point p is found by through the standard form given in Chapter 3 as

$$\overline{P} = \left[\ell_3\ 0\ 0\right]\left[S^{30}\right]\{\bar{n}\} + \left[\ell_4 + \ell_5\ 0\ 0\right]\left[S^{40}\right]\{\bar{n}\} \tag{7.292}$$

which leads to an explicit form

$$\overline{P} = [\ell_3 c3 + (\ell_4 + \ell_5)\, c\,(3+4)]\bar{n}_1 + [\ell_3 s3 + (\ell_4 + \ell_5) s(3+4)]\bar{n}_2$$

The velocity of point $\overline{P}$ is

$$\begin{aligned}
\bar{v}_p &= \frac{d}{dt}\bar{P} \\
&= \left(\left[\ell_3\ 0\ 0\right]\left[\Omega^{30}\right]\left[S^{30}\right]\{\bar{n}\} + \left[\ell_4 + \ell_5\ 0\ 0\right]\left[\Omega^{40}\right]\left[S^{40}\right]\right)\{\bar{n}\} \qquad (7.293)\\
&= -[(\ell_3\dot{\theta}_3 s3) + (\ell_4 + \ell_5)(\dot{\theta}_3 + \dot{\theta}_4)s(3+4)]\bar{n}_1 \\
&\quad + [(\ell_3\dot{\theta}_3 c3) + (\ell_4 + \ell_5)\left(\dot{\theta}_3 + \dot{\theta}_4\right)c(3+4)]\bar{n}_2
\end{aligned}$$

and finally, we find the acceleration of point $\overline{P}$

$$\bar{a}_p = \frac{d^2}{dt^2}\bar{P} \tag{7.294}$$

making use of the definitions in Chapter 2 and Chapter 3 we can express $\bar{a}$ as

$$\begin{aligned}
\bar{a} &= \left[\ell_3\ 0\ 0\right]\left(\left[\Omega^{30}\right]\left[\Omega^{30}\right]\left[S^{30}\right] + \left[\dot{\Omega}^{30}\right]\left[S^{30}\right]\right)\{\bar{n}\} \\
&\quad + \left[\ell_4 + \ell_5\ 0\ 0\right]\left(\left[\Omega^{40}\right]\left[\Omega^{40}\right]\left[S^{40}\right] + \left[\dot{\Omega}^{40}\right]\left[S^{40}\right]\right)\{n\} \\
&= -\left[\ell_3\ddot{\theta}_3 s3 + \ell_3\dot{\theta}_3^2 c3 + (\ell_4 + \ell_5)(\ddot{\theta}_3 + \ddot{\theta}_4)s(3+4)\right. \\
&\quad \left. + \ (\ell_4 + \ell_5)^2\left(\dot{\theta}_3 + \dot{\theta}_4\right)c(3+4)\ \right]\bar{n}_1 \\
&\quad + \left[\ell_3\ddot{\theta}_3 c3 - \ell_3\dot{\theta}_3^2 s3 + (\ell_4 + \ell_5)\left(\ddot{\theta}_3 + \ddot{\theta}_4\right)c\,(3+4)\right. \\
&\quad \left. - (\ell_4 + \ell_5)\left(\dot{\theta}_3 + \dot{\theta}_4\right)^2 s(3+4)\right]\bar{n}_2
\end{aligned} \tag{7.295}$$

where $\left\{\ddot{\theta}\right\}$, $\left\{\dot{\theta}\right\}$ and $\{\theta\}$ are given by solving equation (7.288) or (7.289).

7.13 The Rolling Coin Problem: Kinematics, Forces and Constraints

The formulation for nonholonomic systems presented in this chapter could best be explained through the classical problem of the rolling coin. This is an example of a nonholonomic system used by a number of dynamicists.

Example 7.13.1
Let D denote the coin that is a sharp-edged circular disk that rolls on a plane support S fixed in a reference frame R (Figure 7.19). Line $\bar{n}_1$ is the tangent to the periphery of D at the point of contact C between D and S. Let $\bar{n}_1^1$, $\bar{n}_2^1$ and $\bar{n}_3^1$ be three mutually

perpendicular unit vectors, as shown, $\bar{n}_1^1$ is along $\bar{n}_1$ and $\bar{n}_2^1$ passes through the center G of D. Finally, $\bar{n}_x, \bar{n}_y$ and $\bar{n}_z$ are fixed in R and are mutually perpendicular.

(a) Find the governing equations of motion for the system

(b) Solve for the constraint forces at point C.

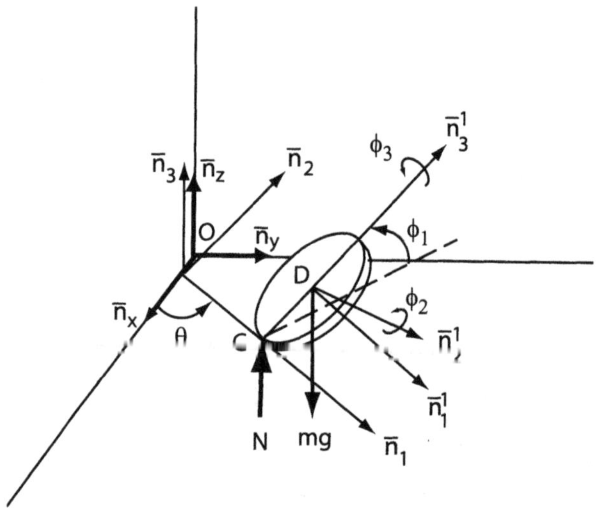

Fig. 7.19. Rolling sharp-edged circular disk.

Solution

Let the rolling coin be body B_1 and the following are the kinematics needed in the analysis. As usual we derive all kinematical quantities in an explicit form, where we define the angular velocity of body B, with respect to R as

$$\bar{\omega} = \{\dot{\phi}\}^T \left\{\bar{n}^1\right\} = \{\dot{\phi}\}^T \left[S^{1R}\right] \left\{\bar{n}^R\right\} \tag{7.296}$$

where

$$\{\dot{\phi}\}^T = [\dot{\phi}_1 \ \dot{\phi}_2 \ \dot{\phi}_3]$$

$$\left\{\bar{n}^R\right\}^T = [\bar{n}_x \ \bar{n}_y \ \bar{n}_z]$$

First we note that

$$\left\{\bar{n}^1\right\} = \begin{bmatrix} 1 & 0 & 0 \\ 0 & \cos(\phi_1 - 90°) & \sin(\phi_1 - 90°) \\ 0 & -\sin(\phi_1 - 90°) & -\cos(\phi_1 - 90°) \end{bmatrix} \{\bar{n}\}$$

expressing $\{\bar{n}\}$ with respect to R we get

$$\left\{\bar{n}^1\right\} = \begin{bmatrix} 1 & 0 & 0 \\ 0 & \sin\phi_1 & -\cos\phi_1 \\ 0 & \cos\phi_1 & \sin\phi_1 \end{bmatrix} \begin{bmatrix} \cos\theta & \sin\theta & 0 \\ -\sin\theta & \cos\theta & 0 \\ 0 & 0 & 1 \end{bmatrix} \left\{\bar{n}^R\right\}$$

Therefore, the shifter is simply

$$[S^{1R}] = \begin{bmatrix} \cos\theta & \sin\theta & 0 \\ -\sin\phi_1\sin\theta & \sin\phi_1\sin\theta & -\cos\phi_1 \\ -\cos\phi_1\sin\theta & \cos\phi_1\cos\theta & \sin\phi_1 \end{bmatrix} \tag{7.297}$$

with

$$\dot{\phi}_3\sin\phi_1 = \dot{\theta} \tag{7.298}$$

Hence the angular velocity components are

$$\{\omega\} = \begin{bmatrix} \dot{\phi}_1\cos\theta - \dot{\phi}_2\sin\phi_1\sin\theta - \dot{\phi}_3\cos\phi_1\sin\theta \\ \dot{\phi}_1\sin\theta + \dot{\phi}_2\sin\phi_1\sin\theta + \dot{\phi}_3\cos\phi_1\cos\theta \\ -\dot{\phi}_2\cos\phi_1 + \dot{\phi}_3\sin\phi_1 \end{bmatrix} \tag{7.299}$$

The coin has six possible degrees of freedom defined by

$$x^T = \left[\phi_1\ \phi_2\ \phi_3\ x\ y\ z\right] \tag{7.300}$$

where ϕ_1, ϕ_2 and ϕ_3 are defined as the rotations shown along the unit vectors given in Figure 7.19, and x, y and z represent the position of the center mass of the coin in R.

In a matrix form the angular velocity can be expressed as

$$\bar{\omega} = \{\dot{x}\}^T [\omega] \left\{\bar{n}^R\right\} \tag{7.301}$$

where $[\omega]$ is the partial angular velocity array defined by

$$[\omega] = \begin{bmatrix} [S^{1R}] \\ 0_{3\times 3} \end{bmatrix} \tag{7.302}$$

which is found through the simple differentiation

$$\frac{\partial\bar{\omega}}{\partial\dot{x}_\ell} = [\omega] \left\{\bar{n}^R\right\} \tag{7.303}$$

The angular acceleration is

$$\bar{\alpha} = \frac{d}{dt}\bar{\omega} = \left(\{\ddot{x}\}^T [\omega] + \{\dot{x}\}^T [\dot{\omega}]\right) \left\{\bar{n}^R\right\} \tag{7.304}$$

and

$$[\dot{\omega}] = \begin{bmatrix} [\dot{S}^{1R}] \\ 0_{3\times 3} \end{bmatrix} \tag{7.305}$$

Differentiation of $\left[S^{1R}\right]$ can be explicitly found

$$\left[\dot{S}^{1R}\right] = \begin{bmatrix} \dot{\phi}_3 s\phi_1 s\theta & \dot{\phi}_3 s\phi_1 c\theta & 0 \\ \dot{\phi}_1 c\phi_1 s\theta - \dot{\phi}_3 s\phi_1 s\phi_1 c\theta & \dot{\phi}_1 c\phi_1 c\theta - \dot{\phi}_3 s\phi_1 c\theta & \dot{\phi}_1 s\phi_1 \\ -\dot{\phi}_2 s\phi_1 s\theta - \dot{\phi}_3 s\phi_1 c\phi_1 c\theta & -\dot{\phi}_2 s\phi_1 c\theta - \dot{\phi}_3 s\dot{\phi}_1 c\phi_1 s\phi & \dot{\phi}_1 c\phi_1 \end{bmatrix} \tag{7.306}$$

The velocity of the center mass of the rolling coin is defined by

$$\bar{v} = \{\dot{x}\}^T [V] \left\{\bar{n}^R\right\} = \dot{x}\bar{n}_x + \dot{y}\bar{n}_y + \dot{z}\bar{n}_z \tag{7.307}$$

where $[V]$ is the partial velocity array which is seen to take a simple form

$$[V] = \begin{bmatrix} 0 \\ I \end{bmatrix} \tag{7.308}$$

where I denotes the identity matrix and 0 a zero matrix. The partial velocity is

$$\frac{\partial \bar{v}}{\partial \dot{x}_\ell} = [V] \left\{\bar{n}^R\right\} \tag{7.309}$$

The acceleration of the center mass is found through differentiation of equation (7.307).

$$\bar{a} = \{\ddot{x}\}^T [V] \left\{\bar{n}^R\right\} = \ddot{x}\bar{n}_x + \ddot{y}\bar{n}_y + \ddot{z}\bar{n}_z \tag{7.310}$$

The equations of motion are found through a, b, c formulation given in Chapter 5 which states

$$\{f\} = a\{\ddot{x}\} + b\{\dot{x}\} + c\{\dot{x}\} \tag{7.311}$$

where, as before, the coefficient of the equations above are

$$[a] = m[V][V]^T + [\omega][I_{1R}][\omega]^T$$
$$[b] = [\omega][I_{1R}][\dot{\omega}]^T$$
$$[c] = [\omega]\left[\Omega_x^{R1}\right][I_{1R}][\omega]^T$$

where $[I_{1R}]$ denotes the moment of inertia of the disk with respect to $\left\{\bar{n}^R\right\}$ in R and is given by

$$[I_{1R}] = \left[S^{R1}\right][I_1]\left[S^{1R}\right] \tag{7.312}$$

where I_1 is the moment of inertia with respect to the center axis.

$$[I_1] = \begin{bmatrix} I_{11} & & 0 \\ & I_{12} & \\ 0 & & I_{13} \end{bmatrix} \tag{7.313}$$

The values I_{11}, I_{12} and I_{13} are the moments of inertia with respect to $\bar{n}_1^1$, $\bar{n}_2^1$ and $\bar{n}_3^1$, respectively. In this problem, $I_{11} = I_{13}$ are assumed to be I_0, then $I_{12} = 2I_0$; hence

$$[I_1] = I_0 \begin{bmatrix} 1 & 0 & 0 \\ 0 & 2 & 0 \\ 0 & 0 & 1 \end{bmatrix} \tag{7.314}$$

From equations (7.312) and (7.297), we obtain the moment of inertia of the disk with respect to R.

$$[I_{1R}] = I_0 \begin{bmatrix} 1 + s^2\phi_1 s^2\theta & & \text{symmetric} \\ -s^2\phi_1 s\theta c\theta & 1 + s^2\phi_1 c^2\theta & \\ c\phi_1 s\theta_1 s\theta & -c\phi_1 s\phi_1 c\theta & 1 + c^2\phi_1 \end{bmatrix} \tag{7.315}$$

The expression $\left[\Omega_x^{R1}\right]$ is a skew matrix composed of $\{\dot{x}\}^T [\omega]$.

Note that

$$\left[\Omega^{R1}\right] = \left[\dot{S}^{R1}\right]\left[S^{1R}\right] = \begin{bmatrix} 0 & -\dot{\theta} s\phi_1 & -\dot{\theta} c\phi_1 \\ \dot{\theta} s\phi_1 & 0 & -\dot{\phi}_1 \\ \dot{\theta} c\phi_1 & \dot{\phi}_1 & 0 \end{bmatrix} \tag{7.316}$$

$$= \begin{bmatrix} 0 & -\dot{\phi}_3 s^2\phi_1 & -\dot{\phi}^3 s\phi_1 c\phi_1 \\ \phi_3 s^2\phi_1 & 0 & -\dot{\phi}_1 \\ \dot{\phi}_3 s\phi_1 c\phi_1 & \dot{\phi}_1 & 0 \end{bmatrix}$$

To obtain the equations we now can evaluate the coefficients $[a]$, $[b]$ and $[c]$. The first term written explicitly after substitution yields

$$[a] = m \begin{bmatrix} 0_{3\times3} \\ I \end{bmatrix} [0_{3\times3} | I] + \begin{bmatrix} [S^{1R}] \\ 0_{3\times3} \end{bmatrix} \left[S^{R1}\right] [I_1] \left[S^{1R}\right] \left[S^{R1} | 0_{3\times3}\right] \tag{7.317}$$

$$= \begin{bmatrix} I_1 & 0_{3\times3} \\ 0_{3\times3} & mI \end{bmatrix}_{6\times6}$$

Let us combine the other two terms in the equations of motion

$$\left[b^*\right] = [b] + [c] \tag{7.318}$$

Substituting the terms of $[b]$ and $[c]$ we get

$$b^* = \begin{bmatrix} [S^{1R}] \\ 0 \end{bmatrix} \left[S^{R1}\right] [I_1] \left[S^{1R}\right] \left[S^{R1} | 0\right] + \begin{bmatrix} [S^{1R}] \\ 0 \end{bmatrix} \left[\Omega_x^{R1}\right] \left[S^{R1}\right] [I_1] \left[S^{1R}\right] \left[S^{R1} | 0\right] \tag{7.319}$$

Since

$$\left[S^{1R}\right]\left[\dot{S}^{R1}\right] = \left[\dot{S}^{R1}\right]\left[S^{1R}\right] = \left[\Omega^{R1}\right] \tag{7.320}$$

Equation (7.319) reduces to

$$[b^*] = \begin{bmatrix} b_{11} & 0 \\ 0 & 0 \end{bmatrix}_{6\times 6} \tag{7.321}$$

where

$$[b_{11}] = \left[S^{1R}\right]\left[\Omega_x^{R1}\right]\left[S^{R1}\right][I_1] + [I_1]\left[\Omega^{R1}\right]$$

which is shown to take the following form

$$[b_{11}] = I_0 \begin{bmatrix} 0 & \begin{matrix} -2(\omega_3 s\phi_1 + \omega_2 c\phi_1 c\theta - \omega_1 s\theta c\phi_i) \\ -\omega_3 c\phi_1 + \omega_2 c\theta s\phi_1 - \omega_1 s\theta s\phi_1 \end{matrix} & \\ \begin{matrix} \omega_3 s\phi_1 + \omega_2 c\phi_1 c\theta \\ -\omega_1 s\theta c\phi_1 \end{matrix} & 0 & \omega_3 s\theta - \omega_1 c\theta \\ \begin{matrix} \omega_3 c\phi_1 - \omega_2 c\theta_s \phi_1 \\ +\omega_1 s\theta s\phi_1 \end{matrix} & 2(\omega_2 s\theta + \omega c\theta) & 0 \end{bmatrix}$$

$$+ I_0 \begin{bmatrix} 0 & -\dot\phi_3 s^2\phi_1 & -\dot\phi^3 s\phi_1 c\phi_1 \\ 2\dot\phi_3 s^2\dot\phi_1 & 0 & -2\dot\phi_1 \\ \dot\phi_3 s\phi_1 c\phi_1 & \dot\phi_1 & 0 \end{bmatrix}$$

$$= I_0 \begin{bmatrix} 0 & -2\dot\phi_3 - \dot\phi_2 & \begin{matrix} +\dot\phi_3 c\phi_1 \\ s\phi_1 \end{matrix} \\ \dot\phi_3 & 0 & -\dot\phi_1 \\ \begin{matrix} -\dot\phi_2 - \dot\phi_3 \\ c\phi_1 s\phi_1 \end{matrix} & 2\dot\phi_1 & 0 \end{bmatrix} + I_0 \begin{bmatrix} 0 & -\dot\phi_3^2 s\phi_1 & \begin{matrix} -\dot\phi^3 s\phi_1 \\ c\phi_1 \end{matrix} \\ 2\dot\phi_3 s^2\dot\phi_1 & 0 & -2\dot\phi_1 \\ \dot\phi_3 s\phi_1 c\phi_1 & \dot\phi_1 & 0 \end{bmatrix}$$

$$= I_0 \begin{bmatrix} 0 & -\dot\phi_3(2 + s^2\phi_1) & \dot\phi_2 \\ \dot\phi_3(1 + 2s^2\phi_1) & 0 & -3\dot\phi_1 \\ -\dot\phi_2 & 3\dot\phi_1 & 0 \end{bmatrix} \tag{7.322}$$

Note how equations (7.317) and (7.321) combined with equation (7.322) form the equations of motion of the rolling disk. What remains is the evaluation of the generalized active forces and any constraint conditions imposed on the system.

The vector of the center mass of the disk is $\overline{OD}$, as shown in Figure 7.19, which is equal to

$$\overline{OD} = \overline{OC} + \overline{CD} \tag{7.323}$$

Since we are assuming no slip, differentiation of the equation above yields

$$\dot{\overline{OD}} = \dot{\overline{OC}} + \dot{\overline{CD}} = \dot{\overline{CD}}$$

The velocity of point C is

$$\bar{v}_c = \dot{\overline{OD}} = \dot{\overline{CD}} = \bar{\omega} \times r\bar{n}_3^1$$

From equation (7.297) and (7.298), we get the velocity of contact C as

$$\bar{\upsilon}_c = r\begin{bmatrix} \bar{n}_x & \bar{n}_y & \bar{n}_z \\ \omega_1 & \omega_2 & \omega_3 \\ -\cos\phi_1\sin\theta & \cos\dot{\phi}_1\cos\theta & \sin\dot{\phi}_1 \end{bmatrix}$$

$$= (\omega_2\sin\phi_1 - \omega_3\cos\phi_1\cos\theta)\,\bar{n}_x - (\omega_3\cos\phi_1\sin\theta + \omega_1\sin\phi_1)\bar{n}_y$$
$$+ (\omega_2\cos\phi_1\cos\theta - \omega_2\cos\phi_1\sin\theta)\,\bar{n}_z\big]\,r$$
$$= r\left(\dot{\phi}_1 s\theta s\phi_1 + \dot{\phi}_2 c\theta\right)\bar{n}_x - r\left(\dot{\phi}_1 c\phi_1 - \dot{\phi}_2 s\theta\right)\bar{n}_y + r\dot{\phi}_1 c\phi_1\bar{n}_z \qquad (7.324)$$

Expressing the center mass velocity in a matrix form, we get

$$\begin{bmatrix} \upsilon_{cx} \\ \upsilon_{cy} \\ \upsilon_{cz} \end{bmatrix} = r\begin{bmatrix} \dot{\phi}_1 c\theta s\phi_1 + \dot{\phi}_2 s\theta \\ -\dot{\phi}_1 c\theta s\phi_1 + \dot{\phi}_2 s\phi \\ \dot{\phi}_1 c\phi_1 \end{bmatrix}$$

We need to impose the condition that υ_c is zero which can be expressed as

$$[B]\{\dot{x}\} = 0 \qquad (7.325)$$

where

$$[B] = \begin{bmatrix} s\theta s\phi_1 & c\theta & 0 & -1 & 0 & 0 \\ -c\theta s\phi_1 & s\theta & 0 & 0 & -1 & 0 \\ c\phi_1 & 0 & 0 & 0 & 0 & 1 \end{bmatrix} \qquad (7.326)$$

The generalized active forces follow the derivations stated in Chapter 4, where using the definition of those forces given by

$$\{f\} = \bar{F}.\frac{\partial\bar{\upsilon}}{\partial x_\ell} + \bar{M}.\frac{\partial\bar{\omega}}{\partial\dot{x}_\ell}$$
$$= [V]\{F\} + [\omega]\{M\} \qquad (7.327)$$

where $\bar{F}$ denotes the resultant external force in R and $\bar{M}$ the corresponding external moment in R acting on the body. Due to gravity, we have

$$\bar{F} = -mg\bar{n}_Z \qquad (7.328)$$

which can be written further as

$$\{F\} = \begin{bmatrix} 0 & 0 & -mg \end{bmatrix}^T \qquad (7.329)$$

Similarly,

$$\bar{M} = -mgrs\phi_1\begin{bmatrix} 1 & 0 & 0 \end{bmatrix}\left\{n^{-1}\right\} = -mgrs\phi_1\begin{bmatrix} 1 & 0 & 0 \end{bmatrix}\left[S^{1R}\right]\left\{\bar{n}^R\right\} \qquad (7.330)$$

$$\{M\} = -mgrs\phi_1\left[S^{R1}\right]\begin{Bmatrix} 1 \\ 0 \\ 0 \end{Bmatrix} \qquad (7.331)$$

Therefore, the generalized active forces array becomes

$$\{f\} = \begin{bmatrix} 0 \\ \hline I \end{bmatrix} \{F\} + \begin{bmatrix} S^{1R} \\ \hline I \end{bmatrix} \{M\}$$

$$= -mg \begin{bmatrix} rs\phi_1 \\ 0 \\ 0 \\ 0 \\ 0 \\ 1 \end{bmatrix} \tag{7.332}$$

Reaction Force N at Point c

$$\bar{N} = mg + m\ddot{z} \quad (N \text{ is in } \bar{n}_Z \text{ direction}) \tag{7.333}$$

The friction force at c has no contribution to the generalized force because $\bar{\upsilon}_c = 0$.

The equations of motion of the rolling disk or rolling coin are found making use of Kane's equations.

$$f + f^* = f^c \tag{7.334}$$

where the generalized external force array f is given by equation (7.332), the generalized inertia force array f^* is given by (7.319), and the constraint force array f^c is

$$f^c = [B]^T \{\lambda\} \tag{7.335}$$

where $\{\lambda\}$ is the array of Lagrange undetermined multipliers

$$\{\lambda\} = \begin{bmatrix} \lambda_1 \\ \lambda_2 \\ \lambda_3 \end{bmatrix} \tag{7.336}$$

Combining equation (7.298), (7.317), (7.321) and (7.322),we obtain the equations of motion of the system:

$$[a]\{\ddot{x}\} + [b*]\{\dot{x}\} + [B]^T \{\lambda\} = \{f\} \tag{7.337}$$

$$\dot{\theta} = \dot{\phi}_3 \sin \phi_1 \tag{7.338}$$

or in a compact form by an augmented matrix form of the equations above. That is, taking θ as a generalized coordinate and equation (7.338) as constraint equation and embedding it into constraint matrix $[B]$, we arrive at

$$\begin{bmatrix} I_0 & & & & & & 0 \\ & 2I_0 & & & & & \\ & & I_0 & & & & \\ & & & m & & & \\ & & & & m & & \\ & & & & & m & \\ 0 & & & & & & 0 \end{bmatrix} \begin{bmatrix} \ddot{\phi}_1 \\ \ddot{\phi}_2 \\ \ddot{\phi}_3 \\ \ddot{x} \\ \ddot{y} \\ \ddot{z} \\ \ddot{\theta} \end{bmatrix}$$

$$+ I_0 \begin{bmatrix} 0 & -\dot{\phi}_3(2+s^2\phi_1) & \dot{\phi}_2 & 0_{3\times 4} \\ \dot{\phi}_3(1+2s^2\phi_1) & 0 & -3\dot{\phi}_1 & \\ \dot{\phi}_2 & 3\dot{\phi}_1 & 0 & \\ & 0_{4\times 3} & & 0_{4\times 4} \end{bmatrix} \begin{bmatrix} \dot{\phi}_1 \\ \dot{\phi}_2 \\ \dot{\phi}_3 \\ \dot{x} \\ \dot{y} \\ \dot{z} \\ \dot{\theta} \end{bmatrix}$$

$$+ \begin{bmatrix} s\theta s\phi_1 & -c\theta s\phi_1 & c\phi_1 & 0 \\ c\theta & s\theta & 0 & 0 \\ 0 & 0 & 0 & s\phi_1 \\ -1 & 0 & 0 & 0 \\ 0 & -1 & 0 & 0 \\ 0 & 0 & -1 & 0 \\ 0 & 0 & 0 & -1 \end{bmatrix} \begin{bmatrix} \lambda_1 \\ \lambda_2 \\ \lambda_3 \\ \lambda_4 \end{bmatrix} = -mg \begin{bmatrix} rs\phi_1 \\ 0 \\ 0 \\ 0 \\ 0 \\ 1 \\ 0 \end{bmatrix} \tag{7.339}$$

We can easily find an orthogonal matrix T numerically in time to make $[T]\,[B]^T = 0$; then we reduce the equations of motion to

$$[T]\,[a]\,\{\ddot{x}\} + [T]\left[b^*\right]\{\dot{x}\} = [T]\,\{f\} \tag{7.340}$$

Numerical Simulation

The problem above was simulated to investigate its motion for 1.6 s. The following data was used.

Input Data

$$m = 1, \qquad r = 1$$

$$I_0 = I_{11} = I_{13} = \frac{mr^3}{4} = 0.25, \qquad I_{12} = \frac{mr_2}{2} = 0.5$$

$$[I_1] = \begin{bmatrix} 0.25 & 0 & 0 \\ 0 & 0.5 & 0 \\ 0 & 0 & 0.25 \end{bmatrix}$$

The following initial conditions were used

$$\begin{bmatrix} \phi_1 \\ \phi_2 \\ \phi_3 \\ x \\ y \\ z \\ \theta \end{bmatrix}_{t=0} = \begin{bmatrix} \frac{\pi}{2} \\ 0 \\ 0 \\ 0 \\ 0 \\ 1 \\ \frac{\pi}{2} \end{bmatrix} \begin{bmatrix} \dot{\phi}_1 \\ \dot{\phi}_2 \\ \dot{\phi}_3 \\ \dot{x} \\ \dot{y} \\ \dot{z} \\ \dot{\theta} \end{bmatrix}_{t=0} = \begin{bmatrix} -0.1 \\ -1 \\ 0.1 \\ -\in \\ 0.1 \\ 0 \\ 0.1 \end{bmatrix}$$

where $\in$ is a very small positive quantity. The time step $\Delta t = 0.1$; the total number of time steps NTT = 17. The following results were found.

Discussion

1. Mass center motion is shown in Figure 7.20.

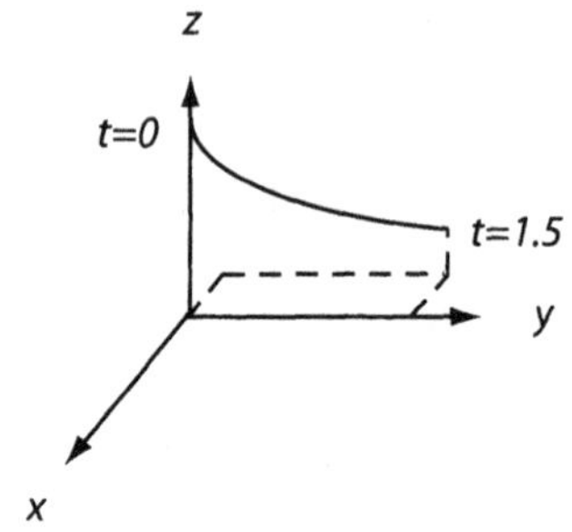

Fig. 7.20. Mass center motion

2. The value of θ increases as t increases; this means that the disk is dropping downward while ϕ increases and ψ decreases (i.e., the disk moves rightward against the $x - z$ plane, as shown in Figure 7.21).

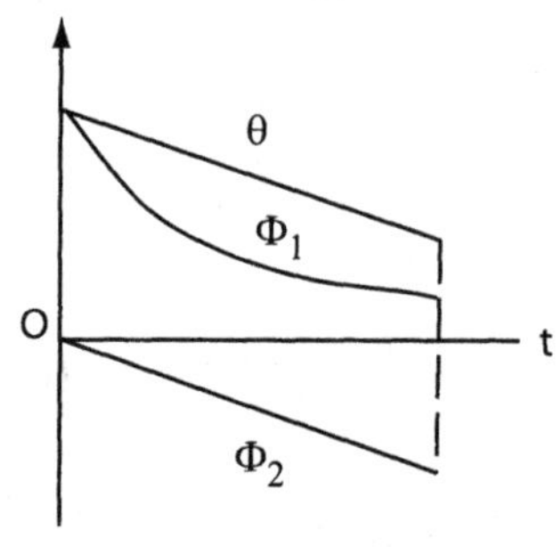

Fig. 7.21. Angular displacements versus time

3. As $\ddot{z}$ decreases, the reaction force at c, N, decreases.

7.14 Analysis and Simulation of Human Locomotion

For years linkage modeling has been the basis of analysis of the mechanics of human locomotion. In particular, the model presented herein puts in perspective the importance of constrained multibody systems and the handling of constraints in particular, after the governing equations of motion have been derived. We examine how the basis assumptions used in the model are converted into constraint conditions imposed on the mechanism system. For simplicity, the human body is modeled by five segments with two-segment lower limbs and the upper body, including hand, arms, and trunk, as one segment. The motion is assumed to be in the sagittal plane. This example is extracted from reference [15].

pc There are five joints, two each at the hips and knees and one between the ankle of the stand leg and the ground. To keep the model simple, the feet are not included in the model as separate segments. The human model is shown in Figure 7.22. The bodies B_1 and B_2 represent the stand leg, while the bodies B_4 and B_5 form the swing leg.

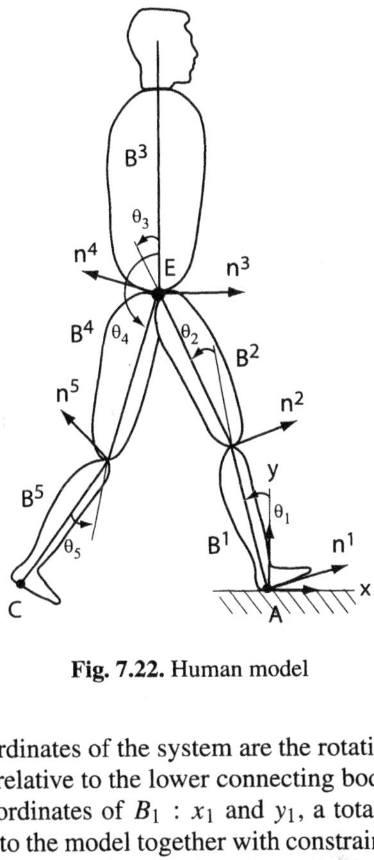

Fig. 7.22. Human model

The generalized coordinates of the system are the rotations of each of the bodies B_1, B_2, B_3, B_4 and B_5, relative to the lower connecting bodies, θ_1, θ_2, θ_3, θ_4 and θ_5; and the translational coordinates of B_1 : x_1 and y_1, a total of seven variables. The inclusion of x_1 and y_1 into the model together with constraining them to be zero, will be solely for the sake of determining the foot forces in the x and y directions. The generalized coordinates of the system can then be expressed as

$$x = [\theta_1, \theta_2, \theta_3, \theta_4, \theta_5, x_1, y_1]^T \tag{7.341}$$

The constraint equations should be selected to yield a unique description of the kinematics of the system.

This model involves seven constraints that describe the kinematics completely. First, the condition that the ankle (point C) has a specified swing trajectory is represented by the following two equations:

$$(L_1c_1 + L_2c_{1-2} + L_2c_{1-4} + L_1c_{1-5})\,\dot{\theta}_1 + (L_2c_{1-2} + L_2c_{1-4} + L_1c_{1-5})\,\dot{\theta}_2 + (L_2c_{1-4} + L_1c_{1-5})\,\dot{\theta}_3 + (L_2c_{1-4} + L_1c_{1-5})\,\dot{\theta}_4 + L_1c_{1-5}\dot{\theta}_5 = g_x(t) \tag{7.342}$$

$$(L_1s_1 + L_2s_{1-2} + L_2s_{1-4} + L_1s_{1-5})\,\dot{\theta}_1 + (L_2s_{1-2} + L_2s_{1-4} + L_1s_{1-5})\,\dot{\theta}_2 + (L_2s_{1-4} + L_1s_{1-5})\,\dot{\theta}_3 + (L_2s_{1-4} + L_1s_{1-5})\,\dot{\theta}_4 + L_1s_{1-5}\dot{\theta}_5 = -g_y(t) \tag{7.343}$$

where $g_x(t)$ and $g_y(t)$ are the prescribed velocities of point C in the x and y directions, respectively. Here L_1 and L_2 are the lengths of the lower leg and upper leg, respectively, and $c_1 = \cos\theta_1$, $c_{i-1} = \cos(\theta_1 + \theta_2 + \cdots + \theta_i)$, $s_1 = \sin\theta_1$, $s_{i-1} = \sin(\theta_1 + \theta_2 + \cdots + \theta_i)$.

Point E must have a specified motion in the x-direction consistent with the motion of point C of the swing leg. This can be expressed as

$$(L_1c_1 + L_2c_{1-2})\,\dot{\theta}_1 + L_2c_{1-2}\dot{\theta}_2 = h_x(t) \tag{7.344}$$

where $h_x(t)$ is the prescribed velocity of point E in the x-direction.

Since the stand ankle (point A) is stationary during the swing, we have the conditions that $\dot{x}_1 = 0$ and $\dot{y}_1 = 0$. The knee joint of the support leg is stiff; therefore, the angular velocities of B_1 and B_2 can be constrained to be equal. This condition is represented as $\dot{\theta}_2 = 0$. Finally, since the upper body B_3 remains in the vertical position throughout the motion, the absolute angular velocity of B_3 will be constrained to be zero, (i.e., $\dot{\theta}_1 + \dot{\theta}_2 + \dot{\theta}_3 = 0$).

Combining equations (7.342) to (7.344) and what we state above, we form constraint equations

$$[B]\,\{y\} = \{g\} \tag{7.345}$$

where

$$[B] = \begin{bmatrix} B_{11} & B_{12} & B_{13} & B_{14} & B_{15} & 0 & 0 \\ B_{21} & B_{22} & B_{23} & B_{24} & B_{25} & 0 & 0 \\ B_{31} & B_{32} & 0 & 0 & 0 & 0 & 0 \\ 0 & 0 & 0 & 0 & 0 & 1 & 0 \\ 0 & 0 & 0 & 0 & 0 & 0 & 1 \\ 0 & 1 & 0 & 0 & 0 & 0 & 0 \\ 1 & 1 & 1 & 0 & 0 & 0 & 0 \end{bmatrix} \tag{7.346}$$

and the $[B]$ coefficients are

$$B_{11} = L_1\,(c_1 + c_{1-5}) + L_2\,(c_{1-2} + c_{1-4})$$

$$B_{12} = L_1c_{1-5} + L_2\,(c_{1-2} + c_{1-4})$$

$$B_{13} = L_1c_{1-5} + L_2c_{1-4}$$

$$B_{14} = L_1c_{1-5} + L_2c_{1-4}$$

$$B_{15} = L_1c_{1-5}$$

$$B_{21} = L_1\,(s_1 + s_{1-5}) + L_2\,(s_{1-2} + s_{1-4})$$

$$B_{22} = L_1 s_{1-5} + L_2 (s_{1-2} + s_{1-4})$$

$$B_{23} = L_1 s_{1-5} + L_2 s_{1-4}$$

$$B_{24} = L_1 s_{1-5} + L_2 s_{1-4}$$

$$B_{25} = L_1 s_{1-5}$$

$$B_{31} = L_1 c_1 + L_2 c_{1-2}$$

$$B_{32} = L_2 c_{1-2}$$

Since $\dot{\theta}_2 = 0$, we can let $B_{12} = B_{22} = B_{32} = B_{72} = 0$. Therefore,

$$[B] = \begin{bmatrix} B_{11} & 0 & B_{13} & B_{14} & B_{15} & 0 & 0 \\ B_{12} & 0 & B_{23} & B_{24} & B_{25} & 0 & 0 \\ B_{31} & 0 & 0 & 0 & 0 & 0 & 0 \\ 0 & 0 & 0 & 0 & 0 & 1 & 0 \\ 0 & 0 & 0 & 0 & 0 & 0 & 1 \\ 0 & 1 & 0 & 0 & 0 & 0 & 0 \\ 1 & 1 & 1 & 0 & 0 & 0 & 0 \end{bmatrix} \tag{7.347}$$

and

$$g = \left[g_x(t)\ g_y(t)\ h_x(t)\ 0\,0\,0\,0 \right]^T \tag{7.348}$$

The equations of motion require the specification of the acceleration profiles $\dot{g}_x(t)$, $\dot{g}_y(t)$ and $\dot{h}_x(t)$, together with the initial conditions for the displacements and velocities. The swing trajectory is obtained by the following acceleration profiles given in reference [15]:

$$\dot{g}_x = \frac{a\pi^2}{2T^2} \cos \frac{T}{T} t \tag{7.349}$$

$$\dot{g}_y = -\frac{2b}{T^2} \pi^2 \sin \left(\frac{2\pi t}{T} + \frac{3}{2}\pi \right) \tag{7.350}$$

where T is the swing time, a the swing length, and b is the maximum vertical displacement of the swing ankle. The complete simulation of the model is shown in Figure 7.23 for a 62-cm swing length and a 1-s swing time. The latter in this analysis is referred to as normal walking, whereas a 62-cm swing length for 0.5 s is fast walking. The joint torques increase drastically when the walking-fast model is used.

The variation of the y component of the foot force (Figure 7.24) is about 10% of the static weight. The increase occurs when the inertial accelerations of the body segments in general are the positive y direction, which is characterized by the earlier and final stages of the swing. The x component of the foot force (Figure 7.25) is solely due to inertia forces. The joint moments (Figures 7.26 and 7.27) depend on both the configuration changes and have a much larger effect on the ankle, knee, and hip joints of the support leg.

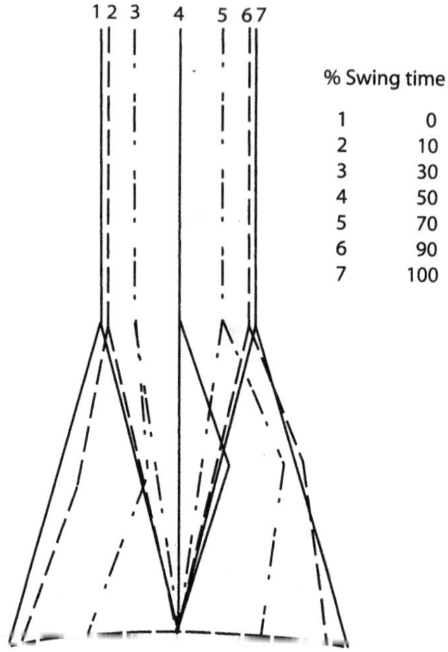

Fig. 7.23. Simulation of walking (as shown in reference [15])

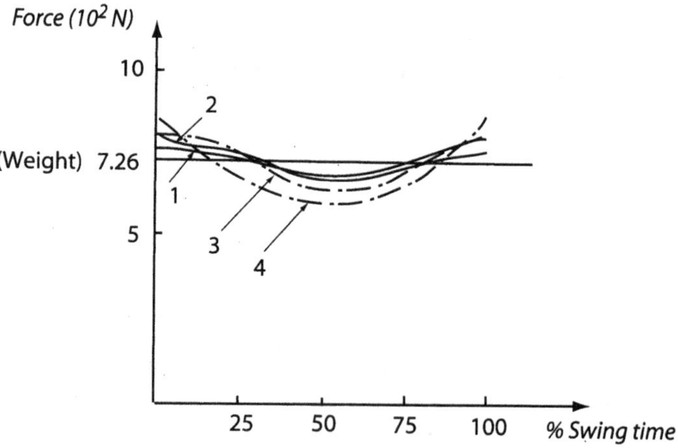

Fig. 7.24. Foot forces in the y direction: (1) normal-level walking; (2) longer swing; (3) fast walking; (4) experimental (Capozzo, Leo, Pedotti) (results obtained by Amirouche and Ider, reference [15])

This example addresses the human walking problem, treating it as a planar linkage with some realistic assumptions. It illustrates how we can handle any configuration of the human model through the use of multibody dynamics modeling. The analysis

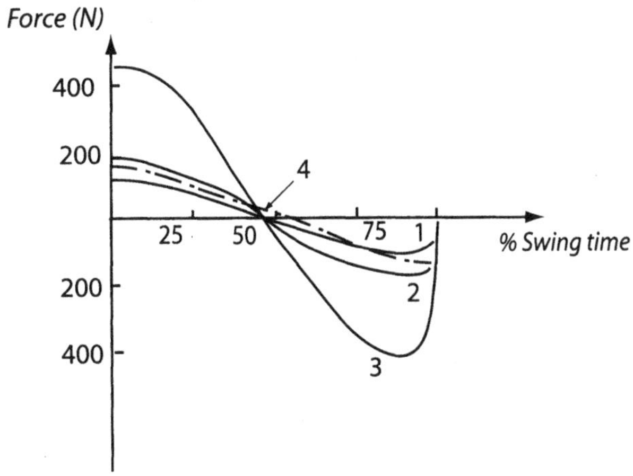

Fig. 7.25. Foot forces in the x direction: (1) normal-level walking; (2) longer swing; (3) fast walking; (4) experimental (Capozzo, Leo, Pedotti)

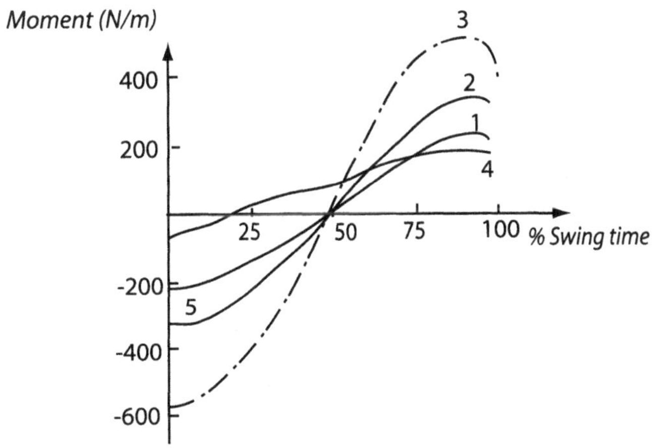

Fig. 7.26. Support ankle moment: (1) normal-level walking; (2) longer swing; (3) fast walking; (4) experimental (Capozzo, Leo, Pedotti) (results obtained by Amirouche and Ider, reference [15])

is further enhanced by introducing the assumptions made on the model in a form of simple nonholonomic constraints. The complete equations of motion can then be given by the planar five-link system together with the constraints; we can then solve for the joint torques (moments) due to the constraints. One can easily see that with proper modeling of the muscles, their contribution to the governing equations of motion can easily be integrated.

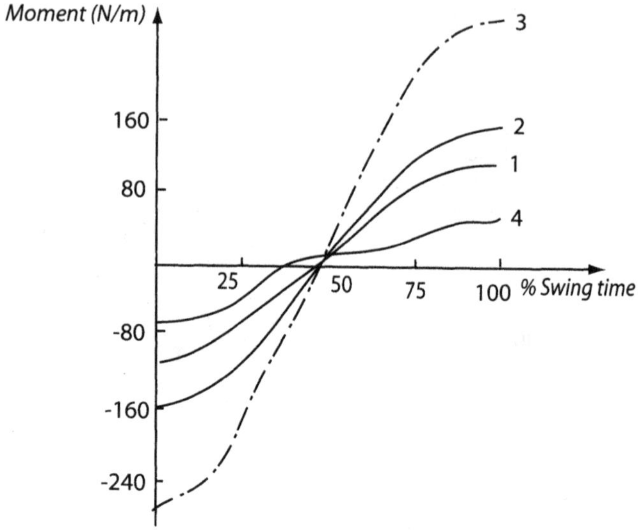

Fig. 7.27. Support knee moment: (1) normal-level walking; (2) longer swing; (3) fast walking; (4) experimental (Capozzo, Leo, Pedotti) (results obtained by Amirouche and Ider, reference [15])

7.15 Problems

7.1. Consider the pendulum shown in Figure 7.28. The mass m moves along a circular path. Determine the equations of motion of m using both the augmented method and the reduced form (eliminating the undetermined multiplies). Solve for the undetermined multipliers (constraint forces). Treat both the bar and the mass as two separate bodies and use the constraint to describe their connection.

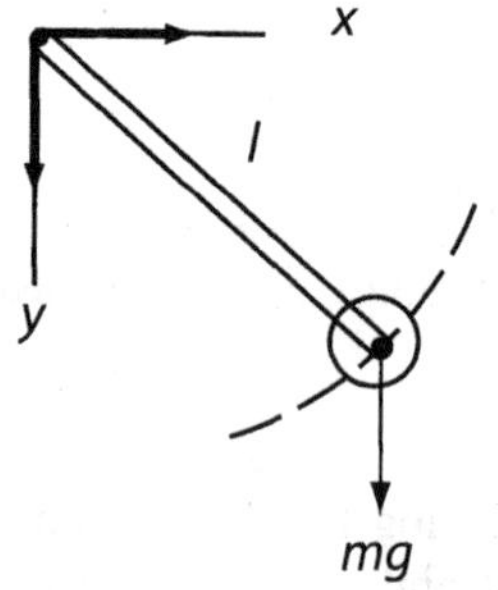

Fig. 7.28. A pendulum with mass m attached to a bar

7.2. Consider the slider–crank mechanism shown in Figure 7.29. Derive the equations of motion of the system if S is to be restricted to move along the x direction only.

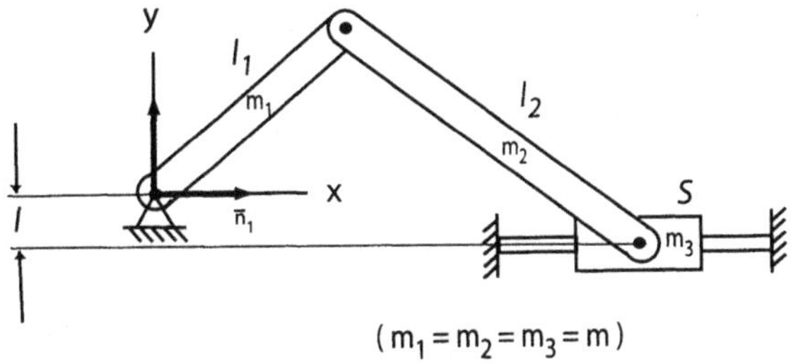

Fig. 7.29. Slider–crank mechanism

7.3. For the deployment of the three link manipulators shown in Figure 7.30, the generalized coordinates ψ_1, ψ_2 and ψ_3 are given by

$$\psi_1(t) = \begin{cases} \pi - \frac{\pi}{2T}\left(t - \frac{T}{2\pi}\sin\frac{2\pi t}{T}\right) & \text{rad} \quad 0 < t \le T \\ \frac{\pi}{2} & \quad t > T \end{cases}$$

$$\psi_2(t) = \begin{cases} \pi - \frac{3\pi}{4T}\left(t - \frac{T}{2\pi}\sin\frac{2\pi t}{T}\right) & \text{rad} \quad 0 < t \le T \\ \frac{\pi}{4} & \quad t > T \end{cases}$$

$$\psi_3(t) = \begin{cases} \pi - \frac{\pi}{T}\left(t - \frac{T}{2\pi}\sin\frac{2\pi t}{T}\right) & \text{rad} \quad 0 < t \le T \\ 0 & \quad t > T \end{cases}$$

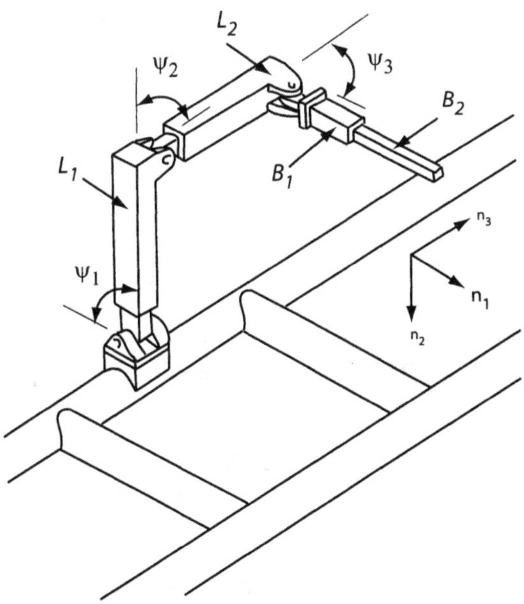

Fig. 7.30. Pace-based manipulator for deployment and maneuvers

(a) Construct the constraint Jacobian matrix resulting from the prescribed motion to those angles.
(b) Find the constraint forces corresponding to the above joint deployment constraints, if $T = 10\,\text{s}$.

7.4. Use the zero-eigenvalue theorem to evaluate the orthogonal complement to the constraint Jacobian matrix given by equation (7.99) for the illustrative Example 7.7.2. Determine the nonsingular G relating the PUTD method and the zero-eigenvalue theorem.

7.5. Solve for the constraint forces resulting from the geometrical constraint of maintaining the loop formation of Example 7.7.2 if the zero-eigenvalue theorem is used in the evaluation of orthogonal complement to B.

7.6. For the system shown in Figure 7.31, find the constraint equations resulting from the closed loops. Assume that all bodies are interconnected by revolute joints.

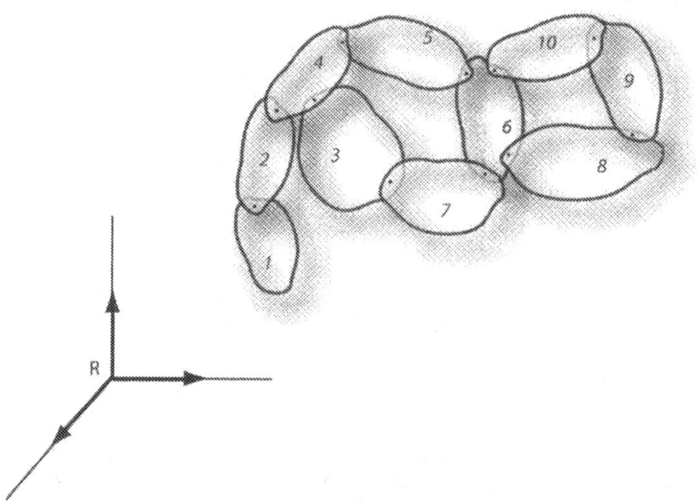

Fig. 7.31. A multibody system with closed loops

7.7. For the robot shown in Figure 7.32, simulate its motion and compute the joint torques that will drive the end effector through the prescribed path (parallel to $\bar{n}_1$)

7.8. If the bar is to remain in contact with the wall at all times, find the corresponding equations of motion, including the constraint conditions. Find the contact force before the bar reaches the ground (see Figure 7.33).

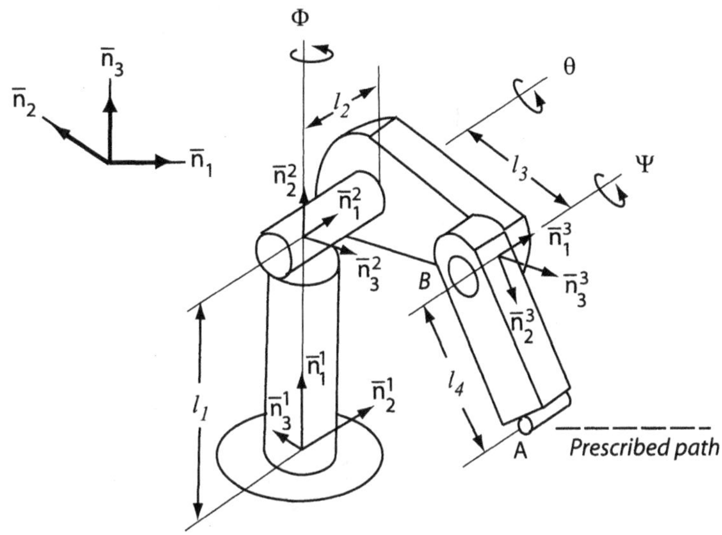

Fig. 7.32. Puma robot constrained to move point A along a prescribed path parallel to the fixed unit vector $\bar{n}_1$

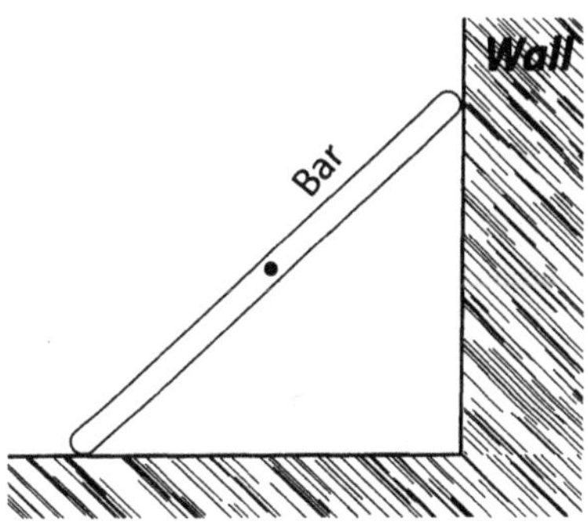

Fig. 7.33. A bar with contact points both at the wall and the ground is to slide until it reaches the ground

7.9. For the mechanism shown in Figure 7.34, find the joint torques $T(t)$ applied at point B such that the velocity of point C is collinear to the line from B to C.

7.10. Use the embedding method to derive the governing equation of motion for the constrained system given in Example 7.10.1.

7.11. Lifting is an activity that requires energy and efforts to be distributed in the proper way to avoid fatal injuries. Assume that the motion for the model shown in Figure 7.35 is in a sagittal plane. Make use of the following assumption to derive the constraint equations and identify the Jacobian matrix.

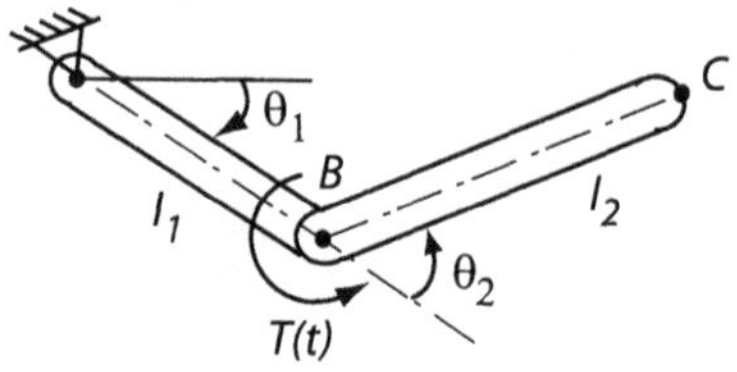

Fig. 7.34. Planar two-link manipulator with a constrained point C

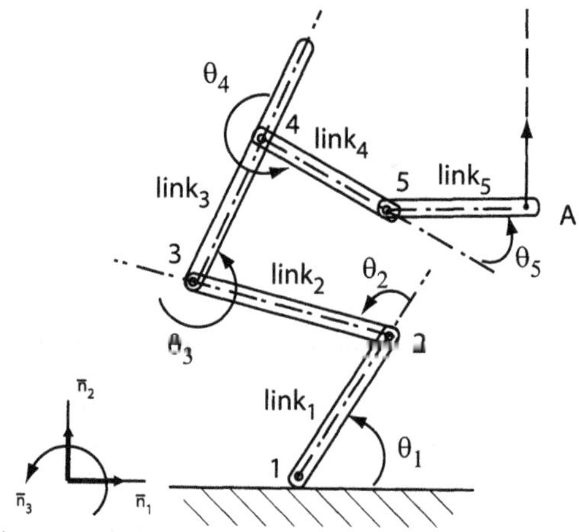

Fig. 7.35. A human model composed of five-link planar linkage

i. Point A is to move in a straight line parallel to $\bar{n}_2$.
ii. The angular velocity of body 5 relative to body 4 is zero.
iii. The angular velocity of body 3 is assumed to be prescribed according to what is known and is equal to $g(t)$.

(*Note*: This is a redundant system for which the solution is obtained by writing the dependent coordinates in terms of the independent ones.)

References

1. Hemani, H. and Weimer, F. C., Modelling of Nonholonomic Dynamic Systems with Applications, *J. Appl. Mech.,* Vol. **48**, 1981, p. 177.
2. Huston, R. L. and Passerello, C. E., On Lagrange's Form of d'Alembert's Principle, *Matrix Tensor Q.,* Vol. **23**, 1973, pp. 109–112.
3. Huston, R. L. and Wang, J. T., Kane's Equations with Undetermined Multipliers—Application with Constrained Multibody Systems, *ARO Symposium on Kinematics and Dynamics of Robot Manipulators,* Rensselaer Polytechnic Institute, Troy, NY, 1987.

4. Kamman, J. W. and Huston, R. L., Dynamic of Constrained Multibody Systems, *J. Appl. Mech.,* Vol. **51**, 1984, pp. 899–903.
5. Kane, T. R., Dynamics of Nonholonomic Systems, *J. Appl. Mech.,* Vol. **28**, 1961, pp. 574–578.
6. Kane, T. R. and Levinson, D. A., Formulation of Equations of Motion for Complex Spacecraft, *J. Guid, Control*, Vol. **3**: No. 2, 1980, pp. 99–112.
7. Kane, T. R. and Levinson, D. A., *Dynamics: Theory and Applications,* McGraw-Hill, New York, 1985.
8. Kirgetov, V. I., The Motion of Controlled Mechanical Systems with Prescribed Constraints (Servo Constraints), *Prikl. Mat. Mekh.,* Vol. **31**: No. 3, 1967, pp. 433–447.
9. Singh, R. P. and Likins, P. W., Singular Value Decomposition for Constrained Dynamical Systems, *J. Appl. Mech.,* Vol. **52**, 1985, pp. 943–948.
10. Walton, W. C., Jr. and Steeves, E. C., 1969, A New Matrix Theorem and Its Application for Establishing Independent Coordinates for Complex Dynamical Systems with Constraints, *NASA Technical Report, TR-327.*
11. Wampler, C., Buffington, K. and Shu-Hui, J., Formulation of Equations of Motion for Systems Subject to Constraints, *J. Apple. Mech.,* Vol. **52**, 1985, pp. 465–470.
12. Wehage, R. A. and Haug, E. J., Generalized Coordinate Partitioning for Dimension Reduction in Analysis of Constrained Dynamics Systems, *J. Mech. Des.,* Vol. **104**: No. 1, 1982, p. 247.
13. Amirouche, F. M. L. and Jyia, T., Automatic Elimination of the Undetermined Multipliers in Kane's Equations Using a Pseudo-uptriangular Decomposition Method, *Comput. Struct.,* Vol. **27**: No. 2, 1987, pp. 203–210.
14. Amirouche, F. M. L. and Ider, S. K., Determination of Constraint Forces in Multibody Systems with Dynamics using Kane's Equation, *Journal de Mécanique Théorique et Appliqué*, Vol. 7, No. 1, pp. 3–20, 1988.
15. Amirouche, F. M. L., Ider, S. K. and Trimble, J., Analytical Method for the Analysis and Simulation of Human locomotion, ASME Trans., Biomechanical Eng., Vol. **112**, Nov. 1990, pp. 379–387.
16. Amirouche, F. M. L., Jyia, T. and Ider, S. K., A Recursive Householder Transformation for Complex Dynamical Systems with Constraints, *J. Appl. Mech.,* Vol. **55**: No. 3, 1988, pp. 729–734.
17. Amirouche F. M. L. and Huston, R. L., Dynamics of Large Constrained Flexible Structures, *J. Dyn. Syst. Meas. Control,* Vol. **110**, Mar. 1988, pp. 78–83.

8

Numerical Stability of Constrained Multibody Systems

8.1 Introduction

Constraints often are represented by their position, velocity and acceleration form. When imposing a set of constraints, holonomic or nonholonomic on dynamical systems it is important that the solution must satisfy the constraints at all levels. To prevent numerical irregularities and divergence of a solution, stability of the equations of motion must be preserved. In this chapter we discuss the numerical stability of the governing equations of motion when integrated to yield the time history of a system. These equations are composed of a set of differential and algebraic equations. Since this is a numerical problem, it will be in the interest of students and practicing engineers to see how the problem is resolved. First, we introduce the Baumgarte technique, used to stabilize and reduce the error due to conventional representation of the constraint equations. Second, a method by Amirouche and Ider, which makes use of a modification of the constraints at singular configurations is introduced to regularize the stability of the dynamical equations. The aforementioned problems are still research topics, and further work is needed.

8.2 Baumgarte Stability Method

In Chapter 7, we presented the numerical form of the governing equations of motion of constrained multibody systems, where the constraints are represented by their form at the acceleration level. For instance, the constraint equations (holonomic or simple nonholonomic) can be represented by

$$h_i\left(q_1, q_2, \ldots, q_n, t\right) = 0 \quad (i = 1, 2, \ldots, m) \tag{8.1}$$

The q's are the generalized coordinates, similar to the x's used in previous chapters, and t denotes time. Equation (8.1) is said to be a representation of the constraints at the position level. Further differentiation will lead its representation at the velocity level, where

$$\dot{h}_i = \sum_{\ell=1}^{n} \frac{\partial h_i}{\partial q_\ell}\frac{dq_\ell}{dt} + \frac{\partial h_i}{\partial t} = \sum_{\ell=1}^{n} b_{i\ell} y_\ell + g_i = 0 \tag{8.2}$$

$$= BY + G$$

where $b_{i\ell}$ is the (i, ℓ) element of the $m \times n$ constraint matrix B defined by

$$B = [b_{i\ell}] = \begin{bmatrix} \frac{\partial h_1}{\partial q_1} & \frac{\partial h_1}{\partial q_2} & \cdots & \frac{\partial h_1}{\partial q_n} \\ \frac{\partial h_2}{\partial q_1} & \frac{\partial h_2}{\partial q_2} & \cdots & \frac{\partial h_2}{\partial q_n} \\ \vdots & & & \\ \frac{\partial h_m}{\partial q_1} & \frac{\partial h_m}{\partial q_2} & \cdots & \frac{\partial h_m}{\partial q_n} \end{bmatrix} \tag{8.3}$$

where y_ℓ is the lth element of the matrix Y, defined by

$$Y = \left[\frac{dq_1}{dt} \; \frac{dq_2}{dt} \cdots \frac{dq_n}{dt} \right]^T, \qquad y_\ell = \frac{dq_\ell}{dt} \tag{8.4}$$

and g_i is the ith element of the $m \times 1$ matrix G given by

$$G = \left[-\frac{\partial h_1}{\partial t} \; -\frac{\partial h_2}{\partial t} \cdots -\frac{\partial h_m}{\partial t} \right]^T \tag{8.5}$$

If we further differentiate equation (8.2), we get the acceleration representation of the constraints as

$$\ddot{h}_i = \sum_{\ell=1}^{n} b_{i\ell} \dot{y}_\ell + \sum_{\ell=1}^{n} \dot{b}_{i\ell} y_\ell - \dot{g}_i = 0 \tag{8.6}$$

One thing we should keep in mind is that $h_i = 0$ implies that $\dot{h}_i = 0$, $\ddot{h}_i = 0$, but $\ddot{h}_i = 0$ [which is what we use in equation (8.6)] does not necessarily yield $\dot{h}_i = 0$ and $h_i = 0$. Therefore, it is important to know that during the integration of the equations of motion, we must try to satisfy all constraint conditions. In the event we do not, it is possible to accumulate numerical errors which might cause serious stability and control problems. To overcome this difficulty, Baumgarte introduced a method to assure *asymptotic stability* by replacing $\ddot{h}_i = 0$ with a more complete representation of the constraints at the acceleration level as

$$\ddot{h}_i + \alpha_i \dot{h}_i + \beta_i h_i = 0 \quad (i = 1, 2, \ldots, m) \tag{8.7}$$

where $\beta_i = 0$ for nonholonomic constraints, and α_i and β_i are arbitrary constants chosen for sufficiently fast decay of the errors [7]. For example, let us consider the constraint equations $\ddot{h}_i = 0$ where $h_i = 0$ is a holonomic constraint. In addition, assume that in the process of the integration of the equations of motion after the nth step, the computer yields the values $h = \delta$, $\dot{h}_i = \epsilon$, which deviate from the exact values $h = 0$, $\dot{h} = 0$. According to the differential equation $\ddot{h}_i = 0$, the computer should produce the value $h = \in t + \delta$. Thus the holonomic constraint is not satisfied in a linearly stable fashion. It is unlikely that subsequent numerical errors will compensate for this behavior.

By virtue of the initial conditions $h(0) = 0$, $\dot{h}(0) = 0$, equation (8.7) leads to $\ddot{h} = 0$ as before, such that the new constraint is analytically equivalent to $\ddot{h} = 0$. But from the numerical point of view, the situation is different. To achieve asymptotical stability

$$\alpha_i > 0 \tag{8.8}$$

and

$$\alpha_i^2 - 4\beta_i \leq 0 \tag{8.9}$$

must be satisfied. We usually choose

$$\alpha_i^2 = 4\beta_i \tag{8.10}$$

Note that from equations (8.8) and (8.9), we have an infinite set of α's and β's which can be used in equation (8.7).

Example 8.2.1

A mass point is moving the $X_1 - X_2$ plane and is to follow circular motion on the unit circle. Assume that no external forces are applied and that the mass m is equal to 1.

Solution

The equations of motion are given by

$$F_i + F_i^* + F_i^c = 0 \tag{8.11}$$

where F_i^* are the inertial forces, F_i^c the constraint forces, and F_i represent the external forces, which are zero in this case.

Since the motion of a mass point is given by a translation only (no rotation is involved), equation (8.11) can be expressed as

$$-m_i\ddot{x}_i - \lambda B_{i\ell} = 0 \tag{8.12}$$

where $B_{i\ell}$ is the constraint Jacobian matrix, and the resulting constraints take the general form

$$\sum_i^m B_{i\ell}\ddot{x}_\ell + G = 0 \tag{8.13}$$

where $B_{i\ell}$ and G are functions of x_i, $\dot{x}_i$ and t. From equation (8.12), we know that

$$\ddot{x}_i = \frac{1}{m_i}[-\lambda B_{i\ell}] \tag{8.14}$$

When we substitute it in equation (8.13), we get an expression for the undetermined multipliers:

$$\lambda = \frac{\sum_{i}^{m} B_{i\ell} F_i / m_i + G}{\sum_{i}^{m} B_{i\ell}^2 / m_i} \tag{8.15}$$

Then equation (8.12) becomes

$$-m_i \ddot{x}_i = \frac{\sum_{i}^{m} B_{i\ell} F_i / m_i + G}{\sum_{i}^{m} B_{i\ell}^2 / m_i} B_i \tag{8.16}$$

In this example, the constraint equation is simply

$$h(x_1, x_2, t) = x_1^2 + x_2^2 - 1 = 0 \tag{8.17}$$

with its velocity representation as

$$\dot{h} = 2x_1\dot{x}_1 + 2x_2\dot{x}_2 - [2x_1 \;\; 2x_2] \begin{bmatrix} \dot{x}_1 \\ \dot{x}_2 \end{bmatrix} \tag{8.18}$$

Its acceleration form is

$$\ddot{h} = 2\dot{x}_1^2 + 2x_1\ddot{x}_1 + 2\dot{x}_2^2 + 2\ddot{x}_2 x_2 \tag{8.19}$$

If we use Baumgarte's method, $\ddot{h}$ should be expressed as (8.7) where

$$\ddot{h} + \alpha\dot{h} + \beta h = x_1\ddot{x} + x_2\ddot{x}_2 + \dot{x}_1^2 + \dot{x}_2^2 + \alpha(x_1\dot{x}_1 + x_2\dot{x}_2) + \frac{\beta}{2}(x_1^2 + x_2^2 - 1) = 0 \tag{8.20}$$

From equation (8.20), we know that

$$B_{11} = x_1, \qquad B_{12} = x_2 \tag{8.21}$$

and

$$G = \dot{x}_1^2 + \dot{x}_2^2 + \alpha(x_1\dot{x}_1 + x_2\dot{x}_2) + \frac{\beta}{2}(x_1^2 + x_2^2 - 1) \tag{8.22}$$

If we substitute in equation (8.16), $m_i = 1$, we obtain

$$\ddot{x}_1 = -\frac{\dot{x}_1^2 + \dot{x}_2^2 + \alpha(x_1\dot{x}_1 + x_2\dot{x}_2) + \frac{\beta}{2}(x_1^2 + x_2^2 - 1)}{x_1^2 + x_2^2} x_1 \tag{8.23}$$

and

$$\ddot{x}_1 = -\frac{\dot{x}_1^2 + \dot{x}_2^2 + \alpha(x_1\dot{x}_1 + x_2\dot{x}_2) + \frac{\beta}{2}(x_1^2 + x_2^2 - 1)}{x_1^2 + x_2^2} x_2 \tag{8.24}$$

Let the initial conditions for $t = 0$ be $x_1(0) = 1$, $\dot{x}_1(0) = 0$, $x_2(0) = 0$, $\dot{x}_2(0) = 1$. Throughout the numerical calculation, we use Runge–Kutta formulas of fourth order, constant step size, 100 steps per revolution. To demonstrate the instability problem, let the initial conditions be slightly modified, such that $x_1(0) = 1$, $\dot{x}_1(0) = 10^{-4}$, $x_2(0) = 0$ and $\dot{x}_2(0) = 1$. After two revolutions, the computed values of the constraint $h(x_1, x_2, t)$ are:

Case 1*:* conventional approach ($\ddot{h} = 0$), for $\alpha = \beta = 0$, $h = 0.00251$.
Case 2*:* Baumgarte approach, for $\alpha = 20$, $\beta = 100$, $h = 0.00006$.

The gain obtained by the stabilization provided by Baumgarte technique is clearly displayed. It is important at this stage to realize that if we are required to perform a simulation of a constrained multibody system for a significant period of time, our results can be hampered by the numerical stability of h, hence yielding an erroneous solution.

8.3 Numerical Solution of a Constrained System's Equations

The most popular formulation of the equations of constrained mechanical motion is the state-space form. In this chapter, we discuss and relate some proposed reformulations of the equations that have improved numerical properties.

Historically, multibody systems have been formulated using several methods: Lagrange equations of the first or second kind, Newton–Euler, and Kane's equations, among others. The equations of motion are a set of differential-algebraic equations (DAE) and their explicit form can be given by the representation developed by Amirouche et al.:

$$\dot{Q} = y \tag{8.25}$$

$$M(q)\dot{y} = F(q, y, t) - B^T(q)\lambda \tag{8.26}$$

$$\Phi(q) = 0 \tag{8.27}$$

where $M(q)$ is an $n \times n$ generalized mass matrix; $F(q, y, t)$ is a vector of applied forces, combined with all the nonlinear terms arising from the generalized inertia forces; and λ represents the m Lagrange multipliers or constraint forces coupled to the system by the $m \times n$ constraint matrix $B = \partial\Phi/\partial q$. Observe how equations (8.25) and (8.26) are first-order differential equations, whereas (8.27) is an algebraic equation. It is possible to rewrite the equations above as a set of ordinary differential equations. First, we differentiate the constraint equations given by (8.27), yielding the m constraint equations at the velocity level:

$$B(q)y = 0 \tag{8.28}$$

and the m constraint equations at the acceleration level are formed by the subsequent differentiation of equation (8.28)

$$B(q)\dot{y} + \dot{B}(q)y = B(q)\dot{y} + Z(q, y) = 0 \tag{8.29}$$

Solving for $\dot{y}$ from equation (8.26), we get

$$\dot{y} = M(q)^{-1}F(q, y, t) - M(q)^{-1}B^T(q)\lambda \tag{8.30}$$

Substituting the result into equation (8.29) results in m constraint equations involving the Lagrange multipliers:

$$B(q)M^{-1}(q)F(q, y, t) - B(q)M^{-1}(q)B^T(q)\lambda + Z(q, y) = 0 \tag{8.31}$$

If the constraint matrix B is assumed to be of full rank m, where each row of the constraint Jacobian matrix is linearly independent, we can write

$$H(q) = B(q)M^{-1}(q)B^T(q) \tag{8.32}$$

where $H(q)$ is a nonsingular $m \times n$ matrix. Equation (8.31) can then be solved for the algebraic variable λ:

$$\lambda = H^{-1}(q)[B(q)M^{-1}(q)F(q, y, t) + Z(q, y)] \tag{8.33}$$

The equations of motion given by equation (8.26) become

$$M(q)\dot{y} = \tilde{F}(q, y, t) \tag{8.34}$$

where

$$\tilde{F}(q, y, t) = F(q, y, t) - B^T(q)H^{-1}(q)[B(q)M^{-1}(q)F(q, y, t) + Z(q, y)] \tag{8.35}$$

Equation (8.34) is called the underlying ordinary differential equation (ODE) of the DAE (see reference [12]). Given initial values for the position and velocity constraints, the solutions to the ODEs are solutions of the original DAEs. It is obvious that the difficulty relies on the fact that accurate initial values are not easy to find, which results in a numerical solution of the problem where the position constraints are not in general satisfied by the numerical solution of equation (8.34). Alternatively, it may be possible to combine and solve the constraints at all levels [equations (8.27) to (8.29)], together with the equations of motion, which will assure that all the constraints are satisfied. The resulting formulation, called the state-space form, is derived in the following sections.

8.4 Effect of Constraint Differentiation

In Section 8.3, we discussed how the use of the constraints in their differentiated form can result in numerical errors that can hamper the performance of the multibody system. Let us show how this method can result in numerical error. Following Fuhrer and Leimkuhler [12], let us assume that the residual of the position constraint is $\Phi(q(t)) = \epsilon_0$ and that the initial conditions satisfy

$$B(q_0)y_0 = 0 \quad \text{and} \quad \Phi(q_0) = \epsilon_0 \tag{8.36}$$

The assumptions above should lead to an increase in the residual, which is linear in time. The example below is taken from reference [12] and is intended to demonstrate the idea.

Example 8.4.1
The governing equations of a constrained system are

$$\dot{x} = y - ax^2 + u(t) \tag{8.37}$$

$$y = ax^2 \tag{8.38}$$

We can think of equation (8.37) as the differential equation of motion and (8.38) as the corresponding constraint at the position level. From these two equations, we can easily see the solution as being

$$x(t) = x_0 + U(t) \text{ with } U(t) = \int_{t_0}^{t} u(\tau)d\tau$$

If the residual of the position constraint is assumed to be $\in_0$, we have

$$y - ax^2 = \in_0 \tag{8.39}$$

To solve this kind of constrained motion problem, we usually make use of the differentiated form of the constraint where we obtain

$$\dot{x} = y - ax^2 + u(t) \tag{8.40}$$

$$\dot{y} = 2ax[y - ax^2 + u(t)] \tag{8.41}$$

Substituting the expression given by equation (8.39), we get the solution:

$$x(t) = \in_0 t + U(t) \tag{8.42}$$

$$y(t) = \in_0 + a[\in_0^2 t^2 + 2 \in_0 tU(t) + U(t)^2] \tag{8.43}$$

The two problems—with the constraint being represented algebraically or in differentiated form—result in different numerical answers.

8.5 Reduction to State-Space Form

A state-space form of a general multibody system can only be established locally, and the reduction process must be carried out in every integration step. It requires the following computations:

- Elimination of constraint forces λ.
- Specification of independent generalized coordinates q_i.
- Elimination of the corresponding dependent variables q_d (the velocity and acceleration must also be represented in terms of the independent coordinates).

The elimination of the λ's result in the ODE given by (8.34).

Next we assume that the vector of generalized coordinates Q is partitioned into $n-m$ independent variables q_i and m dependent variables q_d (i.e., $Q=[q_i^T q_d^T]^T$) (see the embedding method described in Chapter 7). Since the dependent variables can be expressed by the independent ones, we have

$$q_d = H(q_i) \tag{8.44}$$

and the constraint in position level becomes

$$\Phi(q_i, H(q_i)) = 0 \tag{8.45}$$

After further differentiation of generalized coordinates, we get

$$y = \dot{Q} = \begin{bmatrix} I \\ H_{q_i}(q_i) \end{bmatrix} \dot{q}_i = V(q_i)\dot{q}_i \tag{8.46}$$

Similarly, if we differentiate (8.46) with respect to time, we obtain

$$\ddot{Q} = \dot{y} = V(q_i)\ddot{q}_i + \psi(q_i, \dot{q}_i) \tag{8.47}$$

Premultiplying (8.34) by V^T, using the notation $x_1 = q_i$ and $x_2 = \dot{q}_i$, and replacing Q and y by the relations derived above gives

$$\dot{x}_i = x_2 \tag{8.48}$$

$$V(x_1)^T \tilde{M}(x_1)V(x_1)\dot{x}_2 = V(x_1)^T \tilde{F}(x_1, x_2, t) \tag{8.49}$$

with $\tilde{M}(x_1) = M([x_1^T, H(x_1)^T]^T)$ and $\tilde{F}$ being defined similarly from the resulting equations.

The equations above form the local state-space form of a system governed by equations (8.25) to (8.27).

8.6 Modified PUTD Method: A Gaussian Approach

In this section we provide an alternative approach to the PUTD method described in Chapter 6. This modified version will be helpful in studying a different stability problem caused by the vanishing or linear dependency of the constraints. The method presented here is taken from reference [6], which is an Amirouche and Ider technique that uses the PUTD method. We shall formulate a basis for the null space of B, that is, a basis for the set of solution vectors of the homogeneous system

$$B_{i\ell}Y_\ell = 0 \quad (i = 1, \ldots, m;\ \ \ell = 1, \ldots, n) \tag{8.50}$$

The matrix C composed of these basis vectors as columns is then an orthogonal complement of B.

The analysis involves formulating an upper triangular matrix U by utilizing elementary row operations, that is,

$$PB^T = U \tag{8.51}$$

where P is an $n \times n$ nonsingular matrix, and then finding $n - m$ orthogonal vectors, in the n-dimensional space R^n, which are orthogonal to the columns of U. From the definition of B being a full rank m and given P as being a nonsingular matrix, the columns of U are linearly independent with nonzero diagonal matrix.

If the Gaussian elimination method is used to reduce the matrix B^T to upper triangular form, the matrix is premultiplied by a sequence of elementary matrices M_k, $k = 1, \ldots, m$, each involving a sequence of elementary row operations chosen to introduce a column with zeros below the diagonal. At the kth step, let a_k denote the kth column of the matrix. Then M_k is given as

$$M_k = I - \alpha e_k^T \tag{8.52}$$

where e_k denotes the kth column of I and

$$\alpha = [0 \cdots 0 \alpha_{k+1} \alpha_{k+2} \cdots \alpha_n]^T \tag{8.53}$$

with

$$\alpha_i = \frac{a_i^k}{a_k^k} \quad (i = k + 1, \ldots, n) \tag{8.54}$$

Partial pivoting is necessary, since a_k^k may be zero at any step. This must be required at each step, interchanging rows k and ρ_k, $\rho_k \geq k$, so that $a_{pk}^k \neq 0$. Since the rank of B^T is m, m of the n rows must be linearly independent. So row pivoting ensures the completion of the process for the m columns, without encountering a zero pivot.

The matrix can be considered to have a rank of less then m if a sufficiently small pivot is obtained in a step, since rounding errors make remote the probability of encountering an exactly zero pivot.

Denoting the elementary matrix involving row interchange at the kth step by Q_k, the transformation B^T to U can be expressed as

$$M_m Q_m M_{m-1} Q_{m-1} \cdots M_1 Q_1 B^T = U \tag{8.55}$$

Hence the matrix P in equation (8.51) is

$$P = M_m Q_m M_{m-1} \cdots M_1 Q_1 \tag{8.56}$$

If no row interchange is needed at the kth step, then

$$Q_k = I \tag{8.57}$$

Let I_1 and I_2 denote matrices composed of, respectively, the first m columns and the last $(n - m)$ columns of the nth-order identity matrix; therefore, I_2 is orthogonal to I_1. Since the columns of U are linearly combinations of the columns of I_1, I_2 is also orthogonal to U. Then, using equation (8.51), we obtain

$$I_2^T U = I_2^T P B^T = 0 \tag{8.58}$$

or

$$B P^T I_2 = BC = 0 \tag{8.59}$$

where $C = P^T I_2$. Since P is nonsingular, the columns of $P^T I_2$ are linearly independent. Therefore, in R^n-dimensional space, the $n - m$ columns of $P^T I_2$ form a basis of the set of solution vectors to equation (8.50). A linearly combination of these columns is also a solution. Therefore, in general

$$BP^T I_2 q = 0 \tag{8.60}$$

where q is an arbitrary column vector of order $n - m$ (fictitious coordinates that do not have any physical meaning).

Example 8.6.1
For a rectangular matrix given by

$$B = \begin{bmatrix} 0 & 2 & 3 \\ 4 & 0 & 5 \end{bmatrix}$$

find the matrix P to reduce it to an upper triangular matrix U using the Gaussian elimination method.

Solution
Let

$$B^T = \begin{bmatrix} 0 & 4 \\ 2 & 0 \\ 3 & 5 \end{bmatrix} \tag{8.61}$$

Since a_1^1 is zero, we should interchange the first row with the second row; hence

$$Q_1 = \begin{bmatrix} 0 & 1 & 0 \\ 1 & 0 & 0 \\ 0 & 0 & 1 \end{bmatrix}, \qquad Q_1 B^T = \begin{bmatrix} 2 & 0 \\ 0 & 4 \\ 3 & 5 \end{bmatrix} \tag{8.62}$$

Now making use of equation (8.52), we find M_1:

$$M_1 = I_{3\times3} - \alpha e_1^T \tag{8.63}$$

where

$$\alpha = \begin{bmatrix} 0 \\ \frac{0}{2} \\ \frac{3}{2} \end{bmatrix} = \begin{bmatrix} 0 \\ 0 \\ 1.5 \end{bmatrix} \tag{8.64}$$

$$M_1 = \begin{bmatrix} 1 & 0 & 0 \\ 0 & 1 & 0 \\ 0 & 0 & 1 \end{bmatrix} - \begin{bmatrix} 0 \\ 0 \\ 1.5 \end{bmatrix} \begin{bmatrix} 1 & 0 & 0 \end{bmatrix} = \begin{bmatrix} 1 & 0 & 0 \\ 0 & 1 & 0 \\ -1.5 & 0 & 1 \end{bmatrix} \tag{8.65}$$

Checking our first iteration using equation (8.55), we get

$$M_1 Q_1 B^T = \begin{bmatrix} 1 & 0 & 0 \\ 0 & 1 & 0 \\ -1.5 & 0 & 1 \end{bmatrix} \begin{bmatrix} 2 & 0 \\ 0 & 4 \\ 3 & 5 \end{bmatrix} = \begin{bmatrix} 2 & 0 \\ 0 & 4 \\ 0 & 5 \end{bmatrix} \tag{8.66}$$

Since $a_2^2 = 4 \neq 0$, we proceed with the upper triangularization of B^T (no row interchange is needed):

$$Q_2 = I_{3\times 3} \tag{8.67}$$

and

$$\alpha = \begin{bmatrix} 0 & 0 & \frac{5}{4} \end{bmatrix}^T, \qquad e_2^T = \begin{bmatrix} 0 & 1 & 0 \end{bmatrix} \tag{8.68}$$

Then

$$M_2 = \begin{bmatrix} 1 & 0 & 0 \\ 0 & 1 & 0 \\ 0 & 0 & 1 \end{bmatrix} - \begin{bmatrix} 0 \\ 0 \\ 1.25 \end{bmatrix} \begin{bmatrix} 0 & 1 & 0 \end{bmatrix} = \begin{bmatrix} 1 & 0 & 0 \\ 0 & 1 & 0 \\ 0 & -1.25 & 1 \end{bmatrix} \tag{8.69}$$

and our second iteration yields

$$M_2 Q_2 M_1 Q_1 B^T = \begin{bmatrix} 1 & 0 & 0 \\ 0 & 1 & 0 \\ 0 & -1.25 & 1 \end{bmatrix} \begin{bmatrix} 2 & 0 \\ 0 & 4 \\ 0 & 5 \end{bmatrix} = \begin{bmatrix} 2 & 0 \\ 0 & 4 \\ 0 & 0 \end{bmatrix} = U \tag{8.70}$$

Therefore, we can identify the matrix P that upper-triangularizes B^T:

$$P = M_2 Q_2 M_1 Q_1$$

$$= \begin{bmatrix} 1 & 0 & 0 \\ 0 & 1 & 0 \\ 0 & -1.25 & 1 \end{bmatrix} \begin{bmatrix} 1 & 0 & 0 \\ 0 & 1 & 0 \\ -1.5 & 0 & 1 \end{bmatrix} \begin{bmatrix} 0 & 1 & 0 \\ 1 & 0 & 0 \\ 0 & 0 & 1 \end{bmatrix} \tag{8.71}$$

$$= \begin{bmatrix} 0 & 1 & 0 \\ 1 & 0 & 0 \\ -1.25 & -1.5 & 1 \end{bmatrix}$$

Example 8.6.2
For the constraint equations

$$[B] \begin{bmatrix} \dot{x}_1 & 0 & 0 \\ 0 & \dot{x}_2 & 0 \\ 0 & 0 & \dot{x}_3 \end{bmatrix} = \begin{bmatrix} 0 \\ 0 \\ 0 \end{bmatrix}$$

where $[B]$ is as given in Example 8.6.1, find the orthogonal complement to B.

Solution

From equation (8.59) we know that the orthogonal complement matrix $C = P^T I_2$, where I_2 is the second part of the identity matrix $I_{3\times3}$.

$$P^T I_2 = \begin{bmatrix} 0 & 1 & -1.25 \\ 1 & 0 & -1.5 \\ 0 & 0 & 1 \end{bmatrix} \begin{bmatrix} 0 \\ 0 \\ 1 \end{bmatrix} = \begin{bmatrix} -1.25 \\ -1.5 \\ 1 \end{bmatrix} \tag{8.72}$$

The array above is the orthogonal complement matrix to B [equation (8.72)].

8.7 Regularization of the Vanishing Constraints: The Amirouche–Ider Stabilization Method

It is common in the course of a mechanical system's motion that for $B_{s\ell}$, $\ell = 1,\ldots, n$, the Sth constraint equation coefficients may all become zero at some instantaneous special configurations. In this case, the procedures we mentioned in Section 8.6 fail in determining the orthogonal complement array because the rank of B becomes less than m. In addition, the augmented system inertial matrix becomes singular.

Since the generalized constraint forces due to the Sth constraint, $\lambda_S B_{S\ell}$, $\ell = 1, \ldots, n$, are zero, the constraints have no effect on the system at that configuration. One possible procedure is to delete the Sth row of B, find the corresponding orthogonal complement matrix C, and then proceed with the normal coordinate reduction technique. As we mentioned before, the constraint at the acceleration level can be expressed as

$$\dot{h}_i = B_{i\ell}\dot{y}_\ell + \dot{B}_{i\ell}y_\ell - \dot{g}_i = 0 \quad (i = 1,\ldots,m) \tag{8.73}$$

where y denotes the generalized speeds. Suppose that for the $i = s$, $B_{s\ell} = 0$; then we can write

$$\dot{h}s = \dot{B}_{s\ell}\dot{y}_\ell - \dot{g}_s = 0 \tag{8.74}$$

Equation (8.74) must be satisfied by selecting the initial conditions for y_ℓ such that

$$\dot{B}_{s\ell}\dot{y}_\ell - \dot{g}_s = 0 \tag{8.75}$$

in order to achieve the consistency of the equations.

When the Sth constraint equation is removed in this manner, the resulting equations can be used in the neighborhood of the special configuration. The drawback to this approach is that the assumption of zero constraint forces in the neighborhood results in fast deviation of the simulation from the constrained behavior [11].

To capture the effects of the actual constraint forces in the neighborhood of the vanishing constraint, we further differentiate the Sth constraint equation. This will provide additional information on the constraint; hence a modified form of the constraint equation will be used instead of its deletion. To illustrate this approach, consider the time derivative of equation (8.73), when $i = s$:

$$\ddot{h}_s = B_{s\ell}\,\ddot{y}_\ell + 2\dot{B}_{s\ell}\,\dot{y}_\ell + \ddot{B}_{s\ell}\,\dot{y}_\ell - \ddot{g}_s = 0 \quad (\ell = 1,\ldots,n) \tag{8.76}$$

To simplify the representation of equation (8.76), we express $\dot{B}_{s\ell}$ as

$$\dot{B}_{s\ell} = \frac{\partial B_{s\ell}}{\partial q_h}\dot{q}_h + \frac{\partial B_{s\ell}}{\partial t} \tag{8.77}$$

since

$$\dot{q}_h = T_{hj} y_j \tag{8.78}$$

where T is a matrix relating $\dot{q}$ to y. In light of the above, we can write equation (8.77) as

$$\dot{B}_{s\ell} = \phi_{s\ell j}\ y_j + \Psi_{s\ell} \tag{8.79}$$

where

$$\phi_{s\ell j} = \frac{\partial B_{s\ell}}{\partial q_h} T_{hj} \tag{8.80}$$

and

$$\psi_{s\ell} = \frac{\partial B_{s\ell}}{\partial t} \tag{8.81}$$

Similarly, $\dot{g}_s$ can be expressed as

$$\dot{g}_s = \mu_{s\ell} y_\ell + \tau_s \tag{8.82}$$

where

$$\mu_{s\ell} = \frac{\partial g_s}{\partial q_h} T_{h\ell}, \qquad \tau_s = \frac{\partial g_s}{\partial t} \tag{8.83}$$

Further differentiation of equations (8.79) and (8.82) gives the $\ddot{B}$ and $\ddot{g}$ terms as

$$\ddot{B}_{s\ell} = \dot{\phi}_{s\ell j} y_j + \phi_{s\ell j},\ \dot{y}_j + \dot{\psi}_{s\ell} \tag{8.84}$$

$$\ddot{g}_s = \dot{\mu}_{s\ell} y_\ell + \mu_{s\ell}\dot{y}_\ell + \dot{\tau}_s \tag{8.85}$$

Substituting equations (8.79), (8.84) and (8.85) into (8.76), we get

$$B_{s\ell}\ddot{y}_\ell + G_{s\ell}\dot{y}_\ell + J_s = 0 \quad (\ell = 1, \ldots, n) \tag{8.86}$$

where

$$G_{s\ell} = 2\dot{B}_{s\ell} + \phi_{sp\ell} y_p - \mu_{s\ell} \quad (\ell, p = 1, \ldots, n) \tag{8.87}$$

and

$$J_s = \dot{\phi}_{s\ell p} y_p y_\ell + \dot{\psi}_{s\ell} y_\ell - \dot{\mu}_{s\ell} y_\ell - \dot{\tau}_s \tag{8.88}$$

The usefulness of equation (8.86) stems from the assumption that when the elements of the constraint $B_{s\ell}$, $\ell = 1, \ldots, n$, are zero at the special configuration and small at its neighborhood, $B_{s\ell}\ddot{y}_\ell$ is negligible compared to the other terms in equation (8.86), so dropping that term in the equation will yield

$$\ddot{h}'_s = G_{s\ell}\dot{y}_\ell + J_s = 0 \tag{8.89}$$

It is seen that $G_{s\ell}$ and J_s are, in general, functions of q_ℓ, y_ℓ and t. Equation (8.89) is linear in accelerations $\dot{y}_\ell$ and has the form of the constraint at the acceleration level. If one is interested in computing the generalized constraint forces at the neighborhood of singularity, all we have to do is replace the Sth constraint in $F^c = \lambda B^T$ by

$$D_{s\ell} = G_{s\ell} \tag{8.90}$$

In most cases, $G_{s\ell}$ does not vanish where $B_{s\ell}$ vanishes. The governing equations of motion will then be subject to the constraints given by equation (8.89).

To further assure the numerical stability of the constraints, we can make use of the Baumgarte technique, where equation (8.89) takes on the form

$$G_{s\ell}\dot{y}_\ell + J_s + \alpha_s(\dot{B}_{s\ell}y_\ell - \dot{g}_s) + \beta_s(B_{s\ell}y_\ell - g_s) = 0 \tag{8.91}$$

which reduces to

$$G_{s\ell}\dot{y}_\ell + J_s + \alpha_s(\dot{B}_{s\ell}y_\ell - \dot{g}_s) - \beta_s g_s = 0 \tag{8.92}$$

where

$$G_{s\ell}\dot{y}_\ell = -J_s - \alpha_s(\dot{B}_{s\ell}y_\ell - \dot{g}_s) + \beta_s g_s \tag{8.93}$$

and

$$S_s = -J_s - \alpha_s(\dot{B}_{s\ell}y_\ell - \dot{g}_s) + \beta_s g_s \tag{8.94}$$

In addition, $\beta_s = 0$ if the constraint is nonholonomic. Note that $h_s = 0$ is automatically satisfied since $B_{s\ell} = 0$, $\ell = 1, \ldots n$, and for consistency, $g_s = 0$.

Example 8.7.1

Consider a two-link manipulator, $L_1 = \ell$, $L_2 = \ell/2$, $m_2 = m$, $m_1 = 2m_2 = 2m$, as shown in Figure 8.1. Let us assume that point p follows the horizontal line $y = \ell/2$ with a constant velocity along the $-x$-axis. We can expect that for $q_1 = \pi/2$ and $q_2 = \pi$, the system Jacobian matrix becomes singular. This example can be used to check our stability method and can be further compared to Baumgarte's technique. We will show how our stability is completely independent of the asymptotic stability, which is due to the ill representation of the constraints at the acceleration level.

There are two constraint equations for this problem. The first one, the position constraint, keeps point p on the horizontal line $y = \ell/2$. The second one, the velocity constraint, keeps point p at a constant velocity along the $-x$ axis. We can represent the two constraints as follows:

$$h_1(q_1, q_2, t) = \ell \sin q_1 + \frac{\ell}{2}\sin(q_1 + q_2) - \frac{\ell}{2} = 0 \tag{8.95}$$

$$h_2(q_1, q_2, t) = \ell \cos q_1 + \frac{\ell}{2}\cos(q_1 + q_2) + vt - x_0 = 0 \tag{8.96}$$

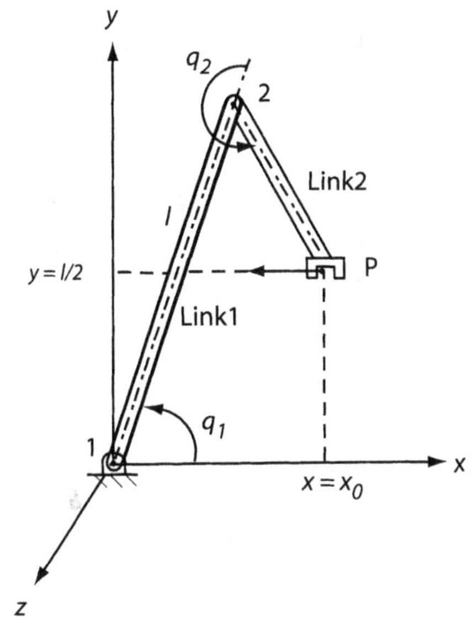

Fig. 8.1. Two-link manipulator undergoes constrained motion

where x_0 is the initial position of x coordinate. Their first-order differentiated forms yield

$$\dot{h}_1 = \ell C_1 y_1 + \frac{\ell}{2} C_{12}(y_1 + y_2) = 0 \tag{8.97}$$

$$\dot{h}_2 = -\ell S_1 y_1 - \frac{\ell}{2} S_{12}(y_1 + y_2) + \upsilon = 0 \tag{8.98}$$

where

$$S_1 = \sin q_1, \qquad S_{12} = \sin(q_1 + q_2)$$

$$C_1 = \cos q_1, \qquad C_{12} = \cos(q_1 + q_2)$$

$$y_1 = \dot{q}_1, \qquad y_2 = \dot{q}_2$$

Using equations (8.97) and (8.98) and expressing the constraint equations in a matrix form, we obtain

$$\begin{bmatrix} \ell C_1 + \frac{\ell}{2} C_{12} & \frac{\ell}{2} C_{12} \\ -\ell S_1 - \frac{\ell}{2} S_{12} & -\frac{\ell}{2} S_{12} \end{bmatrix} \begin{bmatrix} y_1 \\ y_2 \end{bmatrix} = \begin{bmatrix} 0 \\ -\upsilon \end{bmatrix} \tag{8.99}$$

Differentiating equations (8.97) and (8.98) once more, we get the acceleration level of the constraints as

$$\ddot{h}_1 = \left(\ell C_1 + \frac{\ell}{2} C_{12}\right) \dot{y}_1 + \frac{\ell}{2} C_{12} \dot{y}_2 - \left[\ell S_1 y_1^2 + \frac{\ell}{2} S_{12}(y_1 + y_2)^2\right] = 0 \tag{8.100}$$

$$\ddot{h}_2 = \left(-\ell S_1 - \frac{\ell}{2}S_{12}\right)\dot{y}1 - \frac{\ell}{2}S_{12}\dot{y}_2 - \left[\ell C_1 y_1^2 + \frac{\ell}{2}C_{12}(y_1 + y_2)^2\right] = 0 \quad (8.101)$$

which can be written in matrix form as

$$\begin{bmatrix} \ell C_1 + \frac{\ell}{2}C_{12} & \frac{\ell}{2}C_{12} \\ -\ell S_1 - \frac{\ell}{2}S_{12} & -\frac{\ell}{2}S_{12} \end{bmatrix} \begin{bmatrix} \dot{y}_1 \\ \dot{y}_2 \end{bmatrix} = \begin{bmatrix} \ell S_1 y_1^2 + \frac{\ell}{2}S_{12}(y_1 + y_2)^2 \\ \ell C_1 y_1^2 + \frac{\ell}{2}C_{12}(y_1 + y_2)^2 \end{bmatrix} \quad (8.102)$$

Introducing the Baumgarte method, equation (8.102) takes on the form

$$\begin{bmatrix} \ell C_1 + \frac{\ell}{2}C_{12} & \frac{\ell}{2}C_{12} \\ -\ell S_1 - \frac{\ell}{2}S_{12} & -\frac{\ell}{2}S_{12} \end{bmatrix} \begin{bmatrix} \dot{y}_1 \\ \dot{y}_2 \end{bmatrix}$$

$$= \begin{bmatrix} \ell S_1 y_1^2 + \frac{\ell}{2}S_{12}(y_1 + y_2)^2 - \alpha\left[\ell C_1 y_1 + \frac{\ell}{2}C_{12}(y_1 + y_2)\right] \\ -\beta\left(\ell S_1 + \frac{\ell}{2}S_{12} - \frac{\ell}{2}\right) \\ \ell C_1 y_1^2 + \frac{\ell}{2}C_{12}(y_1 + y_2)^2 - \alpha\left[-\ell S_1 y_1 - \frac{\ell}{2}S_{12}(y_1 + y_2) + \upsilon\right] \\ -\beta\left(\ell C_1 + \frac{\ell}{2}C_{12} + \upsilon t - x_0\right) \end{bmatrix} \quad (8.103)$$

From equation (8.99), (8.102), or (8.103), we define the Jacobian constraint matrix as

$$B = \begin{bmatrix} \ell C_1 + \frac{\ell}{2}C_{12} & \frac{\ell}{2}C_{12} \\ -\ell S_1 - \frac{\ell}{2}S_{12} & -\frac{\ell}{2}S_{12} \end{bmatrix} \quad (8.104)$$

From Kane's equation (i.e., $F + F^* + B^T\lambda = 0$), we write the equations of motion for the system as

$$\begin{bmatrix} \frac{3}{2}m\ell^2 + I_1 + I_2 & \frac{1}{4}m\ell^2 C_2 + I_2 \\ \frac{1}{4}m\ell^2 C_2 + I_2 & \frac{1}{16}m\ell^2 + I_2 \end{bmatrix} \begin{bmatrix} \dot{y}_1 \\ \dot{y}_2 \end{bmatrix} + \begin{bmatrix} -\frac{1}{4}m\ell^2 y_2(y_1 + y_2)S_2 + 2mg\ell S_1 \\ \frac{1}{4}m\ell^2 y_1^2 S_2 + \frac{1}{4}mg\ell S_{12} \end{bmatrix}$$

$$= \begin{bmatrix} \ell C_1 + \frac{\ell}{2}C_{12} & -\ell S_1 - \frac{\ell}{2}S_{12} \\ \frac{\ell}{2}C_{12} & -\frac{\ell}{2}S_{12} \end{bmatrix} \begin{bmatrix} \lambda_1 \\ \lambda_2 \end{bmatrix} \quad (8.105)$$

where I_1 and I_2 are moments of inertia of body 1 and body 2, which are measured at the mass center of each body along z-axis, g is the gravitational acceleration, and λ_1 and λ_2 are Lagrange multipliers. The governing equations of motion for the two-link system are given by equations (8.102) and (8.105) for the case in which the Baumgarte method is not used, and by equations (8.103) and (8.105) with Baumgarte representation of the constraints.

We can easily deduce from equations (8.102) and (8.103) that when singularity occurs, (i.e., $q_1 = \pi/2, q_2 = \pi$), the first row of B vanishes. Employing the method proposed, the new representation of the constraints is given below.

$$\begin{bmatrix} -3\ell S_1 y_1 - \frac{3\ell}{2} S_{12}(y_1 + y_2) & -\frac{3\ell}{2} S_{12}(y_1 + y_2) \\ -\ell S_1 - \frac{\ell}{2} S_{12} & -\frac{\ell}{2} S_{12} \end{bmatrix} \begin{bmatrix} \dot{y}_1 \\ \dot{y}_2 \end{bmatrix} = \begin{bmatrix} \ell C_1 y_1^3 + \frac{\ell}{2} C_{12}(y_1 + y_2)^3 \\ \ell C_1 y_1^2 + \frac{\ell}{2} C_{12}(y_1 + y_2)^2 \end{bmatrix} \tag{8.106}$$

Introducing Baumgarte's method, equation (8.106) becomes

$$\begin{bmatrix} -3\ell S_i y_1 - \frac{3\ell}{2} S_{12}(y_1 + y_2) & -\frac{3\ell}{2} S_{12}(y_1 + y_2) \\ -\ell S_1 - \frac{\ell}{2} S_{12} & -\frac{\ell}{2} S_{12} \end{bmatrix} \begin{bmatrix} \dot{y}_1 \\ \dot{y}_2 \end{bmatrix}$$
$$= \begin{bmatrix} \ell C_1 y_1^3 + \frac{\ell}{2} C_{12}(y_1 + y_2)^3 - \alpha \left[\ell C_1 y_1 + \frac{\ell}{2} C_{12}(y_1 + y_2) \right] - \beta \left(\ell S_1 + \frac{\ell}{2} S_{12} - \frac{\ell}{2} \right) \\ \ell C_1 y_1^2 + \frac{\ell}{2} C_{12}(y_1 + y_2)^2 - \alpha \left[-\ell S_1 y_1 - \frac{\ell}{2} S_{12}(y_1 + y_2) + \upsilon \right] - \beta \left(\ell C_1 + \frac{l}{2} C_{12} + \upsilon t - x_0 \right) \end{bmatrix} \tag{8.107}$$

Assume that the two links are slender rods and that

$$m_1 = 2\,\text{kg}, \qquad m_2 = 1\,\text{kg}, \qquad \ell = 1\,\text{m}, \qquad \nu = 0.6\,\text{m/s}$$

Then

$$I_1 = \frac{1}{12} m_1 \ell^2 = 0.167\,\text{kg-m}^2$$
$$I_2 = \frac{1}{12} m_2 \left(\frac{\ell}{2}\right)^2 = 0.208\,\text{kg-m}^2$$

The initial angular positions for links 1 and 2 are, respectively, 1.396263 rad and 3.563268 rad. From these data we can find that $x_0 = 0.295953$ m. The initial angular velocities are chosen to be 0.358620 and −0.867745 rad/s to set the initial error of the velocity constraint close to zero. To compare the effects of our stability methods, we classify the results into the following cases: (1) without Baumgarte (i.e., $\alpha = 0$, $\beta = 0$) and no modification for B; (2) with Baumgarte ($\alpha = 1.0$, $\beta = 0.25$) but no modification for B; (3) without Baumgarte but with modified B; and (4) with Baumgarte and modified B. In each case, we choose 0.05 s as our time step, and the time history yields outputs for q_1 and q_2, the driving torques of links 1 and 2, position error, and velocity error, which are represented as percentages. These results are shown in Tables 8.1 to 8.4. Essentially, from cases 1 and 3, we should be able to test the proposed method and its stability near the singular position. On the other hand, with cases 2 and 4, we can see how the Baumgarte technique affects the stability at singular positions.

T (s)	Q_1 (rad)	Q_2 (rad)	Link 1 Torque (N-m)	Link 2 Torque (N-m)	Position Error (%)	Velocity Error (%)
0.00	1.396	3.563	4.002	0.604	0.119E−04	0.993E−04
0.05	1.414	3.520	3.596	0.543	0.143E−03	0.258E−01
0.10	1.432	3.477	3.190	0.482	0.262E−03	0.508E−01
0.15	1.450	3.434	2.784	0.420	0.334E−03	0.750E−01
0.20	1.467	3.391	2.379	0.359	0.322E−03	0.985E−01
0.25	1.485	3.349	1.974	0.298	0.250E−03	0.122E+00
0.30	1.503	3.306	1.569	0.237	0.179E−03	0.144E+00
0.35	1.520	3.263	1.164	0.176	0.477E−04	0.167E+00
0.40	1.538	3.221	0.761	0.116	0.715E−04	0.189E+00
0.45	1.555	3.179	0.361	0.057	0.143E−03	0.210E+00
0.50	1.573	3.136	−0.214	−0.089	0.715E−04	0.232E+00
0.55	1.590	3.094	−0.173	0.066	0.452E−02	0.255E+00
0.60	1.607	3.051	−0.961	−0.188	0.108E−01	0.273E+00
0.65	1.625	3.009	−1.267	−0.194	0.297E−02	0.296E+00
0.70	1.643	2.967	−1.667	−0.253	0.235E−02	0.319E+00
0.75	1.661	2.924	−2.071	−0.314	0.216E−02	0.341E+00
0.80	1.678	2.882	−2.475	−0.374	0.210E−02	0.364E+00
0.85	1.696	2.839	−2.879	−0.435	0.216E−02	0.387E+00
0.90	1.714	2.796	−3.283	−0.496	0.226E−02	0.411E+00
0.95	1.731	2.753	−3.688	−0.557	0.241E−02	0.435E+00
1.00	1.749	2.710	−4.092	−0.617	0.254E−02	0.460E+00

Table 8.1. Numerical results for case 1 (without Baumgarte, no modification; $\alpha = 0$, $\beta = 0$)

T (s)	Q_1 (rad)	Q_2 (rad)	Link 1 Torque (N-m)	Link 2 Torque (N-m)	Position Error (%)	Velocity Error (%)
0.00	1.396	3.563	4.002	0.604	0.238E−04	0.993E−04
0.05	1.414	3.520	3.596	0.543	0.143E−03	0.557E−02
0.10	1.432	3.477	3.190	0.482	0.238E−03	0.118E−01
0.15	1.450	3.434	2.784	0.420	0.298E−03	0.183E−01
0.20	1.467	3.391	2.378	0.359	0.310E−03	0.251E−01
0.25	1.485	3.348	1.972	0.298	0.274E−03	0.320E−01
0.30	1.503	3.306	1.566	0.237	0.191E−03	0.390E−01
0.35	1.520	3.263	1.161	0.176	0.954E−04	0.459E−01
0.40	1.538	3.221	0.756	0.115	0.238E−04	0.526E−01
0.45	1.556	3.178	0.355	0.056	0.834E−04	0.592E−01
0.50	1.573	3.136	−0.172	−0.066	0.477E−04	0.654E−01
0.55	1.590	3.093	−0.260	0.027	0.317E−02	0.702E−01
0.60	1.608	3.050	−0.947	−0.176	0.722E−02	0.789E−01
0.65	1.626	3.008	−1.274	−0.194	0.138E−02	0.822E−01
0.70	1.644	2.966	−1.679	−0.254	0.126E−02	0.865E−01
0.75	1.661	2.923	−2.084	−0.315	0.135E−02	0.903E−01
0.80	1.679	2.880	−2.490	−0.376	0.150E−02	0.934E−01
0.85	1.697	2.837	−2.896	−0.438	0.168E−02	0.959E−01
0.90	1.715	2.794	−3.303	−0.499	0.193E−02	0.975E−01
0.95	1.732	2.751	−3.709	−0.560	0.217E−02	0.984E−01
1.00	1.750	2.708	−4.116	−0.621	0.232E−02	0.983E−01

Table 8.2. Numerical results for case 2 (without Baumgarte, no modification; $\alpha = 1, \beta = 0.25$)

In this particular case at the singular position, one of the constraints will vanish. As we approach singularity, we should experience the necessary jumps, if we proceed with the standard technique where the constraints are not eliminated. We have run our simulation in two cases, which are labeled as being with and without modification. In Figures 8.2 and 8.3 we can see the curves of the corresponding torques of joints 1 and 2 becoming smoother at the neighborhood of singularity, when our stability method is employed. It is very clear that the stability method proposed (solid line) does not experience the jumps of instability by the constraint forces, when the B is kept unchanged.

This is further illustrated by Figure 8.4, where the jumps in velocity are seen to yield large peaks. The latter could hamper system performance. In Figure 8.5, we can

T (s)	Q_1 (rad)	Q_2 (rad)	Link 1 Torque (N-m)	Link 2 Torque (N-m)	Position Error (%)	Velocity Error (%)
0.00	1.396	3.563	4.002	0.604	0.119E−04	0.993E−04
0.05	1.414	3.520	3.596	0.543	0.143E−03	0.258E−01
0.10	1.432	3.477	3.190	0.482	0.262E−03	0.508E−01
0.15	1.450	3.434	2.784	0.420	0.334E−03	0.750E−01
0.20	1.467	3.391	2.379	0.359	0.322E−03	0.985E−01
0.25	1.485	3.349	1.974	0.298	0.250E−03	0.122E+00
0.30	1.503	3.306	1.569	0.237	0.179E−03	0.144E+00
0.35	1.520	3.263	1.164	0.176	0.834E−04	0.166E+00
0.40	1.538	3.221	0.759	0.115	0.119E−04	0.189E+00
0.45	1.555	3.179	0.355	0.054	0.238E−04	0.210E+00
0.50	1.573	3.136	−0.050	−0.007	0.000E+00	0.232E+00
0.55	1.590	3.094	−0.454	0.069	0.119E−03	0.254E+00
0.60	1.608	3.052	−0.859	−0.130	0.322E−03	0.276E+00
0.65	1.626	3.009	−1.263	−0.191	0.632E−03	0.297E+00
0.70	1.643	2.967	−1.668	−0.252	0.101E−02	0.319E+00
0.75	1.661	2.924	−2.072	−0.313	0.138E−02	0.342E+00
0.80	1.678	2.882	−2.476	−0.374	0.169E−02	0.365E+00
0.85	1.696	2.839	−2.880	−0.435	0.201E−02	0.388E+00
0.90	1.714	2.796	−3.284	−0.496	0.230E−02	0.412E+00
0.95	1.732	2.753	−3.689	−0.556	0.259E−02	0.436E+00
1.00	1.749	2.710	−4.093	−0.617	0.284E−02	0.461E+00

Table 8.3. Numerical results for case 3 (without Baumgarte, with modification; $\alpha = 0, \beta = 0$)

see that the position error stays very small even if B is modified at the singular position. In Figures 8.6 to 8.9 the same simulation is repeated introducing the Baumgarte technique. From this simulation, we can conclude that at the singular position, Baumgarte's technique does not provide stability. This is as expected, since the vanishing of constraint and its reappearance are what cause the instability.

Indeed, this has been our objective in this simulation. There are two major drawbacks to the handling of constraints in the dynamics of multibody systems. As pointed out by Baumgarte, if the constraints are represented by their acceleration form, the accumulation of errors can build up to negate the results. On the other hand, the instantaneous vanishing of constraints can yield a time history where we will experience

T (s)	Q_1 (rad)	Q_2 (rad)	Link 1 Torque (N-m)	Link 2 Torque (N-m)	Position Error (%)	Velocity Error (%)
0.00	1.396	3.563	4.002	0.604	0.238E−04	0.993E−04
0.05	1.414	3.520	3.596	0.543	0.143E−03	0.557E−02
0.10	1.432	3.477	3.190	0.482	0.238E−03	0.118E−01
0.15	1.450	3.434	2.784	0.420	0.298E−03	0.183E−01
0.20	1.467	3.391	2.378	0.359	0.310E−03	0.251E−01
0.25	1.485	3.348	1.972	0.298	0.274E−03	0.320E−01
0.30	1.503	3.306	1.566	0.237	0.191E−03	0.390E−01
0.35	1.520	3.263	1.161	0.176	0.954E−04	0.459E−01
0.40	1.538	3.221	0.755	0.114	0.119E−04	0.526E−01
0.45	1.556	3.178	0.350	0.053	0.119E−04	0.592E−01
0.50	1.573	3.136	−0.056	−0.008	0.119E−04	0.654E−01
0.55	1.591	3.093	−0.462	−0.070	0.834E−04	0.713E−01
0.60	1.608	3.051	−0.867	−0.131	0.250E−03	0.767E−01
0.65	1.626	3.008	−1.273	−0.192	0.465E−03	0.817E−01
0.70	1.644	2.966	−1.679	−0.254	0.727E−03	0.861E−01
0.75	1.661	2.923	−2.085	−0.315	0.978E−03	0.900E−01
0.80	1.679	2.880	−2.491	−0.376	0.112E−02	0.931E−01
0.85	1.697	2.837	−2.897	−0.437	0.126E−02	0.956E−01
0.90	1.715	2.794	−3.303	−0.498	0.139E−02	0.973E−01
0.95	1.732	2.751	−3.710	−0.560	0.149E−02	0.981E−01
1.00	1.750	2.708	−4.117	−0.621	0.153E−02	0.980E−01

Table 8.4. Numerical results for case 4 (with Baumgarte, with modification; $\alpha = 1$, $\beta = 0.25$)

jumps in acceleration that are not suited for actuator design in robotics. The latter can result in serious design problems.

Since this is a numerical type of simulation, we can say that jumps in velocity and acceleration at or near the singular position can be more apparent if the time step of integration is refined. We leave this option for the user to check further. There is no doubt that the proposed approach will carry us further in the analysis and much closer to the exact singular point. The transition is then brief and any instability will be minimized.

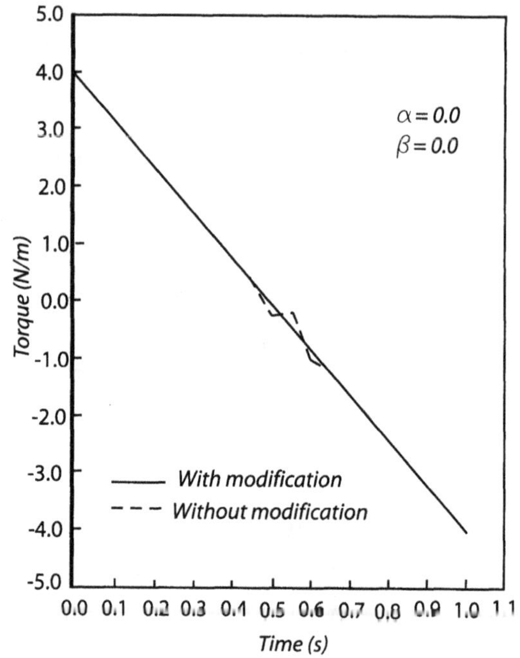

Fig. 8.2. Driving torque for joint 1

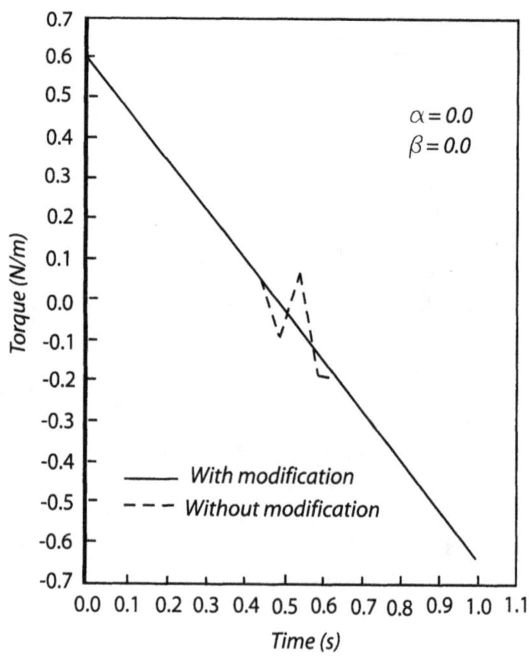

Fig. 8.3. Driving torque for joint 2

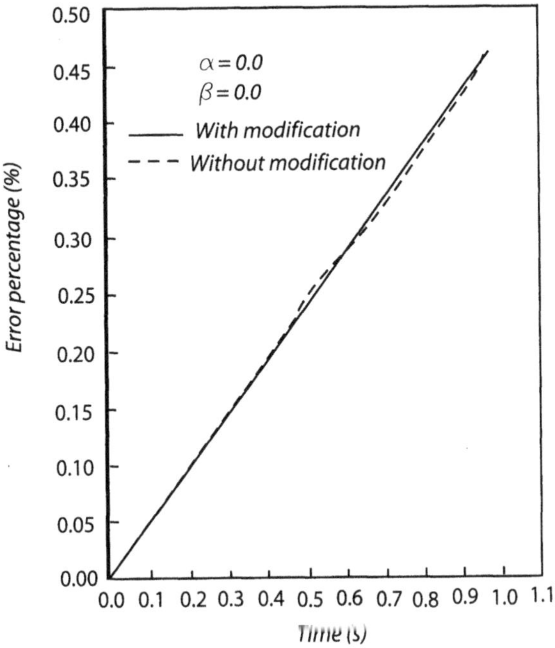

Fig. 8.4. Error of velocity constraint (%)

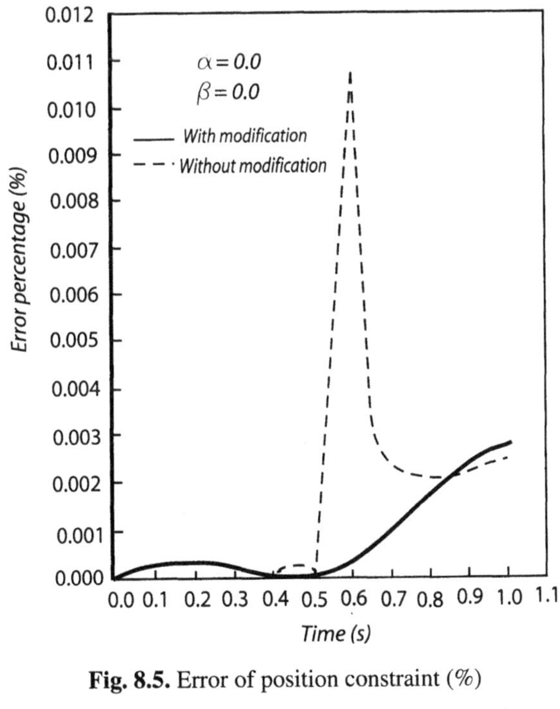

Fig. 8.5. Error of position constraint (%)

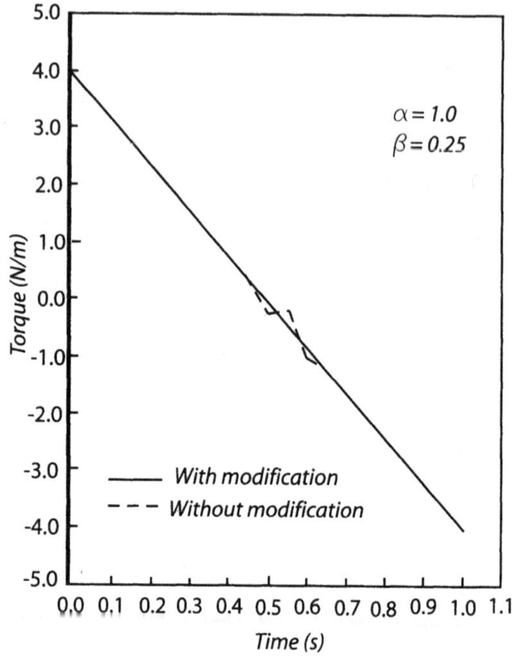

Fig. 8.6. Introducing the Baumgarte technique on torque for joint 1

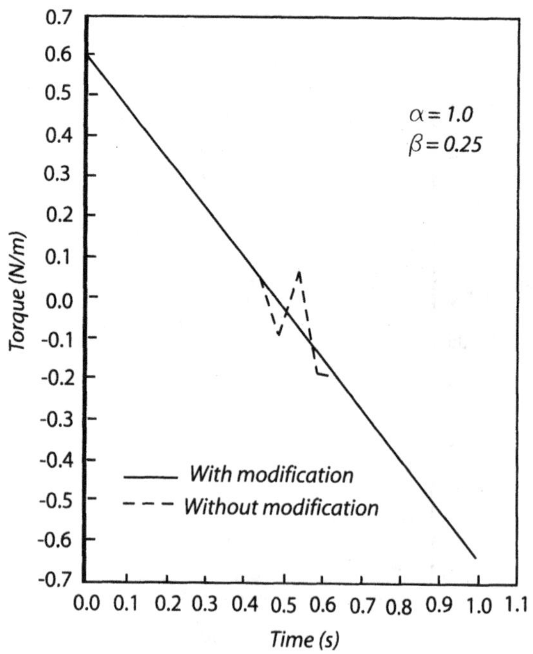

Fig. 8.7. Introducing the Baumgarte technique on torque for joint 2

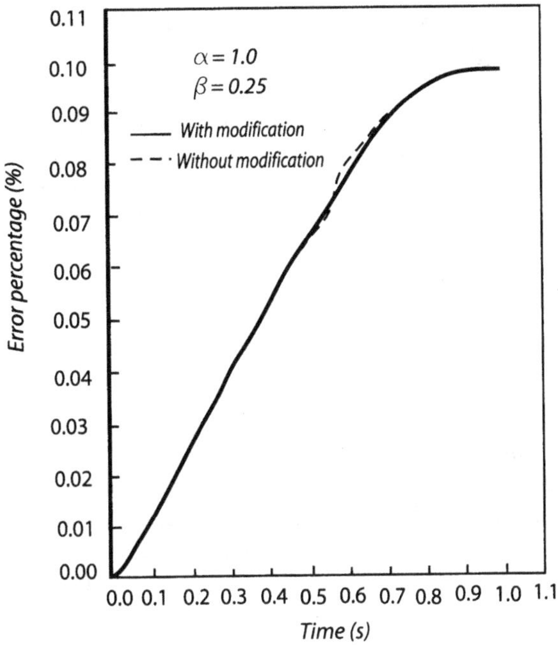

Fig. 8.8. Error of velocity constraint (%)

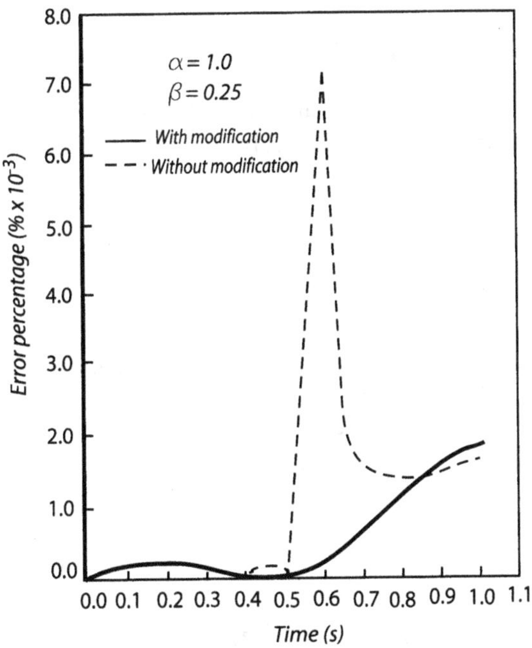

Fig. 8.9. Error of position constraint (%)

8.8 Regularization of Linearly Dependent Constraints: The Amirouche–Ider Method

In the general case, the rank of B may become less than m at some instantaneous special configurations in the course of a dynamic simulation of articulated multibody systems. In this case, it is known that an $n \times m$ matrix B^T can be reduced to Gaussian form U^* by Gaussian elimination row operations, where the rows $1, \ldots, r$ are the nonzero rows of U^*, and the leftmost nonzero entry of row i occurs in column k_i, $i = 1, \ldots, r$, then $k_1 < k_2 < \cdots < k_r$.

The values B^T and U^* are row equivalent with the same rank r, and the row equivalence transformation can be expressed as

$$PB^T = U^* \tag{8.108}$$

where P is the corresponding $n \times n$ nonsingular matrix.

It is seen from the definition of U^* that the columns with indices $k_i, i = 1, \ldots, r$, are linearly independent columns, whereas the remaining $m - r$ columns whose indices are denoted as $d_i, i = 1, \ldots, m - r$, can be expressed as linear combinations of the former set. The (d_i)th columns of B^T can also be written as linear combinations of the (k_i)th columns of B^T. To show this, we can write equation (8.108) as

$$Pb^j = u^j \quad (j = 1, \ldots, m) \tag{8.109}$$

where b^j and u^j denote the jth columns of B^T and U^*, respectively. For the columns of U^*, we have

$$u^{d_j} = Z_{ji} u^{k_i} \quad (i = 1, \ldots, r;\ j = 1, \ldots, m - r) \tag{8.110}$$

where $Z_{ji}, i = 1, \ldots, r$ are the linear combination constants for the (d_j)th column. Substitution of equation (8.109) into (8.110) for u^{d_j} and for each u^{k_i} lead to

$$Pb^{d_j} = u^{d_j} = Z_{ji} u^{k_i} = Z_{ji} Pb^{k_i} \tag{8.111}$$

so

$$b^{d_j} = Z_{ji} b^{k_i} \quad (i = 1, \ldots, r;\ j = 1, \ldots, m - r) \tag{8.112}$$

Representing equation (8.112) in a matrix form, we obtain

$$B_{d_j \ell} - Z_{ji} B_{k_i \ell} = 0 \tag{8.113}$$

We can write the constraint equations in the following forms:

$$\dot{h}_{d_j} = B_{d_j \ell\, y} - g_{d_j} = 0 \tag{8.114}$$

$$\dot{h}_{k_i} = B_{k_i \ell} y - g_{k_i} = 0 \tag{8.115}$$

$$\ddot{h}_{d_j} = \dot{B}_{d_j \ell} y + B_{d_j \ell} \dot{y} - \dot{g}_{d_j} = 0 \tag{8.116}$$

$$\ddot{h}_{k_i} = \dot{B}_{k_i\ell} y + B_{k_i\ell}\dot{y} - \dot{g}_{k_i} = 0 \tag{8.117}$$

We can use equation (8.117) to generate our constraint equations at the acceleration level, but we cannot use equation (8.116) because $B_{d_j\ell}$ is linearly dependent on $B_{k_i\ell}$, as shown in equation (8.113). Since

$$\dot{h}_{d_j} - Z_{ji}\dot{h}_{k_i} = [B_{d_j\ell} - Z_{ji}B_{k_i}]y - [g_{d_j} - Z_{ji}g_{k_i}] = 0 \tag{8.118}$$

and using equation (8.113), equation (8.118) becomes

$$\dot{h}_{d_j} - Z_{ji}\dot{h}_{k_i} = -[g_{d_j} - Z_{ji}g_{k_i}] = 0 \tag{8.119}$$

Its differentiated form is given by

$$\ddot{h}_{d_j} - Z_{ji}\ddot{h}_{k_i} = (B_{d_j\ell} - Z_{ji}B_{k_i})\dot{y} + (\dot{B}_{d_j\ell} - Z_{ji}\dot{B}_{k_i\ell})y - (\dot{g}_{d_j} - Z_{ji}\dot{g}_{k_i}) = 0 \tag{8.120}$$

Making use of equation (8.113) once more, equation (8.120) reduces to

$$\ddot{h}_{d_j} - Z_{ji}\ddot{h}_{k_i} = (\dot{B}_{d_j\ell} - Z_{ji}\dot{B}_{k_i\ell})y - (\dot{g}_{d_j} - Z_{ji}\dot{g}_{k_i}) = 0 \tag{8.121}$$

Note that the $\dot{y}$ term is not explicit in equation (8.111). If we further differentiate equation (8.120), we get

$$\begin{aligned} h'''_{d_j} - Z_{ji}h'''_{k_i} &= (B_{d_j\ell} - Z_{ji}B_{k_i\ell})\ddot{y} + 2(\dot{B}_{d_j\ell} - Z_{ji}\dot{B}_{k_i\ell})\dot{y} \\ &+ (\ddot{B}_{d_j\ell} - Z_{ji}\ddot{B}_{k_i\ell})y - (\ddot{g}_{d_j} - Z_{ji}\ddot{g}_{k_i}) = 0 \end{aligned} \tag{8.122}$$

Equation (8.122) reduces further to

$$\begin{aligned} h'''_{d_j} - Z_{ji}h'''_{ki} &= 2(\dot{B}_{d_j\ell} - Z_{ji}\dot{B}_{k_i\ell})\dot{y} + (\ddot{B}_{d_j\ell} - Z_{ji}\ddot{B}_{k_i\ell})y \\ &- (\ddot{g}_{d_j} - Z_{ji}\ddot{g}_{k_i}) = 0 \end{aligned} \tag{8.123}$$

by dropping the term that multiplies $\ddot{y}$ by virtue of equation (8.113). In essence, the representations of the constraint equations given by equation (8.116) are now represented by equation (8.123), such that

$$\ddot{h}'_{d_j} = 2(\dot{B}_{d_j\ell} - Z_{ji}\dot{B}_{k_i\ell})\dot{y} + (\ddot{B}_{d_j\ell} - Z_{ji}\ddot{B}_{k_i\ell})y - (\ddot{g}_{d_j} - Z_{ji}\ddot{g}_{k_i}) = 0 \tag{8.124}$$

Equations (8.124) and (8.117) become linearly independent and form a consistent set of constraint equations to be used when the rank of the Jacobian matrix is reduced to the linear dependency. This stability procedure can be further enhanced by applying Baugarte's method to stabilize the numerical error due to integration of the equations of motion. The constraint equations in the Baumgarte method become

$$\begin{aligned} &2(\dot{B}_{d_j\ell} - Z_{ji}\dot{B}_{k_i\ell})\dot{y} + (\ddot{B}_{d_j\ell} - Z_{ji}\ddot{B}_{k_i\ell})y - (\ddot{g}_{d_j} - Z_{ji}\ddot{g}_{k_i}) \\ &+ \alpha_{d_j}[(\dot{B}_{d_j\ell} - Z_{ji}\dot{B}_{k_i\ell})y - (\dot{g}_{d_j} - Z_{ji}\dot{g}_{k_i})] + \beta_{d_j}[-(g_{d_j} - Z_{ji}g_{k_i})] = 0 \end{aligned} \tag{8.125}$$

or

$$G_{d_j}\dot{y} + S_{d_j} = 0 \tag{8.126}$$

where

$$G_{d_j} = 2(\dot{B}_{d_j\ell} - Z_{ji}\dot{B}_{k_i\ell}) \tag{8.127}$$

and

$$S_{d_j} = (\ddot{g}_{d_j} - Z_{ji}\ddot{g}_{k_i}) - y[(\ddot{B}_{d_j\ell} + \alpha_{d_j}\dot{B}_{d_j\ell}) - Z_{ji}(\ddot{B}_{k_i\ell} + \alpha_{d_j}\dot{B}_{k_j\ell})] + (\alpha_{d_j}\dot{g}_{d_j} + \beta_{d_j}g_{d_j}) - Z_{ji}(\alpha_{d_j}\dot{g}_{k_i} + \beta_{d_j}g_{k_i}) \tag{8.128}$$

Example 8.8.1

Consider a three-link manipulator with $L_1 = L_2 = L_3 = \ell$ and $m_1 = m_2 = m_3 = m$, as shown in Figure 8.10. Let us assume that point p is fixed in space at the position shown and that the angular velocity of link 1 is considered to be a constant, defined by ω. Link 1 is pinned to the ground, so there are three constraint equations for this problem. The first two define point p fixed along the x- and y-axes; the third, which can be regarded as a velocity constraint, keeps link 1 at a constant angular velocity along the $-z$-axis. We can represent these constraints as follows:

$$h_1(q_1, q_2, q_3, t) = \ell\sin q_1 + \ell\sin(q_1 + q_2) + \ell\sin(q_1 + q_2 + q_3) - p_y = 0 \tag{8.129}$$

$$h_2(q_1, q_2, q_3, t) = \ell\cos q_1 + \ell\cos(q_1 + q_2) + \ell\cos(q_1 + q_2 + q_3) - p_x = 0 \tag{8.130}$$

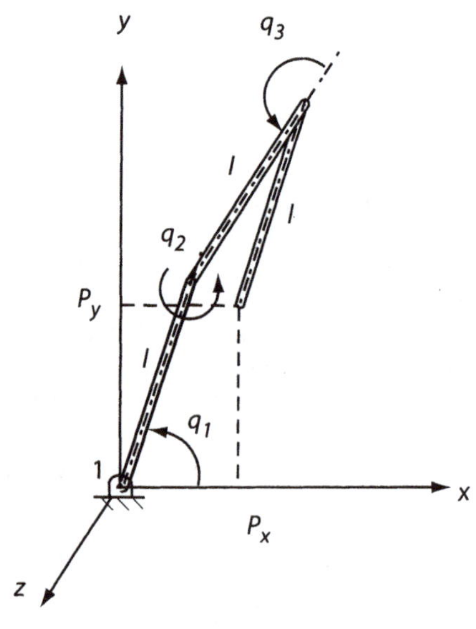

Fig. 8.10. Three-link manipulator with end effector fixed

$$h_3(q_1, q_2, q_3, t) = q_1 - \omega t = 0 \tag{8.131}$$

where p_x and p_y are the initial positions of the x- and y- coordinates. Their first-order differentiated forms yield

$$\dot{h}_1 = \ell C_1 y_1 + \ell C_{12}(y_1 + y_2) + \ell C_{123}(y_1 + y_2 + y_3) = 0 \tag{8.132}$$

$$\dot{h}_2 = -\ell S_1 y_1 - \ell S_{12}(y_1 + y_2) - \ell S_{123}(y_1 + y_2 + y_3) = 0 \tag{8.133}$$

$$\dot{h}_3 = y_1 - \omega = 0 \tag{8.134}$$

where S and C denote sine and cosine functions. Using equations (8.132) to (8.134) and expressing the constraint equations in matrix form, we obtain

$$\begin{bmatrix} \ell C_1 + \ell C_{12} + \ell C_{123} & \ell C_{12} + \ell C_{123} & \ell C_{123} \\ -\ell S_1 - \ell S_{12} - \ell S_{123} & -\ell S_{12} - \ell S_{123} & \ell S_{123} \\ 1 & 0 & 0 \end{bmatrix} \begin{bmatrix} y_1 \\ y_2 \\ y_3 \end{bmatrix} = \begin{bmatrix} 0 \\ 0 \\ \omega \end{bmatrix} \tag{8.135}$$

Differentiating equations (8.132) to (8.134) once more, we get the acceleration level of the constraints as

$$\begin{aligned} \ddot{h}_1 = {} & (\ell C_1 + \ell C_{12} + \ell C_{123})\dot{y}_1 + (\ell C_{12} + \ell C_{123})\dot{y}_2 + \ell C_{123}\dot{y}_3 \\ & - (\ell S_1 y_1^2 + \ell S_{12}(y_1 + y_2)^2 + \ell S_{123}(y_1 + y_2 + y_3)^2) = 0 \end{aligned} \tag{8.136}$$

$$\begin{aligned} \ddot{h}_2 = {} & (-\ell S_1 - \ell S_{12} - \ell S_{123})\dot{y}_1 + (-\ell S_{12} - \ell S_{123})\dot{y}_2 + (\ell S_{123})\dot{y}_3 \\ & - (\ell C_1 y_1^2 + \ell C_{12}(y_1 + y_2)^2 + \ell C_{123}(y_1 + y_2 + y_3)^2) = 0 \end{aligned} \tag{8.137}$$

$$\ddot{h}_3 = \dot{y}_1 = 0 \tag{8.138}$$

which can be written in matrix form as

$$\begin{bmatrix} \ell C_1 + \ell C_{12} + \ell C_{123} & \ell C_{12} + \ell C_{123} & \ell C_{123} \\ -\ell S_1 - \ell S_{12} - \ell S_{123} & -\ell S_{12} - \ell S_{123} & -\ell S_{123} \\ 1 & 0 & 0 \end{bmatrix} \begin{bmatrix} \dot{y}_1 \\ \dot{y}_2 \\ \dot{y}_3 \end{bmatrix}$$
$$- \begin{bmatrix} \ell S_1 y_1^2 + \ell S_{12}(y_1 + y_2)^2 + \ell S_{123}(y_1 + y_2 + y_3)^2 \\ \ell C_1 y_1^2 + \ell C_{12}(y_1 + y_2)^2 + \ell C_{123}(y_1 + y_2 + y_3)^2 \\ 0 \end{bmatrix} \tag{8.139}$$

Introducing the Baumgarte method, equation (8.139) takes the form

$$\begin{bmatrix} \ell C_1 + \ell C_{12} + \ell C_{123} & \ell C_{12} + \ell C_{123} & \ell C_{123} \\ -\ell S_1 - \ell S_{12} - \ell S_{123} & -\ell S_{12} - \ell S_{123} & -\ell S_{123} \\ 1 & 0 & 0 \end{bmatrix} \begin{bmatrix} \dot{y}_1 \\ \dot{y}_2 \\ \dot{y}_3 \end{bmatrix}$$

$$
= \begin{bmatrix} \ell S_1 y_1^2 + \ell S_{12}(y_1+y_2)^2 + \ell S_{123}(y_1+y_2+y_3)^2 - \alpha[\ell C_1 y_1 \\ +\ell C_{12}(y_1+y_2) + \ell C_{123}(y_1+y_2+y_3)] - \beta(\ell S_1 + \ell S_{12} + \ell S_{123} - p_y) \\ \ell C_1 y_1^2 + \ell C_{12}(y_1+y_2)^2 + \ell C_{123}(y_1+y_2+y_3)^2 - \alpha[-\ell S_1 y_1 \\ -\ell S_{12}(y_1+y_2) - \ell S_{123}(y_1+y_2+y_3)] - \beta(\ell C_1 + \ell C_{12} + \ell C_{123} - p_x) \\ -\alpha(y_1 - \omega) - \beta(q_1 - \omega t) \end{bmatrix} \tag{8.140}
$$

From equation (8.135), (8.136), or (8.137), we define the Jacobian constraint matrix as

$$
B = \begin{bmatrix} \ell C_1 + \ell C_{12} + \ell C_{123} & \ell C_{12} + \ell C_{123} & \ell C_{123} \\ -\ell S_1 - \ell S_{12} - \ell S_{123} & -\ell S_{12} - \ell S_{123} & -\ell S_{123} \\ 1 & 0 & 0 \end{bmatrix} \tag{8.141}
$$

If we choose p_x and p_y specifically to satisfy

$$
p_x = \frac{1}{2} \quad p_y = \frac{\sqrt{3}\ell}{2}
$$

we can easily see that when the system is in motion, it can reach a singular position (i.e., $q_1 = 60°$, $q_2 = 0°$ and $q_3 = 180°$). The Jacobian matrix becomes

$$
B = \begin{bmatrix} \frac{l}{2} & 0 & \frac{-l}{2} \\ \frac{-\sqrt{3}l}{2} & 0 & \frac{\sqrt{3}l}{2} \\ 1 & 0 & 0 \end{bmatrix} \tag{8.142}
$$

In equation (8.142) the rows do not vanish (although the second column values are all zero), but the first two rows are linearly dependent on each other. By using the method proposed in Section 8.8, we can modify the first constraint. Then our new B matrix becomes

$$
B = \begin{bmatrix} -3\sqrt{3}(\ell S_1 y_1 + \ell S_{12} & -3\sqrt{3}(\ell S_{12} & -3\sqrt{3}(\ell S_{123} \\ \times(y_1+y_2) & \times(y_1+y_2) & \times(y_1+y_2+y_3)) \\ -3(\ell C_{1y1} + \ell C_{12}(y_1+y_2) & -3(\ell C_{12}(y_1+y_2) & -3(\ell C_{123} \\ +\ell S_{123}(y_1+y_2+y_3)) & +\ell S_{123}(y_1+y_2+y_3)) & \times(y_1+y_2+y_3)) \\ +\ell C_{123}(y_1+y_2+y_3)) & +\ell C_{123}(y_1+y_2+y_3)) & \\ -\ell S_1 - \ell S_{12} - \ell S_{123} & -\ell S_{12} - \ell S_{123} & -\ell S_{123} \\ 1 & 0 & 0 \end{bmatrix} \tag{8.143}
$$

Introducing Baumgarte's method, after substituting the new B matrix above, equation (8.140) takes the form

$$\begin{bmatrix} \begin{array}{c}-3\sqrt{3}(\ell S_1 y_1 + \ell S_{12}\\ \times(y_1+y_2) + \ell S_{123}\\ \times(y_1+y_2+y_3))\end{array} & \begin{array}{c}-3\sqrt{3}(\ell S_{12}\\ \times(y_1+y_2)+\ell S_{123}\\ \times(y_1+y_2+y_3))\end{array} & \begin{array}{c}-3\sqrt{3}(\ell S_{123}(y_1\\ +y_2+y_3)) - 3(\ell C_{123}\\ \times(y_1+y_2+y_3))\end{array} \\ \begin{array}{c}-3(\ell C_1 y_1 + \ell C_{12}\\ \times(y_1+y_2)+\ell C_{123}\\ \times(y_1+y_2+y_3))\\ -\ell S_1 - \ell S_{12} - \ell S_{123}\end{array} & \begin{array}{c}-3(\ell C_{12}\\ \times(y_1+y_2)+\ell C_{123}\\ \times(y_1+y_2+y_3))\\ -\ell S_{12} - \ell S_{123}\end{array} & -\ell S_{123} \\ 1 & 0 & 0 \end{bmatrix}$$

$$\times \begin{bmatrix} \dot{y}_1 \\ \dot{y}_2 \\ \dot{y}_3 \end{bmatrix}$$

$$= \begin{bmatrix} \begin{array}{c}\sqrt{3}(\ell C_1 y_1^3 + \ell C_{12}(y_1+y_2)^3 + \ell C_{123}(y_1+y_2+y_3)^3) - (\ell S_1 y_1^3\\ +\ell S_{12}(y_1+y_2)^3 + \ell S_{123}(y_1+y_2+y_3)^3) - \alpha[\ell C_1 y_1 + \ell C_{12}(y_1+y_2)\\ +\ell C_{123}(y_1+y_2+y_3)] - \beta(\ell S_1 - \ell S_{12} - \ell s_{123} - p_y)\end{array} \\ \begin{array}{c}\ell C_1 y_1^2 + \ell C_{12}(y_1+y_2)^2 + \ell C_{123}(y_1+y_2+y_3)^2\\ -\alpha[-\ell S_1 y_1 - \ell S_{12}(y_1+y_2) - \ell S_{123}(y_1+y_2+y_3)]\\ -\beta(\ell C_1 + \ell C_{12} + \ell C_{123} - p_x)\end{array} \\ -\alpha(y_1 - \omega) - \beta(q_1 - \omega t) \end{bmatrix} \tag{8.144}$$

A numerical solution to the problem above was carried out to show the effectiveness of the proposed method, when the constraints become linearly dependent in the neighborhood of singularity. Even though the selection of the time step during integration of the equations is very critical, the results clearly show how the instability of the constraints can be greatly affected when reaching singularly. On the other hand, the regularization procedure (as demonstrated by the solid lines representing the torques of the planar manipulator, Figures 8.11 to 8.13) can be very effective in going through singularly while maintaining continuity and smoothness of motion. The jerks of the torques at singularity are infinite, and the proposed regularization method can maintain the system stable in the neighborhood of singularity, when the time step is maintained constant through integration of the equations of motion.

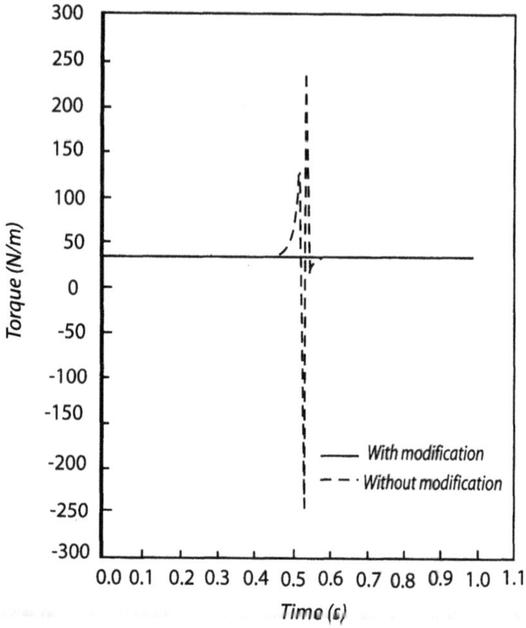

Fig. 8.11. Driving torque for joint 1

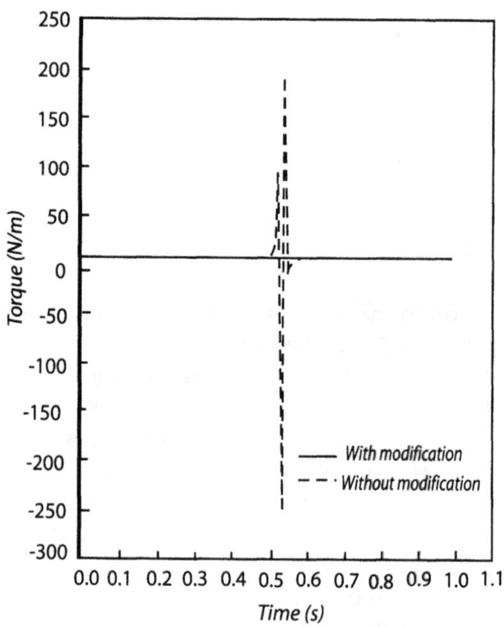

Fig. 8.12. Driving torque for joint 2

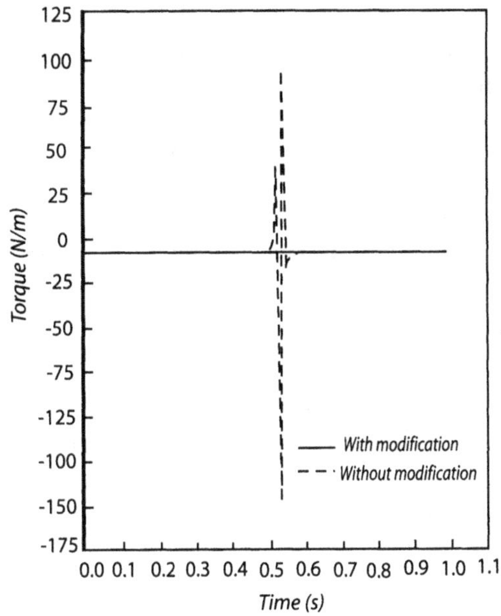

Fig. 8.13. Driving torque for joint 3

References

1. Wehage, R. A. and Haug, E. J., Generalized Coordinate Partitioning for Dimension Reduction in Analysis of Constrained Dynamic Systems, *J. Mech. Des.*, Vol. **104**, 1982, pp. 247–255.
2. Singh, R. P. and Likins, P. W., Singular Value Decomposition for Constrained Dynamical Systems, *J. Appl. Mech.*, Vol. **52**, 1985, pp. 943–948.
3. Kamman, J. W. and Huston, R. L., Dynamics of Constrained Multibody Systems, *J. Appl. Mech.*, Vol. **51**, 1984, pp. 899–903.
4. Amirouche, F. M. L. and Jyia, T., Automatic Elimination of the Undetermined Multipliers in Kane's Equations Using a Pseudo-uptriangular Decomposition Method, *Comput. Struct.*, Vol. **27**, 1987, pp. 203–28.
5. Wampler, C., Buffington, K. and Shu-hui, J., Formulation of Equations of Motion for Systems Subject to Constraints, *J. Appl. Mech.*, Vol. **52**, 1985, pp. 465–470.
6. Ider, S. K. and Amirouche, F. M. L., Coordinate Reduction in Constrained Spatial Dynamic Systems–A New Approach, *J. Appl. Mech.*, Vol. **55**, 1988, pp. 899–905.
7. Baumgarte, J., Stabilization of Constraints and Integrals of Motion in Dynamical Systems, *Comput. Methods Appl. Mech. Eng.*, Vol. **1**, 1972, pp. 1–16.
8. Kane, T. R. and Levinson, D. A., *Dynamics: Theory and Applications*, McGraw-Hill, New York, 1985.
9. Amirouche, F. M. L. and Ider, S. K., Determination of Constraint Forces in Multibody Systems Dynamics Using Kane's Equations, *J. Theor. Appl. Mech.* Vol. **7**, 1988, pp. 3–20.

10. Tung, C. W., Regularization and Stability of the Constraints in the Dynamics of Multibody Systems, Master thesis, Department of Mechanical Engineering, University of Illinois at Chicago, 1991.
11. Amirouche, F. M. L. and Tung, C. W., Regularization and Stability of the Constraints in the Dynamics of Multibody Systems, *J. Nonlin. Dyn.,* 1990, pp. 459–475.
12. Fuhrer, C. and Leimkuhler, B., Numerical Solution of Differential–Algebraic Equations for Constrained Mechanical Motion, *Num. Math.*, DFVLR Report No. DFVLR-FB-89-08, Institut für Dynamik der Flugsystem Oberpfafferhofen, March, 1989.
13. Amirouche, F. M. L., Ider, S. K., Numerical Stability of the Constraints near Singular Positions in the Dynamics of Multibody Systems, *Int. J. Comput. Struct.*, Vol. **31**: No. 1, 1989, pp. 129–138.
14. Fuhrer, C. and Schwertassek, R., Generation and Solution of Multibody System Equations, *Int. J. Nonlinear Mech.,* Vol. **25**: No. 2/3, 1989, pp. 127–141.
15. Gear, C. W., Gupta, G. K. and Leimkuhler, B. J., Automatic Integration of the Euler–Lagrange Equations with Constraints, *J. Comput. Appl. Math.*, Vol. **12**: No. 13, 1985, pp. 77–90.
16. Gear, C. W., Simultaneous Numerical Solution of Differential–Algebraic Equations, IEEE Trans. Circuit Theory, CT-18(1), 89(1971).

9

Linearization and Vibration Analysis of Multibody Systems

9.1 Introduction

In the context of dynamics and control of multibody systems, we are required to linearize the dynamical equations of motion, so that proper control laws can be applied to study the efforts of vibration on the control of flexible structures. The linearization sets the stage for an important topic in the dynamics of flexible (elastic) bodies undergoing large rotations. The objective of this chapter is to highlight the essential elements in linearization and to identify in vibration, the eigenfunctions needed for a description of the elastic deformation associated with the dynamics of multibody systems. In addition to the above, we present a review of the vibration of continuous beams as they are vital to the analysis of flexible bodies presented in the next two chapters.

9.2 Linearization of the Equations of Motion

The linearization of the equation of motion is most useful in the study of stability and control. The linear form of the equations of motion leads to simpler control algorithms, based in most cases on the solution of linear differential equations. The assumption undertaken in the linearization process deals with considering small changes in the generalized coordinates $x_1, x_2, \ldots, x_n$ and in the final governing equations of motion, neglecting the higher-order terms in x, $\dot{x}$ and $\ddot{x}'s$. There are basically two approaches to linearizing the equations of motion. The first is to solve for the full nonlinear equations and motion and then linearize all the terms in the equation by setting

$$x_\ell = x_\ell^* + x_\ell^p \tag{9.1}$$

where x_ℓ^* represents the stable points of the generalized coordinates and x_ℓ^p the perturbation of x_ℓ. By making x^p small and replacing the sine of x_ℓ by x_ℓ and its cosine by unity and dropping all higher terms, we obtain the desired linearized equations of motion. The second approach [1] will be to avoid, in the case of complex systems, the expansion of long expressions that are prone to errors and costly in terms of labor.

Therefore, we proceed in linearizing the equations of motion by first linearizing the kinematical quantities, such as the velocity, angular velocity, acceleration, and angular acceleration. Therefore, we formulate the equations of motion by dropping any higher terms in x_ℓ and $\dot{x}_\ell$. The result will be the linearized equations of motion. The advantages to this approach can be seen in the roles that partial velocity and partial angular velocity play in the formulation of the equations of motion, and how their linearization comprises a large portion of the task required to formulate the equations desired [1–3].

Example 9.2.1

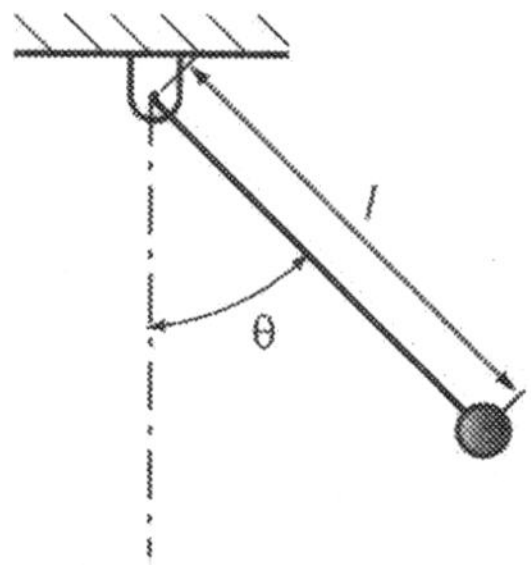

Fig. 9.1. A simple pendulum

Consider the motion of the simple pendulum shown in Figure 9.1. Find the linearized equations of motion about θ^*.

$$\theta = \theta^* + \theta^p \tag{9.2}$$

where θ^* represents the stable points of θ and θ^p the perturbation of θ. Then the equations of motion can be expressed as

$$\ddot{\theta}^p + \frac{g}{L}\sin(\theta^* + \theta^p) = 0 \tag{9.3}$$

Expanding $\sin(\theta^* + \theta^p)$ using Taylor series, and taking only the first two terms of the series by neglecting the higher-order terms, the following equation is obtained:

$$\ddot{\theta}^p + \frac{g}{L}\left[(\theta^* + \theta^p) - \frac{1}{6}(\theta^* + \theta^p)^3\right] = 0 \tag{9.4}$$

Expanding further, we get

$$\ddot{\theta}^p + \frac{g}{L}\left[\theta^p - \frac{1}{2}\theta^p(\theta^*)^2 - \frac{1}{6}(\theta^p)^3 + \theta^* - \frac{1}{2}\theta^*(\theta^p)^2 - \frac{1}{6}(\theta^*)^3\right] = 0 \tag{9.5}$$

If we consider only the first-order terms, equation (9.5) reduces to

$$\ddot{\theta}^p + \frac{g}{L}(\theta^p + \theta^*) = 0 \tag{9.6}$$

If the stable point is such that $\theta^* = 0$, equation (9.3) becomes a homogeneous second-order differential equation in terms of θ^p. In the linearization process, we need to be careful regarding what terms need to be neglected. The approximation of $\sin\theta$ by θ and $\cos\theta$ by 1 are first-order type and sometimes might be erroneous solutions to linearized equations.

Example 9.2.2

Referring to Example 3.9.3 obtain the linearized kinematical expressions needed for the equations of motion (see Figure 9.2).

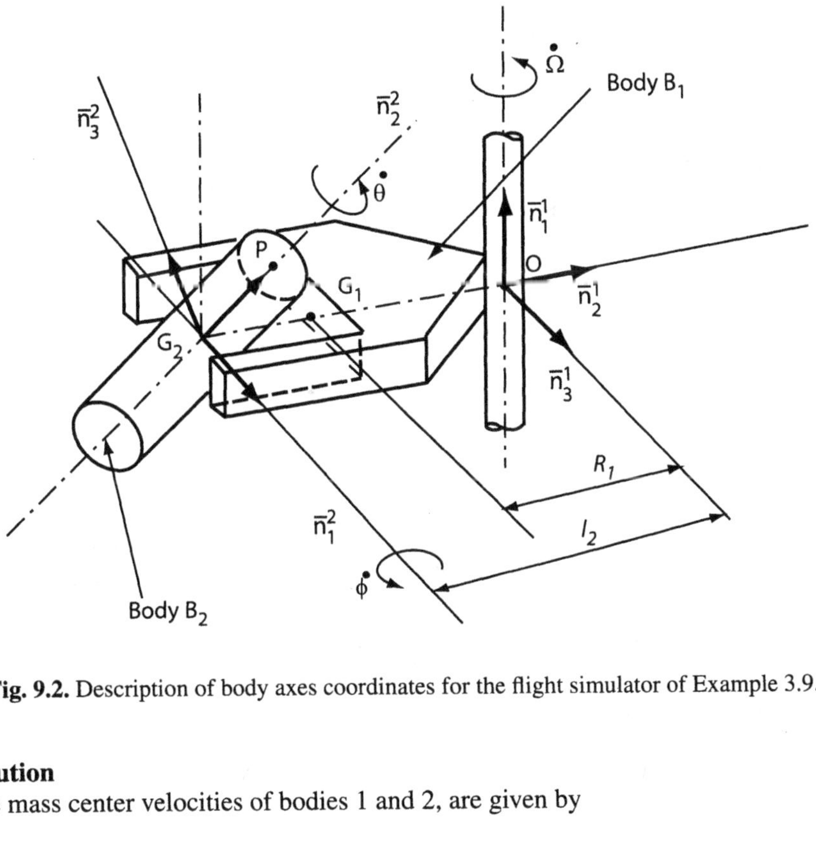

Fig. 9.2. Description of body axes coordinates for the flight simulator of Example 3.9.3

Solution

The mass center velocities of bodies 1 and 2, are given by

$$^R\bar{v}_1^{G_1} = \Omega R_1 \bar{n}_1^1 \tag{9.7}$$

$$^R\bar{v}_2^{G_2} = \Omega L_1 \bar{n}_1^2 \tag{9.8}$$

The angular velocities of B_1 and B_2 in R are found from Example 3.9.3:

$$^R\bar{\omega}_1^B = \Omega \bar{n}_3^1 \tag{9.9}$$

$$^R\bar{\omega}_2^B = (\dot{\phi}c_1 - \dot{\theta}c2s1)\bar{n}_1 + (\dot{\phi}s1 + \dot{\theta}c2c1)\bar{n}_2 + (\Omega + \dot{\theta}s2)\bar{n}_3 \tag{9.10}$$

where

$$s1 = \sin\alpha \quad \text{and} \quad c1 = \cos\alpha$$

$$s2 = \sin\phi \quad \text{and} \quad c2 = \cos\phi$$

Let the generalized coordinate derivatives be such that $\dot{x} = \{\dot{\theta}\ \dot{\phi}\}$, then

$$\frac{\partial\, {}^R\bar{v}^{G_1}}{\partial\dot{\theta}} = \frac{\partial\, {}^R\bar{v}_1^{G}}{\partial\dot{\phi}} = 0 \tag{9.11}$$

and

$$\frac{\partial\, {}^R\bar{v}^{G2}}{\partial\dot{\theta}} = \frac{\partial\, {}^R\bar{v}_2^{G}}{\partial\dot{\phi}} = 0 \tag{9.12}$$

Similarly, we obtain

$$\frac{\partial\, {}^R\bar{\omega}^{B_1}}{\partial\dot{\theta}} = \frac{\partial\, {}^R\bar{\omega}_1^{B}}{\partial\dot{\phi}} = 0 \tag{9.13}$$

and

$$\frac{\partial\, {}^R\bar{\omega}^{B2}}{\partial\dot{\theta}} = -c2s1\bar{n}_1 + c2c1\bar{n}_2 + s2\bar{n}_3 \tag{9.14}$$

$$\frac{\partial\, {}^R\bar{\omega}^{B2}}{\partial\dot{\phi}} = c1\bar{n}_1 + s1\bar{n}_2 \tag{9.15}$$

The next step is to linearize the linear and angular velocities, where we have to consider the generalized coordinates, which are given by

$$x = x^* + x^p \tag{9.16}$$

where x^* is constant and is defined as a stable or reference value of x, and x^p is the perturbation value of x, which is assumed very small. Thus

$$\sin x = \sin(x^* + x^p) = \sin x^* \cos x^p + \cos x^* \sin x^p \tag{9.17}$$

Because x^p is very small,

$$\cos x^p \backsimeq 1 \quad \text{and} \quad \sin x^p \backsimeq x^p$$

Therefore,

$$\sin x = \sin x^* + x^p \cos x^* \tag{9.18}$$

Similarly,

$$\cos x = \cos x^* - x^p \sin x^* \tag{9.19}$$

For our problem, x is identified as either ϕ or θ.

$$\sin\phi = \sin\phi^* + \phi^P \cos\phi^* = s2 \tag{9.20}$$

$$\cos\phi = \cos\phi^* - \phi^P \sin\phi^* = c2 \tag{9.21}$$

Similarly, for θ we obtain

$$\sin\theta = \sin\theta^* + \theta^P \cos\theta^* = s3 \tag{9.22}$$

$$\cos\theta = \cos\theta^* - \theta^P \sin\theta^* = c3 \tag{9.23}$$

Substituting the value of $s2$, $c2$, $s3$ and $c3$ in the expressions of the linear and angular velocities, the only expression that changes is ${}^R\bar{\omega}^{B_2}$, which can be written as

$${}^R\bar{\omega}^{B_2} = [(\dot{\phi}c1 - \dot{\theta}c2sl)]\bar{n}_1 + [(\dot{\phi}s1 + \dot{\theta}c2c1)]\bar{n}_2 + [(\Omega + \dot{\theta}s2)]\bar{n}3 \tag{9.24}$$

If we assume that ϕ^* and θ^* are very small, such that

$$c\phi^* \backsimeq 1, \qquad s\phi^* \backsimeq \phi^*$$

and

$$c\theta^* \backsimeq 1, \qquad s\theta^* \backsimeq \theta^*$$

and that

$$\frac{d\phi^P}{dt} = \frac{d\phi}{dt} = \dot{\phi}$$

$$\frac{d\theta^P}{dt} = \frac{d\theta}{dt} = \dot{\theta}$$

we can obtain the linearized angular velocity of B_2 as

$${}^R\bar{\omega}^{B_2} = (\dot{\phi} - \dot{\theta}\alpha)\bar{n}_1 + (\dot{\phi}\alpha + \dot{\theta})\bar{n}_2 + (\Omega + \dot{\theta}\phi)\bar{n} \tag{9.25}$$

where

$$\theta^* \backsimeq \theta \quad \text{and} \quad \phi^* \backsimeq \phi$$

Thus the angular acceleration is simply obtained from the differentiation of ${}^R\bar{\omega}_2^B$:

$$\begin{aligned}{}^R\bar{\alpha}^{B_2} &= \frac{d}{dt}\left({}^R\bar{\omega}^{B_2}\right) \\ &= (\ddot{\phi} - \dot{\theta}\Omega - \ddot{\theta}\alpha)\bar{n}_1 + (\ddot{\phi}\alpha + \dot{\phi}\Omega + \ddot{\theta})\bar{n}_2 + (0 + \ddot{\theta}\phi + \dot{\theta}\dot{\phi})\bar{n}_3\end{aligned} \tag{9.26}$$

The corresponding linearized mass accelerations for bodies 1 and 2 are

$${}^R\bar{a}^{B_2} = 0 \tag{9.27}$$

and

$${}^R\bar{a}^{B_2} = 0 \tag{9.28}$$

(because Ω = constant).

The partial angular velocity of the linearized equation given by 9.25 is

$$\frac{\partial\, {}^R\bar{\omega}^{B_2}}{\partial \dot{\phi}} = \bar{n}_1 + \alpha\bar{n}_2 \tag{9.29}$$

$$\frac{\partial\, {}^R\bar{\omega}^{B_2}}{\partial \dot{\theta}} = -\alpha\bar{n}_1 + \bar{n}_2 + \phi\bar{n}_3 \tag{9.30}$$

At this point, one can easily formulate the linearized equations of motion of the system by making use of the linearized kinematical expressions. This procedure doesn't guarantee that the final equations are linearized, as it is possible that further terms may need to be simplified or dropped.

Example 9.2.3

Consider the planar pendulum shown in Figure 9.3, which consists of three slender rods. Find the linearized equations of motion of the system. Let us formulate the kinematics of the system. *Note:* Examples 9.2.1 and 9.2.2, however simple, were linearized after the equations of motion had been formulated. This can be a difficult task in multibody systems with more than three or four degrees of freedom. An alternative to this method is the use of Kane's equation.

Let the joint angles in this example be given by the absolute angles measured from the vertical line.

The center mass velocities of G_1, G_2 and G_3 in R are:

$${}^R\bar{\upsilon}^{G_1} = \frac{L}{2}\dot{\theta}_1\bar{n}_1^1 \tag{9.31}$$

$${}^R\bar{\upsilon}^{G_2} = L\dot{\theta}_1\bar{n}_2^1 + \frac{L}{2}\dot{\theta}_2 n_2^2 \tag{9.32}$$

$${}^R\bar{\upsilon}^{G_3} = L\dot{\theta}_1\bar{n}_2^1 + L\dot{\theta}_2\bar{n}_2^2 + \frac{L}{2}\dot{\theta}_3 n_2^3 \tag{9.33}$$

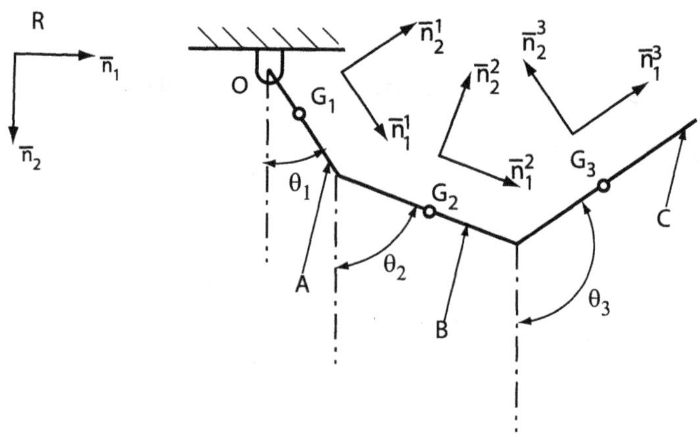

Fig. 9.3. Planar three-link manipulator

The angular velocities of bodies A, B and C in R are obtained as follows:

$$^R\bar{\omega}^A = \dot{\theta}_1\bar{n}_3 \tag{9.34}$$

$$^R\bar{\omega}^B = (\dot{\theta}_2)\bar{n}_3 \tag{9.35}$$

$$^R\bar{\omega}^C = (\dot{\theta}_3)\bar{n}_3 \tag{9.36}$$

Figure 9.3 shows the relationship between the Euler angles and the unit vectors of R.

The mass center acceleration of G_1, G_2 and G_3 corresponding to body A, B and C in R are:

$$^R\bar{a}^{G_1} = \frac{L}{2}\ddot{\theta}_1\bar{n}_2^1 - \frac{L}{2}\dot{\theta}_1^2 n_1^1 \tag{9.37}$$

$$^R\bar{a}^{G_2} = L\ddot{\theta}_1\bar{n}_2^1 - L\dot{\theta}_1^2\bar{n}_1^1 + \frac{L}{2}\ddot{\theta}_2\bar{n}_2^2 - \frac{L}{2}\dot{\theta}_1^2 n_1^2 \tag{9.38}$$

$$^R\bar{a}^{G_3} = L\ddot{\theta}_1\bar{n}_2^1 - L\dot{\theta}_1^2\bar{n}_1^1 + L\ddot{\theta}_2\bar{n}_2^2 - L\dot{\theta}_1^2 n_1^2 + \frac{L}{2}\ddot{\theta}_3\bar{n}_2^3 - \frac{L}{2}\dot{\theta}_3^2 n_1^3 \tag{9.39}$$

The way to proceed at this point is to express both the mass center velocities and accelerations with respect to the orthogonal unit vectors attached to R: namely, $\bar{n}_1$, $\bar{n}_2$ and $\bar{n}_3$. In doing so, let us linearize those expressions as well. Let

$$s\theta \backsimeq \theta$$

$$c\theta \backsimeq 1$$

Then

$$s1 \backsimeq \theta_1 c1 \backsimeq 1$$

$$s2 \backsimeq \theta_2 c2 \backsimeq 1$$

$$s3 \backsimeq \theta_3 c3 \backsimeq 1$$

The mass center velocities become

$$^R\bar{v}^{G_1} = \frac{L}{2}\dot{\theta}_1(-\theta_1\bar{n}_1 + \bar{n}_2) \tag{9.40}$$

$$^R\bar{v}^{G_2} = -L\left(\theta_1\dot{\theta}_1 + \frac{1}{2}\theta_2\dot{\theta}_2\right)\bar{n}_1 + L\left(\dot{\theta}_1 + \frac{\dot{\theta}_2}{2}\right)\bar{n}_2 \tag{9.41}$$

$$^R\bar{v}^{G_3} = -L\left(\theta_1\dot{\theta}_1 + \theta_2\dot{\theta}_2 + \frac{1}{2}\theta_3\dot{\theta}_3\right)\bar{n}_1 + L\left(\dot{\theta}_1 + \dot{\theta}_2 + \frac{1}{2}\dot{\theta}_3\right)\bar{n}_2 \tag{9.42}$$

We can drop the terms that are doomed to be small, and the equations above reduce further to

$$^R\bar{v}^{G_1} = \frac{L}{2}\dot{\theta}_1\bar{n}_2 \tag{9.43}$$

$$^R\bar{v}^{G_2} = L\left(\dot{\theta}_1 + \frac{\dot{\theta}_2}{2}\right)\bar{n}_2 \tag{9.44}$$

$$^R\bar{\upsilon}^{G_3} = L\left(\dot{\theta}_1 + \dot{\theta}_2 + \frac{1}{2}\dot{\theta}_3\right)\bar{n}_2 \tag{9.45}$$

The linearized mass center accelerations are then found to be

$$^R\bar{a}^{G_1} = \frac{L}{2}\ddot{\theta}_1\bar{n}_2 \tag{9.46}$$

$$^R\bar{a}^{G_2} = L\left(\ddot{\theta}_1 + \frac{\ddot{\theta}_2}{2}\right)\bar{n}_2 \tag{9.47}$$

$$^R\bar{a}^{G_3} = L\left(\ddot{\theta}_1 + \ddot{\theta}_2 + \frac{1}{2}\dot{\theta}_3\right)\bar{n}_2 \tag{9.48}$$

These can, of course, be obtained by linearizing equations (9.37) to (9.39) directly. Since we are going to need the partial velocities for the formulation of the equations of motion, then we obtain the following terms

$$\frac{\partial\bar{\upsilon}_1^G}{\partial\dot{\theta}_1} = \frac{L}{2}\bar{n}_2, \qquad \frac{\partial\bar{\upsilon}_2^G}{\partial\dot{\theta}_2} = 0, \qquad \frac{\partial\bar{\upsilon}_1^G}{\partial\dot{\theta}_3} = 0 \tag{9.49}$$

$$\frac{\partial\bar{\upsilon}_2^G}{\partial\dot{\theta}_1} = L\bar{n}_2, \qquad \frac{\partial\bar{\upsilon}_2^G}{\partial\dot{\theta}_2} = \frac{L}{2}\bar{n}_2, \qquad \frac{\partial\bar{\upsilon}_2^G}{\partial\dot{\theta}_3} = 0 \tag{9.50}$$

$$\frac{\partial\bar{\upsilon}^{G3}}{\partial\dot{\theta}_1} = L\bar{n}_2, \qquad \frac{\partial\bar{\upsilon}_3^G}{\partial\dot{\theta}_2} = L\bar{n}_2, \qquad \frac{\partial\bar{\upsilon}_3^G}{\partial\dot{\theta}_3} = \frac{L}{2}\bar{n}_2 \tag{9.51}$$

In a similar fashion we obtain the partial angular velocities.

Formulation of the Linearized Equations of Motion

Using Kane's equation, the generalized inertia and active forces need to be formulated. That is,

$$f_\ell^* + f_\ell = 0(\ell = 1, 3) \tag{9.52}$$

where

$$f_\ell^* = (m_k\bar{a}^{G_k}) \cdot \frac{\partial\bar{\upsilon}^{G_k}}{\partial\dot{\theta}_\ell} + (\bar{T}_k^*) \cdot \frac{\partial\bar{\omega}^k}{\partial\dot{\theta}_\ell} \tag{9.53}$$

The value $\bar{T}_k^*$ represents the inertia torque of body k in R, and for each ℓ we have k that goes from 1 to 3. Similarly, the generalized active forces are obtained making use of equation (9.53) by replacing the inertia and torque forces with the resulting external force and torque acting on each body. Since we only have the gravity force due to the weight, the generalized active forces contribution is simply obtained from

$$f\ell = (m_k g\bar{n}_1) \cdot \frac{\partial\bar{\upsilon}^{G_1}}{\partial\dot{\theta}_\ell} \tag{9.54}$$

Substituting the linearized partial velocities and accelerations into the equations of motion, we obtain automatically the equations desired:

$$\frac{7}{3}\ddot{\theta}_1 + \frac{3}{2}\ddot{\theta}_2 + \frac{1}{2}\ddot{\theta}_3 + \frac{5}{2}\frac{g}{L}\theta_1 = 0 \tag{9.55}$$

$$\frac{3}{2}\ddot{\theta}_1 + \frac{4}{3}\ddot{\theta}_2 + \frac{1}{2}\ddot{\theta}_3 + \frac{3}{2}\frac{g}{L}\theta_2 = 0 \tag{9.56}$$

$$\frac{1}{2}\ddot{\theta}_1 + \frac{1}{2}\ddot{\theta}_2 + \frac{1}{3}\ddot{\theta}_3 + \frac{g}{L}\theta_3 = 0 \tag{9.57}$$

Example 9.2.3 is used primarily to describe the steps to take in linearizing the equations of motion, following Kane's method. We should note, however, that our first assumption was to replace $\sin\theta$ with θ and $\cos\theta$ with 1. Higher-order terms are also dropped. One needs to be cautious as to the assumptions that need to be made and the order of the terms, which are going to have no effect on the equations. It is desirable to use the expansion of $\sin\theta$ and $\cos\theta$ in terms of Taylor series, whenever the generalized coordinates become significant in the analysis.

Example 9.2.4

Consider Example 6.12.3, dealing with the pantograph. Obtain the linearized equations of motion of the mechanical system (follow procedures outlined in the above example).

Solution

The systems has two degrees of freedom, selected to be θ_1 and θ_2. Hence the reduced equations of motion will be in terms of those generalized coordinates. The solution for θ_1 and θ_2 in the time domain can be used in conjunction with the linearized constraint equations to find the time history of the dependent coordinates: namely, θ_2 and θ_4.

The mass center velocities in R of each link are given by

$$^{R}\bar{\upsilon}^{G_1} = \frac{1}{2}L_1\dot{\theta}_1\bar{n}_2^1 \tag{9.58}$$

$$^{R}\bar{\upsilon}^{G_2} = L_1\dot{\theta}_1\bar{n}_2^1 + \frac{1}{2}(\dot{\theta}_1 + \dot{\theta}_2)L_2\bar{n}_2^2 \tag{9.59}$$

$$^{R}\bar{\upsilon}^{G_3} = \frac{1}{2}L_3\dot{\theta}_3\bar{n}_2^3 \tag{9.60}$$

$$^{R}\bar{\upsilon}^{G_4} = L_3\dot{\theta}_3\bar{n}_2^3 + (\dot{\theta}_3 + \dot{\theta}_4)L_4\bar{n}_2^4 \tag{9.61}$$

where G_1, G_2, G_3 and G_4 represent the center mass of each link, respectively. The angular velocities of B_1, B_2 and B_3 with respect to R are:

$$^{R}\bar{\omega}^{B_1} = \dot{\theta}_1 \bar{n}_3^1 \tag{9.62}$$

$$^{R}\bar{\omega}^{B_2} = (\dot{\theta}_1 + \dot{\theta}_2) \bar{n}_3^2 \tag{9.63}$$

$$^{R}\bar{\omega}^{B_3} = \dot{\theta}_3 \bar{n}_3^3 \tag{9.64}$$

$$^{R}\bar{\omega}^{B_4} = (\dot{\theta}_3 + \dot{\theta}_4) \bar{n}_3^4 \tag{9.65}$$

The partial velocities in R for the above linear velocities are:

$$\frac{\partial\ ^{R}\bar{v}_1^{G_1}}{\partial \dot{\theta}_1} = \frac{1}{2} L \bar{n}_2^1 \tag{9.66}$$

$$\frac{\partial\ ^{R}\bar{v}_2^{G_1}}{\partial \dot{\theta}_2} = 0 \tag{9.67}$$

$$\frac{\partial\ ^{R}\bar{v}_1^{G_2}}{\partial \dot{\theta}_1} = L_1 \bar{n}_2^1 + \frac{1}{2} L_2 \bar{n}_2^2 \tag{9.68}$$

$$\frac{\partial\ ^{R}\bar{v}_2^{G_2}}{\partial \dot{\theta}_2} = \frac{1}{2} L_2 \bar{n}_2^2 \tag{9.69}$$

$$\frac{\partial\ ^{R}\bar{v}_1^{G_3}}{\partial \dot{\theta}_1} = 0 \tag{9.70}$$

$$\frac{\partial\ ^{R}\bar{v}_2^{G_3}}{\partial \dot{\theta}_2} = 0 \tag{9.71}$$

$$\frac{\partial\ ^{R}\bar{v}_1^{G_4}}{\partial \dot{\theta}_1} = 0 \tag{9.72}$$

$$\frac{\partial\ ^{R}\bar{v}_2^{G_4}}{\partial \dot{\theta}_2} = 0 \tag{9.73}$$

Similarly, we can obtain the rest of the partial derivatives of the center masses velocities with respect to $\dot{\theta}_3$ and $\dot{\theta}_4$. The partial angular velocities in R for $\dot{\theta}_1$ and $\dot{\theta}_2$ are easily seen to be:

$$\frac{\partial\ ^{R}\bar{\omega}_1^{B_1}}{\partial \dot{\theta}_1} = \bar{n}_3^1 \tag{9.74}$$

$$\frac{\partial\ ^{R}\bar{\omega}_2^{B_1}}{\partial \dot{\theta}_2} = 0 \tag{9.75}$$

$$\frac{\partial\ ^{R}\bar{\omega}_1^{B_2}}{\partial \dot{\theta}_1} = \bar{n}_3^2 \tag{9.76}$$

$$\frac{\partial\ {}^R\bar{\omega}_2^{B_2}}{\partial\dot{\theta}_2} = \bar{n}_3^2 \tag{9.77}$$

$$\frac{\partial\ {}^R\bar{\omega}_1^{B_3}}{\partial\dot{\theta}_1} = 0 \tag{9.78}$$

$$\frac{\partial\ {}^R\bar{\omega}_2^{B_3}}{\partial\dot{\theta}_2} = 0 \tag{9.79}$$

$$\frac{\partial\ {}^R\bar{\omega}_1^{B_4}}{\partial\dot{\theta}_1} = 0 \tag{9.80}$$

$$\frac{\partial\ {}^R\bar{\omega}_2^{B_4}}{\partial\dot{\theta}_2} = 0 \tag{9.81}$$

Since there is no trigonometric function involved in the expressions of linear and angular velocities, the linearized expressions remain the same.

Using the linearized mass center velocity, we can differentiate in time and obtain the linearized mass center accelerations:

$$^R\bar{a}^{G_1} = -\frac{L_1}{2}\dot{\theta}_1^2\bar{n}_1^1 + \frac{L_1}{2}\ddot{\theta}_1\bar{n}_2^1 \tag{9.82}$$

$$^R\bar{a}^{G_2} = -L_1\dot{\theta}_1^2\bar{n}_1^1 + L_1\ddot{\theta}_1\bar{n}_2^1 - \frac{1}{2}(\dot{\theta}_1+\dot{\theta}_2)^2L_2\bar{n}_1^2 + \frac{1}{2}(\ddot{\theta}_1+\ddot{\theta}_2)L_2\bar{n}_2^2 \tag{9.83}$$

$$^R\bar{a}^{G_3} = -\frac{1}{2}L_3\dot{\theta}_3^2\bar{n}_1^3 + \frac{1}{2}L_3\ddot{\theta}_3\bar{n}_2^3 \tag{9.84}$$

$$^R\bar{a}^{G_4} = -L_3\dot{\theta}_3^2\bar{n}_1^3 + L_3\ddot{\theta}_3\bar{n}_2^3 - (\dot{\theta}_3+\dot{\theta}_4)^2L_4\bar{n}_1^4 + (\ddot{\theta}_3+\ddot{\theta}_4)L_4\bar{n}_2^4 \tag{9.85}$$

In a similar fashion the angular accelerations are found:

$$^R\bar{\alpha}^{B_1} = \ddot{\theta}_1\bar{n}_3^1 \tag{9.86}$$

$$^R\bar{\alpha}^{B_2} = (\ddot{\theta}_1+\ddot{\theta}_2)\bar{n}_3^2 \tag{9.87}$$

$$^R\bar{\alpha}^{B_3} = \ddot{\theta}_3\bar{n}_3^3 \tag{9.88}$$

$$^R\bar{\alpha}^{B_4} = (\ddot{\theta}_3+\ddot{\theta}_4)\bar{n}_3^4 \tag{9.89}$$

For the linearization, we assume that

$$\begin{aligned} \cos\theta_1 &\backsimeq 1 \sin\theta_1 \backsimeq \theta_1 \\ \cos\theta_2 &\backsimeq 1 \sin\theta_2 \backsimeq \theta_2 \\ \cos(\theta_1+\theta_2) &\backsimeq 1-\theta_1\theta_2 \backsimeq 1 \end{aligned} \tag{9.90}$$

After linearizing the generalized inertia and active forces, we obtain the following linearized dynamical equations:

$$\left[\left(L_1L_2 + L_1^2 + \frac{1}{4}L_2^2\right)m_{B2} + I_{B1} + I_{B2} + \frac{1}{4}L_1^2 m_{B1}\right]\ddot{\theta}_1$$
$$+\left[\left(\frac{1}{2}L_1L_2 + \frac{1}{4}L_2^2\right)m_{B2} + I_{B2}\right]\ddot{\theta}_2$$
$$+\left[\frac{1}{2}L_1L_2\theta_2\dot{\theta}_1^2 - \frac{1}{2}(\dot{\theta}_1 + \dot{\theta}_2)^2 L_1L_2\theta_2\right]m_{B2}$$
$$+\frac{1}{2}gL_1m_{B1} + gL_1m_{B2} + \frac{1}{2}gL_2m_{B2} = 0 \tag{9.91}$$

$$\left[\left(\frac{1}{2}L_1L_2 + \frac{1}{4}L_2^2\right)m_{B2} + I_{B2}\right]\ddot{\theta}_1 + \left(I_{B2} + \frac{1}{4}L_2^2 m_{B2}\right)\ddot{\theta}_2$$
$$+\frac{1}{2}L_1L_2m_{B2}\theta_2\dot{\theta}_1^2 + \frac{1}{2}gL_2m_{B2} = 0 \tag{9.92}$$

Now the two other angles θ_3 and θ_4 can be determined by the following constraint equations:

$$L_4\cos(\theta_3 + \theta_4) + L_3\cos\theta_3 = L_2\cos(\theta_1 + \theta_2) + L_1\cos\theta_1 \tag{9.93}$$

$$L_4\sin(\theta_3 + \theta_4) + L_3\sin\theta_3 = L_2\sin(\theta_1 + \theta_2) + L_1\sin\theta_1 \tag{9.94}$$

Linearizing equations (9.93) and (9.94) yields

$$L_4\cos(\theta_3 + \theta_4) + L_3\cos\ \theta_3 = L_2 + L_1 \tag{9.95}$$

$$L_4\sin(\theta_3 + \theta_4) + L_3\sin\theta_3 = L_2(\theta_1 + \theta_2) + L_1\theta_1 \tag{9.96}$$

Combining equations (9.91), (9.92), (9.95) and (9.96), we can form a set of linearized equations of motion. Further simplification may be necessary depending on the problem and the solution being sought.

9.3 Free Vibration of Continuous Beams: Natural Mode Shapes and Frequencies

Study of the free vibration of beams and the extraction of their corresponding natural mode shapes and natural frequencies is particularly useful in the dynamics of flexible multibody systems. In this section, we examine three cases—bending, torsion and longitudinal vibration—and derive expressions for their natural mode shapes and natural frequencies. This section is intended to be a review; further information can be found in courses and books dealing with continuous beams and vibration analysis of structures.

9.3.1 Transverse Vibration

Consider the flexure of uniform beam as shown in Fig. 9.4(a). Let the transverse displacement be denoted by $y(x, t)$, where x is the longitudinal position of any point on the beam and t is the time. If we take a free-body diagram of an element of the bar of length dx, we obtain Fig. 9.4(b). Let $Q(x, t)$ and $M(x, t)$ be the corresponding shear forces and bending moments acting on the element, and let m be the mass per unit length, A the cross-sectional area, I the area moment of inertia of the cross section, and J the mass moment of inertia. The relation between the bending moment and the bending deformation along the beam is

$$M(x, t) = EI(x)\frac{\partial \psi(x, t)}{\partial x} \tag{9.97}$$

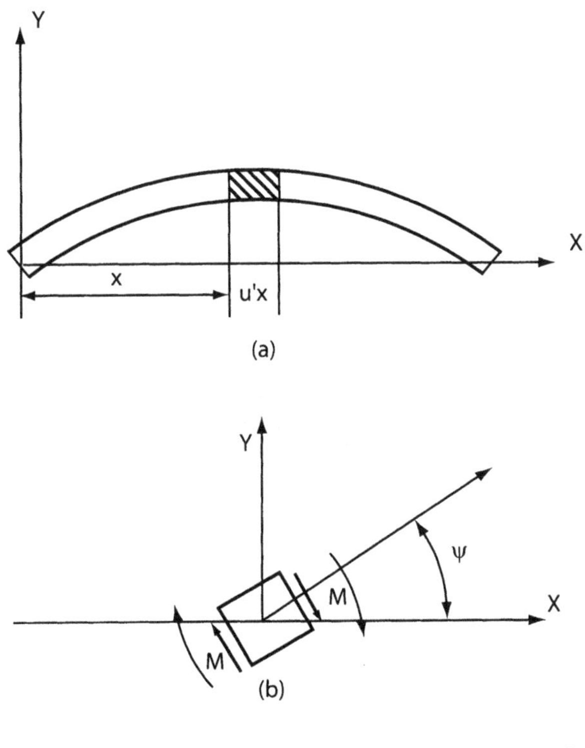

Fig. 9.4. (a) Uniform beam in a flexure mode. (b) Free-body diagram of an element of the beam in flexure

where $\psi(x, t)$ is the angle of rotation due to bending. For $\psi(x, t)$ we have

$$\frac{\partial y(x, t)}{\partial x} = \psi(x, t) \tag{9.98}$$

Equation (9.98) is valid for small deformation where $\tan\psi \approx \psi$ and assuming that the angle of distortion due to the shear force is so small that it can be neglected.

Using the moment equation about the axis normal to x and y, and passing through the center of the element, we get

$$Q = \frac{\partial M}{\partial x} \tag{9.99}$$

Also, we can write

$$W = \frac{\partial Q}{\partial x} \tag{9.100}$$

where W is the distributed load on the beam. Setting W to be the inertia force $-[m(x)(\partial^2 y/\partial t^2)]$, we obtain the equation of motion of the element as

$$\frac{\partial^2}{\partial x^2}\left[EI(x)\frac{\partial^2 y}{\partial x^2}\right] = -m(x)\frac{\partial^2 y}{\partial t^2} \tag{9.101}$$

Equation (9.101) is valid if the rotary inertia force is neglected. On the other hand, if we consider the inertia due to rotation, and we know that the angular acceleration is $(\partial^3 y/\partial x \partial t^2)$; then

$$Q = \frac{\partial M}{\partial x} + J\frac{\partial^3 y}{\partial x \partial t^2} \tag{9.102}$$

and equation (9.101) becomes

$$\frac{\partial^2}{\partial x^2}\left[EI(x)\frac{\partial^2 y}{\partial x^2}\right] = -m(x)\frac{\partial^2 y}{\partial t^2} - J(x)\frac{\partial^4 y}{\partial x^2\,\partial t^2} \tag{9.103}$$

Let us restrict our study to the case where rotary inertia and shear deformation can be neglected. This assumption has been shown by Timoshenko et al. [4] to be accurate for beams with cross-sectional dimensions that are small compared to their length. Therefore, equation (9.101) holds and we can deduce the following:

$$EI\frac{\partial^4 y}{\partial x^2} = -m\frac{\partial^2 y}{\partial t^2} \tag{9.104}$$

Using the separation of variables by setting $y(x, t) = Y(x)f(t)$ for equation (9.104), we obtain

$$f''(t) + \omega^2 f(t) = 0 \tag{9.105}$$

and

$$EI\frac{d^4 Y}{dx^4} - \omega^2 m Y = 0 \tag{9.106}$$

Using the proper boundary conditions, we can extract the natural mode shapes and natural frequencies of the beam in a bending mode. Further, let us consider the case of a fixed–free beam, where the boundary conditions are:

(a) At $x = 0$ the deflection is zero: $Y(0) = 0$

(b) At $x = 0$ the rotation is zero:

$$\frac{dY(0)}{dx} = 0 \tag{9.107}$$

(c) At $x = L$ the bending moment is zero:

$$EI\frac{d^2Y(L)}{dx^2} = 0$$

(d) At $x = L$ the shear force is zero:

$$EI\frac{d^2Y(L)}{dx^2} = 0$$

The solution to equation (9.106) using the boundary conditions given by equations (9.107) yields the natural frequencies obtained from the characteristic equation

$$\cos\beta L \ \cosh\beta L + 1 = 0 \tag{9.108}$$

where

$$\beta_r^4 = \frac{\omega_r^2 m}{EI} \qquad (r = 1, 2, \ldots, \infty)$$

and the eigenfunctions (natural mode shapes) are found to be

$$Y_r(x) = C_r\left[\sin\beta_r x - \sinh\beta_r x + \frac{\cos\beta_r L + \cosh\beta_r L}{\sin\beta_r L - \sinh\beta_r L}(\cos\beta_r x - \cosh\beta_r x)\right] \tag{9.109}$$

The constant C_r can be found through the initial conditions. Similarly, we can obtain the desirable mode shapes for different boundary conditions, such as free–free and fixed–fixed beams. Table 9.3.1 gives the solution for several boundary condition cases.

Table 9.3.1 Transverse vibration of a uniform beam (ROD):

Fixed–Free

$$\cos\beta L \cosh\beta L = -1$$

End 1 (fixed) End 2 (free)

$$Y(0) = Y'(0) = Y''(L) = Y'''(L) = 0$$

$$Y_r(x) = C_r\left[\sin\beta_r x - \sinh\beta_r x + \frac{\cos\beta_r L + \cosh\beta_r L}{\sin\beta_r L - \sinh\beta_r}(\cos\beta_r x - \cosh\beta_r x)\right]$$

End 2 (fixed) End 1 (free)

$$Y''(0) = Y'''(0) = Y(L) = Y'(L) = 0$$

$$Y_r(x) = C_r \left[\sin \beta_r x + \sinh \beta_r x - \frac{\sin \beta_r L + \sinh \beta_r L}{\cos \beta_r L + \cosh \beta_r L} (\cos \beta_r x + \cosh \beta_r x) \right]$$

Free–Free

$$\left[\frac{\sin \beta L - \sinh \beta L}{\cos \beta L - \cosh \beta L} \right]^2 = 1$$

Ends 1 and 2 (free)

$$Y''(0) = Y'''(0) = Y''(L) = Y'''(L) = 0$$

$$Y_r(x) = C_r \left[\sin \beta_r x + \sinh \beta_r x - \frac{\sin \beta_r L - \sinh \beta_r L}{\cos \beta_r L - \cosh \beta_r L} (\cos \beta_r x + \cosh \beta_r x) \right]$$

Pinned–Free

$$\frac{\tan \beta L}{\tanh \beta L} = 1$$

End 1 (pinned) End 2 (free)

$$Y(0) = Y''(0) = Y''(L) = Y'''(L) = 0$$

$$Y_r(x) = C_r \left[\sin \beta_r x + \frac{\sin \beta_r L}{\sinh \beta_r L} \sinh \beta_r x) \right]$$

End 2 (pinned) End 1 (free)

$$Y''(0) = Y'''(0) = Y(L) = Y''(L) = 0$$

$$Y_r(x) = C_r \left[\cos \beta_r x + \cosh \beta_r x - \frac{\cos \beta_r L + \cosh \beta_r L}{\sin \beta_r L + \sinh \beta_r L} (\sin \beta_r x + \sinh \beta_r x) \right]$$

Simply Supported

$$\sin \beta L = 0$$

Ends 1 and 2 (simply supported)

$$Y(0) = Y''(0) = Y(L) = Y''(L) = 0$$

$$Y_r(x) = C_r \left[\sin \beta_r x - \frac{\sin \beta_r L}{\sinh \beta_r L} \sinh \beta_r x) \right]$$

Fixed–Fixed

$$\cos \beta L \ \cosh \beta L = 1$$

Ends 1 and 2 (fixed)

$$Y(0) = Y'(0) = Y(L) = Y'(L) = 0$$

$$Y_r(x) = C_r \left[\sin \beta_r x - \sinh \beta_r x + \frac{\sin \beta_r L - \sinh \beta_r L}{\cos \beta_r L - \cosh \beta_r L} (\cos \beta_r x - \cosh \beta_r x) \right]$$

Fixed–Pinned

$$\frac{\tan \beta L}{\tanh \beta L} = 1$$

End 1 (fixed) End 2 (pinned)

$$Y(0) = Y'(0) = Y(L) = Y''(L) = 0$$

$$Y_r(x) = C_r \left[\sin \beta_r x - \sinh \beta_r x + \frac{\sin \beta_r L - \sinh \beta_r L}{\cos \beta_r L - \cosh \beta_r L} (\cos \beta_r x - \cosh \beta_r x) \right]$$

End 2 (fixed) End 1 (pinned)

$$Y(0) = Y''(0) = Y(L) = Y'(L) = 0$$

$$Y_r(x) = C_r \left[\sin \beta_r x - \frac{\sin \beta_r L}{\sinh \beta_r L} \sinh \beta_r x \right]$$

9.3.2 Longitudinal Vibration

Consider the thin uniform rod shown in Figure 9.5. If the rod is subjected to longitudinal force, there will be a displacement u along the rod that is a function of both the position x and time t. We assume that the place cross-sectional areas remain undeformed during the axial deformation of the rod. If we consider an element of the rod of length dx and denote by u the displacement at x, from Hooke's law we can write

$$\frac{\partial u}{\partial x} = \frac{p}{AE} \tag{9.110}$$

Using Newton's second law of dynamics for the element, we obtain the equation of motion:

$$\frac{\partial p}{\partial x} dx = \rho A \ dx \frac{\partial^2 u}{\partial t^2} \tag{9.111}$$

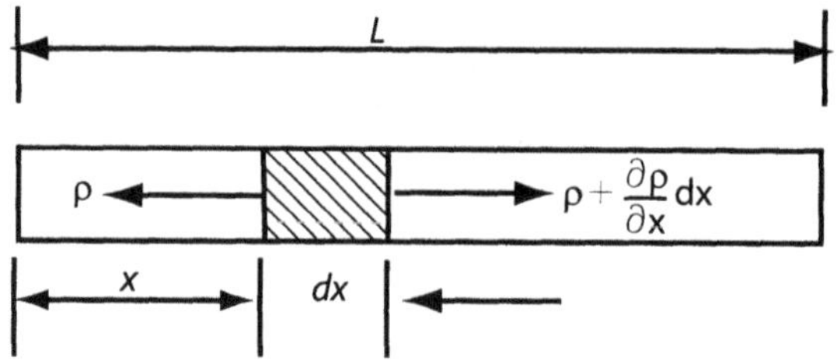

Fig. 9.5. Uniform beam subject to longitudinal forces

Differentiation of equation (9.110) with respect to x and substituting $\partial p/\partial x$ into (9.111), we obtain

$$\frac{\partial^2 u}{\partial x^2} = \frac{1}{c^2}\frac{\partial^2 u}{\partial t^2} \tag{9.112}$$

where $c = \sqrt{E/\rho}$ and ρ is the density of the rod.

A solution to equation (9.112) can be found making use of the separation of variables by setting $u(x, t) = U(x)G(t)$. Then equation (9.112) will result into two equations:

$$\begin{aligned} U(x) &= A \sin \frac{\omega}{c} x + B \cos \frac{\omega}{c} x \\ G(t) &= c \sin \omega t + D \cos \omega t \end{aligned} \tag{9.113}$$

In a similar fashion, using the proper boundary conditions, we can solve for the natural mode shapes and natural frequencies. For example, for the case of fixed–free end conditions, the eigenfunctions are

$$U_r(x) = C_r \sin \beta_r x \tag{9.114}$$

Table 9.3.2 Longitudinal vibration of a uniform beam:

Fixed–Free

$$\beta_r = (2r - 1)\frac{\pi}{2L}$$

End 1 (fixed) End 2 (free)

$$U(0) = U'(L) = 0$$

$$U_r(x) = C_r \sin \beta_r x$$

End 2 (fixed) End 1 (free)

$$U'(0) = U(L) = 0$$

$$U_r(x) = C_r \cos \beta_r x$$

Free–Free

$$\beta_r = \frac{r\pi}{L}$$

End 1 and 2 (free)

$$U'(0) = U'(L) = 0$$

$$U_r(x) = C_r \cos \beta_r x$$

Clamped–Clamped

$$\beta_r = \frac{r\pi}{L}$$

End 1 and 2 (clamped)

$$U(0) = U(L) = 0$$

$$U_r(x) = C_r \sin \beta_r x$$

and

$$\beta_r = (2r - 1)\frac{\pi}{2L} \tag{9.115}$$

Table 9.3.2 gives a summary of eigenfunctions for various end conditions.

9.3.3 Torsional Vibration

The torsional equations of motion for a thin uniform rod are similar to the equations of motion obtained for the longitudinal vibration. Consider an element of the rod subjected to a torque T as shown in Figure 9.6. Assuming that there is no warping and using Hooke's law, the angle twist in any length dx of the rod due to a torque T is

$$d\theta = \frac{T\,dx}{I_p G} \tag{9.116}$$

where T is the torque, I_p is the polar moment of inertia, and G is the shear modulus of elasticity. The differential equation for the element of the rod of motion can be found to take the following form:

$$\frac{\partial^2 \theta}{\partial t^2} = \frac{G}{\rho}\frac{\partial^2 \theta}{\partial x^2} \tag{9.117}$$

A solution to equation (9.117) is found in a similar fashion as equation (9.112) by setting $\theta(x, t) = \Theta(x)G(t)$, or simply

$$\Theta(x) = A \sin \omega\sqrt{\frac{\rho}{G}}x + B \cos \omega\sqrt{\frac{\rho}{G}}x$$

$$G(t) = C \sin \omega t + D \cos \omega t \tag{9.118}$$

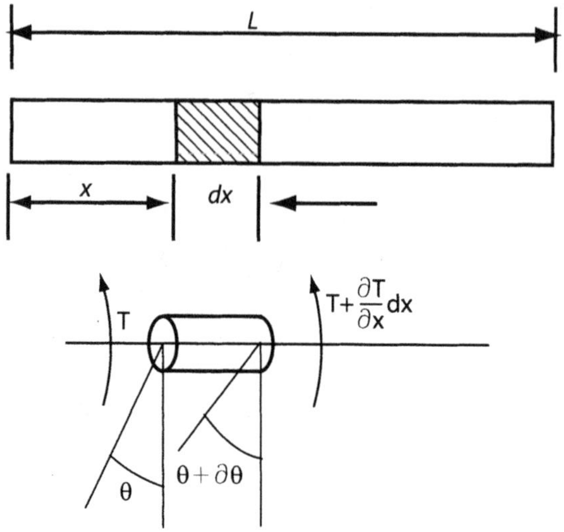

Fig. 9.6. Uniform rod subject to a torsional torque

Similarly, the natural mode shapes and natural frequencies are found by using the proper boundary conditions applied to equation (9.118). For a fixed–free endpoint condition, the eigenfunctions $\Theta(x)$ are found to be

$$\Theta_r(x) = C_r \sin \beta_r \sqrt{\frac{G}{\rho}} x \tag{9.119}$$

where

$$\beta_r = (2r - 1)\frac{\pi}{2L} \tag{9.120}$$

Table 9.3.3 provides a summary for eigenfunction solutions corresponding to various boundary conditions.

Table 9.3.3 Torsional vibration of a uniform rod:

Fixed–Free

$$\beta_r = (2r - 1)\frac{\pi}{2L}$$

End 1 (fixed) End 2 (free)

$$\Theta(0) = \Theta'(L) = 0$$

$$\Theta_r(x) = C_r \sin \beta_r \sqrt{\frac{G}{\rho}} x$$

End 2 (fixed) End 1 (free)

$$\Theta'(0) = \Theta'(L) = 0$$

$$\Theta_r(x) = C_r \cos \beta_r \sqrt{\frac{G}{\rho}} x$$

Free–Free

$$\beta_r = \frac{r\pi}{L}$$

Ends 1 and 2 (free)

$$\Theta'(0) = \Theta'(L) = 0$$

$$\Theta_r(x) = C_r \cos \beta_r \sqrt{\frac{G}{\rho}} x$$

Clamped–Clamped

$$\beta_r = \frac{r\pi}{L}$$

Ends 1 and 2 (clamped)

$$\Theta(0) = \Theta(L) = 0$$

$$\Theta_r(x) = C_r \sin \beta_r \sqrt{\frac{G}{\rho}} x$$

9.4 The Eigenvalue Problem

Vibration analysis is an integrated part of our simulation and study of multibody dynamics performance, safety, and control. In today's technology, most of our articulated structures, mechanisms, and robotic manipulators in general undergo a small vibration during the course of their large rotational motions. To account for such a deformation, in the next two chapters we will be called on to perform eigenvalue problems to extract the mode shapes of the bodies involved. The latter information will be used to describe the modal deformation of the bodies, and hence it will enable us to predict more accurately the dynamics of such systems. Since not all bodies are continuous beams and rods, numerical solutions will be needed to perform the eigenvalue problem. In what follows, we examine the eigenvalue problem and its solution. Since we will only be discussing a few methods pertaining to the solution of continuous and discrete systems, the reader is urged to seek further help if needed in the references provided at the end of the chapter, especially references [5–7].

The eigenvalue problem stems from the solution to linear systems governed by equations of the form

$$M(\ddot{u}) + L(u) = 0 \tag{9.121}$$

where M and L are linear differential operators involving only the space variable x; u is the displacement, which is defined as

$$u(x,\ t) = \phi(x)q(t) \tag{9.122}$$

Substituting equation (9.122) into (9.121) yields

$$\frac{L(\phi)}{M(\phi)} = -\frac{\ddot{q}(t)}{q(t)} = \lambda \tag{9.123}$$

where λ is constant by virtue of the $L(\phi)/M(\phi)$ expression as well as $\ddot{q}/q$. From equation (9.123), we deduce the following:

$$L(\phi) = \lambda M(\phi) \tag{9.124}$$

which defines the eigenvalue problem for continuous systems. In this case, λ is known as the eigenvalue and $\phi(x)$ as its corresponding eigenfunction. One would read equation (9.124) as finding the eigenvalue λ for which there exists $\phi(x)$ satisfying the eigenvalue problem (9.124) and the boundary conditions.

The orthogonality condition is a characteristic of the eigenfunctions. That is, with respect to L and M, we have

$$\int \phi_r M \phi_s dv = 0 \quad \text{for} \quad r \neq s \tag{9.125}$$

and

$$\int \phi_r L \phi_s dv = 0 \quad \text{for} \quad r \neq s \tag{9.126}$$

In engineering applications, M and L are identified from partial differential equations, where u might represent the deflection, whereas in discrete systems, it can easily be proven that in most cases M corresponds to the generalized mass matrix and L to the generalized stiffness. Now, we are ready to solve for λ and ϕ given a second-order differential equation—a partial differential equation of the form (9.121).

Example 9.4.1
Find the eigenvalues and eigenvectors for the lumped mass system shown in Figure 9.7.

Solution
The equations of motion are found to be

$$\begin{aligned} m\ddot{x}_1 &= -kx_1 - kx_1 + kx_2 + kx_3 - kx_1 \\ m\ddot{x}_2 &= -kx_2 - kx_2 - kx_2 + kx_1 - kx_3 \\ m\ddot{x}_3 &= -kx_3 - kx_3 + kx_2 + kx_1 - kx_3 \end{aligned} \tag{9.127}$$

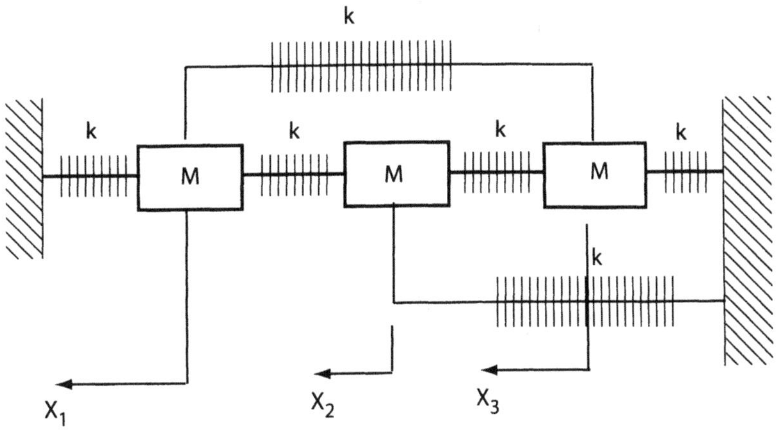

Fig. 9.7. Lumped mass model

In matrix form equations (9.127) can be written as

$$\begin{bmatrix} m & 0 & 0 \\ 0 & m & 0 \\ 0 & 0 & m \end{bmatrix} \begin{bmatrix} \ddot{x}_1 \\ \ddot{x}_2 \\ \ddot{x}_3 \end{bmatrix} + \begin{bmatrix} 3k & -k & -k \\ -k & 3k & -k \\ -k & -k & 3k \end{bmatrix} \begin{bmatrix} x_1 \\ x_2 \\ x_3 \end{bmatrix} = 0 \tag{9.128}$$

Let

$$M = m \begin{bmatrix} 1 & 0 & 0 \\ 0 & 1 & 0 \\ 0 & 0 & 1 \end{bmatrix}, \qquad K = k \begin{bmatrix} 3 & -1 & -1 \\ -1 & 3 & -1 \\ -1 & -1 & 3 \end{bmatrix} \tag{9.129}$$

Then we formulate the eigenvalue problem as follows:

$$[K - \lambda M] = k \begin{bmatrix} 3 & -1 & -1 \\ -1 & 3 & -1 \\ -1 & -1 & 3 \end{bmatrix} - \lambda m \begin{bmatrix} 1 & 0 & 0 \\ 0 & 1 & 0 \\ 0 & 0 & 1 \end{bmatrix} \tag{9.130}$$

Now we compute the determinant of equation (9.130) to extract the eigenvalues corresponding to λ's that makes the determinant zero.

$$|K - \lambda M| = k \begin{vmatrix} 3 - \lambda \dfrac{m}{k} & -1 & -1 \\ -1 & 3 - \lambda \dfrac{m}{k} & -1 \\ -1 & -1 & 3 - \lambda \dfrac{m}{k} \end{vmatrix} = 0 \tag{9.131}$$

Let

$$\omega^2 = \frac{k}{m} \alpha = \frac{\lambda}{\omega^2}$$

The solution to (9.131) yields

$$\alpha_1^2 = 1, \qquad \alpha_2^2 = 4, \qquad \alpha_3^2 = 4 \tag{9.132}$$

The corresponding eigenvalues are

$$\lambda_1 = \omega^2, \qquad \lambda_2 = 4\omega^2, \qquad \lambda_3 = 4\omega^2 \tag{9.133}$$

or

$$\lambda_1 = \frac{k}{m}, \qquad \lambda_2 = 4\frac{k}{m}, \qquad \lambda_3 = 4\frac{k}{m}$$

for

$$\lambda_1 = \frac{k}{m} \quad \text{or} \quad \alpha_1^2 = 1$$

The eigenvector is found by substituting the value of λ_1 into equation (9.131), which yields

$$3(x_1 + x_2 + x_3) = 0 \tag{9.134}$$

resulting in a first eigenvector of

$$\bar{x}_1 = \begin{bmatrix} 1 \\ 1 \\ 1 \end{bmatrix} \tag{9.135}$$

Similarly for $\lambda_2 = 4k/m$, we obtain the second eigenvector,

$$\bar{x}_2 = \begin{bmatrix} 1 \\ 0 \\ -1 \end{bmatrix} \tag{9.136}$$

and finally, for $\lambda_3 = 4k/m$, we get

$$\bar{x}_3 = \begin{bmatrix} 1 \\ -2 \\ 1 \end{bmatrix} \tag{9.137}$$

Note that

$$\bar{x}_2 \cdot \bar{x}_3 = 0 \tag{9.138}$$

9.5 Rayleigh–Ritz Method

The Rayleigh–Ritz method provides a basis for computation of frequencies and mode shapes (eigenfunctions). This method relies on the Rayleigh coefficient, which makes use of the ratio of the maximum kinetic and maximum potential energy. In the evaluation of those kinetic expressions, Rayleigh relied on the assumption that a deflection, as in the case of beams, is expressed by a simple shape function, whereas Ritz extended that concept to the sum of several functions. The deflection could be expressed as

$$y(x) = \sum_{i=1}^{n} c_i \phi_i(x) \tag{9.139}$$

where the ϕ_i are admissible functionss satisfying the geometric boundary conditions of the problem and the c's are constants (to be determined).

Rayleigh's equation is defined as

$$[\lambda T^*_{\max}] - U_{\max} = 0 \tag{9.140}$$

or

$$\lambda = \frac{U_{\max}}{T^*_{\max}} \tag{9.141}$$

where λ is a constant and corresponds to the natural frequencies. For continuous systems, the potential and kinetic energies are expressed as

$$U = \frac{1}{2} \sum \sum k_{ij} c_i c_j \tag{9.142}$$

$$T^* = \frac{1}{2} \sum \sum m_{ij} c_i c_j \tag{9.143}$$

where k_{ij} and m_{ij} are identified from equations (9.142) and (9.143), respectively, for the problem at hand. This is achieved by making use of the Rayleigh–Ritz approximation of the deflection given by equation (9.139). Minimizing equation (9.141) gives

$$\frac{\partial \lambda}{\partial c_i} = 0 \tag{9.144}$$

which yields

$$\frac{\partial U_{\max}}{\partial c_i} - \lambda \frac{\partial T^*_{\max}}{\partial c_i} = 0 \tag{9.145}$$

or simply

$$\sum_{j=1}^{n} (k_{rj} - \lambda m_{rj}) c_j = 0 \tag{9.146}$$

where the c_j are the unknowns. Equation (9.146), known as Galerkin's equation, represents the eigenvalue problem. In matrix form, equation (9.146) can be expressed as

$$[K]\{c\} = \lambda [M]\{c\} \tag{9.147}$$

$[K]$ and $[M]$ are referred to as the global stiffness and mass matrix of the nth-degree-of-freedom system.

In essence, the Rayleigh–Ritz method regards a continuous system as an n-degree-of-freedom system. This makes the derivation of $[M]$ and $[K]$ simple for discrete systems, as is the case for continuous systems, modeled by the finite-element method. One more thought about the method before we go into an example. Additional observations about this method can be summarized as follows:

1. The method is a good approximation for the lower eigenvalues and depends strongly on the trial functions used to approximate the modes of deformation.
2. The coefficients c_i are determined to bring a close resemblance between the eigenfunction $y(x)$ and the natural modes of the system. This is usually achieved by making the Rayleigh quotient stationary.

Example 9.5.1

Consider the system shown in Figure 9.8, where the clamped–free slender bar of a parabolic side is assumed to oscillate longitudinally. Our task is to extract the eigenvalues and the corresponding mode shapes using the Rayleigh–Ritz method discussed above.

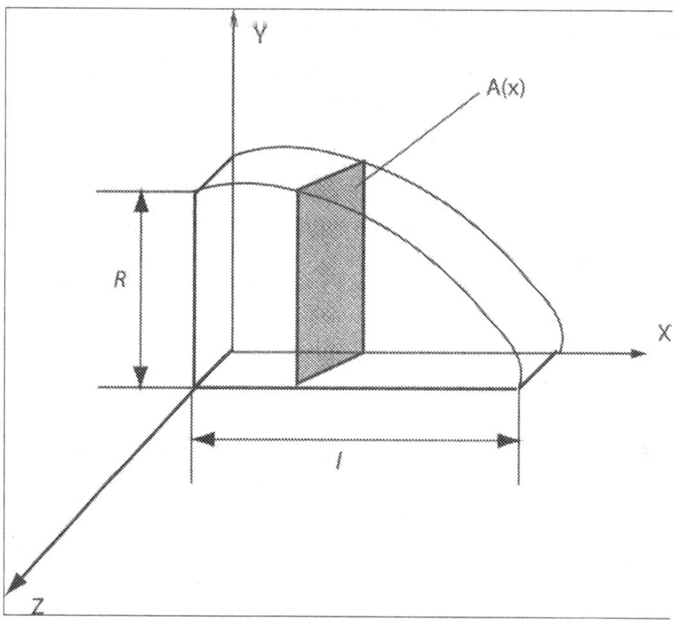

Fig. 9.8. Clamped–free slender bar of a parabolic side

Solution

The mass per unit length and the stiffness along x are

$$m(x) = \frac{R\rho}{\ell^2}(\ell^2 - 2\ell x + x^2) \tag{9.148}$$

$$EA(x) = \frac{ER}{\ell^2}(\ell^2 - 2\ell x + x^2) \tag{9.149}$$

To obtain the approximate first two natural frequencies and mode shapes, we use the displacement as a series of two terms.

$$y(x) = \sum_{i=1}^{2} c_i \phi_i(x) = c_1\phi_1(x) + c_2\phi_2(x) \tag{9.150}$$

Let us assume that ϕ_1 and ϕ_2 can be approximated by sine functions; then we write

$$y(x) = c_1 \sin\frac{\pi x}{2L} + c_2 \sin\frac{3\pi x}{2L} \tag{9.151}$$

The mode shapes will be known once the constants c_1 and c_2 are evaluated. The maximum potential energy is given by

$$U = \frac{1}{2}\int_0^\ell EA(x)\left(\frac{dy}{dx}\right)^2 dx \tag{9.152}$$

or

$$U = \frac{1}{2}\int_0^1 \frac{ER}{\ell^2}(\ell^2 - 2\ell x + x^2)\left(\frac{dy(x)}{dx}\right)^2 dx \tag{9.153}$$

After integration, we obtain

$$U = \frac{ER}{2\ell}\left(\frac{\pi^2}{24}c_1^2 + \frac{3\pi^2}{8}c_2^2 + \frac{1}{4}c_1^2 + \frac{15}{8}c_1c_2 + \frac{1}{4}c_2^2\right) \tag{9.154}$$

This final expression of U is similar to

$$U = \frac{1}{2}\sum_{i=1}^{2}\sum_{i=1}^{2} k_{ij}c_ic_j \tag{9.155}$$

which when expanded yields

$$U = \frac{1}{2}(k_{11}c_1^2 + k_{12}c_1c_2 + k_{21}c_2c_1 + k_{22}c_2^2) \tag{9.156}$$

Similarly, the maximum kinetic energy can be expressed as

$$T = \frac{\omega^2}{2}\int_0^\ell m(x)y^2 dx \tag{9.157}$$

Substituting for $m(x)$ and $y(x)$ from equations (9.148) and (9.151), we get

$$T = \frac{\omega^2}{2}\int_0^\ell \frac{R\rho}{\ell^2}(\ell^2 - 2\ell x + x^2)y^2\, dx \tag{9.158}$$

The above results in

$$T = \frac{\omega^2 R\ell\rho}{2}\left(\frac{1}{6}c_1^2 + \frac{1}{6}c_2^2 - \frac{1}{\pi^2}c_1^2 + \frac{3}{2\pi^2}c_1c_2 - \frac{1}{9\pi^2}c_2^2\right) \tag{9.159}$$

which is similar to

$$T = \frac{\omega^2}{2}\sum_{i=1}^{2}\sum_{j=1}^{2} m_{ij}c_ic_j \tag{9.160}$$

Written in expanded form, we have

$$T = \frac{\omega^2}{2}(m_{11}c_1^2 + m_{12}c_1c_2 + m_{21}c_2c_1 + m_{22}c_2^2) \tag{9.161}$$

where $\omega^2 = \lambda$.

Now, from the potential energy expression, the stiffness matrix can be obtained as

$$K = \begin{bmatrix} \dfrac{ER}{\ell}\left(\dfrac{1}{4} + \dfrac{\pi^2}{24}\right) & \dfrac{15ER}{16\ell} \\ \dfrac{15ER}{16\ell} & \dfrac{ER}{\ell}\left(\dfrac{1}{4} + \dfrac{3\pi^2}{8}\right) \end{bmatrix} \tag{9.162}$$

and the mass matrix is found to be

$$M = \begin{bmatrix} R\ell\rho\left(\dfrac{1}{6} - \dfrac{1}{\pi^2}\right) & \dfrac{3R\ell\rho}{4\pi^2} \\ \dfrac{3R\ell\rho}{4\pi^2} & R\ell\rho\left(\dfrac{1}{6} - \dfrac{1}{9\pi^2}\right) \end{bmatrix} \tag{9.163}$$

Using equation (9.146), we obtain a system of two linear equations, which can be used to solve for the eigenvalues and the constants c_1 and c_2:

$$[K - \lambda M]C = 0 \tag{9.164}$$

where

$$C = [c_1 \quad c_2]^T \tag{9.165}$$

Substituting M and K into (9.164), we obtain

$$\begin{bmatrix} 0.66123\dfrac{ER}{\ell} - 0.06534\lambda R\ell\rho & 0.9375\dfrac{ER}{\ell} - 0.0760\lambda R\ell\rho \\ 0.9375\dfrac{ER}{\ell} - 0.0760\lambda R\ell\rho & 3.9511\dfrac{ER}{\ell} - 0.155409\lambda R\ell\rho \end{bmatrix}\begin{bmatrix} c_1 \\ c_2 \end{bmatrix} = 0 \tag{9.166}$$

If we replace $\lambda \ell^2 \rho / E$ by α, equation (9.166) reduces to

$$\begin{bmatrix} 0.66123 - 0.065364\alpha & 0.9375 - 0.076\alpha \\ 0.9375 - 0.076\alpha & 3.9511 - 0.1554\alpha \end{bmatrix} \begin{bmatrix} c_1 \\ c_2 \end{bmatrix} = 0 \tag{9.167}$$

Setting the determinant of the system (9.167) to zero, we obtain the following characteristic equation:

$$1.73369 - 0.218465\alpha + 0.0043806\alpha^2 = 0 \tag{9.168}$$

The roots are found to be

$$\alpha_1 = 9.90179 \tag{9.169}$$

and

$$\alpha_2 = 39.9688 \tag{9.170}$$

or simply

$$\lambda_1 = 9.90179 \frac{E}{\rho \ell} = \omega_1^2 \tag{9.171}$$

and

$$\lambda_2 = 39.9688 \frac{E}{\rho \ell^2} = \omega_2^2 \tag{9.172}$$

Thus, the first two natural frequencies and mode shapes are

$$\omega_1 = 3.1467 \sqrt{\frac{E}{\rho \ell^2}} \tag{9.173}$$

Substituting α from equation (9.169) into equation (9.167), and solving for c_1 and c_2, we obtain

$$y_1(x) = \sin \frac{\pi}{2L} x - 0.0707143 \ \sin \frac{3\pi}{2L} x \tag{9.174}$$

where $y_1(x)$ is as shown in Figure 9.9. Similarly, we find that

$$\omega_2 = 6.32208 \sqrt{\frac{E}{\rho \ell^2}} \tag{9.175}$$

and for α_2 we get

$$y_2(x) = -\sin \frac{\pi}{2L} x + 0.92894 \ \sin \frac{3\pi}{2L} x \tag{9.176}$$

The second mode, $y_2(x)$, is shown in Figure 9.10.

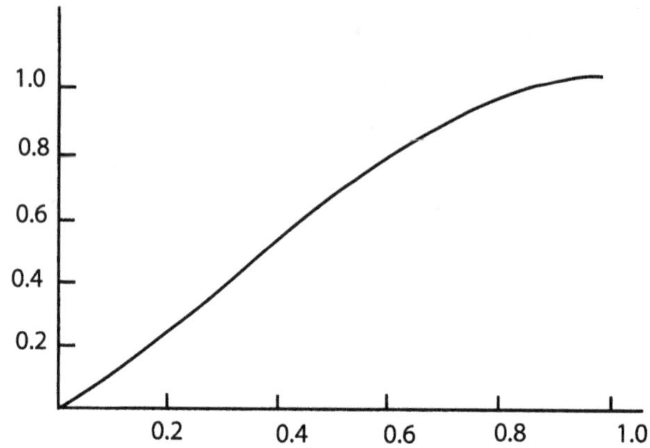

Fig. 9.9. First mode of vibration of the parabolic slender bar in the longitudinal direction

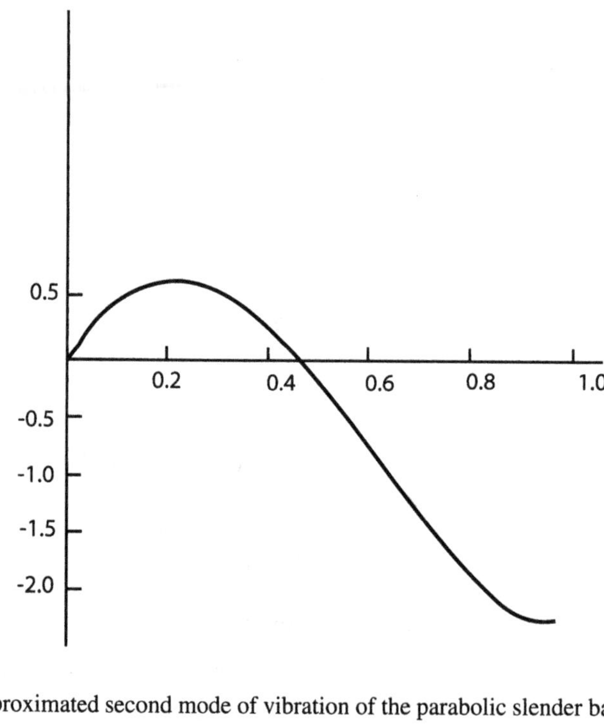

Fig. 9.10. Approximated second mode of vibration of the parabolic slender bar in the longitudinal direction

The accuracy of the mode shapes can be enhanced if a sequence of more terms to the $y(x)$ are used. Consider, for instance, a sequence of three terms:

$$y(x) = c_1 \sin \frac{\pi x}{2\ell} + c_2 \sin \frac{3\pi x}{2\ell} + c_3 \sin \frac{5\pi x}{2\ell} \tag{9.177}$$

Once more, we need to evaluate c_1, c_2 and c_3 and the corresponding eigenvalues of the problem. Substitution of equation (9.177) into (9.153) followed by integration of the entire expression gives

$$U = \frac{ER}{2\ell}\left[\left(\frac{\pi^2}{24}+\frac{1}{4}\right)c_1^2+\left(\frac{3\pi^2}{8}+\frac{1}{4}\right)c_2^2+\left(\frac{25\pi^2}{24}+\frac{1}{4}\right)c_3^2 + \frac{15}{8}c_1c_2+\frac{65}{72}c_1c_3+\frac{255}{321}c_2c_3\right] \tag{9.178}$$

From the potential energy expression U, we can easily deduce the values of the element of the stiffness k_{ij}:

$$k_{11} = \frac{ER}{\ell}\left(\frac{\pi^2}{24}+\frac{1}{4}\right)$$

$$k_{21} = k_{12} = \frac{15}{16}\frac{ER}{2\ell} k_{13} = k_{31} = \frac{65}{144}\frac{ER}{\ell}$$

$$k_{23} = k_{32} = \frac{255}{64}\frac{ER}{2\ell}$$

$$k_{22} = \frac{ER}{\ell}\left(\frac{3\pi^2}{8}+\frac{1}{4}\right)$$

$$k_{33} = \frac{ER}{\ell}\left(\frac{25}{24}\pi^2+\frac{1}{4}\right)$$

Similarly, substituting equation (9.177) into (9.158) and integrating along the bar, we obtain the following closed form of the kinetic energy.

$$T = \frac{\omega^2 R\ell\rho}{2}\left[\left(\frac{1}{6}-\frac{1}{\pi^2}\right)c_1^2+\left(\frac{1}{6}-\frac{1}{9\pi^2}\right)c_2^2 + \left(\frac{1}{6}-\frac{1}{25\pi^2}\right)c_3^2+\frac{3}{2\pi^2}c_1c_2+\frac{15}{8\pi^2}c_2c_3+\frac{5}{18}c_1c_3\right] \tag{9.179}$$

Therefore, from the expression (9.179), we also identify the values of m_{ij}:

$$m_{11} = R\ell\rho\left(\frac{1}{6}-\frac{1}{\pi^2}\right), \qquad m_{12} = m_{21} = \frac{3}{4\pi^2}R\ell\rho$$

$$m_{22} = R\ell\rho\left(\frac{1}{6}-\frac{1}{9\pi^2}\right), \qquad m_{23} = m_{32} = \frac{15}{16\pi^2}R\ell\rho$$

$$m_{33} = R\ell\rho\left(\frac{1}{6}-\frac{1}{25\pi^2}\right), \qquad m_{31} = m_{13} = \frac{5}{36}R\ell\rho$$

Thus the elements k_{ig} and m_{ij} of the matrices K and M, respectively, are known.

Using equation (9.164) and replacing ω^2 by λ and $\lambda\ell^2\rho/E$ by α, we obtain the following system of equations:

$$\begin{bmatrix} 0.66123 - 0.065345\alpha & 0.9375 - 0.076\alpha & 0.451389 - 0.01407\alpha \\ 0.9375 - 0.076\alpha & 3.9511 - 0.1554\alpha & 3.9843 - 0.09498\alpha \\ 0.4513889 - 0.01407\alpha & 3.9843 - 0.09498\alpha & 10.5308 - 0.162614\alpha \end{bmatrix} \begin{bmatrix} c_1 \\ c_2 \\ c_3 \end{bmatrix} = 0 \tag{9.180}$$

Setting the determinant of the system (9.180) to zero, we get

$$10.3272 - 1.42166\alpha + 0.041016\alpha^2 - 0.0002955133\alpha^3 = 0 \tag{9.181}$$

which is found to have the following roots:

$$\alpha_1 = 9.880436$$
$$\alpha_2 = 39.55020$$
$$\alpha_3 = 89.54455$$

or

$$\lambda_1 = 9.880436\frac{E}{\rho\ell^2} = \omega_1^2$$
$$\lambda_2 = 39.55020\frac{E}{\rho\ell^2} = \omega_2^2$$
$$\lambda_3 = 89.54455\frac{E}{\rho\ell^2} = \omega_3^2$$

and consequently, the estimated natural frequencies and natural mode shapes are

$$\omega_1 = 3.14331\sqrt{\frac{E}{\rho\ell^2}} \tag{9.182}$$

and

$$y_1(x) = \sin\frac{\pi x}{2L} - 0.058134\ \sin\frac{3\pi x}{2L} - 0.015158\ \sin\frac{5\pi x}{2L} \tag{9.183}$$

which is displayed in Figure 9.11. For

$$\omega_2 = 6.2889\sqrt{\frac{E}{\rho\ell^2}} \tag{9.184}$$

we have

$$y_2(x) = -\sin\frac{\pi x}{2L} + 0.934\ \sin\frac{3\pi x}{2L} - 0.07749\ \sin\frac{5\pi x}{2L} \tag{9.185}$$

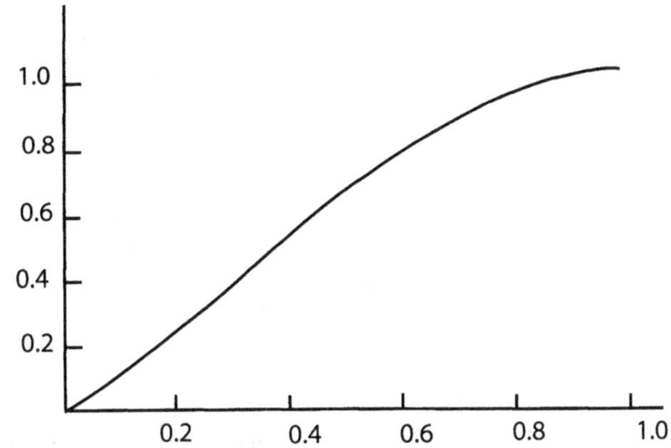

Fig. 9.11. First-mode approximation of the parabolic slender bar longitudinal vibration

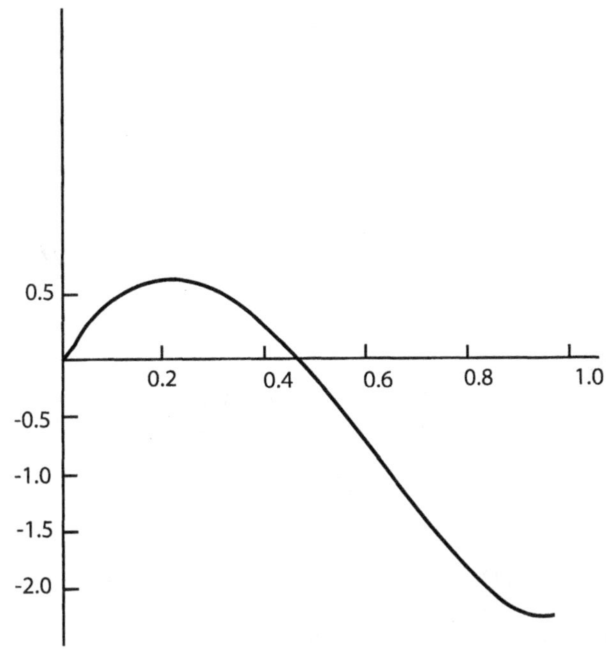

Fig. 9.12. Second-mode approximation for the parabolic slender bar longitudinal vibration

which is shown in Figure 9.12. Finally, for

$$\omega_3 = 9.46279\sqrt{\frac{E}{\rho\ell^2}} \tag{9.186}$$

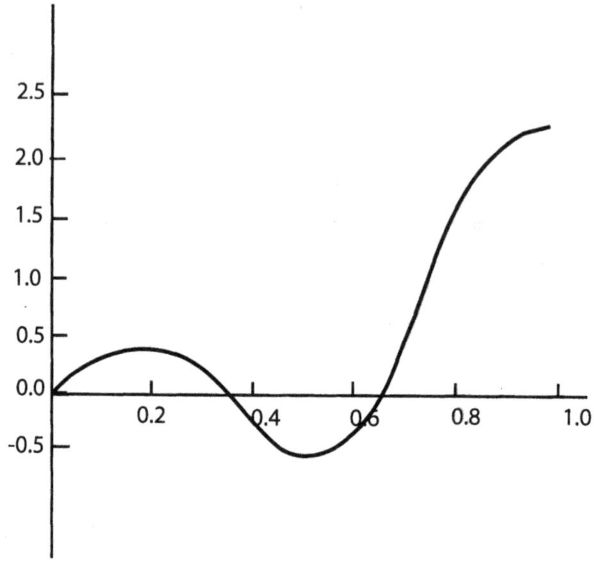

Fig. 9.13 Third mode approximation of the parabolic slender bar longitudinal vibration

we have

$$y_3(x) = \sin\frac{\pi x}{2L} - 1.01371\ \sin\frac{3\pi x}{2L} + 0.93654\ \sin\frac{5\pi x}{2L} \tag{9.187}$$

Figure 9.13 gives the mode shape of ω_3.

Rayleigh–Ritz Method Revisited

In the case where the generalized mass and stiffness matrices of a mechanical system are known, the eigenvalue problem is then given by

$$KX = \lambda MX \tag{9.188}$$

where if we assume that $K$and M are both *positive definite* matrices, the eigenvalues are all positive. In equation (9.188), λ and X correspond to the eigenvalues and eigenvectors respectively.

The Rayleigh quotient is defined as follows.

$$Q = \frac{\{X\}^T[K]\{X\}}{\{X\}^T[M]\{X\}} > 0 \tag{9.189}$$

Let us assume a vector $\bar{w}$ which is a linear combination of trial vector $\bar{x}_1$.

$$\bar{w} = \sum_{i=1}^{w} \alpha_i x_i \tag{9.190}$$

Then

$$Q = \frac{\sum_{i=1}^{n} \alpha_i x_i \sum_{j=1}^{n} K_{ij}\alpha_j x_j}{\sum_{i=1}^{n} \alpha_i x_i \sum_{j=1}^{n} m_{ij}\alpha_j x_j} = \frac{N}{D} \tag{9.191}$$

The minimum of Rayleigh's quotient gives

$$\frac{\partial Q}{\partial \alpha_k} = \frac{D\dfrac{\partial N}{\partial \alpha_k} - N\dfrac{\partial D}{\partial \alpha_k}}{D^2} = 0 \tag{9.192}$$

that is,

$$D\frac{\partial N}{\partial \alpha_k} - N\frac{\partial D}{\partial \alpha_k} = 0 \quad \text{or} \quad \frac{\partial N}{\partial \alpha_k} - \frac{N}{D}\frac{\partial D}{\partial \alpha_k} = 0 \tag{9.193}$$

Performing the partial derivatives of N, we get

$$\begin{aligned} \frac{\partial N}{\partial \alpha_k} &= \frac{\partial}{\partial \alpha_k}\left[\sum_i \sum_j \alpha_i x_i K_{ij}\alpha_j x_j\right] \\ &= \sum_i \sum_j (\delta_{ik} x_i K_{ij}\alpha_j x_j + \alpha_i x_i K_{ij}\delta_{jk} x_j) \\ &= \sum_j x_k K_{kj} x_j \alpha_j + \sum_j x_i K_{ik} x_k \alpha_i \end{aligned} \tag{9.194}$$

So

$$\frac{\partial N}{\partial \alpha_k} = 2\sum_{i=1}^{w} x_k K_{ki} x_i \alpha_i \tag{9.195}$$

Similarly, we obtain

$$\frac{\partial D}{\partial \alpha_k} = 2\sum_{i=1}^{w} x_k m_{ki} x_i \alpha_i \tag{9.196}$$

Let us set

$$K_{ij} = x_i K_{ij} x_j$$

and

$$M_{ij} = x_i m_{ij} x_j$$

Then

$$\frac{\partial N}{\partial \alpha_k} = 2\sum_{i=1}^{w} K_{ij}\alpha_i \tag{9.197}$$

$$\frac{\partial D}{\partial \alpha_k} = 2\sum_{i=1}^{w} M_{ij}\alpha_i$$

Substituting equations (9.196) and (9.197) into (9.193) leads to

$$\sum_{i}^{n} [K_{ki} - \Lambda M_{ki}]\alpha_i = 0 \tag{9.198}$$

where Λ is a diagonal matrix. Therefore, equation (9.198) results into a system of n linear algebraic equations that give n eigenvalues $\lambda_1, \ldots, \lambda_n$ and n eigenvectors α_i. If the stiffness and mass matrices K and M are given, the approximation to the eigenvectors is

$$\bar{W} = \sum_{i=1}^{n} \alpha_i x_i \quad \text{where} \quad n < m \quad (\text{the order of } M \text{ and } K)$$

Let

$$\bar{F} = [\bar{x}_1, \bar{x}_2, \ldots, \bar{x}_n] \tag{9.199}$$

Depending on the choice of F, we can write

$$\bar{K}\bar{X} = \bar{F} \tag{9.200}$$

from which we can solve for the eigenvectors $\bar{X}$. It follows that if we define

$$\bar{\mathbf{K}} = X^T K X$$

and

$$\bar{\mathbf{M}} = X^T M X$$

we can write

$$W = Fa \quad \text{where} \quad a^T = [\alpha_1, \alpha_2, \ldots, \alpha_n]$$

The eigenproblem solution is then defined as

$$(\bar{\mathbf{K}} - \Lambda\bar{\mathbf{M}})\bar{a} = 0 \tag{9.201}$$

where the approximation to the eigenvectors is given by

$$\begin{aligned} W^1 &= Fa^1 \\ W^2 &= Fa^2 \\ &\vdots \\ W^n &= Fa^n \end{aligned} \tag{9.202}$$

Example 9.5.2
Suppose that K and M are given by

$$K = \begin{bmatrix} 3 & -1 & -1 \\ -1 & 3 & -1 \\ -1 & -1 & 3 \end{bmatrix}, \qquad M = \begin{bmatrix} 1 & 0 & 0 \\ 0 & 1 & 0 \\ 0 & 0 & 1 \end{bmatrix}$$

Use the Rayleigh–Ritz method to solve for the eigenvalues and eigenvectors of the system.

Solution
Let

$$W = \alpha_1 \underset{x_1}{\begin{bmatrix} 2 \\ 2 \\ 2 \end{bmatrix}} + \alpha_2 \underset{x_2}{\begin{bmatrix} 1 \\ 0 \\ 2 \end{bmatrix}} \tag{9.203}$$

Then, we write the eigenvectors in a combined matrix as

$$X = \begin{bmatrix} 2 & 1 \\ 2 & 0 \\ 2 & 2 \end{bmatrix} \tag{9.204}$$

The corresponding stiffness and mass matrix for the eigenproblem are

$$\bar{\mathbf{K}} = X^T K X = \begin{bmatrix} 2 & 2 & 2 \\ 1 & 0 & 2 \end{bmatrix} \begin{bmatrix} 3 & -1 & -1 \\ -1 & 3 & -1 \\ -1 & -1 & 3 \end{bmatrix} \begin{bmatrix} 2 & 1 \\ 2 & 0 \\ 2 & 2 \end{bmatrix} = \begin{bmatrix} 12 & 6 \\ 6 & 11 \end{bmatrix} \tag{9.205}$$

$$\bar{\mathbf{M}} = X^T M X = \begin{bmatrix} 2 & 2 & 2 \\ 1 & 0 & 2 \end{bmatrix} \begin{bmatrix} 1 & 0 & 0 \\ 0 & 1 & 0 \\ 0 & 0 & 1 \end{bmatrix} \begin{bmatrix} 2 & 1 \\ 2 & 0 \\ 2 & 2 \end{bmatrix} = \begin{bmatrix} 12 & 6 \\ 6 & 5 \end{bmatrix} \tag{9.206}$$

which results in the following:

$$\bar{\mathbf{K}} - \Lambda \bar{M} = \begin{bmatrix} 12 - 12\lambda & 6 - 6\lambda \\ 6 - 6\lambda & 11 - 5\lambda \end{bmatrix} \tag{9.207}$$

the determinant of which is

$$\left|\bar{\mathbf{K}} - \Lambda \bar{\mathbf{M}}\right| = \lambda^2 - 5\lambda + 4 = 0 \tag{9.208}$$

which gives

$$\lambda_1 = 1 \quad \text{and} \quad \lambda_2 = 4$$

Now we can solve for the constants α_1 and α_2 from

$$(\bar{\mathbf{K}} - \Lambda\bar{\mathbf{M}})\bar{a} = 0 \quad \text{that is} \quad \begin{bmatrix} 12-12\lambda & 6-6\lambda \\ 6-6\lambda & 11-5\lambda \end{bmatrix} \begin{bmatrix} \alpha_1 \\ \alpha_2 \end{bmatrix} = 0 \tag{9.209}$$

for $\lambda_1 = 1$. The first eigenvector is

$$X_1 = \alpha_1 \begin{bmatrix} 2 \\ 2 \\ 2 \end{bmatrix}$$

where α_1 can be any value. Taking $\alpha_1 = \frac{1}{2}$,

$$X_1 = \begin{bmatrix} 1 \\ 1 \\ 1 \end{bmatrix} \tag{9.210}$$

For $\lambda_2 = 4$, we have $\alpha_1 = -1$ and $\alpha_2 = 2$. The second eigenvector is

$$X_2 = -1 \begin{bmatrix} 2 \\ 2 \\ 2 \end{bmatrix} + 2 \begin{bmatrix} 1 \\ 0 \\ 2 \end{bmatrix} = \begin{bmatrix} 0 \\ -2 \\ -2 \end{bmatrix} \tag{9.211}$$

An alternative solution can be obtained through use of the vector F as follows. As before, let the vector W be given by

$$W = \alpha_1 \begin{bmatrix} 2 \\ 2 \\ 2 \end{bmatrix} + \begin{bmatrix} 1 \\ 0 \\ 2 \end{bmatrix} \tag{9.212}$$

Then, we define

$$F = \begin{bmatrix} 2 & 1 \\ 2 & 0 \\ 2 & 2 \end{bmatrix} \tag{9.213}$$

with the columns labelled x_1 and $x2$ respectively.

From $KX = F$, we write

$$\begin{bmatrix} 3 & -1 & -1 \\ -1 & 3 & -1 \\ -1 & -1 & 3 \end{bmatrix} X = \begin{bmatrix} 2 & 1 \\ 2 & 0 \\ 2 & 2 \end{bmatrix} \tag{9.214}$$

Solving for X, we obtain

$$X = \begin{bmatrix} 2 & 1 \\ 2 & \frac{3}{4} \\ 2 & \frac{5}{4} \end{bmatrix} \tag{9.215}$$

Now, we formulate the eigenvalue problem by defining the corresponding stiffness and mass matrices:

$$\bar{\mathbf{K}} = X^T K X = \begin{bmatrix} 2 & 2 & 2 \\ 1 & \frac{3}{4} & \frac{5}{6} \end{bmatrix} \begin{bmatrix} 3 & -1 & -1 \\ -1 & 3 & -1 \\ -1 & -1 & 3 \end{bmatrix} \begin{bmatrix} 2 & 1 \\ 2 & \frac{3}{4} \\ 2 & \frac{5}{4} \end{bmatrix} = \begin{bmatrix} 12 & 6 \\ 6 & \frac{7}{6} \end{bmatrix} \tag{9.216}$$

$$\bar{\mathbf{M}} = X^T K X = \begin{bmatrix} 2 & 2 & 2 \\ 1 & \frac{3}{4} & \frac{5}{6} \end{bmatrix} \begin{bmatrix} 1 & 0 & 0 \\ 0 & 1 & 0 \\ 0 & 0 & 1 \end{bmatrix} \begin{bmatrix} 2 & 1 \\ 2 & \frac{3}{4} \\ 2 & \frac{5}{4} \end{bmatrix} = \begin{bmatrix} 12 & 6 \\ 6 & \frac{25}{8} \end{bmatrix} \tag{9.217}$$

Then the eigenvalue problem becomes

$$(\bar{\mathbf{K}} - \Lambda \bar{\mathbf{M}})\bar{a} = \begin{bmatrix} 12 - 12\lambda & 6 - 6\lambda \\ 6 - 6\lambda & \frac{7}{2} - \frac{25}{8}\lambda \end{bmatrix} \begin{bmatrix} \alpha_1 \\ \alpha_2 \end{bmatrix} = 0 \tag{9.218}$$

Let us evaluate the determinant of (9.218):

$$\left|\bar{\mathbf{K}} - \Lambda \bar{\mathbf{M}}\right| = 12(1-\lambda)\left[\frac{7}{2} - \frac{25}{8}\lambda\right] - 36(1-\lambda)^2 = 0 \tag{9.219}$$

which reduces to

$$\lambda^2 - 5\lambda + 4 = 0$$

The roots are found to be

$$\lambda_1 = 1 \quad \text{and} \quad \lambda_2 = 4 \tag{9.220}$$

For $\lambda_1 = 1$, let us assume that $\alpha_1 = \frac{1}{2}$. Then we get

$$x_1 = \begin{bmatrix} 1 \\ 1 \\ 1 \end{bmatrix} \tag{9.221}$$

For $\lambda_1 = 4$, solving for α_1 and α_2 from equation (9.218), we obtain

$$\alpha_1 = -1 \quad \text{and} \quad \alpha_2 = 2$$

and find the eigenvector to be

$$x_2 = \begin{bmatrix} 0 \\ -2 \\ 2 \end{bmatrix} \tag{9.222}$$

x_1 and x_2 are exactly the same as given by equations (9.210) and (9.211).

9.6 Assumed Modes Method

This method works in a similar fashion to the Rayleigh–Ritz method. The assumed modes assume a solution to the boundary value problem as

$$y(x,t) = \sum_{i=1}^{n} \phi_i q_i(t) \tag{9.223}$$

where ϕ_i are trial functions and $q_i(t)$ are the generalized coordinates. The difference between the two methods is clearly in the representation of $y(x,t)$. The assumed mode method makes use of generalized coordinates, whereas the Rayleigh–Ritz method uses coefficients that have to be determined in the process. The convenience of equation (9.223) as it applies to conversation leads to

$$[M]\{\ddot{q}\} + [K]\{q\} = \{0\} \tag{9.224}$$

where $[M]$ and $[K]$ are the corresponding mass and stiffness matrices, respectively. If we further assume that the solution to equation (9.224) is

$$q = A\sin(\omega t + \psi) \tag{9.225}$$

we get the eigenvalue problem

$$[K]\{A\} = \Lambda[M]\{A\} \tag{9.226}$$

This is the same as predicted by the Rayleigh–Ritz method. The Λ contains the n eigenvalues and the eigenvectors are given by the $\{A\}$ array. The assumed method can also be used to find the forced response to dynamic systems due to external forces or initial impulses.

Since the assumed mode method is popular in the dynamics of flexible multibody systems, it is important to note that its popularity stems from the fact that the admissible functions (ϕ or trial functions) are approximated by taking these from continuous systems such as beams, which in general provide a good approximation. Or simply determine an eigenvalue and make use of the eigenvectors.

9.7 Forced System Response and Selection of Mode Shapes

The dynamic response of a vibrating body due to a forcing function can be obtained by making use of the solution to the corresponding eigenvalue problem. For instance, a solution to the vibration problems discussed in Section 9.6 can be assumed to be written as

$$W(p,\ t) = \sum_{r=1}^{\infty} W_r(p)\eta_r(t) \tag{9.227}$$

where p stands for the spatial independent variables (x in the transverse, longitudinal, and torsional vibration of rods), $W_r(p)$ denotes the mode shapes, and η_r represents the

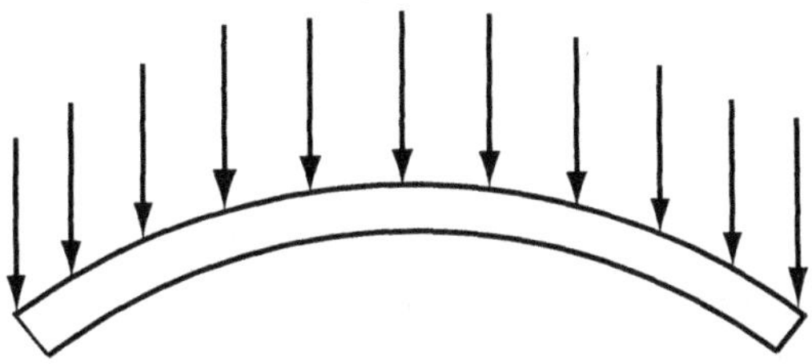

Fig. 9.14. Uniform bar in bending with both ends hinged and subjected to external loads

time-dependent generalized coordinate. Let us examine the solution of a uniform bar in bending with both ends hinged. Let the bar be subjected to external loads $f(x,t)$, as shown in Figure 9.14.

The partial differential equation of motion for the bar is given by

$$\frac{\partial^4 y}{\partial x^4} + \frac{m}{EI}\frac{\partial^2 y}{\partial t^2} = \frac{1}{EI} f(x,\ t) \qquad 0 < x < L \tag{9.228}$$

The boundary conditions are

End 1:

$$x = 0, \quad y(0,\ t) = 0, \quad \frac{\partial^2 y}{\partial x^2} = 0 \tag{9.229}$$

End 2:

$$x = L, \quad y(L,\ t) = 0, \quad \frac{\partial^2 y}{\partial x^2} = 0 \tag{9.230}$$

A solution to the aforementioned problem yields the natural frequencies to be

$$\omega_r = (r\pi)^2 \sqrt{\frac{EI}{mL^4}} \qquad (r = 1,\ 2,\ \ldots) \tag{9.231}$$

and natural mode shapes

$$Y_r(x) = \sqrt{\frac{2}{mL}} \sin\frac{r\pi x}{L} \qquad (r = 1,\ 2,\ \ldots) \tag{9.232}$$

Note that the modes are normalized and satisfy the orthogonality condition. The solution to equation (9.228) can be assumed to be of the form of equation (9.227), or

$$y(x,\ t) = \sum_{r=1}^{\infty} Y_r(x)\eta_r(t) \tag{9.233}$$

which results in

$$\eta_r''(t) + \omega_r^2 \eta_r(t) = F_r(t) \tag{9.234}$$

where

$$F_r(t) = \frac{1}{EI} \int_0^L f(x,\ t) Y_r(x)\, dx (r = 1,\ 2, \ldots, \infty) \tag{9.235}$$

Once more, assuming some initial conditions for $\eta(0)$ and $\dot{\eta}$ (0), we can find a solution $\eta(t)$ for equation (9.234). A general solution to equation (9.228) can then be found by substituting into equation (9.233) the solution to equations (9.234) and (9.232). Furthermore, we can, through the use of computer simulation, examine the dominant role that the mode shapes play in the solution of a particular problem.

9.7.1 Selection of Mode Shapes

In the context of multibody dynamics, we will be looking at the modeling of flexibility effects. This usually occurs, when a rigid body undergoing large amounts of motion can also vibrate at the same time. In the next two chapters, we examine in great detail the kinematics and dynamical formulation of such multibody problems. Consider the body vector q shown in Figure 9.15. If the body undergoes small deformation, it can be represented by

$$p^k = q^k + u^k \tag{9.236}$$

Therefore, the local deformation u^k of body k for the case when it is flexible can be represented by means of equation (9.236), where we write

$$u^k(p,\ t) = \sum_{r=1}^{M_k} \phi_r^k(p)\eta_r^k(t) \tag{9.237}$$

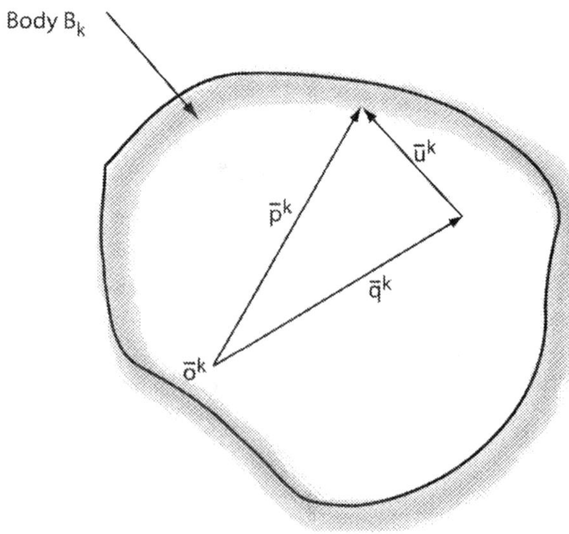

Fig. 9.15. Body vector $\bar{q}^b$ and its representation when subjected to small deformation $\bar{u}^k$

Once more $\phi_r^k(p)$ are the mode shapes for body k, η_r^k the generalized coordinates corresponding to each mode r, and M_k denotes the total number of modes employed.

At this point, it must be noted that since the forced response will be calculated from the solution of the general equations of motion of the whole system, the mode shapes are merely imposed conditions by the use of which we have a small number of generalized coordinates. Although it is not necessary (as will be discussed later) to have exact mode shapes, the accuracy of the solution is improved if the mode shapes selected accurately reflect the actual natural modes.

The following are some of the problems that restrict obtaining exact mode shapes:

1. The geometry of the body may be complex (not uniform), so that it may be impossible to solve the corresponding differential equation for the natural modes.
2. The boundary conditions may be variable during the motion of the multibody system, as is the case if the flexible body is an intermediate body.
3. The flexible body may be subjected to more than one type of loading (e.g., transverse and longitudinal), which is usually the case in a multibody system.

Therefore, one needs to rely on approximate methods such as the Rayleigh–Ritz method and the assumed modes method.

Consider the differential equation of the eigenvalue problem associated with the free vibrations of a body. Let the system be subject to natural and geometric boundary conditions. Then the following classes of functions are distinguished.

1. Eigenfunctions are functions that satisfy the differential equation and all the boundary conditions.
2. Comparison functions are arbitrary functions that do not necessarily satisfy the differential equation, but satisfy both the geometric and natural boundary conditions.
3. Admissible functions are arbitrary functions that do not necessarily satisfy the differential equation and the natural boundary conditions, but satisfy the geometric boundary conditions.

The Rayleigh–Ritz method uses comparison functions in equation (9.139) in place of $W_r(p)$, and the assumed modes method uses admissible functions. Meirovitch [9] suggests using comparison or admissible functions in increasing numbers to observe the convergence numerically. Although convergence to the solution is not shown mathematically, this is illustrated by examples. Clearly, the assumed modes method needs to be employed, if eigenfunctions and even comparison functions are not available. But in many cases it is possible to base the decision, as to the choice of the admissible functions on knowledge of the eigenfunctions of a system with a slightly different configuration. Then for a flexible body in a multibody system, subject to the problems described above, the eigenfunctions for the corresponding simplified beam (uniform fixed–free beam subject to one type of loading) are a good choice of admissible functions.

9.8 Numerical Methods for Eigenvalue Problems

The purpose of this section is to present some effective eigensolution techniques. There are a lot of numerical methods for the solution of eigenvalue problems: the Jacobi method, Householder–QR–inverse iteration method, subspare iteration method, vector iteration method, polynomial iteration method, and so on. Let us introduce the Jacobi method and subspace iteration method and show how they are applied to the numerical solution of eigenvalue problems. For other methods, readers can explore the references.

9.8.1 Jacobi Method

According to Bathe and Wilson [5], the Jacobi method is actually a type of transformation method. The transformation method comprises a group of eigensystem solution procedures that employ the basic properties of the eigenvectors in the Φ matrix, which makes

$$\phi^T K \phi = \Lambda \tag{9.238}$$

and

$$\phi^T M \phi = I \tag{9.239}$$

where Λ denotes a diagonal matrix known as the eigenvalue matrix and I is the identity matrix. Note that ϕ is of order $n \times n$, which diagonalizes K and M, and it is unique. The basic scheme is to reduce K and M into diagonal form using successive pre- and postmultiplication by matrices P_j^T and P_j, respectively ($j = 1, 2, \ldots$). That is, if we let $K_1 = K$ and $M_1 = M$, then

$$\begin{aligned} K_2 &= P_1^T K_1 P_1 \\ K_3 &= P_2^T K_2 P_2 \\ &\vdots \\ K_{j+1} &= P_j^T K_j P_j \end{aligned} \tag{9.240}$$

and

$$\begin{aligned} M_2 &= P_1^T M_1 P_1 \\ M_3 &= P_2^T M_2 P_2 \\ &\vdots \\ M_{j+1} &= P_j M_j P_j \end{aligned} \tag{9.241}$$

where the P_j are selected to bring K_j and M_j closer to diagonal form. When $j \to \infty$, $K_{j+1} \to \Lambda$, $M_{j+1} \to I$, and

$$\phi = P_1 P_2 \cdots P_j \tag{9.242}$$

The basic Jacobi method has been developed for the solution of standard eigenproblems where $M = I$. The generalized Jacobi method can also be used to solve general eigenproblems ($M \mp I$), where the procedures outlined above can be followed. Let $M = I$; then the differential equations for the eigenvalue can be written as

$$\ddot{y} + Ky = 0 \tag{9.243}$$

from which we can deduce the following:

$$K\phi = \lambda\phi \tag{9.244}$$

where λ is the eigenvalue. Then the jth iteration step defined in equations (9.240) reduces to

$$K_{j+1} = P_j^T K_j P_j \tag{9.245}$$

where P_j is an orthogonal matrix which gives

$$P_j^T P_j = I \tag{9.246}$$

The Jacobi method chooses P_j as a rotation matrix selected such that the off-diagonal element in K_j is zeroed. If the element $(k, \ \ell)$ is to be reduced to zero, the corresponding orthogonal matrix P_j is given by

$$P_j = \begin{bmatrix} 1 & & & & \\ & \ddots & & & \\ & & \cos\theta & -\sin\theta & & \\ & & \sin\theta & \cos\theta & & \\ & & & & \ddots & \\ & & & & & 1 \end{bmatrix} \begin{matrix} \\ \\ k \\ l \\ \\ \\ \end{matrix} \tag{9.247}$$

$$\qquad k \quad l \qquad\qquad n \times n$$

where θ is selected from the condition that the element $(k, \ \ell)$ in K_{j+1} be zero; hence

$$\theta = \frac{1}{2}\tan^{-1}\frac{2K_{k\ell}^{(j)}}{K_{kk}^{(j)} - K_{\ell\ell}^{(j)}} \quad \text{for} \quad K_{kk}^{(j)} \neq K_{\ell\ell}^{(j)} \tag{9.248}$$

and

$$\theta = \frac{\pi}{4} \quad \text{for} \quad K_{kk}^{(j)} = K_{\ell\ell}^{(j)} \tag{9.249}$$

Since K is symmetric, we only need to calculate the upper (or lower) triangular part of the matrix. It is important to note that although the transformation in equations (9.240) reduces an off-diagonal element in K_j to zero, that element can again become nonzero during the transformations that follows. A procedure that has been used very effectively is the *threshold Jacobi method,* in which the off-diagonal elements are tested sequentially, row by row (or column by column), and a rotation is applied only if the element is larger than the threshold for that sweep.

Example 9.8.1
Supposing that the mass matrix $M = I$, and the stiffness matrix is given by

$$K = \begin{bmatrix} 3 & -1 & -1 \\ -1 & 3 & -1 \\ -1 & -1 & 3 \end{bmatrix}$$

Use the Jacobi method to find the eigenvalues and eigenvectors for such a system.

Solution
For $i = 1$, $j = 2$, from equation (9.248) we get

$$\cos\theta = 0.7571 \sin\theta = 0.7571$$

and

$$P_1 = \begin{bmatrix} 0.7571 & -0.7571 & 0 \\ 0.7571 & 0.7571 & 0 \\ 0 & 0 & 1 \end{bmatrix} \tag{9.250}$$

Thus

$$P_1^T K_1 P_1 = \begin{bmatrix} 1.999 & 0 & -1.4142 \\ 0 & 3.9999 & 0 \\ -1.4142 & 0 & 3 \end{bmatrix} \tag{9.251}$$

For $i = 1$, $j = 3$ we get

$$\cos\theta = 0.8165 \sin\theta = 0.5773$$

$$P_2 = \begin{bmatrix} 0.8165 & 0 & -0.5773 \\ 0 & 1 & 0 \\ 0.57732 & 0 & 0.8165 \end{bmatrix} \tag{9.252}$$

$$P_2^T P_1^T K P_1 P_2 = \begin{bmatrix} 0.9999 & 0 & 0 \\ 0 & 3.9999 & 0 \\ 0 & 0 & 3.9999 \end{bmatrix} = \Lambda \tag{9.253}$$

Equation (9.253) yields a diagonalization of K after two transformations, the product of which is

$$P_1 P_2 = \begin{bmatrix} 0.5773 & -0.7571 & -0.4082 \\ 0.5773 & 0.7571 & -0.4082 \\ 0.5773 & 0 & 0.8165 \end{bmatrix} = \Phi \tag{9.254}$$

The eigenpairs are:

$$\text{For} \quad \lambda_1 = 1 : x_1 = \begin{bmatrix} 1 \\ 1 \\ 1 \end{bmatrix}$$

$$\text{For} \quad \lambda_2 = 4 : x_2 = \begin{bmatrix} -1 \\ 1 \\ 0 \end{bmatrix}$$

$$\text{For} \quad \lambda_3 = 4 : x_3 = \begin{bmatrix} -1 \\ -1 \\ 2 \end{bmatrix}$$

As we mentioned earlier, we usually need many steps to diagonalize matrix K. As to the case when $M \pm I$, we can use the generalized Jacobi method following procedures described in reference [5].

9.8.2 Subspace Iteration Method

The subspace iteration method is very effective in calculation of the lowest eigenvalues and corresponding eigenvectors of systems with large bandwidth and suitable for large-scale computer computation. The basic objective in the subspace iteration method is to solve for the lowest p eigenvalues and corresponding eigenvectors satisfying

$$K\Phi = M\Phi\Lambda \tag{9.255}$$

where $\Lambda = \text{diag}\,(\lambda_1)$, $\Phi = [\phi_1 \cdots \phi_p]_{n\times p}$, and K and M are of order $n \times n$ matrices. Meanwhile, the eigenvectors also satisfy the orthogonality conditions

$$\Phi^T k\Phi = \Lambda \quad \text{and} \quad \Phi^T M\Phi = I \tag{9.256}$$

In the solution, the iteration with $p(p < n)$ linearly independent vectors can therefore be regarded as an iteration with a subspace. The starting iteration vectors are very important. There is a specific algorithm used to establish the starting iteration vectors. Because iteration is performed with a subspace, convergence of the subspace is all that is required, not that of individual iteration vectors to eigenvectors. In other words, if the iteration vectors are linear combinations of the eigenvectors required, the solution algorithm converges in one step.

Suppose that E_T denotes a target subspace spanned by p eigenvectors that form a base for this subspace, and E_1 denotes a p-dimensional starting subspace. Then, the iteration is performed with the p linearly independent vectors in order with the subspace E_1 until reaching the eigenvector that spans E_T.

The general procedure for subspace iteration can be summarized as follows. Let X_1 store the p starting iteration vectors that span the starting subspace E_1. Simultaneous inverse iteration on the p vectors is then given by

$$KX_{k+1} = MX_k (k = 1,\ 2,\ \ldots) \tag{9.257}$$

where the X_{k+1} span a p-dimensional subspace E_{k+1} and the sequence of subspaces generated converges to E_T, except that E_1 is orthogonal to iterate to E_T. For $k = 1, 2, \ldots$, iterate from E_k to E_{k+1}:

$$K\bar{X}_{k+1} = MX_k \tag{9.258}$$

The projections of the operators K and M on to E_{k+1} are

$$K_{k+1} = \bar{X}_{k+1}^T K\bar{X}_{k+1} \tag{9.259}$$

$$M_{k+1} = \bar{X}_{k+1}^T K\bar{X}_{k+1} \tag{9.260}$$

Making use of equations (9.259) and (9.260), we solve for the eigensystem of the projected operators:

$$K_{k+1}Q_{k+1} = M_{k+1}Q_{k+1}\lambda_{k+1} \tag{9.261}$$

An improved approximation to the eigenvectors can then be found:

$$X_{k+1} = \bar{X}_{k+1}Q_{k+1} \tag{9.262}$$

Then, provided that vectors in $\bar{X}_1$ are not orthogonal to one of the required eigenvectors, we get: as $k \to \infty$,

$$\Lambda_{k+1} \to \Lambda \quad \text{and} \quad X_{k+1} \to \Phi$$

Example 9.8.2
Use the subspace iteration method to solve for the eigenvalue problem given by Example 9.8.1.

Solution
The following stiffness and mass are given:

$$K = \begin{bmatrix} 3 & -1 & -1 \\ -1 & 3 & -1 \\ -1 & -1 & 3 \end{bmatrix} M = I$$

Let us choose

$$X_1 = \begin{bmatrix} 1 & 0 \\ 1 & 1 \\ 0 & 1 \end{bmatrix} \tag{9.263}$$

First Step
From equation (9.258) we get

$$K\bar{X}_2 = MX_1$$

or

$$\begin{bmatrix} 3 & -1 & -1 \\ -1 & 3 & -1 \\ -1 & -1 & 3 \end{bmatrix} \bar{X}_2 = M \begin{bmatrix} 1 & 0 \\ 1 & 1 \\ 0 & 1 \end{bmatrix} \tag{9.264}$$

Solving (9.264) for $\bar{X}_2$, we obtain

$$\bar{X}_2 = \begin{bmatrix} 0.75 & 0.5 \\ 0.75 & 0.75 \\ 0.5 & 0.75 \end{bmatrix} \tag{9.265}$$

From equation (9.259), we find the updated stiffness and mass as

$$\begin{aligned} \bar{\mathbf{K}}_2 = \bar{X}_2^T K \bar{X}_2 &= \begin{bmatrix} \frac{3}{4} & \frac{3}{4} & \frac{1}{2} \\ \frac{1}{2} & \frac{3}{4} & \frac{3}{4} \end{bmatrix} \begin{bmatrix} 3 & -1 & -1 \\ -1 & 3 & -1 \\ -1 & -1 & 3 \end{bmatrix} \begin{bmatrix} \frac{3}{4} & \frac{1}{2} \\ \frac{3}{4} & \frac{3}{4} \\ \frac{1}{2} & \frac{3}{4} \end{bmatrix} \\ &= \frac{1}{4} \begin{bmatrix} 6 & 5 \\ 5 & 6 \end{bmatrix} \end{aligned} \tag{9.266}$$

and

$$\begin{aligned} \bar{\mathbf{M}}_2 &= \bar{X}_2^T M \bar{X}_2 \\ &= \begin{bmatrix} \frac{3}{4} & \frac{3}{4} & \frac{1}{2} \\ \frac{1}{2} & \frac{3}{4} & \frac{3}{4} \end{bmatrix} \begin{bmatrix} 1 & 0 & 0 \\ 0 & 1 & 0 \\ 0 & 0 & 1 \end{bmatrix} \begin{bmatrix} \frac{3}{4} & \frac{1}{2} \\ \frac{3}{4} & \frac{3}{4} \\ \frac{1}{2} & \frac{3}{4} \end{bmatrix} = \frac{1}{16} \begin{bmatrix} 22 & 21 \\ 21 & 22 \end{bmatrix} \end{aligned} \tag{9.267}$$

Using equation (9.261), we formulate the eigenvalue problem

$$\bar{\mathbf{K}}_2 Q_2 = \bar{\mathbf{M}}_2 Q_2 \Lambda_2 \tag{9.268}$$

or

$$(\bar{\mathbf{K}}_2 - \Lambda_2 \bar{\mathbf{M}}_2) Q_2 = 0 \tag{9.269}$$

that is,

$$\frac{1}{4} \begin{bmatrix} 6 & 5 \\ 5 & 6 \end{bmatrix} - \begin{bmatrix} \lambda & 0 \\ 0 & \lambda \end{bmatrix} \frac{1}{16} \begin{bmatrix} 22 & 21 \\ 21 & 22 \end{bmatrix} = 0 \tag{9.270}$$

the determinant of which is

$$\begin{vmatrix} \frac{6}{4} - \lambda \frac{22}{16} & \frac{5}{4} - \lambda \frac{21}{16} \\ \frac{5}{4} - \lambda \frac{21}{16} & \frac{6}{4} - \lambda \frac{22}{16} \end{vmatrix} = 0 \tag{9.271}$$

Solving for the characteristic equation, we find the eigenvalues to be 1.023 and 4.00, which can be written as

$$\Lambda_2 = \begin{bmatrix} 1.023 & 0 \\ 0 & 4.00 \end{bmatrix} \tag{9.272}$$

Their corresponding eigenvectors are found to be

$$Q_2 = \begin{bmatrix} 1 & 1 \\ 1 & -1 \end{bmatrix} \tag{9.273}$$

Then we compute the newly updated eigenvectors from

$$X_2 = \bar{X}_2 Q_2 \begin{bmatrix} \frac{3}{4} & \frac{1}{2} \\ \frac{3}{4} & \frac{3}{4} \\ \frac{1}{2} & \frac{3}{4} \end{bmatrix} \begin{bmatrix} 1 & 1 \\ 1 & -1 \end{bmatrix} = \begin{bmatrix} 1.25 & 0.25 \\ 1.5 & 0 \\ 1.25 & -0.25 \end{bmatrix} \tag{9.274}$$

Second Step

Now that we have an updated X_2, we proceed in the same manner into the second iteration, where $K\bar{X}_3 E = MX_2$ and $(M = I)$ results in

$$\begin{bmatrix} 3 & -1 & -1 \\ -1 & 3 & -1 \\ -1 & -1 & 3 \end{bmatrix} \bar{X}_3 = \begin{bmatrix} 1.25 & 0.25 \\ 1.5 & 0 \\ 1.25 & -0.25 \end{bmatrix} \tag{9.275}$$

with

$$\bar{X}_3 = \begin{bmatrix} 1.3125 & 0.0625 \\ 0.375 & 0 \\ 1.3125 & -0.0625 \end{bmatrix} \tag{9.276}$$

Then

$$\bar{\mathbf{K}}_3 = \bar{X}_3^T K X_3 = \begin{bmatrix} 1.3125 & 1.375 & 1.3125 \\ 0.0625 & 0 & -0.0625 \end{bmatrix} \begin{bmatrix} 3 & -1 & -1 \\ -1 & 3 & -1 \\ -1 & -1 & 3 \end{bmatrix} \begin{bmatrix} 1.3125 & 0.0625 \\ 1.375 & 0 \\ 1.3125 & -0.0625 \end{bmatrix}$$
$$= \begin{bmatrix} 5.34375 & 0 \\ 0 & 0.03125 \end{bmatrix} \tag{9.277}$$

and

$$\bar{\mathbf{M}}_3 = \bar{X}_3^T M \bar{X}_3 = \begin{bmatrix} 5.3359375 & 0 \\ 0 & 0.0078125 \end{bmatrix} \tag{9.278}$$

Then, we obtain

$$\bar{\mathbf{K}}_3 Q_3 = \bar{M}_3 Q_3 \Lambda_3$$

$$\det(\bar{\mathbf{K}}_3 - \Lambda\bar{\mathbf{M}}_3) = \begin{vmatrix} 5.34375 - \lambda 5.3359375 & 0 \\ 0 & 0.03125 - \lambda 0.0078125 \end{vmatrix} = 0 \tag{9.279}$$

which gives

$$\Lambda_3 = \begin{bmatrix} 1.0014 & 0 \\ 0 & 4.0 \end{bmatrix} \tag{9.280}$$

and

$$Q_3 = \begin{bmatrix} 1 & 0 \\ 0 & 1 \end{bmatrix} = I \tag{9.281}$$

The newly updated X_3 are then found to be

$$X_3 = \bar{X}_3 Q_3 = \begin{bmatrix} 1.3125 & 0.0625 \\ 1.375 & 0 \\ 1.3125 & -0.0625 \end{bmatrix} \tag{9.282}$$

or simply

$$X_3 = \begin{bmatrix} 1 & 1 \\ 1.04 & 0 \\ 1 & -1 \end{bmatrix} \tag{9.283}$$

At this point we can easily see that the eigenvalues are converging to $\lambda_1 = 1.0$ and $\lambda_2 = 4.0$, which agree exactly with the answers to Example 9.8.1. The eigenvectors are given by X_3, corresponding to the last updated vector given by equation (9.282).

9.9 Problems

9.1. Give the linearized equations of motion for Problem 4.1. Use the method described in Example 9.2.3. Assume that the angular velocity about the pole indicated describing the swing movement of the bar is given by

$$\Omega(t) = A\sin(\omega t + \psi)$$

where A is a constant defining the initial value of Ω, and ψ is the initial phase angle. (Select A, ω and ψ such that the bar swings between zero and π.)

9.2. Demonstrate the utility of the method of linearizing the equations of motion at the velocity level by comparing the results obtained in Example 9.2.3 with those found through the use of Lagrange equations. What correction is needed to make the latter valid?

9.3. Consider the simple clamped beam shown in Figure 9.16. Use the Rayleigh–Ritz method in approximating the first two eigenvalues and eigenvectors due to the longitudinal vibration. Assume the function ϕ to be

$$\phi(t) = c_1 \sin \frac{\pi t}{\ell} + c_2 \sin \frac{3\pi t}{\ell}$$

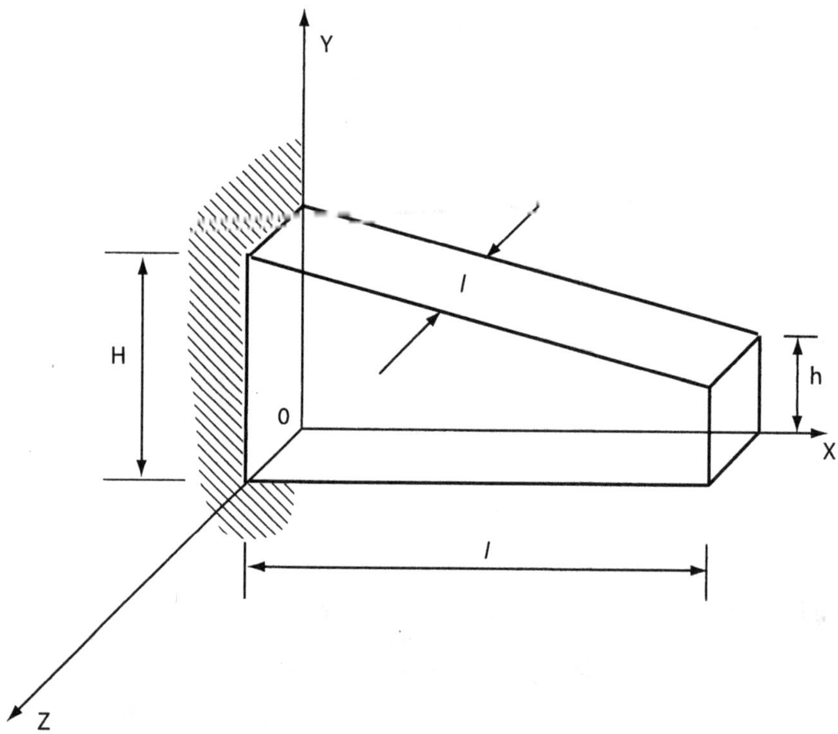

Fig. 9.16. Simple clamped nonuniform beam

9.4. The lumped mass model of the beam shown in Figure 9.17 is used to approximate the eigenvalues and eigenvectors of the system. Assume that the structural stiffness constant is $EA(x)/\ell$ and that the beam is divided into four lumped masses, as shown. Find the eigenvalues and eigenvectors of the lumped mass system and compare your findings to those obtained from the closed-form solution.

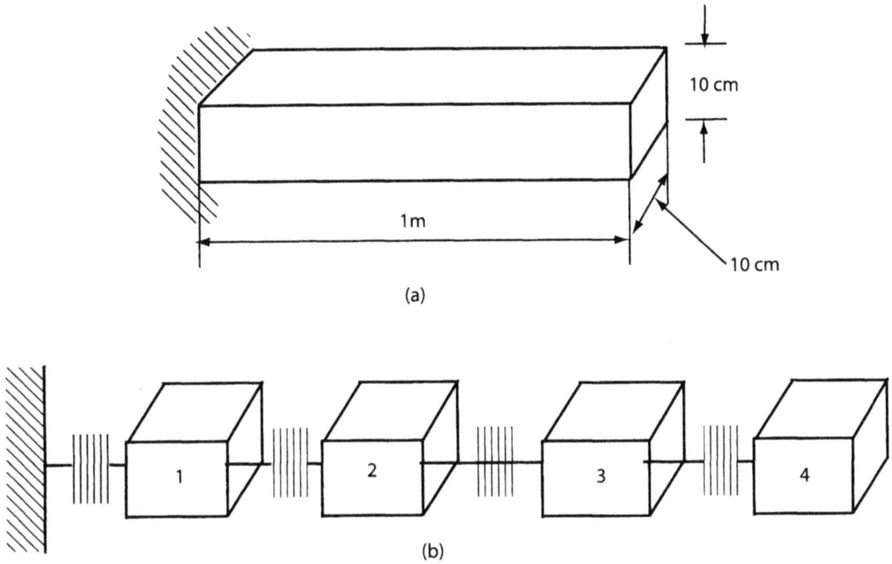

Fig. 9.17. (a) Clamped uniform beam. (b) Lumped mass model equivalent to the one shown in part (a)

Part	Mass (kg)	Stiffness coefficient (N/m)	Damping coefficient (Ns/m)
1	5.3	120,000	1500
2	19.6	105,000	1600
3	2.69	105,000	1800
4	11.76	50,000	1100
5	9.86	50,000	1000
6	4.42	—	—
7	2.22	50,000	8000
8	2.15	50,000	1000
9	2.22	50,000	1000
10	2.15	50,000	1000

Table 9.1. Geometrical data for Problem 9.5

9.5. The human body model shown in Figure 9.18 is used for the study of human body vibration. Give the corresponding eigenvalues and eigenvectors for such a system, making use of the generalized Jacobi method.

9.6. Perform the eigenvalue problem method for a system whose mass and stiffness are

$$M = I \quad \text{and} \quad K = \begin{bmatrix} 1 & 2 & 2 \\ -1 & 3 & 1 \\ 2 & 2 & 1 \end{bmatrix}$$

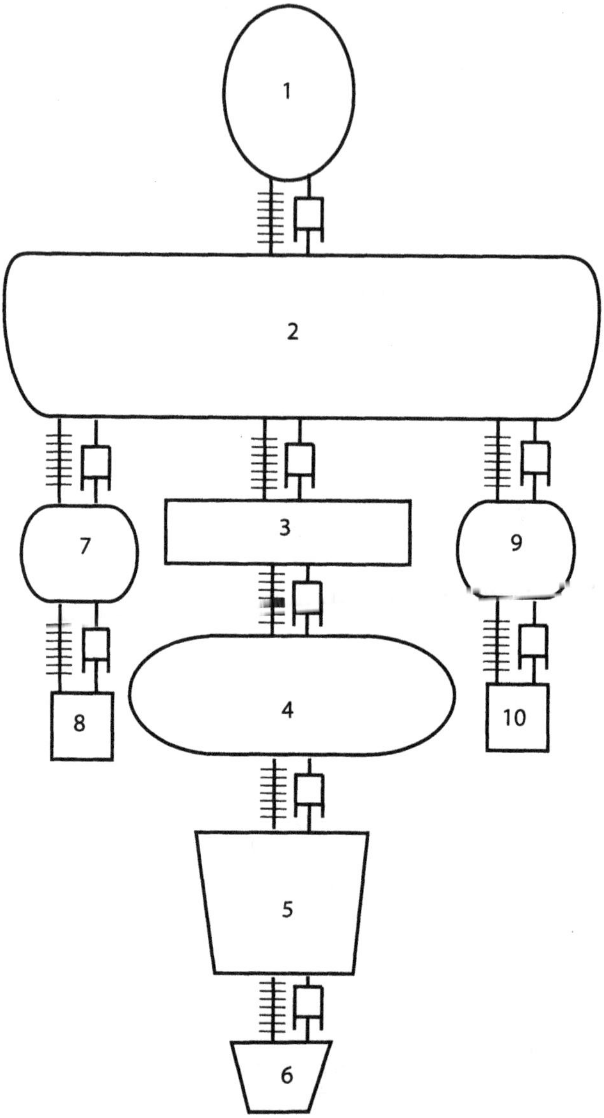

Fig. 9.18. Lumped mass human model

References

1. Kane, T. R. and Levinson, D. A., *Dynamics: Theory and Application,* McGraw-Hill, New York, 1985.
2. Kane, T. R., Likins, P. W. and Levinson, D. A., *Spacecraft Dynamics,* McGraw-Hill, New York, 1983.

3. Amirouche, F. M. L. and Huston, R. L., Dynamics of Large Constrained Flexible Structures, *Dyn. Syst. Meas. Control,* Vol. **110**: No. 1, 1988, pp. 78–83.
4. Timoshenko, S., Young, D. H. and Weaver, W., *Vibration Problems in Engineering,* Wiley, New York, 1974.
5. Bathe, K. J. and Wilson, E. L., *Numerical Methods in Finite-Element Analysis,* Prentice Hall, Englewood Cliffs, NJ, 1976.
6. Jacobi, C. G. J., Über ein leichtes Verfahren die in der Theories der säculärstörungen vorkommenden Gleichungen numerisch aufzulösen, *Crelle's J.,* Vol. **30**, 1846, pp. 51–94.
7. Crandall, S. H., *Engineering Analysis,* McGraw-Hill, New York, 1956.
8. Amirouche, F. M. L., Dynamic Analysis of Treelike Structure Undergoing Large Motions, *J. Eng. Anal. Trans. Comput. Mech.,* Vol. **3**: No. 2, 1986, pp. 117–117
9. Meriovitch, L., *Analytical Methods in Vibrations,* Macmillan, New York, 1969.
10. Meriovitch, L., *Introduction to Dynamics and Control,* Wiley, New York, 1985.
11. Thomson, W. T., *Theory of Vibration with Applications,* 3rd Ed., Prentice Hall, Englewood Cliffs, NJ, 1989.
12. Tse, F. S., Morse, I. E. and Hinkle, R. T., *Mechanical Vibrations: Theory and Applications,* 2nd Ed., Allyn & Bacon, 1979.
13. Amirouche, F. M. L. and Huston, R. L., Collaborative Methods in Modal Analysis, *J. Guid. Dyn. Control,* Vol. **8**: No. 6, 1985, pp. 782–785.
14. Young, D. and Felgar, R. P., Tables of Characteristic Functions Representing Normal Modes of Vibration of a Beam, *University of Texas Publication. 4913,* July, 1949.

10

Dynamics of Multibody Systems with Terminal Flexible Links

10.1 Introduction

In this chapter, we focus our attention on the formulation of the equations of motion of multibody systems with flexible terminal links. It is our interest to see how flexibility (elastic deformation) affects the dynamics, hence providing some insight into how to control it. Vibration of terminal links is more pronounced when high speed of light structure is required. For instance, the accurate positioning of a robotic manipulator requires that we assume a certain vibrating behavior in the control algorithm to be able to predict more precisely the position of its endpoint. In today's technology, it is not possible to foresee any design of a control block for industrial / space manipulators that would not include the effects of flexibility.

In the case of robotics, flexibility effects can be caused by the load carried by the end effectors, the weight of the end effector (grippers) itself, a sudden change in direction of motion, external contact forces, and lighter structures. To understand the actual motion of such systems, we need to formulate equations of motion that incorporate the effects of flexibility. To this end, we introduce two methods to solve such problems: one is based on the method of motion overlapping and the other makes use of the finite-element method to discretize the flexible link.

In essence, in this chapter, we expand the law of motion to derive precise equations of motion of bodies that are flexible and undergoing large rotations.

10.2 Method of Motion Overlapping

To best illustrate the effects of flexibility on the rigid-body motion, consider the general multibody systems depicted in Figure 10.1, which consists of $n - 1$ rigid bodies and a terminal flexible body modeled as a beam. If we assume that the multibody system undergoes small angular velocity and angular acceleration, the following treatment of the problem is proposed.

Let the terminal flexible body be modeled by two parts, as shown in Figure 10.2. The first part is rigid and the second is elastic. Let P be an arbitrary point on the

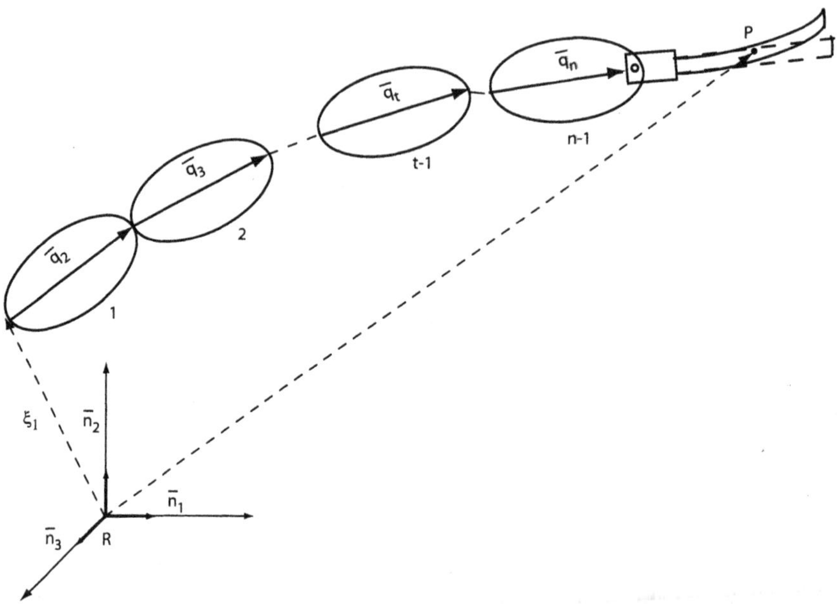

Fig. 10.1. Treelike multibody system with a terminal flexible link

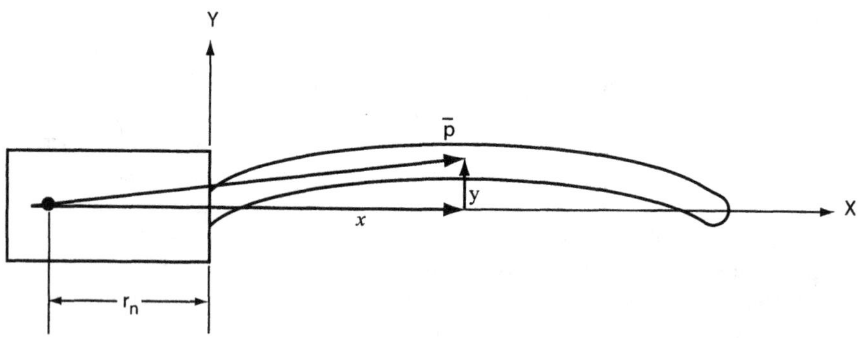

Fig. 10.2. Modeling of the flexible link using a uniform beam attached to a rigid block

flexible link. If we consider only the transverse vibration of the beam (link) in the $x - y$ plane, and make use of the kinematical definitions provided in Chapter 3, we can write the position vector of point P as

$$\bar{p} = \bar{\xi}_1 + \sum_{i=2}^{n} \bar{q}_i + \bar{r}_n \tag{10.1}$$

where the position to their mass center includes the beam deformation and can be expressed as

$$\bar{r}_n = [r_n + x \;\; y \;\; 0] \left\{\bar{n}^n\right\} \quad \text{and} \quad \bar{q}_1 = 0$$

where $\bar{r}_n$ denotes the components of vector $\bar{r}_n$ in $\bar{n}$. Expressing the position $\bar{p}$ with respect to an inertial fixed frame R yields the general expression

$$\bar{p} = \left[\{\xi_1\}^T + \sum_{i=2}^{n} \{q_i\}^T [S^{\upsilon o}] + \{r_n\}^T [S^{no}] \right] \{\bar{n}\} \tag{10.2}$$

where $\upsilon = i - 1$, q_i represents the components of the $\bar{q}$ vector in the body axis, and $[S]$ denotes the shifter matrix.

Let us examine how to derive x and y deformation states in equation (10.1) from beam analysis.

10.2.1 Case 1: Transverse Vibration of the Flexible Body

Let us assume that the terminal elastic link only has a transverse vibrational mode. From Chapter 9, we know that for the cantilever beam shown in Figure 10.2 whose boundary conditions are

$$\begin{aligned} Y(0) &= 0, \qquad Y'(0) = 0 \\ Y''(L) &= 0, \qquad Y'''(L) = 0 \end{aligned} \tag{10.3}$$

the displacement is given by

$$\begin{aligned} y(x,t) &= Y(x)T(t) \\ &= \sum_{r=0}^{\infty} \left[\sin\beta_r x - \sinh\beta_r x + \frac{\cos\beta_r L + \cosh\beta_r L}{\sin\beta_r L - \sinh\beta_r L} (\cos\beta_r x - \cosh\beta_r x) \right] \\ &\quad (a_r \sin\omega_r t + b_r \cos\omega_r t) \end{aligned} \tag{10.4}$$

where

$$\beta_r^2 = \omega_r \sqrt{\frac{m}{EI}} \tag{10.5}$$

and

$$\cos\beta L \cosh\beta L = -1 \tag{10.6}$$

where a_r and b_r are constants that are dependent on the initial conditions, E is the modulus of elasticity, and m is the mass of the link with length ℓ, as shown in Figure 10.2.

Factorizing equation (10.4) further, we get

$$\begin{aligned} y(x,t) = \sum_{r=1}^{\infty} a_r \sin\omega_r t \Bigg[&\sin\beta_r x - \sinh\beta_r x \\ &+ \frac{\cos\beta_r L + \cosh\beta_r L}{\sin\beta_r L - \sinh\beta_r} (\cos\beta_r x - \cosh\beta_r x) \Bigg] \end{aligned} \tag{10.7}$$

Differentiation of $y(x, t)$ with respect to t, we obtain

$$y_t{}'(x,t) = \sum_{r=1}^{\infty} a_r \omega_r \cos \omega_r t \Bigg[\sin \beta_r x - \sinh \beta_r x + \frac{\cos \beta_r L + \cosh \beta_r L}{\sin \beta_r L - \sinh \beta_r} (\cos \beta_r x - \cosh \beta_r x) \Bigg] \tag{10.8}$$

The velocity of an arbitrary point p on the beam is given by differentiation in time of equation (10.2), or

$$\bar{\upsilon}_p = \frac{d}{dt}\bar{p} = \left[\{\dot{\xi}_1\}^T + \sum_{i=2}^{n} \{q_i\}^T [\Omega^{\upsilon o}][S^{\upsilon o}] + \{r_n\}^T [\Omega^{no}][S^{no}] + \{\dot{r}_n\}^T [S^{no}] \right] \{\bar{n}_m\} \tag{10.9}$$

where

$$\{\dot{r}_n\} = \frac{d}{dt}\{r_n\} = \frac{d}{dt}[r_n + x \ \ y(x,t) \ \ 0]^T = [0 \ \ y_t'(x,t) \ \ 0]^T \tag{10.10}$$

Note how the term $r_n + x$ drops out because there is no longitudinal deformation. The acceleration of point p is found by differentiation of equation (10.9), yielding

$$\begin{aligned} \bar{a}_p = \frac{d}{dt}\bar{\upsilon}_p = \{\ddot{\xi}_1\}^T \{\bar{n}\} + \sum_{i=2}^{n} \Big[& \{q_i\}^T [\dot{\Omega}^{\upsilon o}][S^{\upsilon o}] + \{q_i\}^T [\Omega^{\upsilon o}][\Omega^{\upsilon o}][S^{\upsilon o}] \\ & + 2\{\ddot{r}_n\}^T [\Omega^{no}][S^{no}] + \{r_n\}^T [\dot{\Omega}^{no}][S^{no}] \\ & + \{r_n\}^T [\Omega^{no}][\Omega^{no}][S^{no}] + \{\ddot{r}_n\}^T [S^{no}]\{\bar{n}\} \Big] \end{aligned} \tag{10.11}$$

Again, in both equations (10.10) and (10.11) we make use of terms defined in Chapter 3, such as the skew matrix $[\Omega]$ and its rate of change $[\dot{\Omega}]$, plus the shifter matrices $[S]$. In addition, if $r_n = 0$ (see Figure 10.2), the boundary conditions become those of a pinned-free type of beam:

$$Y(0) = 0 \tag{10.12}$$

$$Y''(0) = 0 \tag{10.13}$$

$$Y''(L) = 0 \tag{10.14}$$

$$Y'''(L) = 0 \tag{10.15}$$

and the transverse displacement is

$$\begin{aligned} y(x,t) &= Y(x)T(t) \\ &= \sum_{r=o}^{\infty} \left[\sin \beta_r x + \frac{\sin \beta_r L}{\sinh \beta_r L} \sinh \beta_r x \right] (a_r \sin \omega_r t + b_r \cos \omega_r t) \end{aligned} \tag{10.16}$$

where

$$\beta_r^2 = \omega_r \sqrt{\frac{m}{EI}}$$

for

$$\tan \beta L = \tanh \beta L \tag{10.17}$$

(a_r and b_r are constants dependent on the initial conditions).

The following example highlights the dynamics of a vibrating beam undergoing a transverse motion.

Example 10.2.1
Consider the planar two-bar system shown in Figure 10.3. Find the explicit form of the velocity and acceleration of an arbitrary point p on the flexible link with the boundary conditions given by equations (10.3).

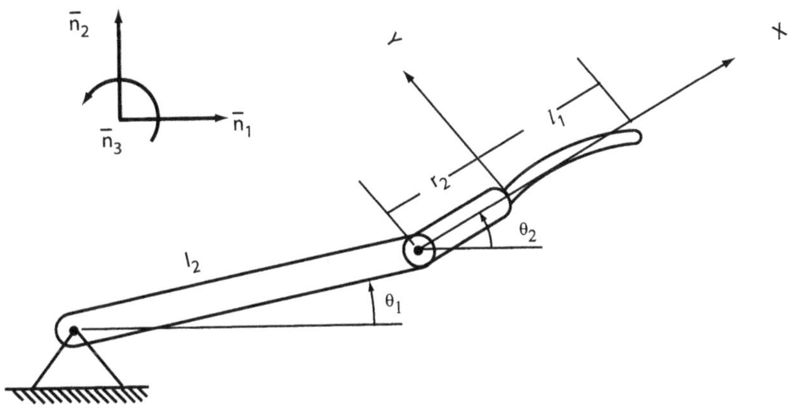

Fig. 10.3. Planar two-link manipulator with a flexible arm

Solution
Let us write the transformation matrices between bodies 1 and 2 and the inertial frame R : The transformation between body 1 and O is

$$[S^{10}] = \begin{bmatrix} \cos\theta_1 & \sin\theta_1 & 0 \\ -\sin\theta_1 & \cos\theta_1 & 0 \\ 0 & 0 & 1 \end{bmatrix} \tag{10.18}$$

and the transformation between body 2 and O is

$$[S^{20}] = \begin{bmatrix} \cos(\theta_1+\theta_2) & \sin(\theta_1+\theta_2) & 0 \\ -\sin(\theta_1+\theta_2) & \cos(\theta_1+\theta_2) & 0 \\ 0 & 0 & 1 \end{bmatrix} \tag{10.19}$$

Let $\bar{p}$ represent the position of an arbitrary point on link 2; then

$$\bar{p} = \ell_1 \bar{n}_1^1 + (r_2 + x)\bar{n}_1^2 + y\bar{n}_2^2 \tag{10.20}$$

which can be written in terms of $\bar{n}_1$, $\bar{n}_2$ and $\bar{n}_3$ as

$$
\begin{aligned}
\bar{p} = {} & [\ell_1 \cos\theta_1 + (r_2 + x)\cos(\theta_1 + \theta_2) - y\sin(\theta_1 + \theta_2)]\,\bar{n}_1 \\
& + [\ell_1 \sin\theta_1 + (r_2 + x)\sin(\theta_1 + \theta_2) + y\cos(\theta_1 + \theta_2)]\,\bar{n}_2
\end{aligned} \tag{10.21}
$$

The velocity of point p is obtained by simple differentiation as

$$
\begin{aligned}
\bar{v}_p = \frac{d\bar{p}}{dt} = {} & [-\ell_1\dot{\theta}_1 \sin\theta_1 - (r_2 + x)(\dot{\theta}_1 + \dot{\theta}_2)\sin(\theta_1 + \theta_2) \\
& - y(\dot{\theta}_1 + \dot{\theta}_2)\cos(\theta_1 + \theta_2) - y_t' \sin(\theta_1 + \theta_2)]\bar{n}_1 \\
& + [\ell_1\dot{\theta}_1 \cos\theta_1 + (r_2 + x)(\dot{\theta}_1 + \dot{\theta}_2)\cos(\theta_1 + \theta_2) \\
& - y(\dot{\theta}_1 + \dot{\theta}_2)\sin(\theta_1 + \theta_2) + y_t' \cos(\theta_1 + \theta_2)]\bar{n}_2
\end{aligned} \tag{10.22}
$$

where y and y_t' are given by equations (10.7) and (10.8) and the acceleration of point p is found to be

$$
\begin{aligned}
\bar{a}_p = \frac{d\bar{v}_p}{dt} = {} & [-\ell_1\ddot{\theta}_1 \sin\theta_1 - \ell_1\dot{\theta}_1^2 \cos\theta_1 - (r_2 + x)(\ddot{\theta}_1 + \ddot{\theta}_2)\sin(\theta_1 + \theta_2) \\
& - (r_2 + x)(\dot{\theta}_2 + \dot{\theta}_2)^2 \cos(\theta_1 + \theta_2) - 2y_t'(\dot{\theta}_1 + \dot{\theta}_2)\cos(\theta_1 + \theta_2) \\
& - y(\ddot{\theta}_1 + \ddot{\theta}_2)\cos(\theta_1 + \theta_2) + y(\dot{\theta}_1 + \dot{\theta}_2)^2 \sin(\theta_1 + \theta_2) \\
& - y_t'' \sin(\theta_1 + \theta_2)]\bar{n}_1 + [\ell_1\ddot{\theta}_1 \cos\theta_1 - \ell_1\dot{\theta}_1^2 \sin\theta_1 \\
& + (r_2 + x)(\ddot{\theta}_1 + \ddot{\theta}_2)\cos(\theta_1 + \theta_2) - (r_2 + x)(\dot{\theta}_1 + \dot{\theta}_2)^2 \sin(\theta_1 + \theta_2) \\
& - 2y_t'(\dot{\theta}_1 + \dot{\theta}_2)\sin(\theta_1 + \theta_2) - y(\ddot{\theta}_1 + \ddot{\theta}_2)\sin(\theta_1 + \theta_2) \\
& - y(\dot{\theta}_1 + \dot{\theta}_2)^2 \cos(\theta_1 + \theta_2) + y_t'' \cos(\theta_1 + \theta_2)]\bar{n}_2
\end{aligned} \tag{10.23}
$$

(**Note:** In the above equations, y, y_t', y_t'' denote the transverse vibration of the beam as given by equation (10.7)).

10.2.2 Case 2: Longitudinal Vibration of the Flexible Beam

Consider the system shown in Figure 10.4. If we restrict the elastic deformation of the terminal link to the longitudinal vibration, then in a fashion similar to that used earlier, we can obtain the kinematics of an arbitrary point. As depicted by Figure 10.5, the position of the arbitrary point p on the flexible body beams is $x + y(x, t)$ after deformation, where $y(x, t)$ is the displacement at x due to the longitudinal vibration. Then $\bar{r}_n$ in equation (10.1) changes to

$$
\bar{r}_n = [x + y \quad 0 \quad 0]\{\bar{n}^n\} \tag{10.24}
$$

where $\{\bar{n}^n\}$ are a set of unit vectors fixed in body N (the elastic body). For the following boundary conditions (fixed-free)

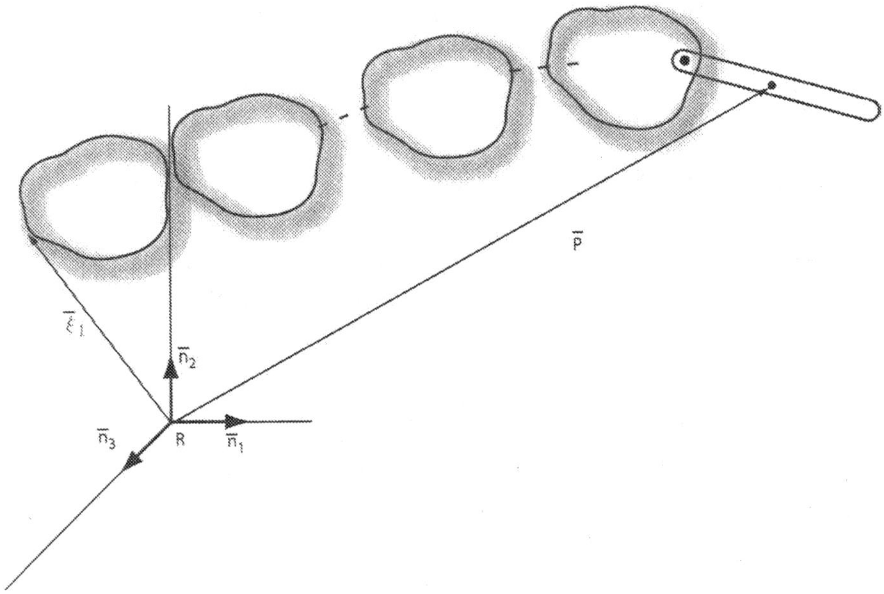

Fig. 10.4. Position of an arbitrary point of the terminal flexible link in R

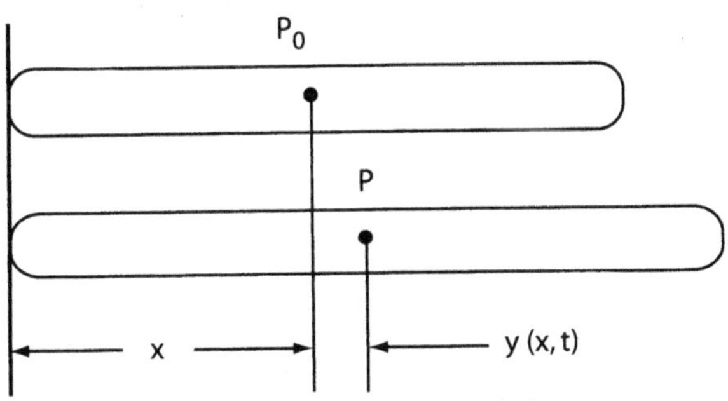

Fig. 10.5. Position components of the flexible link

$$Y(0) = 0, \qquad Y'(L) = 0 \tag{10.25}$$

the longitudinal deformation y is given by

$$y(x,t) = Y(x)T(t) = \sum_{r=n}^{\infty} \sin \beta_r x (a_r \sin \omega_r t + b_r \cos \omega_r t) \tag{10.26}$$

where

$$\beta_r = \frac{2_r - 1}{2L}\pi \quad \omega_r = \beta_r \sqrt{\frac{\rho}{E}}$$

where ρ is the density of the link, E is the modulus of elasticity, and a_r and b_r are constants dependent on the initial conditions.

Substituting equations (10.24) to (10.26) into (10.9) to (10.11), we can obtain the velocity and acceleration on point p.

10.2.3 Case 3: Transverse Vibration of a Flexible Link Considering Inertia Force and Gravity

In the previous two cases we discussed the effect of flexibility of the terminal body by examining how kinematics is derived. Now, let us assume that the flexible body has a transverse vibration in the vertical plane; we can derive the governing equations for an element of the beam by including the effects of gravity force and the inertia force due to the rigid body motion. Hence,

$$EI\frac{\partial^4 y}{\partial x^4} + m\frac{\partial^2 y}{\partial t^2} = f^*(x,t) \tag{10.27}$$

where $f^*(x,t)$ is the inertia force due to rigid-body movement and gravity of the flexible body, which is defined as

$$f^*(x,t) = \int_0^x m\left[-\bar{a}_p \cdot \bar{n}_2^n - \bar{g} \cdot \bar{n}_2^n\right]\frac{\xi}{L}d\xi \tag{10.28}$$

where $\bar{a}_p$ is the acceleration of the arbitrary point p on the link, and $\bar{g}$ is the gravitational acceleration, as shown in Figure 10.6. (Note that we assume that the flexible link

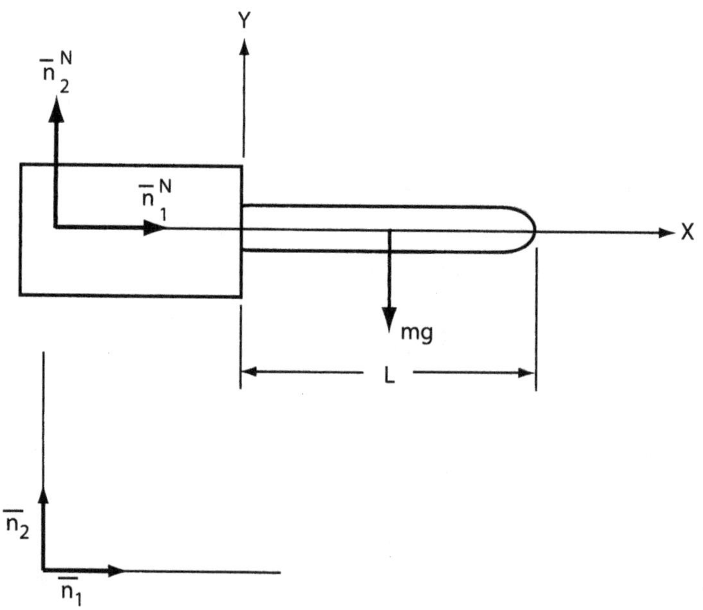

Fig. 10.6. Local axis definition of the flexible link

transverse vibration is in the $\bar{n}_1\bar{n}_2$ plane and that the $\bar{n}_1^n\bar{n}_2^n$ plane is a vertical plane as well.) $\bar{a}_p$ is given by equation (10.11), which is coupled with equation (10.27). Next, let us consider y, y_t' and y_t'' to be small compared to the large angular velocity and acceleration of the system. Therefore, in equation (10.28) the system dynamics between rigid-body motion and induced transverse vibration can be treated as uncoupled, the solution of which is obtained as follows. In the first step, we evaluate $f(x, t)$ from equation (10.7) with equation (10.11). The second step is to solve for $f(x, t)$ using equation (10.27). Then substituting y, y_t' and y_t'' in equation (10.11), we obtain the velocity and acceleration of the arbitrary point p on the flexible link.

Furthermore, if we assume that the beam is uniform and undergoing a large rotation where $\theta = \theta(t)$, then $f^*(x, t)$ from equation (10.28) becomes

$$f^*(x, t) = -\frac{1}{3}mL^2\ddot{\theta} - \frac{1}{2}mgL\cos\theta \tag{10.29}$$

Equation (10.27) can then be solved numerically considering the coupling between the rigid and flexible body motion. On the other hand, if $\ddot{\theta} = 0$, as might be the case when $\dot{\theta}(t)$ is constant, and neglecting the coupling, the governing equation can be solved analytically.

Of course, we can also consider the inertia and gravity forces in $f(x, t)$ for the longitudinal vibration of the link. The displacement of $y(x, t)$ of arbitrary point p is determined by

$$\rho\frac{\partial^2 y}{\partial t^2} = E\frac{\partial^2 y}{\partial x^2} + f^*(x, t) \tag{10.30}$$

and the problem can be solved similarly by the method discussed above.

10.2.4 Case 4: Torsional Vibration of the Flexible Link

For the system shown in Figure 10.4, let the nth body be a flexible link. Then the angular velocity of a small element with center mass point p is

$$^R\bar{\omega}^P = {}^R\bar{\omega}^n + {}^n\bar{\omega}^P \tag{10.31}$$

where $^R\bar{\omega}^P$ denotes the rigid-body angular velocity of body N in R, and $^n\bar{\omega}^P$ is the angular velocity of element P in body n.

$$^n\bar{\omega}^P = [\dot{\theta} \quad 0 \quad 0][S^{no}]\{\bar{n}\} \tag{10.32}$$

where equation (9.117) gives the torsional vibration partial differential equation

$$\frac{\partial^2\theta}{\partial t^2} = \frac{G}{\rho}\frac{\partial^2\theta}{\partial x^2} \tag{10.33}$$

Let us assume the boundary conditions to be fixed–free, that is,

$$\theta(0, t) = \theta_x'(L, t) = 0 \tag{10.34}$$

Then

$$\theta(x,t) = \sum_{n=1}^{\infty} (C_n \cos \omega_n t + D_n \sin \omega_n t) \sin \beta_n x \tag{10.35}$$

where C_n and D_n are constants determined by the initial conditions, and

$$\beta_n = \frac{2n-1}{2L}\pi \text{ and } \omega_n = \sqrt{\frac{G}{\rho}}\beta_n \tag{10.36}$$

Therefore, one can write the angular velocity of an element P as

$${}^{R}\bar{\omega}^{P} = [\{{}^{J}\bar{\omega}^{N}\}^{T}[S^{J0}] + [\dot{\theta}\ \ 0\ \ 0][S^{N0}]]\{\bar{n}\} \tag{10.37}$$

where

$$\dot{\theta} = \sum_{n=1}^{\infty} \omega_n (C_n \sin \omega_n t + D_n \cos \omega_n t) \sin \beta_n x \tag{10.38}$$

If we are interested in studying the effects of torsional vibration in rotating machinery, we proceed as follows. We derive the governing equations of a flexible beam undergoing large rotation by superimposing the perturbation due to the torsional vibration:

$$\frac{\partial^2 \theta^p}{\partial t^2} + \frac{G}{\rho}\frac{\partial^2 \theta^p}{\partial x^2} = T^*(\theta, t)$$

where T^* represents the generalized inertia torque, θ^p the perturbation, and θ the angular displacement of the beam. If we assume that the angular velocity is known in time as it might be prescribed, the equation above can be integrated to yield the time history of the perturbation θ^p. Of course, you can make use of such values based on the assumed solution to θ^p to predict the stability of the rotating beam.

10.3 Derivation of the Equations of Motion Using the Finite-Element Method

Let us consider the multibody system shown in Figure 10.7 with a flexible terminal body B_n. The system shown in Figure 10.8 is equivalent to the one shown in Figure 10.7. Let m_e denote the mass density of element i in the flexible body, which is divided into n elements. $\bar{u}$ is the displacement vector, which changes from original position $\bar{r}_n$ to $\bar{r}_u$. Then

$$\bar{r}_u = \bar{r}_n + \bar{u} \tag{10.39}$$

where $\bar{u}$ can be expressed as

$$\{u\} = [N]\{\rho\} \tag{10.40}$$

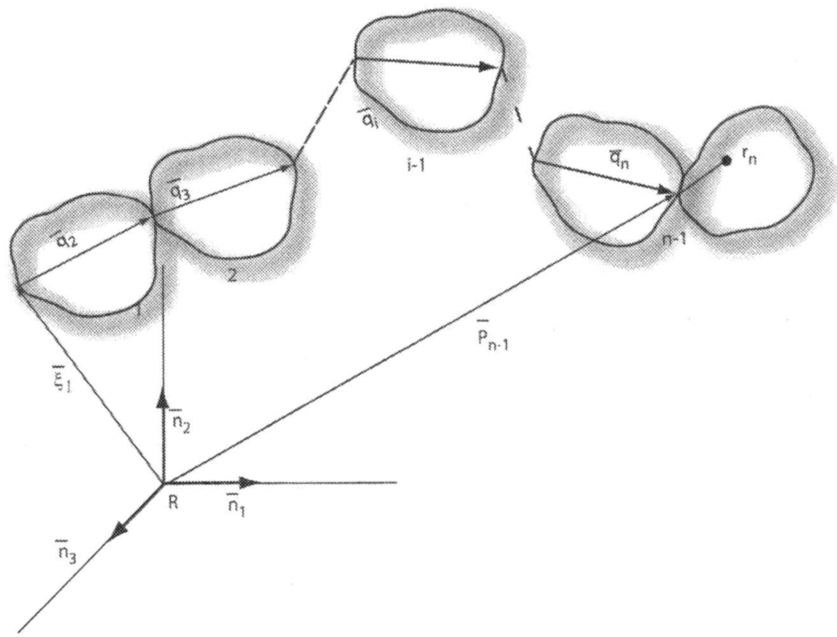

Fig. 10.7. General multibody system with a flexible terminal link

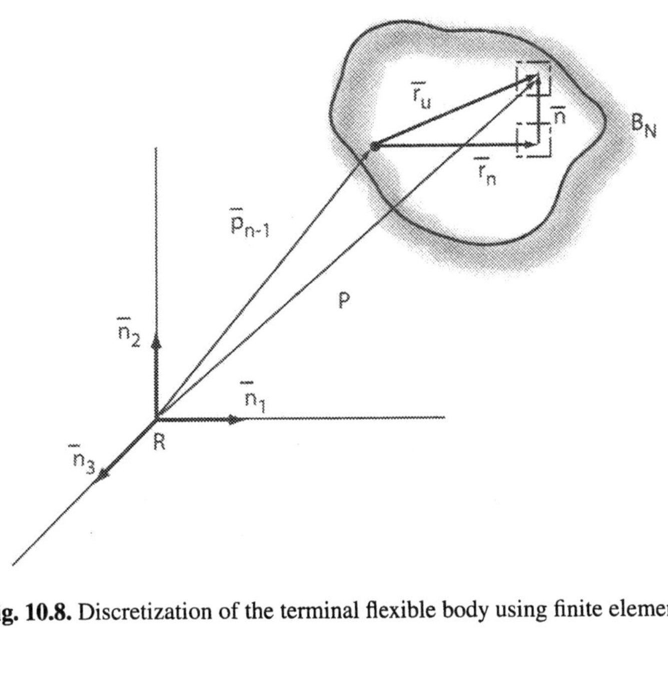

Fig. 10.8. Discretization of the terminal flexible body using finite elements

where $[N]$ denotes the shape function matrix and $\{\rho\}$ the vector of nodal coordinates. Then the position vector of the center mass of an element i in R, $\bar{p}$ can be expressed by

$$\begin{aligned}\bar{p} &= \bar{p}_{n-1} + \bar{r}_u \\ &= \left\{ \{\xi_1\}^T + \sum_{i=2}^{n} \{q_i\}^T [S^{\upsilon o}] + \left[\{r_n\}^T + \{\rho\}^T [N]^T \right] [S^{no}] \right\} \{\bar{n}\}\end{aligned} \tag{10.41}$$

where $\upsilon = i - 1$.

Let $\{\xi_1\} = \{q_1\} = 0$, for convenience; then

$$\bar{p} = \left\{ \sum_{i=2}^{n} \{q_i\}^T [S^{\upsilon o}] + \{r_n\}^T + \{\rho\}^T [N]^T [S^{no}] \right\} \{\bar{n}\} \tag{10.42}$$

Differentiating $\bar{p}$ in time, we obtain the velocity,

$$\begin{aligned}\bar{\upsilon}_p &= \frac{d}{dt}\bar{p} \\ &= \left\{ \sum_{i=2}^{n} \{q_i\}^T [\Omega^{\upsilon o}][S^{\upsilon o}] + [\{r_n\}^T \right. \\ &\left. \quad + \{\rho\}^T [N]^T][\Omega^{no}][S^{no}] + \{\dot{\rho}\}^T [N]^T [S^{no}] \right\} \{\bar{n}\}\end{aligned} \tag{10.43}$$

and the second derivative gives the acceleration:

$$\begin{aligned}\bar{a}_p = \frac{d}{dt}\bar{\upsilon}_p = & \left\{ \sum_{i=2}^{n} [\{q_i\}^T [\dot{\Omega}^{\upsilon o}][S^{\upsilon o}]] + [\{q_i\}^T [\Omega^{\upsilon o}][\Omega^{\upsilon o}][S^{\upsilon o}]] \right. \\ & + [\{r_n\}^T + \{\rho\}^T [N]^T [\dot{\Omega}^{no}][S^{no}]] + [\{r_n\}^T \\ & + \{\rho\}^T [N]^T [\Omega^{no}][\Omega^{no}][S^{no}]] + 2\{\dot{\rho}\}^T [N]^T [\Omega^{no}][S^{no}] \\ & \left. + \{\ddot{\rho}\}^T [N]^T [S^{no}] \right\} \{\bar{n}\}\end{aligned} \tag{10.44}$$

From Kane's equation of motion, we have

$$f + f^* = 0 \tag{10.45}$$

where

$$f^* = \bar{F}^* \cdot \frac{\partial \bar{\upsilon}}{\partial \dot{\rho}} + \bar{T}^* \cdot \frac{\partial \bar{\omega}}{\partial \dot{\rho}} \tag{10.46}$$

where $\bar{F}^*$ and $\bar{T}^*$ represent the inertial force and torque, respectively. For element i they are

$$\bar{F}^* = -m_e \bar{a}_p \tag{10.47}$$

$$\bar{T}^* \cdot \frac{\partial \bar{\omega}}{\partial \dot{\rho}} \simeq 0$$

(where we are assuming that ω of the element is small, hence is negligible). From equation (10.43), we obtain

$$\frac{\partial \bar{\upsilon}}{\partial \dot{\rho}} = [N]^T [S^{no}]\{\bar{n}\} \tag{10.48}$$

From equations (10.44) and (10.48), we get the generalized inertial force

$$\begin{aligned}
f^* = \bar{F}^* \cdot \frac{\partial \bar{\upsilon}}{\partial \dot{\rho}} &= -m_e \bar{a}_p \cdot \frac{\partial \bar{\upsilon}}{\partial \dot{\rho}} \\
&= -m_e \left\{ \sum_{i=2}^{n} (\{q_i\}^T [\dot{\Omega}^{\upsilon o}][S^{\upsilon o}] + \{q_i\}^T [\Omega^{\upsilon o}][\Omega^{\upsilon o}][S^{\upsilon o}]) \right. \\
&\quad + (\{r_n\}^T + \{\rho\}^T [N]^T)[\dot{\Omega}^{no}][S^{no}] + (\{r_n\}^T + \{\rho\}^T [N]^T)[\Omega^{no}] \\
&\quad \cdot [\Omega^{no}][S^{no}] + 2\{\dot{\rho}\}^T [N]^T [\Omega^{no}][S^{no}] + \left. \{\ddot{\rho}\}^T [N]^T [S^{no}] \right\} \{\bar{n}\} \\
&\quad \cdot \{\bar{n}\}^T [S^{no}]^T [N] \\
&= -m_e \{\ddot{\rho}\}^T [N]^T [N] + 2\{\dot{\rho}\}^T [N]^T [\Omega^{no}][N] + \{\rho\}^T ([N]^T [\dot{\Omega}^{no}] \\
&\quad + [\Omega^{no}][\Omega^{no}])[N] + \{r_n\}^T ([\dot{\Omega}^{no}] + [\Omega^{no}][\Omega^{no}])[N] \\
&\quad \left. + \sum_{i=2}^{n} (\{q_i\}^T [\dot{\Omega}^{\upsilon o}]) + \{q_i\}^T [\Omega^{\upsilon o}][\Omega^{\upsilon o}][S^{\upsilon n}][N] \right\}
\end{aligned} \tag{10.49}$$

where $\upsilon = i - 1$.

Now let us discuss f in equation (10.45). f should include stiffness and damping forces and the normalized external forces as well. Let these forces be represented by

$$f = f_e - f_k - f_c \tag{10.50}$$

where the external forces contribution is given by

$$f_e = \bar{F} \cdot \frac{\partial \bar{\upsilon}}{\partial \dot{\rho}} + \bar{T} \cdot \frac{\partial \bar{\omega}}{\partial \dot{\rho}}$$

Since we assume that ω is negligible, $\partial\omega/\partial\dot{\rho} = 0$ and that $\bar{F}$ and $\bar{T}$ denote the resultant external force and moment acting on the element, we write

$$\bar{F} = \{F\}^T \{\bar{n}^n\} = \{F\}^T [S^{no}]\{\bar{n}\} \tag{10.51}$$

which gives

$$f_e = \{F\}^T [S^{no}]\{\bar{n}\}\{\bar{n}\}^T [S^{no}]^T [N] = \{F\}^T [N] \tag{10.52}$$

Let us assume that the stiffness and damping matrices are $[K]$ and $[C]$, respectively, and then we define $[K] = [K]^T$ and $[C] = [C]^T$. From the basic finite-element method the stiffness matrix is

$$[K] = [B]^T [D][B] \tag{10.53}$$

where $[B]$ is the strain–modal displacement matrix, (D) the material property matrix, and the damping coefficient matrix is defined following the procedure above where

$$[C] = [N]^T [\mu][N] \tag{10.54}$$

where $[\mu]$ denotes the damping coefficient matrix. Since f_k, f_c and ρ are all row vectors, we write

$$f_k = \{\rho\}^T [K]$$
$$f_c = \{\dot{\rho}\}^T [C]$$

Combining the terms above, we get

$$f = f_e - f_k - f_c = \{F\}^T [N] - \{\rho\}^T [K] - \{\dot{\rho}\}^T [C] \tag{10.55}$$

Substituting equations (10.55) and (10.49) into (10.45), we obtain

$$\{F\}^T [N] - \{\dot{\rho}\}^T [C] - \{\rho\}^T [K] - m_e \Bigg\{ \{\ddot{\rho}\}^T [N]^T [N] + 2\{\dot{\rho}\}^T [N]^T [\Omega^{no}][N]$$
$$+ \{\rho\}^T [N]^T ([\dot{\Omega}^{no}] + [\Omega^{no}][\Omega^{no}])[N] + \{r_n\}^T ([\dot{\Omega}^{no}] + [\Omega^{no}][\Omega^{no}])[N]$$
$$+ \sum_{i=2}^{n} (\{q_i\}^T [\dot{\Omega}^{vo}] + \{q_i\}^T [\Omega^{vo}][\Omega^{vo}])[S^{vn}][N] \Bigg\} = 0 \tag{10.56}$$

which can be expressed in a compact form as:

$$\{\ddot{\rho}\}^T [M_e] + \{\dot{\rho}\}^T [C_e] + \{\rho\}^T [K_e] = \{F_e\}^T \tag{10.57}$$

where the element mass matrix is

$$[M_e] = m_e [N]^T [N] \tag{10.58}$$

The value $[C_e]$ is composed of two important terms; the first defines the damping matrix and the second is a nonlinear term coupling deformation and rigid-body motion. The equation $[C_e]$ is given by

$$[C_e] = [C] + 2m_e [N]^T [\Omega^{no}][N] = [N]^T [\mu][N] + 2m_e [N]^T [\Omega^{no}][N] \tag{10.59}$$

From equation (10.56), we see that

$$[K_e] = [K] + m_e [N]^T ([\dot{\Omega}^{no}] + [\Omega^{no}][\Omega^{no}])[N]$$
$$= [B]^T [D][B] + m_e [N]^T ([\dot{\Omega}^{no}] + [\Omega^{no}][\Omega^{no}])[N] \tag{10.60}$$

and

$$\{F_e\}^T = \{F\}^T[N] - m_e\left\{\{r_n\}^T([\dot{\Omega}^{no}] + [\Omega^{no}][\Omega^{no}])[N]\right.$$

$$\left. + \sum_{i=2}^{n}(\{q_i\}^T[\dot{\Omega}^{vo}] + \{q_i\}^T[\Omega^{vo}][\Omega^{vo}])[S^{vn}][N]\right\} \tag{10.61}$$

To be consistent with the normal method of expressing the ordinary differential equations, we rewrite the above as

$$[M_e]\{\ddot{\rho}\} + [C_e]\{\dot{\rho}\} + [K_e]\{\rho\} = \{F_e\} \tag{10.62}$$

where

$$[M_e] = m_e[N]^T[N] \tag{10.63}$$

$$[C_e] = [C] + 2m_e[N]^T[\Omega^{on}][N] \tag{10.64}$$

$$[K_e] = [K] + m_e[N]^T([\dot{\Omega}^{on}] + [\Omega^{on}][\Omega^{on}])[N] \tag{10.65}$$

$$\{F_e\} = [N]^T\{F\} - m_e[N]^T\left\{([\dot{\Omega}^{on}] + [\Omega^{on}][\Omega^{on}])\{r_n\}\right.$$

$$\left. + \sum_{i=2}^{n}[S^{nv}]([\dot{\Omega}^{ov}] + [\Omega^{ov}][\Omega^{ov}]\{q_i\})\right\} \tag{10.66}$$

Integrating both sides of equation (10.62), we obtain

$$[M]\{\ddot{\rho}\} + [C]\{\dot{\rho}\} + [K]\{\rho\} = \{F^e\} \tag{10.67}$$

where the matrices $[M]$, $[C]$ and $[K]$ combine both the rigid and the elastic deformation. Their explicit form is given below

$$[M] = \int_v m_e[N]^T[N]dv \tag{10.68}$$

$$[C] = \int_v [N]^T\{\mu\}[N]dv + 2\int_v m_e[N]^T[\Omega^{on}][N]dv = [C_s] + [C_d] \tag{10.69}$$

$$[K] = \int_v [B]^T[D][B]dv + \int_v m_e[N]^T([\dot{\Omega}^{on}] + [\Omega^{on}][\Omega^{on}])[N]dv = [K_s] + [K_d] \tag{10.70}$$

In the same fashion, the generalized forces are divided into two terms

$$\{F\} = \int_v [N]^T\{F\}dv - \int_v m_e[N]^T\left\{([\dot{\Omega}^{on}] + [\Omega^{on}][\Omega^{on}])\{r_n\}\right.$$

$$\left. + \sum_{i=1}^{n}[S^{nv}]([\dot{\Omega}^{ov}] + [\Omega^{ov}][\Omega^{ov}])\{q_i\}\right\} dv = \{F_s\} + \{F_d\} \tag{10.71}$$

where $[C_d]$, $[K_d]$ and $\{F_d\}$ correspond to the dynamic damping, dynamic stiffness and dynamic forces respectively, as a result of the rigid-body motion. Obviously, if the rigid-body motion is omitted, then the following terms are zero.

$$[\Omega^{on}] = 0 \tag{10.72}$$

$$[\dot{\Omega}^{on}] = 0 \tag{10.73}$$

$$[\Omega^{ov}] = 0 \tag{10.74}$$

$$[\dot{\Omega}^{ov}] = 0 \tag{10.75}$$

Therefore,

$$\begin{aligned} [C_d] &= 0 \\ [K_d] &= 0 \\ \{F_d\} &= 0 \end{aligned} \tag{10.76}$$

and equation (10.67) reduces to the ordinary equation of vibration of the body.

Note: The dynamic term associated with C_d, K_d and F_d denotes a unique contribution from the kinematics and they are different from the standard definition of stiffness, as damping due to the structure composition.

Assembling all the elemental matrices given by equation (10.67), we obtain the global differential equations governing the elastic body:

$$[M]\{\ddot{\rho}\} + [C]\{\dot{\rho}\} + [K]\{\rho\} = \{F\} \tag{10.77}$$

The global (generalized) mass, damping, stiffness, and force matrices may overlap due to the common nodal points, depending on the finite-element model of the flexible body, where proper assembly procedures must be used. The damping matrix $[C_s]$ can be estimated by Rayleigh damping as

$$[C_s] = \alpha[M] + \beta[K] \tag{10.78}$$

where α and β are experimentally determined constants.

Example 10.3.1

Consider the bar shown in Figure 10.9 undergoing a large rotation and assumed to be elastic. Assume that E, the modulus of elasticity, m_e, the mass density, and A, the cross-sectional area, are all known constants relating the properties of the bar. Let us assume that the damping coefficient of the bar is zero and that we have zero initial conditions. Derive the equation of motion of the bar using two one-dimensional elements assuming longitudinal deformation.

Solution

As shown in Figure 10.10, the shape function matrix $[N]$ from the standard finite-element method is

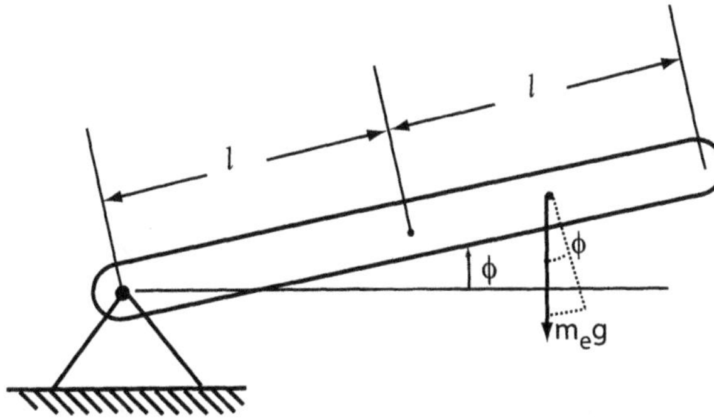

Fig. 10.9. Uniform elastic bar undergoing a large rotation

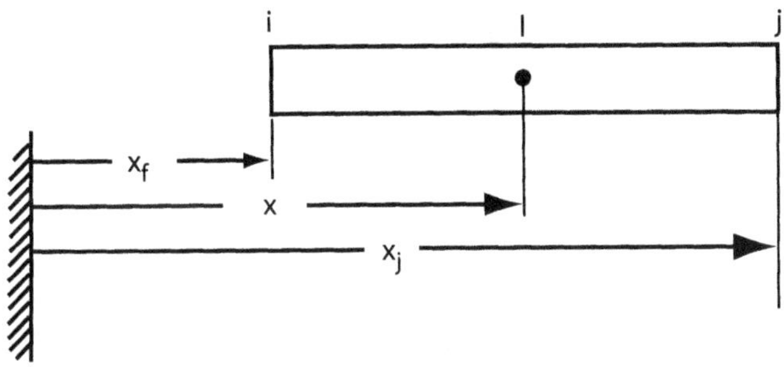

Fig. 10.10. Element node positioning with respect to a fixed-body frame

$$[N] = [N_i N_j] = \begin{bmatrix} \dfrac{x_j - x}{x_j - x_i} & \dfrac{x\_x_i}{x_j - x_i} \end{bmatrix} = \begin{bmatrix} \dfrac{x_j - x}{\ell} & \dfrac{x\_x_i}{\ell} \end{bmatrix} \tag{10.79}$$

The strain–modal displacement matrix

$$[B] = \frac{d[N]}{dx} = \begin{bmatrix} \dfrac{-1}{x_j - x_i} & \dfrac{1}{x_j - x_i} \end{bmatrix} = \begin{bmatrix} -\dfrac{1}{\ell} & \dfrac{1}{\ell} \end{bmatrix} \tag{10.80}$$

The element stiffness matrix

$$[K_s] = \int_{v} [B]^T [D][B] dv \tag{10.81}$$

where D, the material property matrix, is actually the scalar E (the modulus of elasticity); hence

$$[K_s] = \int_{\ell j}^{\ell j} \begin{bmatrix} -\dfrac{1}{\ell} \\ \dfrac{1}{\ell} \end{bmatrix} E \begin{bmatrix} -\dfrac{1}{\ell} & \dfrac{1}{\ell} \end{bmatrix} A dx$$

$$= \frac{EA}{\ell} \begin{bmatrix} 1 & -1 \\ -1 & 1 \end{bmatrix} \tag{10.82}$$

where A is the cross-sectional area of the element. The element mass matrix [see equation (10.68)]

$$[M^e] = \int_v m_e [N]^T [N] dv$$

$$= \frac{m_e A}{\ell^2} \int_{\ell i}^{\ell j} \begin{bmatrix} x_j - x \\ x - x_i \end{bmatrix} [x_j - xx - x_i] \, dx$$

$$= \frac{m_e A}{\ell^2} \int_{\ell i}^{\ell j} \begin{bmatrix} (x_j - x)^2 & (x_j - x)(x - x_i) \\ (x - x_i)(x_j - x) & (x - x_i)^2 \end{bmatrix} dx$$

$$= \frac{1}{6} m_e A \ell \begin{bmatrix} 2 & 1 \\ 1 & 2 \end{bmatrix} \tag{10.83}$$

The nodal force vector [see equation (10.71)]

$$\{F_s\} = \int_v [N]^T \{F\} dv$$

where $\{F\} = -m_e g \sin\varphi$, m_e is the mass density, and g the gravity acceleration. Then

$$\{F_s\} = -\int_{\ell i}^{\ell j} m_e g \sin\varphi [N]^T dx = \frac{1}{\ell} \int_{\ell i}^{\ell j} m_e g \sin\varphi \begin{bmatrix} x_j - x \\ x - x_i \end{bmatrix} dx$$

$$= \frac{1}{2} A m_e g \ell \sin\varphi \begin{Bmatrix} 1 \\ 1 \end{Bmatrix} \tag{10.84}$$

The skew matrix of the angular velocity $[\Omega]$ is defined as follows:

Since we have one body

$$n = 1, \qquad v = n - 1 = 0$$

Then

$$[\Omega^{on}] = [\Omega^{01}] = [\dot{S}^{01}][S^{01}]^{-1} = [\dot{S}^{01}][S^{10}]$$

where

$$[S^{10}] = [S^{01}]^T = \begin{bmatrix} \cos\varphi & \sin\varphi & 0 \\ -\sin\varphi & \cos\varphi & 0 \\ 0 & 0 & 1 \end{bmatrix}$$

The skew matrix is then

$$[\Omega^{01}] = \dot{\varphi} \begin{bmatrix} 0 & -1 & 0 \\ 1 & 0 & 0 \\ 0 & 0 & 0 \end{bmatrix} \tag{10.85}$$

and its product is obtained as follows

$$[\Omega^{01}][\Omega^{01}] = \dot{\varphi}^2 \begin{bmatrix} -1 & 0 & 0 \\ 0 & -1 & 0 \\ 0 & 0 & 0 \end{bmatrix} \tag{10.86}$$

Its derivative is simply

$$[\dot{\Omega}^{01}] = \ddot{\varphi} \begin{bmatrix} 0 & -1 & 0 \\ 1 & 0 & 0 \\ 0 & 0 & 0 \end{bmatrix} \tag{10.87}$$

The element dynamic damping matrix [see equation (10.69)] can be expressed as:

$$[C_d] = 2 \int_{\upsilon} m_e [N]^T [\Omega_{11}^{on}][N] d\upsilon \tag{10.88}$$

Since the element is one-dimensional, we can only consider $\left[\Omega_{11}^{on}\right]$ in the matrix $[\Omega^{on}]$. Therefore,

$$[C_d] = 0 \tag{10.89}$$

The element total damping matrix is found to be

$$[C^e] = [C_s] + [C_d] = 0 + 0 = 0 \tag{10.90}$$

The element dynamic stiffness matrix [see equation (10.70)] after substitution becomes

$$\begin{aligned} [K_d] &= \int_{\upsilon} m_e [N]^T ([\dot{\Omega}_{11}^{on}] + [\Omega_{1t}^{on}][\Omega_{t1}^{on}])[N] d\upsilon \\ &= m_e \int_{\ell i}^{\ell j} \begin{bmatrix} x_j - x \\ x - x_i \end{bmatrix} (0 - \dot{\varphi}^2)[x_j - x \quad x - x_i] A dx \\ &= -\frac{1}{6} m_e A \ell \dot{\varphi}^2 \begin{bmatrix} 2 & 1 \\ 1 & 2 \end{bmatrix} \end{aligned} \tag{10.91}$$

The element total stiffness matrix is

$$[K^e] = [K_s] + [K_d]$$

$$\begin{aligned} \left[K^e\right] &= \frac{EA}{\ell} \begin{bmatrix} 1 & -1 \\ -1 & 1 \end{bmatrix} - \frac{1}{6} m_e A \ell \dot{\varphi}^2 \begin{bmatrix} 2 & 1 \\ 1 & 2 \end{bmatrix} \\ &= A \begin{bmatrix} \frac{E}{\ell} - \frac{1}{3} m_e \ell \dot{\varphi}^2 & -\left(\frac{E}{\ell} + \frac{1}{6} m_e \ell \dot{\varphi}^2\right) \\ -\left(\frac{E}{\ell} + \frac{1}{6} m_e \ell \dot{\varphi}^2\right) & \frac{E}{\ell} - \frac{1}{3} m_e \ell \dot{\varphi}^2 \end{bmatrix} \end{aligned} \tag{10.92}$$

The element dynamic force vector [see equation 10.71)] is expressed as:

$$\{F_d\} = -\int_{\upsilon} m_e[N]^T \left\{([\dot{\Omega}_{11}^{on}] + [\Omega_{1t}^{on}][\Omega_{t1}^{on}])\{r_n\} + \sum_{i=1}^{n}[S^{n\upsilon}]([\dot{\Omega}^{o\upsilon}] + [\Omega^{o\upsilon}][\Omega^{o\upsilon}])\{q_i\}\right\} d\upsilon$$

where $n = 1$, $\upsilon = 0$, and $\{r_n\} = x$; therefore,

$$\begin{aligned}\{F_d\} &= -\int_{\ell i}^{\ell j} m_e \frac{1}{\ell}\begin{bmatrix} x_j - x \\ x - x_i \end{bmatrix}\{(0 - \dot{\varphi}^2)x + 0\}A dx \\ &= \frac{1}{6} m_e A\ell\,\dot{\varphi}^2 \begin{bmatrix} 3x_j - 2\ell \\ 3x_i - 2\ell \end{bmatrix}\end{aligned}$$

or

$$\{F_d\} = \frac{1}{6} m_e A\ell\dot{\varphi}^2 \begin{bmatrix} 3x_i + \ell \\ 3x_j - 5\ell \end{bmatrix} \tag{10.93}$$

The element total force vector

$$\begin{aligned}\{F^e\} &= \{F_s\} + \{F_d\} \\ &= -\frac{1}{2} A m_e g\ell \sin\varphi \begin{bmatrix} 1 \\ 1 \end{bmatrix} + \frac{1}{6} m_e A\ell\dot{\varphi}^2 \begin{bmatrix} 3x_j - 2\ell \\ 3x_i - 2\ell \end{bmatrix} \\ &= \frac{1}{6} A m_e \ell \begin{bmatrix} (3x_j - 2\ell)\dot{\varphi}^2 - 3g\sin\varphi \\ (3x_i - 2\ell)\dot{\varphi}^2 - 3g\sin\varphi \end{bmatrix}\end{aligned} \tag{10.94}$$

From equations (10.67), (10.83), (10.90), (10.92) and (10.94), we obtain the element equations of motion:

$$\frac{1}{6} m_e A\ell \begin{bmatrix} 2 & 1 \\ 1 & 2 \end{bmatrix}\{\ddot{\rho}\} + A \begin{bmatrix} \frac{E}{\ell} - \frac{1}{3} m_e \ell\dot{\varphi}^2 \\ -\left(\frac{E}{\ell} + \frac{1}{6} m_e \ell\dot{\varphi}^2\right) \\ -\left(\frac{E}{\ell} + \frac{1}{6} m_e \ell\dot{\varphi}^2\right) \\ \frac{E}{\ell} - \frac{1}{3} m_e \ell\dot{\varphi}^2 \end{bmatrix}$$

$$= \frac{1}{6} A m_e \ell \begin{bmatrix} (3x_j - 2\ell)\dot{\varphi}^2 - 3g\sin\varphi \\ (3x_i - 2\ell)\dot{\varphi}^2 - 3g\sin\varphi \end{bmatrix} \tag{10.95}$$

If the flexible link is divided into two elements, for element 1 we have $x_j = 2\ell$ and $x_i = \ell$; then

$$\begin{bmatrix} 2 & 1 \\ 1 & 2 \end{bmatrix} \{\ddot{\rho}\} + \begin{bmatrix} \dfrac{6E}{m_e \ell^2} - 2\dot{\varphi}^2 & -\left(\dfrac{6E}{m_e \ell^2} + \dot{\varphi}^2\right) \\ -\left(\dfrac{6E}{m_e \ell^2} + \dot{\varphi}^2\right) & \dfrac{6E}{m_e \ell^2} - 2\dot{\varphi}^2 \end{bmatrix} \{\rho\}$$

$$= \begin{bmatrix} \ell\dot{\varphi}^2 - 3g \sin\varphi + \dfrac{6F_1}{m_e A \ell} \\ -2\ell\dot{\varphi}^2 - 3g \sin\varphi \end{bmatrix} \tag{10.96}$$

For element 2, $x_j = 2\ell$ and $x_i = \ell$; then

$$\begin{bmatrix} 2 & 1 \\ 1 & 2 \end{bmatrix} \{\ddot{\rho}\} + \begin{bmatrix} \dfrac{6E}{m_e \ell^2} - 2\dot{\varphi}^2 & -\left(\dfrac{6E}{m_e \ell^2} + \dot{\varphi}^2\right) \\ \left(\dfrac{6E}{m_e \ell^2} + \dot{\varphi}^2\right) & \dfrac{6E}{m_e \ell^2} - 2\dot{\varphi}^2 \end{bmatrix} \{\rho\}$$

$$= \begin{bmatrix} 4\ell\dot{\varphi}^2 - 3g \sin\varphi \\ \ell\dot{\varphi}^2 - 3g \sin\varphi \end{bmatrix} \tag{10.97}$$

Combining the two elements equations above, we obtain the global representation of the elastic body:

$$\begin{bmatrix} 2 & 1 & 0 \\ 1 & 4 & 1 \\ 0 & 1 & 2 \end{bmatrix} \begin{bmatrix} \ddot{\rho}_1 \\ \ddot{\rho}_2 \\ \ddot{\rho}_3 \end{bmatrix} + \begin{bmatrix} \dfrac{6E}{m_e \ell^2} - 2\dot{\varphi}^2 & -\left(\dfrac{6E}{m_e \ell^2} + \dot{\varphi}^2\right) & 0 \\ -\left(\dfrac{6E}{m_e \ell^2} + \dot{\varphi}^2\right) & \dfrac{12E}{m_e \ell^2} - 4\dot{\varphi}^2 & -\left(\dfrac{6E}{m_e \ell^2} + \dot{\varphi}^2\right) \\ 0 & -\left(\dfrac{6E}{m_e \ell^2} + \dot{\varphi}^2\right) & \dfrac{6E}{m_e \ell^2} - 2\dot{\varphi}^2 \end{bmatrix}$$

$$\cdot \begin{bmatrix} \rho_1 \\ \rho_2 \\ \rho_3 \end{bmatrix} = \begin{bmatrix} \ell\dot{\varphi}^2 - 3g \sin\varphi + \dfrac{6F_1}{m_e A \ell} \\ 2\ell\dot{\varphi}^2 - 6g \sin\varphi \\ \ell\dot{\varphi}^2 - 3g \sin\varphi \end{bmatrix} \tag{10.98}$$

From the boundary conditions we have $\rho_1 = 0$ and $\ddot{\rho}_1 = 0$; then the equations above reduce to

$$\begin{bmatrix} 4 & 1 \\ 1 & 2 \end{bmatrix} \begin{bmatrix} \ddot{\rho}_2 \\ \ddot{\rho}_3 \end{bmatrix} + \begin{bmatrix} \dfrac{12E}{m_e \ell^2} - 4\dot{\varphi}^2 & -\left(\dfrac{6E}{m_e \ell^2} + \dot{\varphi}^2 \right) \\ -\left(\dfrac{6E}{m_e \ell^2} + \dot{\varphi}^2 \right) & \dfrac{6E}{m_e \ell^2} - 2\dot{\varphi}^2 \end{bmatrix} \begin{bmatrix} \rho_2 \\ \rho_3 \end{bmatrix}$$

$$= \begin{bmatrix} 2\ell\dot{\varphi}^2 - 6g \sin\varphi \\ \ell\dot{\varphi}^2 - 3g \sin\varphi \end{bmatrix} \tag{10.99}$$

Let $E/m_e\ell^2 = a$ and $\{\ddot{\rho}\} = \omega^2\{\rho\}$, where ω is the eigenvalue; then for the homogeneous equation (10.99), we obtain

$$\begin{bmatrix} 12a - 4\dot{\varphi}^2 - 4\omega^2 - (6a + \dot{\varphi}^2) - \omega^2 \\ -(6a + \dot{\varphi}^2) - \omega^2 6a - 2\dot{\varphi}^2 - 2\omega^2 \end{bmatrix} \{\rho\} = 0 \tag{10.100}$$

Since $[\rho] \neq 0$, the determinant of the matrix must equal zero; that is,

$$(12a - 4\dot{\varphi}^2 - 4\omega^2)(6a - 2\dot{\varphi}^2 - 2\omega^2) - [(6a + \dot{\varphi}^2) + \omega^2]^2 = 0 \tag{10.101}$$

or

$$7\omega^4 + (14\dot{\varphi}^2 - 60a)\omega^2 + (7\dot{\varphi}^4 - 60a\dot{\varphi}^2 + 36a^2) = 0 \tag{10.102}$$

which yields

$$\omega^2 = \frac{69a - 14\dot{\varphi}^2 \pm \sqrt{(60a - 14\dot{\varphi}^2)^2 - 4 \times 7(7\dot{\varphi}^4 - 60a\dot{\varphi}^2 + 36a^2)}}{2 \times 7}$$

$$= \frac{60a - 14\dot{\varphi}^2 \pm \sqrt{2592}a}{14} \tag{10.103}$$

resulting in

$$\omega_1^2 = 0.649a - \dot{\varphi}^2 \tag{10.104}$$
$$\omega_2^2 = 7.922a - \dot{\varphi}^2$$

Substituting equation (10.104) into equation (10.100), we obtain the modal shape matrix

$$[\phi_1] = \begin{bmatrix} 1 & 1 \\ \sqrt{2} & \sqrt{2} \end{bmatrix} \tag{10.105}$$

The normal modal shape matrix is found to be

$$[\phi] = \begin{bmatrix} \dfrac{1}{\sqrt{8 + 2\sqrt{2}}} & \dfrac{1}{\sqrt{8 - 2\sqrt{2}}} \\ \dfrac{\sqrt{2}}{\sqrt{8 + 2\sqrt{2}}} & \dfrac{-\sqrt{2}}{\sqrt{8 - 2\sqrt{2}}} \end{bmatrix} = \begin{bmatrix} 0.304 & 0.440 \\ 0.430 & -0.622 \end{bmatrix} \tag{10.106}$$

Let the modal coordinates be expressed as functions of the modal coordinates η, where $\{\rho\} = \{\phi\}\{\eta\}$ and premultiply by ϕ^T on both sides of equation (10.99); we get

$$\{\ddot{\eta}\} + \begin{bmatrix} \omega_1^2 & 0 \\ 0 & \omega_2^2 \end{bmatrix} \{\eta\} = \begin{bmatrix} 1.038\ell\dot{\varphi}^2 - 3.114g\sin\varphi \\ 0.258\ell\dot{\varphi}^2 - 0.744g\sin\varphi \end{bmatrix} \tag{10.107}$$

From the initial conditions

$$\{\rho(0)\} = 0 \quad \{\dot{\rho}(0)\} = 0 \tag{10.108}$$

the modal coordinates initial conditions become

$$\{\eta(0)\} = 0 \quad \{\dot{\eta}(0)\} = 0 \tag{10.109}$$

the result of which is

$$\eta_1 = \frac{1}{\omega_1}\int_o^t (1.038\ell\dot{\varphi}^2 - 3.114g\sin\varphi)\sin\omega_1(t-\tau)d\tau \tag{10.110}$$

and

$$\eta_2 = \frac{1}{\omega_2}\int_o^t (0.258\ell\dot{\varphi}^2 - 0.774g\sin\varphi)\sin\omega_2(t-\tau)d\tau \tag{10.111}$$

In matrix form we can express the equations above as

$$\begin{bmatrix} \rho_2 \\ \rho_3 \end{bmatrix} = \{\rho\} = [\phi]\{\eta\}$$

$$= \begin{bmatrix} \dfrac{0.304}{\omega_1}\displaystyle\int_o^t (1.038\ell\dot{\varphi}^2 - 3.114g\sin\varphi)\sin\omega_1(t-\tau)d\tau \\ +\dfrac{0.440}{\omega_2}\displaystyle\int_o^t (0.258\ell\dot{\varphi}^2 - 0.774g\sin\varphi)\sin\omega_2(t-\tau)d\tau \\ \dfrac{0.430}{\omega_1}\displaystyle\int_o^t (1.038\ell\dot{\varphi}^2 - 3.114g\sin\varphi)\sin\omega_1(t-\tau)d\tau \\ -\dfrac{0.662}{\omega_2}\displaystyle\int_o^t (0.258\ell\dot{\varphi}^2 - 0.774g\sin\varphi)\sin\omega_2(t-\tau)d\tau \end{bmatrix} \tag{10.112}$$

where

$$\omega_1 = \sqrt{0.649a - \dot{\varphi}^2} \tag{10.113}$$

$$\omega_2 = \sqrt{7.922a - \dot{\varphi}^2} \tag{10.114}$$

and

$$a = \frac{E}{m_e\ell^2}$$

Let $\dot{\varphi} = 0$; then we obtain the ordinary modal frequencies. Compared with the precise solution obtained by the partial differential equation, for ω_1 (lowest one) the error is only 2.6%, and for ω_2 (second lowest) the error is 19%. If, on the other hand, $\dot{\varphi} = \omega =$ constant, then

$$\begin{aligned}\rho_2(t) = {} & \frac{0.316\ell\omega^2}{\omega_1^2}(1-\cos\omega_1 t) - \frac{0.473g}{\omega_1}\left[\frac{1}{\omega+\omega_1}(\sin\omega t + \sin\omega_1 t)\right. \\ & \left. - \frac{1}{\omega+\omega_1}(\sin\omega t - \sin\omega_1 t)\right] + \frac{0.114\ell\omega^2}{\omega_2^2}(1-\cos\omega_2 t) \\ & - \frac{0.170g}{\omega_2}\left[\frac{1}{\omega+\omega_2}(\sin\omega t + \sin\omega_2 t) - \frac{1}{\omega-\omega_2}(\sin\omega t - \sin\omega_2 t)\right]\end{aligned} \tag{10.115}$$

$$\begin{aligned}\rho_3(t) = {} & \frac{0.446\ell\omega^2}{\omega_1^2}(1-\cos\omega_1 t) - \frac{0.670g}{\omega_1}\left[\frac{1}{\omega+\omega_1}(\sin\omega t + \sin\omega_1 t)\right. \\ & \left. - \frac{1}{\omega-\omega_1}(\sin\omega t - \sin\omega_1 t)\right] - \frac{0.160\ell\omega^2}{\omega_2^2}(1-\cos\omega_2 t) \\ & + \frac{0.241g}{\omega_2}\left[\frac{1}{\omega+\omega_2}(\sin\omega t + \sin\omega_2 t) - \frac{1}{\omega-\omega_2}(\sin\omega t - \sin\omega_2 t)\right]\end{aligned} \tag{10.116}$$

Therefore, from equations (10.79) and (10.116) we can use the following relation

$$[u] = [N]\{\rho\} \tag{10.117}$$

to find the displacement of an arbitrary point on the bar. In addition, we can evaluate the constraint force F_1 at node 1 from equation (10.98). Note that F_1 is just acting along the longitudinal direction of the bar. As shown in Figure 10.11, the total constraint force of the system is

$$\begin{aligned}\{F_c\} &= [F_1 - 2m_e gA\ell\cos\varphi 0]\{\bar{n}^1\} \\ &= [F_1 - 3m_e gA\ell\cos\varphi 0][S^{1o}]\{\bar{n}\} \\ &= [F_1 - 2m_e gA\ell\cos\varphi 0]\begin{bmatrix}\cos\varphi & \sin\varphi & 0 \\ -\sin\varphi & \cos\varphi & 0 \\ 0 & 0 & 1\end{bmatrix}\begin{bmatrix}\bar{n}_1 \\ \bar{n}_2 \\ \bar{n}_3\end{bmatrix} \\ &= (F_1\cos\varphi + 2m_e Ag\ell\cos\varphi\sin\varphi)\bar{n}_1 + (F_1\sin\varphi - 2m_e A\ell g\cos^2\varphi)\bar{n}_2\end{aligned} \tag{10.118}$$

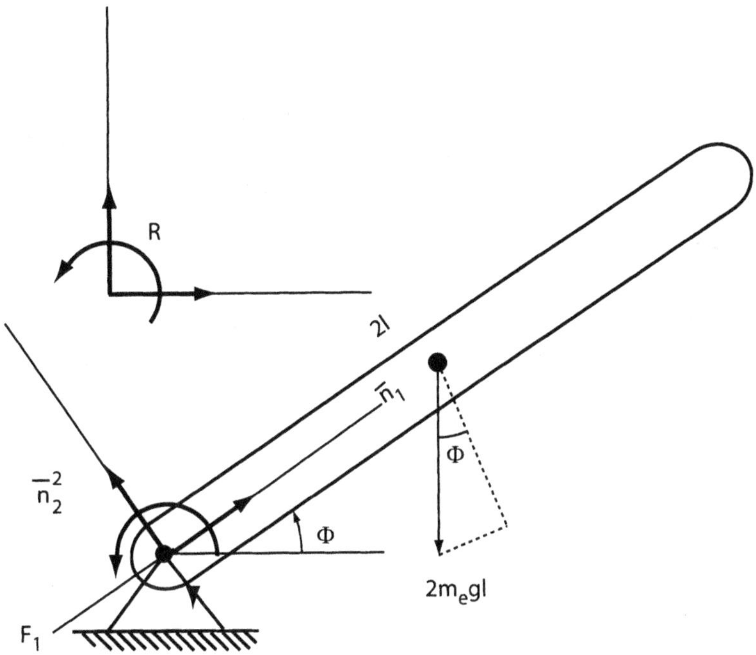

Fig. 10.11. Gravity force modeling and the body axis

Discussion

In this example, for simplicity in carrying out the explicit solution, we selected two one-dimensional elements and discussed the longitudinal vibration of the bar when it is undergoing a large rotation. From equations (10.113) and (10.114), we know that the modal frequencies of the system are decreased due to the angular rotation. The higher the angular velocity, the lower the modal frequencies become. If $\dot{\varphi}$ is time variant, the modal frequencies will vary as well. The modal shapes of the bar have nothing to do with the rotation of the system [see equation (10.106)]. The conclusion one draws is very significant. It means that the modal frequencies of an aircraft moving in space at a high speed are changeable. Therefore, modal frequencies based on laboratory experiments or calculated by material properties and geometric shapes of the system are not accurate enough to avoid resonance in the system. It is apparent from this example that coupling of rigid-body motion with flexibility is important, if the angular velocity $\dot{\varphi}$ takes on significant values.

10.4 Equations of Motion of an Elastic Beam Undergoing Large Rotation: A 2D FE Formulation

In Section 10.3 we discussed the longitudinal vibration of a beam making use of a one-dimensional element. Now, let us choose a two-dimensional element to model the elastic bar (beam) of Example 10.3.1, in order to discuss the effects of transverse

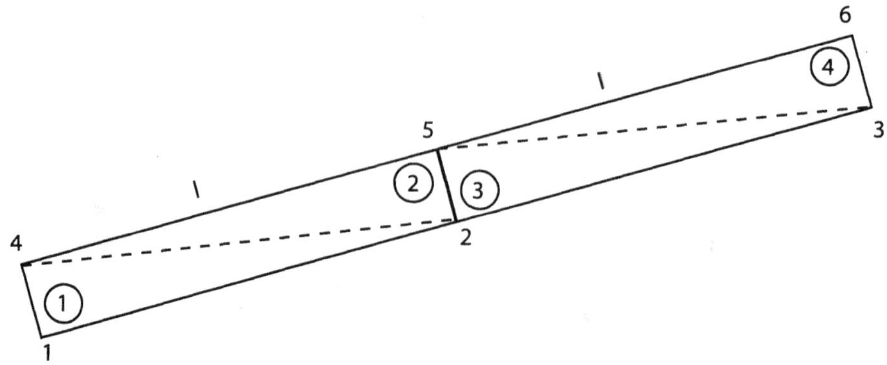

Fig. 10.12. Four-element modeling of the elastic beam

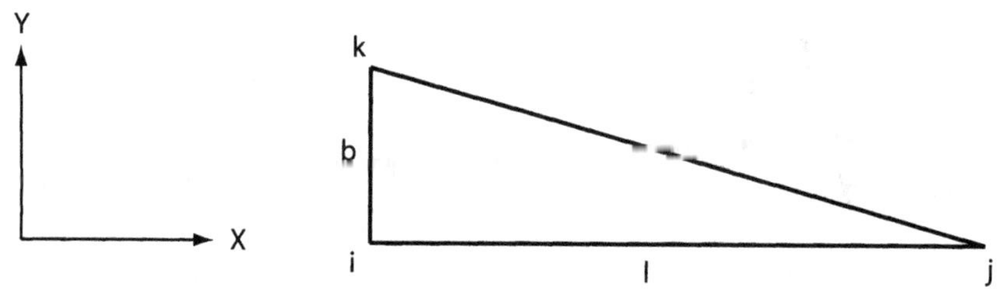

Fig. 10.13. Element shape function description type 1

and longitudinal vibration of a beam undergoing rigid-body motion. For convenience, the beam is divided into four triangular elements, as shown in Figure 10.12.

As shown in Figure 10.13, the shape functions can be expressed as

$$[N_1 \quad N_j \quad N_k] = [1 \quad x \quad y] \begin{bmatrix} 1 & x_i & y_i \\ 1 & x_j & y_j \\ 1 & x_k & y_k \end{bmatrix}^{-1}$$

$$= \frac{1}{2A} \begin{bmatrix} (x_j y_k - x_k y_j) + x(y_j - y_k) + y(x_k - x_j) \\ (x_k y_i - x_i y_k) + x(y_k - y_i) + y(x_i - x_k) \\ (x_i y_j - x_j y_i) + x(y_i - y_j) + y(x_j - x_i) \end{bmatrix}^T \tag{10.119}$$

where A is the area of the triangle $\left(A = \frac{1}{2} b\ell\right)$.

So we may write

$$[N] = \begin{bmatrix} N_i & 0 & N_j & 0 & N_k & 0 \\ 0 & N_i & 0 & N_j & 0 & N_k \end{bmatrix} \tag{10.120}$$

Notice that we have two types of elements from our model in Figure 10.12. For element 1, consider Figure 10.13; then

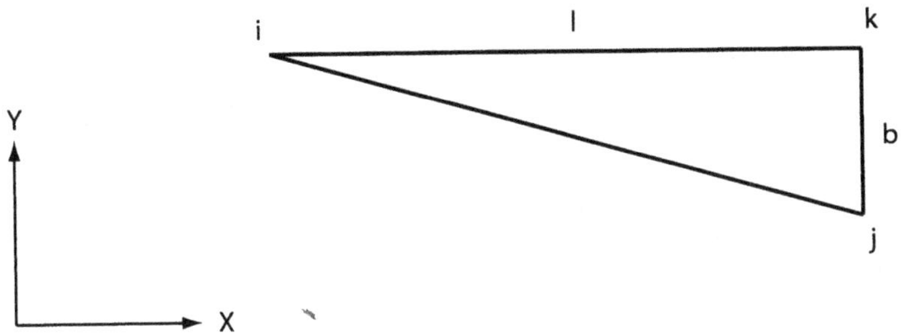

Fig. 10.14. Element shape function description type 2

$$[N] = \frac{1}{b\ell}\begin{bmatrix} b\ell - bx - \ell y & 0 & bx & 0 & \ell y & 0 \\ 0 & b\ell - bx - \ell y & 0 & bx & 0 & \ell y \end{bmatrix} \quad (10.121)$$

For element 2 (see Figure 10.14),

$$[N] = \frac{1}{b\ell}\begin{bmatrix} b\ell - bx & 0 & b\ell - \ell y & 0 & -b\ell + bx + \ell y & 0 \\ 0 & b\ell - bx & 0 & b\ell - \ell y & 0 & -b\ell + bx + \ell y \end{bmatrix} \quad (10.122)$$

Recall that the element displacements are related to the nodal ones:

$$\{u\} = [N]\{\rho\} \quad (10.123)$$

where

$$\{u\} = [u_x \quad u_y]^T \quad (10.124)$$

Since we are assuming transverse and longitudinal displacements

$$\{\rho\} = [u_{ix} \quad u_{iy} \quad u_{jx} \quad u_{jy} \quad u_{kx} \quad u_{ky}]^T \quad (10.125)$$

where each node (i, j, or k) has two displacements. The strain–modal displacement matrix $[B]$ is given by

$$[B] = [L][N] \quad (10.126)$$

where $[L]$ is a linear operator matrix:

$$[L] = \begin{bmatrix} \dfrac{\partial}{\partial x} & 0 \\ 0 & \dfrac{\partial}{\partial y} \\ \dfrac{\partial}{\partial y} & \dfrac{\partial}{\partial x} \end{bmatrix} \quad (10.127)$$

Then equation (10.126) becomes

$$[B] = \frac{1}{2A}\begin{bmatrix} \frac{\partial N_i}{\partial x} & 0 & \frac{\partial N_j}{\partial x} & 0 & \frac{\partial N_k}{\partial x} & 0 \\ 0 & \frac{\partial N_i}{\partial y} & 0 & \frac{\partial N_j}{\partial y} & 0 & \frac{\partial N_k}{\partial y} \\ \frac{\partial N_i}{\partial y} & \frac{\partial N_i}{\partial x} & \frac{\partial N_j}{\partial y} & \frac{\partial N_j}{\partial x} & \frac{\partial N_k}{\partial y} & \frac{\partial N_k}{\partial x} \end{bmatrix}$$

$$= \frac{1}{2A}\begin{bmatrix} y_j - y_k & 0 & y_k - y_i & 0 & y_i - y_j & 0 \\ 0 & x_k - x_j & 0 & x_i - x_k & 0 & x_j - x_i \\ x_k - x_j & y_j - y_k & x_i - x_k & y_k - y_i & x_j - x_i & y_i - y_j \end{bmatrix} \quad (10.128)$$

For elements 1 and 3 of the same type, we obtain the $[B]$ matrix as

$$[B] = \frac{1}{b\ell}\begin{bmatrix} -b & 0 & b & 0 & 0 & 0 \\ 0 & -\ell & 0 & 0 & 0 & \ell \\ -\ell & -b & 0 & b & \ell & 0 \end{bmatrix} \quad (10.129)$$

Similarly, for elements 2 and 4 we derive the following

$$[B] = \frac{1}{b\ell}\begin{bmatrix} -b & 0 & 0 & 0 & b & 0 \\ 0 & 0 & 0 & -\ell & 0 & \ell \\ 0 & -b & -\ell & 0 & \ell & b \end{bmatrix} \quad (10.130)$$

Let us assume that the modulus of elasticity of the material $E = 2.0 \times 10^{11}\text{N/m}^2$ and Poisson's ratio $\mu = 0.3$. Then the material property matrix $[D]$ is found to be

$$[D] = \frac{E}{(1+\mu)(1-2\mu)}\begin{bmatrix} 1-\mu & \mu & 0 \\ \mu & 1-\mu & 0 \\ 0 & 0 & \frac{1-2\mu}{2} \end{bmatrix}$$

$$= 3.85 \times 10^{11}\begin{bmatrix} 0.7 & 0.3 & 0 \\ 0.3 & 0.7 & 0 \\ 0 & 0 & 0.2 \end{bmatrix} \quad (10.131)$$

The element stiffness matrix [see equation (10.81)] is

$$[K_s] = \int_v [B]^T[D][B]dv$$

$$= [B]^T[D][B]\int_v dv$$

$$= [B]^T[D][B]At \quad (10.132)$$

where t is the thickness of the beam and A is the area.

For elements 1 and 3, the element stiffness is explicitly shown below

$$
[K_S] = \frac{3.85 \times 10^{11}}{2b\ell} \begin{bmatrix} -b & 0 & -\ell \\ 0 & -\ell & -b \\ b & 0 & 0 \\ 0 & 0 & b \\ 0 & 0 & \ell \\ 0 & \ell & 0 \end{bmatrix} \begin{bmatrix} 0.7 & 0.3 & 0 \\ 0.3 & 0.7 & 0 \\ 0 & 0 & 0.2 \end{bmatrix} \begin{bmatrix} -b & 0 & b & 0 & 0 & 0 \\ 0 & -\ell & 0 & 0 & 0 & \ell \\ -\ell & -b & 0 & b & \ell & 0 \end{bmatrix}
$$

$$
= \frac{1.925 \times 10^{11} t}{b\ell}
$$

$$
\times \begin{bmatrix} 0.7b^2 + 0.2\ell^2 & & & & & \\ 0.5b\ell & 0.7\ell^2 + 0.2b^2 & & \text{symmetric} & & \\ -0.7b^2 & -0.3b\ell & 0.7b^2 & & & \\ -0.2b\ell & -0.2b^2 & 0 & 0.2b^2 & & \\ -0.2\ell^2 & -0.2b\ell & 0 & 0.2b\ell & 0.2\ell^2 & \\ -0.3b\ell & -0.7\ell^2 & 0.3b\ell & 0 & 0 & 0.7\ell^2 \end{bmatrix} \tag{10.133}
$$

For elements 2 and 4, we derive the corresponding stiffness as

$$
[K_S] = \frac{3.85 \times 10^{11} t}{2b\ell}
$$

$$
\begin{bmatrix} -b & 0 & 0 \\ 0 & 0 & -b \\ b & 0 & -\ell \\ 0 & -\ell & 0 \\ b & 0 & \ell \\ 0 & \ell & b \end{bmatrix} \begin{bmatrix} 0.7 & 0.3 & 0 \\ 0.3 & 0.7 & 0 \\ 0 & 0 & 0.2 \end{bmatrix} \begin{bmatrix} -b & 0 & 0 & 0 & b & 0 \\ 0 & 0 & 0 & -\ell & 0 & \ell \\ 0 & -b & -\ell & 0 & \ell & b \end{bmatrix}
$$

$$
= \frac{1.925 \times 10^{11} t}{b\ell}
$$

$$
\times \begin{bmatrix} 0.7b^2 & & & & & \\ 0 & 0.2b^2 & & & & \\ 0 & 0.2b\ell & 0.2\ell^2 & & \text{symmetric} & \\ 0.3b\ell & 0 & 0 & 0.7\ell^2 & & \\ -0.7b^2 & -0.2b\ell & -0.2\ell^2 & -0.3b\ell & 0.2\ell^2 + 0.7b^2 & \\ -0.3b\ell & -0.2b^2 & 0.2b\ell & -0.7\ell^2 & 0.5b\ell & 0.7\ell^2 + 0.2b^2 \end{bmatrix} \tag{10.134}
$$

The element mass matrix [see equation (10.68)] is defined by

$$[M^e] = \int_v m_e [N]^T [N] dV \tag{10.135}$$

For elements 1 and 3, we perform the necessary integration, to obtain the element mass

$$[M^e] = \frac{tm_e}{b^2\ell^2} \int_0^\ell \int_0^{b(1-x/\ell)} \begin{bmatrix} b\ell - bx - \ell y & 0 \\ 0 & b\ell - bx - \ell y \\ bx & 0 \\ 0 & bx \\ \ell y & 0 \\ 0 & \ell y \end{bmatrix}$$

$$\cdot \begin{bmatrix} bl - bx - ly & 0 & bx & 0 & ly & 0 \\ 0 & bl - bx - ly & 0 & bx & 0 & ly \end{bmatrix} tdydx$$

$$[M^e] = \frac{tm_e}{b^2\ell^2} \int_0^\ell \int_0^{b(1-x/\ell)}$$

$$\times \begin{bmatrix} (b\ell - bx - \ell y)^2 & & & & & \\ 0 & (b\ell - bx - \ell y)^2 & & & \text{symmetric} & \\ bx\,(b\ell - bx - \ell y) & 0 & (bx)^2 & & & \\ 0 & bx\,(b\ell - bx - \ell y) & 0 & (bx)^2 & & \\ \ell y\,(b\ell - bx - \ell y) & 0 & b\ell xy & 0 & (\ell y)^2 & \\ 0 & \ell y\,(b\ell - bx - \ell y) & 0 & b\ell xy & 0 & (\ell y)^2 \end{bmatrix} dxdy$$

$$= \frac{tb\ell m_e}{24} \begin{bmatrix} 2 & & & & & \\ 0 & 2 & & & \text{symmetric} & \\ 1 & 0 & 2 & & & \\ 0 & 1 & 0 & 2 & & \\ 1 & 0 & 1 & 0 & 2 & \\ 0 & 1 & 0 & 1 & 0 & 2 \end{bmatrix} \tag{10.136}$$

For elements 2 and 4 in a similar fashion, the element mass matrix is defined as

$$[M^e] = \frac{m_e}{b^2\ell^2}\int_0^{\ell}\int_{b(1-x/\ell)}^{b}\begin{bmatrix} b\ell - bx & 0 \\ 0 & b\ell - bx \\ b\ell - \ell y & 0 \\ 0 & b\ell - \ell y \\ -b\ell - bx + \ell y & 0 \\ 0 & -b\ell + bx + \ell y \end{bmatrix}$$

$$\cdot \begin{bmatrix} b\ell - bx & 0 & b\ell - by & 0 & \begin{matrix}-b\ell \\ -bx + \ell y\end{matrix} & 0 \\ 0 & b\ell - bx & 0 & b\ell - \ell y & 0 & \begin{matrix}-b\ell \\ +bx + \ell y\end{matrix} \end{bmatrix} t\,dy\,dx$$

$$= \frac{tm_e}{b^2\ell^2}\int_0^{\ell} 1 \int_{b(1-x/\ell)}^{b}$$

$$\cdot \begin{bmatrix} (b\ell - bx)^2 & & & & & \\ 0 & (b\ell - bx)^2 & & & \text{symmetric} & \\ (b\ell - \ell y)(bl - bx) & 0 & (b\ell - by)^2 & & & \\ 0 & (b\ell - \ell y)(b\ell - bx) & 0 & (b\ell - \ell y)^2 & & \\ (-b\ell + bx + \ell y)(b\ell - bx) & 0 & (-b\ell + bx + \ell y)(b\ell - by) & 0 & (-b\ell + bx + \ell y)^2 & \\ 0 & (-b\ell + bx + \ell y)(b\ell - bx) & 0 & (-b\ell + bx + \ell y)(b\ell - \ell y) & 0 & (-b\ell + bx + \ell y)^2 \end{bmatrix} dy\,dx$$

$$= \frac{tb\ell m_e}{24}\begin{bmatrix} 2 & & & & & \\ 0 & 2 & & \text{symmetric} & & \\ 1 & 0 & 2 & & & \\ 0 & 1 & 0 & 2 & & \\ 1 & 0 & 1 & 0 & 2 & \\ 0 & 1 & 0 & 1 & 0 & 2 \end{bmatrix} \qquad (10.137)$$

From equation (10.136) and equation (10.137) we see that, all the elements have the same element mass matrix.

The nodal force vector [see equation (10.71)] is defined by

$$\{F_s\} = \int_v [N]^T \{F\} dv \tag{10.138}$$

where the gravity force contribution at the element level can be expressed as

$$\{F\} = -m_e g \begin{bmatrix} \sin\varphi \\ \cos\varphi \end{bmatrix}$$

Then for elements 1 and 3, the nodal force vector becomes

$$\{F_S\} = -\frac{m_e g}{b\ell} \int_0^\ell \int_0^{b(1-x/\ell)} \begin{bmatrix} b\ell - bx - \ell y & 0 \\ 0 & b\ell - bx - \ell y \\ bx & 0 \\ 0 & bx \\ \ell y & 0 \\ 0 & \ell y \end{bmatrix} \begin{bmatrix} \sin\varphi \\ \cos\varphi \end{bmatrix} t dy dx$$

$$= -\frac{m_e g t b\ell}{6} \begin{bmatrix} 1 & 0 \\ 0 & 1 \\ 1 & 0 \\ 0 & 1 \\ 1 & 0 \\ 0 & 1 \end{bmatrix} \begin{bmatrix} \sin\varphi \\ \cos\varphi \end{bmatrix} = -\frac{1}{6} M_e g t b\ell \begin{bmatrix} \sin\varphi \\ \cos\varphi \\ \sin\varphi \\ \cos\varphi \\ \sin\varphi \\ \cos\varphi \end{bmatrix} \tag{10.139}$$

For elements 2 and 4, the nodal force vector $\{F_s\}$ is the same. The skew matrix of angular velocity $[\Omega^{01}]$ is also the same as the one given by equation (10.85); similarly, $[\dot{\Omega}^{01}]$ is the one provided by equation (10.87).

The element dynamic damping matrix $[C_d]$ matrix is (assuming that all the elements have the same coefficients)

$$[C_d] = 2 \int_v m_e [N]^T [\Omega^{01}][N] dv$$

$$= \frac{2m_e}{b^2\ell^2} \int_0^\ell \int_0^{b(1-x/\ell)} \begin{bmatrix} b\ell - bx - \ell y & 0 \\ 0 & b\ell - bx - \ell y \\ bx & 0 \\ 0 & bx \\ \ell y & 0 \\ 0 & \ell y \end{bmatrix} \dot{\varphi} \begin{bmatrix} 0 & -1 \\ 1 & 0 \end{bmatrix}$$

$$\cdot \begin{bmatrix} b\ell - bx - \ell y & 0 & bx & 0 & \ell y & 0 \\ 0 & b\ell - bx - \ell y & 0 & bx & 0 & \ell y \end{bmatrix} tdxdy$$

$$= \frac{2m_e t\dot{\varphi}}{b^2\ell^2} \int_0^{\ell} \int_0^{b(1-x/\ell)}$$

$$\begin{bmatrix} 0 & & \text{antisymmetric} \\ (b\ell - bx - \ell y)^2 & 0 & 0 \\ 0 & -bx(b\ell - bx - \ell y) & b^2x^2 0 \\ bx(b\ell - bx - \ell y) & 0 & 0 - bx\ell y 0 \\ 0 & -\ell y(b\ell - bx - \ell y) & bx\ell y 0 \ell^2 y^2 0 \\ by(b\ell - bx - \ell y) & 0 & \end{bmatrix} dydx$$

$$= \frac{m_e t b\ell\dot{\varphi}}{12} \begin{bmatrix} 0 & & & & & \\ 2 & 0 & \text{antisymmetric} & & & \\ 0 & -1 & 0 & & & \\ 1 & 0 & 2 & 0 & & \\ 0 & -1 & 0 & -1 & 0 & \\ 1 & 0 & 1 & 0 & 2 & 0 \end{bmatrix} \tag{10.140}$$

The element total damping matrix [see equation (10.69)] is then obtained·

$$[C^e] = [C_s] + [C_d] = [C_d] \tag{10.141}$$

The element dynamic stiffness matrix for all the elements [see equation (10.70)] is defined by

$$[K_d] = \int_v m_e[N]^T([\dot{\Omega}^{o1}] + [\Omega^{o1}][\Omega^{o1}])[N]dv$$

$$= \frac{m_e}{b^2\ell^2} \int_0^{\ell} \int_0^{b(1-x/\ell)} \begin{bmatrix} b\ell - bx - \ell y & 0 \\ 0 & b\ell - bx - \ell y \\ bx & 0 \\ 0 & bx \\ \ell y & 0 \\ 0 & \ell y \end{bmatrix} \left(\ddot{\varphi} \begin{bmatrix} 0 & -1 \\ 1 & 0 \end{bmatrix} \right.$$

$$\left. + \dot{\varphi}^2 \begin{bmatrix} -1 & 0 \\ 0 & -1 \end{bmatrix} \right) \begin{bmatrix} b\ell - bx - \ell y & 0 & bx & 0 & \ell y & 0 \\ 0 & b\ell - bx - \ell y & 0 & bx & 0 & \ell y \end{bmatrix} tdydx$$

$$= \frac{m_e t}{b^2\ell^2} \int_0^{\ell} \int_0^{b(1-x/\ell)}$$

$$
\cdot\begin{bmatrix}
(b\ell - bx - \ell y)^2\dot{\varphi}^2 & \\
-(b\ell - bx - \ell y)^2\ddot{\varphi} & (b\ell - bx - \ell y)^2\dot{\varphi}^2 \\
(b\ell - bx - \ell y)bx\dot{\varphi}^2 & (b\ell - bx - \ell y)bx\ddot{\varphi} \\
-(b\ell - bx - \ell y)bx\ddot{\varphi} & (b\ell - bx - \ell y)bx\,\dot{\varphi}^2 \\
(b\ell - bx - \ell y)\ell y\dot{\varphi}^2 & (b\ell - bx - \ell y)\ell y\,\ddot{\varphi} \\
-(b\ell - bx - \ell y)\ell y\ddot{\varphi} & (b\ell - bx - \ell y)ly\,\dot{\varphi}^2
\end{matrix}
$$

$$
\begin{matrix}
\text{antisymmetric} & & & \\
b^2x^2\dot{\varphi}^2 & & & \\
-b^2x^2\ddot{\varphi} & b^2x^2\dot{\varphi}^2 & & \\
bx\ell y\dot{\varphi}^2 & bx\ell y\ddot{\varphi}^2 & bx\ell y\dot{\varphi}^2 & \\
-bx\ell y\ddot{\varphi} & bx\ell y\dot{\varphi}^2 & -\ell^2y^2\ddot{\varphi} & \ell^2y^2\dot{\varphi}^2
\end{bmatrix} dydx
$$

$$
= \frac{tb\ell m_e}{24}\begin{bmatrix}
-2\dot{\varphi}^2 & -2\ddot{\varphi} & -\dot{\varphi}^2 & -\ddot{\varphi} & -\dot{\varphi}^2 & -\ddot{\varphi} \\
+2\ddot{\varphi} & -2\dot{\varphi}^2 & \ddot{\varphi} & -\dot{\varphi}^2 & +\ddot{\varphi} & -\dot{\varphi}^2 \\
-\dot{\varphi}^2 & -\ddot{\varphi} & -2\dot{\varphi}^2 & -2\ddot{\varphi} & -\dot{\varphi}^2 & -\ddot{\varphi}^2 \\
+\ddot{\varphi} & -\dot{\varphi}^2 & +2\ddot{\varphi} & -2\dot{\varphi}^2 & \ddot{\varphi} & -\dot{\varphi}^2 \\
-\dot{\varphi}^2 & -\ddot{\varphi} & -\dot{\varphi}^2 & -\ddot{\varphi} & -2\dot{\varphi}^2 & -2\ddot{\varphi} \\
+\ddot{\varphi} & -\dot{\varphi}^2 & +\ddot{\varphi} & -\dot{\varphi}^2 & +2\ddot{\varphi} & -2\dot{\varphi}^2
\end{bmatrix} \tag{10.142}
$$

The element total stiffness matrix has both a stiffness and damping contribution where

$$
[K^e] = [K_s] + [K_d] \tag{10.143}
$$

where $[K_s]$ is evaluated by equations (10.133) and (10.134).

The element dynamic force matrix [see equation (10.71)] is computed as follows:

$$
\{F_d\} = -\int_v m_e[N]^T \cdot
$$

$$
\cdot\left\{([\dot{\Omega}^{01}] + [\Omega^{01}][\Omega^{01}])\{r_n\} + \sum_{i=2}^{n}[S^{nv}]([\dot{\Omega}^{ov}] + [\Omega^{ov}][\Omega^{ov}])\{q_i\}\right\} dv \tag{10.144}
$$

where

$$
n = 1, \qquad v = i - 1, \qquad \{r_n\} = \begin{bmatrix} x \\ y \end{bmatrix}
$$

After substitution in the above equation, we get

$$\{F_d\} = -\frac{m_e}{b\ell}\int_0^{\ell}\int_0^{b(1-x/\ell)} \begin{bmatrix} b\ell - bx - \ell y & 0 \\ 0 & b\ell - bx - \ell y \\ bx & 0 \\ 0 & bx \\ \ell y & 0 \\ 0 & \ell y \end{bmatrix}$$

$$\cdot\left(\ddot{\varphi}\begin{bmatrix} 0 & -1 \\ 1 & 0 \end{bmatrix} + \dot{\varphi}^2\begin{bmatrix} -1 & 0 \\ 0 & -1 \end{bmatrix}\right)\begin{bmatrix} x \\ y \end{bmatrix} t\,dx\,dy$$

$$= -\frac{m_e t}{b\ell}\int_0^{\ell}\int_0^{b(1-x/\ell)} \begin{bmatrix} (b\ell - bx - \ell y)(x\dot{\varphi} + y\ddot{\varphi}) \\ (b\ell - bx - \ell y)(-x\ddot{\varphi} + y\dot{\varphi}^2) \\ bx(x\dot{\varphi}^2 + y\ddot{\varphi}\ ) \\ bx(-x\ddot{\varphi} + y\dot{\varphi}^2) \\ \ell y(x\dot{\varphi}^2 + y\ddot{\varphi}) \\ \ell y(-x\ddot{\varphi} + y\dot{\varphi}^2) \end{bmatrix} dy\,dx$$

$$= -\frac{m_e t b\ell}{24}\begin{bmatrix} \ell\dot{\varphi}^2 + b\ddot{\varphi} \\ -\ell\ddot{\varphi} + b\dot{\varphi}^2 \\ 2\ell\dot{\varphi}^2 + b\ddot{\varphi} \\ -\ell\ddot{\varphi} + b\dot{\varphi}^2 \\ \ell\dot{\varphi}^2 + b\ddot{\varphi} \\ -\ell\ddot{\varphi} + 2b\dot{\varphi}^2 \end{bmatrix} \tag{10.145}$$

The element total force vector $\{F^e\}$ is then found as follows:

$$\{F^e\} = \{F_c\} + \{F_d\} = -\frac{1}{6}m_e g t b\ell \begin{bmatrix} \sin\varphi \\ \cos\varphi \\ \sin\varphi \\ \cos\varphi \\ \sin\varphi \\ \cos\varphi \end{bmatrix} - \frac{m_e t b\ell}{24}\begin{bmatrix} \ell\dot{\varphi}^2 + b\ddot{\varphi} \\ -\ell\ddot{\varphi} + b\dot{\varphi}^2 \\ 2\ell\dot{\varphi}^2 + b\ddot{\varphi} \\ -\ell\ddot{\varphi} + b\dot{\varphi}^2 \\ \ell\dot{\varphi}^2 + b\ddot{\varphi} \\ -\ell\ddot{\varphi} + 2b\dot{\varphi}^2 \end{bmatrix}$$

$$= -\frac{m_e t b\ell}{24}\begin{bmatrix} 4g\sin\varphi + \ell\dot{\varphi}^2 + b\ddot{\varphi} \\ 4g\cos\varphi - \ell\ddot{\varphi} + b\dot{\varphi}^2 \\ 4g\sin\varphi + 2\ell\dot{\varphi}^2 + b\ddot{\varphi} \\ 4g\cos\varphi - \ell\ddot{\varphi} + b\dot{\varphi}^2 \\ 4g\sin\varphi + \ell\dot{\varphi}^2 + b\ddot{\varphi} \\ 4g\cos\varphi - \ell\ddot{\varphi} + 2b\dot{\varphi}^2 \end{bmatrix} \tag{10.146}$$

After proper assembly, we obtain the global matrix representation of the equations of motion:

$$[M]\{\ddot{\rho}\} + [C]\{\dot{\rho}\} + [K]\{\rho\} = \{F\} \tag{10.147}$$

where $[M]$ is assembled making use of the element masses obtained by equation (10.136):

$$[M] = \frac{tb\ell m_e}{24}\begin{bmatrix} 2 &&&&&&&&&&& \\ 0 & 2 &&&&&&&&&& \\ 1 & 0 & 6 &&&&&&&&& \\ 0 & 1 & 0 & 6 & \text{symmetric} &&&&&&& \\ 0 & 0 & 1 & 0 & 4 &&&&&&& \\ 0 & 1 & 0 & 1 & 0 & 4 &&&&&& \\ 1 & 0 & 2 & 0 & 0 & 0 & 4 &&&&& \\ 0 & 1 & 0 & 2 & 0 & 0 & 0 & 4 &&&& \\ 0 & 0 & 2 & 0 & 2 & 0 & 1 & 0 & 6 &&& \\ 0 & 0 & 0 & 2 & 0 & 2 & 0 & 1 & 0 & 6 && \\ 0 & 0 & 0 & 0 & 1 & 0 & 0 & 0 & 1 & 0 & 2 & \\ 0 & 0 & 0 & 0 & 0 & 1 & 0 & 0 & 0 & 1 & 0 & 2 \end{bmatrix} \tag{10.148}$$

Similarly, we get

$$[C] = \frac{m_e tb\ell\dot{\varphi}}{24}\begin{bmatrix} 0 &&&&&&&&&&& \\ 2 & 0 &&&&&&&&&& \\ 0 & -1 & 0 &&&&&&&&& \\ 1 & 0 & 6 & 0 &&&&&&&& \\ 0 & 0 & 0 & -1 & 0 & \text{antisymmetric} &&&&&& \\ 0 & 0 & 1 & 0 & 4 & 0 &&&&&& \\ 0 & -1 & 0 & -2 & 0 & 0 & 0 &&&&& \\ 1 & 0 & 2 & 0 & 0 & 0 & 4 & 0 &&&& \\ 0 & 0 & 0 & -2 & 0 & -2 & 0 & -1 & 0 &&& \\ 0 & 0 & 2 & 0 & 2 & 0 & 1 & 0 & 6 & 0 && \\ 0 & 0 & 0 & 0 & 0 & -1 & 0 & 0 & 0 & -1 & 0 & \\ 0 & 0 & 0 & 0 & 1 & 0 & 0 & 0 & 1 & 0 & 2 & 0 \end{bmatrix} \tag{10.149}$$

and the generalized force vector is found to take the following form

$$\{F\} = -\frac{m_e t b \ell}{24} \begin{bmatrix} 4g \sin\varphi + \ell\dot{\varphi}^2 + b\ddot{\varphi} - (24F1/meb\ell) \\ 4g \cos\varphi - \ell\ddot{\varphi} + b\dot{\varphi}^2 - (24F2/meb\ell) \\ 12g \sin\varphi + 3\ell\dot{\varphi}^2 + 3b\ddot{\varphi} \\ 12g \cos\varphi - 3\ell\ddot{\varphi} + 3b\dot{\varphi}^2 \\ 8g \sin\varphi + 2\ell\dot{\varphi}^2 + 2b\ddot{\varphi} \\ 8g \cos\varphi - 2\ell\ddot{\varphi} + 2b\dot{\varphi}^2 \\ 8g \sin\varphi + 2\ell\dot{\varphi}^2 + 2b\ddot{\varphi} \\ 8g \cos\varphi - 2\ell\ddot{\varphi} + 2b\dot{\varphi}^2 \\ 12g \sin\varphi + 3\ell\dot{\varphi}^2 + 3b\ddot{\varphi} \\ 12g \cos\varphi - 3\ell\ddot{\varphi} + 3b\dot{\varphi}^2 \\ 4g \sin\varphi + \ell\dot{\varphi}^2 + b\ddot{\varphi} \\ 4g \cos\varphi - \ell\ddot{\varphi} + b\dot{\varphi}^2 \end{bmatrix} \tag{10.150}$$

where F_1 is the constraint force along the $\bar{n}_1^1$ direction and F_2 is the constraint force along the $\bar{n}_2^1$ direction. The nodal displacement vector showing all the nodal degrees of freedom is given by

$$\{\rho\}^T = [u_{1x} \;\; u_{1y} \;\; u_{2x} \;\; u_{2y} \;\; u_{3x} \;\; u_{3y} \;\; u_{4x} \;\; u_{4y} \;\; u_{5x} \;\; u_{5y} \;\; u_{6x} \;\; u_{6y}] \tag{10.151}$$

where u_{ix} and u_{iy} $(i = 1, 6)$ are the nodal displacements associated with node i. The global stiffness matrix $[K]$ is too long to list here; however, given the element stiffness [see equation (10.143)] we can assemble it without any difficulty. For the problem at hand, the beam is assumed to be pinned to a rotating frame; then the first and second elements of $\{\rho\}^T$ are zero, that is,

$$u_{1x} = 0$$
$$u_{1y} = 0$$

We can delete the first and second rows and columns from the matrices $[M]$, $[C]$ and $[K]$ and the first and second components of $\{\rho\}$ and $\{F\}$. Because of the nature of matrix $[C]$, the equations of motion cannot be decoupled, and the solution to the governing equations of motion can only be carried out by the method of direct integration. Let $\{\ddot{\rho}\} = -\omega^2\{\rho\}$; then from

$$[M]\{\ddot{\rho}\} + [K]\{\rho\} = 0 \tag{10.152}$$

obtain the modal frequencies of the system.

After integration, one can obtain $\{\rho\}$ and solve for F_1 and F_2 from equation (10.147). Then from

$$\{u\} = [N]\{\rho\}$$

we can evaluate the displacement of an arbitrary point on the beam.

10.5 General Equations of Motion of Multibody Systems with Flexible Terminal Links

We discussed in Section 10.4 the development of equations of motion, where the slope of the flexible body was assumed to be small and the angular velocity of the element(s) in the flexible body were negligible. Therefore,

$$\bar{T}^* \cdot \frac{\partial \bar{\omega}}{\partial \dot{\rho}} = 0 \tag{10.153}$$

and the generalized inertia forces reduce to

$$f^* = \bar{F}^* \cdot \frac{\partial \bar{v}}{\partial \dot{\rho}} + \bar{T}^* \cdot \frac{\partial \bar{\omega}}{\partial \dot{\rho}} = \bar{F}^* \cdot \frac{\partial \bar{v}}{\partial \dot{\rho}} \tag{10.154}$$

Now, let us reconsider the multibody system shown in Figure 10.15, where we are going to take into consideration the effects of element rotations due to the elastic deformation of the body. Of course, this is when the elastic body itself is undergoing large rotation. Consider the position vector from the fixed reference frame R to an arbitrary point belonging to an element in body B_N. First we can see that the local position within the body can be expressed as

$$\bar{r}_u = \bar{r}_n + \bar{u} \tag{10.155}$$

and

$$\{u\} = [N]\{\rho\} \tag{10.156}$$

where $[N]$ denotes the shape function matrix and $\{\rho\}$ the vector of nodal unknowns. From Figure 10.15, the position vector $\bar{P}$ is then

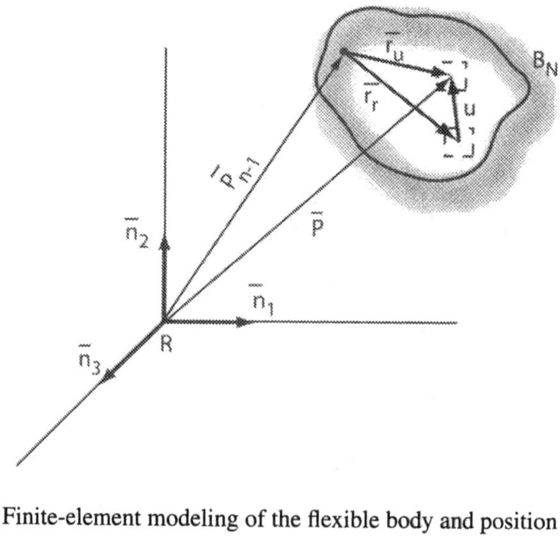

Fig. 10.15. Finite-element modeling of the flexible body and position vector rotations

$$\bar{P} = \bar{P}_{n-1} + \bar{r}_u \tag{10.157}$$

Expressing all the vector components with respect to the unit vectors fixed in R, we get

$$\bar{p} = \left\{ \sum_{i-1}^{n} \{q_i\}^T [S^{VO}] + (\{r_n\}^T + \{\rho\}^T [N]^T)[S^{no}] \right\} \{\bar{n}\} \tag{10.158}$$

Both the velocity and acceleration of the arbitrary point P in R $\bar{\upsilon}_p$ and $\bar{a}_p$ are given by equations (10.43) and (10.44), respectively.

The angular velocity of body n due to rigid-body motion is obtained from equation (3.40), where

$$^R\bar{\omega}^n = \{\dot{x}\}^T [\omega^n]\{\bar{n}\} \tag{10.159}$$

and the angular acceleration

$$^R\bar{\alpha}^n = (\{\ddot{x}\}^T [\omega^n] + \{\dot{x}\}^T [\dot{\omega}^n])\{\bar{n}\} \tag{10.160}$$

Let the slope of the element i be θ; then we can write the element rotation as function of the modal coordinates

$$\{\theta\} = \{\psi\}\{\rho\} \tag{10.161}$$

where $[\psi]$ denotes the shape function matrix and ρ the modal displacements. For the element i, the angular velocity becomes

$$^R\bar{\omega}^i = ^R \bar{\omega}^N + ^N\bar{\omega}^i = (\{\dot{x}\}^T [\omega^n] + \{\dot{\rho}\}[\psi]^T [S^{no}])\{\bar{n}\} \tag{10.162}$$

Notice how the second term denotes the angular velocity of the element i of the elastic body. This corresponds to the contribution due to the flexibility effects of element rotation. The angular acceleration by differentiation of the above yields

$$^R\bar{\omega}^i = (\{\ddot{x}\}^T [\omega^n] + \{\dot{x}\}^T [\dot{\omega}^n] + \{\ddot{\rho}\}^T [\psi]^T [S^{no}] + \{\dot{\rho}\}^T [\psi]^T [\Omega^{no}][S^{no}])\{\bar{n}\}$$

From Kane's equations we can formulate the governing equations of motions. That is, using

$$f + f^* = 0 \tag{10.163}$$

where the generalized inertia forces are

$$f^* = \bar{F}^* \cdot \frac{\partial \bar{\upsilon}}{\partial \dot{\rho}} + \bar{T}^* \cdot \frac{\partial \bar{\omega}}{\partial \dot{\rho}} \tag{10.164}$$

for the element i the inertia force is given by

$$\bar{F}^* = -m_i \bar{a}_p \tag{10.165}$$

where m_i denotes the mass density of element i of the flexible body. From equation (10.43), we obtain the partial velocity

$$\frac{\partial \bar{\upsilon}}{\partial \dot{\rho}} = [N]^T [S^{no}]\{\bar{n}\} \tag{10.166}$$

The first term in equation (10.164) results in a contribution of the inertia forces:

$$\begin{aligned}
f^* &= \bar{F}^* \cdot \frac{\partial \bar{\upsilon}}{\partial \dot{\rho}} = -m_i \bar{a}_P \cdot \frac{\partial \bar{\upsilon}}{\partial \dot{\rho}} \\
&= -m_i \left\{ \sum_{i=2}^{n} (\{q_i\}^T [\dot{\Omega}^{\upsilon o}][S^{\upsilon o}] + \{q_i\}^T [\Omega^{\upsilon o}][\Omega^{\upsilon o}][S^{\upsilon o}]) \right. \\
&\quad + (\{r_n\}^T + \{\rho\}^T [N]^T)[\dot{\Omega}^{no}][S^{no}] + (\{r_n\}^T + \{\rho\}^T [N]^T)[\Omega^{no}][\Omega^{no}][S^{no}] \\
&\quad \left. + 2\{\dot{\rho}\}^T [N]^T [\Omega^{no}][S^{no}] + \{\ddot{\rho}\}^T [N]^T [S^{no}] \right\} \{\bar{n}\} \cdot \{\bar{n}\}^T [S^{no}]^T [N] \qquad (10.167) \\
&= -m_i \left\{ \{\ddot{\rho}\}^T [N]^T [N] + 2\{\dot{\rho}\}^T [N]^T [\Omega^{no}][N] + \{\rho\}^T [N]^T ([\dot{\Omega}^{no}] \right. \\
&\quad + [\Omega^{no}][\Omega^{no}])[N] + \{r_n\}^T ([\dot{\Omega}^{no}] + [\Omega^{no}][\Omega^{no}])[N] \\
&\quad \left. + \sum_{i=2}^{n} (\{q_i\}^T [\dot{\Omega}^{no}] + \{q_i\}^T [\Omega^{\upsilon o}][\Omega^{\upsilon o}])[s^{\upsilon n}][N] \right\}
\end{aligned}$$

where

$$\upsilon = i - 1$$

The moment of inertia needed for the completion of the formulation of the generalized inertia formula is given by

$$T^* = (\bar{\bar{I}}_i \cdot \bar{\omega}) \times \bar{\omega} - \bar{\bar{I}}_i \cdot \bar{\alpha} \tag{10.168}$$

where $\bar{\bar{I}}_i$ denotes the unit volume inertia dyadic of element i, and $\bar{\omega}$ and $\bar{\alpha}$ are its angular velocity and angular acceleration, respectively. Expressing the latter with respect to the unit vectors $\bar{n}$ in R, we get

$$\bar{\bar{I}}_i = \{\bar{n}\}^T [S^{on}][I_i][S^{no}]\{\bar{n}\} = \{\bar{n}\}^T [I_{no}]\{\bar{n}\} \tag{10.169}$$

where $[I_i]$ denotes the unit volume inertia movement of element i:

$$[I_i] = m_i \begin{bmatrix} y^2 + z^2 & xy & xz \\ yx & x^2 + y^2 & yz \\ zx & zy & x^2 + y^2 \end{bmatrix} \tag{10.170}$$

From equation (10.162) the partial angular velocity is found to be

$$\frac{\partial \bar{\omega}}{\partial \dot{\rho}} = [\psi]^T[S^{no}]\{\bar{n}\} \tag{10.171}$$

The second term in the generalized inertia force is then obtained as follows:

$$\begin{aligned} T^* \cdot \frac{\partial \bar{\omega}}{\partial \dot{\rho}} &= -\{\{\dot{x}\}^T[\omega^N][I_{no}]^T[\Omega_x^{no}] + \{\ddot{x}\}^T[\omega^n][I_{no}]^T + \{\dot{x}\}^T[\dot{\omega}^n][I_{no}]^T \\ &\quad + \{\dot{\rho}\}^T[\psi]^T[S^{no}][I_{no}]^T[\Omega_\rho^{no}] + \{\ddot{\rho}\}^T[\psi]^T[S^{no}][I_{no}]^T \\ &\quad + \{\dot{\rho}\}^T[\psi]^T[\Omega^{no}][S^{no}][I_{no}]^T\}[S^{no}]^T[\psi] \\ &= -\{\{\ddot{\rho}\}^T[\psi]^T[I_i][\psi] + \{\dot{\rho}\}^T[\psi]^T([\Omega^{no}][I_i] - [I_i][S^{no}][\Omega_\rho^{no}][S^{no}])[\psi]\} \\ &\quad - \{\{\ddot{x}\}^T[\omega^n][S^{on}][I_i] + \{\dot{x}\}^T[\dot{\omega}^n][S^{on}][I_i] \\ &\quad - [\dot{\omega}^n][S^{on}][I_i][S^{no}][\Omega_x^{no}][S^{on}])\}[\psi] \end{aligned} \tag{10.172}$$

Hence

$$\begin{aligned} f^* &= \bar{F} * \cdot \frac{\partial \bar{v}}{\partial \dot{\rho}} + \bar{T} * \cdot \frac{\partial \bar{\omega}}{\partial \dot{\rho}} \\ &= -\Big\{\{\ddot{\rho}\}^T([N]^T m_i[N] + [\psi]^T[I_i][\psi]) + \{\dot{\rho}\}^T(2[N]^T m_i[\Omega^{no}][N] \\ &\quad + [\psi]^T([\Omega^{no}][I_i] + [I_i][S^{no}][\Omega_\rho^{no}][S^{on}])[\psi]) + m_i\{\rho\}^T[N]^T([\dot{\Omega}^{no}] \\ &\quad + [\Omega^{no}][\Omega^{no}])[N] + m_i\{r_n\}^T([\dot{\Omega}^{no}] + [\Omega^{no}][\Omega^{no}])[N] + m_i\sum_{i=1}^{N-1}(\{q_i\}^T[\dot{\Omega}^{no}] \\ &\quad + \{q_i\}^T[\Omega^{vo}][\Omega^{vo}])[S^{vn}][N] + \{\ddot{x}\}^T[\omega^n][S^{on}][I_i] + \{\dot{x}\}^T([\dot{\omega}^n][S^{on}][I_i] \\ &\quad + [\omega^n][S^{on}][I_i][S^{on}][\Omega_x^{no}][S^{on}])[\psi]\Big\} \end{aligned} \tag{10.173}$$

where $v = i - 1[\Omega_x^{no}]$ is the skew matrix of $\{\dot{x}\}^T$, and $[\omega^n]$ and $[\Omega_\rho^{no}]$ are the skew matrices of $\{\dot{\rho}\}^T[\psi]^T[\Omega^{no}][S^{no}]$.

In the equations of motion given by equation (10.163), the generalized force f can be expressed as

$$f = f_e - f_k - f_c \tag{10.174}$$

where the generalized external forces acting on the element denoted by f_e are given by

$$f_e = \bar{F} \cdot \frac{\partial \bar{v}}{\partial \dot{\rho}} + \bar{T} \cdot \frac{\partial \bar{\omega}}{\partial \dot{\rho}} \tag{10.175}$$

where $\bar{F}$ and $\bar{T}$ are the external forces and moment expressed in R for the element at hand. f_k represents the generalized stiffness force, f_c the generalized damping force. The forces and moments acting on an element can be expressed in R as

$$\bar{F} = \{F\}^T\{\bar{n}^n\} = \{F\}^T[S^{no}]\{\bar{n}\} \quad (10.176)$$

$$\bar{T} = \{T\}^T\{\bar{n}^n\} = \{T\}^T[S^{no}]\{\bar{n}\} \quad (10.177)$$

where $\{F\}^T$ and $\{T\}^T$ denote the resulting forces and moments components with respect to the body axis $\{\bar{n}^N\}$. Then

$$\begin{aligned} f_e &= \{F\}^T[S^{no}]\{\bar{n}\}\cdot\{\bar{n}\}^T[S^{no}]^T[N] + \{T\}^T[S^{no}]\{\bar{n}\}\cdot\{\bar{n}\}^T[S^{no}]^T[\psi] \\ &= \{F\}^T[N] + \{T\}^T[\psi] \end{aligned} \quad (10.178)$$

Assume that the stiffness and damping matrices are $[K]$ and $[C]$, respectively. Note that $[K] = [K]^T$ and $[C] = [C]^T$, and from the finite-element method,

$$[K] = [B]^T[D][B] \quad (10.179)$$

and for a beam element, equation (10.179) reduces to

$$[K] = [N]''^T[D][N]'' \quad (10.180)$$

where $[B]$ denotes the strain–modal displacement matrix, $[D]$ the material property matrix, and $[N]''$ the second derivate of $[N]$. Similarly,

$$[C] = [N]^T[\mu][N] \quad (10.181)$$

where $[\mu]$ denotes the material damping coefficient matrix.

Since f_k, f_c and ρ are all row vectors, then

$$f_k = \{\rho\}^T[K] \quad (10.182)$$

$$f_c = \{\dot{\rho}\}^T[C] \quad (10.183)$$

Therefore, the generalized active force vector becomes

$$f = f_e - f_k - f_c = \{F\}^T[N] + \{T\}^T[\psi] - \{\rho\}^T[K] - \{\dot{\rho}\}^T[C] \quad (10.184)$$

To be consistent with the ordinary fashion of expressing the equations, substituting equations (10.173), (10.178), (10.180) and (10.181) into (10.163) and integrating, we obtain the equations of motion for element i of the terminal elastic body.

$$[M^e]\{\ddot{\rho}\} + [C^e]\{\dot{\rho}\} + [K^e]\{\rho\} = \{F^e\} \quad (10.185)$$

where the element mass matrix contribution from both translation and rotation is

$$[M^e] = \int_v ([N]^T m_i[N] + [\psi]^T[I_i][\psi])dv \quad (10.186)$$

The element damping matrix can be expressed as

$$\begin{aligned} [C^e] = \int_v [N]^T[\mu][N]dv + \int_v \{2[N]^T m_i[\Omega^{on}][N] + [\psi]^T([I_i][\Omega^{on}] \\ + [S^{no}][\Omega_\rho^{on}][S^{on}][I_i])[\psi]\}dv = [C_s^e] + [C_d^e] \end{aligned} \quad (10.187)$$

where $[C_s^e]$ denotes the ordinary damping matrix and $[C_d^e]$ the dynamic damping matrix due to rigid-body motion. The stiffness matrix from equation (10.5.30) is found in a similar fashion as

$$
\begin{aligned}
[K^e] &= \int_v [B]^T[D][B]dv + \int_v [N]^T m_i([\dot{\Omega}^{on}] + [\Omega^{on}][\Omega^{on}])[N]dv \\
&= [K_s^e] + [K_d^e]
\end{aligned} \tag{10.188}
$$

where $[K_s^e]$ denotes ordinary stiffness matrix and $[K_d^e]$ dynamic stiffness matrix due to rigid-body motion. The force vector on the right-hand side of the equations of motion for an arbitrary element in R is

$$
\begin{aligned}
\{F^e\} &= \int_v ([N]^T\{F\} + [\psi]^T\{T\}dv - \int_v \Big\{[N]^T m_i([\dot{\Omega}^{on}] \\
&\quad + [\Omega^{on}][\Omega^{on}])\{r_n\} + [N]^T m_i \sum_{i=1}^{N} [S^{nv}][\dot{\Omega}^{ov}] + [\Omega^{ov}][\Omega^{ov}])\{q_i\} \\
&\quad + [\psi]^T[I_i][S^{no}][\omega^n]^T\{\ddot{x}\} + [\psi]^T([I_i][S^{no}][\dot{\omega}^n]^T \\
&\quad - [S^{no}][\Omega_x^{on}][S^{on}][I_i][S^{no}][\omega^N]^T)\{\dot{x}\}\Big\} dv \\
&= \{F_s^e\} + \{F_d^e\}
\end{aligned} \tag{10.189}
$$

where $\{F_s^e\}$ denotes an ordinary force vector carrying the external applied forces and torques including springs and dampers, and $\{F_d^e\}$ represents the dynamic force vector due to the rigid-body movement.

Assembling all of the element matrices by equation (10.185), we obtain the global equations of motion of the system:

$$
[M]\{\ddot{\rho}\} + [C]\{\dot{\rho}\} + [K]\{\rho\} = \{F\} \tag{10.190}
$$

The equations above represent the governing equations of motion of a multibody system with a flexible terminal link undergoing large motions. The equations take into consideration the element translational and rational vibration (perturbation). Solution of these equations requires computer programming where numerical solutions can be sought.

10.6 Six-Dimensional Beam Element Modeling of Terminal Flexible Links in MBS

In Sections 10.4 and 10.5, we considered the elastic to be a beam discretized into elements, where they were first treated as simple and then as triangular. In both cases we were able to see the effects of longitudinal, transverse, and rotational vibration. In this section we extend our analysis to three dimensions, and a three-dimensional

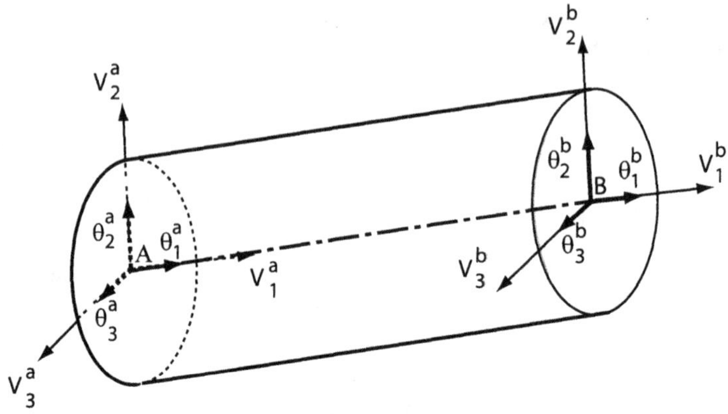

Fig. 10.16. Six-dimensional beam element with nodes A and B

beam element is used (Figure 10.16). Let υ_i and θ_i be expressed in terms of the nodal variables, ρ_j by utilizing polynomials of appropriate order as

$$\begin{aligned} \upsilon_i &= N_{ij}\rho_j \quad (i = 1, 2, 3;\ j = 1, \ldots, 12) \\ \theta_i &= \psi_{ij}\rho_j \end{aligned} \tag{10.191}$$

The nodal coordinates ρ is defined as

$$\{\rho\} = [\upsilon_1^A, V_2^A, \upsilon_3^A, \theta_1^A, \theta_2^A, \theta_3^A, \upsilon_1^B, V_2^B, \upsilon_3^B, \theta_1^B, \theta_2^B, \theta_3^B]^T$$

and the element shape function, including shear deformation, are given by [7]

$$[N_{ij}] = \begin{bmatrix} a_1 & 0 & 0 & 0 & 0 & 0 & a_2 & 0 & 0 & 0 & 0 & 0 \\ 0 & b1 & 0 & 0 & 0 & b2 & 0 & b3 & 0 & 0 & 0 & b4 \\ 0 & 0 & b5 & 0 & b6 & 0 & 0 & 0 & b7 & 0 & b8 & 0 \end{bmatrix} \tag{10.192}$$

and

$$[\psi_{ij}] = \begin{bmatrix} 0 & 0 & 0 & a_1 & 0 & 0 & 0 & 0 & 0 & a_2 & 0 & 0 \\ 0 & 0 & c_5 & 0 & c_6 & 0 & 0 & 0 & c_7 & 0 & c_8 & 0 \\ 0 & c_1 & 0 & 0 & 0 & c_2 & 0 & c_3 & 0 & 0 & 0 & c_4 \end{bmatrix} \tag{10.193}$$

where

$$a_1 = 1 - \zeta \tag{10.194}$$

$$a_2 = \zeta \tag{10.195}$$

$$b_1 = 1 - 3\zeta^2 + 2\zeta^3 \tag{10.196}$$

$$b_2 = L(\zeta - 2\zeta^2 + \zeta^3) \tag{10.197}$$

$$b_3 = 3\zeta^2 - 2\zeta^3 \tag{10.198}$$

$$b_4 = L(-\zeta^2 + \zeta^3) \tag{10.199}$$

$$b_5 = 1 - 3\zeta^2 + 2\zeta^3 \tag{10.200}$$

$$b_6 = L(-\zeta + \zeta^2 - \zeta^3) \tag{10.201}$$

$$b_7 = 3\zeta^2 - 2\zeta^3 \tag{10.202}$$

$$b_8 = L(-\zeta^2 + \zeta^3) \tag{10.203}$$

$$c_1 = \frac{6}{L}(-\zeta + \zeta^2) \tag{10.204}$$

$$c_2 = 1 - 4\zeta + 3\zeta^2 \tag{10.205}$$

$$c_3 = -c_1 \tag{10.206}$$

$$c_4 = -2\zeta + 3\zeta^2 \tag{10.207}$$

$$c_5 = \frac{6}{L}(-\zeta + \zeta^2) \tag{10.208}$$

$$c_6 = 1 - 4\zeta + 3\zeta^2 \tag{10.209}$$

$$c_7 = -c_5 \tag{10.210}$$

$$c_8 = -2\zeta + 3\zeta^2 \tag{10.211}$$

where

$$\zeta = \frac{x - x_i}{x_j - x_i} = \frac{x - x_i}{L} \tag{10.212}$$

Using υ_i and θ_i, the displacement field for an arbitrary point in the beam element, $u_i(x, y, z, t)$ can be derived for small rotations as

$$u_1(x, y, z, t) = \upsilon_1(x, t) + z\theta_2(x,t) - y\theta_3(x, t) \tag{10.213}$$

$$u_2(x, y, z, t) = \upsilon_2(x,t) - z\theta_1(x,t) \tag{10.214}$$

$$u_3(x, y, z, t) = \upsilon_3(x,t) + y\theta_1(x,t) \tag{10.215}$$

Using equations (10.191) and (10.213) to (10.215), we obtain

$$[u] = [N]\{\rho\} \tag{10.216}$$

where

$$[N] = \begin{bmatrix} a_1 & -yc_1 & zc_5 & 0 & zc_6 & -yc_2 & a_2 & -yc_3 & 0 & zc_8 & -yc_4 \\ 0 & b_1 & 0 & -za_1 & 0 & b_2 & 0 & b_3 & -za_2 & 0 & b_4 \\ 0 & 0 & b_5 & ya_1 & b_6 & 0 & 0 & 0 & ya_2 & b_8 & 0 \end{bmatrix} \tag{10.217}$$

The slope shape function $[\psi]$ is the same as $[\psi_{ij}]$ given by equation (10.193).

The element mass matrix $[M^e]$ can be expressed as

$$[M^e] = \int_v [N]^T m_e [N] dv \tag{10.218}$$

Note that in equation (10.217), the shape function $[N]$ has taken into consideration the effects of the slope due to the element rotation; therefore, equation (10.218) is equivalent to equation (10.186). The element mass matrix can be written as

$$[M^e] = m_e LA \begin{bmatrix} M_{11} & M_{12} \\ M_{21} & M_{22} \end{bmatrix} \tag{10.219}$$

where

$$M_{11} = \begin{Bmatrix} \frac{1}{3} & & & & & \\ 0 & \frac{13}{35} + \frac{6Jy}{5AL^2} & & & & \\ 0 & 0 & \frac{13}{35} + \frac{6Jy}{5AL^2} & & \text{symmetric} & \\ 0 & 0 & 0 & \frac{Jx}{3A} & & \\ 0 & 0 & -\frac{11L}{210} - \frac{Jy}{10AL} & 0 & \frac{L^2}{105} + \frac{2Jy}{15A} & \\ 0 & \frac{11L}{210} + \frac{Jz}{10AL} & 0 & 0 & 0 & \frac{L^2}{105} + \frac{2Jz}{15A} \end{Bmatrix}$$

$M_{21} = M_{12}^T$

$$\begin{Bmatrix} \frac{1}{6} & 0 & 0 & 0 & 0 & 0 \\ 0 & \frac{9}{70} - \frac{6Jz}{5AL^2} & 0 & 0 & 0 & \frac{13L}{420} - \frac{Jy}{10AL} \\ 0 & 0 & \frac{9}{70} - \frac{6Jy}{5AL^2} & 0 & -\frac{13L}{420} + \frac{Jy}{10AL} & 0 \\ 0 & 0 & 0 & \frac{Jx}{6A} & 0 & 0 \\ 0 & 0 & \frac{13L}{420} - \frac{Jy}{10AL} & 0 & \frac{L^2}{140} - \frac{Jy}{30A} & 0 \\ 0 & -\frac{13L}{210} + \frac{Jz}{10AL} & 0 & 0 & 0 & \frac{L^2}{140} - \frac{Jz}{30A} \end{Bmatrix} \tag{10.220}$$

$$M_{22} = \begin{Bmatrix} \frac{1}{3} & & & & & \\ 0 & \frac{13}{35} + \frac{6Jz}{5AL^2} & & & & \\ 0 & 0 & \frac{13}{35} + \frac{6Jy}{5AL^2} & & \text{symmetric} & \\ 0 & 0 & 0 & \frac{Jx}{3A} & & \\ 0 & 0 & -\frac{11L}{210} + \frac{Jy}{10AL} & 0 & \frac{L^2}{105} + \frac{2Jy}{15A} & \\ 0 & -\frac{11L}{210} - \frac{Jz}{10AL} & 0 & 0 & 0 & \frac{L^2}{105} + \frac{2Jz}{15A} \end{Bmatrix} \tag{10.221}$$

The ordinary stiffness matrix of element i

$$[K_s^e] = \begin{bmatrix} K_{11} & K_{12} \\ K_{21} & K_{22} \end{bmatrix} \tag{10.222}$$

where

$$K_{11} = \left\{ \begin{array}{cccccc} \dfrac{EA}{L} & & & & & \\ 0 & \dfrac{12EJz}{L^3} & & & & \\ 0 & 0 & \dfrac{12EJy}{L^3} & & \text{symmetric} & \\ 0 & 0 & 0 & \dfrac{GJx}{L} & & \\ 0 & 0 & -\dfrac{6EJy}{L^2} & 0 & \dfrac{4EJy}{L} & \\ 0 & \dfrac{6EJz}{L^2} & 0 & 0 & 0 & \dfrac{4EJz}{L} \end{array} \right\} \tag{10.223}$$

$$K_{21} = K_{12}^T$$

$$= \left\{ \begin{array}{cccccc} -\dfrac{EA}{L} & 0 & 0 & 0 & 0 & 0 \\ 0 & -\dfrac{12EJz}{L^3} & 0 & 0 & 0 & -\dfrac{6EJz}{L^2} \\ 0 & 0 & -\dfrac{12EJy}{L^3} & 0 & \dfrac{6EJy}{L^2} & 0 \\ 0 & 0 & 0 & -\dfrac{GJx}{L} & 0 & 0 \\ 0 & 0 & -\dfrac{6EJy}{L^2} & 0 & \dfrac{2EJy}{L} & 0 \\ 0 & \dfrac{6EJz}{L^2} & 0 & 0 & 0 & \dfrac{2EJz}{L} \\ \dfrac{EA}{L} & & & & & \end{array} \right\} \tag{10.224}$$

$$
K_{22} = \begin{bmatrix}
0 & \frac{12EJz}{L^3} & & & & \\
0 & 0 & \frac{12EJy}{L^3} & & \text{symmetric} & \\
0 & 0 & 0 & \frac{GJx}{L} & & \\
0 & 0 & \frac{6EJy}{L^2} & 0 & \frac{4EJy}{L} & \\
0 & -\frac{6EJz}{L^2} & 0 & 0 & 0 & \frac{4EJz}{L}
\end{bmatrix} \tag{10.225}
$$

where Jx, Jy and Jz are the inertial moments, E is Young's modulus of elasticity, G the shearing modulus of elasticity, and A the cross-sectional area of the beam. What remains is the formulation of the global equations of motion of the elastic beam modeled with a six-dimensional element and undergoing a large amount of motion. The procedure should follow exactly that presented in Sections 10.4 and 10.5; and it is merely a mathematical exercise, since most of the new matrices are given above. We leave this as an exercise.

The derivations above are extremely important, as they take into consideration all possible elements of motion: three for translation and three for rotation. It is essentially an ideal case to study the effects of vibration in all directions and see how they affect the rigid-body motion. A number of papers have been published in this regard, and currently a great deal of effort is being devoted to the development of control laws and algorithms based on such a theory.

10.7 Analysis of Elastic Beams with Time-Variant Boundary Conditions

In our modeling of the elastic body using the finite-element method, it is assumed that the body end conditions do not change during the course of the motion. As we will see in this section, certain applications might have an elastic body whose support is time variant; hence it will affect the elastic behavior of the beam when undergoing large motions. Examples of such beams and their supports and their corresponding derivations of the dynamical equations follow.

Case I

For the system shown in Figure 10.17, we have a uniform beam supported by a moving ball that stays in contact with the elastic beam B_1, and its displacement $x_0 = x_0(t)$ is a function of t. Body B_2 is assumed rigid. The angular velocity of the beam system is $\dot{\phi}$ and the material properties are assumed all known. Further assumptions are that the ball support mass is negligible compared to the mass of the structure, and that it undergoes pure rolling with no friction allowed. It is our intention to solve this problem using a one-dimensional beam element for the modeling of body B_1.

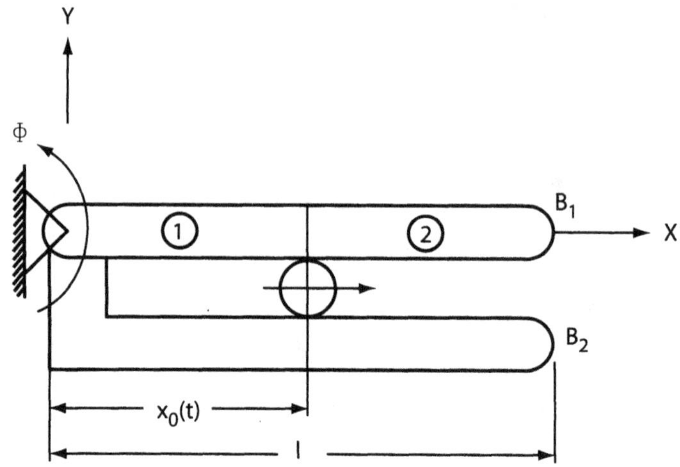

Fig. 10.17. Body B_1 modeled using two one-dimensional beam elements

Let us divide the elastic beam B_1 into two elements to keep the matrices simple. The length of element 1 is assumed to be $x_0 = x_0(t)$ and that of element 2 is $\ell - x_0(t)$. The nodal coordinates ρ for an arbitrary element with nodes A and B is

$$\{\rho\} = \begin{bmatrix} u^A & \theta^A & u^B & \theta^B \end{bmatrix}^T \tag{10.226}$$

where we are allowing only one displacement and one rotation at the nodes. The shape function matrix for the nodal displacement is given by

$$[N] = \begin{bmatrix} 1 - 3\zeta^2 + 2\zeta^3 \\ L(\zeta - 2\zeta^2 + \zeta^3) \\ 3\zeta^2 - 2\zeta^3 \\ L(-\zeta^2 + \zeta^3) \end{bmatrix}^T \tag{10.227}$$

where

$$L = \begin{cases} x_0(t) & \text{for element 1} \\ \ell - x_0(t) & \text{for element 2} \end{cases} \tag{10.228}$$

On the other hand, the shape function matrix for the element nodal rotations is

$$[\varphi] = \begin{bmatrix} \frac{6}{L}(-\zeta + \zeta^2) \\ 1 - 4\zeta + 3\zeta^2 \\ \frac{6}{L}(\zeta - \zeta^2) \\ -2\zeta + 3\zeta^2 \end{bmatrix}^T \tag{10.229}$$

where L is defined as above. Note how the shape function matrices are a function of time due to $x_0(t)$. The mass matrix for a given element is

$$[M^e] = \int_v [N]^T m_e [N] dv + \int_v [\varphi]^T [I_e][\varphi] dv$$

where $[I_e] = m_e z^2$ and m_e is the mass density for both elements ($e = 1, 2$). After integration, we arrive at a simplified element mass given by

$$[M^e] = m_e L A \begin{bmatrix} \frac{13}{35} + \frac{6Jz}{5AL^2} & \frac{11L}{210} + \frac{Jz}{10AL} & \frac{9}{70} - \frac{6Jz}{5AL^2} & \frac{-13L}{420} + \frac{Jz}{10AL} \\ \frac{11L}{210} + \frac{Jz}{10AL} & \frac{L^2}{105} + \frac{2Jz}{15A} & \frac{13L}{420} - \frac{Jz}{10AL} & \frac{-L^2}{140} + \frac{Jz}{30A} \\ \frac{9}{70} - \frac{6Jz}{5AL^2} & \frac{13L}{420} - \frac{Jz}{10AL} & \frac{13}{35} + \frac{6Jz}{5AL^2} & \frac{-11L}{210} - \frac{Jz}{10AL} \\ \frac{-13L}{420} + \frac{Jz}{10AL} & \frac{-L^2}{140} - \frac{Jz}{30A} & \frac{-11L}{210} - \frac{Jz}{10AL} & \frac{L^2}{105} + \frac{2Jz}{15A} \end{bmatrix} \tag{10.230}$$

The stiffness matrix for the

$$[K_s^e] = \int_v [N]^T [D][N]'' dv = \frac{1}{L^3} \int_0^L \frac{d^2[N]^T}{d\zeta^2} EI \frac{d^2[N]}{d\zeta^2} d\zeta$$

$$= \frac{EI}{L^3} \begin{bmatrix} 12 & 6L & -12 & 6L \\ 6L & 4L^2 & -6L & 2L^2 \\ -12 & -6L & 12 & -6L \\ 6L & 2L^2 & -6L & 4L^2 \end{bmatrix} \tag{10.231}$$

The dynamic stiffness matrix is defined as

$$[K_d^e] = \int_v [N]^T m_e([\dot{\Omega}^{on}] + [\Omega^{on}][\Omega^{on}])[N] dv \tag{10.232}$$

where $n = 1$ and

$$[S^{01}] = \begin{bmatrix} \cos\varphi & -\sin\varphi & 0 \\ \sin\varphi & \cos\varphi & 0 \\ 0 & 0 & 1 \end{bmatrix} \tag{10.233}$$

$$[\Omega^{01}] = [\dot{S}^{01}][S^{01}]^T = \dot{\varphi} \begin{bmatrix} 0 & -1 & 0 \\ 1 & 0 & 0 \\ 0 & 0 & 0 \end{bmatrix} \tag{10.234}$$

$$[\Omega^{01}][\Omega^{01}] = \dot{\varphi}^2 \begin{bmatrix} -1 & 0 & 0 \\ 0 & -1 & 0 \\ 0 & 0 & 0 \end{bmatrix} \tag{10.235}$$

and

$$[\Omega^{01}] = \ddot{\varphi} \begin{bmatrix} 0 & -1 & 0 \\ 1 & 0 & 0 \\ 0 & 0 & 0 \end{bmatrix} \tag{10.236}$$

Then

$$\begin{aligned} [K_d^e] &= \int_v [N]^T m_e([\dot{\Omega}_{22}^{01}] + [\Omega_{2t}^{01}][\Omega_{t2}^{01}])[N]dv \\ &= -\dot{\varphi}^2 \int_v [N]^T m_e [N] dv = -\dot{\varphi}^2 \cdot [M^e] \end{aligned} \tag{10.237}$$

Since the displacements of the nodes are just along the y (or $\bar{n}_2$) direction, we can only take the component of the second row and second column in $[\dot{\Omega}^{01}] + [\Omega^{01}][\Omega^{01}]$. The total stiffness matrix for a given element is then found to be

$$[K^e] = [K_s^e] + [K_d^e]$$

$$= \begin{bmatrix} \frac{12EI}{L^3} - m_e LA\dot{\varphi}^2 \left(\frac{13}{35} + \frac{6Jz}{5AL^2}\right) & & & \text{symmetric} \\ \frac{6EI}{L^2} - m_e LA\dot{\varphi}^2 \left(\frac{11L}{210} + \frac{Jz}{10AL}\right) & \frac{4EI}{L} - m_e LA\dot{\varphi}^2 \left(\frac{L^2}{105} + \frac{2Jz}{15A}\right) & & \\ \frac{-12EI}{L^3} - m_e LA\dot{\varphi}^2 \left(\frac{9}{70} - \frac{6Jz}{5AL^2}\right) & \frac{-6EI}{L^2} - m_e LA\dot{\varphi}^2 \left(\frac{13L}{420} - \frac{Jz}{10AL}\right) & \frac{12EI}{L^3} - m_e LA\dot{\varphi}^2 \left(\frac{13}{35} + \frac{6Jz}{5AL^2}\right) & \\ \frac{6EI}{L^2} - m_e LA\dot{\varphi}^2 \left(\frac{-13L}{420} + \frac{Jz}{10AL}\right) & \frac{2EI}{L} - m_e LA\dot{\varphi}^2 \left(\frac{-L^2}{140} - \frac{Jz}{30A}\right) & \frac{-6EI}{L^2} - m_e LA\dot{\varphi}^2 \left(\frac{-11L}{210} - \frac{Jz}{10AL}\right) & \frac{4EI}{L} - m_e LA\dot{\varphi}^2 \left(\frac{L^2}{105} + \frac{2Jz}{15A}\right) \end{bmatrix} \tag{10.238}$$

Let us assume that the damping coefficients for the elastic beam are zero; then the ordinary damping matrix for both elements is

$$[C_s^e] = 0 \tag{10.239}$$

whereas the dynamic damping matrix [from equations (10.221) and (10.234)] is obtained as follows:

$$\begin{aligned}[C_d^e] = 2\int_v m_e[N]^T[\Omega_{22}^{on}][N]dv + \int_v [\psi]^T[I_e][\Omega_{33}^{on}] \\ - [S_{3t}^{no}][\Omega_{t\ell}^{on}][S_{\ell 3}^{on}][I_e][\psi])dv = 0 + 0 = 0\end{aligned} \tag{10.240}$$

Since the rotation coordinate of the system is in the $\bar{n}_3$ direction, we can only take the component of the third row and third column in counting the coefficients relative to $[\psi]$. From equation (10.223) we know the ordinary force vector

$$\{F_s^e\} = \int_v ([N]^T\{F\} + [\psi]^T\{T\})dv \tag{10.241}$$

where

$$\begin{aligned}\{F\} &= m_e g \cos\varphi \\ \{T\} &= 0\end{aligned} \tag{10.242}$$

Then

$$\{F_s^e\} = \frac{1}{2} m_e g A L \cos\varphi \begin{bmatrix} 1 \\ L/6 \\ 1 \\ -L/6 \end{bmatrix} \tag{10.243}$$

In equation (10.223),

$$n = 1, \qquad \{r_1\} = \begin{bmatrix} x \\ 0 \\ 0 \end{bmatrix}, \qquad \{\dot{x}\} = \begin{bmatrix} 0 \\ 0 \\ \dot{\varphi} \end{bmatrix}, \qquad \{\ddot{x}\} = \begin{bmatrix} 0 \\ 0 \\ \dot{\varphi} \end{bmatrix}, \qquad [\omega^1] = \text{identity}$$

and

$$[S^{10}] = \begin{bmatrix} \cos\varphi & \sin\varphi & 0 \\ -\sin\varphi & \cos\varphi & 0 \\ 0 & 0 & 1 \end{bmatrix}$$

In addition, we have

$$\{F_d^e\} = 0 - \int_v [\varphi]^T m_e z^2 \ddot{\varphi} dv = \frac{m_e J z \ddot{\varphi}}{L} \begin{bmatrix} 1 \\ 0 \\ -1 \\ 0 \end{bmatrix} \tag{10.244}$$

Then

$$\{F^e\} = \{F_s^e\} + \{F_d^e\} = m_e \begin{bmatrix} \frac{1}{2} gAl\cos\varphi + Jz\ddot{\varphi}/L \\ \frac{1}{12} gAL^2 + \cos\varphi \\ \frac{1}{2} gAL\cos\varphi - Jz\ddot{\varphi}/L \\ -\frac{1}{12} gAL^2\cos\varphi \end{bmatrix} \tag{10.245}$$

Thus far, we can assemble the element equations of motion

$$[M^e]\{\ddot{\rho}\} + [K^e]\{\rho\} = \{F^e\} \tag{10.246}$$

where $[M^e]$ is obtained from equation (10.230). The expression $[K^e]$ is given by equation (10.238) and $\{F^e\}$ is given by equation (10.245). After appropriately assembling, we can obtain the global equations of motion. Note that for element 1, $L = x_0$, and for element 2, $L = \ell - x_0[x_0 = x_0(t)]$. Hence

$$[M]\{\ddot{\rho}\} + [K]\{\rho\} = \{F\} \tag{10.247}$$

where

$$[M] = \begin{bmatrix} M_{11} & M_{12} \\ M_{21} & M_{22} \end{bmatrix}_{6\times 6} \tag{10.248}$$

and

$$M_{11} = m_e \begin{bmatrix} \frac{13Ax_0}{35} + \frac{6Jz}{5x_0} & & \\ \frac{11Ax_0^2}{210} + \frac{Jz}{10} & \frac{Ax_0^3}{105} + \frac{2Jzx_0}{15} & \\ \frac{9Ax_0}{70} - \frac{6Jz}{5x_0} & \frac{13Ax_0^2}{420} - \frac{Jz}{10} & \frac{13AL}{35} + \frac{6Jz\ell}{5x_0(\ell - x_0)} \end{bmatrix}$$

$$M_{21} = M_{12}^T = m_e \begin{bmatrix} \frac{13Ax_0}{420} + \frac{Jz}{10} & \frac{Ax_0^3}{140} - \frac{Jzx_0}{30} & \frac{11AL(\ell - 2x_0)}{210} \\ 0 & 0 & \frac{9A(\ell - x_0)}{70} - \frac{6Jz}{5(\ell - x_0)} \\ 0 & 0 & \frac{-13A(\ell - x_0)}{420} + \frac{Jz}{10} \end{bmatrix}$$

$$M_{22} = m_e \begin{bmatrix} \dfrac{A\ell(\ell^2 - 3\ell x_0 + 3x_0^2)}{105} + \dfrac{2JzL}{15} & & \text{symmetric} \\ \dfrac{13A(\ell - x_0)^2}{420} - \dfrac{Jz}{10} & \dfrac{13A(\ell - x_0)}{35} + \dfrac{6Jz}{5(\ell - x_0)} & \\ \dfrac{-A(\ell - x_0)^3}{140} - \dfrac{Jz(\ell - x_0)}{30} & \dfrac{-11A\ell(\ell - x_0)^2}{210} - \dfrac{Jz}{10} & \dfrac{A(\ell - x_0)^3}{105} \dfrac{2Jz(\ell - x_0)}{15} \end{bmatrix}$$

The stiffness matrix can be expressed as

$$[K] = \begin{bmatrix} K_{11} & K_{12} & K_{13} \\ K_{21} & K_{22} & K_{23} \\ K_{31} & K_{32} & K_{33} \end{bmatrix}_{6\times 6} \tag{10.249}$$

where

$$K_{11} = \begin{bmatrix} \dfrac{12EI}{x_0^3} - m_e\dot{\varphi}^2\left[\dfrac{13Ax_0}{35} + \dfrac{6Jz}{5x_0}\right] & \text{symmetric} \\ \dfrac{6EI}{x_0^2} - m_e\dot{\varphi}^2\left[\dfrac{11Ax_0^2}{210} + \dfrac{Jz}{10}\right] & \dfrac{4EI}{x_0} - m_e\dot{\varphi}^2\left[\dfrac{Ax_0^3}{105} + \dfrac{2Jzx_0}{15}\right] \end{bmatrix}$$

$$K_{21} = K_{12}^T = \begin{bmatrix} -\dfrac{12EI}{x_0^3} - m_e\dot{\varphi}^2\left[\dfrac{9Ax_0}{70} + \dfrac{6Jz}{5x_0}\right] \\ \dfrac{-6EI}{x_0} - m_e\dot{\varphi}^2\left[\dfrac{13Ax_0^2}{420} - \dfrac{Jz}{10}\right] \\ \dfrac{6EI}{x_0^2} + m_e\dot{\varphi}^2\left[\dfrac{13Ax_0^2}{420} - \dfrac{Jz}{10}\right] \\ \dfrac{2EI}{x_0} + m_e\dot{\varphi}^2\left[\dfrac{Ax_0^3}{440} + \dfrac{Jzx_0}{30}\right] \end{bmatrix}$$

$$K_{31} = K_{13}^T = \begin{bmatrix} 0 & 0 \\ 0 & 0 \end{bmatrix}$$

$$K_{22} = \begin{bmatrix} 12EI\left[\dfrac{1}{x_0^3} + \dfrac{1}{(\ell - x_0)^3}\right] - m_e\dot{\varphi}^2\left[\dfrac{13A\ell}{35} + \dfrac{6Jz\ell}{5x_0(\ell - x_0)}\right] & \text{symmetric} \\ 6EI\left[\dfrac{1}{x_0^2}\dfrac{1}{(\ell - x_0)^2}\right] - m_e\dot{\varphi}^2\dfrac{11A\ell(\ell - 2x_0)}{210} & 4EI\left[\dfrac{1}{x_0} + \dfrac{1}{\ell - x_0}\right] - m_e\dot{\varphi}^2\left[\dfrac{A\ell(\ell - 3\ell x_0 + 3x_0^2)}{105} + \dfrac{2^2Jz\ell}{15}\right] \end{bmatrix}$$

$$K_{23} = K_{32}^T = \begin{bmatrix} -\dfrac{12EI}{(\ell - x_0^3)} - m_e\dot{\varphi}^2\left[\dfrac{9A(\ell - x_0)}{70} + \dfrac{6Jz}{5(\ell - x_0)}\right] & -\dfrac{6EI}{(\ell - x_0)^2} - m_e\dot{\varphi}^2\left[\dfrac{13A(\ell - x_0)^2}{420} - \dfrac{Jz}{10}\right] \\ -\dfrac{6EI}{(\ell - x_0^3)} \quad m_e\dot{\varphi}^2\left[\dfrac{13A(\ell - x_0)}{420} + \dfrac{Jz}{10}\right] & \dfrac{2EI}{\ell - x_0} + m_e\dot{\varphi}^2\left[\dfrac{A(\ell - x_0)^3}{140} - \dfrac{Jz(\ell - x_0)}{30}\right] \end{bmatrix}$$

$$K_{33} = \begin{bmatrix} -\dfrac{12EI}{(\ell - x_0)^3} - m_e\dot{\varphi}^2\left[\dfrac{13A(\ell - x_o)}{35} + \dfrac{6Jz}{5(\ell - x_0)}\right] & \text{symmetric} \\ -\dfrac{6EI}{(\ell - x_o)^2} - m_e\dot{\varphi}^2\left[\dfrac{11A(\ell - x_0)^2}{420} + \dfrac{Jz}{10}\right] & \dfrac{4EI}{\ell - x_0} - m_e\dot{\varphi}^2\left[\dfrac{A(\ell - x_0)^3}{105} + \dfrac{2Jz(\ell - x_0)}{15}\right] \end{bmatrix}$$

and finally,

$$\{F\} = m_e \begin{bmatrix} \frac{1}{2} g A x_0 \cos\varphi + J z \ddot{\varphi}/x_0 \\ \frac{1}{12} g A x_0^2 \cos\varphi \\ \frac{1}{2} g A \ell \cos\varphi - J z \ddot{\varphi} \ell/[x_0(\ell - x_0)] \\ -\frac{1}{12} g A \ell(\ell - 2x_0)\cos\varphi \\ \frac{1}{2} g A(\ell - x_0)\cos\varphi - J z \ddot{\varphi}/(\ell - x_0) \\ -\frac{1}{12} g A(\ell - x_0)^2 \cos\varphi \end{bmatrix} \tag{10.250}$$

From the boundary conditions, where the elastic beam nodal coordinates are

$$\{\rho\} = [u_1 \quad \theta_1 \quad u_2 \quad \theta_2 \quad u_3 \quad \theta_3]^T \tag{10.251}$$

we have node 1 of element 1 pinned to the rotating frame, and node 2 has a support. Then

$$u_1 = 0 \text{ and } u_2 = 0$$

Therefore, deleting the first and third rows and columns in matrices $[M]$, $[K]$, $\{\rho\}$ and $\{F\}$, respectively, we obtain the resulting equations of motion. Furthermore, we can solve for the modal frequencies and mode shapes of the system by letting

$$\{\ddot{\rho}\} = -\omega^2\{\rho\} \tag{10.252}$$

and solving for the eigenvalues and eigenfunctions from

$$\left|-\omega^2[m] + [K]\right| = 0 \tag{10.253}$$

Equation (10.253) then results in the modal frequencies $\omega_i^2 (i = 1, \ldots, 4)$, which can be expressed as

$$\omega_i^2 = \dot{\omega}_i^2 - \dot{\varphi}^2 \tag{10.254}$$

where ωi denote the natural frequencies excluding the rigid-body motion. We can also say that

$$\omega_i = \omega_i\,[x_0(t), \dot{\varphi}] \tag{10.255}$$

and the mode shapes are

$$\phi_i = \phi_i\,[x_0(t)] \tag{10.256}$$

The latter is not a function of the rigid-body motion. Finally, we can obtain the solution to equation (10.247) given some initial conditions and proceeding in a fashion similar to Example 10.3.1. This is quite an interesting problem, which can be investigated further when the elastic body is discretized into more than two elements. The solution can be carried out to find the optimum $x_0(t)$ that would minimize the vibration of elastic beam (see reference [17]).

Case 2

Let us consider the system shown in Figure 10.18. The moving ball in case 1 is changed to the moving spring-damping unit with stiffness coefficient k and damping coefficient c.

The global equations of motion are given by equation (10.247), and the mass matrix and stiffness matrix are given by equations (10.248) and (10.249), respectively, while in the global force vector we should add the spring and damping force due to the spring-damping unit.

$$\{F\} = m_e \begin{bmatrix} \frac{1}{2} g A x_0 \cos\varphi + J z \ddot{\varphi}/x_0 \\ \frac{1}{12} g A x_0^2 \cos\varphi \\ \frac{1}{2} g A \ell \cos\varphi - J z \ddot{\varphi} \ell [x_0(\ell - x_0)] \\ -\frac{1}{12} g A \ell (\ell - 2x_0) \cos\varphi \\ \frac{1}{2} g A (\ell - x_0) \cos\varphi - J z \ddot{\varphi} (\ell - x_0) \\ -\frac{1}{12} g A (\ell - x_0)^2 \cos\varphi \end{bmatrix} + \begin{bmatrix} 0 \\ 0 \\ -c\dot{u}_2 - k u_2 \\ 0 \\ 0 \\ 0 \end{bmatrix} \tag{10.257}$$

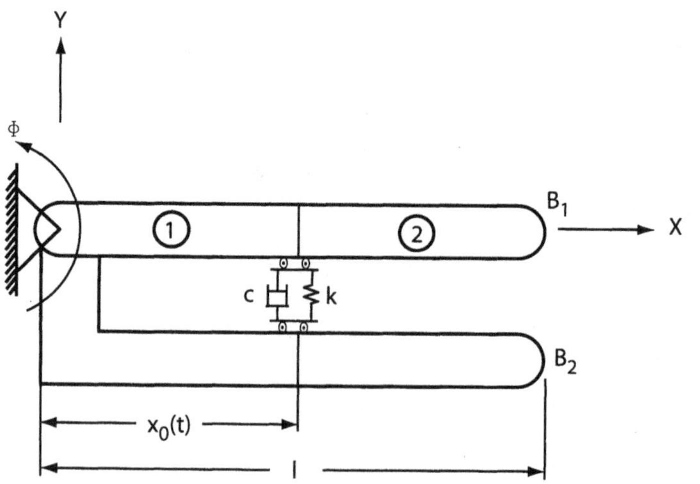

Fig. 10.18. Spring-damper adjustable support between the elastic body B_1 and the rigid body B_2

The boundary condition is simply $u_1 = 0$. The difference between the boundary condition in cases 2 and 1 is that $u_2 \neq 0$. Therefore, deleting the first rows and columns in matrices $[M]$, $[K]$, $\{\rho\}$ and $\{F\}$, respectively, will result in the global equations of motion. They take the form

$$[M_g]\{\ddot{\rho}_g\} + [C_g][\dot{\rho}_g] + [K_g]\{\rho_g\} = \{F_g\} \tag{10.258}$$

where

$$[M_g] = \begin{bmatrix} M_{11} & M_{12} \\ M_{21} & M_{22} \end{bmatrix} \tag{10.259}$$

with

$$M_{11} = m_e \begin{bmatrix} \frac{Ax_0^3}{105} + \frac{2Jzx_0}{15} & & \text{symmetric} \\ \frac{13Ax_0^2}{420} - \frac{Jz}{10}\frac{13A\ell}{15} + \frac{6Jz\ell}{5x_0(\ell - x_0)} & & \\ \frac{Ax_0}{140} - \frac{Jzx_0}{30}\frac{11A\ell(\ell - 2x_0)}{210} & & \frac{A\ell(\ell^2 - 3\ell x_0 + 3x_0^2)}{105} + \frac{2Jz\ell}{15} \end{bmatrix}.$$

$$M_{21} = M_{12}^T = \begin{bmatrix} 0 & \frac{9Ax_0}{70} - \frac{6Jz}{5(\ell - x_0)} & \frac{13A(\ell - x_0)^2}{420} - \frac{Jz}{10} \\ 0 & \frac{-13A(\ell - x_0)}{420} + \frac{Jz}{10} & -\frac{A(\ell - x_0)^3}{440} + \frac{Jz(\ell - x_0)}{30} \end{bmatrix}$$

$$M_{22} = \begin{bmatrix} \frac{13A(\ell - x_0)}{35} + \frac{6Jz}{5(\ell - x_0)} & \text{symmetric} \\ \frac{-11A(\ell - x_0)^2}{210} - \frac{Jz}{10} & \frac{A(\ell - x_0)^3}{105} + \frac{2Jz(\ell - x_0)}{15} \end{bmatrix}$$

The stiffness matrix could be expressed as

$$[K_g] = \begin{bmatrix} K_{11} & K_{12} & K_{13} \\ K_{21} & K_{22} & K_{23} \\ K_{31} & K_{32} & K_{33} \end{bmatrix} \tag{10.260}$$

where

$$K_{11} = \frac{4EI}{x_0} - m_e\dot{\varphi}^2\left(\frac{Ax_0^3}{105} + \frac{2Jzx_0}{15}\right)$$

$$K_{21} = K_{12}^T = \begin{bmatrix} \dfrac{-6EI}{x_0^2} - m_e\dot{\varphi}^2\left(\dfrac{13Ax_0^2}{420} - \dfrac{Jz}{10}\right) \\[2ex] \dfrac{2EI}{x_0} + m_e\dot{\varphi}^2\left(\dfrac{Ax_0^3}{140} - \dfrac{Jzx_0}{30}\right) \end{bmatrix}$$

$$K_{21} = K_{13}^T = \begin{bmatrix} 0 \\ 0 \end{bmatrix}$$

$$K_{21} = \begin{bmatrix} 12EI\left[\dfrac{1}{x_0^3} - \dfrac{1}{(\ell - x_0)^3}\right] - m_e\dot{\varphi}^2\left[\dfrac{13A\ell}{35} + \dfrac{6Jz\ell}{5x_0(\ell - x_0)}\right] + k & \text{symmetric} \\[2ex] 6EI\left[\dfrac{1}{x_0^2} - \dfrac{1}{(\ell - x_0)^2}\right] - m_e\dot{\varphi}^2\dfrac{11A\ell(\ell - 2x_0)}{210} & 4EI\left(\dfrac{1}{x_0} + \dfrac{1}{\ell - x_0}\right) - m_e\dot{\varphi}^2\dfrac{A\ell(\ell^2 - 3\ell x_0 + 3x_0^2)}{105} + \dfrac{2Jz\ell}{15} \end{bmatrix}$$

$$K_{32} = K_{23}^T \begin{bmatrix} \dfrac{-12EI}{(\ell - x_0)^3} - m_e\dot{\varphi}^2\left[\dfrac{9A(\ell - x_0)}{70} - \dfrac{6Jz}{5(\ell - x_0)}\right] \\[2ex] \dfrac{6EI}{(\ell - x_0)^2} - m_e\dot{\varphi}^2\left[\dfrac{13A(\ell - x_0)^2}{420} + \dfrac{Jz}{10}\right] \\[2ex] \dfrac{-6EI}{(\ell - x_0)^2} - m_e\dot{\varphi}^2\left[\dfrac{-13A(\ell - x_0)}{420} + \dfrac{Jz}{10}\right] \\[2ex] \dfrac{2EI}{\ell - x_0} + m_e\dot{\varphi}^2\left[\dfrac{A(\ell - x_0)^3}{140} + \dfrac{Jz(\ell - x_0)}{30}\right] \end{bmatrix} \tag{10.261}$$

$$K_{33} = \begin{bmatrix} \dfrac{12EI}{(\ell - x_0)s3} - m_e\dot{\phi}^2\left[\dfrac{13A(\ell - x_0)}{35} + \dfrac{6Jz}{5(\ell - x_0)}\right] & \text{symmetric} \\[2ex] \dfrac{-6EI}{(\ell - x_0)^2} + m_e\dot{\phi}^2\left[\dfrac{11A\ell(\ell - x_0)}{210} + \dfrac{Jz}{10}\right] & \\[2ex] \dfrac{4EI}{\ell - x_0} - m_e\dot{\phi}^2\left[\dfrac{A(\ell - x_0)^3}{105} + \dfrac{2Jz(\ell - x_0)}{15}\right] & \end{bmatrix}$$

The damping matrix has one coefficient and is given by

$$[Cg] = \begin{bmatrix} 0 & & & & \\ & c & & & \\ & & 0 & & \\ & & & 0 & \\ & & & & 0 \end{bmatrix} \tag{10.262}$$

Finally, we obtain the generalized force array from equation (10.257) and the boundary conditions:

$$\{Fg\} = m_e \begin{bmatrix} \frac{1}{12} g A x_0^2 \cos\phi \\ \frac{1}{2} g A \ell \cos\phi - J z \dot{\phi}^2 \ell / [x_0(\ell - x_0)] \\ -\frac{1}{12} g A \ell (\ell - 2x_0) \cos\phi \\ \frac{1}{12} g A (\ell - x_0) \ \cos\phi - [J z \ddot{\phi} / (\ell - x_0)] \\ -\frac{1}{12} g A (\ell - x_0)^2 \cos\phi \end{bmatrix} \tag{10.263}$$

The nodal displacement vector is

$$\{\rho_g\} = [\theta_1 \quad u_2 \quad \theta_2 \quad u_3 \quad \theta_3]^T \tag{10.264}$$

Let $\{\ddot{\rho}_g\} = -\omega^2\{\rho_g\}$; then the eigenvalue problem can be stated as

$$\left| -\omega^2 [Mg] + [Kg] \right| = 0$$

from which we can obtain the modal frequencies $\omega_i\,(i = 1, \ldots, 5)$ and mode shapes $[\phi]$ needed in the analysis. For the time response of the system, we must use the method of direct numerical integration since the damping matrix $[Cg]$ cannot be decoupled; that is,

$$[\phi]^T [Cg][\phi] \neq \text{diagonal matrix} \tag{10.265}$$

Case 3

In this particular case, the elastic beam has a mass (B_2) that rolls along its edge without slipping, and we are interested in the dynamic response of the beam when undergoing large rotation. It is assumed that the mass is comparably small to that of the beam. Let the system as shown in Figure 10.19 represent such a case, where the beam mass system rotates at point O with an angular velocity $\dot{\phi}$. Although this is not the problem with time-variant boundary conditions, we can use the method used in case 2 to derive the governing equations of motion.

For convenience, let us divide the beam into two elements, where element 1 is of length $x_0(t)$ and element 2 of length $\ell - x_0(t)$. Similar to case 2, we can obtain the force vector $\{F\}$ [see equation (10.257)] with the second term changed to

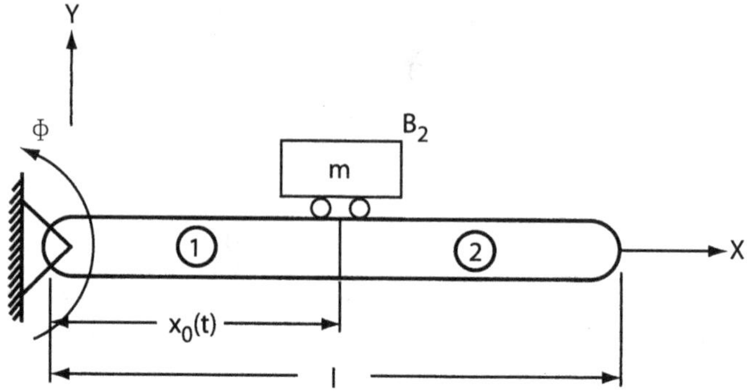

Fig. 10.19. Elastic beam undergoing large rotation with a movable mass

$$\begin{bmatrix} 0 \\ -m\ddot{u}_2 \\ 0 \\ 0 \\ 0 \end{bmatrix}$$

That is, in equation (10.258), $[Cg] = 0$ and Mg_{22} of $[Mg]$ should add m and delete k in Kg_{22} [see equation (10.260)]. Then $\{Fg\}$ remains the same as equation (10.263).

Since $[Cg] = 0$, the equation of motion can be decoupled. Let $\{\ddot{\rho}_g\} = -\omega^2\{\rho_g\}$ and $\left|-\omega^2[Mg] + [Kg]\right| = 0$; we can obtain the modal frequencies ω_i $(i = 1, \ldots, 5)$, mode shapes $[\phi]$ and the solution of the system.

Discussion

In these three cases, we have taken two elements, for convenience; obviously, the precision is not sufficient in practice. We need to divide the beam into n elements with n steps of T [the time period for the analysis (i.e., $T = n\Delta t$)]. The nodal coordinate $\{\rho\}$ is expressed as

$$\{\rho\} = [u_1 \quad \theta_1 \quad u_2 \quad \theta_2 \quad \cdots \quad u_i \quad \theta_i \quad \cdots \quad u_n \quad \theta_n]^T \tag{10.266}$$

If at the beginning, the movable support (for cases 1 and 2) or the movable body (for case 3) is at node i, then we need to change the stiffness matrix, damping matrix, or mass matrix, correspondingly to cases 1 to 3. Then at the first step the movable support is at node $i + 1$ (or $i - 1$). One would simulate the dynamics of such a system to find the optimum displacement of the support to minimize the vibration of the beam [17].

10.8 Problems

10.1. The element mass matrix for all elements 1, 2, 3 and 4 in Section 10.4 are found to be equal. Show how the calculations can have been avoided in proving that for elements 2 and 4 the results would have been the same as those for elements 1 and 3.

10.2. Evaluate the global stiffness matrix for the beam shown in Figure 10.12.

10.3. Consider the rotating beam in Section 10.4 if the angular rotation $\dot{\phi}$ = const.
(a) Find the reduced governing equations of motion.
(b) Find the displacements due to the longitudinal and transverse vibration of the center mass of the beam. (For that you need to program the equations of motion. *Note*: During the simulation you might be interested in seeing how the values of ϕ affect the center mass elastic deformation.)

10.4. In equation (10.41) we assumed that $\xi_1 = 0$ and derived the equations of motion of a multibody system with a terminal flexible link. Find the contribution to equation (10.67) if body 1 is assumed to have a translational motion $\bar{\xi}_1$ in R.

10.5. For the system shown in Figure 10.20, suppose that the nth body is flexible and the rest are rigid. In addition to the general rotations, they also have relative displacements, denoted by $\bar{\xi}_1\,(i = 1, \ldots, n)$, between adjacent bodies. Derive the equations of motion in the form of equation (10.67), assuming that the angular velocity of the elastic body element $\bar{\omega}$ is negligible.

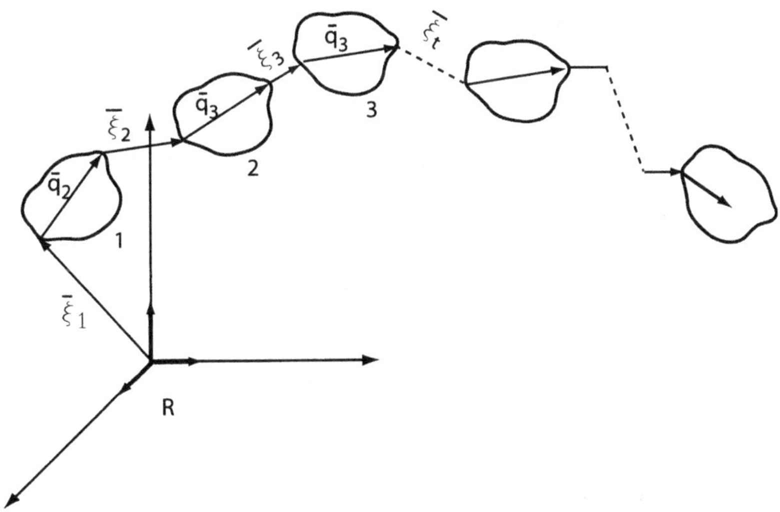

Fig. 10.20. General treelike multibody system with a terminal flexible body

10.6. The robotic system shown in Figure 10.21 was studied in chapter 3, where all the links were assumed rigid. Based on the derivations presented by equation (10.67), give the explicit form of the dynamical equations of motion for the robot if the terminal link is assumed flexible. (Use two elements and model the link as a beam type.) In this problem, we are interested primarily in the effects of transverse vibration, so a simple beam element type will be suitable for modeling the flexible link. (*Note:* This is a perfect example for symbolic manipulation programs. Students are urged to make use of such softwares, if available.)

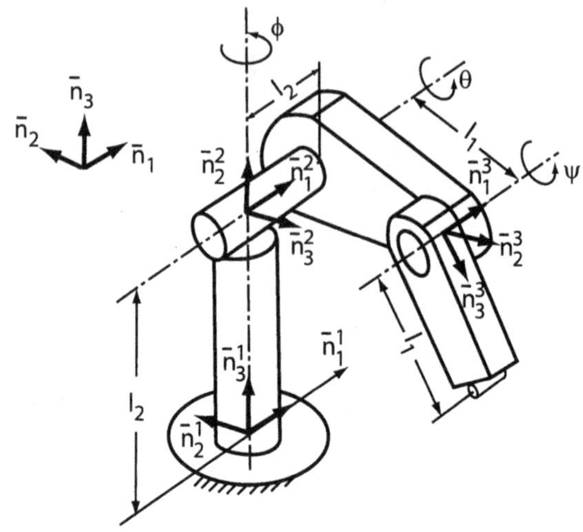

Fig. 10.21. Robot taken from Figure 3.7 where the third link is assumed flexible

References

1. Midha, A., Erdman, A. G. and Frohrib, D. A., Finite Element Approach to Mathematical Modelling of High-Speed Elastic Linkages, *Mech. Mach. Theory*, Vol. **13**, 1978, pp. 603–618.
2. Turcic, D. A. and Midha, A., Generalized Equations of Motion for the Dynamic Analysis of Elastic Mechanism Systems, *J. Dyn. Syst. Meas. Control*, Vol. **106**, 1984.
3. Ho, J. Y. L. and Herber, D. R., Development of Dynamics and Control Simulation of Large Flexible Space Systems, *J. Guid. Control Dyn.*, Vol. **8**: No. 3, 1985.
4. Sunada, W. and Dubowsky, S., The Application of Finite Element Methods to the Dynamic Analysis of Flexible Spatial and Co-planar Linkage Systems, *J. Mech. Des.*, Vol. **103**, July 1981.
5. Singh, R. P., Vandervoort, R. J. and Likins, P. W., Dynamics of Flexible Bodies in Tree Topology—A Computer Oriented Approach, Paper AIAA 84-1024, *25th Structures, Structural Dynamics and Materials Conference*, Palm Springs, Calif., May 1984.
6. Kane, T. R., Ryan, R. R. and Banerjee, A. K., Dynamics of a Cantilever Beam Attached to a Moving Base, *J. Guid. Control Dyn.*, Vol. **10**: No. 2, 1987.
7. Fertis, D. G., *Dynamics and Vibration of Structures*, Wiley, New York, 1973.
8. Nothmann, G. A., Vibration of a Cantilever Beam with Prescribed End Motion, *Appl. Mech.*, Dec. 1948.
9. Hoppmann, W. H., Impact of a Multispan Beam, *J. Appl. Mech.*, Dec. 1950.
10. Hoppmann, W. H., Impact of a Mass on a Damped Elastically Supported Beam, *J. Appl. Mech.*, June 1948.
11. Mindlin, R. D. and Goodman, L. E., Beam Vibrations with Time-Dependent Boundary Conditions, *J. Appl. Mech.*, Dec. 1950.

12. Young, D. Vibration of a Beam with Concentrated Mass, Spring and Dashpot, *J. Appl. Mech.*, Mar. 1948.
13. Hoppmann, W. H., Forced Lateral Vibration of Beam Carrying a Concentrated Mass, *J. Appl. Mech.*, Sept. 1952.
14. Amirouche, F. M. L. and Huston, R. L., Dynamics of Large Constrained Flexible Structures, *J. Dyn. Syst. Meas. Control*, Vol. **110**: No. 1, 1988, pp. 78–83.
15. Bainum, P. M. and Kumar, V. K., Dynamics of Orbiting Flexible Beams and Platforms in the Horizontal Orientation, *Acta Astronant.*, Vol. **9**: No. 3, 1982, pp. 119–127.
16. Huston, R. L., Flexibility Effects in Multibody Systems, *Mech. Res. Commun.*, Vol. **7**: No. 4, 1980, pp. 261–268.
17. Xie, M. and Amirouche, F. M. L., Minimization of Vibration in Elastic Beams with Time Variant Boundary Conditions in the Dynamics of Multibody Systems, *J. Guid. Dyn. Control*, July 1992.
18. Amirouche F. M. L. and Xie, M., Dynamic Analysis of Multibody Systems with Time-Variant Mode Shapes, *13th Biennial ASME Conference on Mechanical Vibration and Noise,* Sept. 22–25, 1991.

11

Dynamic Analysis of Multiple Flexible-Body Systems

11.1 Introduction

Dynamic analysis of multibody systems has gained tremendous popularity in the past two decades. The research has been the subject of numerous publications [1–10]. While most of the significant theory has been developed, some computational issues remain to be resolved. The latter is a result of the strategic development of the kinematics involved and the formalism used in for the development of the equations of motion. In the recursive formulation developed in reference [27–31], relative coordinates were used in conjunction with Boolean matrices and the tensor notation, which made the notation somehow difficult to follow. While the tensor notation adopted kept the equations in compact form, it hid some vital information from interested readers. This chapter focuses on developing an all-purpose algorithm for the dynamic simulation of flexible treelike systems, making use of matrix representation at all levels. The equations developed are applicable to high-speed systems undergoing large amounts of rotation, and a strict finite-element formulation is presented. The inertial moment of the elements, the damping forces of the structure, and geometrical stiffening forces are all included in an explicit form to be able to monitor their contribution and how they influence the dynamics of multibody systems.

An explicit matrix representation of the partial velocities and partial angular velocities for treelike structures is also presented. The governing equations of motion are in a form, which is easy for implementation and the recursive generation of the expressions needed for each element of a particular flexible body can be adopted to any element type. A tree array is used for bookkeeping and the generalized coordinates are partitioned into three groups: the rotational degrees of freedom, followed by the translational and nodal coordinate degrees of freedom. The effects of geometric stiffening for high-speed systems are included automatically via the nonlinear strain energy formulation.

11.2 Topology and Kinematics of Flexible Treelike Systems

For the system in Figure 11.1, assume that the flexible bodies can be divided into N_j $(j = 1, \ldots, n)$ elements. Let $\bar{\zeta}_j (j = 1, \ldots, n)$ denote the displacement between

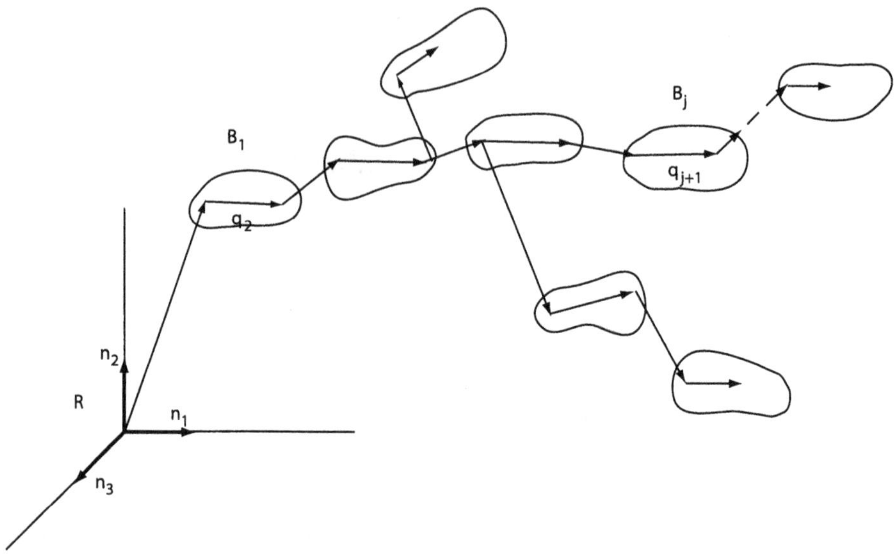

Fig. 11.1. General treelike multibody system with flexible bodies

any two bodies as shown, and $\bar{q}_{j+1}(j = 1, \ldots, n-1)$ the distance vector between the two joints of body j with $\bar{q}_1$ assumed to be zero.

Let the displacement and slope of an arbitrary element of a particular discretized elastic body be u and u_y respectively. Then, following the finite-element method, we can write the nodal displacements and rotation as

$$\{u\} = [N]\{\rho\} \tag{11.1}$$

and

$$\{\theta\} = [\psi]\{\rho\} \tag{11.2}$$

where $[N]$ and $[\psi]$ are the shape functions and $\{\rho\}$ denotes the nodal coordinate vector. Note that the shape functions depend on the kind of element selected; therefore, our equations are written to accommodate any type of element.

The position vector of the mass center of element i of body j can be expressed as

$$\bar{P}_{ji} = \left\{ \sum_{h=0}^{H(j)} \{q_\ell\}^T [S^{\ell-1,0}] + \sum_{h=0}^{H(j)} \{\zeta_\ell\}^T [S^{\ell-1,0}] + (\{r_{ji}\}^T + \{\rho_{ji}\}^T [N]^T)[S^{j0}] \right\} \{\bar{n}\} \tag{11.3}$$

where $\ell = \Gamma_h(j)$ resulting from the tree array, indicating the chain of bodies that are on the path from body j to the reference frame (see Section 3.2). The value $[S^{jo}]$ is the transformation matrix between the set of unit vectors $\{n^j\}$ in body j and $\{\bar{n}\}$ in R, and $\{r_{ji}\}$ represents the coordinate of the undeformed position vector of element i of body j. Equation (11.3) follows from the notation adapted in particular in Chapter 3. The velocity of the mass center of element i of body j is then found to be

$$\bar{v}_{ji} = \frac{d}{dy}\bar{P}_{ji} = \left\{ \sum_{h=0}^{H(j)} (\{\dot{q}_\ell\}^T[S^{\ell-1,0}] + \{q_\ell\}^T[\dot{S}^{\ell-1,0}] + \{\dot{\zeta}_\ell\}^T[S^{\ell-1,0}] \right.$$
$$\left. +\{\zeta_\ell\}^T[\dot{S}^{\ell-1,0}]) + (\{\dot{\rho}_{ji}\}^T[N]^T)[S^{jo}] + \{r_{ji}\}^T + (\{\rho_{ji}\}^T[N]^T)[\dot{S}^{jo}] \right\}\{\bar{n}\} \tag{11.4}$$

$$= \left\{ \sum_{h=0}^{H(j)} (\{q_\ell\}^T + \{q_\ell\}^T[\dot{S}^{\ell-1,0}] + \{r_{ji}\}^T + \{\rho_{ji}\}^T[N]^T)\,[\dot{S}^{j0}] + \sum_{h=0}^{H(j)} (\{\dot{q}_\ell\}^T \right.$$
$$\left. +\{\dot{q}_\ell\}^T[S^{\ell-1,0}] + \{\dot{\rho}_{ji}\}^T[N]^T)[S^{jo}] \right\}\{\bar{n}\}$$

Keep in mind that $H(j)$ denotes the exponent of body j resulting from the tree array of the multibody system being studied. It essentially defines the path to take from body j to fixed inertial frame R.

Recall from Sections 3.2 and 3.6 that we can isolate matrices that are associated with the generalized coordinates derivatives. Making use of such a concept, equation (11.4) becomes

$$\bar{v}_{ji} = \frac{d}{dt}\bar{P}_{ji} = \{\dot{x}\}^T\{V^{ji}\} + \{\dot{q}\}^T\{V_s^j\} + \{\dot{\zeta}\}^T\{V_s^j\} + \{\dot{\rho}\}^T\{V_e^{ji}\} \tag{11.5}$$

Equation (11.5) represents an excellent form to identify matrices that will become important in equations of motion development. The q's are additional generalized coordinates introduced for convenience only, and will be eliminated later in the formulation of the equations of motion. An explicit form and representation of the matrices associated with the generalized coordinates derivatives are given below: From equation (11.5) we have

$$\{V^{ji}\} = [V^{ji}]\{\bar{n}\} \tag{11.6}$$

$[V^{ji}]$ follows the generalized coordinates pattern and it is an $3n \times 3$ matrix, the components of which are

$$[V^{ji}] = [W] \begin{bmatrix} ([Sq_2] + [S\zeta_2])[S^{10}] \\ ([Sq_3] + [S\zeta_3])[S^{20}] \\ \vdots \\ ([Sq_j] + [S\zeta_j])[S^{j-1,0}] \\ ([Sr_{ij}] + [S_{ij}])[S^{jo}] \\ \vdots \\ 0 \end{bmatrix} \tag{11.7}$$

The matrix $[W]$ is used to transform the velocity vector $\{y\}$ (see Sections 3.3 to 3.5), which is a combination of rigid-body motion generalized coordinates derivatives $\{\dot{x}\}$ and trigonometric functions composed of $\{x\}$. The skew matrices are $[Sq_j]$, $[S\zeta_j]$ and $[Sr_{ij}]$ associated with the body vector q_j, the translation vector between adjacent bodies, ζ_j, and the position to the center mass r_j. Those matrices are

$$[Sq_j] = \begin{bmatrix} 0 & -q_{j3} & q_{j2} \\ q_{j3} & 0 & -q_{j1} \\ -q_{j2} & q_{j1} & 0 \end{bmatrix} \tag{11.8}$$

$$[S\zeta_j] = \begin{bmatrix} 0 & -\zeta_{j3} & \zeta_{j2} \\ \zeta_{j3} & 0 & -\zeta_{j1} \\ -\zeta_{j2} & \zeta_{j1} & 0 \end{bmatrix} \tag{11.9}$$

$$[Sr_{ij}] = \begin{bmatrix} 0 & -r_{ji3} & r_{ji2} \\ r_{ji3} & 0 & -r_{ji1} \\ -r_{ji2} & r_{ji1} & 0 \end{bmatrix} \tag{11.10}$$

and the elements of the skew matrix $[S_{ji}]$ come from the vector $\{\rho_{ji}\}^T[N]^T$. We also note that the nodal coordinate array is given by

$$\{\rho\} = [\rho_2^1\rho_2^1 \cdots \rho_{N1}^1, \ldots, \ \rho_1^j\rho_2^j\rho_{Nj}^j, \ldots, \rho_1^n\rho_2^n \cdots \rho_{Nn}^n]^T \tag{11.11}$$

where j denotes the body number and N^j the total number of elements in body j. The coefficients associated with $\{\rho\}$ are given by

$$\{V_e^{ji}\} = [V_e^{ji}]\{\bar{n}\} \tag{11.12}$$

with the matrix $[V_e^{ji}]$ being

$$[V_e^{ji}] = \begin{bmatrix} 0 \\ 0 \\ \vdots \\ [N]^T[S^{jo}] \\ \vdots \\ 0 \end{bmatrix} \begin{matrix} 1,1 \\ 1,2 \\ \vdots \\ j,i \\ \vdots \\ n,N^n \end{matrix} \tag{11.13}$$

The last columns are used to help the reader understand the set-up of the matrix above, where j refers to the body and i to the element (as we have N^j elements in each body). In addition, we have the matrix $[V_s^j]$ associated with the body vector q, which is composed of the transformation matrices between the bodies representing j and R. This matrix represents the orientation of the bodies due to the large rotations.

$$[V_s^j] = \begin{bmatrix} [I] \\ [S^{1,0}] \\ \vdots \\ [S^{j-1,0}] \\ \vdots \\ 0 \end{bmatrix} \begin{matrix} 1 \\ 2 \\ \vdots \\ j \\ \vdots \\ n \end{matrix} \tag{11.14}$$

The acceleration of the center mass of element i of body j is found by differentiating in time equation (11.5), which yields

$$\begin{aligned} \bar{a}_{ji} = \frac{d}{dt}\bar{v}_{ji} &= \{\ddot{x}\}^T\{V^{ji}\} + \{\ddot{q}\}^T\{V_s^j\} + \{\ddot{\zeta}\}^T\{V_s^j\} + \{\ddot{\rho}\}^T\{V_e^{ji}\} \\ &+ \{\dot{x}\}^T\{\dot{V}^{ji}\} + \{\dot{q}\}^T\{\dot{V}_s^j\} + \{\dot{\zeta}\}^T\{\dot{V}_s^j\} + \{\dot{\rho}\}^T\{\dot{V}_e^{ji}\} \end{aligned} \tag{11.15}$$

Once more, equation (11.15) shows the simplicity of expressing the acceleration of the center mass of a particular element of an elastic body, in a multibody structure.

The matrices associated with the second derivatives of the generalized coordinates are a function of $[\dot{S}]$, which can easily be computed using $[\Omega][S]$. For the development of general-purpose codes, the formulation of those matrices is computationally expensive; hence, making use of the predefined quantities saves time and space. The angular velocity of body $j (j = 1, \ldots, n)$ in R can be expressed as

$$^R\bar{\omega}^j = \{\dot{x}\}^T[\omega^j]\{\bar{n}\} = \{\dot{x}\}^T\{\omega^j\} \tag{11.16}$$

where the angular velocity is divided into two parts with $[\omega^j]\{\bar{n}\}$ equal to $\{\omega^j\}$.

$$[\omega^j] = \begin{bmatrix} [I] \\ [S^{10}] \\ [S^{20}] \\ \vdots \\ [S^{j-1,0}] \\ \vdots \\ 0 \end{bmatrix} \begin{matrix} 1 \\ 2 \\ 3 \\ \vdots \\ j \\ \vdots \\ n \end{matrix} \tag{11.17}$$

The angular acceleration of body j in R is obtained by simple differentiation of equation (11.16):

$$^R\bar{\alpha}^j = \frac{d}{dt}\bar{\omega}^j = \{\ddot{x}\}^T\{\omega^j\} + \{\dot{x}\}^T\{\dot{\omega}^j\} \tag{11.18}$$

Similarly, the angular velocity of element i of body j expressed in R is composed of two parts:

$$^R\bar{\omega}^{ji} = {}^R\bar{\omega}^j + {}^j\bar{\omega}^i = \{\dot{x}\}^T\{\omega^j\} + \{\dot{\rho}\}^T\{\omega^{ji}\} \tag{11.19}$$

where the first part is associated with the rigid-body motion and the second part represents the contribution from the flexible motion. The latter is expressed as

$$\{\omega^{ji}\} = [\omega^{ji}]\{\bar{n}\} \tag{11.20}$$

and the matrix $[\omega^{ji}]$ is obtained as follows:

$$[\omega^{ji}] = \begin{bmatrix} 0 \\ 0 \\ \vdots \\ [\psi]^T[S^{jo}] \\ \vdots \\ 0 \end{bmatrix} \begin{matrix} 1,1 \\ 1,2 \\ \vdots \\ j,i \\ \vdots \\ n, N^n \end{matrix} \tag{11.21}$$

The last column indicates the body and the element number, so that one can follow the matrix description given. On the other hand, the angular acceleration of an arbitrary element i in body j expressed in R takes the following form:

$$^R\bar{\alpha}^{ji} = \frac{d}{dt}{}^R\bar{\omega}^{ji} = \{\ddot{x}\}^T\{\omega^j\} + \{\ddot{\rho}\}T\{\omega^{ji}\} + \{\dot{x}\}^T\{\dot{\omega}^j\} + \{\dot{\rho}\}^T\{\dot{\omega}^{ji}\} \tag{11.22}$$

Let the generalized speeds be defined by

$$\{\dot{y}\} = [\dot{x}^T \quad \dot{q}^T \quad \dot{\zeta}^T \quad \dot{\rho}^T]^T \tag{11.23}$$

Note that this y is different from the one we discussed earlier (Sections 3.2 to 3.4). From equations (11.5), (11.16) and (11.19), we obtain the partial velocities and partial angular velocities as follows: for $\ell = 1, \ldots, n$,

$$\frac{\partial \bar{v}_{ii}}{\partial \dot{y}_\ell} = \frac{\partial \bar{v}_{ii}}{\partial \dot{x}_\ell} = \{V^{ji}\} \tag{11.24}$$

$$\frac{\partial \bar{\omega}^{ji}}{\partial \dot{y}_\ell} = \frac{\partial \bar{\omega}^{ji}}{\partial \dot{x}_\ell} = \{\omega^j\} \tag{11.25}$$

for $\ell = n+1, \ldots, 2n$,

$$\frac{\partial \bar{v}_{ji}}{\partial \dot{y}_\ell} = \frac{\partial \bar{v}_{ji}}{\partial \dot{q}_\ell} = \{V_s^j\} \tag{11.26}$$

$$\frac{\partial \bar{\omega}^{ji}}{\partial \dot{y}_\ell} = \frac{\partial \bar{\omega}^{ji}}{\partial \bar{q}_\ell} = 0 \tag{11.27}$$

for $\ell = 2n+1, \ldots, 3n$,

$$\frac{\partial \bar{v}_{ji}}{\partial \dot{y}_\ell} = \frac{\partial \bar{v}_{ji}}{\partial \dot{\zeta}_\ell} = \{V_s^j\} \tag{11.28}$$

$$\frac{\partial \bar{\omega}^{ji}}{\partial \dot{y}_\ell} = \frac{\partial \bar{\omega}^{ji}}{\partial \dot{\zeta}_\ell} = 0 \tag{11.29}$$

and finally, for $\ell = 3n+1, \ldots, 3n+N$,

$$\frac{\partial \bar{\upsilon}_{ji}}{\partial \dot{y}_\ell} = \frac{\partial \bar{\upsilon}_{ji}}{\partial \dot{\rho}_\ell} = \{V_e^{ji}\} \tag{11.30}$$

$$\frac{\partial \bar{\omega}^{ji}}{\partial \dot{y}_\ell} = \frac{\partial \bar{\omega}^{ji}}{\partial \dot{\rho}_\ell} = \{\omega^{ji}\} \tag{11.31}$$

with

$$N = \sum_{j=1}^{n} N^j \tag{11.32}$$

So far we have developed expressions in matrix form for an arbitrary element of a flexible body of a treelike multibody system. What follows will be the formulation of the equations of motion. Our intention is to minimize our efforts in the derivation of additional equations and have a clear path to the contribution of the different kinematical expressions.

11.3 Kinetics

In the dynamics of flexible multibody systems if we make use of the finite-element method, then we need to write the corresponding equations of motion for each element.

Let m_{ji} denote the mass density of element i of body j; then the generalized inertia force can be obtained by dividing it into four parts.

1. For $\dot{y}_\ell = \dot{x}_\ell$,

$$\begin{aligned}\{f_x^*\} &= -\sum_{j=1}^{n}\sum_{j=1}^{Nj}\int_{V_{ji}} m_{ji}\bar{a}_{ji}.\frac{\partial \bar{\upsilon}_{ji}}{\partial \dot{x}_\ell} dV \\ &= -\sum_j\sum_i\int_{V_{ji}} m_{ji}[V^{ji}]\{[V^{ji}]^T\{\ddot{x}\} + [V_s^j]^T(\{\ddot{q}\} + \{\ddot{\zeta}\}) + [V_e^{ji}]^T\{\ddot{\rho}\} \\ &\quad + [\dot{V}^{ji}]^T\{\dot{x}\} + [\dot{V}_s^j]^T(\{\dot{q}\} + \{\dot{\zeta}\}) + [\dot{V}_e^{ji}]^T\{\dot{\rho}\}\}dV\end{aligned} \tag{11.33}$$

2. For $\dot{y}_\ell = \dot{q}_\ell$,

$$\begin{aligned}\{f_q^*\} &= \sum_j\sum_i\int_{V_{ji}} m_{ji}\bar{a}_{ji}.\frac{\partial \bar{\upsilon}_{ji}}{\partial \dot{q}_\ell} dV \\ &= -\sum_j\sum_i\int_{V_{ji}} m_{ji}[V_s^j]\{[V^{ji}]^T\{\ddot{x}\} + [V_s^j]^T(\{\ddot{q}\} + \{\ddot{\zeta}\}) + [V_e^{ji}]^T\{\ddot{\rho}\} \\ &\quad + [\dot{V}^{ji}]^T\{\dot{x}\} + [\dot{V}_s^j]^T(\{\dot{q}\}] + \{\dot{\zeta}\}) + [\dot{V}_e^{ji}]^T\{\dot{\rho}\}\}dV\end{aligned} \tag{11.34}$$

3. For $\dot{y}_\ell = \dot{\zeta}_\ell$,

$$\{f_\zeta^*\} = \{f_q^*\} \tag{11.35}$$

4. For $\dot{y}_\ell = \dot{\rho}_\ell$,

$$\begin{aligned}
\{f_q^*\} &= -\sum_j \sum_i \int_{V_{ji}} m_{ji} \bar{a}_{ji} \cdot \frac{\partial \bar{v}_{ji}}{\partial \dot{\rho}_\ell} dV \\
&= -\sum_j \sum_i \int_{V_{ji}} m_{ji} [V_e^{ji}] \{[V^{ji}]^T \{\ddot{x}\} + [V_s^j]^T (\{\ddot{q}\} + \{\ddot{\zeta}\}) + [V_e^{ji}]^T \{\ddot{\rho}\} \\
&\quad + [\dot{V}^{ji}]^T \{\dot{x}\} + [\dot{V}_s^j] (\{\dot{q}\} + \{\dot{\zeta}\}) + [\dot{V}_e^{ji}]^T \{\dot{\rho}\}\} dV
\end{aligned}$$

Let $[I_{ji}]$ represent the moment density of the inertia of element i of body j in R. Then we write

$$[I_{ji}] = [I_{ji}]^T = m_{ji} [S^{oj}] \begin{bmatrix} y^2 + z^2 & xy & xz \\ yx & x^2 + z^2 & yz \\ zx & zy & y^2 + x^2 \end{bmatrix} [S^{jo}] \tag{11.36}$$

where as before S^{oj} is the shifter between the unit vectors $\bar{n}^j$ and R. The inertia moment (torque) is given by

$$\bar{T}_{ji}^* = -\bar{\omega}^{ji} \times (\bar{\bar{I}}_{ji} \cdot \bar{\omega}^{ji}) - \bar{\bar{I}}_{ji} \cdot \bar{\alpha}^{ji} \tag{11.37}$$

where $\bar{\bar{I}}_{ji}$ is the inertia dyadic of element i of body j. Then after substitution, equation (11.37) becomes

$$\begin{aligned}
\bar{T}_{ji}^* &= -(\{\bar{n}\}^T (\omega^j]^T \{\dot{x}\} + \{\bar{n}\}^T [\omega^{ji}]^T \{\dot{\rho}\}) \times \{\{\bar{n}\}^T [I_{ji}] \{\bar{n}\} \\
&\quad \cdot (\{\bar{n}\}^T [\omega^j]^T \{\dot{x}\} + \{\bar{n}^T [\omega^{ji}]^T \{\dot{\rho}\})\} + \{\bar{n}\}^T [I_{ji}] \{\bar{n}\} \cdot \{\bar{n}\}^T ([\omega^j]^T \{\ddot{x}\} \\
&\quad + [\omega^{ji}]^T \{\ddot{\rho}\} + [\dot{\omega}^j]^T \{\dot{x}\} + [\dot{\omega}^{ji}]^T \{\dot{\rho}\})
\end{aligned} \tag{11.38}$$

After some laborious work, we arrive at the desired matrix representation in terms of the generalized coordinate derivatives selected:

$$\begin{aligned}
\bar{T}_{ji}^* &= -(\{\dot{x}\}^T [\omega^j][I_{ji}][\Omega_x^{jo}] + \{\dot{\rho}\}^T [\omega^{ji}][I_{ji}][\Omega_\rho^{jo}] + \{\ddot{x}\}^T [\omega^j][I_{ji}] \\
&\quad + \{\bar{x}\}^T [\dot{\omega}^j][I_{ji}] + \{\ddot{\rho}\}^T [\omega^{ji}][I_{ji}] + \{\dot{\rho}\}^T \{\dot{\omega}^{ji}][I_{ji}])
\end{aligned} \tag{11.39}$$

where $[\Omega_x^{jo}]$ is the skew matrix of angular velocity components of the bodies and $[\Omega_\rho^{jo}]$ is the skew matrix of angular velocity associated with the elements. Note that $[\Omega_x^{jo}]T = [\Omega_x^{oj}] = -[\Omega_x^{jo}]$, $[\Omega_\rho^{jo}]^T = [\Omega_\rho^{oj}] = -[\Omega_\rho^{jo}]$.

The generalized inertia moment contribution for the four sets of generalized coordinates is obtained by integrating over the volume of the element (V_{ji}), then summing the contribution of all the elements for each body. This is obtained as follows:

1. For $\dot{y}_\ell = \dot{x}_\ell$,

$$\{T_x^*\} = \sum_j \sum_i \int_{V_{ji}} \bar{T}^{*ji} \cdot \frac{\partial \bar{\omega}^{ji}}{\partial \dot{x}_\ell} dV$$

$$= -\sum_j \sum_i \int_{V_{ji}} [\omega^j]([\Omega_x^{oj}][I_{ji}][\omega^j]^T\{\dot{x}\} + [\Omega_\rho^{oj}][I_{ji}][\omega^{ji}]^T\{\dot{\rho}\}$$

$$+ [I_{ji}][\omega^j]^T\{\ddot{x}\} + [I_{ji}][\dot{\omega}^j]^T\{\dot{x}\} + [I_{ji}][\omega_{ji}]^T\{\ddot{\rho}\} + [I_{ji}][\dot{\omega}^{ji}]^T\{\dot{\rho}\})dV \tag{11.40}$$

2. For $\dot{y}_\ell = \dot{q}_\ell$,

$$\{T_q^*\} = \sum_j \sum_i \int_{V_{ji}} \bar{T}^{*ji} \cdot \frac{\partial \bar{\omega}^{ji}}{\partial \dot{q}_\ell} dV = 0 \tag{11.41}$$

3. For $\dot{y}_\ell = \dot{\zeta}_\ell$,

$$\{T_\zeta^*\} = \sum_j \sum_i \int_{V_{ji}} \bar{T}^{*ji} \cdot \frac{\partial \bar{\omega}^{ji}}{\partial \dot{\zeta}_\ell} dV = 0 \tag{11.42}$$

4. For $\dot{y}_\ell = \dot{\rho}_\ell$,

$$\{T_p^*\} = \sum_j \sum_i \int_{V_{ji}} \bar{T}^{*ji} \cdot \frac{\partial \bar{\omega}^{ji}}{\partial \dot{\rho}_\ell} dV$$

$$= -\sum_j \sum_i \int_{V_{ji}} [\omega^{ji}]([\Omega_x^{jo}][I_{ji}]^T[\omega^j]\{\dot{x}\} + [\Omega_\rho^{jo}][I_{ji}][\omega^{ji}]^T\{\dot{\rho}\}$$

$$+ [I_{ji}][\omega^j]^T\{\ddot{x}\} + [I_{ji}][\dot{\omega}^j]^T\{\dot{x}\} + [I_{ji}][\omega^{ji}]^T\{\ddot{\rho}\} + [I_{ji}][\dot{\omega}^{ji}]^T\{\dot{\rho}\})dV \tag{11.43}$$

The equations of motion are obtained by means of Kane's method, which states that

$$f + f^* = 0 \tag{11.44}$$

where

$$f = f_e - f_s - f_c \tag{11.45}$$

where f_e denotes the generalized external forces, f_s the generalized stiffness force array of the elastic bodies, and f_c the generalized damping force of the bodies.

The generalized inertia force f^* is composed of two parts: the contribution from the inertia forces and the inertia torques.

$$f^* = \{f^*\} + \{T^*\} \tag{11.46}$$

Combining equations (11.33) to (11.43), the equations of motion above take the following unique form

$$[M]\{\ddot{y}\} + [C]\{\dot{y}\} + [K]\{y\} = \{F\} \tag{11.47}$$

or

$$\begin{bmatrix} M_{11} & & \text{symmetric} & \\ M_{21} & M_{22} & & \\ M_{31} & M_{32} & M_{33} & \\ M_{41} & M_{42} & M_{43} & M_{44} \end{bmatrix} \begin{Bmatrix} \ddot{x} \\ \ddot{q} \\ \ddot{\zeta} \\ \ddot{\rho} \end{Bmatrix} + \begin{bmatrix} C_{11} & C_{12} & C_{13} & C_{14} \\ C_{21} & C_{22} & C_{23} & C_{24} \\ C_{31} & C_{32} & C_{33} & C_{34} \\ C_{41} & C_{42} & C_{43} & C_{44} \end{bmatrix} \begin{Bmatrix} \dot{x} \\ \dot{q} \\ \dot{\zeta} \\ \dot{\rho} \end{Bmatrix}$$
$$+ \begin{bmatrix} 0 & & & \\ & 0 & & \\ & & 0 & \\ & & & K_{44} \end{bmatrix} \begin{Bmatrix} x \\ q \\ \zeta \\ \rho \end{Bmatrix} = \begin{Bmatrix} F_1 \\ F_2 \\ F_3 \\ F_4 \end{Bmatrix} \tag{11.48}$$

where the coefficients of the generalized mass, dynamic damping and stiffness M, C and K are explicitly given by

$$[M_{11}] = \sum_j \sum_i \int_{V_{ji}} (m_{ji}[V^{ji}][V^{ji}]^T + [\omega j][I_{ji}][\omega j]^T) dV \tag{11.49}$$

$$[M_{21}] = [M_{31}] = [M_{12}]^T = [M_{13}]^T = \sum_j \sum_i \int_{V_{ji}} m_{ji}[V^{ji}][V_s^j]^T dV \tag{11.50}$$

$$[M_{41}] = [M_{14}] =^T \sum_j \sum_i \int_{V_{ji}} (m_{ji}[V_e^{ji}][V^{ji}]^T + [\omega^{ji}][I_{ji}][\omega j]^T) dV \tag{11.51}$$

$$[M_{22}] = [M_{23}] = [M_{32}] = [M_{33}] = \sum_j \sum_i \int_{V_{ji}} m_{ji}[V_e^{ji}][V_s^j]^T dV \tag{11.52}$$

$$[M_{42}] = [M_{43}] = [M_{24}]^T = [M_{34}]^T = \sum_j \sum_i \int_{V_{ji}} m_{ji}[V_e^{ji}][V_s^j]^T dV \tag{11.53}$$

$$[M_{44}] = \sum_j \sum_i \int_{V_{ji}} (m_{ji}[V_e^{ji}][V_e^{ji}]^T + [\omega^{ji}][I_{ji}][\omega^{ji}]^T) dV \tag{11.54}$$

$$[C_{11}] = \sum_j \sum_i \int_{V_{ji}} \{m_{ji}[V_{ji}][\dot{V}_{ji}]^T + [\omega^j][I_{ji}][\dot{\omega}^j]^T + [\Omega_x^{oj}][I_{ji}][\omega^j]^T\} dV \tag{11.55}$$

$$[C_{12}] = [C_{13}] = \sum_j \sum_i \int_{V_{ji}} m_{ji}[V^{ji}][\dot{V}_s^j]^T dV \tag{11.56}$$

$$[C_{14}] = \sum_j \sum_i \int_{V_{ji}} \{m_{ji}[V_{ji}][\dot{V}_e^{ji}]^T + [\omega^j]([I_{ji}][\dot{\omega}^{ji}]^T + [\Omega_{-}\rho^{oj}][I_{ji}][\omega^{ji}]^T\} dV \tag{11.57}$$

$$[C_{21}] = [C_{31}] = \sum_j \sum_i \int_{V_{ji}} m_{ji}[V_s^j [\dot{V}^{ji}]^T dV \tag{11.58}$$

$$[C_{22}] = [C_{23}] = [C_{32}] = [C_{33}] = \sum_j \sum_i \int_{V_{ji}} m_{ji}[V_s^j][\dot{V}_s^j]^T dV \tag{11.59}$$

$$[C_{24}] = [C_{34}] = \sum_j \sum_i \int_{V_{ji}} m_{ji}[V_s^j][\dot{V}_e^{ji}]^T dV \tag{11.60}$$

$$[C_{41}] = \sum_j \sum_i \int_{V_{ji}} \{m_{ji}[V_e^{ji}][\dot{V}_{ji}]^T + [\omega^{ji}][I_{ji}][\dot{\omega}^j]^T$$

$$+ [\Omega_x^{oj}][I_{ji}][\omega^j]^T\}dV \tag{11.61}$$

$$[C_{42}] = [C_{43}] = \sum_j \sum_i \int_{V_{ji}} m_{ji}[V_e^{ji}][\dot{V}_s^j]^T dV \tag{11.62}$$

$$[C_{44}] = \sum_j \sum_i \int_{V_{ji}} \{m_{ji}[V_e^{ji}][\dot{V}_e^{ji}]^T + [\omega^{ji}][I_{ji}][\dot{\omega}^{ji}]^T$$

$$+ [\Omega_\rho^{oj}][I_{ji}][\omega^{ji}]^T\}dV + [C_s] \tag{11.63}$$

where $[C_s]$ is the ordinary damping matrix of the bodies.

$$[K_{44}] = [K_s] + [K_G] \tag{11.64}$$

where $[K_s]$ is the ordinary stiffness matrix of the bodies, and $[K_G]$ is the geometrical stiffness contribution, which we discuss later. The generalized force vector is given by

$$\{F_\ell\} = \sum_j \sum_i \int_{S_{ji}} \frac{\partial \bar{\upsilon}_{ji}}{\partial \dot{y}_\ell} \cdot \bar{f}_{ji} ds + \sum_j \sum_i \int_{V_{ji}} \frac{\partial \upsilon_{ji}}{\partial \dot{y}_\ell} \cdot \bar{b}_{ji} dV$$

$$+ \sum_j \sum_j \int_{S_{ji}} \frac{\partial \bar{\omega}_{ji}}{\partial \dot{y}_\ell} . \bar{M}_{ji} ds \tag{11.65}$$

where $\bar{f}_{ji}$ denotes external surface traction, $\bar{b}_{ji}$ denotes external body force, and $\bar{M}_{ji}$ is the external surface distributed moment. The subscripts i and j are such that i is the element and j the body, so b_{ji} reads as the external force exerted on element i of body j. Substituting $\dot{y}$ with $\dot{x}_\ell$, $\dot{q}_\ell$, $\dot{\zeta}_\ell$ and $\dot{\rho}_\ell$, respectively, into equation (11.65), we get

$$\{F_1\} = \sum_j \sum_i \int_{S_{ji}} [V^{ji}]\{f_{ji}\} dS + \sum_j \sum_i \int_{V_{ji}} [V^{ji}]\{b_{ji}\} dV$$

$$+ \sum_j \sum_i \int_{S_{ji}} [\omega^j]\{M_{ji}\} dS \tag{11.66}$$

$$\{F_2\} = \{F_3\} = \sum_j \sum_i \int_{S_{ji}} [V_s^j]\{f_{ji}\}dS + \sum_j \sum_i \int_{V_{ji}} [V_s^j]\{b_{ji}\}dV \tag{11.67}$$

$$\{F_4\} = \sum_j \sum_i \int_{V_{ji}} [V_e^{ji}]\{f_{ji}\}ds + \sum_j \sum_i \int_{V_{ji}} [V_e^{ji}]\{b_{ji}\}dV$$
$$+ \sum_j \sum_i \int_{S_{ji}} [\omega^{ji}]\{M_{ji}\}dS \tag{11.68}$$

Let $\bar{P}$ be a concentrated force acting on a point A in the element i of body j, and $\bar{M}$ be a moment acting on the element. Then the equations above should have the following terms at appropriate positions:

$$\frac{\partial \bar{v}^A}{\partial \dot{y}_\ell} \cdot \bar{P} + \frac{\partial \bar{\omega}^{ji}}{\partial \dot{y}_\ell} \cdot \bar{M} \tag{11.69}$$

Let us discuss equation (11.47). Since the elements of the second row are the same as those of the third row, we can, for convenience, delete the third row. This is due to the fact that q is redundant, as it can be expressed as function of the ρ's. So we can further simplify the equations of motion by omitting the q vector as a generalized coordinate. Let q_{j+1}^o $(j = 0, \ldots, n-1)$ be the original undeformed vector of $\bar{q}_{j+1}$, which is the distance vector between two joints, which is a function of time t. Assume that one nodal displacement coordinate of the element that is closest to the next joint of body j is $\{\rho_{jb}\}$, as shown in Figure 11.2. Then

$$\{q_{j+1}(t)\} = \{q_{j+1}^o\} + \{\rho_{jb}\} \tag{11.70}$$

$$\{\dot{q}_{j+1}(t)\} = \{\dot{\rho}_{jb}\} \tag{11.71}$$

$$\{\ddot{q}_{j+1}(t)\} = \{\ddot{\rho}_{jb}\} \tag{11.72}$$

Substituting equations (11.71) and (11.72) into (11.47), we obtain the equations of motion that combine both rigid-body motion and flexibility. They are expressed in

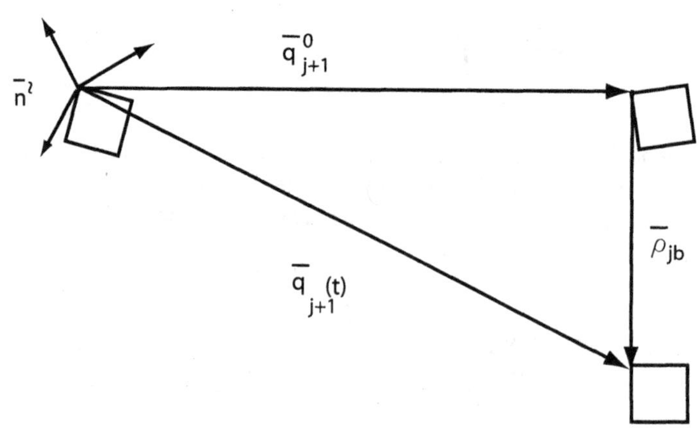

Fig. 11.2. Distance between two elements of a flexible body before and after deformation is assumed

a matrix form and represent the general form of the dynamics of flexible multibody systems using the finite-element method.

$$\begin{bmatrix} M_{11} & M_{12} & M_{14}^* \\ M_{21} & M_{22} & M_{24}^* \\ M_{41} & M_{42} & M_{44}^* \end{bmatrix} \begin{Bmatrix} \ddot{x} \\ \ddot{\zeta} \\ \ddot{\rho} \end{Bmatrix} + \begin{bmatrix} C_{11} & C_{12} & C_{14}^* \\ C_{21} & C_{22} & C_{24}^* \\ C_{42} & C_{42} & C_{44}^* \end{bmatrix} \begin{Bmatrix} \dot{x} \\ \dot{\zeta} \\ \dot{\rho} \end{Bmatrix} + \begin{bmatrix} 0 & & \\ & 0 & \\ & & K_{44} \end{bmatrix} \begin{Bmatrix} x \\ \zeta \\ \rho \end{Bmatrix} = \begin{Bmatrix} F_1 \\ F_2 \\ F_4 \end{Bmatrix} \tag{11.73}$$

where the coefficients of $[M]$ and $[C]$ are such that

$$[M_{14}^*] = [M_{14}] + [M_{12}][Q] \tag{11.74}$$

$$[M_{24}^*] = [M_{24}] + [M_{22}][Q] \tag{11.75}$$

$$[M_{44}^*] = [M_{44}] + [M_{42}][Q] \tag{11.76}$$

$$[C_{14}^*] = [C_{14}] + [C_{12}][Q] \tag{11.77}$$

$$[C_{24}^*] = [C_{14}] + [C_{22}][Q] \tag{11.78}$$

$$[C_{44}^*] = [C_{44}] + [C_{42}][Q] \tag{11.79}$$

The $[Q]$ matrix is an identifier of the modal coordinates and is expressed as:

$$[Q] = \begin{array}{c} \begin{array}{ccccc} 1,1 & 1,b & j+1,b & n,b & n,N^n \end{array} \\ \left[\begin{array}{ccccccc} 0 & \cdots & 1 & & & & \\ & & & \ddots & & 0 & \\ & & & & 1 & & \\ & 0 & & & & & \\ & & & & & \ddots & \\ & & & & & 1 \;\cdots & 0 \end{array} \right] \end{array} \begin{array}{c} 1 \\ \vdots \\ \\ j+1 \\ \vdots \\ n \end{array} \tag{11.80}$$

In matrix form the body vectors $\{q\}$ can be written as

$$\{q\} = \{q^o\} + [Q]\{\rho\} \tag{11.81}$$

with $\{q\}$ being the body vector of all bodies such that

$$\{q\} = [q_2^T, q_3^T, \ldots, q_n^T]^T \tag{11.82}$$

Substituting equation (11.81) with the appropriate element into equation (11.7) or (11.8), the coordinate $\{q\}$ disappears in equation (11.73). The numbers of the equations and their corresponding unknowns should then be the same.

If body $j(j = 1, \ldots, n)$ is a rigid body, we can simply let $\dot{q}_j = \ddot{q}_j = 0$, and the whole body is considered as an element in the integration of the coefficients in equation (11.73). Then substitute the constant $\{q_j\}$ into equation (11.7). The same is the case when several bodies are rigid.

Note that in equations (11.74) to (11.79) one can avoid the multiplication with the $\{Q\}$ matrices and develop an index notation for modifying the related matrices to implement in a "do loop" to get the $[M^*]$ and $[C^*]$ matrices. This procedure will reduce the computational time in the simulation of large systems.

11.4 An Illustrative Example

To illustrate the use of the method developed in this chapter, we opt to show how the exact equations of motion can be found instead of giving numerical answers. Let us consider the link manipulator system shown in Figure 11.3. Assume that body 1 is a rigid body with length ℓ_1 and body 2 is flexible with length ℓ_2. All properties of the materials of both bodies are assumed to be known. As shown in Figure 11.4, body 2 is divided into two one-dimensional elements for simplicity and the system rotates in the horizontal plate. The joints are one-dimensional revolute joints.

The generalized coordinates adopted for the system are

$$\{y\} = \begin{bmatrix} x_1 & x_2 & \rho_1 & \rho_2 & \rho_3 \end{bmatrix}^T \tag{11.83}$$

where the joint angles are $x_1 = x_3^1$ and $x_2 = x_3^2$, and the nodal coordinates are $\rho_1 = \rho_1^2$, $\rho_2 = \rho_2^2$, and $\rho_3 = \rho_3^2$. The shape function for the elements of body 2 is

$$[N] = \begin{bmatrix} 1 - \xi & \xi \end{bmatrix} \tag{11.84}$$

Since the elements used to model the flexibility of body 2 are nondimensional, their slopes are zero and their corresponding shape function $[\psi] = 0$. In equation (11.3), the components of the body 1 and the translation are

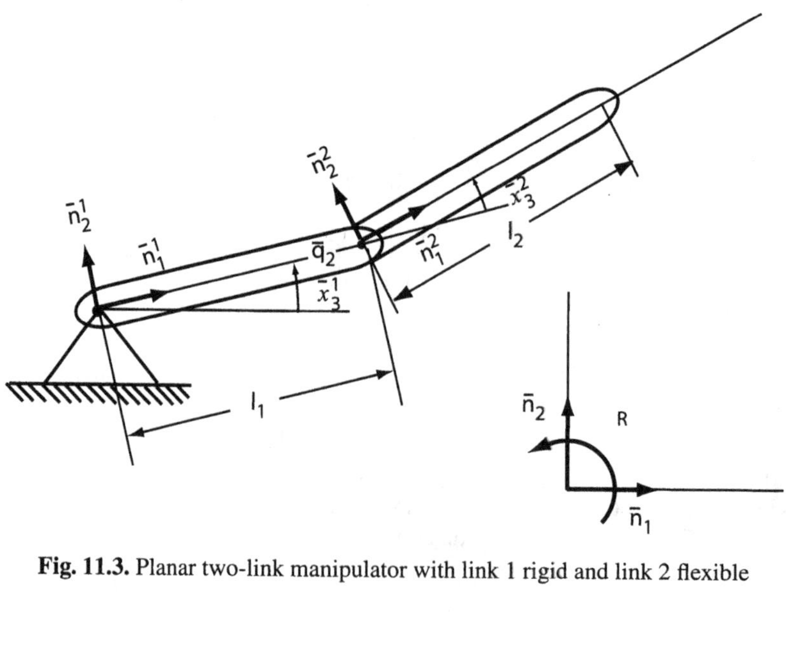

Fig. 11.3. Planar two-link manipulator with link 1 rigid and link 2 flexible

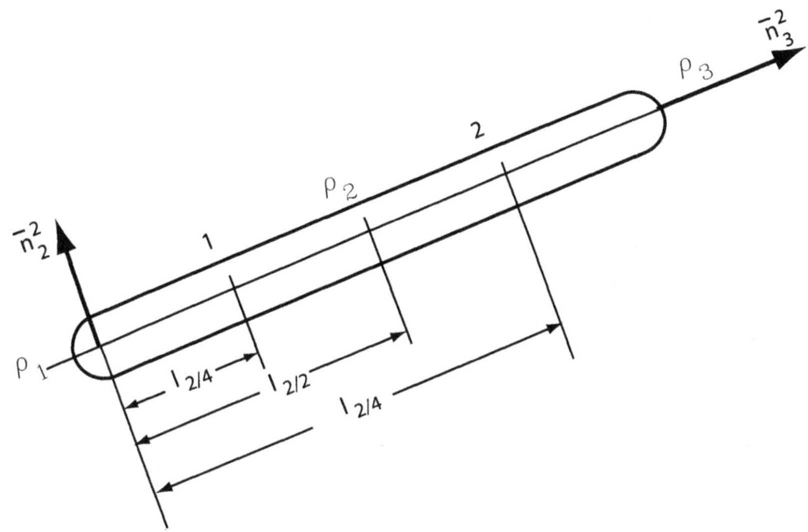

Fig. 11.4. Discretization of the flexible link and generalized coordinates rotation

$$\{q_2\} = \begin{Bmatrix} \ell_1 \\ 0 \\ 0 \end{Bmatrix} \quad \text{and} \quad \{\zeta_1\} = \{\zeta_2\} = 0$$

The position vectors to the center masses in the body coordinates are

$$\{r_{11}\} = \begin{Bmatrix} \frac{1}{2}\ell_1 \\ 0 \\ 0 \end{Bmatrix}, \qquad \{r_{21}\} = \begin{Bmatrix} \frac{1}{4}\ell_2 \\ 0 \\ 0 \end{Bmatrix}, \qquad \{r_{22}\} = \begin{Bmatrix} \frac{3}{4}\ell_2 \\ 0 \\ 0 \end{Bmatrix}$$

The skew matrices associated with the position vectors needed in the analysis are obtained as follows

$$[Sq_2] = \begin{bmatrix} 0 & 0 & 0 \\ 0 & 0 & -\ell_1 \\ 0 & \ell_1 & 0 \end{bmatrix} \tag{11.85}$$

$$[S\zeta_1] = [S\zeta_2] = 0 \tag{11.86}$$

$$[Sr_{11}] = \begin{bmatrix} 0 & 0 & 0 \\ 0 & 0 & -\frac{1}{2}\ell_1 \\ 0 & \frac{1}{2}\ell_1 & 0 \end{bmatrix} \tag{11.87}$$

where $[S_{11}] = 0$ (because of the rigid body)

$$[Sr_{21}] = \begin{bmatrix} 0 & 0 & 0 \\ 0 & 0 & -\frac{1}{4}\ell_2 \\ 0 & \frac{1}{4}\ell_2 & 0 \end{bmatrix} \tag{11.88}$$

$$[Sr_{22}] = \begin{bmatrix} 0 & 0 & 0 \\ 0 & 0 & -\frac{3}{4}\ell_2 \\ 0 & \frac{3}{4}\ell_2 & 0 \end{bmatrix} \tag{11.89}$$

$$[S_{21}] = \begin{bmatrix} 0 & 0 & 0 \\ 0 & 0 & -[(1-\xi)\rho_1 + \xi\rho_2] \\ 0 & (1-\xi)\rho_1 + \xi\rho_2 & 0 \end{bmatrix} \tag{11.90}$$

$$[S_{22}] = \begin{bmatrix} 0 & 0 & 0 \\ 0 & 0 & -[(1-\xi)\rho_2 + \xi\rho_3] \\ 0 & (1-\xi)\rho_2 + \xi\rho_3 & 0 \end{bmatrix} \tag{11.91}$$

The transformation matrices between body 1 and O and body 2 and O are found to be:

$$[S^{10}] = \begin{bmatrix} \cos\ x_1 & \sin\ x_1 & 0 \\ -\sin\ x_1 & \cos\ x_1 & 0 \\ 0 & 0 & 1 \end{bmatrix} \tag{11.92}$$

$$[S^{21}] = \begin{bmatrix} \cos\ x_2 & \sin\ x_2 & 0 \\ -\sin\ x_2 & \cos\ x_2 & 0 \\ 0 & 0 & 1 \end{bmatrix}$$

$$[S^{20}] = [S^{21}][S^{10}] \begin{bmatrix} \cos\ (x_2\ +\ x_1) & \sin\ (x_2\ +\ x_1) & 0 \\ -\sin\ (x_2\ +\ x_1) & \cos\ (x_2\ +\ x_1) & 0 \\ 0 & 0 & 1 \end{bmatrix} \tag{11.93}$$

The skew matrices of the angular velocities are derived from the shifters, where

$$[\Omega^{01}] = [\dot{S}^{01}][S^{10}] = \dot{x}_1 \begin{bmatrix} 0 & -1 & 0 \\ 1 & 0 & 0 \\ 0 & 0 & 0 \end{bmatrix} \tag{11.94}$$

$$[\Omega^{02}] = [\dot{S}^{02}][S^{20}] = (\dot{x}_1\ +\ \dot{x}_2) \begin{bmatrix} 0 & -1 & 0 \\ 1 & 0 & 0 \\ 0 & 0 & 0 \end{bmatrix} \tag{11.95}$$

The matrix $[W]$ which is used in transforming $\begin{bmatrix} \dot{x}_1 & \dot{x}_1\ +\ \dot{x}_2 \end{bmatrix}$ to $\begin{bmatrix} \dot{x}_1 & \dot{x}_2 \end{bmatrix}$ is obtained by

$$[W] = \begin{bmatrix} 1 & 1 \\ 0 & 1 \end{bmatrix} \tag{11.96}$$

The velocity coefficient matrices associated with $\dot{\rho}$ for both element 1 and element 2 in body 2 are:

$$[V_e^{21}] = \begin{bmatrix} (1-\xi)\cos(x_2+x_1) & (1-\xi)\sin(x_2+x_1) & 0 \\ \xi\cos(x_2+x_1) & \xi\sin(x_2+x_1) & 0 \\ 0 & 0 & 0 \end{bmatrix} \tag{11.97}$$

$$[V_e^{22}] = \begin{bmatrix} 0 & 0 & 0 \\ (1-\xi)\cos(x_2+x_1) & (1-\xi)\sin(x_2+x_1) & 0 \\ \xi\cos(x_2+x_1) & \xi\sin(x_2+x_1) & 0 \end{bmatrix} \tag{11.98}$$

Note that because of the one-dimensional element, we can only take the first row in the transformation matrix $[S^{20}]$. Since $x_1 = x_3^1$, which is the $\bar{n}_3^1$ direction, we can take only the third row from equation (11.7). Similarly, following equation (11.5), we obtain the remaining partial velocity matrices:

$$[V^{11}] = \frac{1}{2}\ell_1 \begin{bmatrix} \sin x_1 & -\cos x_1 & 0 \\ 0 & 0 & 0 \end{bmatrix} \tag{11.99}$$

$$[V^{21}] = \begin{bmatrix} -\ell_1\sin x_1 - \left[\frac{1}{4}\ell_2 + (1-\xi)\rho + \xi\rho_2\right]\sin(x_2+x_1) & \ell_1\cos x_1 + \left[\frac{1}{4}\ell_2 + (1-\xi)\rho_1 + \xi\rho_2\right]\cos(x_2+x_1) & 0 \\ -\left[\frac{1}{4}\ell_2 + (1-\xi)\rho_1 + \xi\rho_2\right]\sin(x_2+x_1) & \left[\frac{1}{4}\ell_2 + (1-\xi)\rho_1 + \xi\rho_2\right]\cos(x_2+x_1) & 0 \end{bmatrix} \tag{11.100}$$

$$[V^{22}] = \begin{bmatrix} -\ell_1\sin x_1 - \left[\frac{3}{4}\ell_2 + (1-\xi)\rho_2 + \xi\rho_3\right]\sin(x_2+x_1) & \ell_1\cos x_2 + \left[\frac{3}{4}\ell_2 + (1-\xi)\rho_2 + \xi\rho_3\right]\cos(x_2+x_1) & 0 \\ -\left[\frac{3}{4}\ell_2 + (1-\xi)\rho_2 + \xi\rho_3\right]\sin(x_2+x_1) & \left[\frac{3}{4}\ell_2 + (1-\xi)\rho_2 + \xi\rho_3\right]\cos(x_2+x_1) & 0 \end{bmatrix} \tag{11.101}$$

From equation (11.14), by taking the first row, we obtain the partial velocity arrays

$$[V_s^1] = \begin{bmatrix} 1 & 0 & 0 \\ 0 & 0 & 0 \end{bmatrix} \tag{11.102}$$

$$[V_s^2] = \begin{bmatrix} 1 & 0 & 0 \\ \cos\ x_1 & \sin\ x_1 & 0 \end{bmatrix} \tag{11.103}$$

Taking the third row, from equation (11.17), we derive the partial angular velocity for body 1 and 2 as:

$$[\omega^1] = \begin{bmatrix} 0 & 0 & 1 \\ 0 & 0 & 0 \end{bmatrix} \tag{11.104}$$

$$[\omega^2] = \begin{bmatrix} 0 & 0 & 1 \\ 0 & 0 & 1 \end{bmatrix} \tag{11.105}$$

Let us not consider the elements rotation in this example, then we set the following partial angular velocity arrays to zero

$$[\omega^{11}] = [\omega^{21}] = [\omega^{22}] = 0 \tag{11.106}$$

The moment of inertia matrix can be computed from the definition given in Section 4.5, where:

$$[I_{ji}] = m_{ji}[S^{oj}] \begin{bmatrix} 0 & & \\ & 0 & \\ & & (\xi\ell)^2 \end{bmatrix} [S^{jo}] \tag{11.107}$$

$j = 1,\ i = 1$, and $j = 2,\ i = 1, 2$.

Since $q_1 = 0$ and $\dot{q}_2 = \ddot{q}_2 = 0$ (rigid body), we get

$$[Q] = 0 \tag{11.108}$$

The mass matrix coefficients can then be obtained by substituting the kinematical quantities above and carrying out the integration as defined

$$\begin{aligned}
[M_{11}] &= \sum_j \sum_i \int_{V_{ji}} (m_{ji}[V^{ji}][V^{ji}]^T + [\omega^j][I_{ji}][\omega^j]^T)\ dV \\
&= \int_0^1 m_1[V^{11}][V^{11}]^T A_1\ell_1 d\xi + \sum_{i=1}^{2} \int_0^1 \frac{1}{2} m_2[V^{2i}][V^{2i}]^T A_2\ell_2\ d\xi \\
&\quad + \int_0^1 m_1[\omega^1][I_{11}][\omega^1]^T A_1\ell_1\ d\xi + \sum_{i=1}^{2} \int_0^1 \frac{1}{2} m_2[\omega^2][I_{2i}][\omega^2]^T A_2\ell_2\ d\xi \\
&= \begin{bmatrix} \frac{1}{3} m_1 A_1 \ell_1^3 + \frac{1}{24} m_2 A_2 \ell_2^3 + m_{11} & \frac{1}{24} m_2 A_2 \ell_2^3 + m_{12} \\ \frac{1}{24} m_2 A_2 \ell_2^2 + m_{21} & \frac{1}{24} m_2 A_2 \ell_2^3 + m_{22} \end{bmatrix}
\end{aligned} \tag{11.109}$$

where m_1 and m_2 denote the mass densities of bodies 1 and 2, and A_1 and A_2 are their respective cross-sectional areas. The coefficients m_{11}, m_{12}, $(m_{12} = m_{21})$ and

m_{22} in equation (11.109) are functions of $\rho_k (k = 1, 2, 3)$ and $x_k (k = 1, 2)$ and they are left as an exercise for the reader.

$$[M_{14}^*] = [M_{14}] = [M_{41}]^T = \sum_j \sum_i \int_{V_{ji}} m_{ji}[V^{ji}][V_e^{ji}]\, dV$$
$$= 0 + \int_0^1 \frac{1}{2} m_2([V^{21}][V_e^{21}]^T + [V^{22}][V_e^{22}]^T) A_2 \ell_2 d\xi = \begin{bmatrix} m_{13} & m_{14} & m_{15} \\ m_{23} & m_{24} & m_{25} \end{bmatrix} \tag{11.110}$$

where $m_{ji} (i = 1, 2, \; j = 3, 4, 5; \; m_{ij} = m_{ji})$ are functions of ρ_k $(k = 1, 2, 3)$ and $x_k = (k = 1, 2)$.

$$[M_{44}^*] = [M_{44}] = \sum_j \sum_i \int_{V_{ji}} [V_e^{ji}][V_e^{ji}]^T dV$$
$$= 0 + \int_0^1 \frac{1}{2} m_2([V_e^{21}][V_e^{21}]^T + [V_e^{22}][V_e^{22}]^T) A_2 \ell_2 d\xi \tag{11.111}$$
$$= \frac{1}{12} m_2 A_2 \ell_2 \begin{bmatrix} 2 & 1 & 0 \\ 1 & 4 & 1 \\ 0 & 1 & 2 \end{bmatrix}$$

Note that $[M_{44}]$ is exactly the same as the ordinary mass matrix and is automatically overlapping. The generalized mass matrix is then obtained by

$$[M] = \begin{bmatrix} M_{11} & M_{14} \\ M_{41} & M_{44} \end{bmatrix} \tag{11.112}$$

or more explicitly as

$$[M] = \begin{bmatrix} \frac{1}{3} m_1 A_1 \ell \frac{3}{1} + \frac{1}{24} m_2 A_2 \ell \frac{3}{2} + m_{11} & & & & \\ \frac{1}{24} m_2 A_2 \ell \frac{3}{2} + m_{21} & \frac{1}{24} m_2 A_2 \ell \frac{3}{2} + m_{22} & & \text{symmetric} & \\ m_{31} & m_{32} & \frac{1}{6} m_2 A_2 \ell_2 & & \\ m_{41} & m_{42} & \frac{1}{12} m_2 A_2 \ell_2 & \frac{1}{3} m_2 A_2 \ell_2 & \\ m_{51} & m_{52} & 0 & \frac{1}{12} m_2 A_2 \ell_2 & \frac{1}{6} m_2 A_2 \ell_2 \end{bmatrix} \tag{11.113}$$

where $m_{ij}(i = 1, \dots, 5;\ j = 1, 2)$ are given by equations (11.109) and (11.110). The dynamic damping matrix coefficients are found following the definitions defined previously. First let us substitute the matrices for C_{11} and perform the necessary integration to evaluate the coefficients of C.

$$[C_{11}] = \sum_j \sum_i \int_{V_{ji}} m_{ji}[V^{ji}][V^{ji}]^T + [\omega^j]([I_{ji}][\dot{\omega}^j]^T + [\Omega_x^{oj}][I_{ji}][\omega_j]^T dV$$

$$= \int_0^1 m_1[V^{11}][\dot{V}^{11}]^T A_1\ell_1 d\xi + \int_0^1 \frac{1}{2} m_2([V^{21}][\dot{V}^{21}]^T$$

$$+ [V^{22}][\dot{V}^{22}]^T)A_2\ell_2\, d\xi + \int_0^1 [\omega^1]([I_{11}][\dot{\omega}^1]^T$$

$$+ [\Omega_x^{01}][I_{11}][\omega^1]^T A_1\ell_1 d\xi + \int_0^1 \frac{1}{2}\{[\omega^2]([I_{21}] + [I_{22}])[\dot{5}\omega^2]^T$$

$$+ [\omega^2][\Omega_x^{02}]([I_{21}] + [I_{22}])[\omega^2]^T\}A_2\ell_2\, d\xi \tag{11.114}$$

where the skew matrices defining the angular velocities of body 1 and 2 are:

$$[\Omega_x^{01}] = \dot{x}_1 \begin{bmatrix} 0 & -1 & 0 \\ 1 & 0 & 0 \\ 0 & 0 & 0 \end{bmatrix} \tag{11.115}$$

$$[\Omega_x^{02}] = (\dot{x}_1 + \dot{x}_2) \begin{bmatrix} 0 & -1 & 0 \\ 1 & 0 & 0 \\ 0 & 0 & 0 \end{bmatrix} \tag{11.116}$$

From equation (11.83) we get the partial velocity derivative

$$[\dot{V}^{11}] = -\frac{1}{2}\ell_1\dot{x}_1 \begin{bmatrix} \cos x_1 & \sin x_1 & 0 \\ 0 & 0 & 0 \end{bmatrix} \tag{11.117}$$

The expressions $[\dot{V}^{21}]$ and $[\dot{V}^{22}]$ can be obtained from equations (11.100) and (11.101) by differentiation. From equations (11.104) and (11.105), it follows that the partial angular velocities derivatives for body 1 and body 2 are

$$[\dot{\omega}^1] = [\dot{\omega}^2] = 0$$

The other coefficients for the matrix $[C]$ are obtained following their integral definition

$$[C_{14}^*] = [C_{14}] = \sum_j \sum_i \int_{V_{ji}} m_{ji}[V^{ji}][V_e^{ji}]^T dV$$

$$= 0 + \frac{1}{2} m_2 A_2 \ell_2 \int_0^1 ([V^{21}][\dot{V}^{21}]^T + [V^{22}][\dot{V}^{22}]^T)\, d\xi \tag{11.118}$$

$$[C_{41}] = \sum_j \sum_i \int_{V_{ji}} m_{ji} [V_e^{ji}][\dot{V}^{ji}]^T \, dV$$

$$= 0 + \frac{1}{2} m_2 A_2 \ell_2 \int_0^1 ([V^{21}][\dot{V}^{21}]^T + [V^{22}][\dot{V}^{22}]^T) \, A_2 \ell_2 \, d\xi \quad (11.119)$$

Note that $[C_{14}]^T \neq [C_{41}]$. This simply states that the dynamic damping matrix is not symmetric. One needs to be reminded that the dynamic damping is a contribution of generalized velocities, viewed as the terms that are responsible for such things as Coriolis effects, centrifugal terms and other nonlinearities. The term C that contains the ordinary damping force coefficient is given by.

$$[C_{44}^*] = [C_{44}] = \sum_i \sum_i \int_{V_{ji}} m_{ji} [V_e^{ji}][\dot{V}_e^{ji}]^T \, dV + [C_s]$$

$$= 0 + \frac{1}{2} \int_0^1 m_2 ([V_e^{21}][\dot{V}_e^{21}]^T + [V_e^{22}][\dot{V}_e^{22}]^T) \, A_2 \ell_2 \, d\xi + [C_s]$$

$$= 0 + [C_s] = [C_s]$$

If we do not consider the ordinary damping force from the flexible body, then

$$[C_{44}^*] = 0 \quad (11.120)$$

The generalized damping matrix reduces to

$$[C] = \begin{bmatrix} C_{11} & C_{14} \\ C_{41} & C_{44} \end{bmatrix} = \begin{bmatrix} C_{11} & C_{14} \\ C_{41} & 0 \end{bmatrix} \quad (11.121)$$

where C_{11}, C_{14} and C_{41} are given by equations (11.114), (11.118) and (11.119). The stiffness matrix, on the other hand, is given by

$$[K_{44}] = [K_s] = \sum_j \sum_i \int_{V_{ji}} \frac{d}{d\xi}[V_e^{ji}] \frac{d}{d\xi}[V_e^{ji}]^T \frac{E \, dV}{\ell_{ji}^2}$$

$$= 0 + \frac{1}{2} \int_0^1 \left(\frac{d}{d\xi}[V_e^{21}] \frac{d}{d\xi}[V_e^{21}]^T + \frac{d}{d\xi}[V_e^{22}] \frac{d}{d\xi}[V_e^{22}]^T \right) \frac{E A_2 \ell_2 d\xi}{\left(\frac{1}{2}\ell_2\right)^2} \quad (11.122)$$

where E denotes Young's modulus of elasticity. From equations (11.97) and (11.98), we obtain

$$\frac{d}{d\xi}[V_e^{21}] = \begin{bmatrix} -\cos(x_2 + x_1) & -\sin(x_2 + x_1) & 0 \\ \cos(x_2 + x_1) & \sin(x_2 + x_1) & 0 \\ 0 & 0 & 0 \end{bmatrix} \quad (11.123)$$

$$\frac{d}{d\xi}[V_e^{22}] = \begin{bmatrix} 0 & 0 & 0 \\ -\xi \cos(x_2 + x_1) & -\sin(x_2 + x_1) & 0 \\ \cos(x_2 + x_1) & \sin(x_2 + x_1) & 0 \end{bmatrix} \quad (11.124)$$

Substituting of equation (11.123) and (11.124) into (11.122) results in to

$$[K_{44}] = 2\int_0^1 \begin{bmatrix} 0 & -1 & 0 \\ -1 & 1 & 0 \\ 0 & 0 & 0 \end{bmatrix} \frac{EA_2}{\ell_2} d\xi + 2\int_0^1 \begin{bmatrix} 0 & 0 & 0 \\ 0 & 1 & -1 \\ 0 & -1 & 1 \end{bmatrix} \frac{EA_2}{\ell_2} d\xi$$

$$= \frac{2EA_2}{\ell_2} \begin{bmatrix} 0 & -1 & 0 \\ -1 & 2 & -1 \\ 0 & -1 & 1 \end{bmatrix} \tag{11.125}$$

where $[K_{44}]$ is the ordinary stiffness, which is automatically overlapping. Because node 1 of body 2 is connected to the joint where the local relative coordinate $\{\bar{n}^2\}$ lies, we can assume that $\rho_1 = 0$. In addition, if we assume that no external force and moment are acting on the system, we obtain the following governing equations of motion:

$$[M]\{\ddot{y}\} + [C]\{\dot{y}\} + [K]\{y\} = 0 \tag{11.126}$$

where

$$[M] = \begin{bmatrix} \frac{1}{3}m_1 A_1 \ell\frac{3}{1} + \frac{1}{24}m_2 A_2 \ell\frac{3}{2} + m_{11} & & \text{symmetric} \\ \frac{1}{24}m_2 A_2 \ell\frac{3}{2} + m_{21} & \frac{1}{24}m_2 A_2 \ell\frac{3}{2} + m_{22} & \\ m_{41} & m_{42} & \frac{1}{3}m_2 A_2 \ell_2 \\ m_{51} & m_{52} & \frac{1}{12}m_2 A_2 \ell_2 \quad \times\frac{1}{6}m_2 A_2 \ell_2 \end{bmatrix} \tag{11.127}$$

m_{ij} $(i = 1,\ 2,\ 4,\ 5;\ j = 1,\ 2)$ are given by equations (11.109) and (11.110) by letting $\rho_1 = 0$.

$$[C] = \begin{bmatrix} C_{11} & C_{12} & C_{14} & C_{15} \\ C_{21} & C_{22} & C_{24} & C_{25} \\ C_{41} & C_{42} & 0 & 0 \\ C_{51} & C_{52} & 0 & 0 \end{bmatrix} \tag{11.128}$$

where $C_{ij}(i, j = 1,\ 2,\ 4,\ 5)$ are given by equations (11.114), (11.118) and (11.119), and

$$[K] = \frac{2EA_2}{\ell_2} \begin{bmatrix} 0 & 0 & 0 & 0 \\ 0 & 0 & 0 & 0 \\ 0 & 0 & 2 & -1 \\ 0 & 0 & -1 & 1 \end{bmatrix} \tag{11.129}$$

with the generalized coordinates being

$$\{y\} = [x_1 \quad x_2 \quad \rho_1 \quad \rho_3]^T \tag{11.130}$$

If the initial conditions $\{y_o\}$ and $\{\dot{y}_o\}$ are given, we can solve the differential equations of motion. The numerical solution will be as follows. Let

$$\{z\} = \begin{Bmatrix} y \\ \dot{y} \end{Bmatrix} \tag{11.131}$$

Then equation (11.126) becomes

$$\{\dot{z}\} = [A]\{z\} + \{F^*\} \tag{11.132}$$

where

$$[A] = \begin{bmatrix} 0 & 1 \\ M^{-1}K & M^{-1}C \end{bmatrix} \quad \text{and} \quad \{F^*\} = \begin{Bmatrix} 0 \\ M^{-1}F \end{Bmatrix} = 0$$

At $t = 0$, we have

$$\{z_0\} = \begin{Bmatrix} \dot{y}_0 \\ y_0 \end{Bmatrix} \tag{11.133}$$

Then from equation (11.132) we can obtain $\{\dot{z}_0\}$. The solution to equation (11.132) can be obtained through several integration schemes, one of which is based on Newton's method, where we can write

$$\{z_1\} = \{z_0\} + \{\dot{z}_0\}\ \Delta t \tag{11.134}$$

From equation (11.132), we then find $\{\dot{z}_1\}$. If we are interested in a simulation for a given period T, we let

$$T = m\ \Delta t \tag{11.135}$$

and $Ti = i\Delta t, i = i, \ldots, m$. Equation (11.132) can then be solved step-by-step from $i = 0$ to $i = m$. Integration routines to handle such problems are available through several mathematical tools such as MATLAB and IMSL routines, among others.

11.5 Reduction of the Equations of Motion Through Modal Analysis

From vibration theory, we know that for an arbitrary body, the equations of motion without damping are given by

$$[M_j]\{\ddot{\rho}_j\} + [K_j]\{\rho_j\} = 0 \tag{11.136}$$

This is an ordinary eigenvalue and eigenvector problem, where $[M_j]$ is the mass matrix of body j and $[K_j]$ its corresponding stiffness matrix. Let $\{\rho_j\} = [\Phi_j]\{\eta_j\}$ $[\Phi_j]$

denote the shape matrix of body j and $[\eta_j]$ the modal coordinates. In general, $[\Phi_j]$ $(j = 1, \ldots, n)$ is a square matrix. To save memory space and computing time in the simulation of large-scale multibody systems, we can select the mode shapes which are sufficiently significant in engineering analysis. Let us write $[\Phi_j]$ into its global form.

$$[\Phi] = \begin{bmatrix} [\Phi_1] & & & & & 0 \\ & [\Phi_2] & & & & \\ & & \ddots & & & \\ & & & [\Phi_2] & & \\ & & & & \ddots & \\ & & & & & [\Phi_n] \end{bmatrix} \tag{11.137}$$

Substituting the nodal coordinates $\{\rho\}$ by $\{\eta_j\}$ into equation (11.73) using equation (11.137), and after some work, we obtain

$$[M][\ddot{y}\} + [C]\{\dot{y}\} + [K]\{y\} = \{F\} \tag{11.138}$$

where

$$\{y\} = [x^T \quad \xi^T \quad \eta^T]^T \tag{11.139}$$

Observe how the equations are now in terms of the generalized coordinates x associated with the rigid-body rotation, ξ is the rigid-body translation, and η is associated with modal coordinates of the mode shapes selected. The generalized mass matrix takes the following form:

$$[M] = \begin{bmatrix} M_{11} & M_{12} & M_{13} \\ M_{21} & M_{22} & M_{23} \\ M_{31} & M_{32} & M_{33} \end{bmatrix} \tag{11.140}$$

where

$$[M_{11}] = \sum_j \sum_i \int_{V_{ji}} (m_{ji}[V^{ji}][V^{ji}]^T + [\omega^j][I_{ji}][\omega^j]^T)\ dV \tag{11.141}$$

$$[M_{12}] = \sum_j \sum_i \int_{V_{ji}} m_{ji}[V^{ji}][V_s^j]^T\ dV \tag{11.142}$$

$$[M_{13}] = \sum_j \sum_i \int_{V_{ji}} \{m_{ji}[V^{ji}]([V_e^{ji}]^T + [V_s^j]^T[Q]) + [\omega^j][I_{ji}][\omega^{ji}]^T\}\ dV[\Phi] \tag{11.143}$$

$$[M_{21}] = [M_{12}]^T \tag{11.144}$$

$$[M_{22}] = \sum_j \sum_i \int_{V_{ji}} m_{ji}[V_s^j][V_s^j]^T\ dV \tag{11.145}$$

$$[M_{23}] = \sum_j \sum_i \int_{V^{ji}} m_{ji}[V_s^j][V_e^{ji}]^T + [V_s^j]^T[Q]^T \, dV[\Phi] \tag{11.146}$$

$$[M_{31}] = \sum_j \sum_i \int_{V^{ji}} (m_{ji}[V_e^{ji}][V^{ji}]^T + [\omega_{ji}][I_{ji}][\omega^j]^T) \, dV \tag{11.147}$$

$$[M_{32}] = \sum_j \sum_i \int_{V^{ji}} m_{ji}[V_e^{ji}][V_s^{ji}]^T \, dV \tag{11.148}$$

$$[M_{33}] = \sum_j \sum_i \int_{V^{ji}} \{m_{ji}[V_e^{ji}]([V_e^{ji}]^T + [V_s^j]^T[Q]) + [\omega^{ji}][I_{ji}][\omega^{ji}]^T\} \, dV[\Phi] \tag{11.149}$$

Similarly,

$$[C] = \begin{bmatrix} C_{11} & C_{12} & C_{13} \\ C_{21} & C_{22} & C_{23} \\ C_{31} & C_{32} & C_{33} \end{bmatrix} \tag{11.150}$$

where

$$[C_{11}] = \sum_j \sum_i \int_{V^{ji}} \{m_{ji}[V^{ji}][\dot{V}^{ji}]^T + [\omega^j][I_{ji}][\dot{\omega}^j] + [\Omega_x^{oj}][I_{ji}][\omega^j]^T\} \, dV \tag{11.151}$$

$$[C_{12}] = \sum_j \sum_i \int_{V^{ji}} m_{ji}[V^{ji}][\dot{V}_s^j]^T \, dV \tag{11.152}$$

$$[C_{13}] = \sum_j \sum_i \int_{V^{ji}} \{m_{ji}[V^{ji}]([\dot{V}^{ji}]^T + [\dot{V}_s^j]^T[Q])$$

$$+ [\omega^j][I_{ji}][\dot{\omega}^{ji}]^T + [\Omega_\rho^{oj}][I_{ji}][\omega^{ji}]^T\} \, dV[\Phi] \tag{11.153}$$

$$[C_{21}] = \sum_j \sum_i \int_{V^{ji}} m_{ji}[V_s^j][\dot{V}^{ji}]^T \, dV \tag{11.154}$$

$$[C_{22}] = \sum_j \sum_i \int_{V^{ji}} m_{ji}[V_s^j][\dot{V}_s^j]^T \, dV \tag{11.155}$$

$$[C_{23}] = \sum_j \sum_i \int_{V^{ji}} m_{ji}[V_s^{ji}]([\dot{V}_e^{ji}]^T + [\dot{V}_s^j]^T[Q]) \, dV[\Phi] \tag{11.156}$$

$$[C_{31}] = \sum_j \sum_i \int_{V^{ji}} \{m_{ji}[V_e^{ji}][\dot{V}^{ji}]^T + [\omega^j][I_{ji}][\dot{\omega}^j]^T + [\Omega_x^{oj}][I_{ji}][\omega^j]^T\} \, dV \tag{11.157}$$

$$[C_{32}] = \sum_j \sum_i \int_{V^{ji}} m_{ji} [V_e^{ji}][V_s^j]^T \, dV \tag{11.158}$$

$$[C_{33}] = \sum_j \sum_i \int_{V^{ji}} \{m_{ji} [V_e^{ji}]([\dot{V}_e^{ji}]^T + [\dot{V}_s^j]^T [Q])$$

$$+ [\omega^{ji}][I_{ji}][\dot{\omega}^{ji}]^T + [\Omega_\rho^{oj}][I_{ji}][\omega^{ji}]^T\} \, dV[\Phi] + [C_s][\Phi] \tag{11.159}$$

The generalized stiffness matrix is

$$[K] = \begin{bmatrix} 0 & 0 & 0 \\ 0 & 0 & 0 \\ 0 & 0 & K_{33} \end{bmatrix} \tag{11.160}$$

where

$$[K_{33}] = ([K_S] + [K_G])[\Phi] \tag{11.161}$$

Equation (11.161) shows how the stiffness is affected when the modal coordinates are introduced. Finally, we write the generalized force vector as

$$\{F\} = [F_1^T \;\; F_2^T \;\; F_3^T]^T \tag{11.162}$$

where each component contribution is defined based on the rigid-body motion and elastic deformation forces.

$$\{F_1\} = \sum_j \sum_i \int_{S_{ji}} [V^{ji}]\{f_{ji}\} \, dS + \sum_j \sum_i \int_{V^{ji}} [V^{ji}]\{b_{ji}\} \, dV$$

$$+ \sum_j \sum_i \int_{S_{ji}} [\omega^i]\{M_{ji}\} \, dS \tag{11.163}$$

$$\{F_2\} = \sum_j \sum_i \int_{S_{ji}} [V_s^j]\{f_{ji}\} \, dS + \sum_j \sum_i \int_{V_{ji}} [V_s^j]\{b_{ji}\} \, dV \tag{11.164}$$

$$\{F_3\} = \sum_j \sum_i \int_{S_{ji}} [V_e^{ji}]\{f_{ji}\} \, dS + \sum_j \sum_i \int_{V_{ji}} [V_e^{ji}]\{b_{ji}\} \, dV$$

$$+ \sum_j \sum_i \int_{S_{ji}} [\omega^{ji}]\{M_{ji}\} \, dS \tag{11.165}$$

For $[C_S]$, $[K_S]$, $[K_G]$, $\{f_{ji}\}$, $\{b_{ji}\}$ and $\{M_{ji}\}$, see equations (11.62) to (11.68) for details and formulation.

11.6 Effect of Geometric Stiffening

The stiffness matrices that we discussed before are all based on small deformation and material elastic linearity. In this case, the solution can be obtained by deleting quadratic terms in the strain–displacement relations. But when the deformation is large enough to cause significant change in the geometry of the deformable body, we have to consider the nonlinearity between strain and displacement, that is, geometric elastic nonlinearity. The material nonlinearity due to a nonlinear constitution, such as plastic or nonlinear viscoelastic behavior of the deformable bodies, is not discussed here. The linear strain energy theory assumes that the deformation components are independent. This assumption does not hold when a deformable body is subjected to a high-speed overall motion. Therefore, higher-order terms in the strain energy need to be considered.

Consider a one-dimensional beam element, where the element axis is located at the centroidal line (see Figure 11.5). Let $\upsilon_i(x,\ t)$ and $\theta_i(x,\ t)$ represent, respectively, the deformation displacement and rotation components of axis frames, located along the centroidal line x distance away from the element axis in the undeformed state. Then the displacement field for arbitrary points in the beam element, $u_i(x, y, z, t)$, can be expressed for small deformation rotations in terms of the undeformed state coordinates.

$$u_1 = \upsilon_1 + z\theta_2 - y\theta_3 \tag{11.166}$$

$$u_1 = \upsilon_2 + z\theta_1 \tag{11.167}$$

$$u_3 = \upsilon_3 + y\theta_1 \tag{11.168}$$

The general nonlinear equation for strain components $\in_{ij}$ is

$$\in_{ji} = \frac{1}{2}\left(\frac{\partial u_i}{\partial x_j} \quad \frac{\partial u_j}{\partial x_i} \quad \frac{\partial u_r}{\partial x_i} \quad \frac{\partial u_r}{\partial x_j}\right) \quad (i,\ j,\ r = 1,\ 2,\ 3) \tag{11.169}$$

where $x_i,\ i = 1,\ 2,\ 3$, denote the undeformed coordinates x, y and z.

If the change in volume of an infinitesimal element in the deformed body is negligible, the strain energy is then given by

$$U = \int_{\upsilon_k} \in^T \sigma \, d\upsilon = \int_{\upsilon_k} \in^T E \in \, d\upsilon \tag{11.170}$$

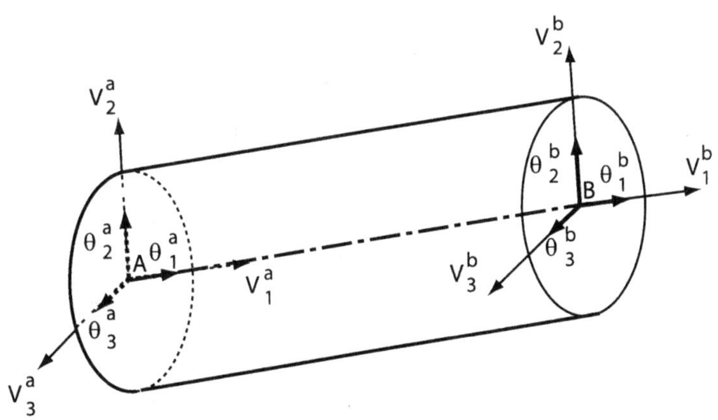

Fig. 11.5. Beam element model with end nodes A and B

where $\in$ and σ are the vectors of strain and stress components respectively.

$$\in = [\in_{11},\ \in_{22},\ \in_{33},\ \in_{12},\ \in_{13},\ \in_{23}]^T \tag{11.171}$$

$$\sigma = [\sigma_{11},\ \sigma_{22},\ \sigma_{33},\ \sigma_{12},\ \sigma_{13},\ \sigma_{23}]^T \tag{11.172}$$

and E is the matrix of Hooke's elastic coefficients.

From equations (11.166) to (11.172), we can obtain the strain energy equation

$$\begin{aligned} U = \frac{E}{A}\int_L \left[\left(\frac{\partial \upsilon_1}{\partial x}\right)^2 A + \left(\frac{\partial \theta_2}{\partial x}\right)^2 I_2 + \left(\frac{\partial \theta_3}{\partial x}\right)^2 I_3\right] dx \\ + \frac{G}{2}\int_L \left[\left(\frac{\partial \theta_1}{\partial x}\right)^2 k_1 + \left(\frac{\partial \upsilon_2}{\partial x} - \theta_3\right)^2 k_2 A + \left(\frac{\partial \upsilon_3}{\partial x} - \theta_2\right)^2 k_3 A\right] dx \\ + \frac{E}{A} A \int_L \left[\frac{\partial \upsilon_1}{\partial x}\left(\frac{\partial \upsilon_2}{\partial x}\right)^2 + \frac{\partial \upsilon_1}{\partial x}\left(\frac{\partial \upsilon_3}{\partial x}\right)\right]^2 dx \end{aligned} \tag{11.173}$$

where A is the cross-sectional area, E the modulus of elasticity, G the shear modulus, I_2 and I_3 the central second moments of area, k_1 the torsional constant for the cross section, and k_2 and k_3 the shear correction factors. Note that in equation (11.173), only the first two integral expressions would be obtained if the linear strain-displacement relations were used. The last integral expression contains the geometric stiffening terms due to the coupling of the axial and transverse displacements. As we know, υ_i and θ_i $(i = 1, 2, 3$: see Figure 11.5) can be expressed in terms of the nodal variables ρ_j by utilizing polynomials of appropriate order as

$$\upsilon_i = N_i \rho_j \tag{11.174}$$

$$\theta_i = \psi_{ij} \rho_j \tag{11.175}$$

where ρ_j are the displacements and rotations at the nodes,

$$\{\rho\} = [\upsilon_1^A, \upsilon_2^A, \upsilon_3^A, \theta_1^A, \theta_2^A, \theta_3^A, \upsilon_1^B, \upsilon_2^B, \upsilon_3^B, \theta_1^B, \theta_2^B, \theta_3^B)^T \tag{11.176}$$

and the element shape functions including shear deformation, are given by

$$[N] = \begin{bmatrix} a_1 & 0 & 0 & 0 & 0 & 0 & a_2 & 0 & 0 & 0 & 0 & 0 \\ 0 & b_1 & 0 & 0 & 0 & b_2 & 0 & b_3 & 0 & 0 & 0 & b_4 \\ 0 & 0 & b_5 & 0 & b_6 & 0 & 0 & 0 & b_7 & 0 & b_8 & 0 \end{bmatrix} \tag{11.177}$$

$$[\psi] = \begin{bmatrix} 0 & 0 & 0 & a_1 & 0 & 0 & 0 & 0 & 0 & a_2 & 0 & 0 \\ 0 & 0 & c_5 & 0 & c_6 & 0 & 0 & 0 & c_7 & 0 & c_8 & 0 \\ 0 & c_1 & 0 & 0 & 0 & c_2 & 0 & c_3 & 0 & 0 & 0 & c_4 \end{bmatrix} \tag{11.178}$$

where

$$a_1 = 1 - \xi \tag{11.179}$$

$$a_2 = \xi \tag{11.180}$$

$$b_1 = \frac{1}{1+\lambda_2}[1 - 3\xi^2 + 2\xi^3 + (1-\xi)\lambda_2] \tag{11.181}$$

$$b_2 = \frac{L}{1+\lambda_2}\left[\xi - 2\xi^2 + \xi^3 + \frac{1}{2}\lambda_2(\xi - \xi^2)\right] \tag{11.182}$$

$$b_3 = \frac{1}{1+\lambda_2}[3\xi^2 - 2\xi^3 + \lambda_2\xi] \tag{11.183}$$

$$b_4 = \frac{L}{1+\lambda_2}\left[-\xi^2 - \xi^3 - \frac{1}{2}\lambda_2(\xi - \xi^2)\right] \tag{11.184}$$

$$b_5 = \frac{1}{1+\lambda_3}[1 - 3\xi^2 + 2\xi^3 + \lambda_3(1-\xi)] \tag{11.185}$$

$$b_6 = \frac{1}{1+\lambda_3}\left[-\xi + \xi^2 - \xi^3 - \frac{1}{2}\lambda_3(\xi - \xi^2)\right] \tag{11.186}$$

$$b_7 = \frac{1}{1+\lambda_3}[3\xi^2 - 2\xi^3 + \lambda_3\xi] \tag{11.187}$$

$$b_8 = \frac{1}{1+\lambda_3}\left[-\xi^2 + \xi^3 - \frac{1}{2}\lambda_3(\xi - \xi^2)\right] \tag{11.188}$$

$$c_1 = \frac{b}{(1+\lambda_2)L}(-\xi + \xi^2) \tag{11.189}$$

$$c_2 = \frac{1}{1+\lambda_2}[1 - 4\xi + 3\xi^2 + \lambda_2(1-\xi)] \tag{11.190}$$

$$c_3 = -c_1 \tag{11.191}$$

$$c_4 = \frac{1}{1+\lambda_2}[-2\xi + 3\xi^2 + \lambda_2\xi] \tag{11.192}$$

$$c_5 = \frac{b}{(1+\lambda_3)L}(-\xi + \xi^2) \tag{11.193}$$

$$c_6 = \frac{1}{1+\lambda_3}[1 - 4\xi + 3\xi^2 + \lambda_3(1-\xi)] \tag{11.194}$$

$$c_7 = -c_5 \tag{11.195}$$

$$c_8 = \frac{1}{1+\lambda_3}[-2\xi + 3\xi^2 + \lambda_3\xi] \tag{11.196}$$

where

$$\xi = \frac{x}{L} \quad \lambda_2 = \frac{12EI_3}{k_2GAL^2} \quad \lambda_3 = \frac{12EI_2}{k_3GAL^2} \tag{11.197}$$

Using equations (8.44) to (8.46) and

$$u_i = \phi_i\rho_j \quad (i = 1,\ 2,\ 3;\ j = 1, \ldots, 12) \tag{11.198}$$

then shape function matrix can be expressed as

$$[\phi] = \begin{bmatrix} a_1 & -yc_1 & zc_5 & 0 & zc_6 & -yc_2 & a_2 & -yc_3 & zc_7 & 0 & zc_8 & -yc_4 \\ 0 & b_1 & 0 & -za_1 & 0 & b_2 & 0 & b_3 & 0 & -za_2 & 0 & b_4 \\ 0 & 0 & b_5 & ya_1 & b_6 & 0 & 0 & 0 & b_7 & ya_2 & b_8 & 0 \end{bmatrix} \tag{11.199}$$

From equation (11.173), we know that the strain energy due to geometric stiffening is given by

$$U^G = \frac{E}{2}A \int_L \frac{\partial v_1}{\partial x}\left[\left(\frac{\partial v_2}{\partial x}\right)^2 + \left(\frac{\partial v_3}{\partial x}\right)^2\right] dx \tag{11.200}$$

From equation (11.174), we get

$$U^G = F \int_L \left[\left(\frac{\partial v_2}{\partial x}\right)^2 + \left(\frac{\partial v_3}{\partial x}\right)^2\right] dx \tag{11.201}$$

where

$$F = \frac{EA}{2L}(v_1^B - v_1^A) \tag{11.202}$$

Substituting equation (11.202) into (11.201), we obtain the geometrical stiffness matrix:

$$[K_G] = \frac{F}{L}\begin{bmatrix} 0 & & & & & & & & & & & \\ 0 & g_1 & & & & & & & & & & \\ 0 & 0 & g_3 & & & & & & & & & \\ 0 & 0 & 0 & 0 & & & & & & & & \\ 0 & 0 & g_4 & 0 & g_5 & & & & & & & \\ 0 & g_2 & 0 & 0 & 0 & g_7 & & & & & & \\ 0 & 0 & 0 & 0 & 0 & 0 & 0 & & & & & \\ 0 & -g_1 & 0 & 0 & 0 & -g_2 & 0 & g_1 & & & & \\ 0 & 0 & -g_3 & 0 & -g_4 & 0 & 0 & 0 & g_3 & & & \\ 0 & 0 & 0 & 0 & 0 & 0 & 0 & 0 & 0 & 0 & & \\ 0 & 0 & g_4 & 0 & g_6 & 0 & 0 & 0 & -g_4 & 0 & g_5 & \\ 0 & g_2 & 0 & 0 & 0 & g_8 & 0 & -g_2 & 0 & 0 & 0 & g_7 \end{bmatrix} \tag{11.203}$$

where

$$g_1 = \frac{1}{(1+\lambda_2)^2}\left(\frac{6}{5} + 2\lambda_2 + \lambda_2^2\right) \tag{11.204}$$

$$g_2 = \frac{L}{10(1+\lambda_2)^2} \tag{11.205}$$

$$g_3 = \frac{1}{(1+\lambda_2)}\left(\frac{6}{5} + 2\lambda_3 + \lambda_3^2\right) \tag{11.206}$$

$$g_4 = -\frac{L}{10(1+\lambda_2)^2} \tag{11.207}$$

$$g_5 = \frac{L^2}{(1+\lambda_3)^2}\left(\frac{2}{15} + \frac{\lambda_3}{6} + \frac{\lambda_3^2}{12}\right) \tag{11.208}$$

$$g_6 = -\frac{L^2}{(1+\lambda_3)^2}\left(\frac{1}{30} + \frac{\lambda_3^2}{3}\right) \tag{11.209}$$

$$g_7 = \frac{L^2}{(1+\lambda_2)^2}\left(\frac{2}{15} + \frac{\lambda_2}{6} + \frac{\lambda_3^2}{12}\right) \tag{11.210}$$

$$g_8 = \frac{L^2}{(1+\lambda_2)^2}\left(\frac{1}{30} + \frac{\lambda_2^2}{3}\right) \tag{11.211}$$

11.6.1 Geometric Stiffness Matrix of an Isoparametric Brick Element

Based on the procedure presented earlier, for the derivation of geometric stiffness matrix for the beam element, we can also derive a similar geometric stiffness matrix for an isoparametric brick element [28]. Let us consider the eight-noded brick element shown in Figure 11.6.

The computation of the geometric stiffness matrix z^{ki} for the ith element of body k can be carried out numerically by using the equation

$$\frac{\partial U^{ki}}{\partial \Delta^{ki}} = Z^{ki}\Delta^{ki} \tag{11.212}$$

where U^{ki} is the strain energy component for the ith element of body k and Δ is the associated vector of elastic nodal coordinates. Equation (11.212) represents the relationship between nodal elastic coordinates and the associated force vector, through the use of the geometrical stiffness matrix.

This can be written by using the nonlinear theory of elasticity for linear–elastic materials:

$$U^{ki} = \int_{V^{ki}} (2\gamma_{xy}^s)^2 + (2\gamma_{yz}^s)^2 + (2\gamma_{zx}^s)^2 \det\ [J(x,\ \ y,\ \ z)]\ dV \tag{11.213}$$

where γ_{xy}^s, γ_{yz}^s and γ_{zx}^s are the components of shear strain, det $[J]$ is the Jacobian of the transformation from the body coordinates, $\bar{n}_1$, $\bar{n}_2$ and $\bar{n}_3$, are the natural or dimensionless coordinates $(\mu,\ \omega,\ \upsilon)$, and V^{ki} is the element domain. Transforming to local coordinates of the finite element, equation (11.213) becomes

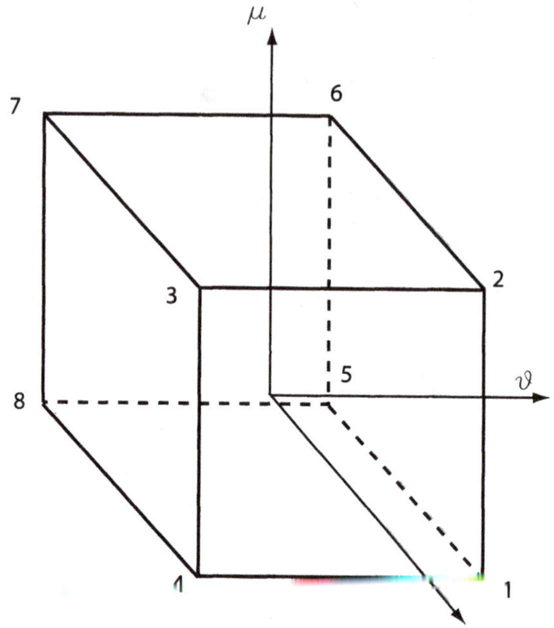

Fig. 11.6. Three-dimensional eight-noded isoparametric element

$$U^{ki} = \int_{-1}^{+1}\int_{-1}^{+1}\int_{-1}^{+1} (2\gamma_{xy}^{s})^2 + (2\gamma_{yz}^{s})^2 + (2\gamma_{zx}^{s})^2 \ \det\left[J(\mu,\omega,\upsilon)\right] d\mu\, d\omega\, d\upsilon \tag{11.214}$$

Using the residual weights W_i, W_j and W_k, corresponding to the Gauss points (μ_{gp}, ω_{gp}, υ_{gp}), we can write the equation for the strain energy for the element as

$$U^{ki} = \sum_{i=1}^{\mu_{gp}}\sum_{j=1}^{\omega_{gp}}\sum_{i=1}^{\upsilon_{gp}} (2y_{xy}^{s})^2 + (2y_{yz}^{s})^2 + (2y_{zx}^{s})^2 \ \det\left[J(\mu,\ \omega,\ \upsilon)\right] W_i W_j W_k$$

Using for the sake of convenience u_1, u_2, u_3 for and x_1, x_2, x_3 for x, y, z and $\gamma_{12}, \gamma_{23}, \gamma_{31}$ for $\gamma_{xy}, \gamma_{yz}, \gamma_{zx}$, we can write the nonlinear strain displacement relation as

$$\gamma_{ab}^{s} = \frac{1}{2}\left(\frac{\partial u_a}{\partial x_b} + \frac{\partial u_b}{\partial x_a}\right) + \sum_{c=1}^{3} \frac{\partial u_c}{\partial x_a} . \frac{\partial u_c}{\partial x_b} \quad (a,\ b\ =\ 1,\ 2,\ 3) \tag{11.215}$$

The linear elastic displacements δ, ρ, σ with respect to the body reference coordinates, x, y, z can be written in terms of the elastic nodal displacements of three-dimensional isoparametric Brick elements $\rho_1, \rho_2, \ldots, \rho_{48}$ through the corresponding shape functions $T_1, T_2, \ldots, T_8$ as

$$\delta = \Gamma_1\rho_1 + \Gamma_2\rho_7 + \Gamma_3\rho_{13} + \Gamma_4\rho_{19} + \cdots + \ \Gamma_8\rho_{43} \tag{11.216}$$

$$\rho = \Gamma_1 \rho_2 + \Gamma_2 \rho_8 + \Gamma_3 \rho_{14} + \Gamma_4 \rho_{20} + \cdots + \Gamma_8 \rho_{44} \tag{11.217}$$

$$\sigma = \Gamma_1 \rho_3 + \Gamma_2 \rho_9 + \Gamma_3 \rho_{15} + \Gamma_4 \rho_{21} + \cdots + \Gamma_8 \rho_{45} \tag{11.218}$$

The components of shear strain γ^s needed in the evaluation of the geometrical stiffness can be written in terms of the derivatives of the linear elastic displacements. The derivatives of linear elastic displacements with respect to x can also be written in terms of the derivatives of the dimensionless natural coordinates of the isoparametric elements. Square terms in equation (11.213) can be achieved after simplifying, and it can be written in the following fashion:

$$(\gamma_{xy}^s)^2 = \frac{1}{4}\left(\frac{\partial \delta}{\partial y} + \frac{\partial \rho}{\partial x} + \frac{\partial \delta}{\partial x}\cdot\frac{\partial \delta}{\partial y} + \frac{\partial \rho}{\partial x}\cdot\frac{\partial \rho}{\partial y} + \frac{\partial \sigma}{\partial x}\cdot\frac{\partial \sigma}{\partial y}\right)^2 \tag{11.219}$$

Expanding the expression (11.219) and simplifying by neglecting the higher-order terms and retaining the terms up to the second order, we get

$$(\gamma_{xy}^s)^2 = \frac{1}{4}\left[\left(\frac{\partial \delta}{\partial y}\right)^2 + \left(\frac{\partial \rho}{\partial x}\right)\right]^2 \tag{11.220}$$

The square terms on the right-hand side of equation (11.220) can be simplified using the partial derivatives discussed earlier. Similarly, the other two terms in equation (11.213), $(2 \cdot \gamma_{yz}^s)^2$ and $(2 \cdot \gamma_{zx}^s)^2$, in equation (11.213), can be evaluated.

Subsequent substitution of these terms in equation (11.214), followed by required numerical integration, gives the geometrical stiffness matrix. For more details, the reader is encouraged to see reference [29].

11.7 Dynamic Simulations: Applications

A general-purpose computer program for multibody systems called DYAMUS [32] has been developed utilizing the procedures presented in this chapter. The program automatically eliminates the constrained joint degrees of freedom by deleting the corresponding rows in the partial velocity vectors, hence eliminating constraint equations for joint connections.

11.7.1 Dynamic Simulation of a Space-Based Robotic Manipulator Using Beam Elements

As discussed in earlier sections of this chapter, geometric stiffening plays an important role in the accurate dynamic prediction of flexible-body systems. To show the significance of geometric stiffening effects, the space-based robotic manipulator used by Kane et al. and discussed in reference [30] by Ider and Amirouche will be considered (Figure 11.7). Kane et al. developed the equations of motion for a cantilever beam attached to a moving base and made use of the explicit constraint equation between the axial and transverse deflections in the derivation of the generalized inertia forces to obtain the related geometric stiffening terms.

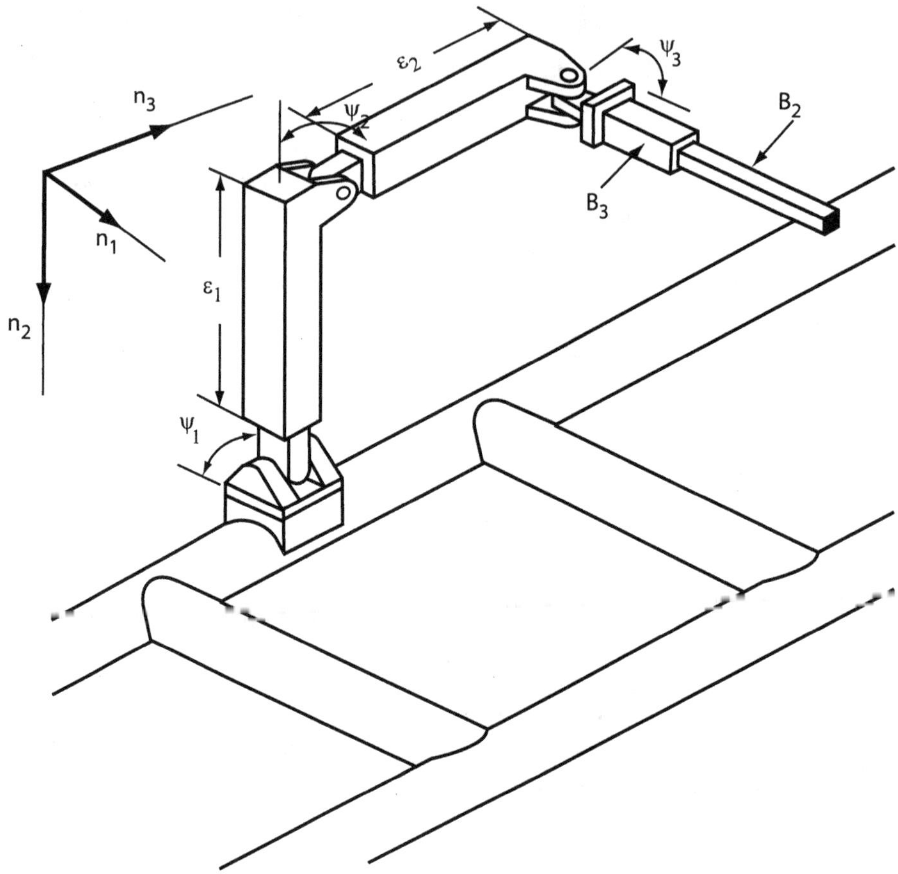

Fig. 11.7. Space-based manipulator used for deployment and maneuvers

Here, the same data for the manipulator will be used. Links 1 and 2 are rigid, each with a length of 8 m. The flexible link contains two distinct segments, B_1 and B_2, with material properties $E = 6.895 \times 10^{10}\,\text{N/m}^2$, $G = 2.6519 \times 10^{10}\,\text{N/m}^2$, and mass density $m_e = 2766.67\,\text{kg/m}^3$. The value B_1 is $2\frac{2}{3}$ m long and has the following section properties: $A = 3.84 \times 10^{-4}\,\text{m}^2$, $I_2 = 1.50 \times 10^{-7}\,\text{m}^4$, $I_3 = 1.50 \times 10^{-7}\,\text{m}^4$, $k_2 = 2.09$, $k_3 = 2.09$, and $\kappa 1 = 2.2 \times 10^{-7}\,\text{m}^4$. The value B_2 is $5\frac{1}{3}$m long and has the properties $A = 7.3 \times 10^{-5}\,\text{m}^2$, $I_2 = 4.8746 \times 10^{-9}\,\text{m}^4$, $I_3 = 8.2181 \times 10^{-9}\,\text{m}^4$, $k_2 = 3.174$, $k_3 = 1.52$, and $\kappa_1 = 2.433 \times 10^{-11}\,\text{m}^4$ (see reference [30]).

First, the deployment maneuver lasting $T = 15$ s, during which the angles ψ_1, ψ_2 and ψ_3 change from 180° to 90°, 180° to 45°, and 180° to 0°, respectively, is simulated. The deployment motion is given by

$$\psi_1(1) = \begin{cases} \pi - \dfrac{\pi}{2T}\left(t - \dfrac{T}{2\pi}\sin\dfrac{2\pi t}{T}\right) & \text{rad} \quad 0 < t \le T \\ \pi/2 & \qquad\quad t > T \end{cases} \tag{11.221}$$

$$\psi_2(1) = \begin{cases} \pi - \dfrac{3\pi}{4T}\left(t - \dfrac{T}{2\pi}\sin\dfrac{2\pi t}{T}\right) \text{ rad} & 0 < t \leq T \\ \pi/4 & t > T \end{cases} \tag{11.222}$$

$$\psi_3(1) = \begin{cases} \pi - \dfrac{\pi}{T}\left(t - \dfrac{T}{2\pi}\sin\dfrac{2\pi t}{T}\right) \text{ rad} & 0 < t \leq T \\ 0 & t > T \end{cases} \tag{11.223}$$

The tip deflections along n_2 and n_3, respectively, are given in Figure 11.8.

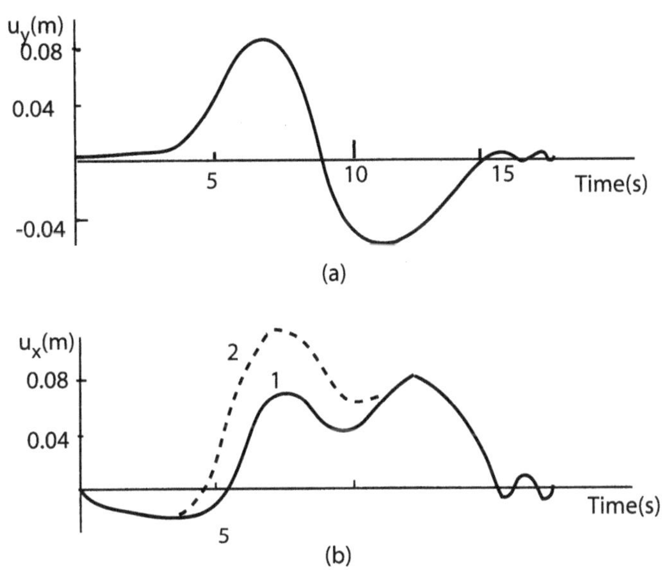

Fig. 11.8. (a) Tip deflection u_y. (b) Tip deflection u_z : (1) Simulation using equation (11.138) (also presented by Ider and Amirouche [30]); (2) Simulation by Kane et al. [13], including eccentricity

Second, the spin-up maneuver is simulated, for which the initial configuration is given by Figure 11.9. The ψ_1 and ψ_3 are fixed 90° and 0°, respectively, and ψ_2 increases from 0 to 6 rad/s in a time interval of $T = 15$ s. This motion is given as

$$\psi_2(1) = \begin{cases} \dfrac{6}{T} - \dfrac{t^2}{2} + \dfrac{T^2}{4\pi^2}\left(\cos\dfrac{2\pi t}{T} - 1\right) \text{ rad} & 0 < t \leq T \\ (6t - 45) \text{ rad} & t > T \end{cases} \tag{11.224}$$

Again, the tip transverse deflection is plotted in Figure 11.10, and it correlates with results by reference [20]. This simulation shows clearly how the geometric stiffening affects the spin-up motions.

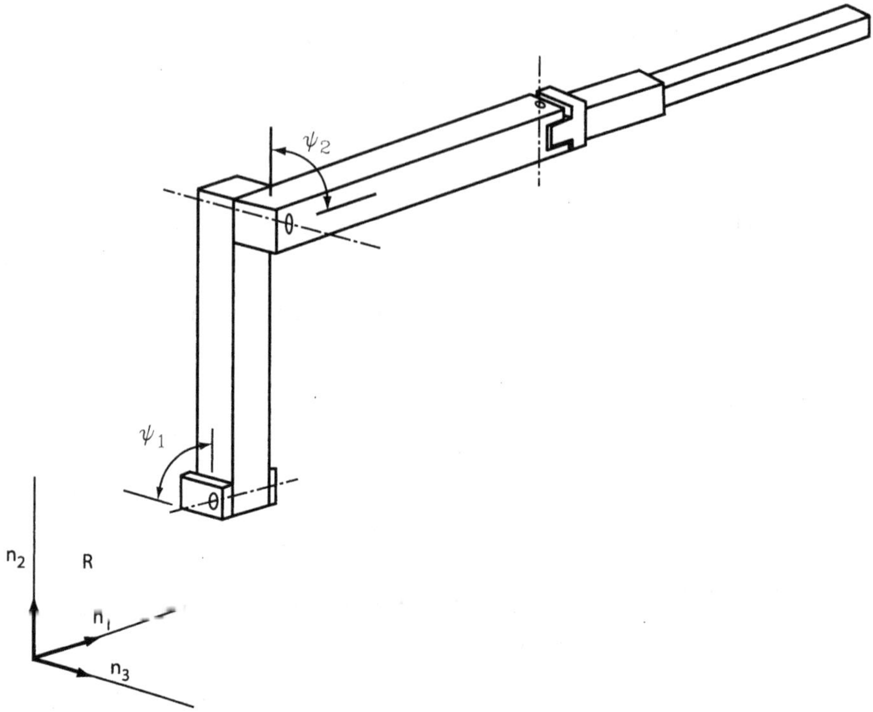

Fig. 11.9. Initial configuration for spin-up motion for the space-based manipulator

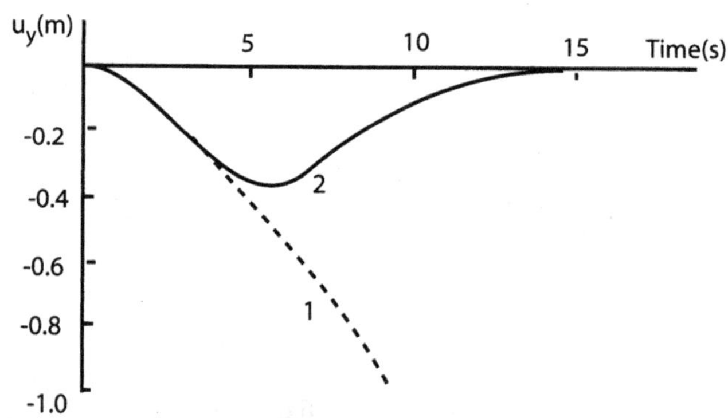

Fig. 11.10. Tip deflection u_y for spin-up motion: (1) without geometric stiffening; (2) with geometric stiffening (from reference [30])

Spin-up motion together with deployment of other parts, such as helicopter rotor blades and satellites with flexible appendages, are of practical interest. For this reason, an additional simulation is performed. Starting from the initial configuration given in Figure 11.9, in addition to the spin-up motion of link 2 given in equation (11.224), link 1 will be deployed from 90° and 40° as

$$\psi_2(1) = \begin{cases} \dfrac{\pi}{T} - \dfrac{5}{9}\dfrac{\pi}{2T}\left(t - \dfrac{T}{2\pi}\sin\dfrac{2\pi t}{T}\right) \text{ rad} & 0 < t \leq T \\ 2\pi/9 & t > T \end{cases} \tag{11.225}$$

The tip deflections along n_2 and n_3 for this simulation are given in Figures 11.11 and 11.12. It can easily be seen that the displacement along n_2 is affected by the additional centrifugal forces while following the same pattern. The deflection in the n_3 direction diverges when the geometric stiffening is not included. This is because centrifugal softening becomes the dominant effect stemming from high $\dot{\psi}_2$. Otherwise, the deflection tends to stabilize with time, when the geometric stiffening forces become dominant. The effect of the deployment acceleration on the displacement along n_3 changes sign depending on ψ_2, which undergoes approximately seven revolutions in 15 s.

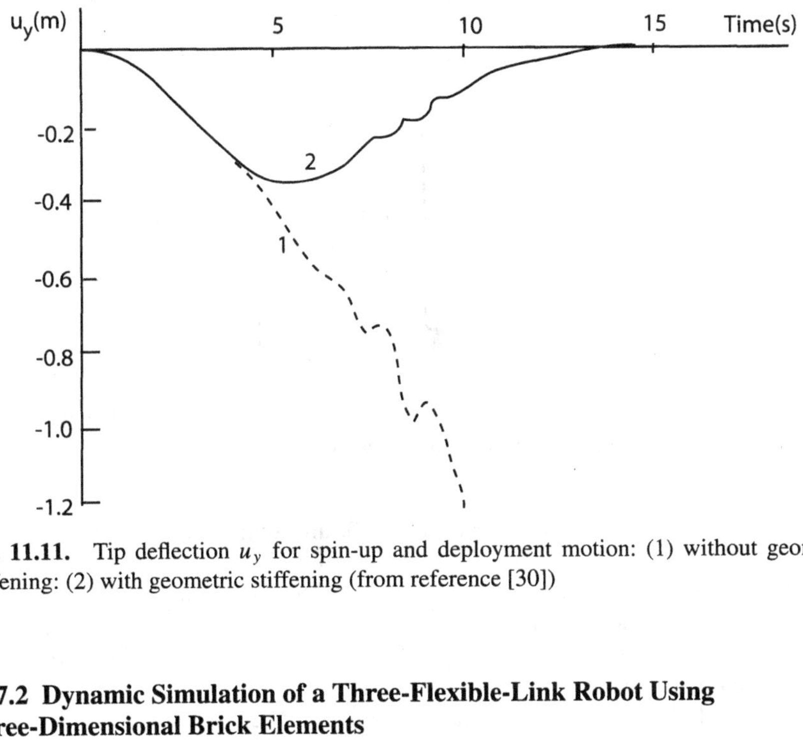

Fig. 11.11. Tip deflection u_y for spin-up and deployment motion: (1) without geometric stiffening: (2) with geometric stiffening (from reference [30])

11.7.2 Dynamic Simulation of a Three-Flexible-Link Robot Using Three-Dimensional Brick Elements

In this example, a simulation similar to that presented earlier will be discussed. The discretization of the links will be carried out using three-dimensional isoparametric Brick elements [28]. The robot will be made to perform in a constrained environment. The effect of coupling between the longitudinal and transverse elastic displacements on the overall dynamical behavior of the system is described. Subsequent development of internal elastic forces in the flexible links on the midnode elastic displacements of one of the links is examined. A detailed description of the simulation is given below.

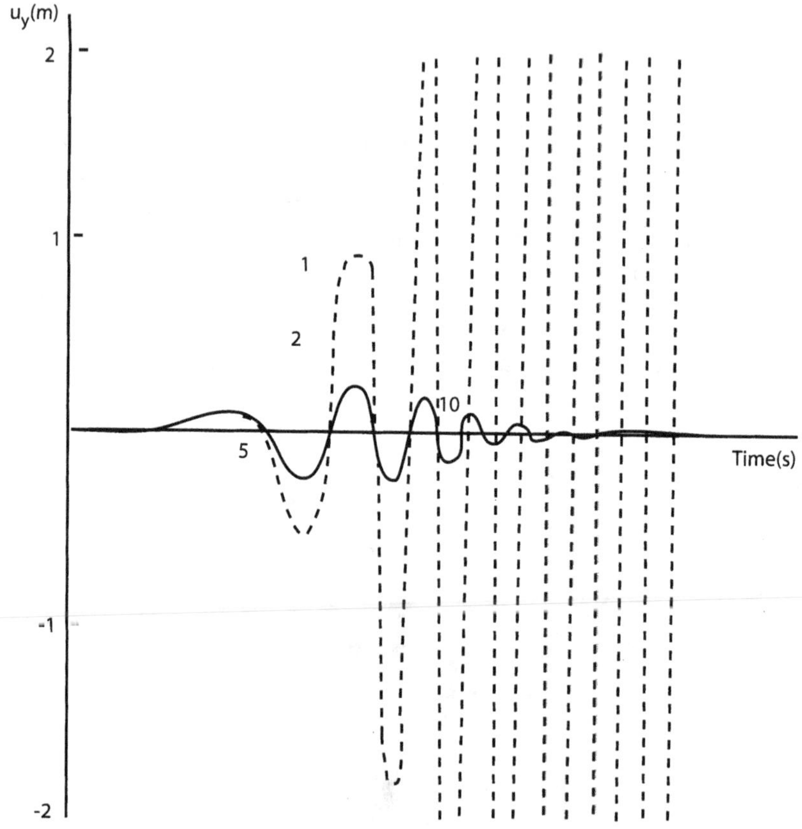

Fig. 11.12. Tip deflection u_z for spin-up and deployment motion: (1) without geometric stiffening; (2) with geometric stiffening (from reference [30])

The first two links, shown in Figure 11.13, are long cylindrical structures discretized using two Brick elements for each. The third link is the plate carrying solar panels used in space applications, which is discretized using four elements with a total of 20 nodes. Thus, the total number of nodes for the present system is 44, making the size of the problem 264 degrees of freedom. The length of the first two links is 200 cm each, with a square cross section of 10 cm by 10 cm. The solar panel has a surface area of 200 cm by 200 cm with a thickness of 1 cm.

The details of discretization are shown in Figure 11.14. The first revolute joint connection is assumed between the fixed reference frame (lower body) and node 4 of the first link (upper body). In a similar fashion, the second and third revolute joint connections are assumed to be between node numbers 12 of links 1 and 2 (constituting the lower body) and node numbers 4 for links 2 and 3 (constituting the upper body).

Each body has its own local reference frame N_j^i, $i = 1, 3$ and $j = 1, 3$. The fixed reference frame is represented as N_j^o. The local reference frames N_j^i are assumed to be located at node numbers 4 for all links. N_j^{2*} and N_j^{3*} are located at node numbers

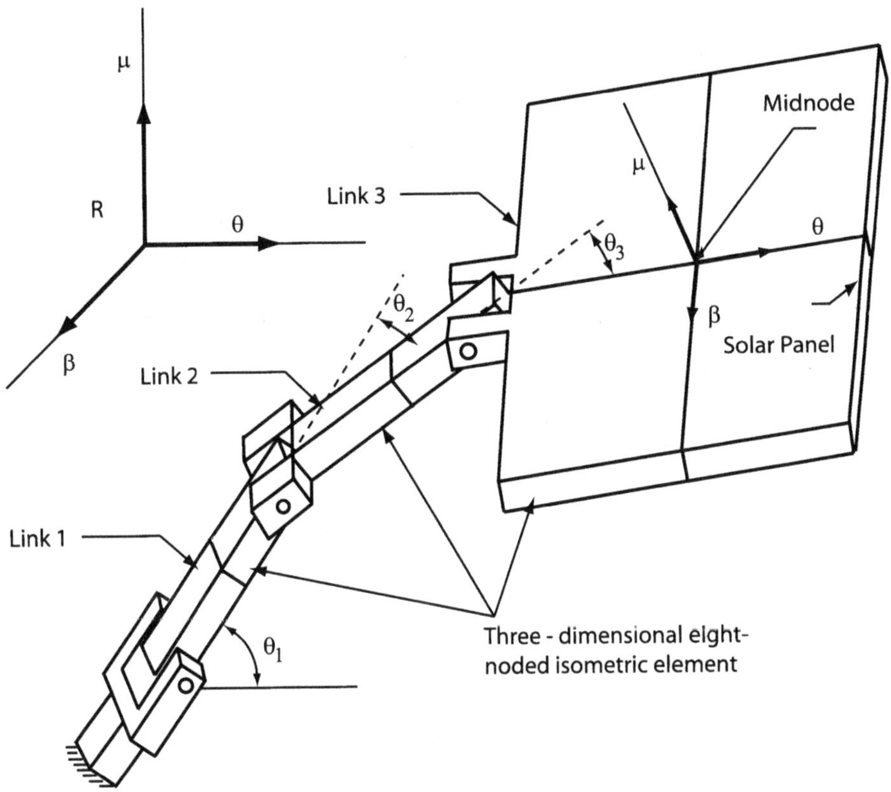

Fig. 11.13. Three-link flexible space robot

12 on bodies 1 and 2, respectively. The element numbering is shown separately for each body. Thus, bodies 1 and 2 have only two elements each, numbered 1 and 2, whereas body 3 has four elements, numbered 1, 2, 3 and 4. All the elastic degrees of freedom at the origins of the reference frames N_j^i and N_j^{i*} are suppressed but the elastic degrees of freedom of all other nodes are free. One rigid-body degree of freedom for each body (i.e., θ_1, θ_2, and θ_3) is kept free, whereas all other rigid-body degrees of the rotations and three translations are suppressed. The material properties of the bodies (links) of the robot are taken as follows:

$$E = 2.1 \times 10^{10}\ \text{kg/m}^2$$

$$E = 0.8 \times 10^{10}\ \text{kg/m}^2$$

$$\rho = 7.825 \times 10^3\ \text{kg/m}^3$$

Two mode shapes, first and second, were assumed for the purpose of present analysis. Thus the generalized coordinates of the system would be

$$\theta_1, \theta_2, \theta_3, \eta_{11}, \eta_{12}, \eta_{21}, \eta_{22}, \eta_{31}, \eta_{32}$$

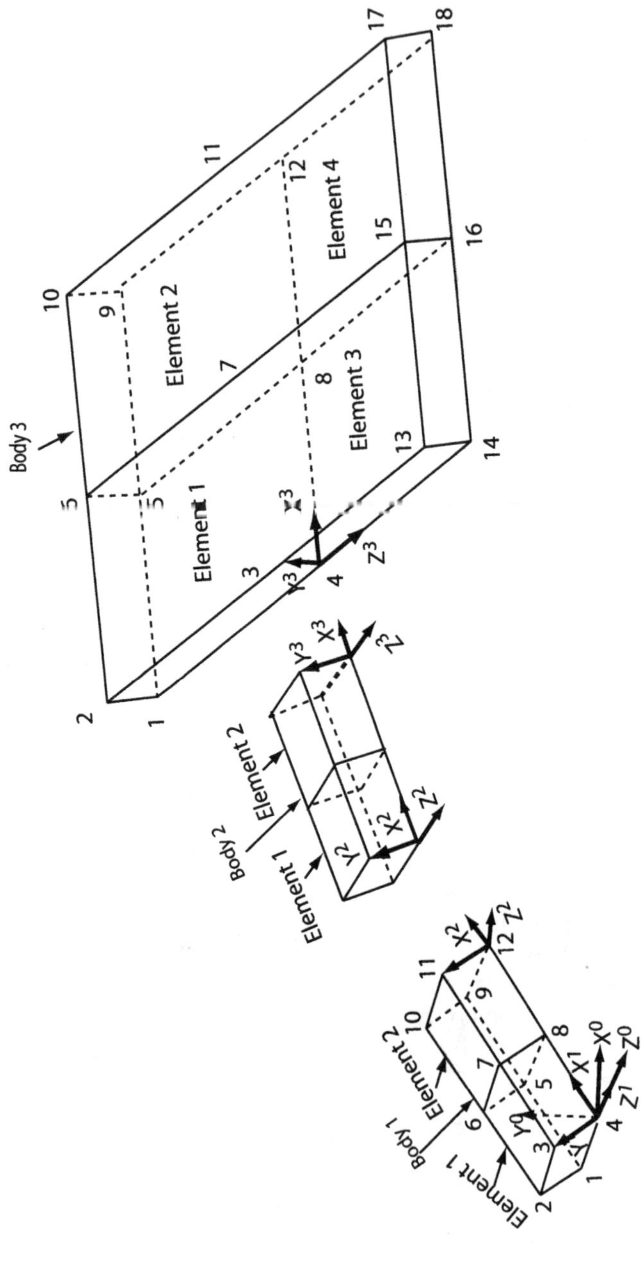

Fig. 11.14. Details of finite-element discretization

The first three correspond to the rigid-body rotations, whereas the last six correspond to the modal coordinates, two for each body. The generalized speeds are the derivatives of the generalized coordinates and are given by

$$\dot{\theta}_1, \dot{\theta}_2, \dot{\theta}_3, \dot{\eta}_{11}, \dot{\eta}_{12}, \dot{\eta}_{21}, \dot{\eta}_{22}, \dot{\eta}_{31}, \dot{\eta}_{32}$$

The initial values of the generalized speeds are assumed to be zero.

$$\{\dot{\theta}_1, \dot{\theta}_2, \dot{\theta}_3, \dot{\eta}_{11}, \dot{\eta}_{12}, \dot{\eta}_{21}, \dot{\eta}_{22}, \dot{\eta}_{31}, \dot{\eta}_{32}\} = \{0\}$$

The space robot under consideration is assumed to perform a deployment maneuver at a high speed, such that links from the initial configuration reach a final position of $\theta_1 = 60°$, $\theta_2 = -20°$, $\theta_3 = -20°$. The time during which this takes place is $T = 1.0$ s. The deployment motion is given by

$$\theta_1(t) = \frac{3\pi - 8\pi}{3t}\left(t - \frac{T}{4\pi}\sin\frac{4\pi t}{T}\right) \qquad 0 < t \leq T \tag{11.226}$$

$$\theta_1(t) = \frac{\pi}{3} \qquad t \geq T \tag{11.227}$$

$$\theta_3(t) = -\pi + \frac{8\pi}{3t}\left(t - \frac{T}{4\pi}\sin\frac{4\pi t}{T}\right) \qquad 0 < t \leq T \tag{11.228}$$

$$\theta_2(t) = \frac{\pi}{9} \qquad t \geq T \tag{11.229}$$

$$\theta_3(t) = -\pi + \frac{8\pi}{3t}\left(t - \frac{T}{4\pi}\sin\frac{4\pi t}{T}\right) \qquad 0 < t \leq T \tag{11.230}$$

$$\theta_3(t) = \frac{-\pi}{9} \qquad t \geq T \tag{11.231}$$

where t and T are the time at any instant and the period for which the maneuver lasts.

The constraint equation for the prescribed angular motion takes the following form:

$$\ddot{\theta}_1(t) = -\left(\frac{32\pi^2}{3t^2}\right)\sin\frac{4\pi t}{T} \qquad 0 < t \leq T \tag{11.232}$$

$$\ddot{\theta}_1(t) = 0 \qquad t \geq T \tag{11.233}$$

$$\ddot{\theta}_2(t) = -\left(\frac{32\pi^2}{9t^2}\right)\sin\frac{4\pi t}{T} \qquad 0 < t \leq T \tag{11.234}$$

$$\ddot{\theta}_2(t) = 0 \qquad t \geq T \tag{11.235}$$

$$\ddot{\theta}_3(t) = -\left(\frac{32\pi^2}{9t^2}\right)\sin\frac{4\pi t}{T} \qquad 0 < t \leq T \tag{11.236}$$

$$\ddot{\theta}_3(t) = 0 \qquad t \geq T \tag{11.237}$$

The elastic displacement for the midnode on the surface of the solar panel represented by node 7 is plotted as shown in Figure 11.15 by the dashed lines for the case with a consideration of the geometric stiffening. Under these circumstances, the midnode must experience a deflection Δ_m in the θ direction, as shown in Figure 11.13, and from which one can predict the deflection in qualitative terms on physical grounds. It can be seen from the plot that at the initial stage of motion, the deflection Δ_m takes a positive value, which is reduced subsequently by the restoring force due to the elastic property of the link material. Eventually, these displacements overshoot the original position of the midnode and take negative values. This process continues as long as the rigid-body motion of the link is in progress. Once the links of the robot reach their final positions, the expected residual "ringing" takes place because of the oscillatory character of the elastic displacement of the midnode. In a similar simulation, when the geometric stiffening effect due to the elasticity of the material is not considered, the absence of the elastic restoring forces results in a continuous increase in the displacement, as shown by the solid lines. Thus, consideration of the geometric stiffening effect is an essential requirement for an accurate prediction of the dynamic behavior of a system with flexible bodies.

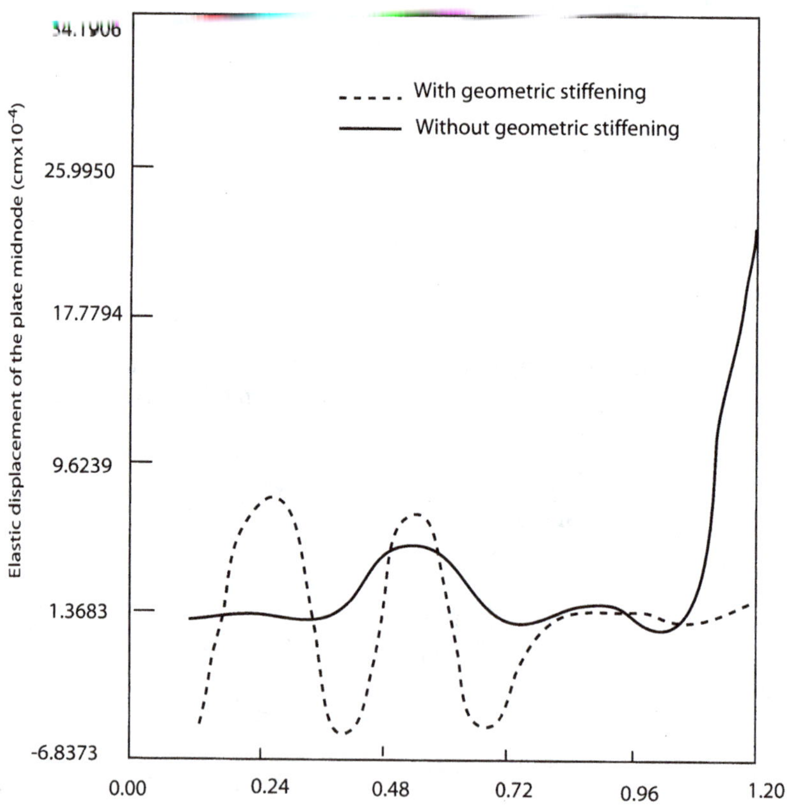

Fig. 11.15. Variation of elastic displacements with time

11.8 Summary

An explicit matrix formulation of the dynamical equations for flexible treelike multi-body systems is presented in this chapter. The method highlights the simplicity of the approach taken when a least set of relative coordinates is used and a matrix representation is adopted right at the kinematics level, by identifying the skew matrices associated with the position vectors and the angular velocities. The example used in Section 11.4 describes further the procedures outlined and their utility in generating the equations of motion. In the final sections, two simulations were carried out pertaining to high-speed spin-up maneuvers performed by a flexible space-based robot under a constrained environment. Two different types of finite elements, the beam and brick elements, have been used for the purpose of discretization. The effect of geometric stiffness consideration on nodal elastic displacement, subsequent to the development of internal elastic restoring forces in the flexible bodies, has been developed. This shows the importance of a consideration of the geometrical stiffness effects in accurate prediction of the dynamic behavior of multiple elastic body systems, particularly those that encounter high-speed deployments.

11.9 Problems

11.1. Evaluate the coefficients m_{11}, m_{12} and m_{12} in equation (11.109).

11.2. For the two-link planar manipulator discussed in Section 11.4, the first link is assumed rigid, whereas the second one is flexible. Two elements were used to model the flexible link. Derive the equations of motion if four elements were used. Solve for the eigenvalue of the flexible link and consider the first mode only where $(i,\ j\ =\ 1,\ 2,\ 3)$

$$\{\rho\} = [\o]\{\eta\}$$

Show how the equations of motion can be reduced. Discuss the utility of the modal analysis reduction.

11.3. Derive the geometrical stiffness matrix given by equation (11.203).

11.4. Consider the two-arm manipulator in Problem 11.2. If link 1 is flexible and modeled with two elements as in Section 11.4 and link 2 is treated as rigid, show how the equations of motion are affected. Discuss the perturbation of the tip of link 2 under both circumstances.

11.5. Derive the constraint conditions for the planar robot in Problem 11.2, if it is restricted to move along the horizontal direction only at a speed $v(t)$. Let the position of the tip with the ground be $h = \text{const}$. In addition, find an explicit form of the generalized constraint forces.

11.6. Identify the Jacobian matrix in Problem 11.5 and find its corresponding orthogonal complement at time $t = 0$ if $\theta_1 = 30^\circ$ and $\theta_2 = 45^\circ$. (You need to use a computer for that.)

11.7. Simulate the dynamics of the planar under normal conditions for the assumption given in Problem 11.5.

11.8. Based on the procedure explained in Section 11.6.1 for the geometric stiffness matrix, derive the next two terms in equation (11.213): $(2\gamma_{yz}^s)^2$ and $(2\gamma_{zx}^s)^2$.

11.9. Discretize the third link of the flexible space robot shown in Figure 11.13 by using nine brick elements, and show the element and node numbering details. Formulate the eigenvalue of the link and give the frequencies and mode shapes.

11.10. Write down the constraint conditions for the robot shown in Figure 11.16 such that all three links stand vertical.

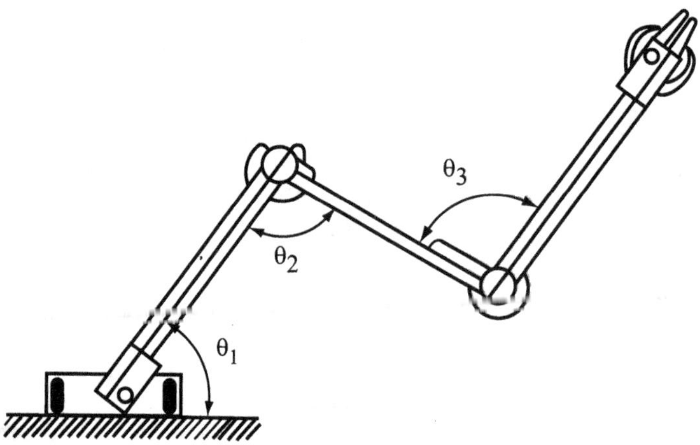

Fig. 11.16. Three-link manipulator undergoing a prescribed motion

References

1. Bodley, C. S., Devers, A. D., Park, A. C. and Frisch, H. P., A Digital Computer Program for the Dynamic Interaction Simulation of Controls and Structure (DISCOS), Vols. **1** and **2**, *NASA TP-1219*, May 1978.
2. Frisch, H. P., A Vector-Dyadic Development of the Equations of Motion for N-Coupled Flexible Bodies and Point Masses, *NASA TN D-8047*, Aug. 1975.
3. Ho, J. Y. L., Direct Path Method for Flexible Multibody Spacecraft Dynamics, *J. Spacecraft Rockets*, Vol. **14**, Jan.–Feb. 1977, pp. 102–111.
4. Ho, J. Y. L. and Herber D. R., Development of Dynamics and Control Simulation of Large Flexible Space Systems, *J. Guide Control Dyn.*, Vol. **8**, May–June 1985, pp. 374–384.
5. Singh, R. P., Vandervoort, R. J. and Likins, P. W., Dynamics of Flexible Bodies in Tree Topology—A Computer Oriented Approach, *AIAA Paper 84-1024,* 1984.
6. Huston, R. L., Multibody Dynamics including the Effects of Flexibility and Compliance, *Comput. Struct.*, Vol. **14**: No. 5–6, 1981, pp. 443–451.
7. Singh, R. P. and Likins, P. W., Manipulator Interactive Design with Interconnected Flexible Elements, Paper TAG-11:00, *Proceedings of the American Control Conference*, American Automatic Control Council, IEEE Service Center, 1983, pp. 505–512.

8. Sunada, W. and Dubowsky, S., The Application of Finite-Element Methods to the Dynamic Analysis of Flexible Spatial and Co-planar Linkage Systems, *J. Mech. Des.*, Vol. **103**: No. 3, 1981, pp. 643–651.
9. Book, W. J. and Majett, M., Controller Design for Flexible Distributed Parameter Mechanical Arms via Combined State Space and Frequency Domain Techniques, *J. Mech. Des.*, Vol. **103**: No. 3, 1981, pp. 101–120.
10. Bejczy, A. K., ad Paul, R. P., Simplified Robot Arm Dynamics for Control, *Proceedings of the 20th IEEE Conference on Decision and Control,* Institute for Electrical and Electronics Engineers Control System Society, IEEE Service Center.
11. Geradin, M., Robter, G. and Bernardin, C., Dynamic Modeling of Manipulators with Flexible Members, in *Advance Software in Robotics* (A. Danthine and M. Geradin, eds.), North-Holland, Amsterdam, 1983, pp. 27–42.
12. Cannon, R. H. and Schmitz, E., Initial Experiments on the End-Point Control of a Flexible One Link Robot, *Int. J. Robot. Res.*, Vol. **3**: No. 3, 1984, pp. 62–75.
13. Kane, T. R., Ryan, R. R. and Banerjee, A. K., Dynamics of a Cantilever Beam Attached to a Moving Base, *J. Guid. Control Dyn.*, Vol. **10**: No. 2, 1987.
14. Zienkiewicz, O. C., *The Finite-Element Method,* McGraw-Hill, New York, 1977.
15. Chen, W. F. and Han, D. J., *Plasticity for Structural Engineers,* Springer-Verlag, New York, 1988.
16. Kraus, H., *Creep Analysis,* Wiley, New York, 1980.
17. Andrade, C., *Creep and Recovery,* American Society for Metals, Metals Park, OH, 1957, pp. 176–198.
18. Garofalo, F., Properties of Crystalline Solids, *ASTM Special Technical Publication 283,* 1960, p. 82.
19. Gandhi, M. V. and Thompson, B. S., The Finite-Element Analysis of Flexible Components of Mechanical Systems Using a Mixed Variational Principle, *ASME Paper 80-DET-64.*
20. Kane, T. R., Ryan, R. R. and Banerjee, A. K., Dynamics of a Cantilever Beam Attached to a Moving Base, *J. Guid. Control Dyn.*, Vol. **10**: No. 2, 1987.
21. Kane, T. R. and Levinson, D. A., *Dynamics: Theory and Applications,* McGraw-Hill, New York, 1985.
22. Huston, R. L. and Passerello, C., Multibody Structural Dynamics Including Translation between the Bodies, *Comput. Struct.,* Vol. **11**, 1980, pp. 715–720.
23. Amirouche, F. M. L. and Ider, S. K., Determination of the Generalized Constraint Forces in Multibody Systems Dynamics Using Kane's Equations, *J. Theor. Appl. Mech.,* Vol. **7**: No. 1, 1988.
24. Przemieniecki, J. S., *Theory of Matrix Structural Analysis,* McGraw-Hill, New York, 1968.
25. Dym, C. L. and Shames, I. M., *Solid Mechanics: A Variational Approach,* McGraw-Hill, New York, 1973.
26. Hurty, W. C., Dynamic Analysis of Structural Systems Using Component Modes, *AIAA J.,* Vol. **3**, Apr. 1965, pp. 678–685.
27. Ider, S. K. and Amirouche, F. M. L., A Recursive Formulation of the Equations of Motion for Articulated Structures with Closed-Loops—An Automated Approach, *Int. J. Comput. Struct.,* Vol. **30**: No. 5, 1988.

28. Shareef, N. H. and Amirouche, F. M. L., Implementation of 3D-Isoparametric Finite Elements on a Supercomputer for this Formulation of Recursive Dynamical Equations of Multibody Systems, AIAA guidance, navigation and control conference, New Orleans, LA, Technical Papers. V, 1991, pp. 12–14.
29. Shareef, N. H., Dynamic Analysis of Time Variant Constraints in Multibody Systems—An Exploitation of Vector Processors, Ph.D. thesis submitted at the University of Illinois at Chicago, 1991.
30. Ider, S. K. and Amirouche, F. M. L., Nonlinear Modelling of Flexible Multibody Systems Dynamics Subject to Variable Constraints, *J. Appl. Mech.,* Vol. **56**: No. 1, 1989.
31. Ider, S. K. and Amirouche, F. M. L., The Influence of Geometric Nonlinearities in the Dynamics of Flexible Treelike Structures, *J. Guid. Control Dyn.,* Vol. **12**: No. 6, 1989, pp. 830–837.
32. Amirouche, F. M. L., *DYAMUS (Dynamic Analysis of Multibody Systems): A General-Purpose Program.* Department of Mechanical Engineering, University of Illinois at Chicago, 1990.
33. Amirouche, F. M. L. and Shareef, N., DARS: Dynamic Analysis of Rotorcraft Systems—A General-Purpose Code, *NASA Lewis Technical Report,* 1991.

12

Modeling of Flexibility Effects Using the Boundary-Element Method

12.1 Introduction

The dynamic simulation of flexible multibody systems has been a real challenge to dynamists and researchers for the past two decades. Areas of interest include robotics, aircraft dynamics, vehicle dynamics and rotorcraft systems. Over the years, numerous papers have been published, tackling a broad range of theoretical problems. More precisely, the inclusion of geometric nonlinearities in the dynamics has proven to be a major contributing factor in predicting accurately the dynamics of flexible (elastic) beams and plates and optimization, which is a recurrence in variational design.

In the formulations proposed in Chapter 10 and 11 the computational methods used in the dynamic simulation of flexible multibody systems were based on either the finite-element method (FEM) or the assumed (component) mode method (AMM) (CMM). In the first method, the flexible body is discretized into a finite number of elements; the equations of motion are then generated in terms of physical nodal coordinates of the elements. In the second method, the body is modeled by assuming that its deformation is approximated in terms of its assumed modes of vibration obtained from modal analysis of a finite-element method. The method can also be used with FEM in order to reduce the number of elastic coordinates of eliminating the undesired frequency modes of vibration. The approximation methods used in the dynamic simulation of multibody systems need to be computationally efficient, and a reliable tool in predicting accurate system motion and dynamic stresses. In fact, the improvement and application of these methods have been the subject of a multitude of publications (see the appendix) contributing tremendously in understanding and solving various problems. From the abundant available publications, we will refer to a limited number of references that will put in perspective the proposed method by Kherdjoudj, et. al. [1].

It is worthwhile to shed some light on the advantages of this method when applied in multibody dynamics. The BEM is a powerful alternative to the FEM suited for solving many engineering problems such as robotics and mechanisms design. The effectiveness of this method in its accurate mathematical description of the elastic body stems in part from the use of Green's functions, which give rise to diagonally dominant influence matrices, and from the use of intrinsic approximations, and is limited only to the surface of the domain. In other words, the computation is restricted to points of interest and does not involve all the mesh points as in the FEM. This feature may qualify the BEM to be computationally more efficient than the FEM, because of the reduction in dimension of the problem by one, therefore leading to a minimum number of the governing equations and reducing the modeling time considerably compared to the finite-element methods. This characteristic may permit the BEM to be suitable for control design purposes, a task that the FEM has failed. Mukherjee and Morjaria [2] showed that the results from the BEM are more accurate than those from the FEM, when the solution is needed at the boundary and few of the internal points. Radaj, et al. [3] applied the BEM with the substructure technique and the FEM to the analysis of notch stresses. They concluded that the BEM displayed much better results than the FEM in terms of CPU time and convergence. Error estimators have also been developed for both methods by Rencis, et al. [4] who demonstrated through some examples that the BEM error estimator is computationally more efficient than the FEM one.

The advantages that the BEM enjoys are related to its more complex mathematical foundations which have been developed throughout the century. This method has successfully been used to solve problems such as elastostatics [5–7], elastodynamics [8–11], potential theory [12], acoustics [12], plasticity [12–14], fracture mechanics [12], thermoelasticity [25] and much more. However, its application to free vibration analysis problems has been limited to eigenfrequency formulations, in which the influence matrices are implicitly in terms of the frequencies. The method used to solve such problems is known as the determinant searching method, for which the computational cost is very high. Nardini and Brebbia [8–11] formulated a method known as the dual reciprocity method (DRM), where the volume integral arising from the inertial term is transformed into DRM integrals by approximating the displacement in terms of fictitious coefficients and known functions. After the particular solutions and their associated tractions are obtained from the equilibrium equations, the mass matrix can then be formulated. Thus, this method can be used to cast easily a generalized eigenvalue problem, a transient dynamic analysis problem, or transform the volume integrals due to body forces into DRM integrals. However, the main drawback of this method is that an inversion of a matrix of the size of the total number degrees of freedom of the boundary and few internal nodes in needed. This makes the method lose some of its computational efficiency for large-scale problems. Ahmad and Banerjee [13] proposed a similar approach based on complementary function and a particular integral. Their method is known as the particular integral method (PIM); it does not require matrix inversion, but it can only be used for eigenvalue problems.

The finite element has been used extensively for the numerical solution of large-displacement and large-deformation problems. Various formulations have been presented. Bathe [15] presented in detail the total and updated Lagrangian formulations and demonstrated the applicability and the differences in the two formulations. Chandra and Mukherjee (see Reference 12) derived and updated Lagrangian formulation for the boundary-element method. Their approach was based on an appropriate hypoelastics law and Betti's reciprocal theorem. Mattiasson, et al. [13] applied the updated Lagrangian formulation for the boundary-element method in large strain and large displacement problems of elastoplasticity. Their formulation was also based on Betti's reciprocal theorem. The resulting equations are similar to that obtained by Chandra and Mukherjee (see Reference 12). Okada, et. al. [14] formulated a full tangent stiffness field boundary element which takes the initial stress-velocity gradient (displacement gradient) coupling terms accurately into account. These terms include the large strain plasticity using the concept of material director triad and the associated plastic spin. In geometric and material nonlinear problems of solid mechanics, the fundamental solutions to the linear differential operators of the problem are used, as test integrals for which the domain has to be discretized into internal cells. This leads to an increase in the data needed to run the program and hence the boundary-element method loses some of its attraction.

12.2 Model Description and Notation

The model as depicted in Figure 12.1 consists of continuum body B attached at point O to frame B. The motion of B consists of a coupling between the elastic deformation with respect to the frame B and the large rigid-body rotations and translations of B with respect to the inertial frame R. In this analysis, the Lagrangian description is used, where the configuration at time $t + \Delta t$ denoted by $C^{t+\Delta t}$ is the current and

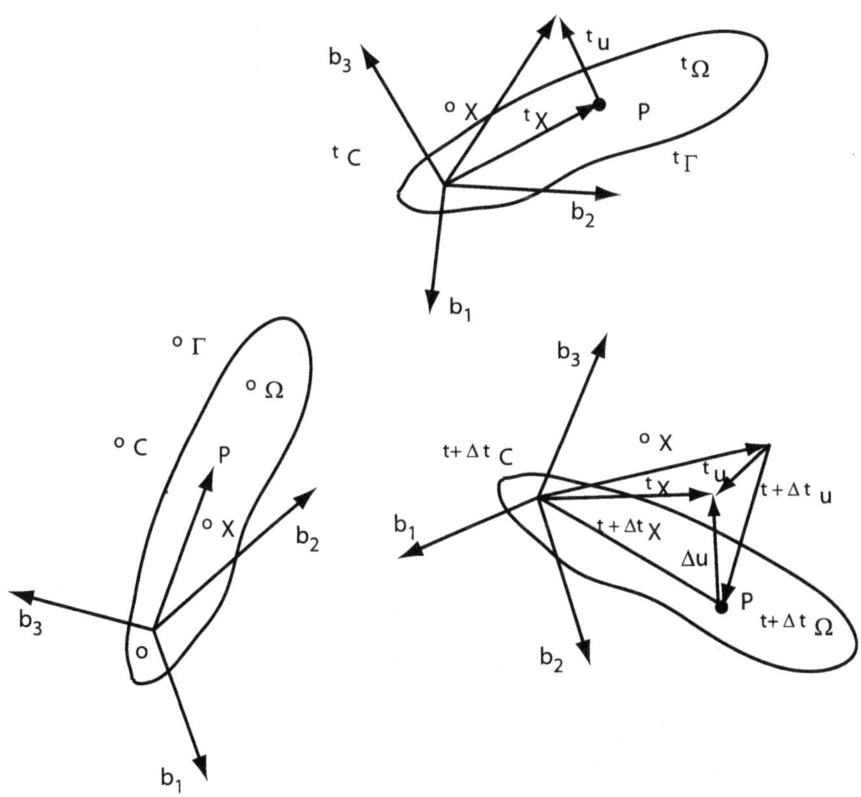

Fig. 12.1. Motion of a continuum body

unknown configuration in which the equations of motion have to be satisfied, and the configuration at time t denoted by C^t is the previous known configuration.

The large motion is characterized by six generalized speeds $\{\upsilon_i | i = 1, 2, 3\}$ and $\{\omega_i | i = 1, 2, 3\}$, where the first set measures the inertial translational velocity ${}^R\upsilon^O$ of point O such that

$$^{t+\Delta t}\upsilon_i = {}^N\upsilon^O \cdot \mathbf{b}_i \quad (i = 1, 2, 3) \tag{12.1}$$

and the second set measures the inertial angular velocity ${}^R\omega^O$ of body B at the configuration $\tau^{t+\Delta t}$, such that

$$^{t+\Delta t}\omega_i = {}^N\omega^O \cdot \mathbf{b}_i \quad (i = 1, 2, 3) \tag{12.2}$$

where $\{\mathbf{b}_i | i = 1, 2, 3\}$ is a dextral set of mutually perpendicular unit vectors rigidly attached to body B at point O as shown in Figure 12.1.

The position of a generic point P in body B is defined by the set $\{^{t+\Delta t}x_i | i = 1, 2, 3\}$ for which the components are the $\{\mathbf{b}_i | i = 1, 2, 3\}$ measure numbers of the position vector $\mathbf{P}^{\overline{OP}}$ defined from point O to point P in the configuration $C^{t+\Delta t}$, i.e.,

$$^{t+\Delta t}x_i = \mathbf{P}^{\overline{OP}}.\mathbf{b}_i \quad (i = 1, 2, 3) \tag{12.3}$$

where the left superscript refers to the configuration of the body B. The position vector $P^{\overline{OP}}$ at configuration $C^{t+\Delta t}$ can be expressed in a vectorial form as

$$\mathbf{P}^{\overline{OP}} = {}^{t+\Delta t}\mathbf{x}_i = {}^0\mathbf{x} + {}^{t+\Delta t}\mathbf{u} \tag{12.4}$$

or

$$\mathbf{P}^{\overline{OP}} = {}^t\mathbf{x} + {}_t\Delta\mathbf{u} \tag{12.5}$$

The components $\{^{t+\Delta t}u_i | i = 1, 2, 3\}$ are the $\{\mathbf{b}_i | i = 1, 2, 3\}$ measure numbers of the displacement vector ${}^{t+\Delta t}\mathbf{u}$ of body B from the initial configuration C^0 to the current configuration $C^{t+\Delta t}$, and $\{{}_t\Delta u_i | i = 1, 2, 3\}$ are the $\{\mathbf{b}_i | i = 1, 2, 3\}$ measure numbers of the incremental displacement vector ${}_t\Delta\mathbf{u}$ of body B from the updated configuration C^t to the current configuration $C^{t+\Delta t}$, such that

$${}_t\Delta u_i = {}^{t+\Delta t}u_i - {}^t u_i \quad (i = 1, 2, 3) \tag{12.6}$$

12.3 Partial Velocities

The inertial velocity vector of a generic point P of body B at time $t + \Delta t$ is given by

$${}^R\upsilon^P = {}^R\upsilon^O + {}^R\omega^B \times \mathbf{P}^{\overline{OP}} + {}^B\upsilon^P \tag{12.7}$$

or in component form as

$$^{t+\Delta t}\bar{\upsilon}_i \doteq {}^{t+\Delta t}\upsilon_i + e_{imn}\,{}^{t+\Delta t}\omega_m({}^t x_n + {}_t\Delta u_n) + {}^{t+\Delta t}\dot{u}_i \quad (i, m, n = 1, 2, 3) \tag{12.8}$$

The components $\{{}^{t+\Delta t}\bar{v}_i | i = 1, 2, 3\}$ are the $\{\mathbf{b}_i | i = 1, 2, 3\}$ measure numbers of the velocity vector ${}^R v^P$ at the configuration $C^{t+\Delta t}$. e_{imn} defines the permutation symbol. Similarly, the inertial acceleration vector of a generic point P of body B at time $t + \Delta t$ can be obtained by taking the time derivative of the velocity vector in the inertial frame R:

$$ {}^R\mathbf{a}^P = \frac{{}^B d\, {}^R v^P}{dt} + {}^R\omega^B \times {}^R v^P \tag{12.9} $$

which results in the following expression:

$$ \begin{aligned} {}^R\mathbf{a}^P &= {}^R\mathbf{a}^O + {}^R\omega^B \times ({}^R\omega^B \times \mathbf{P}^{\overline{OP}}) + {}^R\alpha^B \times \mathbf{P}^{\overline{OP}} \\ &\quad + {}^B\mathbf{a}^P + 2\,{}^R\omega^B \times {}^B v^P + {}^R\omega^B \times {}^R v^O \end{aligned} \tag{12.10} $$

Its components form in tensor notation is defined by

$$ \begin{aligned} {}^{t+\Delta t}\bar{a}_i &= {}^{t+\Delta t}\dot{v}_i + e_{imn}e_{npq}\,{}^{t+\Delta t}\omega_m\,{}^{t+\Delta t}\omega_p({}^t x_q + {}_t\Delta u_q) \\ &\quad + e_{imn}\,{}^{t+\Delta t}\dot{\omega}_m({}^t x_n + {}_t\Delta u_n) + e_{imn}\,{}^{t+\Delta t}\omega_m\,{}^{t+\Delta t}v_n \\ &\quad + {}^t\ddot{u}_i + {}_t\Delta\ddot{u}_i + 2e_{imn}\,{}^{t+\Delta t}\omega_m({}^t\dot{u}_n + {}_t\Delta\dot{u}_n) \qquad (i, m, n, q = 1, 2, 3) \end{aligned} \tag{12.11} $$

where e_{ijk} defines the Levi–Civita permutation symbol *.

The generalized speeds s$\{u_r^* | r = 1, 2, \ldots, v + 6\}$ are given by

$$ u_r^* = \left\{ \begin{array}{cc} \Delta\dot{q}_r & (r = 1, \ldots, v) \\ {}^{t+\Delta t}v_{(r-v)} & (r = v + 1, \ldots, v + 3) \\ {}^{t+\Delta t}\omega_{(r-v-3)} & (r = v + 4, \ldots, v + 6) \end{array} \right\} \tag{12.12} $$

where $\{u_r^* = {}_t\Delta\dot{q}_r | r = 1, 2, \ldots, v\}$ are the temporal derivatives of the incremental generalized coordinates $\{{}_t\Delta\dot{q}_r | r = 1, 2, \ldots, v\}$, which characterize the deformation of the body B in the frame B between the configurations C^t and $C^{t+\Delta t}$ and $\{u_r^* | r = v + 1, 2, \ldots, v + 6\}$ are the generalized speeds for the arbitrary large overall motion of the frame B in the inertial frame N.

During the motion, the configuration of the body B undergoes a continuous change of shape; this elastic deformation is depicted by the displacement vector ${}^{t+\Delta t}\mathbf{u}$ in the current configuration $C^{t+\Delta t}$. According to the Rayleigh–Ritz approach, the displacement vector can be approximated as an expansion series of a set of v linearly independent functions $\{\phi_{ir} | i = 1, 2, 3$ and $r = 1, \ldots, v\}$ and generalized coordinates $\{q_r | r = 1, \ldots, v\}$ such that

$$ {}^{t+\Delta t}u_i = \sum_{r=1}^{v} \phi_{ir}^{t+\Delta t} q_r = \sum_{r=1}^{v} \phi_{ir}({}^t q_r + {}_t\Delta q_r) \tag{12.13} $$

The partial velocity ${}^{N}\upsilon_r^P$ can be obtained by differentiation of the inertial velocity ${}^{N}\upsilon^P$ with respect to the generalized speeds. Consequently, the $\{\mathbf{b}_i | i = 1, 2, 3\}$ measure numbers of the partial velocity can be expressed as

$$ {}^{t+\Delta t}\upsilon_{ir} = \frac{\partial^{t+\Delta t}\bar{\upsilon}_i}{\partial u_r^*} \tag{12.14} $$

or

$$ {}^{t+\Delta t}\upsilon_{ir} = \frac{\partial^{t+\Delta t}\upsilon_i}{\partial u_r^*} + e_{imn}\frac{\partial^{t+\Delta t}\omega_m}{\partial u_r^*}({}^{t}x_n + {}_{t}\Delta u_n) + \frac{\partial^{t+\Delta t}\dot{u}_i}{\partial u_r^*} \tag{12.15} $$

12.4 Generalized Inertia Forces

The generalized inertia forces corresponding to the rth generalized speed at the configuration $C^{t+\Delta t}$ and referred to $C^{t+\Delta t}$ can be expressed in the following form:

$$ {}^{t+\Delta t}_{t+\Delta t}F_r^* = -\int_{{}^{t+\Delta t}\Omega} {}^{R}\mathbf{a}^P \cdot {}^{R}\upsilon_r^{P\,t+\Delta t} dm \tag{12.16} $$

In terms of components of the inertial acceleration and the partial velocities, we write

$$ {}^{t+\Delta t}_{t+\Delta t}F_r^* = -\int_{{}^{t+\Delta t}\Omega} {}^{t+\Delta t}\bar{a}_i \, {}^{t+\Delta t}\upsilon_{ir} \, {}^{t+\Delta t}\rho^{t+\Delta t} d\Omega \quad (r = 1, \ldots, \upsilon + 6) \tag{12.17} $$

The term ${}^{t+\Delta t}\rho^{t+\Delta t}d\Omega$ is the mass of the differential element dB of the flexible-body B that occupies the domain ${}^{t+\Delta t}d\Omega$ in the current configuration $C^{t+\Delta t}$, ${}^{t+\Delta t}\upsilon_{ir}$ is the partial velocity and $\bar{a}_i$ are $\{\mathbf{b}_i | i = 1, 2, 3\}$, the measure numbers of the inertial acceleration of point P given by expression (12.11). Because the current configuration $C^{t+\Delta t}$ is not known, we can assume that the direction and magnitude of the inertial forces ${}^{t+\Delta t}\rho^{t+\Delta t}\bar{a}_i$ are independent of the specific configuration and that the specific mass is conserved throughout the motion, hence

$$ {}^{t+\Delta t}\rho^{t+\Delta t}\bar{a}_i^{t+\Delta t}d\Omega = {}^{t}\rho^{t+\Delta t}\bar{a}_i^{t}d\Omega \tag{12.18} $$

for which

$$ {}^{t}\rho^{t}d\Omega = {}^{t+\Delta t}\rho^{t+\Delta t}d\Omega \tag{12.19} $$

Therefore, the expression of the generalized inertia forces at the current configuration $C^{t+\Delta t}$, but referred to the updated configuration C^t becomes:

$$ {}^{t+\Delta t}_{t}F_r^* = -\int_{{}^{t}\Omega} {}^{t+\Delta t}\bar{a}^{t+\Delta t}\upsilon_{ir} \, {}^{t}\rho^{t} d\Omega \tag{12.20} $$

12.5 Generalized Active Forces

From the Lagrangian formulation, Kane's form of generalized active forces at the configuration and expressed in the geometry of the same configuration can be represented in the following expression:

$$
\begin{aligned}
{}^{t+\Delta t}_{t+\Delta t}F_r = & -\int_{^{t+\Delta t}\Omega} \sigma \cdot \frac{\partial \Delta \mathbf{e}}{\partial_t \Delta q_r} {}^{t+\Delta t}d\Omega + \int_{^{t+\Delta t}\Gamma} \mathbf{p} \cdot {}^{N}\upsilon_r^{P\,t+\Delta t} d\Gamma \\
& + \int_{^{t+\Delta t}\Omega} \gamma \cdot {}^{N}\upsilon_r^{P\,t+\Delta t} d\Omega
\end{aligned}
\tag{12.21}
$$

where σ and $\Delta\mathbf{e}$ represent the Cauchy stress and the infinitesimal strain dyadics, respectively. They are expressed in the frame B as

$$\sigma = {}^{t+\Delta t}\sigma_{ij}\mathbf{b}_i\mathbf{b}_j \tag{12.22}$$

$$\Delta\mathbf{e} = {}_t\Delta e_{ij}\mathbf{b}_i\mathbf{b}_j \tag{12.23}$$

The vectors ρ and γ define, respectively, the traction and the body forces. Using the components of the vectorial forces quantities defined above, the generalized active forces can be expressed further as

$$
\begin{aligned}
{}^{t+\Delta t}_{t+\Delta t}F_r = & -\int_{^{t+\Delta t}\Omega} {}^{t+\Delta t}\sigma_{ij} \frac{\partial \Delta e_{ij}}{\partial_t} \Delta q_r {}^{t+\Delta t}d\Omega \\
& + \int_{^{t+\Delta t}\Gamma} {}^{t+\Delta t}_{t+\Delta t}p_i {}^{t+\Delta t}\upsilon_{ir}^{t+\Delta t} d\Gamma + \int_{^{t+\Delta t}\Omega} {}^{t+\Delta t}_{t+\Delta t}\gamma_i {}^{t+\Delta t}\upsilon_{ir}^{t+\Delta t} d\Omega
\end{aligned}
\tag{12.24}
$$

where ${}_t\Delta e_{ij}$ are the components of the infinitesimal strain tensor given by

$${}_t\Delta e_{ij} = \frac{1}{2}\left(\frac{\partial_t \Delta u_i}{\partial^{t+\Delta t}x_j} + \frac{\partial_t \Delta u_i}{\partial^{t+\Delta t}x_i}\right) \tag{12.25}$$

The expression ${}^{t+\Delta t}\sigma_{ij}$ are the components of the Cauchy stress tensor, ${}^{t+\Delta t}_{t+\Delta t}p_i$ and ${}^{t+\Delta t}_{t+\Delta t}\gamma_i$ define the components of the traction and body forces per unit mass, respectively, at the configuration $C^{t+\Delta t}$ and referred to in the configuration $C^{t+\Delta t}$ and represent the incremental components of the displacement and the generalized coordinates, respectively.

Since the configuration $C^{t+\Delta t}$ is not known, the stresses ${}^{t+\Delta t}\sigma_{ij}$ cannot be computed using the current geometry of the elastic body. Consequently, the presence of this difficulty brings about a phenomenon known as geometric nonlinearity. To solve this problem, consider the solution known at the configurations $C^0, C^{\Delta t}, C^{2\Delta t}, \ldots, C^t$, but the solution at the current configuration $C^{t+\Delta t}$ needs to be evaluated. Therefore, we can choose any of the above known configurations to serve as a reference configuration. In this work, C^t is used as the reference configuration. This leads to the updated Lagrangian formulation, where the second Piola–Kirchhoff stress tensor components ${}^{t+\Delta t}_{t}S_{ij}$ are used. The components ${}^{t+\Delta t}_{t}S_{ij}$ refer to the unit areas in the updated configuration C^t but are resolved with respect to axes in the current configuration $C^{t+\Delta t}$. Therefore, the Green–Lagrange strain tensor, which is energetically conjugate to ${}^{t+\Delta t}_{t}S_{ij}$, must be used in place of the infinitesimal strain tensor.

When the updated Lagrangian formulation is considered, the expression for the generalized active forces becomes

$$\begin{aligned} {}^{t+\Delta t}_{t}F_r = &- \int_{{}^t\Omega} {}^{t+\Delta t}_{t}S_{ij} \frac{\partial_t^{t+\Delta t}\varepsilon_{ij}}{\partial_t \Delta q_r} {}^t d\Omega \\ &+ \int_{{}^t\Gamma} {}^{t+\Delta t}_{t+\Delta t}p_i {}^{t+\Delta t}\upsilon^t_{ir} d\Gamma + \int_{{}^t\Omega} {}^{t+\Delta t}_{t}\gamma_i {}^{t+\Delta t}\upsilon^t_{ir} d\Omega \end{aligned} \tag{12.26}$$

with the condition that

$$\frac{\partial_t^{t+\Delta t}\varepsilon_{ij}}{\partial_t \Delta q_r} = 0 \quad \text{when} \quad (r = \upsilon + 1, \ldots, \upsilon + 6) \tag{12.27}$$

The Green–Lagrange strain tensor is given by

$${}^{t+\Delta t}_{t}\varepsilon_{ij} = \frac{1}{2}\left({}^{t+\Delta t}_{t}u_{i,j} + {}^{t+\Delta t}_{t}u_{j,i} + {}^{t+\Delta t}_{t}u_{k,i}\,{}^{t+\Delta t}_{t}u_{k,j}\right) \quad (i, j, k = 1, 2, 3) \tag{12.28}$$

Because the configuration is chosen as reference, the Green–Lagrange strain tensor can be expressed as the increment:

$${}^{t+\Delta t}_{t}\varepsilon_{ij} = {}_t\Delta\varepsilon_{ij} \quad (i, j = 1, 2, 3) \tag{12.29}$$

which can be decomposed into two parts:

$${}_t\Delta\varepsilon_{ij} = {}_t\Delta\varepsilon^L_{ij} + {}_t\Delta\varepsilon^{NL}_{ij} \ (i, j = 1, 2, 3) \tag{12.30}$$

The expression ${}_t\Delta\varepsilon^L_{ij}$ is the linear part which contains the terms involving the derivatives of the displacement components, and ${}_t\Delta\varepsilon^{NL}_{ij}$ is the nonlinear part containing the quadratic terms. The latter generates the geometric stiffening effects.

The linear part of the strain tensor can be written as

$${}_t\Delta\varepsilon^L_{ij} = \frac{1}{2}\left({}^{t+\Delta t}_{t}u_{i,j} + {}^{t+\Delta t}_{t}u_{j,i}\right) \tag{12.31}$$

and the nonlinear part takes the following form:

$${}_t\Delta\varepsilon^{NL}_{ij} = \frac{1}{2}\left({}^{t+\Delta t}_{t}u_{k,i}\,{}^{t+\Delta t}_{t}u_{k,j}\right) \tag{12.32}$$

Integrating twice by parts and using the divergence theorem, the expression for the generalized active forces becomes

$$
\begin{aligned}
{}_t^{t+\Delta t}F_r = &-\int_{{}^t\Omega} [{}_t^{t+\Delta t}S_{ij,j} + ({}_t\Delta u_{i,j}{}_t^{t+\Delta t}S_{kj})_k]\frac{\partial_t \Delta u_i}{\partial_t \Delta q_r}{}^t d\Omega \\
&+ \int_{{}^t\Omega} {}_t^{t+\Delta t}\gamma_i{}^{t+\Delta t}\upsilon_{ir}^t d\Omega \\
&- \int_{{}^t\Gamma} [{}_t^{t+\Delta t}S_{ij} + ({}_t\Delta u_{i,k}{}_t^{t+\Delta t}S_{kj})^t n_j]\frac{\partial_t \Delta u_i}{\partial_t \Delta q_r}{}^t d\Gamma \\
&+ \int_{{}^t\Gamma} {}_t^{t+\Delta t}p_i{}^{t+\Delta t}\upsilon_{ir}^t d\Gamma
\end{aligned}
\tag{12.33}
$$

where ${}_t^{t+\Delta t}p_i$ $(i = 1, 2, 3)$ are the components of the traction vector ${}_t^{t+\Delta t}\mathbf{p}$ acting on the boundary with unit outer normal ${}^t\mathbf{n}$, whose components are the direction cosines ${}^t n_j (j = 1, 2, 3)$, and ${}_t^{t+\Delta t}\gamma_i$ $(i = 1, 2, 3)$ are the components of the body force per unit mass referred to the reference configuration.

12.6 Generalized Constraint Forces

The constraint equations at the velocity level can be expressed as

$$
{}_{t+\Delta t}^{t+\Delta t}[Q_{ij}]^{T\,t+\Delta t}\{\bar{\upsilon}_i\} = 0 \tag{12.34}
$$

After the linearization of equation (12.34) with respect to the configuration C^t, the generalized constraint forces at the configuration $C^{t+\Delta t}$ can be written as

$$
{}_t^{t+\Delta t}F_r^C = -{}^{t+\Delta t}\upsilon_{ir}{}_t^{t+\Delta t}Q_{ij}{}^{t+\Delta t}\lambda_j \quad (j = 1, \ldots, \upsilon_c) \tag{12.35}
$$

where the tensor ${}_t^{t+\Delta t}Q_{ij}$ contains all the influences of the joints at the configuration $C^{t+\Delta t}$, v_c is the number of constraint equations and ${}^{t+\Delta t}\lambda_j$ are the undetermined Lagrange multipliers representing the constraint forces.

12.7 Equations of Motion of a Continuum Body

Using Kane's dynamical equations, we write the following

$$
{}_t^{t+\Delta t}F_r^* + {}_t^{t+\Delta t}F_r + {}_t^{t+\Delta t}F_r^C = 0 \qquad (r = 1, \ldots, \upsilon + 6) \tag{12.36}
$$

and substituting equations (12.20), (12.33), (12.35) and (12.36), the general equations of motion of the continuum body B under consideration take the following form:

$$
\begin{aligned}
&\int_{{}^t\Omega} [{}_t^{t+\Delta t}S_{ij,j} + ({}_t\Delta u_{i,j}{}_t^{t+\Delta t}S_{kj})_k]\frac{\partial_t \Delta u_i}{\partial_t \Delta q_r}{}^t d\Omega + \int_{{}^t\Omega} {}_t^{t+\Delta t}\gamma_i{}^{t+\Delta t}\upsilon_{ir}^t d\Omega \\
&- \int_{{}^t\Gamma} [{}_t^{t+\Delta t}S_{ij} + ({}_t\Delta u_{i,k}{}_t^{t+\Delta t}S_{kj})^t n_j]\frac{\partial_t \Delta u_i}{\partial_t \Delta q_r}{}^t d\Gamma + \int_{{}^t\Gamma} {}_t^{t+\Delta t}p_i{}^{t+\Delta t}\upsilon_{ir}^t d\Gamma \\
&= \int_{{}^t\Omega} {}^{t+\Delta t}\bar{a}_i{}^{t+\Delta t}\upsilon_{ir}^t \rho\, {}^t d\Omega + {}^{t+\Delta t}\upsilon_{ir}{}_t^{t+\Delta t}Q_{ij}{}^{t+\Delta t}\lambda_j
\end{aligned}
\tag{12.37}
$$

Considering the relationship

$$\frac{\partial_t \Delta \dot{u}_i}{\partial u_r^*} = \frac{\partial_t \Delta u_i}{\partial_t \Delta q_r} \quad \text{when} \quad (r = 1, \ldots, v) \tag{12.38}$$

the following three equations of motion can be defined independently as:

Deformation equations:

$$\int_{^t\Omega} [{}_t^{t+\Delta t}S_{ij,j} + ({}_t\Delta u_{i,j}{}_t^{t+\Delta t}S_{kj})_k] \frac{\partial_t \Delta u_i}{\partial_t \Delta q_r} {}^t d\Omega + \int_{^t\Omega} {}_t^{t+\Delta t}\gamma_i \frac{\partial_t \Delta u_i}{\partial_t \Delta q_r} {}^t d\Omega$$

$$- \int_{^t\Gamma} [{}_t^{t+\Delta t}S_{ij} + ({}_t\Delta u_{i,k}{}_t^{t+\Delta t}S_{kj})^t n_j] \frac{\partial_t \Delta u_i}{\partial_t \Delta q_r} {}^t d\Gamma + \int_{^t\Gamma} {}_t^{t+\Delta t}p_i \frac{\partial_t \Delta u_i}{\partial_t \Delta q_r} {}^t d\Gamma$$

$$= \int_{^t\Omega} {}^{t+\Delta t}\bar{a}_i \frac{\partial_t \Delta u_i}{\partial_t \Delta q_r} {}^t\rho \, {}^t d\Omega + \frac{\partial_t \Delta u_i}{\partial_t \Delta q_r} {}_t^{t+\Delta t}Q_{ij} {}^{t+\Delta t}\lambda_j \tag{12.39}$$

Translation equations:

$$\int_{^t\Omega} {}_t^{t+\Delta t}\gamma_i {}^t d\Omega + \int_{^t\Gamma} {}_t^{t+\Delta t}p_i {}^t d\Gamma = \int_{^t\Omega} {}^{t+\Delta t}\bar{a}_i {}^t\rho \, {}^t d\Omega + {}_t^{t+\Delta t}Q_{ij} {}^{t+\Delta t}\lambda_j \tag{12.40}$$

Rotation equations:

$$\int_{^t\Omega} {}_t^{t+\Delta t}\gamma_i e_{imn}({}^t x_n + {}_t\Delta u_n) {}^t d\Omega + \int_{^t\Gamma} {}_t^{t+\Delta t}p_i e_{imn}({}^t x_n + {}_t\Delta u_n) {}^t d\Gamma$$

$$= \int_{^t\Omega} {}^{t+\Delta t}\bar{a}_i {}^t e_{imn}({}^t x_n + {}_t\Delta u_n) \rho \, {}^t d\Omega + e_{imn}({}^t x_n + {}_t\Delta u_n) {}_t^{t+\Delta t}Q_{ij} {}^{t+\Delta t}\lambda_j \tag{12.41}$$

The dynamical boundary conditions can be deduced from the above equations of motion for the deformation of the continuum body as

$$[{}_t^{t+\Delta t}S_{ij} + {}_t\Delta u_{i,k}{}_t^{t+\Delta t}S_{kj}]^t n_j = {}^t\bar{p}_i + {}_t\Delta\bar{p} \quad \text{on} \quad {}^t\Gamma_2 \tag{12.42}$$

whereas the kinematical boundary conditions can be incorporated by means of the following relation:

$${}^{t+\Delta t}u_i = {}^t u_i + {}_t\Delta\bar{u} \quad \text{on} \quad {}^t\Gamma_1 \tag{12.43}$$

The total boundary of the elastic body at the configuration forms a direct sum of the boundaries such that

$${}^t\Gamma = {}^t\Gamma_1 \oplus {}^t\Gamma_2 \tag{12.44}$$

This implies that the intersection of ${}^t\Gamma_1$ and ${}^t\Gamma_2$ is empty, i.e., ${}^t\Gamma_1 \cap {}^t\Gamma_2 = \phi$.

The initial conditions are

$${}^0u_i({}^0x) = {}^0u_i \quad \text{on} \quad {}^0\Omega \tag{12.45}$$

$${}^0\dot{u}_i({}^0x) = {}^0\dot{u}_i \quad \text{on} \quad {}^0\Omega \tag{12.46}$$

where ${}^0\Omega$ is the domain of the body at the initial configuration.

12.8 Linearization of the Equations of Motion

Once the formulation of the equations of motion is complete, we proceed with their linearization. Consider the second Piola–Kirchhoff stress tensor, which can be decomposed incrementally as

$$ {}_{t}^{t+\Delta t}S_{ij} = {}^{t}\sigma_{ij} + {}_{t}\Delta S_{ij} \tag{12.47} $$

where ${}^{t}\sigma_{ij}$ are the components of the Cauchy stress tensor at reference configuration C^t, and ${}_{t}\Delta S_{ij}$ are the components of the second Piola–Kirchhoff stress increment tensor referred to the configuration C^t expressed as

$$ {}_{t}^{t+\Delta t}S_{ij} = E_{ijkl}\,{}_{t}\Delta\varepsilon_{kl} \tag{12.48} $$

where E_{ijkl} is the matrix of the elastic coefficients.

The linearization of the equations of motion is accomplished by setting

$$ {}_{t}\Delta\varepsilon_{kl} \cong {}_{t}\Delta\varepsilon_{kl}^{L} \tag{12.49} $$

Consequently, the second Piola–Kirchhoff stress increment tensor becomes

$$ {}_{t}^{t+\Delta t}S_{ij} \cong E_{ijkl}\,{}_{t}\Delta\varepsilon_{kl}^{L} \tag{12.50} $$

From the linearized quantities defined above, the linearized equations of motion for the deformation of the elastic body takes the following form:

$$ \begin{aligned} &\int_{{}^{t}\Omega} \{{}^{t}\sigma_{ij,j} + {}_{t}^{\Delta}S_{ij,j} + [{}_{t}\Delta u_{i,j}({}^{t}\sigma_{kj} + {}_{t}^{\Delta}S_{kj})]_{k}\}\phi_{ir}\,{}^{t}d\Omega + \int_{{}^{t}\Omega} {}_{t}^{t+\Delta t}\gamma_{i}\phi_{ir}\,{}^{t}d\Omega \\ &\quad - \int_{{}^{t}\Gamma} [{}_{t}^{t+\Delta t}S_{ij} + {}_{t}\Delta u_{i,k}\,{}_{t}^{t+\Delta t}S_{kj}]\,{}^{t}n_{j}\phi_{ir}\,{}^{t}d\Gamma + \int_{{}^{t}\Gamma} {}_{t}^{t+\Delta t}p_{i}\phi_{ir}\,{}^{t}d\Gamma \\ &= \int_{{}^{t}\Omega} {}^{t+\Delta t}\bar{a}_{i}\phi_{ir}\,{}^{t}\rho\,{}^{t}d\Omega + \phi_{ir}\,{}_{t}^{t+\Delta t}Q_{ij}\,{}^{t+\Delta t}\lambda_{j} \end{aligned} \tag{12.51} $$

where ϕ_{ir} is given by

$$ \phi_{ir} = \frac{\partial_{t}\Delta u_{i}}{\partial_{t}\Delta q_{r}} \quad \text{with} \quad (r = 1, \ldots, v) \tag{12.52} $$

12.9 Weighted Residual Statement

Previously, the displacement was approximated by the series $u_i = \sum_{r=1}^{v}\phi_{ir}q_r$, where $\{\phi_{ir}|r = 1, \ldots, v\}$ is a set of linearly independent functions and q_r generalized coefficients. In Galerkin's approach, the weighting field belongs to the same space as the displacement field. Thus, w_i are chosen in such a way that $w_i = \sum_{r=1}^{v}\psi_{ir}\alpha_r$ and $\phi_{ir} = \psi_{ir}$. Therefore, we can construct kinematically admissible weighting functions

w_i, that can be written as a series of the set of functions ϕ_{ir} multiplied by arbitrary coefficients α_r, such that

$$w_i = \sum\nolimits_{r=1}^{v} \psi_{ir}\alpha_r \tag{12.53}$$

where we consider only the incremental equations of motion that describe the elastic deformation of the continuum body B. Once these equations are weighted using the coefficients α_r and the relations, we obtain the following expression:

$$\sum\nolimits_{r=1}^{v} ({}_t^{t+\Delta t}F_r^* + {}_t^{t+\Delta t}F_r^{+}{}_t^{t+\Delta t}F_r^C)\alpha_r = 0 \quad (r = 1, \ldots, v) \tag{12.54}$$

Therefore, the equations of motion weighted over the domain at reference configuration C^t can be written as

$$\begin{aligned}&\int_{{}^t\Omega} \{{}^t\sigma_{ij,j} + {}_t^{\Delta}S_{ij,j} + [{}_t\Delta u_{i,j}({}^t\sigma_{kj} + {}_t^{\Delta}S_{kj})]_k\} w_i{}^t d\Omega + \int_{{}^t\Omega} {}_t^{t+\Delta t}\gamma_i w_i{}^t d\Omega \\ &- \int_{{}^t\Gamma} [{}_t^{t+\Delta t}S_{ij} + {}_t\Delta u_{i,k}{}_t^{t+\Delta t}S_{kj}]{}^t n_j w_i{}^t d\Gamma + \int_{{}^t\Gamma} {}_t^{t+\Delta t}p_i w_i{}^t d\Gamma \\ &- \int_{{}^t\Omega} {}^{t+\Delta t}\ddot{u}_i w_i{}^t\rho\,{}^t d\Omega + w_i{}_t^{t+\Delta t}Q_{ij}{}^{t+\Delta t}\lambda_j\end{aligned} \tag{12.55}$$

where w_i denotes the weighting field. If functions w_i are regarded as virtual displacements, equation (12.55) would be similar to the concept of virtual work.

For boundary problems the approximate solution should satisfy the field equation, but not the boundary conditions for which the solution is to be determined. Thus the weighting function are chosen to satisfy the equation without any boundary conditions. In other words, the weighting function are chosen as the fundamental solution of the following structural field defined over the domain at reference configuration C^t:

$$\begin{aligned}&{}^t\sigma_{ij,j}^* + \delta_{ik}\delta({}^t\mathbf{x} - {}^t\chi){}^t e_k = 0 \quad \text{in} \quad {}^t\Omega_\infty \\ &{}^t\varepsilon_{ij}^* = \frac{1}{2}({}^t u_{i,j}^* + {}^t u_{j,i}^*) \quad \text{in} \quad {}^t\Omega_\infty \\ &{}^t\sigma_{ij} = E_{ijkl}{}^t\varepsilon_{kl}^* \quad \text{in} \quad {}^t\Omega_\infty\end{aligned} \tag{12.56}$$

where $\delta({}^t\mathbf{x} - {}^t\chi)$ is the Dirac delta function and ${}^t\sigma_{ij,j}^*, {}^t\varepsilon_{ij}^*, {}^t u_{i,j}^*$ are the components of the stresses, the strains and displacements respectively, defined over the infinite domain ${}^t\Omega_\infty$ at the field point ${}^t\mathbf{x}$ due to a unit load acting at the source point ${}^t\chi$ in the kth direction corresponding to the reference configuration C^t. Therefore, the equation leads to the following important property:

$$\begin{aligned}\int_{{}^t\Omega} {}^t\sigma_{ij,j}^*{}^t u_i^t d\Omega &= -\int_{{}^t\Omega} \delta_{ik}\delta({}^t\mathbf{x} - {}^t\chi){}^t e_k{}^t u_i^t({}^t\mathbf{x})d\Omega({}^t\mathbf{x}) \\ &= \left\{\begin{array}{c} -{}^t u_i({}^t\chi){}^t e_k \quad \text{if} \quad {}^t\chi \in {}^t\Omega \\ 0 \quad \text{if} \quad {}^t\chi \in {}^t\Omega_c \end{array}\right\}\end{aligned} \tag{12.57}$$

where ${}^t\Omega_c$ is the complement of ${}^t\Omega$. Equation (12.55) can be recognized as the weighted residual statement. Using the fundamental solution ${}^tu_i^*$ of the field defined by the equation (12.56), and incorporating the kinematical boundary conditions (12.45) into equation (12.55), we obtain the following equation:

$$\int_{{}^t\Omega} \{{}^t\sigma_{ij,j} + {}^{\Delta}_{t}S_{ij,j} + [{}_t\Delta u_{i,k}({}^t\sigma_{kj} + {}^{\Delta}_{t}S_{kj})]_j\}^t {}^tu_i^{*t} d\Omega$$

$$= -\int_{{}^t\Gamma_1} [{}^tu_i^+ {}_t\Delta u_i - {}^t\bar{u}_i^- {}_t\Delta\bar{u}_i]^t p_i^{*t} d\Gamma + \int_{{}^t\Gamma_2} [{}^tp_i^+ {}_t\Delta p_i - {}^t\bar{p}_i^- {}_t\Delta\bar{p}_i]^{tt} u_i^{*t} d\Gamma$$

$$- \int_{{}^t\Omega} {}^{t+\Delta t}_{t}\gamma_i {}^tu_i^{*t} d\Omega + \int_{{}^t\Omega} {}^{t+\Delta t}\bar{a}_i {}^tu_i^{*t}\rho\, {}^td\Omega + {}^tu_{i\,t}^{*t+\Delta t} Q_{ij} {}^{t+\Delta t}\lambda_j \qquad (12.58)$$

where ${}^tp_i^*$ are the components of the traction corresponding to the fundamental solution. The boundary element formulation is based on the application of the Macwell–Betti reciprocal theorem and the fundamental solution. Given the linearization of the symmetric second Piola–Kirchhoff increment stress tensor, it is possible to apply the Maxwell–Betti reciprocal theorem between the field defined by ${}^{\Delta}_{t}S_{ij}, {}^{\Delta}_{t}\varepsilon_{ij}$, and the fundamental structural field defined by ${}^t\sigma^*_{ij,j}, {}^t\varepsilon^*_{ij}$ at the reference configuration C^t. Therefore, we can write the following relation:

$$\int_{t\Omega} {}^t\sigma_{ij}^{*t}\varepsilon_{ij} {}^td\Omega = \int_{t\Omega} {}_t\Delta S_{ij} {}^t\varepsilon_{ij}^{*t} d\Omega \qquad (12.59)$$

Similarly, the Maxwell–Betti theorem can be applied between the field defined by the Cauchy stress and the strain tensor ${}^t\sigma_{ij}; {}^t\varepsilon_{ij}$ defined in C^t and the structural field as

$$\int_{t\Omega} {}^t\sigma_{ij}^{*t}\varepsilon_{ij} {}^td\Omega = \int_{t\Omega} {}^t\sigma_{ij} {}^t\varepsilon_{ij}^{*t} d\Omega \qquad (12.60)$$

After carrying out twice the integration by parts and employing the divergence theorem, using equations (12.59), (12.60) and adding and subtracting the nonlinear term $\int [{}_t\Delta u_{i,m}({}^t\sigma_{mj} + {}_t\Delta S_{mj})]^t n_j {}^tu_i^{*t} d\Gamma$, the underlined integrals in equation (12.58) result in the following expression:

$$\int_{{}^t\Omega} {}^t\sigma^*_{ij,j} {}^tu_i {}^td\Omega + \int_{{}^t\Gamma} {}^tp_i {}^tu_i^{*t} d\Gamma - \int_{{}^t\Gamma} {}^tu_i {}^tp_i^{*t} d\Gamma$$

$$+ \int_{{}^t\Omega} {}^t\sigma^*_{ij,jt}\Delta u_i {}^td\Omega + \int_{{}^t\Gamma} {}_t\Delta p_i {}^tu_i^{*t} d\Gamma - \int_{{}^t\Gamma} {}_t\Delta u_i {}^tp_i^{*t} d\Gamma$$

$$- \int_{{}^t\Gamma} [{}_t\Delta u_{i,m}(t\sigma_{mj} + {}_t\Delta S_{mj})]^t n_j u_i^{*t} d\Gamma + \int_{{}^t\Omega} [{}_t\Delta u_{i,m}(t\sigma_{mj} + {}_t\Delta S_{mj})]^t_{,j} u_i^{*t} d\Omega$$

$$= -\int_{{}^t\Omega} {}^{t+\Delta t}_{t}\gamma_i^t u_i^{*t} d\Omega + \int_{{}^t\Omega} {}^{t+\Delta t}a_i^t u_i^{*t}\rho\, {}^td\Omega + {}^tu_{i\,t}^{*t+\Delta t} Q_{ij}^{t+\Delta t}\lambda_j \qquad (12.61)$$

where ${}_t\Delta p_i$ are the components of the incremental pseudo-traction vector $\Delta \mathbf{p}$ defined on the boundary ${}^t\Gamma$ of the reference configuration

$$
{}_t\Delta p_i = {}_t\Delta S_{ij}{}^t n_j + {}_t\Delta u_{i,m}({}^t\sigma_{mj} + {}_t\Delta S_{mj}){}^t n_j \tag{12.62}
$$

and ${}^t n_j$ are the components in the jth direction of the normal ${}^t n$ at the field point defined by the vector ${}^t\mathbf{x}$ at the reference configuration. The fundamental solution components can be expressed as

$$
\begin{aligned}
{}^t u_i^*({}^t\mathbf{x}) &= {}^t u_{ki}^*({}^t\chi, \mathbf{x}){}^t e_k({}^t\mathbf{x}) \\
{}^t p_i^*({}^t\mathbf{x}) &= {}^t p_{ki}^*({}^t\chi, \mathbf{x}){}^t e_k({}^t\mathbf{x})
\end{aligned} \tag{12.63}
$$

where ${}^t u_{ki}^*({}^t\chi, \mathbf{x}), {}^t p_{ki}^*({}^t\chi, \mathbf{x})$ are the displacement and traction kernels in the ith direction due to a unit load acting in the kth direction.

Using (12.62), (12.63) and the property of the Dirac delta function in equation (12.57), the incremental integral equations of motion (12.61) become

$$
\begin{aligned}
&-{}^t u_k({}^t\chi) + \int_{{}^t\Gamma} {}^t u_{ki}^*({}^t\chi, \mathbf{x}){}^t p_i({}^t\mathbf{x}){}^t d\Gamma({}^t\mathbf{x}) - \int_{{}^t\Gamma} {}^t p_{ki}^*({}^t\chi, \mathbf{x}){}^t u_i({}^t\mathbf{x}){}^t d\Gamma({}^t\mathbf{x}) \\
&-{}_t\Delta u_k({}^t\chi) + \int_{{}^t\Gamma} {}^t u_{ki}^*({}^t\chi, \mathbf{x}){}_t\Delta p_i({}^t\mathbf{x}){}^t d\Gamma({}^t\mathbf{x}) - \int_{{}^t\Gamma} {}^t p_{ki}^*({}^t\chi, \mathbf{x}){}_t\Delta u_i({}^t\mathbf{x}){}^t d\Gamma({}^t\mathbf{x}) \\
&-\int_{{}^t\Gamma} {}^t u_{ki}^*({}^t\chi, \mathbf{x})[{}_t\Delta u_{i,m}(t\sigma_{mj} + {}_t\Delta S_{mj})]({}^t\mathbf{x}){}^t n_j{}^t d\Gamma({}^t\mathbf{x}) \\
&+\int_{{}^t\Omega} {}^t u_{ki}^*({}^t\chi, \mathbf{x})[{}_t\Delta u_{i,m}(t\sigma_{mj} + {}_t\Delta S_{mj})]_{,j}({}^t\mathbf{x}){}^t d\Omega({}^t\mathbf{x}) \\
&= -\int_{{}^t\Omega} {}^t u_{ki}^*({}^t\chi, \mathbf{x}){}_t^{t+\Delta t}\gamma_i({}^t\mathbf{x}){}^t d\Omega({}^t\mathbf{x}) + \int_{{}^t\Omega} {}^t u_{ki}^*({}^t\chi\mathbf{x}){}^{t+\Delta t}a_i\,({}^t\mathbf{x}){}^t\rho{}^t d\Omega({}^t\mathbf{x}) \\
&+{}^t u_{ki}^*({}^t\chi, \mathbf{x}){}_t^{t+\Delta t}Q_{ij}^{t+\Delta t}\lambda_j
\end{aligned} \tag{12.64}
$$

The above equation gives the values of the components of displacement ${}^t u_k\,({}^t\chi)$ and the incremental displacement ${}_t\Delta u_k\,({}^t\chi)$ at any internal point ${}^t p_i({}^t\mathbf{x}), {}^t u_i({}^t\mathbf{x})$ in terms of the boundary values ${}_t\Delta p_i({}^t\mathbf{x})$, ${}_t\Delta u_i({}^t\mathbf{x})$ the fundamental solutions ${}^t u_{ki}^*({}^t\chi, \mathbf{x})$, ${}^t p_{ki}^*({}^t\chi, \mathbf{x})$, the body and the inertia forces. The fundamental solution due to Kelvin for elastostatics in two-dimensional (2D) plane strain can be found in references [29] and [30]

$$
{}^t u_{ij}^*({}^t\chi, {}^t\mathbf{x}) = \frac{1}{8\pi(1-v)\mu}\left[(3-4v)\delta_{ij}\ln\left(\frac{1}{t_r}\right) + {}^t r_{,i}{}^t r_{,j}\right] \tag{12.65}
$$

For a three-dimensional (3D) case,

$$ {}^{t}u_{ij}^{*}({}^{t}\chi, {}^{t}\mathbf{x}) = \frac{1}{16\pi(1-v)\mu\, {}^{t}r}\left[(3-4v)\delta_{ij} + {}^{t}r_{,i}{}^{t}r_{,j}\right] \tag{12.66} $$

$$ {}^{t}p_{ij}^{*}({}^{t}\chi, {}^{t}\mathbf{x}) = \frac{-1}{4\alpha\,\pi(1-v){}^{t}r^{\alpha}}\left[\left\{(1-2v)\delta_{ij} + 2\beta\, {}^{t}r_{,i}{}^{t}r_{,j}\right\}\frac{\partial^{t}r}{\partial^{t}n} - (1-2v)({}^{t}r_{,i}{}^{t}n_{j} - {}^{t}r_{,j}{}^{t}n_{i})\right] \tag{12.67} $$

where $\alpha = 1$, $\beta = 2$ for 2D plane strain and $\alpha = 2$, $\beta = 3$ for 3D problems.

The ${}^{t}u_{ij}^{*}({}^{t}\chi, {}^{t}\mathbf{x})$ and ${}^{t}p_{ij}^{*}({}^{t}\chi, {}^{t}\mathbf{x})$ are the displacements and tractions, respectively, at point ${}^{t}\mathbf{x}$ in the jth direction due to a unit load at point ${}^{t}\chi$ in the ith direction at the reference configuration. The distance between the points ${}^{t}\chi$ and ${}^{t}\mathbf{x}$ is denoted by ${}^{t}r$ and is given by

$$ {}^{t}r = \left|{}^{t}\mathbf{x} - {}^{t}\chi\right| \tag{12.68} $$

The kernels ${}^{t}u_{ij}^{*}({}^{t}\chi, {}^{t}\mathbf{x})$ and ${}^{t}p_{ij}^{*}({}^{t}\chi, {}^{t}\mathbf{x})$ display singularities when the distance approaches zero. Weak singularity such as $O(\ln(1/{}^{t}r))$ can be integrated, whereas strong singularity of the form $O(1/{}^{t}r)$, which exists only in the Cauchy principal value sense, is nonintegrable.

After using the limiting process [22,30], where the source point is taken to the boundary ${}^{t}\chi \in {}^{t}\Omega \rightarrow {}^{t}\chi \in {}^{t}\Gamma$, the integrals $\int_{{}^{t}\Gamma} {}^{t}p_{ki}^{*}({}^{t}\chi, \mathbf{x}){}^{t}u_{i}({}^{t}\mathbf{x}){}^{t}d\Gamma({}^{t}\mathbf{x})$, $\int_{{}^{t}\Gamma} {}^{t}p_{ki}^{*}({}^{t}\chi, \mathbf{x})_{t}\Delta u_{i}({}^{t}\mathbf{x}){}^{t}d\Gamma({}^{t}\mathbf{x})$ are obtained (existing only in the sense of the Cauchy principal value for a smooth boundary ${}^{t}\Gamma$ at the source point ${}^{t}\chi$). Then, the general incremental integral equations of motion for the elastic deformation of the continuum body become

$$ \begin{aligned} &{}^{t}c_{ki}({}^{t}\chi){}^{t}u_{k}({}^{t}\chi) - \int_{{}^{t}\Gamma} {}^{t}u_{ki}^{*}({}^{t}\chi, \mathbf{x}){}^{t}p_{i}({}^{t}\mathbf{x}){}^{t}d\Gamma({}^{t}\mathbf{x}) + \int_{{}^{t}\Gamma} {}^{t}p_{ki}^{*}({}^{t}\chi, \mathbf{x}){}^{t}u_{i}({}^{t}\mathbf{x}){}^{t}d\Gamma({}^{t}\mathbf{x}) \\ &+ {}^{t}c_{ki}({}^{t}\chi)_{t}\Delta u_{k}({}^{t}\chi) - \int_{{}^{t}\Gamma} {}^{t}u_{ki}^{*}({}^{t}\chi, \mathbf{x})_{t}\Delta p_{i}({}^{t}\mathbf{x}){}^{t}d\Gamma({}^{t}\mathbf{x}) \\ &+ \int_{{}^{t}\Gamma} {}^{t}p_{ki}^{*}({}^{t}\chi, \mathbf{x})_{t}\Delta u_{i}({}^{t}\mathbf{x}){}^{t}d\Gamma({}^{t}\mathbf{x}) \\ &+ \int_{{}^{t}\Gamma} {}^{t}u_{ki}^{*}({}^{t}\chi, \mathbf{x})[{}_{t}\Delta u_{i,m}(t\sigma_{mj} + {}_{t}\Delta S_{mj})]({}^{t}\mathbf{x}){}^{t}n_{j}{}^{t}d\Gamma({}^{t}\mathbf{x}){}^{t}\Omega \\ &- \int_{{}^{t}\Omega} {}^{t}u_{ki}^{*}({}^{t}\chi, \mathbf{x})[{}_{t}\Delta u_{i,m}(t\sigma_{mj} + {}_{t}\Delta S_{mj})]_{,j}({}^{t}\mathbf{x}){}^{t}d\Omega({}^{t}\mathbf{x}) \\ &= \int_{{}^{t}\Omega} {}^{t}u_{ki}^{*}({}^{t}\chi, \mathbf{x})_{t}^{t+\Delta t}\gamma_{i}({}^{t}\mathbf{x}){}^{t}d\Omega({}^{t}\mathbf{x}) \end{aligned} $$

$$- \int_{{}^t\Omega} {}^t\rho\, {}^t u_{ki}^*({}^t\chi, \mathbf{x})[{}^{t+\Delta t}\dot{v}_i + e_{imn} e_{npq}\, {}^{t+\Delta t}\omega_m\, {}^{t+\Delta t}\omega_p({}^t x_q + {}_t\Delta u_q)$$

$$+ e_{imn}\, {}^{t+\Delta t}\dot{\omega}_m({}^t x_n + {}_t\Delta u_n) + e_{imn}\, {}^{t+\Delta t}\omega_m\, {}^{t+\Delta t}v_n + {}^t\ddot{u}_i + {}_t\Delta\ddot{u}_i$$

$$+ 2e_{imn}\, {}^{t+\Delta t}\omega_m({}^t\dot{u}_n + {}_t\Delta\dot{u}_n)]({}^t\mathbf{x})^t d\Omega({}^t\mathbf{x}) - {}_t^{t+\Delta t}C_{kj}\, {}^{t+\Delta t}\lambda_j \tag{12.69}$$

where ${}_t^{t+\Delta t}C_{kj}$ denotes the terms due to the constraint equations

$${}_t^{t+\Delta t}C_{kj} = {}^t u_{ki}^*({}^t\chi, {}^t\mathbf{x})_t^{t+\Delta t}Q_{ij} \tag{12.70}$$

The expression ${}^t c_{ki}({}^t\chi)$ depends on the geometry at the neighborhood of the boundary point ${}^t\chi$ (Figure 12.2); it is given by

$${}^t c_{ki}\,({}^t\chi) = \delta_{ki} + \lim_{\varepsilon\to 0} \int_{\Gamma_\varepsilon} {}^t p_{ki}^*({}^t\chi, {}^t\mathbf{x})^t d\Gamma({}^t\mathbf{x}) \tag{12.71}$$

${}^t c_{ki} = \delta_{ki}$ for an interior point ${}^t\chi \in {}^t\Omega$, and ${}^t c_{ki} = \frac{1}{2}\delta_{ki}$ for ${}^t\chi \in {}^t\Gamma$, where the boundary is smooth at the point ${}^t\chi$. However, for a corner point, ${}^t c_{ki}$ must be evaluated explicitly using complicated expressions or implicitly by using rigid-body modes.

To obtain the numerical solution of the foregoing integral equations of motion, the boundary of the continuum body has to be discretized into a number of elements.

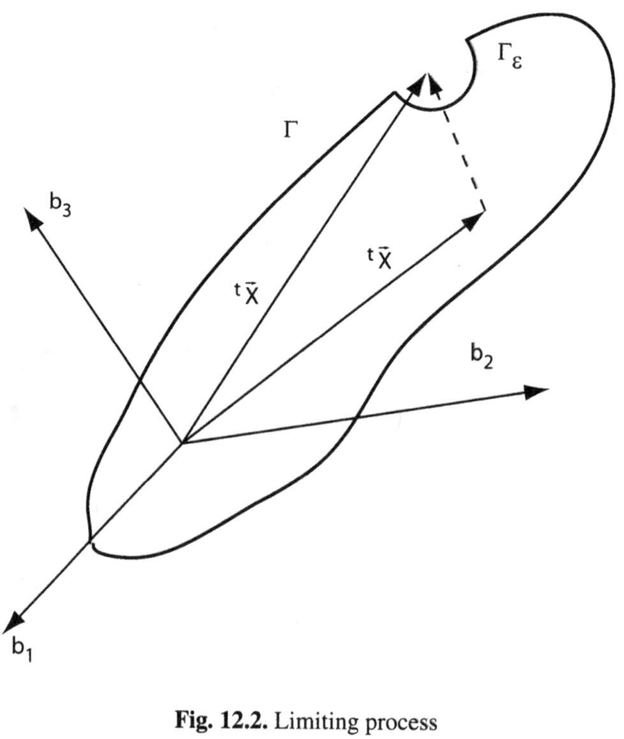

Fig. 12.2. Limiting process

The profiles of the components of displacements ${}^t u_i$ and tractions ${}^t p_i$, and their corresponding incremental values ${}_t\Delta u_i$, ${}_t\Delta p_i$ within each element are described in terms of their nodal points using a set of interpolation functions

$$ {}^t u_i({}^t\mathbf{x}) = \sum_{n=1}^{\upsilon_n} N^{nt} u_i^n({}^t\mathbf{x}) \quad \text{and} \quad {}_t\Delta u_i({}^t\mathbf{x}) = \sum_{n=1}^{\upsilon_n} N^{nt}{}_t\Delta u_i^n({}^t\mathbf{x}) \tag{12.72} $$

$$ {}^t p_i({}^t\mathbf{x}) = \sum_{n=1}^{\upsilon_n} N^{nt} p_i^n({}^t\mathbf{x}) \quad \text{and} \quad {}_t\Delta p_i({}^t\mathbf{x}) = \sum_{n=1}^{\upsilon_n} N^{nt}{}_t\Delta p_i^n({}^t\mathbf{x}) \tag{12.73} $$

where υ_n is the number of nodes within the isoparametric element and N^n are the interpolation functions. Then, the first six terms of boundary integrals in equation (12.69) can be discretized as

$$ \begin{aligned} &{}^t c_{ki}({}^t\chi^\alpha){}^t u_k({}^t\chi^\alpha) - \sum_{e=1}^{\upsilon_e}\sum_{n=1}^{\upsilon_n}\left\{\int_{{}^t\Gamma_e} {}^t u_{ki}^*({}^t\chi^\alpha, \mathbf{x}) N^{nt} d\Gamma_e\right\}{}^t p_i({}^t\chi^{K(e,n)}) \\ &+ \sum_{e=1}^{\upsilon_e}\sum_{n=1}^{\upsilon_n}\left\{\int_{{}^t\Gamma_e} {}^t p_{ki}^*({}^t\chi^\alpha, \mathbf{x}) N^{nt} d\Gamma_e\right\}{}^t u_i({}^t\chi^{K(e,n)}) \\ &+ {}^t c_{ki}({}^t\chi^\alpha){}_t\Delta u_k({}^t\chi^\alpha) - \sum_{e=1}^{\upsilon_e}\sum_{n=1}^{\upsilon_n}\left\{\int_{{}^t\Gamma_e} {}^t u_{ki}^*({}^t\chi^\alpha, \mathbf{x}) N^{nt} d\Gamma_e\right\}{}_t\Delta p_i({}^t\chi^{K(e,n)}) \\ &+ \sum_{e=1}^{\upsilon_e}\sum_{n=1}^{\upsilon_n}\left\{\int_{{}^t\Gamma_e} {}^t p_{ki}^*({}^t\chi^\alpha, \mathbf{x}) N^{nt} d\Gamma_e\right\}{}_t\Delta u_i({}^t\chi^{K(e,n)}) \end{aligned} \tag{12.74} $$

where ${}^t\chi^\alpha$ is the node at which the integral equation is written and the number $K(e, n)$ represents the boundary node n of element e in the global co-ordinates and υ_e is the number of the elements along the boundary. The above expression can be put in a compact form as

$$ \begin{aligned} &{}^t H_{ki}({}^t\chi^\alpha, {}^t\chi^{K(e,n)}){}^t u_i({}^t\chi^{K(e,n)}) - {}^t T_{ki}({}^t\chi^\alpha, {}^t\chi^{K(e,n)}){}^t p_i({}^t\chi^{K(e,n)}) \\ &+ {}^t H_{ki}({}^t\chi^\alpha, {}^t\chi^{K(e,n)}){}_t\Delta u_i({}^t\chi^{K(e,n)}) - {}^t T_{ki}({}^t\chi^\alpha, {}^t\chi^{K(e,n)}){}_t\Delta p_i({}^t\chi^{K(e,n)}) \end{aligned} \tag{12.75} $$

in which the term ${}^t H_{ki}({}^t\chi^\alpha, {}^t\chi^{K(e,n)})$ is given by

$$ {}^t H_{ki}({}^t\chi^\alpha, {}^t\chi^{K(e,n)}) = c_{ki}({}^t\chi^\alpha) + {}^t\tilde{H}_{ki}({}^t\chi^\alpha, {}^t\chi^{K(e,n)}) \tag{12.76} $$

Consequently, expression (12.74) can be cast in a matrix form

$$ {}^t\mathbf{H}^t\{\mathbf{u}\} - {}^t\mathbf{T}^t\{\mathbf{p}\} + {}^t\mathbf{H}\{\Delta\mathbf{u}\} - {}^t\mathbf{T}\{\Delta\mathbf{p}\} \tag{12.77} $$

The matrices ${}^t\mathbf{H}$ and ${}^t\mathbf{T}$ are evaluated using the known geometry of the reference configuration; see Figure 12.2. These matrices are fully populated and are non-symmetric if the body under consideration is not subdivided.

12.10 Body Forces

The domain integrals occurring in the integral equations of motion are a nuisance when evaluated numerically, because the whole body needs to be discretized into cells. This increases the amount of computation and inhibits the advantage of the boundary element method. To avoid the discretization of the whole body when domain integrals are present, a suitable technique known as the dual reciprocity method (DRM) is used to transform these integrals from their domain into DRM integrals.

Let I_b denote the term involving the domain integrals due to the body forces, expressed over the space of the reference configuration as

$$I_b = \int_{{}^t\Omega} \rho\, {}^t u^*_{ki}({}^t\chi, \mathbf{x})_t^{t+\Delta t}\gamma_i({}^t\mathbf{x})^t d\Omega({}^t\mathbf{x})$$

$$- \int_{{}^t\Omega} {}^t u^*_{ki}({}^t\chi, \mathbf{x})[{}^{t+\Delta t}\dot{v}_i + e_{imn}{}^{t+\Delta t}\omega_m{}^{t+\Delta t}v_n$$

$$+ e_{imn}e_{npq}{}^{t+\Delta t}\omega_m{}^{t+\Delta t}\omega_p{}^t x_q + e_{imn}{}^{t+\Delta t}\dot{\omega}_m{}^t x_n]({}^t\mathbf{x})^t d\Omega({}^t\mathbf{x}) \qquad (12.78)$$

To transform the above domain integrals into DRM integrals, the terms due to body forces within the domain ${}^t\Omega$ are approximated, using a set of known spatial functions $f^\mu({}^t\mathbf{x})(\mu = 1, \ldots, v_d)$ and unknown functions ϕ_i^n, where v_d is the number of DRM points, given by $v_d = v_b + v_i$, for which v_b is the number of boundary nodes and v_i is the number of few internal nodes. If the body consists of several regions, v_d would be equal to the number of boundary nodes of the region.

The choice of the class of functions $f^\mu({}^t\mathbf{x})$ used for the approximation of the domain integrals will determine the accuracy of the solution. According to [18, 20], even simple interpolation functions will be sufficient for computation of the dynamic analysis for lower modes of vibrations. According to the above references, an adequate choice of the approximating functions $f^\mu({}^t\mathbf{x})$ that is sufficient for dynamic analysis can be written as a conical function of type distance and constant that guarantees the completeness, which can then be incorporated through the following mapping:

$$f^\mu({}^t\mathbf{x}) = 1 + {}^t r({}^t\mathbf{x}, {}^t\mathbf{x}^\mu) \qquad (12.79)$$

where ${}^t r$ is the Euclidean distance between the domain point ${}^t\mathbf{x}$ and the DRM point ${}^t\mathbf{x}^\mu$. The best results were always reported using conical functions. When the function $f^\mu({}^t\mathbf{x})$ is evaluated at each DRM point at the reference configuration, we obtain a square matrix ${}^t F$ of order $(2v_d \times 2v_d)$ or $(3v_d \times 3v_d)$ for 2D plain strain or 3D problems, respectively. Consequently, the body forces due to gravitational loading at configuration $C^{t+\Delta t}$ but referred to C^t can be represented as

$${}_t^{t+\Delta t}\gamma_i = \sum_{\mu=1}^{v_d} f^\mu({}^t\mathbf{x})^{t+\Delta t}\phi^\mu_{1i} \qquad (12.80)$$

and the body forces due to the effect of large overall motion are

$$-{}^t\rho\, {}^{t+\Delta t}\dot{v}_i = \sum_{\mu=1}^{v_d} f^\mu({}^t\mathbf{x})\, {}^{t+\Delta t}\phi^\mu_{2i} \tag{12.81}$$

$$-{}^t\rho\, e_{imn}\, {}^{t+\Delta t}\omega_m\, {}^{t+\Delta t}v_n = \sum_{\mu=1}^{v_d} f^\mu({}^t\mathbf{x})\, {}^{t+\Delta t}\phi^\mu_{3i} \tag{12.82}$$

$$-{}^t\rho\, e_{imn}\, {}^{t+\Delta t}\alpha_m\, {}^t x_n = -{}^t\rho\, {}^{t+\Delta t}\dot{\tilde{\omega}}_{ij}\, {}^t x_j = \sum_{\mu=1}^{v_d} f^\mu({}^t\mathbf{x})\, {}^{t+\Delta t}\phi^\mu_{4i} \tag{12.83}$$

$$-{}^t\rho\, e_{imn} e_{npq}\, {}^{t+\Delta t}\omega_m\, {}^{t+\Delta t}\omega_p\, {}^t x_q = -{}^t\rho\, {}^{t+\Delta t}\tilde{\omega}^2_{ij}\, {}^t x_j = \sum_{\mu=1}^{v_d} f^\mu({}^t\mathbf{x})\, {}^{t+\Delta t}\phi^\mu_{5i} \tag{12.84}$$

where ${}^{t+\Delta t}\omega_i$ are the components of the angular velocity of frame B at the current configuration $C^{t+\Delta t}$ and $\dot{\tilde{\omega}}_{ij}$ are the components of the matrix $[\dot{\tilde{\omega}}_{ij}]$, which is obtained by the time differentiation of the following skew-symmetric angular velocity matrix:

$$^{t+\Delta t}\omega = {}^{t+\Delta t}[\dot{\tilde{\omega}}_{ij}] = \begin{bmatrix} 0 & -\omega_3 & \omega_2 \\ \omega_3 & 0 & -\omega_1 \\ -\omega_2 & \omega_1 & 0 \end{bmatrix} t + \Delta t \tag{12.85}$$

and $[{}^{t+\Delta t}\tilde{\omega}^2_{ij}]$ the product of the matrix $[{}^{t+\Delta t}\tilde{\omega}_{ij}]$ by itself. ω_1, ω_2 and ω_3 are the components of the inertial angular velocity ${}^N\omega^B$, where the left superscript is taken out for the sake of notation simplicity.

Using expressions (12.80), (12.82), (12.83) and (12.84) in equation (12.78), the domain integral I_{23} is expressed as product of two separate terms

$$I_b = \sum_{\mu=1}^{v_d} {}^t I^\mu_{kl}({}^{t+\Delta t}\phi^\mu_{1l} + {}^{t+\Delta t}\phi^\mu_{2l} + {}^{t+\Delta t}\phi^\mu_{3l} + {}^{t+\Delta t}\phi^\mu_{4l} + {}^{t+\Delta t}\phi^\mu_{3l} + {}^{t+\Delta t}\phi^\mu_{5l}) \tag{12.86}$$

where ${}^t I^\mu_{kl}$ is a domain integral over the geometry of the reference configuration:

$${}^t I^\mu_{kl} = \int_{{}^t\Omega} \delta_{li} f^\mu({}^t\mathbf{x})\, {}^t u^*_{ki}({}^t\chi, \mathbf{x})\, {}^t d\Omega({}^t\mathbf{x}) \tag{12.87}$$

and ${}^{t+\Delta t}\phi^\mu_{1l}$, ${}^{t+\Delta t}\phi^\mu_{2l}$, ${}^{t+\Delta t}\phi^\mu_{3l}$, ${}^{t+\Delta t}\phi^\mu_{4l}$, ${}^{t+\Delta t}\phi^\mu_{5l}$ are functions due to the loading at time $t + \Delta t$. The domain integral ${}^t I^\mu_{kl}$ is transformed to a DRM integral, using the fact that the functions $\delta_{li} f^\mu({}^t\mathbf{x}) (\mu = 1, \ldots, v)$ constitute a set of body forces acting in the lth direction and in equilibrium with a system of stresses ${}^t\hat{\sigma}^\mu_{ij,j}$ at the configuration C^t

$$ {}^{t}\hat{\sigma}^{\mu}_{ij,j} + \delta_{li} f^{\mu}({}^{t}\mathbf{x})^{t}e_{l} = 0 \tag{12.88} $$

Multiplying both sides by the fundamental solution and integrating over the domain of the body at the configuration C^t,

$$ \int_{{}^{t}\Omega} {}^{t}u^{*}_{ki}({}^{t}\chi, {}^{t}\mathbf{x})\delta_{li} f^{\mu}({}^{t}\mathbf{x})^{t}e_{l}{}^{t}d\Omega({}^{t}\mathbf{x}) = \int_{{}^{t}\Omega} {}^{t}u^{*}_{ki}({}^{t}\chi, {}^{t}\mathbf{x})^{t}\hat{\sigma}^{\mu}_{ij,j}({}^{t}\mathbf{x})^{t}d\Omega({}^{t}\mathbf{x}) \tag{12.89} $$

The right-hand side of the above equation (12.89) can be treated in a similar manner as in equation (12.58) by using the Maxwell–Betti reciprocal theorem and the property of the fundamental solution:

$$ \begin{aligned} {}^{t}I_{kl}({}^{t}\chi^{\alpha}, {}^{t}\mathbf{x}^{\mu}) = {} & c_{ki}({}^{t}\chi^{\alpha})^{t}\hat{u}_{li}({}^{t}\chi^{\alpha}, {}^{t}\mathbf{x}^{\mu}) - \int_{{}^{t}\Gamma} {}^{t}u^{*}_{ki}({}^{t}\chi, {}^{t}\mathbf{x})^{t}\hat{p}_{li}({}^{t}\mathbf{x}, {}^{t}\mathbf{x}^{\mu})^{t}d\Gamma({}^{t}\mathbf{x}) \\ & + \int_{{}^{t}\Gamma} {}^{t}p^{*}_{ki}({}^{t}\chi, {}^{t}\mathbf{x})^{t}\hat{u}_{li}({}^{t}\mathbf{x}, {}^{t}\mathbf{x}^{\mu})^{t}d\Gamma({}^{t}\mathbf{x}) \end{aligned} \tag{12.90} $$

where ${}^{t}\hat{u}_{li}({}^{t}\mathbf{x}, {}^{t}\mathbf{x}^{\mu})$ are the particular solutions of equation (12.88) and ${}^{t}\hat{p}_{li}({}^{t}\mathbf{x}, {}^{t}\mathbf{x}^{\mu})$ are their corresponding tractions. For $f^{\mu}({}^{t}\mathbf{x})$, the particular solutions can be found in [30].

The components of the particular solution and their corresponding tractions can be expressed in terms of their nodal values along the element as

$$ {}^{t}\hat{u}_{li}({}^{t}\mathbf{x}, {}^{t}\mathbf{x}^{\mu}) = \sum_{n=1}^{\upsilon_n} N^{n\,t}\hat{u}^{n}_{li}({}^{t}\mathbf{x}^{K(e,n)}, {}^{t}\mathbf{x}^{\mu}) \tag{12.91} $$

$$ {}^{t}\hat{p}_{li}({}^{t}\mathbf{x}, {}^{t}\mathbf{x}^{\mu}) = \sum_{n=1}^{\upsilon_n} N^{n\,t}\hat{p}^{n}_{li}({}^{t}\mathbf{x}^{K(e,n)}, {}^{t}\mathbf{x}^{\mu}) \tag{12.92} $$

After discretization of the integrals in (12.90), we obtain the system of equations

$$ \begin{aligned} {}^{t}I_{kl}({}^{t}\chi^{\alpha}, {}^{t}\mathbf{x}^{\mu}) = {} & -\sum_{e=1}^{\upsilon_e}\sum_{n=1}^{\upsilon_n} {}^{t}T_{ki}({}^{t}\chi^{\alpha}, {}^{t}\mathbf{x}^{K(e,n)})\hat{p}^{n}_{li}({}^{t}\mathbf{x}^{K(e,n)}, {}^{t}\mathbf{x}^{\mu}) \\ & + [c_{ki}({}^{t}\chi^{\alpha}) + \sum_{e=1}^{\upsilon_e}\sum_{n=1}^{\upsilon_n} {}^{t}H_{ki}({}^{t}\chi^{\alpha}, {}^{t}\mathbf{x}^{K(e,n)})]^{t}\hat{u}^{n}_{li}({}^{t}\mathbf{x}^{K(e,n)}, {}^{t}\mathbf{x}^{\mu}) \end{aligned} \tag{12.93} $$

or in matrix form as

$$ {}^{t}\mathbf{M} = {}^{t}\mathbf{H}^{t}\mathbf{U} - {}^{t}\mathbf{T}^{t}\hat{\mathbf{P}} \tag{12.94} $$

where the matrices ${}^{t}\mathbf{U}$ and ${}^{t}\hat{\mathbf{P}}$ contain simply the values of the components of the particular solutions ${}^{t}\hat{u}_{li}({}^{t}\mathbf{x}^{\alpha}, {}^{t}\mathbf{x}^{\mu})$ and ${}^{t}\hat{p}_{li}({}^{t}\mathbf{x}^{\alpha}, {}^{t}\mathbf{x}^{\mu})$, respectively.

The integral equation (12.93) corresponding to the body forces and evaluated at the DRM points can be represented in a matrix form as

$$ {}^{t+\Delta t}_{t}\{\mathbf{b}\} = {}^{t}\mathbf{M}({}^{t+\Delta t}\{\phi_1\} + {}^{t+\Delta t}\{\phi_2\} + {}^{t+\Delta t}\{\phi_3\} + {}^{t+\Delta t}\{\phi_4\} + {}^{t+\Delta t}\{\phi_5\}) \quad (12.95) $$

where the vectors ${}^{t+\Delta t}\{\phi_1\}$, ${}^{t+\Delta t}\{\phi_2\}$, ${}^{t+\Delta t}\{\phi_3\}$, ${}^{t+\Delta t}\{\phi_4\}$ and ${}^{t+\Delta t}\{\phi_5\}$which consist, respectively, of unknowns ${}^{t+\Delta t}\phi^{\mu}_{1l}$, ${}^{t+\Delta t}\phi^{\mu}_{2l}$, ${}^{t+\Delta t}\phi^{\mu}_{3l}$, ${}^{t+\Delta t}\phi^{\mu}_{4l}$, ${}^{t+\Delta t}\phi^{\mu}_{5l}$ are obtained from the set of linear equations given by

$$ {}^{t+\Delta t}\{\gamma\} = {}^{t}F^{t+\Delta t}\{\phi_1\} \quad (12.96) $$

$$ {}^{t+\Delta t}\{\dot{\upsilon}\} = {}^{t}F^{t+\Delta t}\{\phi_2\} \quad (12.97) $$

$$ {}^{t+\Delta t}\Omega^{t+\Delta t}\{\upsilon\} = {}^{t}F^{t+\Delta t}\{\phi_3\} \quad (12.98) $$

$$ {}^{t+\Delta t}\dot{\Omega}^{t}\{\mathbf{x}\} = {}^{t}F^{t+\Delta t}\{\phi_4\} \quad (12.99) $$

$$ {}^{t+\Delta t}\Omega^{2t}\{\mathbf{x}\} = {}^{t}F^{t+\Delta t}\{\phi_5\} \quad (12.100) $$

where the matrices Ω and $\Omega^{\mathbf{2}}$ consist of the submatrices ω and $\omega^{\mathbf{2}}$, respectively. The order of theses matrices is $(2\upsilon_d \times 2\upsilon_d)$ or $(3\upsilon_d \times 3\upsilon_d)$ for 2D plain strain or 3D problems, respectively.

$$ {}^{t+\Delta t}\Omega = \begin{bmatrix} {}^{t+\Delta t}\omega & & & \cdots \\ & {}^{t+\Delta t}\omega & & \cdots \\ & & {}^{t+\Delta t}\omega & \cdots \\ \vdots & \vdots & \vdots & \cdots \end{bmatrix} \quad (12.101) $$

$$ {}^{t+\Delta t}\Omega^2 = \begin{bmatrix} {}^{t+\Delta t}\omega^2 & & & \cdots \\ & {}^{t+\Delta t}\omega^2 & & \cdots \\ & & {}^{t+\Delta t}\omega^2 & \cdots \\ \vdots & \vdots & \vdots & \cdots \end{bmatrix} \quad (12.102) $$

Let $I_{\mathfrak{Z}}$ denote the terms involving the domain integrals due to the inertia loading:

$$ I_{\mathfrak{Z}} = \int_{{}^{t}\Omega} \rho\, {}^{t}u^{*}_{ki}({}^{t}\chi, {}^{t}\mathbf{x})[{}^{t}\ddot{u}_i + {}_{t}\Delta\ddot{u}_i + 2e_{imn}{}^{t+\Delta t}\omega_m({}^{t}\dot{u}_n + {}_{t}\Delta\dot{u}_n) $$

$$ + e_{imn}e_{npq}{}^{t+\Delta t}\omega_m{}^{t+\Delta t}\omega_{pt}\Delta u_q + e_{imn}{}^{t+\Delta t}\dot{\omega}_{mt}\Delta u_n]({}^{t}\mathbf{x})^{t}d\Omega({}^{t}\mathbf{x}) \quad (12.103) $$

To transform the domain integral $I_{\mathfrak{Z}}$ into a DRM one, the displacement components ${}^{t+\Delta t}u_i$ need to be approximated by the same spatial functions $f^{\mu}({}^{t}\mathbf{x})$ defined by (12.79) and unknown functions ${}^{t+\Delta t}\eta^{\mu}_i$ due to the inertia loading at the current configuration.

Therefore, the displacement components and their first and second time derivatives can be written as

$$^{t+\Delta t}u_i = {}^t u_i + {}_t\Delta u_i = \sum_{\mu=1}^{v_d} f^{\mu}({}^t\mathbf{x})^t \eta_i^{\mu} + \sum_{\mu=1}^{v_d} f^{\mu}({}^t\mathbf{x})_t \Delta\eta_i^{\mu} \tag{12.104}$$

$$^{t+\Delta t}\dot{u}_i = {}^t \dot{u}_i + {}_t\Delta \dot{u}_i = \sum_{\mu=1}^{v_d} f^{\mu}({}^t\mathbf{x})^t \dot{\eta}_i^{\mu} + \sum_{\mu=1}^{v_d} f^{\mu}({}^t\mathbf{x})_t \Delta\dot{\eta}_i^{\mu} \tag{12.105}$$

$$^{t+\Delta t}\ddot{u}_i = {}^t u_i + {}_t\Delta \ddot{u}_i = \sum_{\mu=1}^{v_d} f^{\mu}({}^t\mathbf{x})^t \ddot{\eta}_i^{\mu} + \sum_{\mu=1}^{v_d} f^{\mu}({}^t\mathbf{x})_t \Delta\ddot{\eta}_i^{\mu} \tag{12.106}$$

Using the DRM principle in a similar manner as I_B the term due to the inertial forces can be expressed as

$$\begin{aligned} I_{\Im} = & \sum_{\mu=1}^{v_d} {}^t I_{kl}({}^t\chi^{\alpha}, {}^t\mathbf{x}^{\mu})({}^t\ddot{\eta}_1^{\mu} + 2^{t+\Delta t}\tilde{\omega}_{lp}{}^t\dot{\eta}_P^{\mu}) \\ & + \sum_{\mu=1}^{v_d} {}^t I_{kl}({}^t\chi^{\alpha}, {}^t\mathbf{x}^{\mu})({}_t\Delta\ddot{\eta}_1^{\mu} + 2^{t+\Delta t}\tilde{\omega}_{lp\,t}\Delta\dot{\eta}_P^{\mu} + {}^{t+\Delta t}\dot{\tilde{\omega}}_{lp\,t}\Delta\eta_P^{\mu} + {}^{t+\Delta t}\tilde{\omega}_{lp\,t}^2\Delta\eta_P^{\mu}) \end{aligned} \tag{12.107}$$

or in matrix form as

$$\begin{aligned} {}_t^{t+\Delta t}\{\mathbf{i}\} = & -[\rho^t\mathbf{M}^t\{\ddot{\eta}\} + 2\rho^t\mathbf{M}^{t+\Delta t}\Omega^t\{\dot{\eta}\} \\ & + \rho^t\mathbf{M}^t\{\Delta\ddot{\eta}\} + 2\rho^t\mathbf{M}^{t+\Delta t}\Omega^t\{\Delta\dot{\eta}\} \\ & + \rho^t\mathbf{M}({}^{t+\Delta t}\dot{\Omega} + {}^{t+\Delta t}\Omega^2)\{\Delta\eta\}] \end{aligned} \tag{12.108}$$

For which ${}^t\{\dot{\eta}\}$ and ${}^t\{\ddot{\eta}\}$ are obtained from the equations

$$^t\{\dot{\mathbf{u}}\} = {}^t F^t\{\dot{\eta}\} \tag{12.109}$$

$$^t\{\ddot{\mathbf{u}}\} = {}^t F^t\{\ddot{\eta}\} \tag{12.110}$$

where the array $\{\eta\}$ consist of the values of components ${}^t\eta_i^{\mu}$ at each DRM point. Similar relations can be applied for incremental displacement, velocity and acceleration vectors.

12.11 Formulation of the Nonlinear Stiffness Matrix

From the divergence theorem, the integrals due to the nonlinear terms in the equations of motion can be rewritten as

$$\int_{{}^t\Gamma} {}^t u_{ki}^*({}^t\chi, {}^t\mathbf{x})[{}_t\Delta u_{i,m}({}^t\sigma_{mj} + {}_t\Delta S_{mj})]({}^t\mathbf{x})^t n_j({}^t\mathbf{x})^t d\Gamma({}^t\mathbf{x})$$

$$- \int_{{}^t\Omega} {}^t u_{ki}^*({}^t\chi, {}^t\mathbf{x})[{}_t\Delta u_{i,m}({}^t\sigma_{mj} + {}_t\Delta S_{mj})]_{,j}({}^t\mathbf{x})^t d\Omega({}^t\mathbf{x}) \tag{12.111}$$

$$= \int_{{}^t\Omega} {}^t u_{ki}^*({}^t\chi, {}^t\mathbf{x})_t\Delta u_{i,m}({}^t\sigma_{mj} + {}_t\Delta S_{mj})^t d\Omega({}^t\mathbf{x})$$

This integral equation plays an important role in modeling the geometric nonlinearity, which is a function of the stresses at the reference configuration.

The right-hand side of equation (12.111) requires the discretization of the whole domain of the body continuum. Furthermore, the presence of the gradients of ${}^t u_{ki}^*({}^t\chi, {}^t\mathbf{x})$ creates difficulty for applying the DRM. Therefore, it is preferable to use the left-hand side of equation (12.111) because it consists of a boundary integral and a domain integral which can be transformed easily to a DRM integral.

12.11.1 Nonlinear Stiffness Matrix due to Boundary Integral

The nonlinear boundary integral in equation (12.69) can be expressed as

$$I_{\Xi}^{\Gamma} = \int_{{}^t\Gamma} \breve{p}_m({}^t\mathbf{x})_t\Delta u_{i,m}{}^t u_{ki}^*({}^t\chi, {}^t\mathbf{x})^t d\Gamma({}^t\mathbf{x}) \tag{12.112}$$

For which the components $\breve{p}_m({}^t\mathbf{x})$ are given by:

$$\breve{p}_m({}^t\mathbf{x}) = ({}^t\sigma_{mj} + {}_t\Delta S_{mj})^t n_j({}^t\mathbf{x}) = {}^t p_m({}^t\mathbf{x}) + {}_t\Delta S_{mj}({}^t\mathbf{x})^t n_j({}^t\mathbf{x}) \tag{12.113}$$

Within each element, the components of the incremental displacements ${}_t\Delta u_i$ and the tractions $\breve{p}_i$ can be expressed in terms of their nodal points using a set of interpolation functions. Then, the gradient of the components of the incremental displacements can be written as

$${}_t\Delta u_{i,m} = \sum_{n=1}^{v_n} \frac{\partial N^n}{\partial^t x_m} {}_t\Delta u_i^m \tag{12.114}$$

Substituting (12.72), (12.73) and (12.114) into equation (12.112), we can obtain the following integral:

$$I_{\Xi}^{\Gamma} = \int_{{}^t\Gamma} [{}_t\Delta u_{i,m}({}^t\sigma_{mj} + {}_t\Delta S_{mj})]({}^t\mathbf{x})^t n_j({}^t\mathbf{x})^t u_{ki}^*({}^t\chi, {}^t\mathbf{x})^t d\Gamma({}^t\mathbf{x})$$

$$= \sum_{e=1}^{v_e} \sum_{n=1}^{v_n} K_{ki}^{\Gamma}({}^t\chi^{\alpha}, {}^t\mathbf{x}^{k(e,n)})_t\Delta u_i({}^t\chi^{k(e,n)}) \tag{12.115}$$

where $K^{\Gamma}_{ki}({}^t\chi^{\alpha}, {}^t\mathbf{x}^{k(e,n)})$ are the elements of the nonlinear matrix due to boundary integrals:

$$K^{\Gamma}_{ki}({}^t\chi^{\alpha}, {}^t\mathbf{x}^{k(e,n)}) = \sum_{l=1}^{\upsilon_n} {}^t p_m^{k(e,l)} \left\{ \int_{{}^t\Gamma_e} {}^t u^*_{ki}({}^t\chi, {}^t\mathbf{x}) N^l \frac{\partial N^n}{\partial^t x_m} {}^t d\Gamma_e \right\}$$

$$+ \sum_{l=1}^{\upsilon_n} E_{mjzr\,t}\Delta u_z^{K(e.l)} \left\{ \int_{{}^t\Gamma_e} {}^t u^*_{ki}({}^t\chi, {}^t\mathbf{x}) \frac{\partial N^n}{\partial^t x_r} {}^t n_j({}^t\mathbf{x}) \frac{\partial N^n}{\partial^t x_m} {}^t d\Gamma_e \right\} \quad (12.116)$$

or in matrix form as

$$\mathbf{K}^{\Gamma} = \mathbf{K}^{\Gamma}_S({}^t\sigma^{\mu}) + \mathbf{K}^{\Gamma}_G({}_t\Delta\mathbf{u}^{\mu}) \quad (12.117)$$

where ${}^t\sigma^{\mu}$ and ${}_t\Delta\mathbf{u}^{\mu}$ are the stress dyadic and the incremental displacement vector evaluated at each nodal boundary point.

12.11.2 Nonlinear Stiffness Matrix due to DRM Integral

The nonlinear domain integral is rewritten as

$$I^{\Gamma}_{\Xi} = \int_{{}^t\Omega} {}^t u^*_{ki}({}^t\chi, {}^t\mathbf{x}) \breve{S}_{ij,j}({}^t\mathbf{x}) {}^t d\Omega({}^t\mathbf{x}) \quad (12.118)$$

where $\breve{S}_{ij}$ is a nonsymmetric tensor given by

$$\breve{S}_{ij}({}^t\mathbf{x}) = {}_t\Delta u_{i,k}({}^t\sigma_{kj} + {}_t\Delta S_{kj})({}^t\mathbf{x}) \quad (12.119)$$

We can assume that the tensor $\breve{S}_{ij}$ can be approximated as a sum of the products of υ_d spatial functions $f^{\mu}({}^t\mathbf{x})$, corresponding to the updated configuration and υ_d unknown functions β^{μ}_{ij}. Hence,

$$\breve{S}_{ij} = \sum_{\mu=1}^{\upsilon_d} f^{\mu}({}^t\mathbf{x}) \beta^{\mu}_{ij} \quad (12.120)$$

The spatial derivatives of $\breve{S}_{ij}$ with respect to the updated configuration are approximated as

$$\breve{S}_{ij,j} = \sum_{\mu=1}^{\upsilon_d} f^{\mu}({}^t\mathbf{x}) \psi^{\mu}_i \quad (12.121)$$

Consequently, the domain integral I_{Ξ} can be expressed as the DRM integral

$$I^{\Omega}_{\Xi} = \sum_{\mu=1}^{\upsilon_d} {}^t I_{kl}({}^t\chi^{\alpha}, {}^t\mathbf{x}^{\mu}) \psi^{\mu}_i \quad (12.122)$$

Taking the derivatives with respect to ${}^t x_j$, equation (12.121) results in

$$\breve{S}_{ij,j} = \sum_{\mu=1}^{\upsilon_d} f^{\mu}_{,j}({}^t\mathbf{x}) \beta^{\mu}_{ij} \quad (12.123)$$

or

$$\frac{\partial \breve{S}_{ij}}{\partial^t x_j} = \sum_{\mu=1}^{v_d} \left[\frac{\partial f^\mu({}^t\mathbf{x})}{\partial^t x_1} \beta_{i1}^\mu + \frac{\partial f^\mu({}^t\mathbf{x})}{\partial^t x_2} \beta_{i2}^\mu + \frac{\partial f^\mu({}^t\mathbf{x})}{\partial^t x_3} \beta_{i3}^\mu \right] \tag{12.124}$$

The expressions (12.121), (12.122), (12.124) can be cast in a matrix form in the following manner. Consider the square matrix ${}^tF$ which consists of the values of the functions $f^\mu({}^t\mathbf{x})$ at every DRM point at the reference configuration C^t.

Then, equation (12.121) can be written as

$$\{\breve{S}_d\} = {}^tF\{\psi\} \tag{12.125}$$

or

$$\{\psi\} = {}^tF^{-1}\{\breve{S}_d\} \tag{12.126}$$

where ${}^tF^{-1}$ is the inverse matrix of ${}^tF$ and the terms $\breve{S}_{ij,j}$ are the elements of the column matrix $\{\breve{S}_d\}$, whose order is $(2v_d \times 1)$ or $(3v_d \times 1)$ for 2D plane strain of 3D problems, respectively. From equation (12.123), $\{\breve{S}_d\}$ can be written in matrix form as:

$$\{\breve{\tau}_d\} = {}^tF_1\{{}^t\breve{B}_1\} + {}^tF_2\{{}^t\breve{B}_2\} + {}^tF_3\{{}^t\breve{B}_3\} \tag{12.127}$$

where the matrices ${}^tF_1, {}^tF_2, {}^tF_3$ are the derivatives of ${}^tF$ with respect to the spatial coordinates ${}^tx_1$, ${}^tx_2$ and ${}^tx_3$, respectively. The expressions $\{\breve{B}_1\}$, $\{\breve{B}_2\}$ and $\{\breve{B}_3\}$ are the column matrices whose elements are, respectively, β_{i1}^μ, β_{i2}^μ and β_{i3}^μ. These vectors are obtained from equation (12.114) as

$$\{\breve{B}_1\} = {}^tF^{-1}\{\breve{S}_1\} \tag{12.128}$$

$$\{\breve{B}_2\} = {}^tF^{-1}\{\breve{S}_2\} \tag{12.129}$$

$$\{\breve{B}_3\} = {}^tF^{-1}\{\breve{S}_3\} \tag{12.130}$$

The substitution of equations (12.128), (12.129) and (12.130) in equation (12.127) results in

$$\{\breve{S}_d\} = {}^t\breve{F}_1{}^tF^{-1}\{\breve{S}_1\} + {}^t\breve{F}_2{}^tF^{-1}\{\breve{S}_2\} + {}^t\breve{F}_3{}^tF^{-1}\{\breve{S}_3\} \tag{12.131}$$

After substituting (12.131) in equation (12.124), we obtain the following expression:

$$\{\psi\} = \Im_1\{\breve{S}_1\} + \Im_2\{\breve{S}_2\} + \Im_3\{\breve{S}_3\} \tag{12.132}$$

where

$$\Im_1 = {}^tF^{-1t}F_1{}^tF^{-1} \tag{12.133}$$

$$\Im_2 = {}^tF^{-1t}F_2{}^tF^{-1} \tag{12.134}$$

$$\Im_3 = {}^tF^{-1t}F_3{}^tF^{-1} \tag{12.135}$$

and $\{\breve{S}_1\}$, $\{\breve{S}_2\}$ and $\{\breve{S}_3\}$ are column matrices each of order $(2v_d \times 1)$ or $(3v_d \times 1)$ for 2D plane strain and for 3D problems:

$$\{\breve{\tau}_j\} = \left[\underbrace{\breve{S}_{1j}\breve{S}_{2j}\breve{S}_{3j}}_{\mu=1} \ldots \underbrace{\breve{S}_{1j}\breve{S}_{2j}\breve{S}_{3j}}_{\mu=v_d} \right]^T \tag{12.136}$$

Using the finite differences, the incremental displacement gradient ${}_t\Delta u_{m,k}$ can be computed as

$$\frac{\partial_t \Delta u_m}{\partial^t x_k} = \sum_{l=1}^{N} L_l({}^t x_k)_t \Delta u_m({}^t x_l^{\mu}) \tag{12.137}$$

where $L_l({}^t x_k)$ are the Lagrange coefficient polynomials. Hence, equation (12.120) becomes

$$\breve{S}_{ij} = A_{ijmk} L_{kl}({}^t x_k)_t \Delta u_m({}^t x_l^{\mu}) \tag{12.138}$$

for which the tensor $L_{kl}({}^t x_k)$ consists of the coefficient polynomials $L_l({}^t x_k)$ and the tensor A_{ijmk} is given by

$$A_{ijmk} = \delta_{im}({}^t\sigma_{kj} + {}_t\Delta S_{kj})({}^t\mathbf{x}^{\mu}) \tag{12.139}$$

Therefore, the equation (12.138) can be cast in matrix form as

$$\breve{S} = AL\{\Delta\mathbf{u}\} \tag{12.140}$$

where the column matrices corresponding to the matrix $\breve{S}$ can be expressed as

$$\{\breve{S}_1\} = A_1 L\{\Delta\mathbf{u}\} \tag{12.141}$$

$$\{\breve{S}_2\} = A_2 L\{\Delta\mathbf{u}\} \tag{12.142}$$

$$\{\breve{S}_3\} = A_3 L\{\Delta\mathbf{u}\} \tag{12.143}$$

The substitution of (12.141), (12.141), (12.141) into equation (12.131) results in

$$\{\psi\} = [\Im_1 A_1 + \Im_2 A_2 + \Im_3 A_3] L\{\Delta u\} \tag{12.144}$$

Then, the nonlinear stiffness matrix due to the DRM integral can be cast in the form

$$K_S^{\Omega}({}^t\sigma^{\mu}) + K_G^{\Omega}({}_t\Delta^{\mu}) = {}^t M[\Im_1 A_1 + \Im_2 A_2 + \Im_3 A_3] L \tag{12.145}$$

Consequently, the total nonlinear stiffness matrices due to boundary and DRM integrals can be represented by

$$\mathbf{K}_S({}^t\sigma^{\mu}) = \mathbf{K}_S^{\Gamma}({}^t\sigma^{\mu}) - \mathbf{K}_S^{\Omega}({}^t\sigma^{\mu}) \tag{12.146}$$

$$\mathbf{K}_{Gt}(\Delta\mathbf{u}^{\mu}) = \mathbf{K}_G^{\Gamma}({}_t\Delta\mathbf{u}^{\mu}) - \mathbf{K}_G^{\Omega}({}_t\Delta\mathbf{u}^{\mu}) \tag{12.147}$$

where the superscripts Γ and Ω refer to the boundary and DRM points, respectively. When a substructuring technique is utilized, the number of DRM points is taken to be equal to the number of boundary points, $v_d = v_b$. Therefore, the subtraction of the matrices is possible because the matrices have the same order.

12.12 General Equations of Motion

After collecting all the formulations developed above and carrying out the necessary substitutions, we arrive at the following incremental dynamical equations of motion expressed in matrix form as

$$ {}^{t}\mathbf{M}\{\Delta\ddot{\mathbf{u}}\} + {}_{t}^{t+\Delta t}\mathbf{G}\{\Delta\dot{\mathbf{u}}\} + [{}_{t}^{t+\Delta t}\mathbf{K}_{D} + \mathbf{K}_{\tau}]\{\Delta\mathbf{u}\} + {}_{t}^{t+\Delta t}\mathbf{C}^{t+\Delta t}\{\lambda\} = \{\Delta\mathbf{F}\} \tag{12.148} $$

where the matrix $\mathbf{K}_{\tau}$ is the tangent stiffness matrix given by

$$ \mathbf{K}_{\tau} = {}^{t}\mathbf{H} + \mathbf{K}_{S}\,({}^{t}\sigma^{\mu}) + \mathbf{K}_{G}\,({}_{t}\Delta\mathbf{u}^{\mu}) \tag{12.149} $$

${}^{t}\mathbf{M}$ is the matrix given by

$$ {}^{t}\mathbf{M} = {}^{t}\rho\,{}^{t}M^{t}\,F^{-1} \tag{12.150} $$

${}_{t}^{t+\Delta t}\mathbf{G}$ is the coriolis matrix:

$$ {}_{t}^{t+\Delta t}\mathbf{G} = 2{}^{t}\mathbf{M}^{t+\Delta t}\Omega \tag{12.151} $$

${}_{t}^{t+\Delta t}\mathbf{K}_{D}$ defines the stiffness matrix related to the overall rotation:

$$ {}_{t}^{t+\Delta t}\mathbf{K}_{D} = {}^{t}\mathbf{M}({}^{t+\Delta t}\dot{\Omega} + {}^{t+\Delta t}\Omega^{2}) \tag{12.152} $$

${}_{t}^{t+\Delta t}\mathbf{C}$ is the constraint Jacobian matrix and ${}^{t+\Delta t}\{\lambda\}$ represents the undetermined vector of Lagrange multipliers. $\{\Delta\mathbf{F}\}$ is the incremental forcing vector given by

$$ \begin{aligned} \{\Delta\mathbf{F}\} &= {}^{t}\mathbf{T}^{t+\Delta t}\{\mathbf{p}\} \\ &\quad + {}^{t}\mathbf{M}({}^{t+\Delta t}\{\gamma\} - {}^{t+\Delta t}\{\dot{\upsilon}\} - {}^{t+\Delta t}\Omega^{t+\Delta t}\{\upsilon\}) - {}_{t}^{t+\Delta t}\mathbf{K}_{D}{}^{t}\{\mathbf{x}\} \\ &\quad - [{}^{t}\mathbf{M}\{\ddot{\mathbf{u}}\} + {}_{t}^{t+\Delta t}\mathbf{G}\{\dot{\mathbf{u}}\} + {}^{t}\mathbf{H}^{t}\{\mathbf{u}\}] \end{aligned} \tag{12.153} $$

The matrices defined above are evaluated using the geometry of reference configuration, where the order of each matrix is $(2\upsilon_d \times 2\upsilon_d)$ or $(3\upsilon_d \times 3\upsilon_d)$ for two of three dimension problems, respectively. The equations of motion are formulated using the fact that the number of DRM points is the same as the number of boundary points.

The contribution of the stress or the geometric stiffening in the stiffness matrix is due to the internal forces based on the reference stresses. These forces must balance the external applied loading due to the large overall motion of the elastic body. The internal nodal forces at the reference configuration C^{t} can be expressed as

$$ F_{\text{int}}({}^{t}\mathbf{u}^{\mu}) = {}^{t}\mathbf{H}^{t}\{\mathbf{u}\} \tag{12.154} $$

At the current and unknown configuration $C^{t+\Delta t}$ the internal nodal forces can be written as

$$F_{\text{int}}({}^{t+\Delta t}\mathbf{u}^{\mu}) = {}^{t+\Delta t}\mathbf{H}^{t+\Delta t}\{\mathbf{u}\} \tag{12.155}$$

Hence, the tangent stiffness matrix can be formulated as

$$\mathbf{K}_{\tau}\{\Delta\mathbf{u}\} = \mathbf{F}_{\text{int}}({}^{t+\Delta t}\mathbf{u}^{\mu}) - \mathbf{F}_{\text{int}}({}^{t}\mathbf{u}^{\mu}) \tag{12.156}$$

The internal nodal forces at the current configuration $\mathbf{F}_{\text{int}}({}^{t+\Delta t}\mathbf{u}^{\mu})$ contain implicitly the geometric stiffening nonlinearity which arise when the geometry of reference configuration C^t is adapted in the analysis. The tangent stiffness matrix is obtained after the linearization is performed about the reference configuration. This term can be obtained also by using the Taylor series to expand the function $\mathbf{F}_{\text{int}}(\mathbf{u}^{\mu})$ with respect to the deformation at the reference configuration and neglecting the quadratic terms, hence

$$\mathbf{F}_{\text{int}}({}^{t+\Delta t}\mathbf{u}^{\mu}) = \mathbf{F}_{\text{int}}({}^{t}\mathbf{u}^{\mu}) + \frac{\partial \mathbf{F}_{\text{int}}(\mathbf{u}^{\mu})}{\partial u_i^{\mu}}\bigg|_{\mathbf{u}^{\mu}={}^{t}\mathbf{u}^{\mu}}({}^{t+\Delta t}u_i^{\mu} - {}^{t}u_i^{\mu}) \tag{12.157}$$

From which the term $\frac{\partial \mathbf{F}_{\text{int}}(\mathbf{u}^{\mu})}{\partial u_i^{\mu}}|_{\mathbf{u}^{\mu}={}^{t}\mathbf{u}^{\mu}}$ represents the tangent stiffness matrix given by equation (12.149).

Equations (12.157) are only approximations to the actual equations to be solved in each time step. The linearization of equations (12.148) may introduce errors which ultimately result in solution instability. For this reason, it may be necessary to iterate in each load step until, within the necessary assumptions on the numerical time integration, equations (12.153) are satisfied to a required tolerance. In order to ensure the stability and the accuracy of the integration of the equations of motion, the time step must be small. The third term of the tangent stiffness matrix arising from the incremental components of the displacement $\mathbf{K}_G({}_t\Delta\mathbf{u}^{\mu})$ can be neglected on the basis of the linearization. Therefore, the tangent stiffness matrix reduces to the following expression:

$$\mathbf{K}_{\tau} = {}^{t}\mathbf{H} + \mathbf{K}_S({}^{t}\sigma^{\mu}) \tag{12.158}$$

Using the nodal displacement components ${}^{t}u_i^{\mu}$, the nodal stresses components ${}^{t}\sigma_{ij}^{\mu}$ relative to the configuration C^t and the nodal applied loading corresponding to the configuration $C^{t+\Delta t}$ by referring to C^t, the incremental equations of motion (1.50) can be used to solve for the nodal boundary unknowns, which are the incremental components ${}_t\Delta u_i^{\mu}$ and ${}_t\Delta p_i^{\mu}$. The nodal values of ${}_t\Delta u_i^{\mu}$ are utilized to approximate the nodal displacement components ${}^{t+\Delta t}u_i^{\mu}$ corresponding to the configuration $C^{t+\Delta t}$ simply by using the kinematical relations given by equations (12.96–12.100). The linearized equations of motion result in an error known as the out-of-balance forces between the applied loading and the internal resisting forces of the structure at the current configuration $C^{t+\Delta t}$. If these forces are not small, iterations must be performed in order to reduce them further. Hence, a Newton–Raphson scheme is employed to solve the linearized equations of motion, where an iterative increment vector $\{\delta\,\mathbf{u}\}_i$ is computed from the following equation:

$$[a_0{}^{t}\mathbf{M} + a_1{}_t^{t+\Delta t}\mathbf{G} + {}_t^{t+\Delta t}\mathbf{K}_D + \mathbf{K}_{\tau}]\{\delta\,\mathbf{u}\}_i$$

$$= \left\langle {}^{t}\mathbf{T}_t^{t+\Delta t}\{\mathbf{p}\} + {}^{t}\mathbf{M}\left[{}^{t+\Delta t}\{\gamma\} - {}^{t+\Delta t}\{\dot{\upsilon}\} - {}^{t+\Delta t}\Omega^{t+\Delta t}\{\upsilon\}\right]\right\rangle \tag{12.159}$$

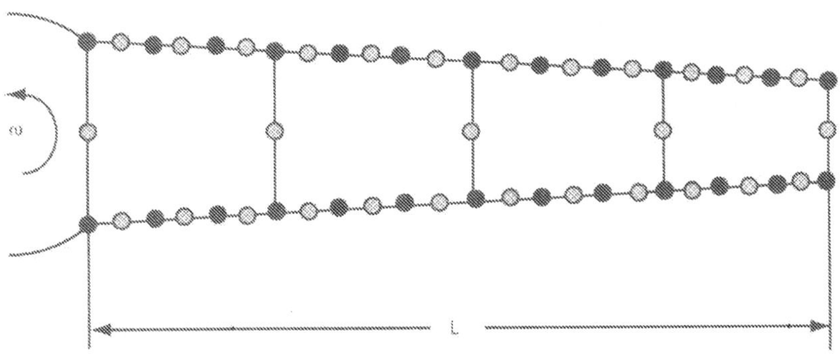

Fig. 12.3. Rotating tapered cantilever beam

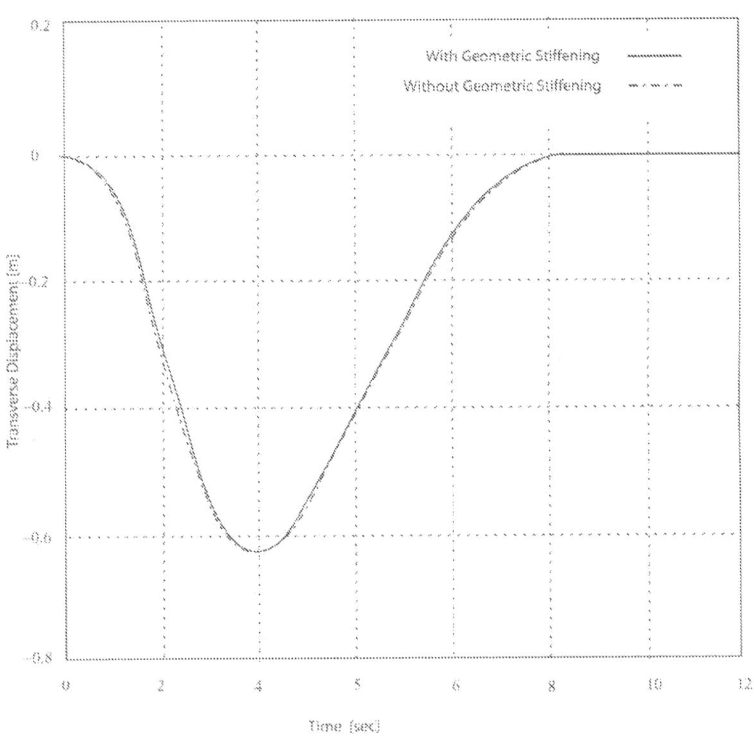

Fig. 12.4. Time history of the tip deflection u_2

and used to update the vectors ${}^{t+\Delta t}\{\mathbf{u}\}_i$, ${}^{t+\Delta t}\{\dot{\mathbf{u}}\}_i$ and ${}^{t+\Delta t}\{\ddot{\mathbf{u}}\}_i$ by using the relations

$$ {}^{t+\Delta t}\{\mathbf{u}\}_i = {}^{t+\Delta t}\{\mathbf{u}\}_{i-1} + \{\delta\,\mathbf{u}\}_i \tag{12.160} $$

$$ {}^{t+\Delta t}\{\dot{\mathbf{u}}\}_i = {}^{t+\Delta t}\{\dot{\mathbf{u}}\}_{i-1} + a_1\{\delta\,\mathbf{u}\}_i \tag{12.161} $$

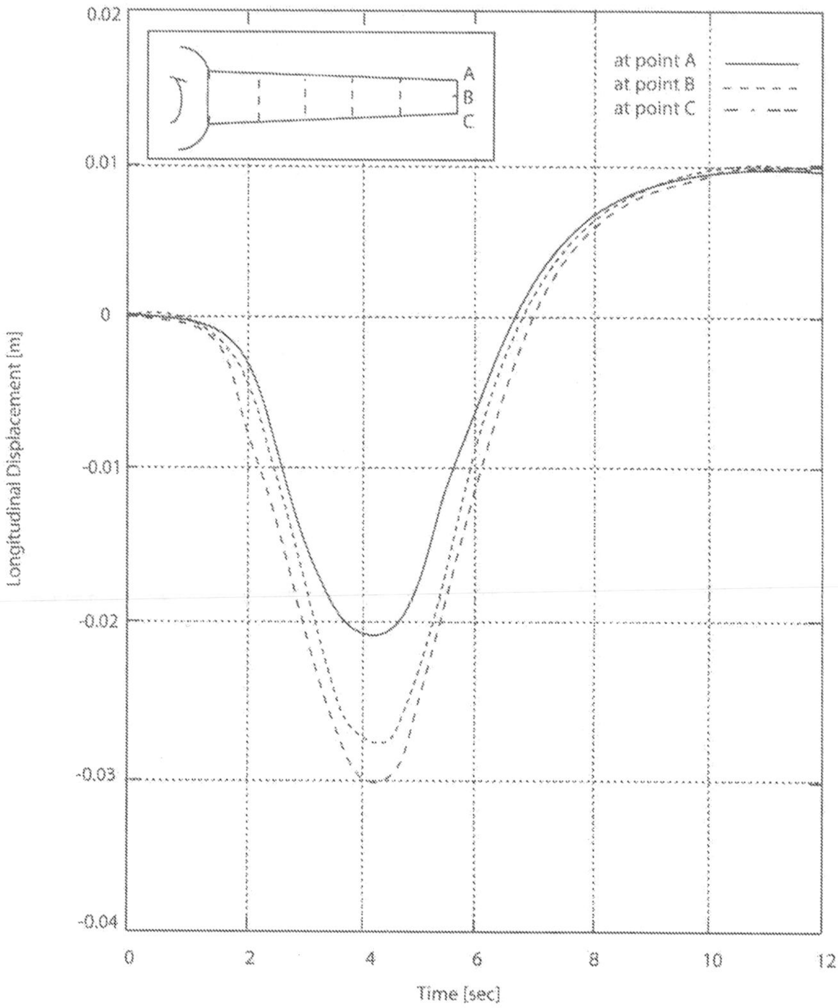

Fig. 12.5. Time history of the tip deflection u_1

$$^{t+\Delta t}\{\ddot{\mathbf{u}}\}_i = {}^{t+\Delta t}\{\ddot{\mathbf{u}}\}_{i-1} + a_0\{\delta\,\mathbf{u}\}_i \tag{12.162}$$

$$^{t+\Delta t}\{\mathbf{x}\}_i = {}^{t+\Delta t}\{\mathbf{x}\}_{i-1} + \{\delta\,\mathbf{u}\}_i \tag{12.163}$$

where a_0 and a_1 are integration constants due to the integration scheme used (Newman or Huboult).

It should be mentioned that when the body is divided into several substructures, one can notice that the overall mass matrix is formulated by assembling all the mass matrices from each substructure. Therefore, the computation of the mass matrix requires only the inverse of the matrix ${}^tF$, for which the size is equal to the total

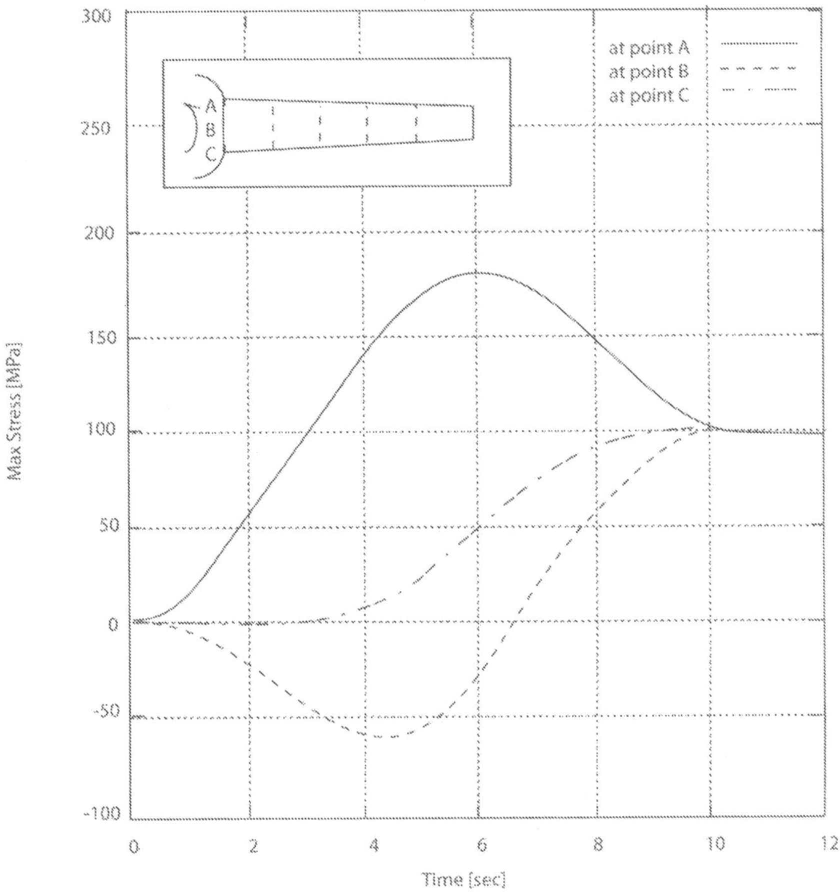

Fig. 12.6. Time history of the maximum stress $\sigma_{\max}$ at the base of a beam

number of the nodal degrees of freedom of the boundary of the substructure and not the whole domain of the body. As a matter of fact, when a substructuring technique is used, the overall matrices related to the entire body are block-banded. This feature improves remarkably the computational efficiency of the method.

12.13 Numerical Example

The problem consists of a 2D (plane strain) body in the form of a long tapered beam attached to a moving base under a prescribed spin-up maneuver. The maneuver starts at time $t = 0$ and ends at time $t = T_s$ during which time the measure numbers of the inertial angular velocity are $\omega_1 = 0$, $\omega_2 = 0$ and $\omega_3 = \omega(t)$, given by the expression

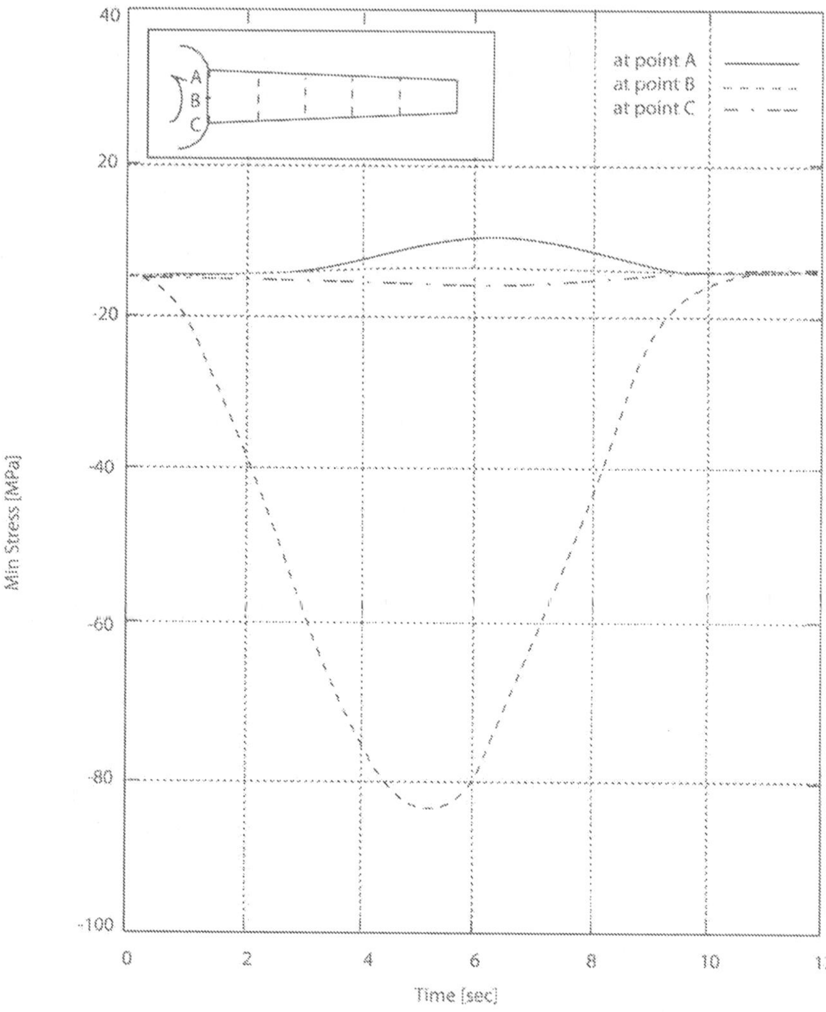

Fig. 12.7. Time history of the minimum stress σ_{min} at the base of a beam

$$\omega(t) = \left\{ \begin{array}{c} \dfrac{\omega_s}{T_s}\left(t - \dfrac{T_s}{2\pi}\sin\left(2\pi\, t\, T_s\right)\right) \text{ if } t < T_s \\ \\ \omega_s \text{ if } t \geq T_s \end{array} \right\} \tag{12.164}$$

If the spin-up time is 10 s and ω_s is the steady-state, the spin-up speed is equal to 30 rad/s. The inertial linear velocity is taken to be zero and body forces due to gravitation and the external traction are assumed to be zero.

The physical properties of the beam are Young's modulus $E = 7 \times 10^{10}$ Pa. Poisson's ratio $\upsilon = 0.26$, the density of the material $\rho = 2767\,\text{kg/m}^3$ the length of

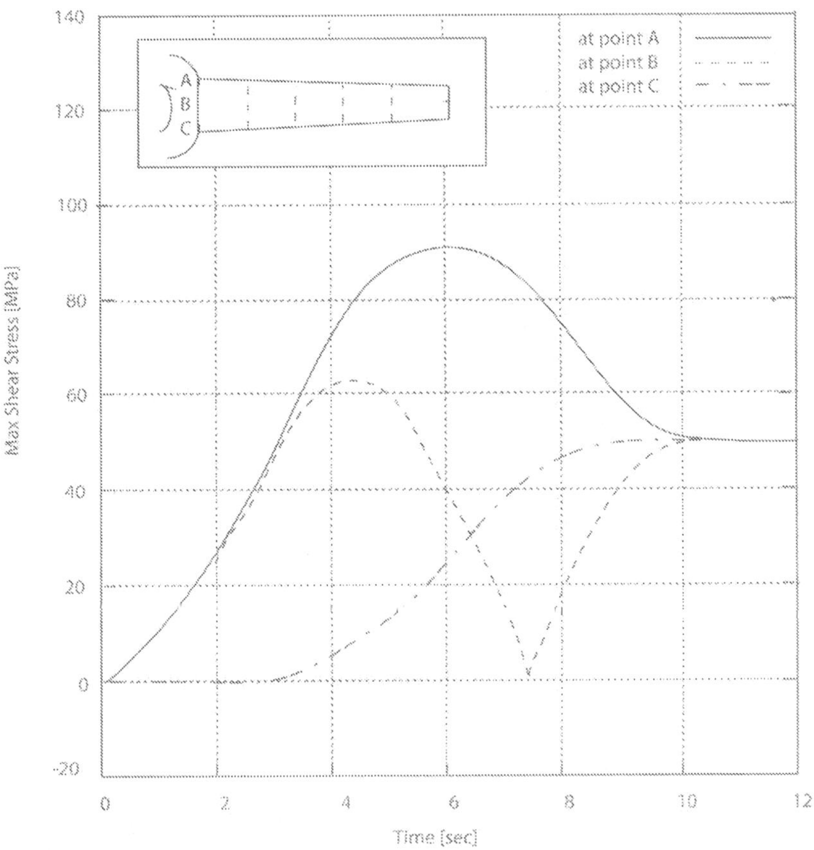

Fig. 12.8. Time history of the maximum shear stress $\tau_{\max}$ at the base of a beam

the beam $L = 10\,\mathrm{m}$, the root and the tip thickness of the beam $h_r = 0.15\,\mathrm{m}$ and $h_t = 0.1\,\mathrm{m}$, respectively. The beam is modeled using five substructures as shown in Figure 12.3. Each substructure is discretized to ten quadratic boundary elements.

Shown in Figure 12.4 is the transient behavior of the $\mathbf{b}_2$-direction tip deflection for a spin-up time of 10 s, where the effects of stress stiffening are investigated. In both cases, with (solid line) and without (dashed line) the inclusion of the stiffening effects, the responses are almost similar. This is due to the fact that the beam does not have enough flexibility features to show the effects of geometric stiffening to influence the tip deflection dynamics. The accuracy of the BEM for problems where the body has a large aspect ratio of 100/1, such as for thin beams, depends on the mesh and the substructure technique used. The latter can improve drastically the accuracy of the solutions in such problems.

Figure 12.5 shows the $\mathbf{b}_1$-direction deflection at three points of the tip cross-section of the beam. Figures 12.6, 12.7, 12.8 and 12.9 show, respectively, the behavior of the principal stresses and the maximum shear stress at three points of the base

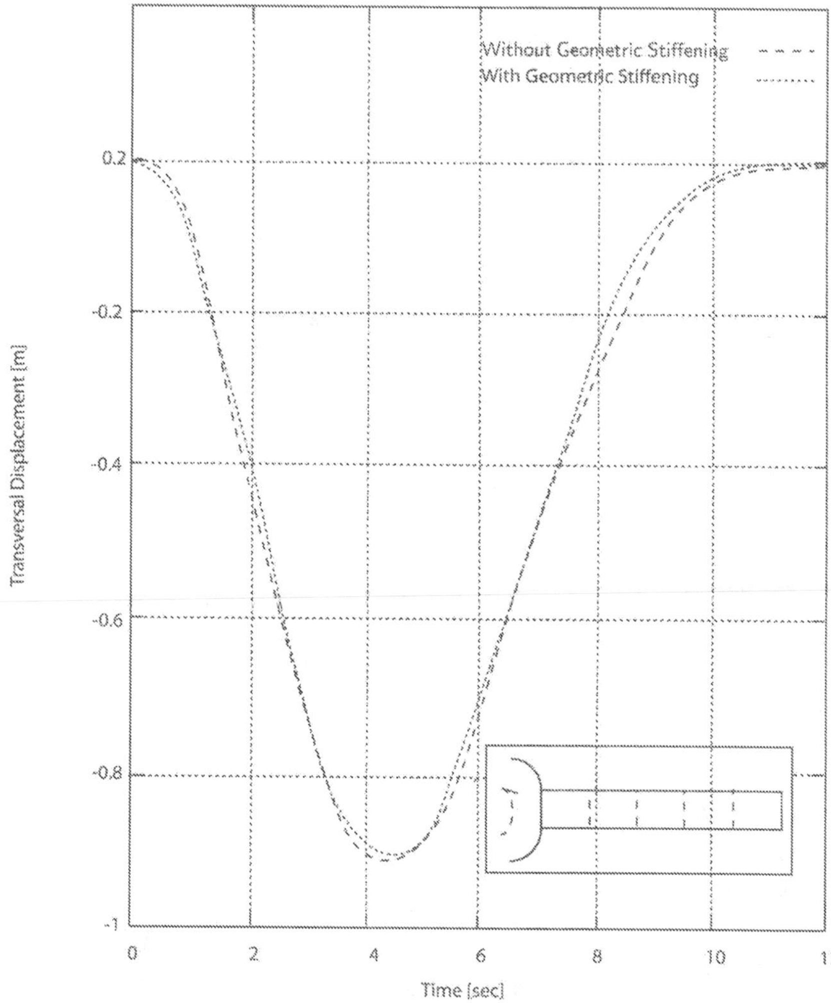

Fig. 12.9. Time history of the tip deflection u_2 when $\varpi_s = 20$ rad/s

cross-section. The maximum shear stress response at point C displays a cusp at time $t = 7.35$ s, which is due to the fact that point C becomes under principal stresses at that instant. These stresses are induced by the large overall motion. The steady-state responses were validated by the beam theory analysis.

A straight beam where the length was kept the same and its thickness changed to $h = 0.1$ m is subjected first to a spin-up of $\omega_s = 20$ rad/s (Figure 12.9) and then $\omega_s = 30$ rad/s (Figure 12.10). These figures demonstrate how the increase in the spin-up speed affects the dynamic response of the tip deflection where the curves are no longer superimposed. However, we see a clear distinction between the two cases; when the geometric stiffening is not present, the curve deepens, allowing a larger

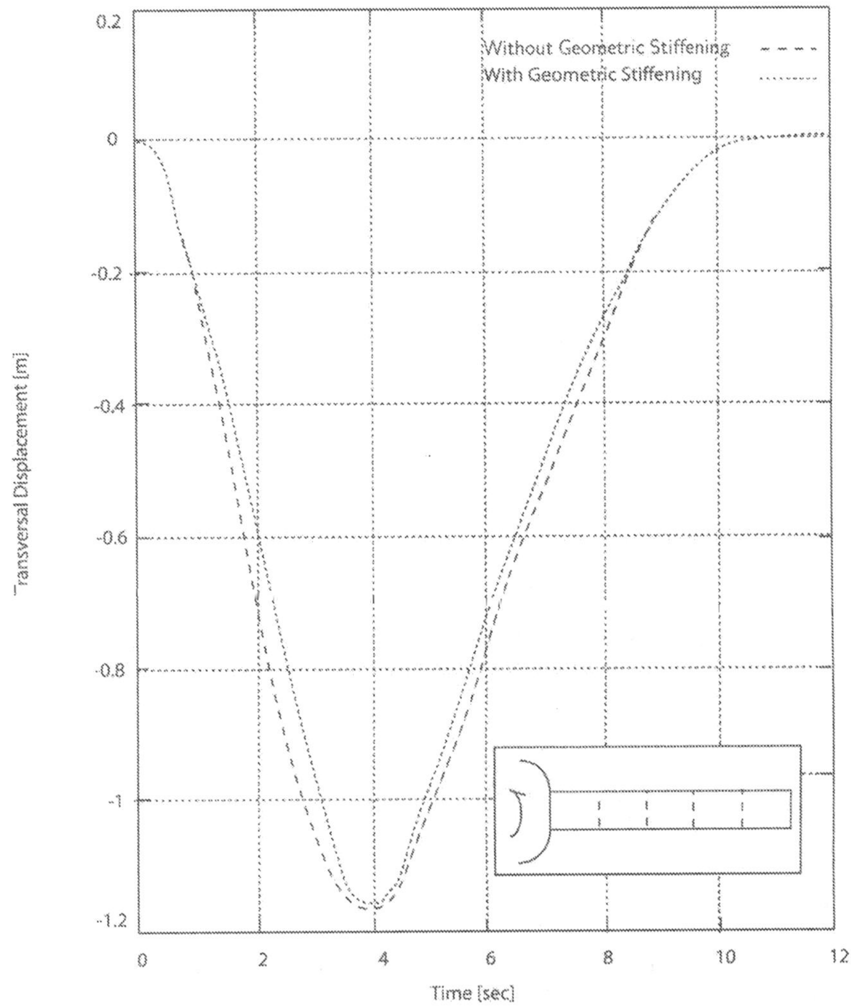

Fig. 12.10. Time history of the tip deflection u_2 when $\omega_s = 30$ rad/s

deflection. Therefore, when the stress stiffening is considered, the tip deflection shows less deformation indicating a stiffer structure.

No iterations were needed in the solution of the system of equations of motion, except for the case of Figure 12.11. During the numerical integration of the equations, a Newton–Raphson iterative scheme was used when the Euclidean norm of out-of-balance forces were greater than 10^{-6}. The norm for the convergence is the Euclidean norm of the incremental displacements which is less than 10^{-5} times the current norm of the displacement. The number of iterations required to meet the criterion was about 3–5 iterations for each time step starting from $t \approx 6$ s to $t \approx 10$ s.

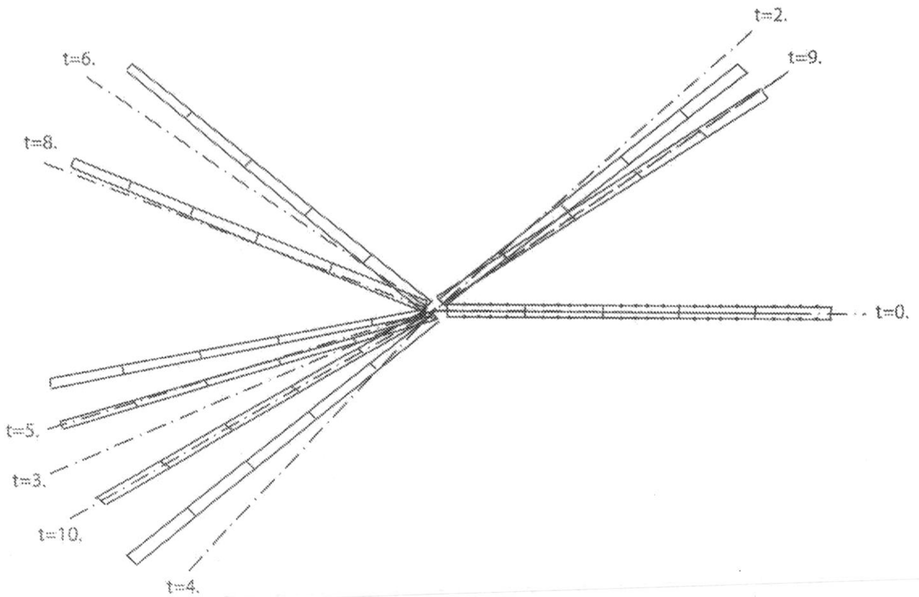

Fig. 12.11. Deformed shapes during the spin-up motion

To take advantage of the sparseness characteristic of matrices involved, routines for sparse LU decomposition and solution are used. All matrix–matrix and matrix–vector multiplication are performed taking into account the sparseness of the matrices.

12.14 Conclusion

A general formalism using the incremental boundary-element method is presented for the derivation of the general equations of 3D elastic body undergoing large overall base rotation. The method stems from use of the continuum principles and boundary integral equations. The updated Lagrangian formulation was found to be suitable for the derivation of the linearized boundary integral equations. The effect of stiffening due to the presence of reference stresses are used to balance the inertial forces from the gross base motion. The effects of these stresses are included in the tangent stiffness matrix. All the domain integrals occurring in the formulation are expressed as DRM integrals. For numerical simulation, a Houboult time integration scheme was used to establish a system of quasi-static equilibrium and a Newton–Raphson type iteration scheme was used to solve these equations. A substructure technique was utilized in the computer algorithm, in order to improve the accuracy for bodies with a large aspect ratio and increase the computational efficiency. One can conclude that the formulation presented is capable of capturing the phenomenon of dynamic stiffening within reasonable accuracy. The boundary-element method proves to be a reliable method of analysis in the dynamic simulation of flexible bodies.

References

1. Kerdjoudj, M., Amirouche F., Implementation of the Boundary-Element Method in the Dynamics of Flexible Bodies, *International Journal for Numerical Methods in Engineering*, Vol. **39**, 1996, pp. 321–354.
2. Mukherjee, G. and Morjaria, M., On the Efficiency and Accuracy of the Boundary-Element Method and the Finite-Element Method, *International Journal for Numerical Methods in Engineering*, Vol. **20**, 1984, pp. 515–522.
3. Radaj, D., Mohrmann, W. and Schilberth, G., Economy and Convergence of Notch Stress Analysis Using Boundary and the Finite-Element Method, *International Journal for Numerical Methods in Engineering*, Vol. **20**, 1984, pp. 565–572.
4. Rencis, J. J., Urekew, T. J., Jong, K., Kirk, R. and Frederico, P., A Posteriori Error Estimation for the Finite-Element and Boundary-Element Method, *Computer Structure*, Vol. **37**, No. 1, 1990, pp. 103–117.
5. Banerjee, P. K., Wilson, R. B. and Miller, N., Advanced Elastic and Inelastic Three-Dimensionl Analysis of Gas Turbine Engine Structure by the BEM, *International Journal for Numerical Methods in Engineering*, Vol. **26**, 1988, pp. 393–411.
6. Danson, D. J., A Boundary-Element Formulation of Problems in Linear Isotropic Elasticity with Body Forces, in: *Boundary-Element Methods*, (Brebbia, et al., eds.), Springer-Verlag, Berlin, 1981, pp. 105–122.
7. Ahmad, S. and Banerjee, P. K., Inelastic Transient Dynamic Analysis of Three-Dimensional Problems by the BEM, *International Journal for Numerical Methods in Engineering*, Vol. **29**, 1990, pp. 371–390.
8. Nardini, D. and Brebbia, C. A., Boundary Integral Formulation of Mass Matrices for Dynamic Analysis, in: *Topics in Boundary-Element Research*, Vol. 2, Springer-Verlag, New York, 1985, pp. 191–208.
9. Nardini, D. and Brebbia, C. A., Transient Dynamic Analysis by the Boundary-Element Method, in: *Boundary-Elements Computational Mechanics Publications*, Southampton, and Springer-Verlag, New York, 1983, pp. 719–730.
10. Brebbia, C. A. and Nardini, D., Dynamic Analysis in Solid Mechanics by the Alternative Boundary-Element Procedure, *International Journal of Soil Dynamics and Earthquake Engineering*, Vol. **2**, No. 4, 1 1983, pp. 228–233.
11. Nardini, D. and Brebbia, C. A., A New Approach to Free Vibration Analysis Using Boundary Elements, Computational Mechanics Publications, Southampton, and Springer-Verlag, New York, 1982, pp. 313–326.
12. Khun, G. and Möhrmann, W., Boundary-Element Method in Elastostatics: Theory and Applications, *Applied Mathematical Modelling*, Vol. **7**, No. 2, 1983, pp. 97–105.
13. Mattiasson, H., J., K., K., Runesson, K. and Samuelsson, A., On the Use of the Boundary-Element Method for Elastoplastic, Large Deformation Problems, *International Journal for Numerical Methods in Engineering*, Vol. **25**, 1988, pp. 165–176.
14. Okada, H., Rajiyah, H. and Atluri, S., A Full Tangent Field Boundary Element for Geometric and Material Nonlinear Problems of Solid Mechanics, *International Journal for Numerical Methods in Engineering*, Vol. **29**, 1990, pp. 15–35.

15. Bathe, K. J., Finite-Element Procedures in Engineering Analysis, Prentice-Hall, Englewood Cliffs, NJ, 1982.
16. Benerjee, P. K. and Butterfield, R., Boundary-Element Methods in Engineering, McGraw-Hill, London, 1981.
17. Brebbia, C. A. and Dominguez, J. *Boundary Element: An Introductory Course*, Computational Mechanics Publications, Southampton and McGraw-Hill, New York, 1989.
18. Partridge. P. W., Brebbia, C. A. and Wrobel, L. C., *The Dual Reciprocity Boundary-Element Method*, Computational Mechanics Publications, Southampton, Boston, 1992.
19. Ali A., Rajakumar, C. and Yunus, S. M., On the Formulation of the Acoustic Boundary-Element Eigenvalue Problems, *International Journal for Numerical Methods in Engineering*, Vol. **31**, 1991, pp. 1271–1282.
20. Kleiber, M., *Incremental Finite-Element Modeling in Nonlinear Solid Mechanics*, Wiley, New York, 1989.
21. Washizu, K., *Variation Methods in Elasticity and Plasticity*, 2nd ed., Pergamon Press, 1975.
22. Malvern, L. E., *Introduction to the Mechanics of a Continuous Medium*, Prentice-Hall, Englewood Cliffs, NJ, 1969.
23. Ahmed, S. and Banerjee, P. K., Free Vibration Analysis by BEM Using Particular Integrals, *J. Engng. Mech.*, ASCE 112, 1986, pp. 682–695.
24. Beskos, D. E., *Boundary-Element Methods in Mechanics: Computational Methods in Mechanics*, First Series, Vol. **3**, Elsevier, 1987.
25. Leipholz, H., H., E., On Galerkin's Method Interpreted as a Generalized Integral Transformation, *Computer Methods in Applied Mechanics and Engineering*, 65, 1987, pp. 177–189.
26. Guiggiani, M. and Casalini, P., Direct Computation of Cauchy Principal Value Integrals an Advanced Boundary Elements, *International Journal for Numerical Methods in Engineering*, Vol. **24**, 1987, pp. 1711–1720.
27. Cerrolaza, M. and Alarcon, E., A Bi-Cubic Transformation for the Numerical Evaluation for the Cauchy Principal Value Integrals in Boundary Methods, *International Journal for Numerical Methods in Engineering*, Vol. **29**, 1990, pp. 371–390.
28. Kane, J. H. and Saigal, S., Numerical Integration and Sparse Blocked Equation Solution Techniques for Large-Scale Boundary-Element Analysis, in *Solution of Superlarge Problems in Computational Mechanics* (Kane, J. H., Carlson, A. D., and Cox, D. L., eds.), Plenum Press, New York, 1989.

Appendix A

Multibody Dynamics Flowchart for the Construction of the Equations of Motion with Constraints

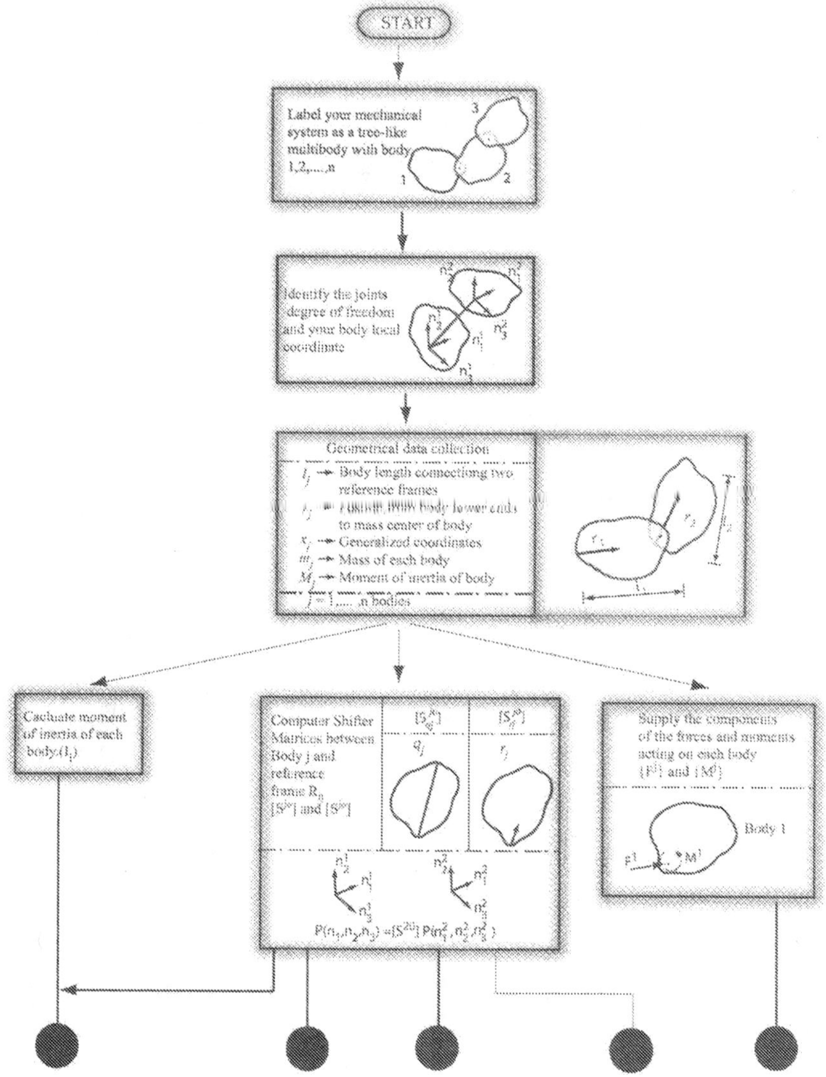
START
Label your mechanical system as a tree-like multibody with body 1,2,....,n
Identify the joints degree of freedom and your body local coordinate
Geometrical data collection
Body length connectiong two reference frames
to mass center of body
Generalized coordinates
Mass of each body
Moment of inertia of body
n bodies
Caclulate moment of inertia of each body
Computer Shifter Matrices between Body j and reference
Supply the components of the forces and moments acting on each body
Body 1

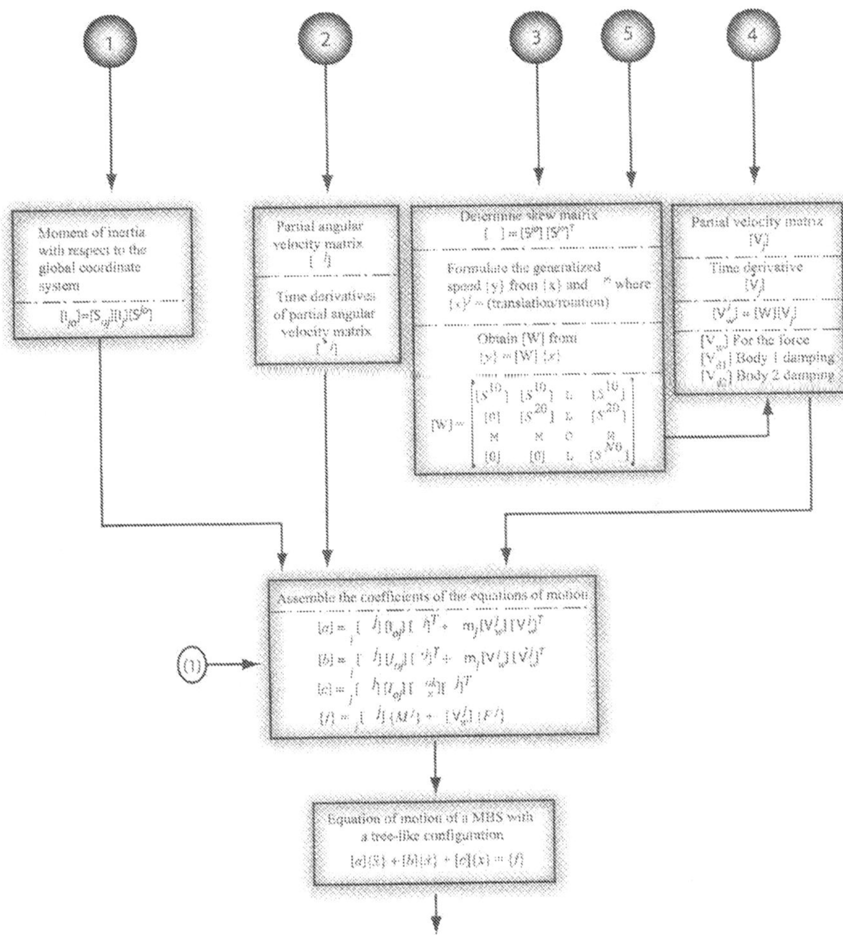
1
2
3
5
4
Moment of inertia with respect to the global coordinate system
Partial angular velocity matrix
Time derivatives of partial angular velocity matrix
Determine skew matrix
Formulate the generalized speed {y} from {x} and
where
(translation/rotation)
Obtain [W] from
{y} = [W] {x}
Partial velocity matrix
Time derivative
For the force
Body 1 damping
Body 2 damping
Assemble the coefficients of the equations of motion
(1)
Equation of motion of a MBS with a tree-like configuration

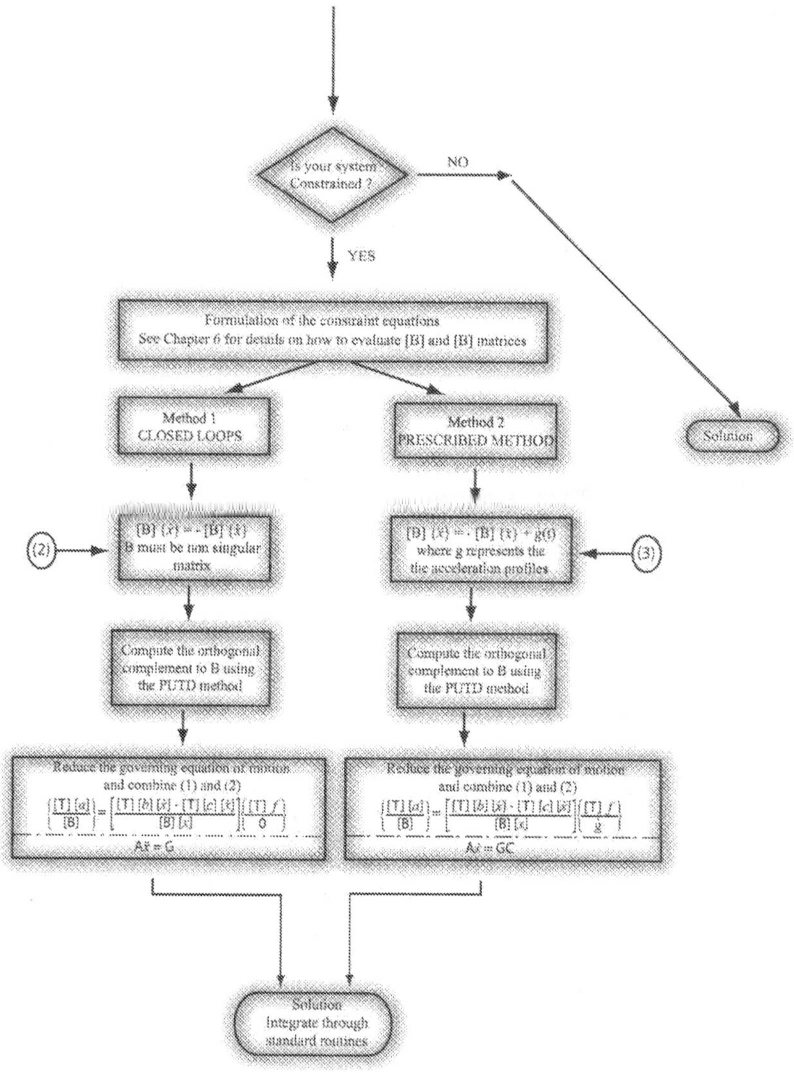
Is your system Constrained ?
NO
YES
Formulation of the constraint equations
See Chapter 6 for details on how to evaluate [B] and [Ḃ] matrices
Method 1
CLOSED LOOPS
Method 2
PRESCRIBED METHOD
Solution
(2)
B must be non singular matrix
where g represents the the acceleration profiles
(3)
Compute the orthogonal complement to B using the PUTD method
Compute the orthogonal complement to B using the PUTD method
Reduce the governing equation of motion and combine (1) and (2)
Reduce the governing equation of motion and combine (1) and (2)
Solution
Integrate through standard routines

Appendix B

Centroid Location and Area Moment of Inertia

Centroid Location	Centroid Location	Area Moment of Inertia
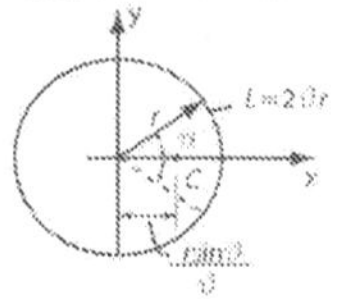 Circular arc segment	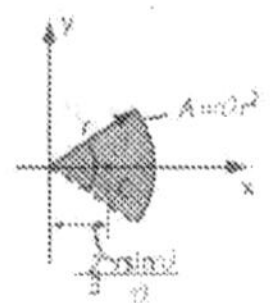	$I_x = \frac{1}{4}r^4(\theta - \frac{1}{2}\sin 2\theta)$ $I_y = \frac{1}{4}r^4(\theta + \frac{1}{2}\sin 2\theta)$
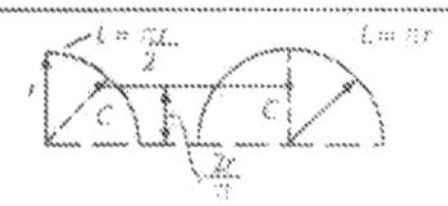 Quarter and semicircular arcs	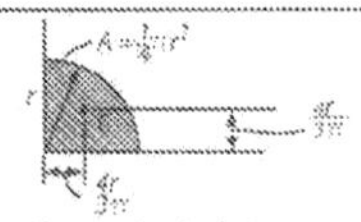 Quarter circular area	$I_x = \frac{1}{16}\pi r^4$ $I_y = \frac{1}{16}\pi r^4$
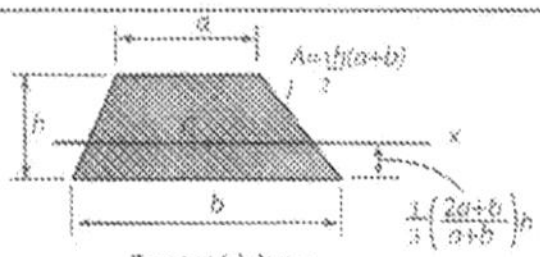 Trapezoidal area	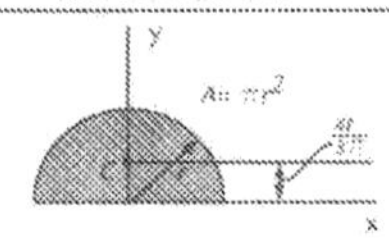 Semicircular area	$I_x = \frac{1}{8}\pi r^4$ $I_y = \frac{1}{8}\pi r^4$
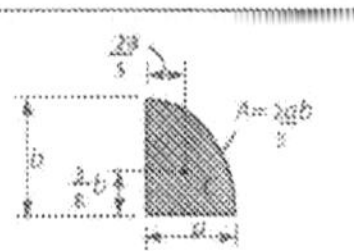 Semiparabolic area	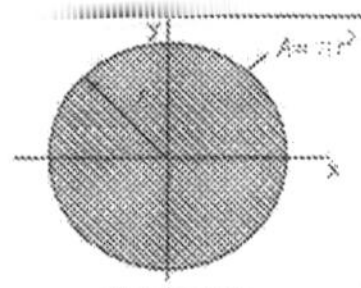 Circular area	$I_x = \frac{1}{4}\pi r^4$ $I_y = \frac{1}{4}\pi r^4$
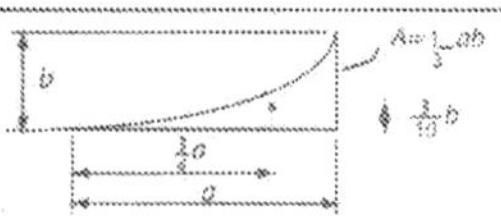 Exparabolic area	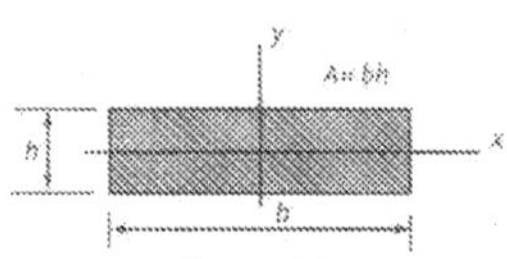 Rectangular area	$I_x = \frac{1}{12}bh^3$ $I_y = \frac{1}{12}hb^3$
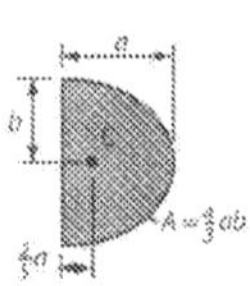 Parabolic area	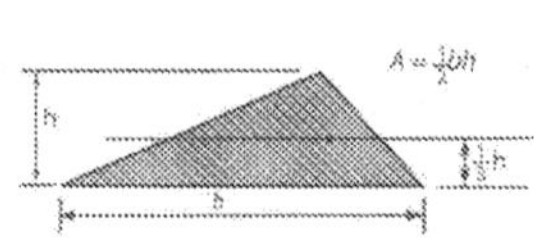 Triangular area	$I_x = \frac{1}{36}bh^3$

Appendix C

Center of Gravity and Mass Moment of Inertia of Homogeneous Solids

Center of Gravity and Mass Moment of Inertia of Homogenous Solids

Sphere

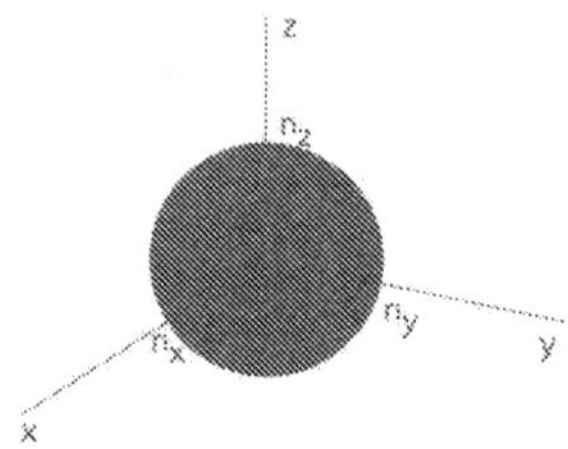

$I_{xx} = I_{yy} = I_{zz} = \frac{2}{5} mr^2$

$V = \frac{4}{3} \pi r^3$

Hemisphere

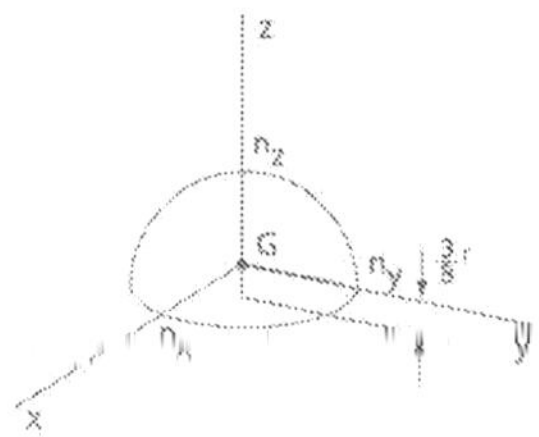

$I_{xx} = I_{yy} = \frac{83}{320} mr^2$

$I_{zz} = \frac{2}{5} mr^2$

$V = \frac{2}{3} \pi r^3$

Thin Circular Disk

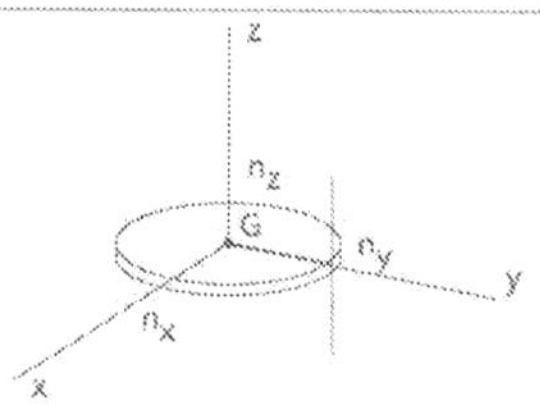

$I_{xx} = I_{yy} = \frac{1}{4} mr^2$

$I_{zz} = \frac{1}{2} mr^2 \quad I_{z'z'} = \frac{3}{2} mr^2$

Thin Ring

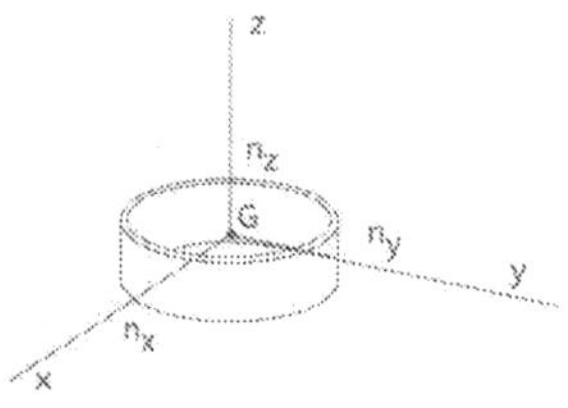

$I_{xx} = I_{yy} = \frac{1}{2} mr^2 \quad I_{zz} = mr^2$

Cylinder

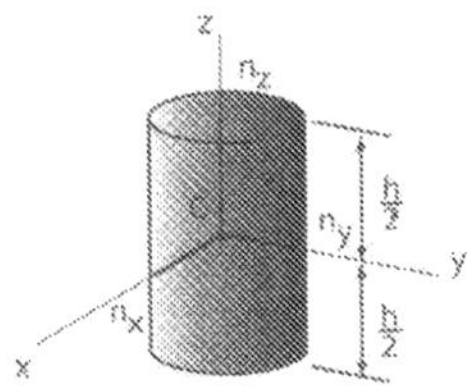

$I_{xx}=I_{yy}=\frac{1}{12}m(3r^2+h^2)$ $I_{zz}=\frac{1}{2}mr^2$

$V=\pi r^2 h$

Cone

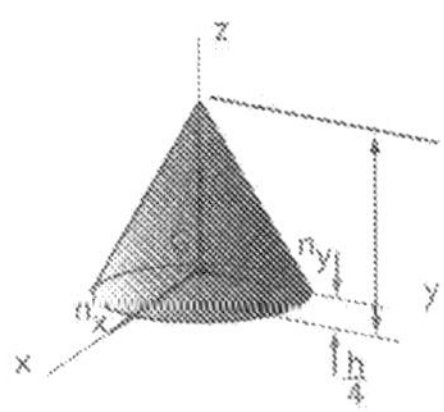

$I_{xx}=I_{yy}=\frac{3}{80}m(4r^2+h^2)$ $I_{zz}=\frac{3}{10}mr^2$

$V=\frac{1}{3}\pi r^2 h$

Thin Plate

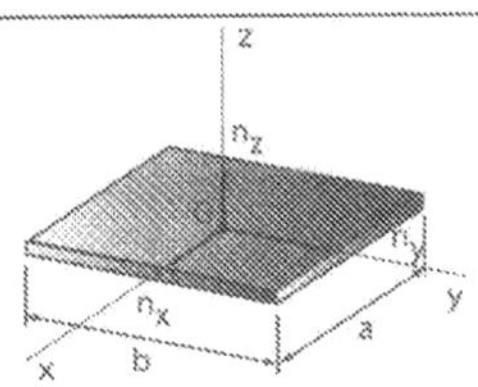

$I_{xx}=\frac{1}{12}mb^2$ $I_{yy}=\frac{1}{12}ma^2$ $I_{zz}=\frac{1}{12}m(a^2+b^2)$

Slender Rod

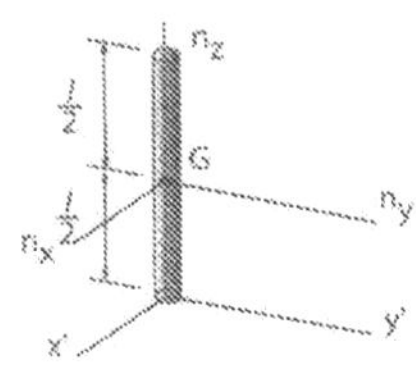

$I_{xx}=I_{yy}=\frac{1}{12}ml^2$ $I_{x'x'}=I_{y'y'}=\frac{1}{3}ml^2$ $I_{zz}=0$

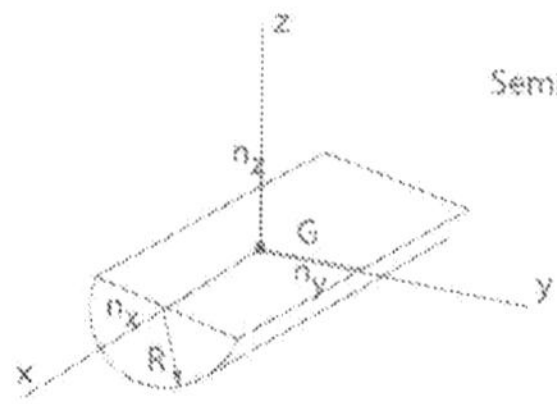

Semicylinder

$I_{xx} = \frac{9\pi^2 - 64}{36\pi^2} mR^2 + \frac{1}{12} mL^2$ $I_{zz} = \frac{9\pi^2 - 32}{18\pi^2} mR^2$

$I_{yy} = \frac{1}{4} mR^2 + \frac{1}{12} mL^2$

$V = \frac{1}{2}\pi R^2 L$

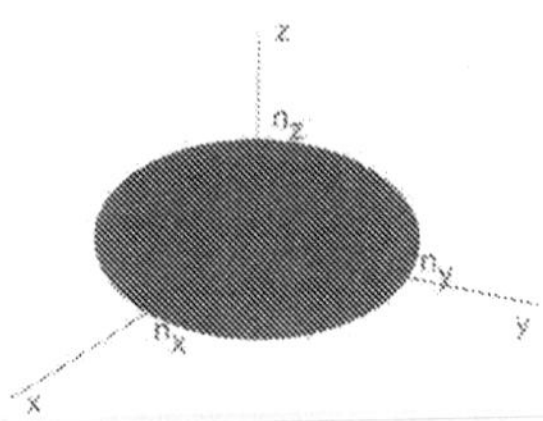

Ellipsoid

$I_{xx} = \frac{1}{5} m(b^2 + c^2)$ $I_{zz} = \frac{1}{5} m(b^2 + a^2)$

$I_{yy} = \frac{1}{5} m(a^2 + c^2)$

$V = \frac{4}{3}\pi abc$

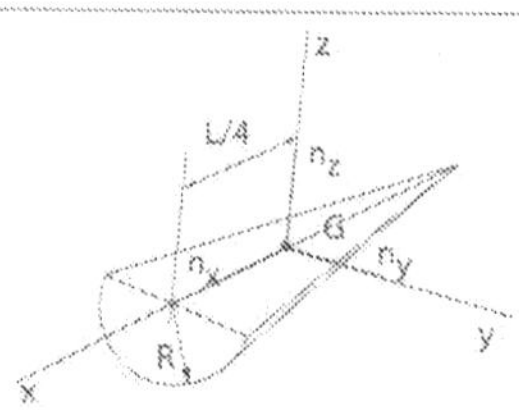

Half Cone

$I_{xx} = \frac{3}{20} mR^2 + \frac{3}{80} mL^2$ $I_{zz} = (\frac{3}{20} - \frac{1}{\pi^2}) mR^2 = \frac{3}{80} mL^2$

$I_{yy} = (\frac{3}{10} - \frac{1}{\pi^2}) mR^2$

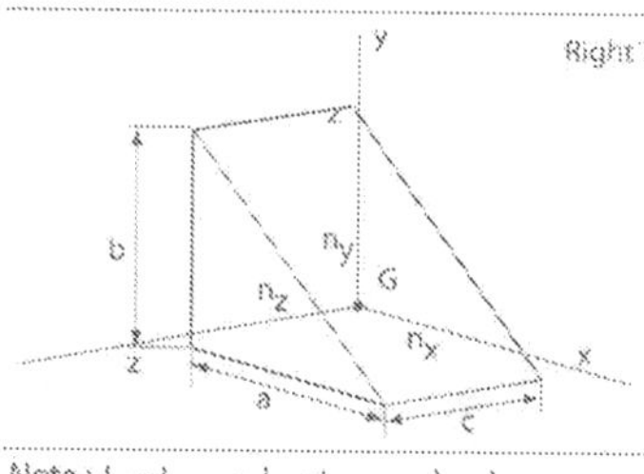

Right Triangular Prism

$I_{xx} = \frac{1}{18} mb^2 + \frac{1}{12} mc^2$ $I_{zz} = \frac{1}{18} m(b^2 + c^2)$

$I_{yy} = \frac{1}{18} ma^2 + \frac{1}{12} mc^2$

$V = \frac{1}{2} abc$

Note : $I_{xx} = I_{n_x n_x}$; $I_{yy} = I_{n_y n_y}$; $I_{zz} = I_{n_z n_z}$

Appendix D

Symbols Description

Symbols	Description
${}^R\bar{n}^r$	Unit vector of axis frame 'r' with respect to the fixed inertial frame 'R'
S_{ij}^{rs}	Direction cosine or transformation matrix between unit vectors in body s and r
$\{\bar{n}^r\}$	Matrix representation of the vector arrays
$[S^{rs}] = \{\bar{n}^r\}\{\bar{n}^s\}^T$	Matrix representation of the direction cosine of the transformation from r to s
δ_{ij}	Kronecker delta
$S_{ij}^{rs} = \begin{bmatrix} \bar{n}_1^s \cdot \bar{n}_1^r & \bar{n}_2^s \cdot \bar{n}_1^r & \bar{n}_3^s \cdot \bar{n}_1^r \\ \bar{n}_1^s \cdot \bar{n}_2^r & \bar{n}_2^s \cdot \bar{n}_2^r & \bar{n}_3^s \cdot \bar{n}_2^r \\ \bar{n}_1^s \cdot \bar{n}_3^r & \bar{n}_2^s \cdot \bar{n}_3^r & \bar{n}_3^s \cdot \bar{n}_3^r \end{bmatrix}$	Direction cosine matrix using unit vectors defining frames r and s
$\bar{n}_i^s$	Unit vector of the frame S
V_i^s	Represents the coordinates (components) of $\bar{v}$ in $\bar{n}^s$
$\bar{\bar{d}} = d_{ij}\bar{n}_i\bar{n}_j$	A second-order tensor (called dyad) (also dyadic)
$\bar{\bar{d}} = \bar{a}\bar{b}$	Tensor product of any two vectors
d_{ij}	The component matrix of $\bar{\bar{d}}$
T_N	Nth-order tensor
$T_{i_1,\ldots,i_n}^r$	Components of $\bar{T}$ in $\bar{n}^r$ and $\bar{n}^s$
e_{ijt}	Permutation symbol when $i = \lvert 3, j = \lvert 3, t = \lvert 3$
$\Gamma_0(k)$	Defines the body array and is composed basically of all the body numbers
$\Gamma_1(k)$	Defines the lower body array, i.e., the lower adjacent connecting body to each of the bodies described by $\Gamma_0(k)$
$[a]^{sk}$	Denotes the skew matrix of tensor $\bar{a}$, e.g., $[a]^{sk} = \begin{bmatrix} 0 & -a_3 & a_2 \\ a_3 & 0 & -a_1 \\ -a_2 & a_1 & 0 \end{bmatrix}$
${}^R\underline{\omega}^n$	Angular velocity of body n, w.r.t., R, the reference frame
$H(k)$	The degree of connectivity for each body K
$[X_\ell]^T$	Generalized coordinates for the entire system
S_{ij}^{KO}	Shifter matrix (also called as transformation matrix) of body k with respect to the origin 'O'

Symbols	**Description**
B_k	Any body; $k = 0, 1, 2, \ldots k$
$[\Omega^{KJ}]$	Skew matrix from K to reference frame J; $[\Omega^{KJ}] = [\dot{S}^{KJ}]\,[S^{KJ}]^T$
$\bar{F}_n$	External forces acting on a rigid body; $n = 1, 3$
$M_{pp'}$	Moment of a force about a given line PP'
$\bar{\lambda}$	Unit vector acting along a line PP'
$\bar{r}$	Position vector
y_{kn}	The components of the angular velocity of body k and R
$[\omega^n]$	Partial angular velocity of body n in matrix form
${}^m\bar{\omega}^n$	Relative angular velocity of body n with respect to body m
$\dot{\omega}^k$	Rate of change of ω^k
$[w]$	The mapping function, (Gimball matrix) e.g., $[w] = \begin{bmatrix} [S^{10}] & [S^{10}] & [S^{10}] & \ldots & [S^{10}] \\ [0_{3\times3}] & [S^{20}] & [S^{20}] & \ldots & [S^{20}] \\ [0_{3\times3}] & [0_{3\times3}] & [S^{30}] & \ldots & [S^{30}] \\ & & & \ddots & \vdots \\ [0_{3\times3}] & [0_{3\times3}] & [0_{3\times3}] & \cdots & [S^{n0}] \end{bmatrix}$
$\bar{R}$	Resultant of forces acting on a rigid body; i.e., $\bar{R} = R_x\bar{n}_x + R_y\bar{n}_y + R_z\bar{n}_z$
$\bar{\mu}$	Displacement
$\bar{f}_n$	Force acting on a body 'n', expressed as $\bar{f} = f(\mu, \dot{\mu}\bar{\delta})$
$\bar{v}^{P1}$	Velocity of point P_1
$\bar{\eta}$	Unit vector directed along a particular axis
$\bar{T}j$	External torque exerted on body j
$\bar{F}_{1c}$	The force due to the damper on body '1'
C	Damping coefficient
$\{f_c\}$	Generalized active force due to the dampers
$\{r_n\}$	Mass center coordinates of body 'n' w.r.t., R
$\bar{T}^*$	The inertia torque, defined as $\bar{T}^* = \sum_{i=1}^{n} m_i\bar{r}_i \times \bar{a}_i$
$\bar{F}^*$	The inertia force

Symbols	Description
$\bar{I}_a$	The second moment of a set S (comprised of N particles) relative to a fixed point O for a unit vector $\bar{n}_a$, defined as $\bar{I}_a = \sum_{i=1}^{n} m_i \bar{r}_i \times (\bar{n}_a \times \bar{r}_i)$
${}^{O}\bar{I}_a^s$	Second moment of 's' relative to O, the origin
δW	Virtual work
δr	Virtual displacement
G_k	Mass center of a body k
$\bar{F}_k^c$	Constraint forces applied to body K and passing through G_k
$\bar{M}_\ell^*$	Generalized inertial moments ($l = 1, 6N$), where N is the number of bodies
$\bar{\alpha}^k$	The angular acceleration of body k in R
$\bar{\alpha}_a^k$	Scalar components of the angular acceleration of $\bar{\alpha}^k$ in R along the unit vector $\bar{n}_a$
R^n	n dimensional real space
B_{il}	Constraint Jacobian matrix
$\left\Vert B_1^{(1)} \right\Vert$	Denotes the norm of vector $\bar{B}_1^{(1)}$.
$H^{(n)}$.	The Householder transformation matrix
λ_i	i^{th} eigenvalue element
$\hat{T}$	Transformation matrix
$\bar{d}_i$	Orthonormal set of vectors
ϕ_i	Eigenfunctions or eigenvectors
V	Potential energy of the system
T	Kinetic energy of the system
$\{u\}$	Nodal displacement array
$\{\rho\}$	Nodal coordinate array
$[N]$	Shape function
$[\Psi]$	Shape function
ξ_j	The position vector of the center of the mass
$\{x\}.[Sq_j], [S\xi_j]$	Skew matrices associated with the body vector q_j
$[I_{ji}]$	The moment density of the inertia of element i of body j in R
$[C_s]$	The ordinary damping matrix of the bodies

Symbols	Description
$[K_s]$	The ordinary stiffness matrix
$[K_G]$	The geometrical stiffness contribution
$\bar{f}_{ji}$	Denotes external surface traction
$\bar{b}_{ji}$	External body force acting on element I in body j
$\bar{M}_{ji}$	The external surface distributed moment acting on element I in body j
$\bar{P}_A$	A concentrated force acting on a point A
$[K_G]$	Geometrical stiffness contribution
f_e	Generalized external force array
f_s	Generalized external stiffness force array
E	Young's modulus of elasticity
E_{ji}	Strain components
U	Strain energy
σ	Vector of stress components
G	Shear modulus
K_n	Shear correction factor; $n = 1, 2, \ldots, n$
$[\emptyset]$	Shape function
Z^{ki}	Geometric stiffness matrix for the ith element of body k
Δ_m	Deflection in θ direction
e_{ijk}	Defines the Levi–Civita permutation symbol $(i, j, k = 1, 2, 3 \ldots)$
σ	Cauchy stress
Δe	The infinitesimal strain dyadics
${}^{t+\Delta t}\sigma_{ij}$	Components of the Cauchy stress tensor
${}^{t+\Delta t}_{t+\Delta t}p_i$	Define the components of the traction
${}^{t+\Delta t}_{t+\Delta t}\gamma_i$	Define the components of body forces
${}^{t+\Delta t}_{t}\varepsilon_{ij}$	The Green–Lagrange strain tensor
${}^{t+\Delta t}_{t}S_{ij}$	The second Piola–Kirchhoff stress tensor components
E	Matrix of elastic coefficients
α_r	Arbitrary coefficients
w_i	Kinematically admissible weighting function

Symbols	Description
${}^t\sigma^*_{ij,j}$	Components of the stress defined over the infinite domain ${}^t\Omega_\infty$
${}^t\varepsilon_{ij}{}^*$	The components of the strain defined over the infinite domain ${}^t\Omega_\infty$
${}^tu^*_{i,j}$	The components of the displacement defined over the infinite domain ${}^t\Omega_\infty$
${}^tp^*_i$	The components of the traction corresponding to the fundamental solution
${}^t\mathbf{x}$	Field point
${}^t\chi$	Unit load acting at the source point
$\Delta\mathbf{p}$	Pseudo-traction vector
${}_t\Delta\mathbf{p}_i$	Components of the incremental pseudo-traction vector $\Delta\mathbf{p}$
${}^t\Gamma$	Boundary of the reference configuration
${}^tc^{(}_{ki}{}^{t}\chi)$	Depends on the geometry at the neighborhood of the boundary point ${}^t\chi$
I_{23}	Denotes the term involving the domain integrals due to the body forces expressed over the space of the inference configuration
$f^\mu({}^t\mathbf{x})$	Spatial functions, $(\mu = 1, \ldots, v_d)$
V_d	Number of drm points
${}^{t+\Delta t}_{t}G$	Coriolis matrix
${}^{t+\Delta t}_{t}K_d$	Stiffness matrix related to overall rotation
${}^{t+\Delta t}\lambda_j$	The undetermined Lagrange multipliers representing the constraint forces
$\{\Delta F\}$	Incremental forcing function
${}^tu^\mu$	Tangent stiffness matrix

Appendix E

Units and Conversion

LENGTH	**VOLUME**
1 in = 2.54 c	1 imp. gal = 4.546090 litres = 4 quarts
1 ft = 12 in = 30.48 cm	1 imp. quart = 40 fluid ounces
1 yd = 3 ft	1 bushel = 8 gallons
1 mi = 5280 ft	1 U. S. gal (liq) = 3.785306 litres
1 nautical mile = 6080 ft	1 Hogshead = 238.47427 litres
1 mi = 1.609344 km	1 Noggin = 142.0652 cm
VELOCITY	**FORCE**
1 ft/sec = 3.048×10^{-1} m/sec	1 Newton = 0.22480894 lb
1 knot (nautical mile/hr) = 5.1444×10^{-1} m/sec	1 Newton = 10^5 dynes (g cm/s^2)
1 mile/hr(mi/hr) = 0.44704 m/sec	1 lb = 4.448 N
1 rad/sec = 9.5493 rpm	1 kip = 1000 lb = 4448 N
	1 ton = 2000 lb = 8896 N
ACCELERATION	
1 ft/sec^2 = 0.3048 m/sec^2	**POWER**
1 in/sec^2 = 2.54×10^{-2} m/sec^2	1 horse power = 745.700 W
	1 horse power = 550 ft lb/s
ANGLE	1 foot-pound/min = 0.022597 W
1 deg = 0.0174 radian	1 Nm/sec = 8.7873 in-lb/sec
1 radian = 57.325 deg	
	MOMENT OF FORCE
WEIGHT/MASS	1 pound-foot (lb-ft) = 1.3558 Newton-meter (N.m)
1 lb = 16 oz	1 pound-inch (lb-in) = 0.11298 Newton-meter (N.m)
1 lb = 453.59237 g	
1 metric ton = 1000 kg	**MOMENT OF INERTIA**
1 carat = 200 mg	1 $inch^4$ = 41.623×10^{-8} $meter^4$ (m^4)
1 slug = 14.5939 kg	
	PRESSURE, STRESS
AREA	1 atmosphere (std) = 1.0133 N/m^2 (Pascal)
1 mi^2 = 640 acre	1 $pound/foot^2$(lb/ft^2) = 47.880 N/m^2 (Pascal)
1 hec = 10000 m^2	1 $pound/inch^2$(lb/ft^2) = 47.880 N/m^2 (Pascal)
1 $foot^2$(ft^2) = 9.2903×10^{-2} m^2	
	SPRING CONSTANT
MOMENT& ENERGY	1 pound/inch (lb/in) = 175.13 N/m
1 Btu = 1055.6 J (N m = kg m^2/s^2)	1 pound/foot (lb/ft) = 0.08333 lb/in.
1 calorie = 4.1868 J	1 pound/foot (lb/ft) = 14.594 N/m
1 foot-pound force(ft-lbf) = 1.3558 J	
1 kilowatt-hour (kw-h) = 3.6×10^6 J	
1 in.-lb = 0.1138 N-m	
1 ft-lb = 12 in-lb	

Alphabetic list of Greek symbols

A	α	alpha	I	ι	iota	P	ρ	rho
B	β	beta	K	κ	kappa	Σ	σ	sigma
Γ	γ	gamma	Λ	λ	lambda	T	τ	tau
Δ	δ	delta	M	μ	mu	Υ	υ	upsilon
E	ε	epsilon	N	ν	nu	Φ	ϕ	phi
Z	ζ	zeta	Ξ	ξ	xi	X	χ	chi
H	η	eta	O	o	omicron	Ψ	ψ	psi
Θ	θ	theta	Π	π	pi	Ω	ω	omega

Prefixes for decimal multiples and submultiples of SI units.

Symbol	Prefix	Factor
T	tera	10^{12}
G	giga	10^{9}
M	mega	1,000,000
k	kilo	1,000
h	hector	100
d	deci	0.1
c	centi	0.01
m	milli	0.001
m	micro	$1. \times 10^{-6}$
n	nano	$1. \times 10^{-9}$
p	pico	$1. \times 10^{-12}$
f	femto	$1. \times 10^{-15}$
a	atto	$1. \times 10^{-18}$

Physical Properies of Common Metals

Metal	Modulus of Elasticity, E		Modulus of Rigidity, G		Poisson's Ratio, v	Unit Wieght,w	Density, ρ	Coefficient of Thermal Expansion,α		Thermal Conductivity		Specific Heat	
	Mpsi	Gpa	Mpsi	Gpa		(lb/in.^{3})	(mg/m^3)	$10^{-6}/^{\circ}F$	$10^{-6}/^{\circ}C$	Btu/h-ft-$^{\circ}$F	W/m-$^{\circ}$C	Btu/lbm-$^{\circ}$F	J/kg-$^{\circ}$C
Aluminium alloy	10.4[a]	72	3.9	27	0.32	0.1	2.8	12.0	22	100	173	0.22	920
Beryl. Copper	18.5	127	7.2	50	0.29	0.3	8.3	9.3	17	85	147	0.10	420
Brass, Bronze	16.0	110	6.0	41	0.33	0.31	8.7	10.5	19	45	78	0.10	420
Copper	17.5	121	6.6	46	0.33	0.32	8.9	9.4	17	220	381	0.10	420
Iron, gray cast	15.0	103	6.0	41	0.26	0.26	7.2	6.4	12	29	50	0.13	540
Magnesium alloy	6.5	45	2.4	17	0.35	0.065	1.8	14.5	26	55	95	0.28	1170
Nickel alloy	30.0	207	11.5	79	0.30	0.3	8.3	7.0	13	12	21	0.12	500
Steel carbon	30.0	207	11.5	79	0.30	0.28	7.7	6.7	12	27	47	0.11	460
Steel, alloy	30.0	207	11.5	79	0.30	0.28	7.7	6.3	11	22	38	0.11	460
Steel, stainless	27.5	190	10.6	73	0.30	0.28	7.7	8.0	14	12	21	0.11	460
Titanium alloy	16.5	114	6.2	43	0.33	0.16	4.4	4.9	9	7	12	0.12	500
Zinc alloy	12.0	83	4.5	31	0.33	0.24	6.6	15.0	27	64	111	0.11	460

[a]Values given are representative. Exact values may vary with composition and processing,sometimes greatly

References

1. Ambrosio J, Pereira M and Dias J, Distributed and discrete nonlinear deformations on multibody dynamics, *Nonlinear Dynamics*, Vol. 10, 1996, pp. 359–379.
2. Ambrosio J, Pereira M and Silva F, Crashworthiness of transportation systems: Structural impact and occupant protection, *NATO-ASI Series E*, Vol. 332, Kluwer Academic Publishers, Dordrecht, 1997.
3. Amirouche F and Huston R L, Collaborative techniques in modal analysis, *Journal of Guidance, Control and Dynamics (AIAA)*, Vol. 8, No. 6, 1985, pp. 783–786.
4. Amirouche F and Huston R L, Dynamics of constrained tree-like structures, *Journal of Dynamic Systems, Measurement and Control (ASME)*, Vol. 110, No. 1, 1988, pp. 78–83.
5. Amirouche F and Ider S K, A look at the constraint violations in constrained multibody dynamics, *Journal of Dynamic Systems, Measurement and Control (ASME)*, Vol. 111, No. 2, 1989, pp. 238–243.
6. Amirouche F and Ider S K, A recursive formulation of the equations of motion for articulated structures with closed loops – An automated approach, *Computers & Structures*, Vol. 30, No. 5, 1988, pp. 1135–1145.
7. Amirouche F and Ider S K, Coordinate reduction in the dynamics of constrained multibody systems – A new approach, *Journal of Applied Mechanics (ASME)*, Vol. 55, No. 2, 1988, pp. 889–904.
8. Amirouche F and Ider S K, Influence of geometric nonlinearities in the dynamics of flexible tree-like structures, *Journal of Guidance, Control and Dynamics (AIAA)*, Vol. 12, No. 6, 1989, pp. 830–837.
9. Amirouche F and Ider S K, Numerical stability of the constraints near singular positions in the dynamics of multibody systems, *Computers & Structures*, Vol. 33, No. 1, 1989, pp. 129–138.
10. Amirouche F and Ider S K, Simulation and analysis of a biodynamic human model subjected to low accelerations – A correlation study, *Journal of Sound and Vibration*, Vol. 123, No. 2, 1988, pp. 281–292.
11. Amirouche F and Khang J, Impact force analysis in mechanical hand design, Part I, *International Journal of Robotics and Automation*, Vol. 3, No. 3, 1988, pp. 158–164.
12. Amirouche F and Shareef N H, Gain in computational efficiency by vectorization of general purpose code for multibody dynamic simulations, *Journal of Computers and Structures*, Vol. 41, No. 2, 1991, pp. 292–302.

13. Amirouche F and Tongyi J, A pseudo-uptriangular decomposition method for constrained multibody systems using Kane's equations, *Journal of Guidance, Control and Dynamics (AIAA)*, Vol. 10, No. 7, 1988, pp. 39–46.
14. Amirouche F and Tongyi J, Automatic elimination of the undetermined multipliers in Kane's equations using a pseudo-uptriangular decomposition (PUTD) method, *Journal of Computers and Structures*, Vol. 27, No. 2, 1987, pp. 203–210.
15. Amirouche F and Tongyi J, Modeling of clearances and joint flexibility effects in multibody systems dynamics, *Computers & Structures*, Vol. 29, No. 6, 1988, pp. 983–991.
16. Amirouche F and Tongyi J, Natural dynamics and control of manipulators in presence of obstacles, *Journal of Dynamic Systems, Measurement and Control (ASME)*, Vol. 113, 1991, pp. 171–174.
17. Amirouche F and Tongyi J, Spatial coordination of a two-arm robot subjected to geometrical and kinematical constraints, *Advances in Engineering Software*, Vol. 12, No. 1, 1990, pp. 21–26.
18. Amirouche F and Tung C W, On-line collision avoidance and constraint violation stabilization, *Journal of Nonlinear Dynamics*, Vol. 5, 1994, pp. 1–17.
19. Amirouche F and Vashi T R, Accessible region and synthesis of a two-arm robot, *International Journal of Robotics and Automation*, Vol. 3, No. 3, 1988, pp. 133–139.
20. Amirouche F and Xie M, An explicit matrix formulation of the dynamical equations for flexible multibody systems: A recursive approach, *Computers & Structures*, Vol. 46, No. 2, 1993, pp. 311–321.
21. Amirouche F and Xie M, Minimization of vibration in elastic beams with time variant boundary conditions in the multibody systems, *Journal of Guidance, Control and Dynamics (AIAA)*, Vol. 15, No. 5, 1992, pp. 1265–1271.
22. Amirouche F, Biodynamic analysis of the human body subjected to vibration, *IEEE Engineering in Medicine and Biology*, 1987, pp. 22–26.
23. Amirouche F, Determination of the generalized constraint forces in multibody systems using Kane's equations, *Journal of Theoretical and Applied Mechanics*, Vol. 7, No. 1, 1988, pp. 3–20.
24. Amirouche F, Dynamics analysis of tree-like structure undergoing large motions – A finite segment approach, *Journal of Engineering Analysis, Computational Mechanics*, Vol. 3, No. 2, 1986, pp. 111–117.
25. Amirouche F, Equations of motion of a two-arm robot manipulator including the flexibility effects, *International Journal of Robotics and Automation*, Vol. 2, No. 2, 1987, pp. 77–85.
26. Amirouche F, Estimation of the stiffness matrix in the study of flexibility effects in large structures, *Journal of Vibration, Acoustics, Stress and Reliability in Design (ASME)*, Vol. 109, 1987, pp. 283–288.
27. Amirouche F, Ider S K and Trimble J, Analytical method for the analysis and simulation of human locomotion, *Journal of Biomechanical Engineering (ASME)*, Vol. 112, No. 4, 1990, pp. 379–385.
28. Amirouche F, Modeling of human reactions to whole-body vibration, *Journal of Biomechanical Engineering (ASME)*, Vol. 109, No. 211, 1987, pp. 210–216.
29. Amirouche F, Palkovic L and Woodrrooffe J, Optimal driver seat suspension design for heavy-trucks, *International Journal of Heavy Vehicle Systems*, Vol. 2, No. 1, 1995, pp. 19–45.
30. Amirouche F, Palkovics L and Vida G, Vibration analysis of a straight truck using DAVS (Dynamic Analysis of Vehicle Systems), Zobory I (ed.) 4th Mini Conference on Vehicle System Dynamics, Identification and Anomalies, Budapest: TU Budapest, 1994, pp. 103–124.

31. Amirouche F, Shareef N H and Xie M, Dynamic analysis of flexible gear trains/ transmissions – An automated approach, *Journal of Applied Mechanics (ASME)*, Vol. 59, No. 4, 1992, pp. 976–982.
32. Amirouche F, Shareef N H and Xie M, Time variant analysis of rotorcraft systems dynamics – An exploitation of vector-processors, *Journal of Guidance, Control and Dynamics (AIAA)*, Vol. 16, No. 1, 1993, pp. 96–110.
33. Amirouche F, Tajiri G and Valco M, Dynamic analysis of gear meshing teeth using modified Timoshenko beam theory, *International Journal for Numerical Methods in Engineering*, Vol. 39, 1996, pp. 1–22.
34. Amirouche F, Tongyi J and Ider S K, A recursive householder transformation for complex dynamical systems with constraints, *Journal of Applied Mechanics (ASME)*, Vol. 55, No. 3, 1988, pp. 729–734.
35. Amirouche F, Xie M and Patwardhan A, Optimization of the contact damping and stiffness coefficients to minimize human body vibration, *Journal of Biomechanical Engineering (ASME)*, Vol. 116, No. 4, 1994, pp. 413–420.
36. Amirouche F, Xie M, Shareef N H and Valco M, Finite element modeling of contact conditions in multibody dynamics, *Journal of Nonlinear Dynamics*, Vol. 4, 1993, pp. 83–102.
37. Anantharaman M and Hiller M, Dynamic analysis of complex multibody systems using methods for differential-algebraic equations. Solid mechanics and its applications; advanced multibody system dynamics; Kluwer Academic Publishers, 1993, pp. 173–194.
38. Anantharaman M and Hiller M, Numerical simulation of mechanical systems using methods for differential-algebraic equations, *International Journal for Numerical Methods in Engineering*, Vol. 32, 1991, pp. 1531–1542.
39. Anderson K S and Duan S, A hybrid parallelizable low order algorithm for dynamics of multi-rigid-body systems: Part I, Chain systems, *Mathematical and Computer Modelling*, Vol. 30, 1999, pp. 193–215.
40. Anderson K S and Duan S, A new order-*N* hybrid parallelizable algorithm for multi-rigid-body dynamics, submitted to *Journal of Applied Mechanics (ASME)*, in review.
41. Anderson K S and Duan S, Highly parallelizable low order dynamics algorithm for complex multi-rigid-body systems, *Journal of Guidance, Control and Dynamics (AIAA)*, Vol. 23, No. 2, 2000, pp. 355–364.
42. Anderson K S and Hsu Y H, Analytical full-recursive sensitivity analysis for multibody chain systems, *Multibody Systems Dynamics*, Vol. 8, No. 1, 2002, pp. 1–27.
43. Anderson K S, An efficient modeling of constrained multibody systems for application with parallel computing, *Zeitschrift für Angewandte Mathematik und Mechanik*, Vol. 73, No. 6, 1993, pp. 935–939.
44. Anderson K S, An order-*N* formulation for the motion simulation of general constrained multi-rigid-body systems, *Computers & Structures*, Vol. 43, No. 3, 1992, pp. 565–579.
45. Anderson K S, Efficient modeling of general multibody dynamic systems with flexible components, *Computational Dynamics in Multibody Systems*, M S Pereira and J A C Ambrosia (eds), Kluwer Academic Publishers, Dordrecht, The Netherlands, 1995.
46. Andrew G C and Kesavan H K, Simulation of multibody systems using the vector-network model. In: Dynamics of multibody systems, K Magnus (ed.), Springer-Verlag, New York, 1977, pp. 1–13.
47. Angeles J and Kecskemethy A, Kinematics and dynamics of multibody systems, Kluwer Academic Publishers, Dordrecht, 1995.
48. Argyris J H, An excursion into large rotations, *Computer methods in Applied Mechanics and Engineering*, Vol. 32, 1982, pp. 85–155.

49. Armstrong B, Dynamics for robot control: Friction modeling and ensuring excitation during parameter identification, Ph.D. Thesis, Stanford University, Stanford, CA, 1988.
50. Ascher U and Petzold L, Projected implicit Runge–Kutta methods for differential algebraic equations, *SIAM Journal on Numerical Analysis*, Vol. 28, 1991, pp. 1097–1120.
51. Atluri S N and Fukuchi N, Finite deformation analysis of shells – A hybrid finite element method based on assumed stress-function vector and rotation tensor, *AIAA Journal*-1983-861.
52. Atluri S N and Nakagaki M, An elastic-plastic analysis of fatigue crack closure in modes I and II, *AIAA Journal*-1979-758.
53. Atluri S N, A new assumed stress hybrid finite element model for solid continua, *AIAA Journal*, 0001-1452, Vol. 9, No. 9, 1971, pp. 1647–1649.
54. Atluri S N, Computational nonlinear mechanics in aerospace engineering, *Progress in Astronautics and Aeronautics*, 1992.
55. Banerjee A K, Block-diagonal equations for multibody elastodynamics with geometric stiffness and constraints, *Journal of Guidance, Control and Dynamics*, 0731-5090, Vol. 16, No. 6, 1993, pp. 1092–1100.
56. Banerjee A K, Contributions of multibody dynamics to space flight: A brief review, *Journal of Guidance, Control and Dynamics*, 0731-5090, Vol. 26, No. 3, 2003, pp. 385–394.
57. Banerjee A K and Dickens J M, Dynamics of an arbitrary flexible body in large rotation and translation, (ASME, ASCE, and AHS, Structures, Structural Dynamics, and Materials Conference, Mobile, AL, April 3–5, 1989) *Journal of Guidance, Control and Dynamics*, 0731-5090.
58. Banerjee A K, Dynamics of a variable-mass, flexible-body system, *Journal of Guidance, Control and Dynamics*, 0731-5090, Vol. 23, No. 3, 2000, pp. 501–508.
59. Banerjee A K, Efficient generalized speeds and elastic mutibody dynamics equations with analytic Jacobian, AIAA/AAS Astrodynamics Specialist Conference, Denver, CO, August 14–17, 2000.
60. Banerjee A K, Recursive formulation with geometric stiffness in multibody elastodynamics (AAS 95-398), 1589.
61. Banichuk N V, Klimov D M and Schiehlen W (eds), Dynamical problems of rigid-elastic systems and structures, Springer-Verlag, Berlin, 1991.
62. Bates L and Sniatycki J, Nonholonomic reduction, *Reports on Mathematical Physics*, Vol. 32, No. 1, 1993, pp. 99–115.
63. Battle J A, On Newton's and Poisson's rules of recursive dynamics, *Transactions of the ASME Journal of Applied Mechanics*, Vol. 60, 1993, pp. 376–381.
64. Bauchau O A and Theron N J, Energy decaying scheme for nonlinear elastic multi-body systems, *Computers & Structures*, Vol. 59, No. 2, 1996, pp. 317–331.
65. Bauchau O A, Analysis of flexible multi-body systems with intermittent contacts, *Multibody System Dynamics*, Vol. 4, 2000, pp. 23–54.
66. Bauchau O A, Bottasso C L and Nikishkov Y G, Modeling rotorcraft dynamics with finite element multibody procedures, *Mathematical and Computer Modelling*, Vol. 33, No. 10–11, 2001, pp. 1113–1137.
67. Bauchau O A, On the modeling of prismatic joints in flexible multi-body systems, *Computer Methods in Applied Mechanics and Engineering*, Vol. 181, No. 1–3, 2000, pp. 87–105.
68. Bauchau O A, Rodriguez J and Bottasso C L, Modeling of unilateral contact conditions with application to aerospace systems involving backlash, freeplay and friction, *Mechanics Research Communications*, 2002, pp. 193–214.

69. Baumgarte J, Stabilization of constraints and integrals of motion in dynamical systems, *Computer Methods in Applied Mechanics and Engineering*, Vol. 1, 1972, pp. 1–16.
70. Bayo E, de Jalón G J and Serna M A, A modified Lagrangian formulation for the dynamic analysis of constrained mechanical systems, *Computer Methods in Applied Mechanics and Engineering*, Vol. 71, 1988, pp. 183–195.
71. Bayo J, de Jalón G, Avello A and Cuadrado J, An efficient computational method for real time multibody dynamic simulation in fully cartesian coordinates, *Computer Methods in Applied Mechanics and Engineering*, Vol. 92, 1991, pp. 377–395.
72. Bestle D and Eberhard P, Multicriteria multimodel design optimization. In: *IUTAM Symposium on Optimization of Mechanical Systems*, D Bestle and W Schiehlen (eds), Kluwer Academic Publishers, Dordrecht, 1996, pp. 33–40.
73. Bianchi G and Schiehlen W (eds), Dynamics of multibody systems, Springer-Verlag, Berlin, 1986.
74. Blajer W, Index of differential-algebraic equations governing the dynamics of constrained mechanical systems, *Applied Mathematical Modelling*, Vol. 16, No. 2, 1992, pp. 70–77.
75. Blajer W, State and energy stabilization in multibody system simulations, *Mechanics Research Communications*, Vol. 26, No. 3, 1999, pp. 261–268.
76. Bottasso C L and Trainelli L, Implementation of effective procedures for unilateral contact modeling in multibody dynamics, *Mechanics Research Communications*, Vol. 28, No. 3, 2001.
77. Brogliato B, Nonsmooth impact mechanics, Springer-Verlag, London, 1996.
78. Burton W S and Noor A K, Structural analysis of the adhesive bond in a honeycomb core sandwich panel, *Finite Elements in Analysis and Design*, Vol. 26, No. 3, 1997, pp. 213–227.
79. Canudas de Wit C, Olson H, Astrom K and Lischinsky P, A new model for control of systems with friction, *IEEE Transaction on Automatic Control*, Vol. 40, No. 3, 1995, pp. 419–425.
80. Casey J and Lam V C, A tensor method for the kinematic analysis of systems of rigid bodies, *Mechanism and Machine Theory*, Vol. 21, No. 1, 1986, pp. 87–97.
81. Ceccarelli M, Cuadrado J and Dopico D, An optimum synthesis for gripping mechanisms by using natural coordinates, *Proc. Inst. Mech. Eng. Part C: J. Mech. Eng. Science*, Vol. 216, 2002, pp. 643–653.
82. Chace M A and Smith D A, DAMN—A digital computer program for the dynamic analysis of generalized mechanical systems, *SAE Paper*, 710244, 1971.
83. Chaffin D B, A computerized biomechanical model-development for the use in studying gross body actions, *Journal of Biomechanics*, Vol. 2, No. 4, 1969, pp. 429–441.
84. Chen K-H and Pletcher R H, Primitive variable, strongly implicit calculation procedure for viscous flows at all speeds, *AIAA Journal*, Vol. 29, No. 0, 1993, pp. 1241–1249.
85. Chen S and Tortorelli D A, an energy-conserving and filtering method for stiff nonlinear multibody dynamics, *Multibody System Dynamics*, Vol. 10, No. 4, 2003, pp. 341–362.
86. CISM courses and lectures, Vol. 360, Springer-Verlag, Vienna, 1995.
87. Conway B A and Widham J W, Equations of motion for an N-body satellite with moving joints, *Journal of Guidance and Control*, Vol. 8, No. 4, 1985, pp. 537–539.
88. Crespo and Da Silva M R M, The Hamiltonian formulation and integrals of the motion for non-conservative and time-varying dynamical systems, *Report ADC* 73-1-1, October 1973, Dept. of Aerospace Engineering, University of Cincinnati, Ohio.
89. Cuadrado J, Cardenal J and de Jalón J G, Flexible mechanisms through natural coordinates and component synthesis: An approach fully compatible with the rigid case, *Int. Journal for Numerical Methods in Engineering*, Vol. 39, 1996, pp. 3535–3551.

90. Cuadrado J, Gutiérrez R, Naya M A and González I, Experimental validation of a flexible MBS dynamic formulation through comparison between measured and calculated stresses on a prototype car, *Multibody System Dynamics*, Vol. 11, No. 2, 2004, pp. 147–166.
91. Cuadrado J, Gutiérrez R, Naya M A and Morer P, Comparison in terms of accuracy and efficiency between a MBS dynamic formulation with stress analysis and a non-linear FEA code, *Int. J. for Numerical Methods in Engineering*, Vol. 51, No. 9, 2001, pp. 1033–1052.
92. Daberkow A and Schiehlen W, Concept, development and implementation of DAMOSC: The object oriented approach to multibody systems. In: *Computers in Engineering*, 1991, Proceedings ASME International Computers in Engineering Conference, K Ishii (ed.), ASME, New York, 1994, pp. 937–951.
93. Danielson K T and Noor A K, Three-dimensional finite element analysis in cylindrical coordinates for nonlinear solid mechanics problems, *Finite Elements in Analysis and Design*, Vol. 27, No. 3, 1997, pp. 225–249.
94. de Carufel J, Martin E and Piedboeuf J-C, Control strategies for hardware-in-the-loop simulation of flexible space robots, *IEEE Proceedings-D: Control Theory and Applications*, Vol. 147, No. 6, 2000, pp. 569–579.
95. de Jalón G J and Bayo E, Kinematic and dynamic simulation of multibody systems, Springer-Verlag, New York, 1994.
96. Denavit J and Hartenberg R S, A kinematic motion for lower pair mechanisms based on matrices, *Journal of Applied Mechanics*, Vol. 22, 1955, pp. 215–221.
97. Dürrbaum A, Klier W and Hahn H, Comparison of automatic and symbolic differentiation in mathematical modeling and computer simulation of rigid-body systems, *Multibody System Dynamics*, Vol. 7, No. 4, 2002, pp. 331–355.
98. El-Gindy M and Ilosvai L, Computer simulation study on a vehicle's directional response in some severe manoeuvres – Part 1: Rapid lane-change manoeuvres, *Int. Journal of Vehicle Design*, Vol. 4, No. 4, 1983, pp. 386–401.
99. El-Gindy M and Palkovics L, Possible application of artificial neural networks to vehicle dynamics and control: A literature review, *Int. Journal of Vehicle Design*, Vol. 14, No. 5, 1993.
100. El-Gindy M, Thomas P J and Palkovics L, Computer simulation modelling and field tests of a Canadian logging truck: Tractor/pole-trailer. In: *Proc. of 12th Symposium on Engineering Application of Mechanics*, MacGill University, 1994.
101. Enos M J (ed.), Dynamics and control of mechanical systems, American Mathematical Society, Providence, 1993.
102. Essen H, On the geometry of nonholonomic dynamics, *Journal of Applied Mechanics*, Vol. 61, No. 3, 1994, pp. 689–690.
103. Euler L, Nova methods motum corporum rigidarum determinandi, Novi Commentarii Academiae Scientiarum Petropolitanae 20, 1776, 208–238. See also Euler L., Opera Omnia, Series II, Vol. 9, pp. 99–125.
104. Feng Z-Q, Joli P and Seguy N, FER/Mech–a software with interactive graphics for dynamic analysis of multibody system, *Advances in Engineering Software*, Vol. 35, No. 1, 2004.
105. Fleischer G E and Likins P W, Attitude dynamics simulation subroutines for systems of hinge-connected rigid bodies, *JPL Technical Report*, 32-1592, Pasadena, CA, 1974.
106. Fuhrer C and Schwertassek R, Generation and solution of multibody system equations, *International Journal of Non-Linear Mechanics*, Vol. 25, 1990, pp. 127–141.
107. Gaspar P, Bokor J and Palkovics L, Closed-loop identification schemes for active suspension design, Zobory I (ed.) 4th Mini Conference on Vehicle System Dynamics, Identification and Anomalies, Budapest: TU Budapest, 1994, pp. 535–545.

108. Gear C W, Differential-algebraic equation index transformations, *SIAM Journal of Scientific and Statistical Computing*, Vol. 9, 1988, pp. 39–47.
109. Gear C W, Numerical initial value problems in ordinary differential equations, Automatic Computation, Prentice-Hall, Englewood Cliffs, NJ, 1971.
110. Glocker C and Pfeiffer F, Complementarity problems in multibody systems with planar friction, *Archive of Applied Mechanics*, Vol. 63, 1993, pp. 452–463.
111. Glocker C and Pfeiffer F, Dynamical systems with unilateral contacts, *Nonlinear Dynamics*, Vol. 3, 1992, pp. 107–115.
112. Goldsmith W, Impact: The theory and physical behavior of colliding solids, Edward Arnold, London, 1960.
113. Gonthier Y, Mcphee2 J, Lange C and Piedboeuf J-C, A regularized contact model with asymmetric damping and dwell-time dependent friction, *Multibody System Dynamics*, Vol. 11, No. 3, 2004, pp. 209–233.
114. Grotjahn M, Daemi M and Heimann B, Friction and rigid body identification of robot dynamics, *Internat. J. Solids Structures*, Vol. 38, No. 10–13, 2001, pp. 1889–1902.
115. Grotjahn M, Heimann B and Abdellatif H, Identification of friction and rigid-body dynamics of parallel kinematic structures for model-based control, *Multibody System Dynamics*, Vol. 11, No. 3, 2004, pp. 273–294.
116. Hairer E and Wanner G, Solving ordinary differential equations II, Stiff and Differential-Algebraic Problems, Springer-Verlag, Berlin, 1991.
117. Han H and Park T, Robust optimal design of multi-body systems, *Multibody System Dynamics*, Vol. 11, No. 2, 2004, pp. 167–183.
118. Han I and Gilmore B J, Multi-body impact motion with friction: Analysis, simulation and experimental validation, *Journal of Mechanical Design*, Vol. 115, 1993, pp. 412–422.
119. Hansen J M and Tortorelli D A, An efficient method for synthesis of mechanisms using an optimization method. In: Optimization of Mechanical System, D Bestle and W Schiehlen (eds), Kluwer Academic Publishers, Dordrecht, 1996, pp. 129–138.
120. Haug E J, Computer-aided kinematics and dynamics of mechanical systems, Allyn and Bacon, Boston, 1989.
121. Hemani H, A state space model for interconnected rigid bodies, *IEEE Trans. Autom. Control*, Vol. AC-27, No. 2, 1982, pp. 376–382.
122. Hemani H, Some aspects of Euler–Newton equations of motion, *Ing. Arch.*, Vol. 52, 1985, pp. 167–176.
123. Hiller M and Kecskemethy A, Dynamics of multibody systems with minimal coordinates, *Computer-Aided Analysis of Rigid and Flexible Mechanical Systems*, 1994, pp. 61–100.
124. Hiller M, Modeling and simualtion of complex multibody systems in mechatronic systems, Eurotech Direct '91 – Machine Systems, *Proceedings Eur. Eng. Res. and Tech. Transfer Congress*, Birmingham, 1991, pp. 159–163.
125. Hollerback J M, A recursive Lagrangian formulation of manipulator dynamics and a comparative study of dynamics of formulation complexity, *IEEE Trans. Syst. Man Cybern.*, Vol. SMC-10, No. 11, 1980, pp. 730–736.
126. Hooker W W and Margulies G, The dynamical attitude equations for n-body satellite, *Journal on Astronomical Science*, Vol. 12, No. 4, 1965, pp. 123–128.
127. Hunt K H and Crossley F R E, Coefficient of restitution interpreted as damping in vibroimpact, *Transactions of the ASME Journal of Applied Mechanics*, Series E 42, 1975, pp. 440–445.
128. Huston R L and Kamman J W, A representation of fluid forces in finite segment cable models, *Computers & Structures*, Vol. 14, 1981, pp. 281–287.
129. Huston R L and Kamman J W, Testing the accuracy of numerical simulations of multibody systems, *Mechanics Research Communications*, Vol. 28, No. 1, 2001, pp. 1–6.

130. Huston R L and Kamman J W, Validation of finite segment cable models, *Computers & Structures*, Vol. 15, No. 6, 1982, pp. 653–660.
131. Huston R L and Passerello C E, Eliminating singularities in governing equations of mechanical systems, *Mech. Res. Commun.*, Vol. 3, No. 5, 1976, pp. 361–365.
132. Huston R L and Passerello C E, Multibody structure dynamics including translation between the bodies, *Comput. Struct.*, Vol. 11, 1980, pp. 713–720.
133. Huston R L and Passerello C E, On multi-rigid-body system dynamics, *Comput. Struct.*, Vol. 10, 1979, pp. 439–446.
134. Huston R L and Passerello C E, On the dynamics of a human body model, *Journal on Biomechanics*, Vol. 4, No. 5, 1971, pp. 369–378.
135. Huston R L, Multibody dynamics – Modeling and analysis methods, *Applied Mechanics Review*, Vol. 44, 1991, pp. 109–117.
136. Huston R L, Multibody Dynamics, Butterworth-Heinemann, Boston, 1990.
137. Huston R L, Passerello C E and Harlow M W, Dynamics of multi-rigid-body systems, *Journal of Applied Mechanics*, Vol. 45, No. 4, 1978, pp. 889–894.
138. Ibrahimbegovic A and Mamouri S, On rigid components and joint constraints in nonlinear dynamics of flexible multibody systems employing 3D geometrically exact beam model, *Computer Methods in Applied Mechanics and Engineering*, Vol. 188, No. 4, 2000, pp. 805–831.
139. Ibrahimbegovic A, Taylor R L and Lim H, Non linear dynamics of flexible multibody systems, *Computers & Structures*, Vol. 81, No. 12, 2003, pp. 1113–1132.
140. Ider S K and Amirouche F M, Coordinate reduction in the dynamics of constrained multibody systems – A new approach, *Journal of Applied Mechanics (ASME)*, Vol. 55, No. 4, 1988, pp. 899–904.
141. Ider S K and Amirouche F M, Determination of constraint forces in multibody systems dynamics using Kane's equations, *Journal de Mecanique Theorique et Applique*, Vol. 7, No. 1, 1988, pp. 3–20.
142. Ider S K and Amirouche F M, Influence of geometric nonlinearities in the dynamics of flexible tree-like structures, *Journal of Guidance, Control and Dynamics (AIAA)*, Vol. 12, No. 6, 1989, pp. 830–837.
143. Ider S K and Amirouche F M, Nonlinear modeling of flexible multibody systems dynamics subjected to variable constraints, *Journal of Applied Mechanics (ASME)*, Vol. 56, No. 2, 1989, pp. 444–451.
144. Ider S K and Amirouche F M, Numerical stability of the constraints near singular positions in the dynamics of multibody systems, *Computers & Structures*, Vol. 33, 1989, pp. 129–137.
145. Ider S K and Amirouche F M, Stability analysis of constrained multibody systems, *Journal of Computational Mechanics*, Vol. 6, No. 5/6, 1990, pp. 327–340.
146. Ider S K, Özgören M K and Ay V, Trajectory tracking control of robots with flexible links, *Mechanism and Machine Theory*, Vol. 37, No. 11, 2002, pp. 1377–1394.
147. Ider S K and Amirouche F M, On the constraint violations in the dynamic simulations of multibody systems, *Journal of Dynamic Systems, Measurement and Control (ASME)*, Vol. 111, No. 2, 1989, pp. 238–244.
148. Ider S K, Finite element based recursive formulation for real time dynamic simulation of flexible multibody systems, *Computers & Structures*, Vol. 40, 1991, pp. 939–945.
149. Ider S K, Force and motion trajectory tracking control of flexible-joint robots, *Mechanism and Machine Theory*, Vol. 35, No. 3, 2000, pp. 363–378.
150. Ider S K, Inverse dynamics control of constrained robots in the presence of joint flexibility, *Journal of Sound and Vibration*, Vol. 224, No. 5, 1999, pp. 879–895.

151. Ider S K, Open-loop flexibility control in multibody systems dynamics, *Mechanism and Machine Theory*, Vol. 30, No. 6, 1995, pp. 861–869.
152. Ider S K, Stability of control forces in redundant multibody systems, *Computers & Structures*, Vol. 58, 1996, pp. 115–121.
153. Jerkovsky W, The transformation operator approach to multi-systems dynamics, Part I: The general approach, *Matrix Tensor Q.*, Vol. 27, No. 2, 1976, pp. 48–50.
154. Jiang Z, Streit D A and El-Gindy M, Heavy vehicle ride comfort: Literature survey, *Heavy Vehicle Systems, special series of Int. J. of Vehicle Design*, Vol. 8, No. 3, 2001, pp. 258–284.
155. Johnson K L, Contact Mechanics, Cambridge University Press, London, 1985.
156. Jorge A C, Ambrosio J, Nikravesh P E and Pereira M S, Crashworthiness analysis of a truck, *Mathematical and Computer Modelling*, Vol. 14, 1990, pp. 959–964.
157. Joseph F, McGrath, Rampalli R and Ryan R R, Integration methods for multibody dynamics, (AAS 95-455), 1603.
158. Jourdain P E B, Note on an analogue at Gauss' principle of least constraints, Quarterly Journal 2, 1995, 12–19. *Journal on Pure Applied Mathematics*, Vol. 40, 1909, pp. 153–197.
159. Junker F, Modular-hierarchisch strukturierte modellbildung mechanischer systeme, *Archive of Applied Mechanics*, Vol. 65, 1995, pp. 227–245.
160. Kalaba R E and Udwadia F E, Equations of motion for nonholonomic, constrained dynamical systems via Gauss's principle, *Journal of Applied Mechanics*, Vol. 60, 1993, pp. 662–668.
161. Kamman J W and Huston R L, Dynamics of constrained multibody systems, *J. Appl. Mech.*, Vol. 51, 1984, pp. 899–903.
162. Kamman J W and Huston R L, Modeling of submerged cable dynamics, *Computers & Structures*, Vol. 20, No. 103, 1985, pp. 623–629.
163. Kane T R and Levinson D A, Dynamics: Theory and Applications, McGraw-Hill, New York, 1985.
164. Kane T R and Levinson D A, Formulation of equations of motion for complex spacecraft, *Journal of Guidance and Control*, Vol. 3, No. 2, 1980, pp. 99–112.
165. Kane T R and Scher M P, A dynamical explanation of the falling cat phenomena, *J. Guidance*, Vol. 10, No. 2, 1969, pp. 119–131.
166. Kane T R and Wang C F, On the derivation of equations of motion, *J. Soc. Ind. App. Math.*, Vol. 13, No. 2, 1965, pp. 487–492.
167. Kane T R and Wang C F, On the derivation of equations of motion, *J. Soc. Ind. App. Math.*, Vol. 13, No. 2, 1965, pp. 487–492.
168. Kane T R, Dynamics of nonholonomic systems, *Journal of Applied Mechanics*, 1961, pp. 574–578.
169. Kane T R, Ryan R R and Banerjee A K, Dynamics of a cantilever beam attached to a moving base, *J. Guidance*, Vol. 10, No. 2, 1987, pp. 139–151.
170. Keller J B, Impact with friction, *Journal of Applied Mechanics (ASME)*, Vol. 53, 1986, pp. 1–4.
171. Khulief Y A and Mohiuddin M, On the dynamic analysis of rotors using modal reduction, *Internat. J. Finite Element Anal. Design*, Vol. 26, No. 1, 1997, pp. 41–55.
172. Khulief Y A, Finite element dynamic analysis of multibody systems with impulsive constraints, Sabbatical Leave Research Report, College of Engineering Sciences, King Fahd University of Petroleum and Minerals, Dhahran, Saudi Arabia, 1998.
173. Khulief Y A, Spatial formulation of elastic multibody systems with impulsive constraints, *Dynamics Multibody System*, Vol. 4, 2000, pp. 383–406.

174. Khulief Y A, Spatial formulation of elastic multibody systems with impulsive constraints, *Multibody System Dynamics*, Vol. 4, No. 4, 2000, pp. 383–406.
175. Kielau G and Maišer P, Nonholonomic multibody dynamics, *Multibody System Dynamics*, Vol. 9, No. 3, 2003, pp. 213–236.
176. Kim H W, Bae D S and Choi K K, Configuration design sensitivity analysis of dynamics for constrained mechanical systems, *Computer Methods in Applied Mechanics and Engineering*, Vol. 190, No. 40–41, 2001, pp. 5271–5282.
177. Kim S S and Haug E J, Recursive formulation for flexible multibody dynamics, Part II, Closed loop systems, *Computer Methods in Applied Mechanics and Engineering*, Vol. 74, 1989, pp. 251–269.
178. Kobayashi N and Watanabe M, Dynamics and stability of spaghetti and reverse spaghetti problems coupled with fluid force, *Multibody System Dynamics*, Vol. 11, No. 2, 2004, pp. 111–125.
179. Kortum W and Sharp R S, Multibody computer codes in vehicle system dynamics, Swet & Zeitlinger, Amsterdam, 1993.
180. Kortum W and Lugner P, Systemdynamik und Regelung von Fahrzeugen, Springer-Verlag, Berlin, 1996.
181. Kortum W and Schiehlen W, General purpose vehicle system dynamics software based on multibody formalisms, *Vehicle System Dynamics*, Vol. 14, 1985, pp. 229–263.
182. Kuang J, Meehan P A, Leung A Y T and Tan S, Nonlinear dynamics of a satellite with deployable solar panel arrays, *International Journal of Non-Linear Mechanics*, Vol. 39, No. 7, 2004.
183. Laananen D, Bolokbasi A and Coltman J, Computer simulation of an aircraft seat and occupant in a crash environment–Volume I: Technical Report, US Department of Transportation, Federal Aviation Administration. Report No. DOT/FAA/CT-82/33-I, 1983.
184. Lagrange J L, Mecanique Analytique, L'Acad'emie Royal des Sciences, Paris, 1788.
185. Lankarani H M and Nikravesh P E, Canonical impulse-momentum equations for impact analysis of multibody systems, *Journal of Mechanical Design*, Vol. 114, 1992, pp. 180–186.
186. Latchézar G, Paul F and Jean-Claude S, Piecewise over actuation of parallel mechanisms following singular trajectories: Modeling, simulation and control, ECCOMAS Thematic Conference Multibody 2003, *Advances in Computational Multibody Dynamics*, July 1–4, 2003, Lisbon, Portugal, MB2003-043.
187. Lawson C L and Hanson R J, Solving least squares problems, *Classics in Applied Mathematics*, SIAM, Philadelphia, PA, 1995.
188. Leamy M J, Noor A K and Wasfy T M, Dynamic simulation of a tethered satellite system using finite elements and fuzzy sets, *Computer Methods in Applied Mechanics and Engineering*, Vol. 190, No. 22, 2001, pp. 37–38.
189. Leister G, Beschreibung und Simulation von Mehrkörpersystemen mit geschlossenen kinematischen.
190. Lens E V, Cardona A and Géradin M, Energy preserving time integration for constrained multibody systems, *Multibody System Dynamics*, Vol. 11, No. 1, 2004, pp. 41–61.
191. Levinson D A, Equations of motion for multi-rigid-body systems via symbolic manipulation, *Journal of Spacecraft and Rockets*, Vol. 14, 1977, pp. 479–487.
192. Lichtenberg A and Liebermann M, Regular and chaotic dynamics, Springer-Verlag, New York, 1992.
193. Li-Fu L, Variational problem of general displacement in nonholonomic systems, *Acta. Mechanica Solida Sinica*, Vol. 6, 1993, pp. 357–364.
194. Likins P W and Wirsching P H, Use of synthetic modes in hybrid coordinate dynamic analysis, *AIAA Journal*, Vol. 6, 1968.

195. Lilov L and Chirikov V A, On the dynamics equations of systems of interconnected bodies, *Prikl. Mat. Mekh. (J. Appl. Math. Mech.)*, Vol. 45, 1982, pp. 383–390.
196. Lilov L and Lorer M, Dynamic analysis for multi-rigid-body systems based on the gauss principle, *Z. Angew. Math. Mech.*, Vol. 72, 1982, pp. 539–545.
197. Lin M and Gottschalk S, Collision detection between geometric models: A survey. In: *The Proceedings of IMA Conference on Mathematics of Surfaces*, R Cripps (ed.), University of Birmingham, 1998, pp. 37–56.
198. Lo J, Huang G and Metaxas D, Human motion planning based on recursive dynamics and optimal control techniques, *Multibody System Dynamics*, Vol. 8, No. 4, 2002, pp. 433–458.
199. Lubich Ch, Extrapolation integrators for constrained multibody systems, IMPACT *Comp. Sci. Eng.*, Vol. 3, 1991, pp. 213–234.
200. Magnus K, Kreisel-Theorie und Anwendungen, Springer-Verlag, Berlin, 1971.
201. Mariot J-P, Knevez J-Y and Barbedette B, Tripod and ball joint automotive transmission kinetostatic model including friction, *Multibody System Dynamics*, Vol. 11, No. 2, 2004, pp. 127–145.
202. Marshall R N, Jensen R K and Wood G A, A general Newtonian simulation of an *N*-segment open chain model, *J. Biomech.*, Vol. 18, No. 5, 1985, pp. 359–367.
203. McPhee J J and Anderson R J, A model reduction procedure for the dynamic analysis of rail vehicles subjected to linear creep forces, *Vehicle System Dynamics*, Vol. 25, 1996, pp. 349–367.
204. McPhee J J and Dubey R N, Dynamics of multibody systems with known configuration changes, *Journal of Applied Mechanics*, Vol. 58, No. 1, 1991, pp. 215–221.
205. Meirovitch L and Kwak M K, Rayleigh-Ritz based substructure synthesis for flexible multibody systems, *AIAA Journal*, Vol. 29, 1991, pp. 1709–1719.
206. Meirovitch L and Stemple T J, Dynamics and control of flexible multibody systems, *Proceedings of the First Joint Conference of International Simulation Societies*, Zurich, Switzerland, 1994, pp. 268–272.
207. Meirovitch L and Stemple T J, Hybrid equations of motion for flexible multibody systems using quasi-coordinates, *Journal of Guidance, Control and Dynamics*, Vol. 18, 1995, pp. 678–688.
208. Meirovitch L, Computational methods in structural dynamics, Sijthoff & Noordhoff, USA, 1980.
209. Meirovitch L, Hybrid state equations of motion for flexible bodies in terms of quasi-coordinates, *Journal of Guidance, Control and Dynamics*, Vol. 14, 1991, pp. 1008–1013.
210. Mikkola A M and Shabana A A, A non-incremental finite element procedure for the analysis of large deformation of plates and shells in mechanical system applications, *Multibody System Dynamics*, Vol. 9, No. 3, 2003, pp. 283–309.
211. Mladenova C D, Kinematics and dynamics of a rigid body in nonholonomic coordinates, *Systems & Control Letters*, Vol. 22, 1994, pp. 257–265.
212. Morecki A, Biomechanics of engineering, modeling, simulation, control, Springer-Verlag, Vienna, 1987.
213. Muller P C, Descriptor systems: A new way to model mechatronic systems. In: *Proceedings 3rd European Control Conference*, Vol. 3, Part 2, 1995, pp. 2725–2729.
214. Muller P C, Rentrop P, Kortüm W and Führer C, Constrained mechanical systems in descripton form: Identification, simulation and control, *Advanced Multibody System Dynamics*, W Schiehlen (ed.), Kluwer Academic Publishers, Dordrecht, 1993, pp. 451–456.
215. Muller P C, Stabilität und Matrizen, Springer-Verlag, Berlin, 1977.
216. Multibody mechanical systems, Part I: Systems with tree configuration, *Acta Mech.*, Vol. 50, 1984, pp. 141–161.

217. Munteanu M Gh, Ray P and Gogu G, Method for a fast and simple dynamic analysis of 2D and 3D mechanisms, *Multibody System Dynamics*, Vol. 11, No. 1, 2004, pp. 63–85.
218. Nagata T, Modi V J and Matsuo H, Dynamics and control of flexible multibody systems, Part II: Simulation code and parametric studies with nonlinear control, *Acta Astronautica*, Vol. 49, No. 11, 2001, pp. 595–610.
219. Nagata T, Modi V J and Matsuo H, Dynamics and control of flexible multibody systems, Part I: General formulation with an order-N forward dynamics, *Acta Astronautica*, Vol. 49, No. 11, 2001, pp. 581–594.
220. Nagata T, Modi V J and Matsuo H, Dynamics and control of flexible multibody systems, Part II: Simulation code and parametric studies with nonlinear control, *Acta Astronautica*, Vol. 49, No. 11, 2001, pp. 595–610.
221. Nayfeh A H and Lacarbonara W, On the discretization of distributed-parameter systems with quadratic and cubic nonlinearities, *Nonlinear Dynamics*, Vol. 13, 1997, pp. 203–220.
222. Nayfeh A H and Mook D T, Nonlinear oscillations, Wiley, New York, 1979.
223. Nayfeh A H and Nayfeh S A, On nonlinear modes of continuous systems, *Journal of Vibration and Acoustics*, Vol. 116, 1994, pp. 129–136.
224. Nayfeh A H, Nayfeh J F and Mook D T, On methods for continuous systems with quadratic and cubic nonlinearities, *Nonlinear Dynamics*, Vol. 3, 1992, pp. 145–162.
225. Nayfeh A H, Nayfeh S A and Pakdemirli M, On the discretization of weakly nonlinear spatially continuous systems, In: *Nonlinear Dynamics and Stochastic Mechanics*, W Kliemann and N Sri Namachchivaya (eds), CRC Press, Boca Raton, FL, 1995, pp. 175–200.
226. Nayfeh A H, Nonlinear Interactions, Wiley, New York, 2000.
227. Nayfeh A H, On direct methods for constructing nonlinear normal modes of continuous systems, *Journal of Vibration and Control*, Vol. 1, 1995, pp. 389–430.
228. Newton I, Philosophiae Naturalis Principia Mathematica, Royal Society, London, 1687.
229. Nigg B M and Herzog W, Biomechanics of the Musculo-Skeletal System, John Wiley & Sons, Toronto, 1994.
230. Nikravesh P E and Affifi H A, Construction of the equations of motion for multibody dynamics using point and joint coordinates, computer-aided analysis of rigid and flexible mechanical systems, NATO ASI Series E: Applied Sciences, Vol. 268, 1994, pp. 31–60 Kluwer Academic Publishers.
231. Nikravesh P E and Gim G, Ride and stability analysis of a sports car using multibody dynamic simulation, *Mathematical and Computer Modelling*, Vol. 14, 1990, pp. 953–958.
232. Nikravesh P E, Computer-aided analysis of mechanical systems, Prentice-Hall, Englewood Cliffs, NJ, 1988.
233. Noor A K and Atluri S N, Advances and trends in computational structural mechanics, *AIAA Journal*, 0001-1452, Vol. 25, No. 7, 1987, pp. 977–995.
234. Noor A K and Peters J M, Penalty finite element models for nonlinear dynamic analysis, 26th Conference Structures, Structural Dynamics and Materials , Orlando, FL, April 15–17, 1985, Technical Papers, Part 2 (A85-30226 13–39), New York, American Institute of Aeronautics and Astronautics, 1985, pp. 369–378.
235. Noor A K and Wasfy T M, Multibody dynamic simulation of the next generation space telescope using finite elements and fuzzy sets, *Computer Methods in Applied Mechanics and Engineering*, Vol. 190, No. 5–7, 2000, pp. 803–824.
236. Noor A K, Recent advances in reduction methods for nonlinear problems, *Computers & Structures*, Vol. 13, 1981, pp. 31–44.
237. Noor A K, Starnes J H Jr. and Peters J M, Thermomechanical buckling and postbuckling responses of composite panels with skewed stiffeners, *Finite Elements in Analysis and Design*, Vol. 27, No. 2, 1997, pp. 193–214.

238. Oheron P, Nikravesh P E and Arabyan A, A multibody model simulating tilt wing conversion, *Technical Report CAEL-91-5*, Computer Aided Engineering Laboratory, University of Arizona, 1991.
239. Oriordain K, Thomas P M, Phillips J P and Gilchrist M D, Reconstruction of real world head injury accidents resulting from falls using multibody dynamics, *Clinical Biomechanics*, Vol. 18, No. 7, 2003, pp. 590–600.
240. Palkovics L, Semsey A and Gerum E, Roll-over prevention system for commercial vehicles – Additional sensorless function of the electronic brake system, *Vehicle System Dynamics*, Vol. 32, No. 248, 1999, pp. 285–297.
241. Pan W and Haug E J, Flexible multibody dynamic simulation using optimal lumped inertia matrices, *Computer Methods in Applied Mechanics and Engineering*, Vol. 173, No. 1–2, 1999, pp. 189–200.
242. Park J H, Singh R, Pyo C R and Atluri S N, Structural integrity of panels with multisite fatigue damage, AIAA/ASME/ASCE/AHS/ASC Structures, Structural Dynamics, and Materials Conference 35th, Hilton Head, SC, April 18–20, 1994, Technical Papers. Pt. 2 (A94-23876 06-39), Washington, DC, American Institute of Aeronautics and Astronautics, 1994, pp. 1191–1200.
243. Pennestrì E, Renzi A and Santonocito P, Dynamic modeling of the human arm with video-based experimental analysis, *Multibody System Dynamics*, Vol. 7, No. 4, 2002, pp. 389–406.
244. Pennock G R and Yang A T, Dynamic analysis of a multi-rigid-body open-chain system, *J. Mech. Transm. Autom. Des.*, Vol. 105, No. 1, 1983, pp. 28–38.
245. Pereira M F O S and Ambrosio J A C (eds), Computer-aided analysis of rigid and flexible mechanical systems, Kluwer Academic Publishers, Dordrecht, 1994.
246. Pfeiffer F and Gebler B, A multistage-approach to the dynamics and control of elastic robots. In: 1988 *IEEE International Conference on Robotics and Automation*, IEEE Comput Soc. Press, Washington, Vol. 1, 1988, pp. 2–8.
247. Pfeiffer F and Glocker C, Multibody Dynamics with Unilateral Contacts, Wiley, New York, 1996.
248. Piedbœuf J-C, de Carufel J and Hurteau R, Friction and stick-slip in robots: Simulation and experimentation, *Multibody System Dynamics*, Vol. 4, No. 4, 2000, pp. 341–354.
249. Piedboeuf J C, de Carufel J, Aghili F and Dupuis E, Task verification facility for the Canadian special purpose dextrous manipulator. In: 1999 *IEEE International Conference on Robotics and Automation*, Detroit, MI, 1999, pp. 1077–1083.
250. Popov G, Sankar S, Sankar T S and Vatistas G H, Dynamics of liquid sloshing in horizontal cylindrical road containers, *Proceedings of the Institute of Mechanical Engineers*, Part C 207, pp. 399–406.
251. Popov G, Sankar S, Sankar T S and Vatistas G H, Liquid sloshing in rectangular road containers, *Computers & Fluids*, Vol. 21, 1992, pp. 551–559.
252. Popp K and Stelter P, Stick-slip vibrations and chaos, *Philosophical Transactions of the Royal Society of London*, A336, 1990, pp. 89–105.
253. Prasad P and Chou C C, A review of mathematical occupant simulation models. In: Crashworthiness and Occupant Protection in Transportation Systems, *ASME AMD*, Vol. 106/BED, Vol. 13, T Khalil and A King (eds), 1989, pp. 36–112.
254. Preprints of Symposium on Robot Control 1994, L Sciavicco, C Bonivento and F Nicolo (eds), Napoli, Vol. 1, 1994, pp. 329–334.
255. Presse C and Gautier M, New criteria of exciting trajectories for robot identification. In: *Proceedings of the IEEE International Conference on Robotics and Automation*, Atlanta, GA, IEEE, New York, 1993, pp. 907–912.

256. Quaranta' G, Mantegazza P and Masarati P, Assessing the local stability of periodic motions for large multibody non-linear systems using proper orthogonal decomposition, *Journal of Sound and Vibration*, Vol. 271, No. 3–5, pp. 1015–1038.
257. Rabinowicz E, Stick and slip, *Scientific American*, Vol. 194, No. 5, 1956, pp. 109–118.
258. Rakheja S, Ranganathan R and Sankar S, Field testing and validation of directional dynamics model of a tank truck, *International Journal of Vehicle Design*, Vol. 13, 1992, pp. 251–275.
259. Rakheja S, Sankar S and Ranganathan R, Roll plane analysis of articulated tank vehicles during steady turning, *Vehicle System Dynamics*, Vol. 17, 1988, pp. 81–104.
260. Ranganathan R, Rakheja S and Sankar S, Steady turning stability of partially filled tank vehicles with arbitrary tank geometry, *Journal of Dynamic Systems, Measurement and Control*, Vol. 111, 1989, pp. 481–489.
261. Rismantab-Sany J and Shabana A A, Impulsive motion of nonholonomic deformable multibody systems, Part I, Kinematic and dynamic equations, *Journal of Sound and Vibration*, Vol. 127, 1988, pp. 193–204.
262. Roberson R E and Schwertassek R, Dynamics of multibody systems, Springer-Verlag, Berlin, 1988.
263. Roberson R E and Wittenburg J, A dynamical formalism for an arbitrary number of interconnected rigid bodies with reference to the problem of attitude control, *Proceedings of the 3rd International Congress of Automated Control*, London, 1966
264. Roberson R E and Wittenburg J, A dynamical formalism for an arbitrary number of interconnected rigid bodies with reference to the problem of satellite attitude control, *Proceedings 3rd Congr. Int. Fed. Autom. Control*, Butterworth, Vol. 1, Book 3, Paper 46D, London, 1967.
265. Roberson R E, On recursive determination of body frames for multibody state dynamic simulation, *J. Appl. Mech.*, Vol. 52, No. 3, 1985, pp. 698–700.
266. Roberson R E, On the practical use of Euler-Rodrigues parameters in multibody system dynamic simulation, *Ing. Arch.*, Vol. 54, 1985, pp. 114–123.
267. Roberson R E, On the ways to use normalized Euler parameters as kinetic state variable in dynamic simulation, *Z. Angew. Math. Mech.*, Vol. 718, 1985, pp. 380–382.
268. Santini P and Gasbarri P, Dynamics of multibody systems in space environment; Lagrangian vs. Eulerian approach, *Acta Astronautica*, Vol. 54, No. 1, 2004.
269. Schaller C, Fuhrer C and Simeon B, Simulation of multi-body systems by a parallel extrapolation method, *Mechanics of Structures and Machines*, Vol. 22, 1994, pp. 473–486.
270. Schiehlen W (ed.), Multibody Systems Handbook, Springer-Verlag, Berlin, 1990.
271. Schiehlen W and Kallenbach R G, Identification of robot system parameters. In: Theory of Robots, *Proceedings IFAC/IFIP/IMACS Symposium*, P Kopacek, I Troch and K Desoyer (eds), Pergamon Press, Oxford, 1988, pp. 119–124.
272. Schiehlen W and Schäfer P, Modeling of vehicles with controlled components. In: The dynamics of vehicles on roads and on tracks, *Proceedings 11th IAVSD Symposium*, R J Anderson (ed.), Swets & Zeitlinger, Amsterdam, 1989, pp. 488–501.
273. Schiehlen W, Computational aspects in multibody system dynamics, *Computer Methods in Applied Mechanics and Engineering*, Vol. 90, 1991, pp. 569–582.
274. Schiehlen W, Dynamics of Satellites, Springer-Verlag, Wien, 1970.
275. Schneider E, Popp K and Irretier H, Noise generation in railway wheels due to rail-wheel contact forces, *Journal of Sound and Vibration*, Vol. 120, 1989, pp. 227–244.
276. Schwertassek R and Roberson R E, A perspective on computer-oriented multibody dynamical formalisms and their implementations. In: *Dynamics of Multibody Systems*, G Bianchi and W Schiehlen (eds), Springer-Verlag, Berlin, 1986, pp. 263–273.

277. Shabana A A and Rismantab-Sany J, Impulsive motion of nonholonomic deformable multibody systems, Part II: Impact analysis, *Journal of Sound and Vibration*, Vol. 127, 1988, pp. 205–219.
278. Shabana A A, Dynamics of multibody systems, Wiley, New York, 1989.
279. Shareef N H and Amiroche F M L, Implementation of 3-D isoparametric finite elements on supercomputer for the formulation of recursive dynamical equations of multi-body systems, AIAA guidance, navigation and control conference, New Orleans, LA, *Technical Papers. V*, 1991, pp. 12–14.
280. Shi P and McPhee J, Dynamics of flexible multibody systems using virtual work and linear graph theory, *Multibody System Dynamics*, Vol. 4, 2000, pp. 355–381.
281. Shi P and McPhee J, Dynamics of flexible multibody systems using virtual work and linear graph theory, *Multibody System Dynamics*, Vol. 4, No. 4, 2000, pp. 355–381.
282. Shi P, McPhee J and Heppler G, A deformation field for Euler-Bernoulli beams with application to flexible multibody dynamics, *Multibody System Dynamics*, Vol. 5, 2001, pp. 79–104.
283. Shiiba T and Suda Y, Development of driving simulator with full vehicle model of multi-body dynamics, *JSAE Review*, Vol. 23, No. 2, 2002, pp. 223–230.
284. Silva M P T and Ambrosio J A C, Sensitivity of the results produced by the inverse dynamic analysis of a human stride to perturbed input data, *Gait and Posture*, Vol. 19, No. 1, 2004.
285. Silver W M, On the equivalence of Lagrangian and Newton–Euler dynamics for manipulators, *Int. J. Robot. Res.*, Vol. 1, No. 2, 1982, pp. 60–70.
286. Simeon B, Numerical analysis of flexible multibody systems, *Multibody System Dynamics*, Vol. 6, No. 4, 2001, pp. 305–325.
287. Simeon B, Numerische integration mechanischer Mehrkörpersysteme: Projizierende Deskriptorformen, Algorithmen und Rechenprogramme, Fortschritt-Berichte, Reihe 20 Nr. 130, VDIVerlag Düsseldorf, 1994.
288. Slibar A and Troger H, Das stationäre Fahrverhalten des Tank-Sattelaufliegers. In: Proceedings IUTAM-Symposium, H B Pacejka (ed.), Swets & Zeitlinger, Amsterdam, 1976, pp. 152–172.
289. Stadler W, (ed.), Multicriteria optimization in engineering and in the sciences, Plenum Press, New York, 1988.
290. Stadler W, Analytical Robotics and Mechatronics, McGraw-Hill, London, 1995.
291. Stejskal V and Valasek M, Kinematics and dynamics of machinery, Marcel Dekker, New York, 1996.
292. Stelzle W, Kecskemethy A and Hiller M, A comparative study of recursive methods, *Archive of Applied Mechanics*, Vol. 66, 1995, pp. 9–19.
293. Stronge W J, Impact Mechanics, Cambridge University Press, Cambridge, 2000.
294. Sun Q, Chang K-H, Dormer K J, Dyer R K Jr and Gan R Z, An advanced computer-aided geometric modeling and fabrication method for human middle ear, *Medical Engineering & Physics*, Vol. 24, No. 9, 2002, pp. 595–606.
295. Tenaglia C A, Orin D E, La Farge R A and Lewis C, Toward development of generalized contact algorithm for polyhedral objects. In: *Proceedings of the 1999 IEEE International Conference on Robotics and Automation*, 1999, pp. 2887–2892.
296. Tong R T, Amirouche F and Palkovics L, Ride control - A two-state suspension design for cabs and seats, 16th IAVSD symposium, *Pretoria*, 1999, pp. 139–141.
297. Tongyi J and Amirouche F, Optimum impact force in motion control of multibody systems subject to inequality constraints, *Mechanics Research Communications*, Vol. 16, No. 3, 1989, pp. 163–173.

298. Tongyi J and Amirouche F, Optimum impact force in motion control of multibody systems subject to intermittent constraints, *Computers & Structures*, Vol. 33, No. 5, 1990, pp. 1243–1249.
299. Trindade M A E and Sampaio R, Influence of the flexibility in the dynamics of multibody systems, III APCOM/96 – Third Asian-Pacific Conference on Computational Mechanics, 1996b, Setembro, Seoul, Korea.
300. Troger H and Steindl A, Nonlinear Stability and Bifurcation Theory, Springer-Verlag, Wien, 1991.
301. Tsai F F and Haug E J, Real-time multibody system dynamic simulation, Part I: A parallel algorithm and numerical results, *Mechanics of Structures and Machines*, Vol. 19, 1991, pp. 99–127.
302. Tsai F F and Haug E J, Real-time multibody system dynamic simulation, Part II: A modified recursive formulation and topological analysis, *Mechanics of Structures and Machines*, Vol. 19, 1991, pp. 129–162.
303. Uicker J J Jr, On the dynamic analysis of spatial linkages using 4 by 4 matrices, Ph D Thesis, Northwestern University, Evanston, 1965.
304. van Campen D, (ed.), Interaction between dynamics and control in advanced mechanical systems, Kluwer Academic Publishers, Dordrecht, 1997.
305. Van Staeyen K, Tijskens E and Ramon H, A multibody dynamic approach for colliding particles, *Journal of Sound and Vibration*, Vol. 266, No. 3, 2003, pp. 481–491.
306. von Dombrowski S, Analysis of large flexible body deformation in multibody systems using absolute coordinates, *Multibody System Dynamics*, Vol. 8, No. 4, 2002, pp. 409–432.
307. Vukobratovic M and Stokic D, Control of manipulation robots, Springer-Verlag, Berlin, 1982.
308. Vukobratovic M, Computer method for dynamic model construction of active articulated mechanisms using kinetostatic approach, *Mechanism and Machine Theory*, Vol. 13, 1978, pp. 19–39.
309. Vukobratovic M, Dynamics of articulated mechanisms and synthesis of artificial motion, *Mechanism and Machine Theory*, Vol. 13, 1978, pp. 1–18.
310. Vukobratovic M, Frank A A and Juricic D, On the stability of biped locomotion, *IEEE Transactions on Biomedical Engineering* BME-17, 1970, pp. 25–36.
311. Wagner J F, Adapting the principle of integrated navigation systems to measuring the motion of rigid multibody systems, *Multibody System Dynamics*, Vol. 11, No. 1, 2004, pp. 87–110.
312. Wallentowitz H, (ed.), AVEC'96, Proceedings international symposium on advanced vehicle control, Aachen University of Technology, Aachen, 1996.
313. Wang J T and Huston R L, Computational methods in constrained multibody dynamics: Matrix formalisms, *Computers & Structures*, Vol. 29, No. 2, 1988, pp. 331–338.
314. Wasfy T M and Noor A K, Application of fuzzy sets to transient analysis of space structures, *Finite Elements in Analysis and Design*, Vol. 29, No. 3–4, 1998, pp. 153–171.
315. Watson P and Dabell B J, Cycle counting and fatigue damage, *J. Soc. Env. Eng.*, Vol. 9, 1976, pp. 3–8.
316. Watt A, Fundamentals of three-dimensional computer graphics, Addison-Wesley, Reading, MA, 1989.
317. Whittaker E T, Analytical dynamics, 4th edition, Cambridge University Press, London, 1937.
318. Wittenbauer F, Graphische Dynamik, Springer-Verlag, Berlin, 1923.
319. Wittenburg J, Dynamics of systems of rigid bodies, Teubner, Stuttgart, 1977.

320. Wylie C R E and Barrett L C, Advanced engineering mathematics, McGraw-Hill, Singapore, 1985.
321. Xie M and Amirouche F, Treatment of creep and material nonlinearities in flexible multibody dynamics, *AIAA Journal*, Vol. 32, No. 1, 1994, pp. 190–197.
322. Yamakawa J and Watanabe K, A spatial motion analysis model of tracked vehicles with torsion bar type suspension, *Journal of Terramechanics*, 2004.
323. Yao-huang L and Yong-da Z, Routh's equations for general non holonomic mechanical systems of variable mass, *Applied Mathematics and Mechanics*, Vol. 14, 1993, pp. 285–298.
324. Yong K H and Noor A K, Buckling and postbuckling of composite panels with cutouts subjected to combined loads, *Finite Elements in Analysis and Design*, Vol. 22, No. 2, 1996, pp. 163–185.
325. Yu H, Zhang J and Xu Y, Noether's theory for nonholonomic systems relative to noninertial reference frame, *Applied Mathematics and Mechanics*, Vol. 14, 1993, pp. 527–536.

List of Figures

Index